Lecture Notes in Computer Science 3017

Commenced Publication in 1973
Founding and Former Series Editors:
Gerhard Goos, Juris Hartmanis, and Jan van Leeuwen

Springer
Berlin
Heidelberg
New York
Hong Kong
London
Milan
Paris
Tokyo

Bimal Roy   Willi Meier (Eds.)

Fast Software Encryption

11th International Workshop, FSE 2004
Delhi, India, February 5-7, 2004
Revised Papers

Volume Editors

Bimal Roy
Applied Statistics Unit, Indian Statistical Institute
203, B.T. Road, Calcutta 700 108, India
E-mail: bimal@isical.ac.in

Willi Meier
FH Aargau, 5210 Windisch, P.O. Box , Switzerland
E-mail: meierw@fh-aargau.ch

Library of Congress Control Number: 2004107501

CR Subject Classification (1998): E.3, F.2.1, E.4, G.2, G.4

ISSN 0302-9743
ISBN 3-540-22171-9 Springer-Verlag Berlin Heidelberg New York

Springer-Verlag is a part of Springer Science+Business Media

springeronline.com

Printed in Germany

Typesetting: Camera-ready by author, data conversion by DA-TeX Gerd Blumenstein
Printed on acid-free paper SPIN: 11011880 06/3142 5 4 3 2 1 0

Preface

Fast Software Encryption is a now eleven years old workshop on symmetric cryptography, including the design and analysis of block ciphers and stream ciphers as well as hash functions and message authentication codes. The FSE workshop was first held in Cambridge in 1993, followed by Leuven in 1994, Cambridge in 1996, Haifa in 1997, Paris in 1998, Rome in 1999, New York in 2000, Yokohama in 2001, Leuven in 2002, and Lund in 2003.

This Fast Software Encryption Workshop, FSE 2004, was held February 5–7, 2004 in Delhi, India. The workshop was sponsored by IACR (the International Association for Cryptologic Research) and organized in cooperation with the Indian Statistical Institute, Delhi, and the Cryptology Research Society of India (CRSI).

This year a total of 75 papers were submitted to FSE 2004. After a seven week reviewing process, 28 papers were accepted for presentation at the workshop. In addition, we were fortunate to have in the program two invited talks by Adi Shamir and David Wagner.

During the workshop a rump section was held. Seven presentations were made and all the presenters were given the option of submitting their presentations for possible inclusion in the proceedings. Only one paper from this session was submitted, which was refereed and accepted. This paper appears at the end of these proceedings.

We would like to thank the following people. First Springer-Verlag for publishing the proceedings in the Lecture Notes in Computer Science series. Next the submitting authors, the committee members, the external reviewers, the general co-chairs Subhamoy Maitra and R.L. Karandikar, and the local organizing committee, for their hard work. Bart Preneel for letting us use COSIC's Webreview software in the review process and Thomas Herlea for his support. We are indebted to Lund University, especially Thomas Johansson, Bijit Roy and Sugata Gangopadhyay for hosting the Webreview site. Additionally we would like to thank Partha Mukhopadhyay, Sourav Mukhopadhyay, Malapati Raja Sekhar, and Chandan Biswas for handling all the submissions and Madhusudan Karan for putting together the pre-proceedings. We would also like to thank the sponsors: Infosys Technology Ltd., Honeywell Corporation and Via Technology.

The organizing committee consisted of Sanjay Burman (CAIR, Bangalore), Ramendra S. Baoni (Bisecure Technologies Pvt. Ltd., Delhi), Hiranmoy Ghosh (Tata Infotech Ltd. Delhi), Abdul Sakib Mondal (Infosys Technologies Ltd., Bangalore), Arup Pal (ISI, Delhi), N.R. Pillai (SAG, Delhi), P.K.Saxena (SAG, Delhi), and Amitabha Sinha (ISI, Kolkata), who served as Treasurer. Thank you to them all.

February 2004 Bimal Roy and Willi Meier

Fast Software Encryption 2004

February 5–7, 2004, Delhi, India

Sponsored by the
International Association for Cryptologic Research

in cooperation with the
Indian Statistical Institute, Delhi
and
Cryptology Research Society of India

General Co-chairs
Subhamoy Maitra, Indian Statistical Institute, Kolkata
and
R.L. Karandikar, Indian Statistical Institute, Delhi

Program Co-chairs
Bimal Roy, Indian Statistical Institute, Kolkata
and
Willi Meier, Fachhochschule Aargau, Switzerland

Program Committee

Eli Biham	Technion Israel
Claude Carlet	INRIA, France
Don Coppersmith	IBM, USA
Cunsheng Ding	Hong Kong University of Science and Technology
Helena Handschuh	Gemplus, France
Thomas Johansson	Lund University, Sweden
Charanjit S. Jutla	IBM Research, USA
Lars R. Knudsen	Technical University of Denmark
Kaoru Kurosawa	Ibaraki University, Japan
Kaisa Nyberg	Nokia, Finland
C. Pandu Rangan	Indian Institute of Technology, Chennai
Dingyi Pei	Chinese Academy of Sciences
Bart Preneel	K.U. Leuven, Belgium
Matt Robshaw	Royal Holloway, University of London, U.K.
Serge Vaudenay	EPFL, Switzerland
C.E. Veni Madhavan	Indian Institute of Science, Bangalore
Xian-Mo Zhang	Macquarie University, Australia

External Reviewers

Gildas Avoine
Thomas Baignères
Elad Barkan
Thierry Berger
Qingjun Cai
Carlos Cid
Christophe Clavier
Nicolas Courtois
Christophe De Cannière
Orr Dunkelman
Eric Filiol
Henri Gilbert
Marc Girault
Philippe Guillot
Louis Guillou
Shoichi Hirose
Tetsu Iwata
Thomas Jakobsen
Fredrik Jönsson
Jakob Johnsson
Pascal Junod
Ju-Sung Kang
Tania Lange
Joseph Lano
Yong Li
Yi Lu
Miodrag Mihaljevic
Marine Minier
Jean Monnerat
Sean Murphy
Enes Pasalic
Souradyuti Paul
Michael Quisquater
Haavard Raddum
Palash Sarkar
Takeshi Shimoyama
Xiaojian Tian
Xuesong Wang
Guohua Xiong
Akihiro Yamamura
Julien Brouchier
Ju-Sung Kang

Table of Contents

New Cryptographic Primitives Based on Multiword T-Functions

Alexander Klimov and Adi Shamir

Computer Science department
The Weizmann Institute of Science, Rehovot 76100, Israel
{ask,shamir}@weizmann.ac.il

Abstract. A *T-function* is a mapping from n-bit words to n-bit words in which for each $0 \le i < n$ bit i of the output can depend only on bits $0, 1, \ldots, i$ of the input. All the boolean operations and most of the numeric operations in modern processors are T-functions, and their compositions are also T-functions. In earlier papers we considered 'crazy' T-functions such as $f(x) = x + (x^2 \vee 5)$, proved that they are invertible mappings which contain all the 2^n possible states on a single cycle for any word size n, and proposed to use them as primitive building blocks in a new class of software-oriented cryptographic schemes. The main practical drawback of this approach is that most processors have either 32 or 64 bit words, and thus even a maximal length cycle (of size 2^{32} or 2^{64}) may be too short. In this paper we develop new ways to construct invertible T-functions on multiword states whose iteration is guaranteed to yield a single cycle of arbitrary length (say, 2^{256}). Such mappings can lead to stream ciphers whose software implementation on a standard Pentium 4 processor can encrypt more than 5 gigabits of data per second, which is an order of magnitude faster than previous designs such as RC4.

1 Introduction

There are two basic approaches to the design of secret key cryptographic schemes, which we can call 'tame' and 'wild'. In the tame approach we try to use only simple primitives (such as linear feedback shift registers) with well understood behaviour, and try to prove mathematical theorems about their cryptographic properties. Unfortunately, the clean mathematical structure of such schemes can also help the cryptanalyst in his attempt to find an attack which is faster than exhaustive search. In the wild approach we use crazy compositions of operations (which mix a variety of domains in a nonlinear and nonalgebraic way), hoping that neither the designer nor the attacker will be able to analyse the mathematical behaviour of the scheme. The first approach is typically preferred in textbooks and toy schemes, but real world designs often use the second approach.

In several papers published in the last few years [5, 6], we tried to bridge this gap by considering 'semi-wild' constructions which look like crazy combinations of boolean and arithmetic operations, but have many analyzable mathematical properties. In particular, we defined the class of T-functions which contains arbitrary compositions of plus, minus, times, or, and, xor operations on n-bit words,

B. Roy and W. Meier (Eds.): FSE 2004, LNCS 3017, pp. 1–15, 2004.

and showed that it is easy to analyse their invertibility and cycle structure for arbitrary word sizes. Such constructions can replace LFSRs and linear congruential mappings (which are vulnerable to correlation and algebraic attacks) in a new class of stream ciphers and pseudo random generators.

The paper is organized in the following way. In section 2 we recall the basic definitions from [5] for single word mappings, and consider several ways in which they can be extended to the multiword case. In section 3 we extend our bit-slice technique to analyse the invertibility of multiword T-functions. In section 4 we extend our technique from [6] to analyse the cycle structure of multiword T-functions. Finally, in section 5 we provide experimental data on the speed of several possible implementations of our functions on a PC.

2 Multiword T-Functions

Invertible mappings with a single cycle have many cryptographic applications. The main context in which we study them in this paper is pseudo random generation and stream ciphers. Modern microprocessors can directly operate on up to 64-bit words in a single clock cycle, and thus a univariate mapping can go through at most 2^{64} different states before entering a cycle. In some cryptographic applications this cycle length may be too short, and in addition the cryptanalyst can guess a 64 bit state in a feasible computation. A common way to increase the size of the state and extend the period of a generator is to run in parallel and combine the outputs of several generators with different periods. The overall period is determined by the least common multiple of their individual periods. This works well with LFSRs, whose periods $2^{n_1}-1, 2^{n_2}-1, \ldots$ can be relatively prime, and thus the overall period can be their product. However, our univariate mappings have periods of $2^{n_1}, 2^{n_2}, \ldots$ whose least common multiple is just $2^{\max(n_1, n_2, \ldots)}$.

A partial solution to this problem is to cyclically use a large number of different state update functions, starting from a secret state and a secret index. For example, we can use 64-bit words and $2^{16}-1$ different constants C_k to get a guaranteed cycle length of almost 2^{80} from the following simple generator:

Theorem 1. *Consider the sequence $\{(x_i, k_i)\}$ defined by iterating*

$$x_{i+1} = x_i + (x_i^2 \vee C_{k_i}) \bmod 2^n,$$
$$k_{i+1} = k_i + 1 \bmod m,$$

where each x is an n-bit word and C_k is some n-bit constant for each $k = 0, \ldots, m-1$. Then the sequence of pairs (x_i, k_i) has a maximal period (of size $m2^n$) if and only if m is odd, and for all k, $[C_k]_0 = 1$ and $\bigoplus_{k=0}^{m-1} [C_k]_2 = 1$.

A special case of this theorem for $m = 1$ is that the function $f(x) = x + (x^2 \vee C)$ is invertible with a single cycle if and only if both the least significant bit and the third least significant bit in C are 1, and the smallest such C is 5.

Unfortunately, the cyclic change of state update functions is inconvenient, and it cannot yield really large cycles (e.g., of 2^{256} possible states). We can try to solve the problem by using a single high precision variable x (say, with 256 bits), but the multiplication of such long variables can become prohibitively expensive. What we would like to do is to define the mapping by operating separately on the various input words, without trying to interpret the result as a natural mathematical operation on multi-precision words.

Let us first review the definitions from [5] in the case of univariate mappings. Let x be an n-bit word. We can view x as a vector of bits denoted by $([x]_{n-1}, \ldots, [x]_0)$, where the least significant bit has number 0. In this bit notation, the univariate function $f(x) = x + 1 \pmod{2^n}$ can be expressed in the following way:

$$
\begin{aligned}
[f(x)]_0 &= f_0([x]_0) && = [x]_0 \oplus 1 \\
[f(x)]_1 &= f_1([x]_1 ; [x]_0) && = [x]_1 \oplus \alpha_1([x]_0) \\
[f(x)]_2 &= f_2([x]_2 ; [x]_1 , [x]_0) && = [x]_2 \oplus \alpha_2([x]_1 , [x]_0) \\
&\vdots && \vdots \\
[f(x)]_{n-1} &= f_{n-1}([x]_{n-1} ; [x]_{n-2}, \ldots, [x]_0) \\
&= [x]_{n-1} \oplus \alpha_{n-1}([x]_{n-2}, \ldots, [x]_0),
\end{aligned}
\tag{1}
$$

where each α_i denotes one of the carry bits. Note that for any bit position i, $[f(x)]_i$ depends only on $[x]_i, \ldots, [x]_0$ and does not depend on $[x]_{n-1}, \ldots, [x]_{i+1}$. We call any univariate function f which has this property a *T-function* (where 'T' is short for *triangular*). Note further that each carry bit α_i depends only on strictly earlier input bits $[x]_{i-1}, \ldots, [x]_0$ but not on $[x]_i$. This is a special type of a T-function, which we call a *parameter*. To provide some intuition from the theory of linear transformations on n-dimensional spaces, we can say that T-functions roughly correspond to lower triangular matrices, parameters roughly correspond to lower triangular matrices with zeroes on the diagonal, and a T-function can be roughly represented as a diagonal matrix plus a parameter.

Let us now define these notions for functions which map several input words into one output word. The natural extension of the notion of a T-function in this case is to allow bit i of the output to depend only on bits 0 to i of each one of the inputs. The observation which makes this notion interesting is that all the boolean operations and most of the arithmetic operations available on modern processors are T-functions. In particular, addition ('$+$'), subtraction ('binary $-$'), negation ('unary $-$'), multiplication ('$*$'), or ('$\vee$'), and ('$\wedge$'), exclusive or ('$\oplus$'), and complementation ('$\neg$') (where the boolean operations are performed on all the n bits in parallel and the arithmetic operations are performed modulo 2^n) are T-functions with one or two inputs. We call these eight functions *primitive operations*. Note that circular rotations and right shifts are not T-functions, but left shifts can be expressed as multiplication by a power of 2 and thus they are T-functions. Since the composition of T-functions is also a T-function, any 'crazy' function which contains arbitrarily many primitive operations is always a T-function.

In order to define multiword mappings f which can be iterated, we have to further extend the notion to functions with the same number m of input and output words. We can represent the multiword input as the following $n \times m$ bit matrix $B^{n \times m}$:

$$x = \begin{pmatrix} x_0 \\ x_1 \\ \vdots \\ x_{m-1} \end{pmatrix} = \begin{pmatrix} [x]_{0,n-1} & \cdots & [x]_{0,1} & [x]_{0,0} \\ [x]_{1,n-1} & \cdots & [x]_{1,1} & [x]_{1,0} \\ \vdots & \ddots & \vdots & \vdots \\ [x]_{m-1,n-1} & \cdots & [x]_{m-1,1} & [x]_{m-1,0} \end{pmatrix}. \quad (2)$$

We can now consider the columns of the bit matrix as parallel bit slices with no internal bit order, and say that a multiword mapping is a T-function if all the bits in column i of the matrix of output words can depend only on bits in columns 0 to i of the matrix of input words. In this interpretation it is still true that any composition of primitive operations is a multiword T-function, but some of the proven properties of univariate T-functions (e.g., that all the cycle lengths are powers of 2) are no longer true.

An alternative definition of multiword T-functions is to concatenate all the input words into one long word, to concatenate all the output words into one long word, and then to use the standard univariate definition of a T-function in order to limit which input bits can affect which output bits. If we denote the l input words by $x_u, x_v, \ldots$, then we define the single logical variable x by

$$x = (x_u, \ldots, x_w) = ([x]_{n(l-1)+(n-1)}, \ldots, [x]_{n(l-1)}, \cdots [x]_{n-1}, \ldots, [x]_0). \quad (3)$$

Note that in this interpretation $f(x) = (f_u, f_v) = (x_u + x_v, x_v)$ is a T-function, but the very similar $f(x) = (f_u, f_v) = (x_u, x_u + x_v)$ is not a T-function, and thus we cannot compose primitive operations in an arbitrary way. On the other hand, we can obtain many new types of T-functions in which low-order words can be manipulated by non-primitive operations (such as cyclic rotation) before we use them to compute higher order output words.

Our actual definition of multiword T-functions combines and generalizes these two possible interpretations. Let x be an $nl \times m$ bit matrix ($\mathbb{B}^{nl \times m}$):

$$\begin{pmatrix} [x]_{0,n(l-1)+(n-1)} & \cdots & [x]_{0,n(l-1)+1} & [x]_{0,n(l-1)} & \cdots & [x]_{0,n-1} & \cdots & [x]_{0,0} \\ [x]_{1,n(l-1)+(n-1)} & \cdots & [x]_{1,n(l-1)+1} & [x]_{1,n(l-1)} & \cdots & [x]_{1,n-1} & \cdots & [x]_{1,0} \\ \vdots & \ddots & \vdots & \vdots & \cdots & \vdots & \ddots & \vdots \\ [x]_{m-1,n(l-1)+(n-1)} & \cdots & [x]_{m-1,n(l-1)+1} & [x]_{m-1,n(l-1)} & \cdots & [x]_{m-1,n-1} & \cdots & [x]_{m-1,0} \end{pmatrix}. \quad (4)$$

We consider it as an $m \times l$ matrix of n bits words:

$$x = \begin{pmatrix} x_{0,u} & \cdots & x_{0,w} \\ \vdots & \ddots & \vdots \\ x_{m-1,u} & \cdots & x_{m-1,w} \end{pmatrix}.$$

We concatenate the l words in each row into a single logical variable, and then consider the collection of the m long variables as the inputs to the T-function.

Finally, we allow the bits in column i of the output matrix to depend only on bits in columns $0, \ldots, i$ in the input matrix.

To demonstrate this notion, consider the following mapping over 4-tuples of words:

$$f(x) = \begin{pmatrix} x_{0,u} + x_{1,u}x_{0,v} & x_{0,v} + x_{1,v} \\ x_{0,u} - x_{1,u}(x_{1,v} \bullet 1) & x_{0,v} \oplus x_{1,v} \end{pmatrix}.$$

This is a valid T-function under our general multiword definition even though it contains the non-primitive right shift operation $\bullet\ 1$.

3 Bit Slice Analysis and Invertibility

The main tool we use in order to study the invertibility of T-functions is bit slice analysis. Its basic idea is to define the mapping from $[x]_i$ to $[f(x)]_i$ by abstracting out the complicated dependency on $[x]_{0\ldots i-1}$ via the notion of parameters. For example, the size of the explicit description of the mapping $[f(x)]_i = \phi([x]_0, \ldots, [x]_i)$ in the function $f(x) = x + (x^2 \vee 5)$ grows exponentially with i, but it can be written as $[f(x)]_i = [x]_i \oplus \alpha_i$, where α_i is some function of $[x]_0, \ldots, [x]_{i-1}$ (that is, a parameter). By using this parametric representation we can easily prove the invertibility of the mapping by induction on i, since if we already know bits 0 to $i-1$ of the input x and bit i of the output $f(x)$, we can (in principle) calculate the value of α_i and thus derive in a unique way bit i of the input. Intuitively, this is the same technique we use in order to solve a triangular system of linear equations, except that in our case the explicit description of α_i can be extremely complicated, and thus we do not use this technique as a real inversion algorithm for $f(x)$, but only in order to prove that this inverse is uniquely defined.

The main observation in [5] was that such an abstract parametric representation can be easily derived for any composition of primitive operations by the following recursive definition, in which i can be any bit position except zero:

$$\begin{aligned}
[xy]_0 &= [x]_0 \wedge [y]_0 \\
\left[x \overset{+}{\underset{\oplus}{-}} y\right]_0 &= [x]_0 \oplus [y]_0 \\
\left[x \overset{\wedge}{\vee} y\right]_0 &= [x]_0 \overset{\wedge}{\vee} [y]_0 \\
[xy]_i &= [x]_i\, \alpha_{[y]_0} \oplus \alpha_{[x]_0}\, [y]_i \oplus \alpha_{xy} \\
\left[x \overset{+}{-} y\right]_i &= [x]_i \oplus [y]_i \oplus \alpha_{x \pm y} \\
\left[x \overset{\oplus}{\underset{\vee}{\wedge}} y\right]_i &= [x]_i \overset{\oplus}{\underset{\vee}{\wedge}} [y]_i
\end{aligned} \tag{5}$$

To demonstrate this technique, consider our running example: $\left[x + (x^2 \vee 5)\right]_0 = [x]_0 \oplus \left[x^2 \vee 5\right]_0 = [x]_0 \oplus 1$ and, for $i > 0$, $\left[x + (x^2 \vee 5)\right]_i = [x]_i \oplus (\left[x^2\right]_i \vee [5]_i) \oplus \alpha_{x+(x^2 \vee 5)} = [x]_i \oplus (([x]_i\, \alpha_{[x]_0} \oplus \alpha_{[x]_0}\, [x]_i \oplus \alpha_{x^2}) \vee [5]_i) \oplus \alpha_{x+(x^2 \vee 5)} = [x]_i \oplus \alpha$.

This invertibility test can be easily generalized to the multivariate case (2). Let us show an example of such a construction. We start from an arbitrary non singular matrix which denotes a possible bit slice mapping, such as:

$$\begin{matrix} 1 & \alpha \\ 0 & 1 \end{matrix}$$

We can add to this linear mapping an affine part (where α, β, and γ are arbitrary parameters) and get the following bit slice structure:

$$\begin{pmatrix} [x_0]_i \\ [x_1]_i \end{pmatrix} \to \begin{pmatrix} [x_0]_i \oplus \alpha\,[x_1]_i \oplus \beta \\ [x_1]_i \oplus \gamma. \end{pmatrix} \tag{6}$$

It is easy to check that for $i > 0$ the i-th bit slice of the following mapping matches (6):

$$\begin{pmatrix} x_0 \\ x_1 \end{pmatrix} \to \begin{pmatrix} x_0 + (x_0^2 \wedge x_1) \\ x_1 + x_0^2. \end{pmatrix}$$

Unfortunately, the least significant bit slice of this mapping is not invertible:

$$\begin{pmatrix} [x_0]_0 \\ [x_1]_0 \end{pmatrix} \to \begin{pmatrix} [x_0]_0 \oplus [x_0]_0\,[x_1]_0 \\ [x_1]_0 \oplus [x_0]_0\,. \end{pmatrix}$$

So we have to apply a little tweak to fix it:

$$\begin{pmatrix} x_0 \\ x_1 \end{pmatrix} \to \begin{pmatrix} x_0 + ((x_0^2 \wedge x_1) \vee 1) \\ x_1 + x_0^2. \end{pmatrix}$$

The reader may get the impression that the bit slice mappings of invertible functions are always linear. From (5) it is easy to see that every expression which uses only $\oplus$, $+$, $-$ and $\times$ has linear i-th bit slice, but in general this is not true.

4 The Single Cycle Property

A T-function has the single cycle property if its repeated application to any initial state goes through all the possible states. Let us recall the basic results from [6] in the univariate case. Invertibility is a prerequisite of the single cycle property. If a T-function has a single cycle modulo 2^k then it has a single cycle modulo 2^{k-1}. If a T-function has a cycle of length l modulo 2^{k-1} then modulo 2^k it has either a cycle of length $2l$ or two cycles of length l. Taking into account the fact that modulo 2^1 a function has either one cycle of length two or two cycles of length one we can conclude that the size of any cycle of a T-function is always a power of 2.

In the univariate case a bit slice of an invertible T-function has the form $[f(x)]_i = [x]_i \oplus \alpha$. From (5) it follows that $f(x)$ has one of the following forms: $f_1(x) = x \oplus r_1(x)$, $f_2(x) = x + r_2(x)$ or $f_3(x) = x r_3(x)$, where the r_i are parameters (in the case of multiplication additionally we need $[r_3]_0 = 1$). It is easy to see that $[f_3(x)]_0 = [x]_0\,[r_3(x)]_0 = [x]_0$, that is it has two cycles modulo 2 and so it can not form a single cycle modulo 2^n. So, a single cycle function has either[1] the first or the second form. In order to analyse the cycle structure of these forms the following definitions of *even* and *odd* parameters

[1] Note that there is no *exclusive* or here since every function can be represented in both forms, for example $x + 1 = x \oplus (x \oplus (x + 1))$.

were introduced. Suppose that $r(x)$ is a parameter, that is $r(x) = r(x + 2^{n-1}) \pmod{2^n}$. So, $r(x) = r(x + 2^{n-1}) + 2^n b(x) \pmod{2^{n+1}}$. Consider

$$B[r,n] = 2^{-n} \sum_{i=0}^{2^{n-1}-1} (r(i + 2^{n-1}) - r(i)) \pmod 2 = \bigoplus_{i=0}^{2^{n-1}-1} b(i). \quad (7)$$

The parameter is called *even* if $B[r,n]$ is always zero, and *odd* if $B[r,n]$ is always one[2].

Let us give several examples of even parameters:

- $r(x) = C$, where C is an arbitrary constant ($r(x) = r(x + 2^{n-1})$ and so, $b = 0$ and $B = 0$);
- $r(x) = 2x$ ($r(x + 2^{n-1}) = r(x) + 2^n \pmod{2^{n+1}}$, so $b(x) = 1$ and B is even as long as 2^{n-1} is even, that is for $n \geq 2$);
- $r(x) = x^2$ ($r(x + 2^{n-1}) = r(x) + 2^n x + 2^{2(n-1)}$, so $b(x) = [x]_0$ and B is even for $n \geq 3$);
- $r(x) = 4g(x)$, where $g(x)$ is an arbitrary T-function ($r(x + 2^{n-1}) - r(x) = 4(g(x + 2^{n-1}) - g(x)) = 0 \pmod{2^n}$);
- $r(x) = r'(x) \underset{\oplus}{\overset{+}{-}} r''(x)$, where r' and r'' are simultaneously even or odd parameters ($B = B' \oplus B''$).
- $r(x) = r'(x) \vee C$, where C is an arbitrary constant and $r'(x)$ is an even parameter (if $[C]_i = 0$ then $[r(x)]_i = [r'(x)]_i$, and if $[C]_i = 1$ then $[r(x)]_i = [C]_i$, so in both cases $[r(x)]_i$ is the same as for some even parameter.)

The following theorem was proved in [6]:

Theorem 2. *Let N_0 be such that $x \to x + r(x) \bmod 2^{N_0}$ defines a single cycle and for $n > N_0$ the function $r(x)$ is an* even *parameter. Then the mapping $x \to x + r(x) \bmod 2^n$ defines a single cycle for all n.*

We can use our running example of $f(x) = x + (x^2 \vee C)$ to demonstrate this theorem. If the binary form of C ends with $\ldots 1\overset{0}{\underset{1}{}}1$, then $C = 5, 7 \pmod 8$, and $x^2 \vee C = C \pmod 8$ is an odd constant modulo 2^3 so $x + C$ has a single cycle modulo 2^3. In addition, x^2 is an even parameter for $n \geq 3$, and this is not affected by 'or'ing it with an arbitrary constant. In [6] it was shown that $x \to x + (x^2 \vee C)$ is the smallest nonlinear expression which defines a single cycle, in other words there is no nonlinear expression which defines a single cycle and consists of less than three operations.

Another important class of single cycle mappings is $f(x) = 1 + x + 4g(x)$ for an arbitrary T-function $g(x)$. It turns out that x86 microprocessors have an instruction which allows us to calculate $1 + x + 4y$ with a single instruction[3] and

[2] Note that in the general case B is a function of n, and thus the parameter can be neither even nor odd. We often relax these definitions by allowing exceptions for small n such as 1 or 2.

[3] The `lea` (load effective address) instruction makes it possible to calculate any expression of the form $C + R_1 + kR_2$, where C is a constant, R_1 and R_2 are registers and k is a power of two. Its original purpose was to simplify the calculation of addresses in arrays.

thus the single cycle mapping $f(x) = 1 + x + 4x^2$ can be calculated using only two instructions.

Odd parameters are less common and harder to construct. Their main application is in mappings of the form $x \oplus r(x)$:

Theorem 3. *Let N_0 be such that $x \to x \oplus r(x) \bmod 2^{N_0}$ defines a single cycle and for $n > N_0$ the function $r(x)$ is an* odd *parameter. Then the mapping $x \to x \oplus r(x) \bmod 2^n$ defines a single cycle for all n.*

Proof. Let us prove this by induction: suppose that the mapping defines a single cycle modulo 2^n and we are going to prove that it defines a single cycle modulo 2^{n+1}. Since there are only two possible cycle structures modulo 2^{n+1} (a single cycle of size 2^{n+1} or two cycles of size 2^n) we will prove that the second case is impossible, that is $\left[x^{(0)}\right]_n \neq \left[x^{(2^n)}\right]_n$ at least for one x. Recall that $[x]_n$ is the most significant bit of x modulo 2^{n+1}. Let $x^{(0)} = 0$, since r is a parameter it follows that $r(i) = r(i + 2^n) \pmod{2^{n+1}}$, and so $\left[x^{(2^n)}\right]_n = \left[r(x^{(0)}) \oplus \ldots \oplus r(x^{(2^n-1)})\right]_n = \bigoplus_{i=0}^{2^n-1} [r(i)]_n =$
$$\bigoplus_{i=0}^{2^{n-1}-1} \left(\left[r(i + 2^{n-1})\right]_n \oplus [r(i)]_n\right) = \bigoplus_{i=0}^{2^{n-1}-1} \left(\left[r(i + 2^{n-1}) - r(i)\right]_n\right) = 1.$$
Here we use the fact that $\left[r(i + 2^{n-1})\right]_n - [r(i)]_n = \left[r(i + 2^{n-1}) - r(i)\right]_n$.

Recently the related notions of measure preservation and ergodicity of compatible functions over p-adic numbers were independently studied by Anashin [1].[4] His motivation was mathematical rather than cryptographic, and he used different techniques. In order to study if a T-function is invertible (respectively, has a single cycle property) he tried to represent it as $f(x) = d + cx + pv(x)$ (respectively, $f(x) = c + rx + p(v(x + 1) - v(x))$), or to represent it as Mahler interpolation series, or to use the notion of uniform differentiability. The first characterization is the most general (that is $v(x)$ can be any T-function) and complete (he proved that every invertible (respectively, a single cycle function) can be represented in this form. For example, it follows that there exists $v_f(x)$, such that $f(x) = x + (x^2 \vee 5) = 1 + x + 2(v_f(x + 1) - v_f(x))$ and in order to prove that $f(x)$ defines a single cycle it is enough to find such a function $v_f(x)$. This example shows that this criterion is not that good in practice in checking properties of a given function but it allows us to construct arbitrary complex functions with needed properties. For the second approach any function f can be represented as a Mahler interpolation series $\sum_{i=0}^{\infty} a_i \left(\frac{x(x-1)\cdots(x-i+1)}{i!}\right)$. It turns out, for example, that a T-function is invertible if and only if $\|a_1\|_2 = 1$ and $\|a_i\|_2 \leq 2^{-\lfloor \log_2 i \rfloor - 1}$, for $i = 2, 3, \ldots$. This is used to prove theoretical results similar to the previous one, but once again for practical purposes it is usually hard to represent a given function as a Mahler series. The uniformly differentiable[5] func-

[4] This could be translated to our terminology as follows: *measure preservation* — invertibility, *ergodicity* — a single cycle property, *compatible* — T-function. To simplify reading we will continue to use our terminology, but an interested reader who will refer to his paper should keep in mind this "dictionary".

[5] It is exactly the same concept as the usual notion of uniform differentiability of real functions, but with respect to the p-adic distance.

tions allow us to use Hensel lifting. However, there are several obstacles to the application of this theorem in practice. First of all, it is not always easy to tell if a given function is uniformly differentiable. Consider, for example, $x + (x^2 \vee 5)$: $\vee$ is not a differentiable operation but according to [2] in this particular case $(x+(x^2 \vee 5))' = (x+x^2+5-(x^2 \wedge 5))' = 1+2x+2x(u \wedge 5)|'_{u=x^2}$, but $(u \wedge 5)' = 0$ for $\|h\|_2 \leq \frac{1}{8}$, and so the whole expression is uniformly differentiable. Unfortunately, this trick does not work for the general case $x+(x^2 \vee C)$. Moreover, there are invertible mappings which are not uniformly differentiable. The second obstacle is how to find N_0. In the very restricted case of polynomials (expressions which use only $+$, $-$ and $\times$) it is possible to calculate this number in advance, but this is not possible even if we add only $\oplus$: we found a family of functions $f_i(x)$ (which use only $+$, $-$, and $\oplus$) such that for every N there is $N_0 \geq N$, i_{N_0} such that $f_{i_{N_0}}$ has a single cycle modulo 2^{N_0} but not modulo 2^{N_0+1}. In particular, if $f_0 = x - 1$, $f_i = ((f(i-1, x) \oplus x) + x) \oplus (x + x)$ then $N_0 = i + 2$, $i = 2^r$.

Both Anashin's techniques and our techniques can be used to completely characterize all the univariate polynomial mappings modulo 2^n which are invertible with a single cycle:

Theorem 4. *A polynomial $P(x) = \sum_{i=0}^{d} a_i x^i$ is invertible modulo any 2^n if and only if it is invertible modulo 4, and it has a single cycle modulo any 2^n if and only if it has a single cycle modulo 8.*

Proof. From (5) it follows[6] that $[P(x)]_0 = [a_0]_0 + [(a_1 + \cdots + a_d)]_0 [x]_0$ and $[P(x)]_i = [a_1]_0 [x]_i \oplus \bigoplus_{\text{odd } k \geq 3} [a_k]_0 [x]_0 [x]_i \oplus \alpha$. Since a T-function is invertible if and only if each bit slice is invertible, the following conditions are necessary and sufficient for the invertibility of a polynomial: $[(a_1 + \cdots + a_d)]_0 = 1$, $[a_1]_0 = 1$ and $\bigoplus_{\text{odd } k \geq 3} [a_k]_0 = 0$. In order to prove the single cycle property let us represent the polynomial in the following form: $P(x) = x + P'(x)$. Since in order to generate a single cycle $P(x)$ should be invertible it follows that $[(a'_1 + \cdots + a'_d)]_0 = 0$, $[a'_1]_0 = 0$ and $\bigoplus_{\text{odd } k \geq 3} [a'_k]_0 = 0$. Now we need to prove that $P'(x)$ is an even parameter, that is, that $\bigoplus_{i=0}^{2^{n-1}-1} b(i) = 2^{-n} \sum_{i=0}^{2^{n-1}-1} (P'(i + 2^{n-1}) - P'(i)) \pmod 2 = 0$. Let us do it separately for different parts of P'. It is easy to show that a_0, $a_1 x$ with even a_1 and $a_k x^k$ for even k are even parameters for $n \geq 3$. Let us prove that the sum of odd powers is also an even parameter if $\sum_k [a_k]_0 = 0 \bmod 2$ and $n \geq 3$: $2^{-n} \sum_{i=0}^{2^{n-1}-1} \sum_{\text{odd } k \geq 3} (a_k (i + 2^{n-1})^k - a_k i^k) = 2^{-n} \sum_i \sum_k a_k k i^{k-1} 2^{n-1} + \sum_i \sum_k a_k (\frac{k(k-1)}{2} i^{k-2} 2^{2(n-1)-n} + \cdots) = 0 \pmod 2$. So, if $P(x)$ defines a single cycle modulo 8 then it defines a single cycle modulo any 2^n. On the other hand if $P(x)$ does not define a single cycle modulo 8 then it does not define a single cycle modulo any $2^n \geq 8$.

It is easy to verify that exactly 1/8 of all the polynomials are invertible, and 1/64 of all the polynomials have a single cycle, and thus it is easy to pick random polynomials with these properties. Typical examples of quadratic single cycle polynomials are $f(x) = (x + 1)(2x + 1)$ and $f(x) = 6x^2 - x + 1$.

[6] See [5] for details.

Our goal now is to construct invertible multiword T-functions whose iteration defines a single cycle. Let us start with T-functions of type (2), which map m parallel input words to m parallel output words. We would like to construct a mapping

$$\begin{pmatrix} x_0 \\ \vdots \\ x_{m-1} \end{pmatrix} \to \begin{pmatrix} f_0(x_0, \ldots, x_{m-1}) \\ \vdots \\ f_{m-1}(x_0, \ldots, x_{m-1}) \end{pmatrix}$$

which is invertible and has the maximum possible period of 2^{mn}. Several simple constructions can be easily shown to be impossible. For example:

Theorem 5. *No T-mapping of the form* $(x_0, x_1) \to (f_0(x_0), f_1(x_0, x_1))$ *can have a period of* 2^{2n}.

Proof. Suppose there is a mapping $(x_0, x_1) \to (f_0(x_0), f_1(x_0, x_1))$, where f_0 and f_1 are T-functions, such that it has a period of size $P = 2^{2i}$ modulo 2^i. This means that, for example, $f^{(P)}(0,0) = (0,0)$ and $\forall p < P$, $f^{(p)}(0,0) \neq (0,0)$. Let $p = 2^{2(i-1)}$. Since the period of the mapping modulo 2^{i-1} is at most $2^{2(i-1)}$ it follows that $f^{(p)}(0,0) \neq (0,0)$ if and only if the same holds for the most significant bits, but since f_0 depends only on x_0 modulo 2^i the period of f_0 is at most 2^i and so (for sufficiently large i) the most significant bit of $f_0^{(p)} = f_0^{(2p)} = f_0^{(3p)}$ and since the most significant bit of f_1 can assume only two possible values it follows that either $f^{(p)} = f^{(2p)}$ or $f^{(p)} = f^{(3p)}$ which is a contradiction.

Also it can be shown that it is impossible to obtain a single cycle function of this type from slice-linear mappings, which are T-functions which have the following form: $[f_j]_i = \bigoplus_{k=0}^{m-1} [C_{j,k}]_i [x_k]_i \oplus \alpha_{j,i}$, where $(C_{i,j})$ is a constant invertible matrix and $m > 2$.

Note that the construction is trivial in the two extreme cases of $m = 1$ and $n = 1$. In the univariate case ($m = 1$) the answer is given by theorem 3: $x \to x \oplus \alpha$, where α is an odd parameter. The case of $n = 1$ is also simple because every function is then a T-function, and it is easy to define a counting transformation such as $f_i = x_i \oplus (x_0 \wedge \cdots \wedge x_{i-1})$ which goes through all the states in the natural order $0 \ldots 00 \to 0 \ldots 01 \to 0 \ldots 10 \to \cdots \to 1 \ldots 11 \to 0 \ldots 00$. Let us combine these two cases:

$$f_i(x_0, \ldots, x_{m-1}) = x_i \oplus (\alpha_i(x_0, \ldots, x_{m-1}) \wedge x_0 \wedge \cdots \wedge x_{i-1}), \tag{8}$$

where each α_i is an odd parameter, that is

$$\bigoplus_{(x_0, \ldots, x_{m-1}) = (0, \ldots, 0)}^{(2^n-1, \ldots, 2^n-1)} [\alpha_i(x_0, \ldots, x_{m-1})]_n = 1$$

and $[\alpha_i]_0 = 1$. We can now prove:

Theorem 6. *The mapping defined by (8) defines a single cycle of length* 2^{mn}.

Proof. Let us prove it by induction. For $n = 1$ we know that it is true. Suppose that it is a single cycle modulo 2^n and let us prove this for 2^{n+1}. Suppose without loss of generality that $(x_0, \ldots, x_{m-1})^{(0)} = (0, \ldots, 0)$. From the induction hypothesis it follows that $\{(x_0, \ldots, x_{m-1})^{(i)} \bmod 2^n\}_{i=0}^{2^{mn}}$ contains all the possible tuples, so $\left[x_0^{(2^{mn})}\right]_n = x_0^{(0)} \oplus \bigoplus \alpha_0 = 1$, more generally, $\left[x_0^{(l+2^{mn})}\right]_n = \left[x_0^{(l)}\right]_n \oplus 1$, so

$$\bigoplus_{(x_0,\ldots,x_{m-1})=(0,\ldots,0)}^{(2^{n+1}-1,2^n-1,\ldots,2^n-1)} [\alpha_1(x_0, \ldots, x_{m-1})]_n \wedge [x_0]_n = \bigoplus_{(x_0,\ldots,x_{m-1})=(0,\ldots,0)}^{(2^n-1,2^n-1,\ldots,2^n-1)} [\alpha_1(x_0, \ldots, x_{m-1})]_n .$$

So, if we consider the next variable x_1: $\left[x_1^{(2^{mn+1})}\right]_n = x_1^{(0)} \oplus \bigoplus \alpha_1 \wedge x_0 = x_1^{(0)} \oplus 1$. Using similar arguments, we can prove that $\left[x_{m-1}^{(2^{mn+(m-1)})}\right]_n = \left[x_{m-1}^0\right]_n \oplus 1$, that is modulo 2^{n+1} the period is $2^{mn+m} = 2^{m(n+1)}$.

In order to use this theorem we need an odd parameter. We know that $f(x) = x + r(x)$ defines a single cycle if $r(x)$ is an even parameter. We also know that every such function can be represented as $f(x) = x \oplus s(x)$, where $s(x)$ is an odd parameter. So, for any even parameter $r(x)$ the following expression $s(x) = (x + r(x)) \oplus x$ is an odd parameter. For example, if $r(x) = x^2 \vee 5$, then $s(x) = (x + (x^2 \vee 5)) \oplus x$ is an odd parameter. Note that if $f(x)$ is invertible then $[r(x)]_0 = 1$ and so $[s(x)]_0 = [r(x)]_0 = 1$. To construct an odd parameter with m variables we need the following lemma:

Lemma 1.

$$\bigoplus_{t=0}^{2^n} \alpha(t) = \bigoplus_{(x_0,\ldots,x_{m-1})=(0,\ldots,0)}^{(2^n-1,\ldots,2^n-1)} \alpha(x_0 \wedge \cdots \wedge x_{m-1})$$

Proof. In order to prove the lemma it is sufficient to prove that for every t there is an *odd* number of tuples $(x_0, \ldots, x_{m-1})$ such that $x_0 \wedge \cdots \wedge x_{m-1} = t$. In fact, we can directly calculate this number: if t contains k zeros then there are $(2^m - 1)^k$ such tuples, which is an odd number.

For example, the following mapping defines a single cycle for any m and n:

$$f_i(x_0, \ldots, x_{m-1}) = x_i \oplus (s(x_0 \wedge \cdots \wedge x_{m-1}) \wedge x_0 \wedge \cdots \wedge x_{i-1}), \tag{9}$$

where $s(t) = (t + (t^2 \vee 5)) \oplus t$. Note that to evaluate this mapping for each particular $(x_0, \ldots, x_{m-1})$ we need one squaring, one addition and $3m$ bitwise operations, or $3m + 2$ operations in total. Unfortunately, this mapping has the property that during update each bit of x_{m-1} is changed with probability $2^{-(m+1)}$ instead of $\frac{1}{2}$ as expected for a random mapping. To avoid this we can add any even parameter[7] with zero in the least significant bit, simplify $s(t)$ and obtain, for example,

[7] Note that in the multivariate case every parameter which does not use all the variables is even.

the following mapping:

$$\begin{pmatrix} x_0 \\ x_1 \\ x_2 \\ x_3 \end{pmatrix} \rightarrow \begin{pmatrix} x_0 \oplus s & \oplus (x_1^2 \wedge M) \\ x_1 \oplus (s \wedge a_0) & \oplus (x_2^2 \wedge M) \\ x_2 \oplus (s \wedge a_1) & \oplus (x_3^2 \wedge M) \\ x_3 \oplus (s \wedge a_2) & \oplus (x_0^2 \wedge M) \end{pmatrix}, \tag{10}$$

where $a_0 = x_0$, $a_1 = a_0 \wedge x_1$, $a_2 = a_1 \wedge x_2$, $a_3 = a_2 \wedge x_3$, $s = (a_3 + C) \oplus a_3$, C is any odd constant and $M = 1 \ldots 1110_2$ or we can enhance the inter-variable mixing with the following mapping:

$$\begin{pmatrix} x_0 \\ x_1 \\ x_2 \\ x_3 \end{pmatrix} \rightarrow \begin{pmatrix} x_0 \oplus s & \oplus 2x_1x_2 \\ x_1 \oplus (s \wedge a_0) & \oplus 2x_2x_3 \\ x_2 \oplus (s \wedge a_1) & \oplus 2x_3x_0 \\ x_3 \oplus (s \wedge a_2) & \oplus 2x_0x_1 \end{pmatrix} \tag{11}$$

During the iteration of these mappings each bit is changed in approximately one half of the cases. Each mapping requires 24 operations to obtain a cycle of size 2^{256}.

We next analyse mappings of type (3), which consider multiple words as concatenated parts of a single multi-precision logical variable x, for example, $x = 2^{3n}x_a + 2^{2n}x_b + 2^n x_c + x_d$. In this case our running example $x \rightarrow x + (x^2 \vee C)$ can be represented as $(x_a, x_b, x_c, x_d) \rightarrow (f_a(x_a; x_b, x_c, x_d), \ldots, f_d(x_d))$ with appropriate f_a, f_b, f_c and f_d, but the squaring of a $4n$-bit word would be rather inefficient on an n-bit processor. Note that there is no requirement that the whole mapping $x \rightarrow f(x)$ has a simple interpretation as a mapping of a $4n$ bit word although we need this interpretation to prove the single cycle property.

Let us start with the simplest single cycle mapping $x \rightarrow x + 1$. It can be implemented as follows: $(x_a, x_b, x_c, x_d) \rightarrow (x_a + \kappa_b, x_b + \kappa_c, x_c + \kappa_d, x_d + 1)$, where κ_d is the carry (overflow) from x_d, κ_c is the carry from x_c, et cetera. Many contemporary processors (including x86 and SPARC) have an operation adc (addition with carry) $(x, y, c) \rightarrow (z, c')$, where $z = (x + y + c) \bmod 2^n$ and $c' = (x + y + c \geq 2^n)$, so the addition of κ can be done at no additional cost[8] if the mapping has the following form: $(x_a, x_b, x_c, x_d) \rightarrow (x_a + f_a + \kappa_b, x_b + f_b + \kappa_c, x_c + f_c + \kappa_d, x_d + f_d + 1)$. The only restriction on each f is that it should be an even parameter. For example,

$$\begin{pmatrix} x_d \\ x_c \\ x_b \\ x_a \end{pmatrix} \rightarrow \begin{pmatrix} x_d + (s_d^2 \vee C_d) \\ x_c + (s_c^2 \vee C_c) + \kappa_d \\ x_b + (s_b^2 \vee C_b) + \kappa_c \\ x_a + (s_a^2 \vee C_a) + \kappa_b, \end{pmatrix} \tag{12}$$

[8] This is not always true. For example, on `Intel Pentium 4` processor the ordinary `add` instruction has the latency (the number of clock cycles that are required for the execution core to complete the execution of all of the μops that form a IA-32 instruction) 0.5 and the throughput (the number of clock cycles required to wait before the issue ports are free to accept the same instruction again) 0.5, but `adc` has the latency 8 and the throughput 3 [4].

where $s_d = x_d$, $s_c = s_d \oplus x_c$, $s_b = s_c + x_b$, $s_a = s_b \oplus x_a$, C_a, C_b, C_c are odd constants[9] and C_d ends with $\dots 1\,{}^{0}_{1}\,1_2$.

This mapping uses 15 operations to obtain a cycle of size 2^{256}. So, it can be much faster than the previous example but only on processors that have the adc instruction. If this instruction has to be emulated[10] this mapping can be slower. Note that almost all high level programming languages do not have such an instruction so it has to be emulated or to be written in assembly language.

The number of operations is not usually a good predictor of the speed of the implementation, since on modern microprocessors several operations can be done in parallel, use different number of clocks, etc. We should take into account that an update function is not the complete generator. There is also an output function which produces the output bits given the internal state. Since the least significant bits of the state are repeated in small cycles the simplest output function is the one that gives away the most significant bits of the state. It seems that it is easier to implement such a function for the second mapping since x_a and x_b form the most significant part of the state, but their least significant bits are also weak since they depend only of the least significant bits of x_i and a single bit carry.[11] One solution is to give away the most significant part of x_a (32 bits) or use a more sophisticated output function and give away more bits thus raising the ratio of the output bits per operation. For example, $O_1 = ((x_a \bullet \frac{n}{2}) \oplus x_b)(((x_c \bullet \frac{n}{2}) \oplus x_d) \vee 1)$, where $\bullet$ could be substituted with circular rotation if the target microprocessor supports it, uses only five operations and doubles the size of the output. It can be shown that O_1 has maximal period and produces each possible value with the same probability (since it is an invertible mapping of x_b for fixed x_a, x_c and x_d). Note that we give here only examples of the possible mappings and output functions, but in order to construct a secure stream cipher they have to be subjected to a lengthy and thorough cryptanalysis.

5 Experimental Results

We tested the actual execution speed of our mappings on an IA-32 machine. We used a standard PC with a 1.7 GHz Pentium 4 processor with 256KB of cache. In each experiment we encrypted one gigabyte of data by encrypting 10^4 times a buffer which is 10^5 bytes long (this ensured that the data was prefetched into L2 cache). The tests were done on an otherwise idle Linux system. To calculate

[9] Note that the only purpose of C_a, C_b and C_c is to make parameters out of $x_\cdot^2$, that is mask the least significant bit. The evenness of the parameter is guaranteed since n from (7) is larger than 3.

[10] There are basically two ways to emulate it: check if $a + b \geq a$ or $a + b \geq b$ for unsigned a and b or, usually faster since conditions break pipeline, to set $n = 63$ and use the most significant bit of the sum as $\kappa_\cdot$. The second approach adds two operation for each addition of $\kappa_\cdot$ and gives 21 operations overall with reduced word size.

[11] It is possible to change the definitions of s_a,s_b,s_c,s_d to incorporate right shifts or rotations, but this will also increase the calculation time.

the overhead produced by memory access we "encrypted" the buffer by xoring it with a 32 bit constant. It took 0.42 seconds, and we did *not* subtract it from our actual results.

In this paper we propose a general approach rather than a concrete stream cipher, and thus in our performance tests we experimented with many different state sizes, update mappings and output functions. As a typical example, we used 256 bit states defined as four words of length $n = 64$ and updated them by (11). We wanted to use a simple output function which is just the top half of each x_i, but we discovered that this variant is vulnerable to a (theoretical) attack in which the attacker guesses the 17 least significant bits of each x_i (i.e., a total of 68 guessed bits) and searches in 2^{34} bytes of data for places where two of the x_i end with 17 zeroes and thus their product ends with 34 zeros. To protect against this attack, we slightly modified the state update function to:

$$\begin{pmatrix} x_0 \\ x_1 \\ x_2 \\ x_3 \end{pmatrix} \to \begin{pmatrix} x_0 \oplus s & \oplus\, (2(x_1 \vee C_1)x_2) \\ x_1 \oplus (s \wedge a_0) & \oplus\, (2x_2(x_3 \vee C_3)) \\ x_2 \oplus (s \wedge a_1) & \oplus\, (2(x_3 \vee C_3)x_0) \\ x_3 \oplus (s \wedge a_2) & \oplus\, (2x_0(x_1 \vee C_1)) \end{pmatrix}, \tag{13}$$

where C_1 and C_2 are constants with several ones in the least significant half, for example, $C_1 = \mathsf{12481248}_{16}$ and $C_3 = \mathsf{48124812}_{16}$. To take advantage of the SSE2 instruction set of Pentium 4 processors (which contains 128-bit integer instructions that allow us to operate on two 64-bit integers simultaneously), we run two generators $((x_0, x_1, x_2, x_3)$ and $(x'_0, x'_1, x'_2, x'_3))$ in parallel and the output is generated by $(x_0 \bullet 32) \oplus x_1$, $(x_2 \bullet 32) \oplus x_3$, $(x'_0 \bullet 32) \oplus x'_1$ and $(x'_2 \bullet 32) \oplus x'_3$. Since each command operates on a pair (x_i, x'_i) we can run two generators instead of one at no additional cost. With this optimization, the complete encryption operation required only 1.56 seconds per gigabyte, and thus the experimentally verified encryption speed was approximately 5.13 gigabits/second.

To put our results in perspective, the reader should consider the RC4 stream cipher, which is one of the fastest software oriented ciphers available today. The web site of Crypto++ contains benchmarks for highly optimized implementations of many well known cryptographic algorithms [3]. For RC4 it quotes a speed of 110 megabyte per second on a 2.1 GHz Pentium 4 processor. This can be scaled to $1000/110 \times 1.7/2.1 = 11.2$ seconds required to encrypt a gigabyte of data on a 1.7GHz processor, which is almost an order of magnitude slower than the speed we obtain with our approach.

References

[1] V. Anashin, "Uniformly Distributed Sequences of p-adic integers, II". Available from http://www.arxiv.org/ps/math.NT/0209407. 8
[2] V. Anashin, private communication. 9
[3] Crypto++ 5.1 Benchmarks:
http://www.eskimo.com/~weidai/benchmarks.html. 14
[4] "IA-32 Intel Architecture Optimization Reference Manual". Available from http://www.intel.com/design/pentium4/manuals/248966.htm. 12

[5] A. Klimov and A. Shamir, "*A New Class of Invertible Mappings*", CHES 2002. 1, 2, 3, 5, 9
[6] A. Klimov and A. Shamir, "*Cryptographic Applications of T-functions*", SAC 2003. 1, 2, 6, 7

Towards a Unifying View of Block Cipher Cryptanalysis

David Wagner*

University of California, Berkeley

Abstract. We introduce *commutative diagram cryptanalysis*, a framework for expressing certain kinds of attacks on product ciphers. We show that many familiar attacks, including linear cryptanalysis, differential cryptanalysis, differential-linear cryptanalysis, mod n attacks, truncated differential cryptanalysis, impossible differential cryptanalysis, higher-order differential cryptanalysis, and interpolation attacks can be expressed within this framework. Thus, we show that commutative diagram attacks provide a unifying view into the field of block cipher cryptanalysis. Then, we use the language of commutative diagram cryptanalysis to compare the power of many previously known attacks. Finally, we introduce two new attacks, *generalized truncated differential cryptanalysis* and *bivariate interpolation*, and we show how these new techniques generalize and unify many previous attack methods.

1 Introduction

How do we tell if a block cipher is secure? How do we design good ciphers? These two questions are central to the study of block ciphers, and yet, after decades of research, definitive answers remain elusive. For the moment, the art of cipher evaluation boils down to two key tasks: we strive to identify as many novel cryptanalytic attacks on block ciphers as we can, and we evaluate new designs by how well they resist known attacks.

The research community has been very successful at this task. We have accumulated a large variety of different attack techniques: differential cryptanalysis, linear cryptanalysis, differential-linear attacks, truncated differential cryptanalysis, higher-order differentials, impossible differentials, mod n attacks, integrals, boomerangs, sliding, interpolation, the yo-yo game, and so on. The list continues to grow. Yet, how do we make sense of this list? Are there any common threads tying these different attacks together?

In this paper, we seek unifying themes that can put these attacks on a common foundation. We have by no means accomplished such an ambitious goal; rather, this paper is intended as a first step in that direction. In this paper, we show how a small set of ideas can be used to generate many of today's known attacks. Then, we show how this viewpoint allows us to compare the strength of different types of attacks, and possibly to discover new attack techniques. We

* This work is supposed by NSF ITR CNS-0113941.

B. Roy and W. Meier (Eds.): FSE 2004, LNCS 3017, pp. 16–33, 2004.

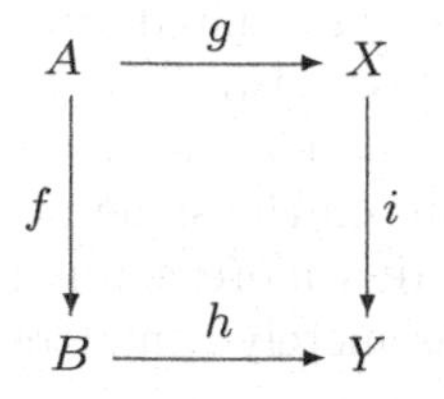

Fig. 1. An example of a commutative diagram. The intended meaning of this diagram is that $h \circ f = i \circ g$

hope this perspective will be of some interest, if only to see a different way to think about the known cryptanalytic attacks on block ciphers.

2 Background

What is a block cipher? A block cipher is a map $E : \mathcal{K} \times \mathcal{M} \to \mathcal{M}$ so that E_k is invertible for all keys $k \in \mathcal{K}$, and both E_k and E_k^{-1} can be efficiently computed. The set $\mathcal{M}$ is the space of texts; for instance, for AES, it is $\mathcal{M} = \{0,1\}^{128}$.

When is a block cipher secure? A block cipher is secure if it behaves as a pseudorandom permutation. In other words, it must be secure against *distinguishing attacks*: no efficient algorithm A given interactive access to encryption and decryption black boxes should be to distinguish the real cipher (i.e., E_k and E_k^{-1}) from a truly random permutation (i.e., π and π^{-1}, where π is uniformly distributed on the set of all permutations on $\mathcal{M}$) with non-negligible advantage. The distinguishing advantage of an attack A is given by Adv $A = \Pr[A^{E_k, E_k^{-1}} = 1] - \Pr[A^{\pi, \pi^{-1}} = 1]$.

In this paper, we focus exclusively on distinguishing attacks. Usually, once a distinguishing attack is found, a key-recovery attack soon follows; the hard part is in finding a distinguishing attack in the first place, or in building a cipher secure against distinguishing attacks.

How are block ciphers built? Most block ciphers are *product ciphers.* In other words, the cipher is built as the composition of individual round transformations: we choose a round function $f : \mathcal{M} \to \mathcal{M}$, compute a sequence of round keys $k_1, \ldots, k_n$ as a function of the key k, and set $E_k = f_{k_n} \circ \cdots \circ f_{k_1}$. The function f computes one round of the cipher.

Commutative Diagrams. In the discussion to follow, it will be useful to introduce a concept from abstract algebra: that of *commutative diagrams.* Commutative diagrams are a concise notation for expressing functional composition properties. An example of a commutative diagram can be found in Fig. 1. In this example, the symbols A, B, X, Y represent sets, and the symbols f, g, h, i are functions with signatures $f : A \to B$, $g : A \to X$, $h : B \to Y$, and $i : X \to Y$. We say that the diagram "commutes" if $h \circ f = i \circ g$, or in other words, if

$h(f(a)) = i(g(a))$ for all $a \in A$. Notice how paths correspond to functions, obtained by composing the maps associated with each edge in the path. In this diagram, there are two paths from A to Y, corresponding to two functions with signature $A \to Y$. Informally, the idea is that it doesn't matter which path we follow from A to Y; we will obtain the same map either way. More complicated diagrams can be used to express more complex relationships, and identifying the set of implied identities is merely a matter of chasing arrows through the diagram.

Markov Processes. Also, we recall the notion of Markov processes. A Markov process is a pair of random variables I, J, and to it we associate a transition matrix M given by $M_{i,j} = \Pr[J = j | I = i]$. We call the sequence of random variables $I - J - K$ a Markov chain if K is conditionally independent of I given J. We can associate transition matrices M, M', M'' to the Markov processes $I - J$, $J - K$, and $I - K$, respectively, and if $I - J - K$ forms a Markov chain, we will have $M'' = M' \cdot M$. In other words, composition of Markov processes corresponds to multiplication of their associated transition matrices.

The maximum advantage of an adversary at distinguishing one Markov process from another can be calculated using decorrelation theory [14]. Let $||M||_\infty$ denote the ℓ_∞ norm of the matrix M, i.e., $||M||_\infty = \max_i \sum_j |M_{i,j}|$. We can consider an adversary A who is allowed to choose a single input, feed it through the Markov process, and observe the corresponding output. The maximum advantage of any such adversary at distinguishing M from M' will then be exactly $\frac{1}{2}||M - M'||_\infty$. If U denotes the uniform $m \times n$ transition matrix, i.e., $U_{i,j} = 1/n$ for all i, j, then $||M_1 M_2 - U||_\infty \leq ||M_1 - U||_\infty \cdot ||M_2 - U||_\infty$.

The above calculations can be extended to calculate the advantage $\text{Adv}\, A = \Pr[A^M = 1] - \Pr[A^{M'} = 1]$ of an adversary A who can interact repeatedly with the Markov process. First, if $\frac{1}{2}||M - M'||_\infty = \epsilon$ denotes the advantage of a single-query adversary, then an adversary making q queries has advantage at most $q \cdot \epsilon$. In practice, when ϵ is small, the advantage of a q-query adversary often scales roughly as $\sim \sqrt{q} \cdot \epsilon$. Hence, as a rough rule of thumb, $\Theta(1/\epsilon^2)$ queries often are necessary and sufficient to distinguish M from M' with non-trivial probability [3]. We emphasize, though, that this heuristic is not always valid; there are many important exceptions.

The advantage of a q-query adversary can be computed more precisely. If M is a $m \times n$ matrix, let $[M]^q$ denote the $m^q \times n^q$ matrix given by $([M]^q)_{i,j} = M_{i_1,j_1} \times \cdots \times M_{i_q,j_q}$. Define the matrix norm $||M||_a = \max_{i_1} \sum_{j_1} \cdots \max_{i_q} \sum_{j_q} |M_{i,j}|$. Then, in an adaptive attack, the maximum advantage of any q-query adversary is exactly $\max_A \text{Adv}\, A = \frac{1}{2}||[M]^q - [M']^q||_a$. In a non-adaptive attack, the maximum advantage of any q-query non-adaptive adversary is exactly $\frac{1}{2}||[M]^q - [M']^q||_\infty$.

Organization. The rest of this paper studies cryptanalysis of product ciphers. First, we describe commutative diagrams and their relevance to cryptanalysis.

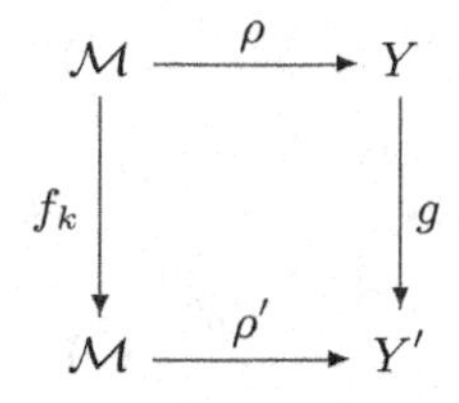

Fig. 2. A local property of the round function f_k

Then, we explore statistical attacks, a probabilistic generalization of commutative diagram attacks, and then we further generalize by introducing the notion of higher-order attacks. Finally, we explore algebraic attacks.

3 Commutative Diagram Attacks

The basic recipe for analyzing a product cipher is simple:

1. Identify local properties of the cipher's round functions.
2. Piece these together to obtain a global property of the cipher as a whole.

In this way, we seek to exploit the structure of a product cipher—namely, its construction as a composition of round functions—to simplify the cryptanalyst's task.

How do we identify local properties of the round function that are both non-trivial and can be spliced together suitably? This is where commutative diagrams can help. Let $f_k : \mathcal{M} \to \mathcal{M}$ denote a round function. Suppose we can find a property of the input that is preserved by the round function; then this would suffice as a local property. If there is some partial information about x that allows to predict part of the value of $f_k(x)$, this indicates a pattern of non-randomness in the round function that might be exploitable.

One way to formalize this is using *projections.* A projection is a function $\rho : \mathcal{M} \to Y$ from the text space to a smaller set Y. If we have two projections ρ, ρ' so that $\rho'(f_k(x))$ can be predicted from $\rho(x)$, then we have a local property of the round function. To make this more precise, we look for projections $\rho : \mathcal{M} \to Y$, $\rho' : \mathcal{M} \to Y'$ and a function $g : Y \to Y'$ so that $\rho' \circ f_k = g \circ \rho$ for all $k \in \mathcal{K}$, or in other words, so that the diagram in Fig. 2 commutes for all k.

A commutative diagram is *trivial* if it remains satisfied if we replace f_k by any random permutation π. Each non-trivial commutative diagram for f is an interesting local property of the round function.

Such local properties can be pieced together to obtain global properties by exploiting the compositional behavior of commutative diagrams. Refer to Fig. 3. If both small squares commute (i.e., if $\rho' \circ f_{k_1} = g \circ \rho$ and $\rho'' \circ f_{k_2} = g' \circ \rho'$), then whole diagram commutes (e.g., $\rho'' \circ f_{k_2} \circ f_{k_1} = g' \circ g \circ \rho$). In other words, if ρ, ρ' form a local property of the first round f_{k_1} and if ρ', ρ'' form a local

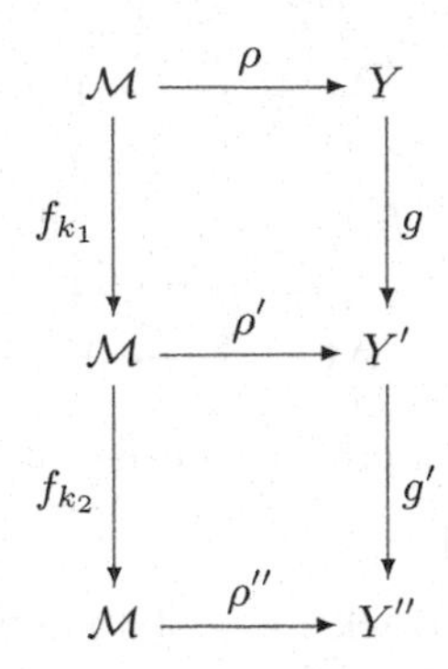

Fig. 3. Splicing together a local property for f_{k_1} and a local property for f_{k_2} to obtain a property of their composition

property of the second round f_{k_1}, then ρ, ρ'' form a global property of the first two rounds $f_{k_2} \circ f_{k_1}$.

Note that the requirement is that we have a local property of each round, *and* that the local properties match up appropriately. If ρ, ρ' is a local property of the first round and φ, φ' is a local property of the second round, these two can only be composed if $\rho' = \varphi$. Thus there is a compatibility requirement that must be satisfied before two local properties can be composed.

The same kind of reasoning can be extended inductively to obtain a global property of the cipher as a whole. The requirement is that we obtain local properties for the rounds that are compatible. See Fig. 4. If each local property is non-trivial, then the global property so obtained will be non-trivial.

Any non-trivial global property for the cipher as a whole immediately leads to a distinguishing attack. Suppose we have ρ, ρ', g so that $\rho' \circ E_k = g \circ \rho$ holds for all k. Then our distinguishing attack is straightforward: we obtain a few known-plaintext/ciphertext pairs (x_i, y_i) and we check whether $\rho'(y_i) \stackrel{?}{=} g(\rho(x_i))$ holds for all of them. When the known texts are obtained from the real cipher (E_k), these equalities will always hold. However, since our property is non-trivial, the equalities will not always hold if the pairs (x_i, y_i) were obtained from an ideal cipher (π, a random permutation). The distinguishing advantage of such an attack depends on the details of the projections chosen, but we can typically expect to obtain a significant attack.

Example: Madryga. As a concrete example of a commutative diagram attack, let us examine Madryga, an early cipher design. Eli Biham discovered that the Madryga round function preserves the parity of its input. Hence, we may choose the parity function as our projection $\rho : \{0,1\}^{64} \to \{0,1\}$, i.e., $\rho(x_1, \ldots, x_{64}) = x_1 \oplus \cdots \oplus x_{64}$. When taken with the identity function, we obtain a global property for Madryga, as depicted in Fig. 5. This yields a dis-

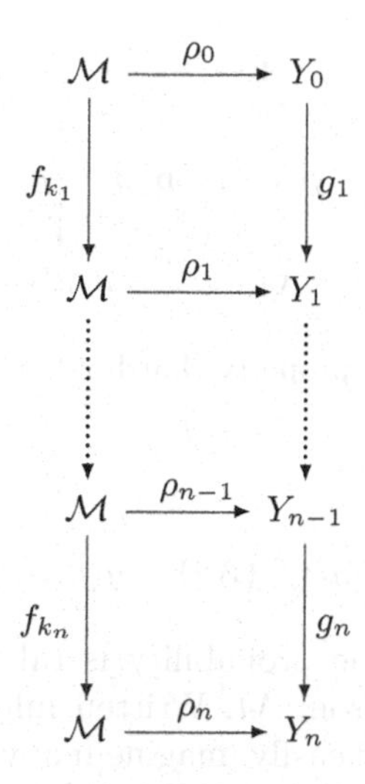

Fig. 4. Splicing together local properties for each round to obtain a global property for E_k, the cipher as a whole

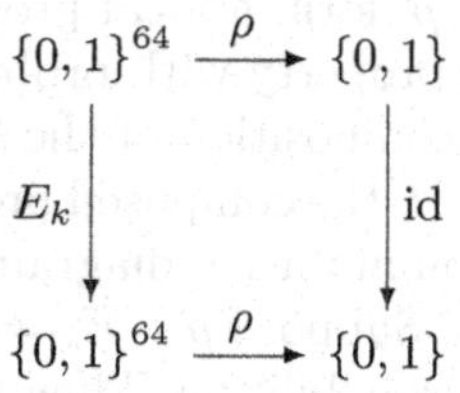

$$\begin{array}{ccc} \{0,1\}^{64} & \xrightarrow{\rho} & \{0,1\} \\ \downarrow E_k & & \downarrow \text{id} \\ \{0,1\}^{64} & \xrightarrow{\rho} & \{0,1\} \end{array}$$

Fig. 5. A global property of the Madryga block cipher. Here ρ is the parity function

tinguishing attack on Madryga with advantage 1/2, as $\Pr[\rho(E_k(x)) = \rho(x)] = 1$ yet $\Pr[\rho(\pi(x)) = \rho(x)] = 1/2$.

Commutative diagrams are mathematically elegant. However, they are not, on their own, powerful enough to successfully attack many ciphers; another idea is needed. We shall describe next how these ideas may be extended to model statistical attacks, which turn out to be significantly more powerful.

4 Statistical Attacks

We now turn our attention to statistical attacks. The natural idea is to look at diagrams that only commute with some probability.

A reasonable first attempt might be to introduce the notion of *probabilistic commutative diagrams.* See Fig. 6, which is intended to show a diagram that commutes with probability p. In other words, though the relation $\rho' \circ f_k = g \circ \rho$

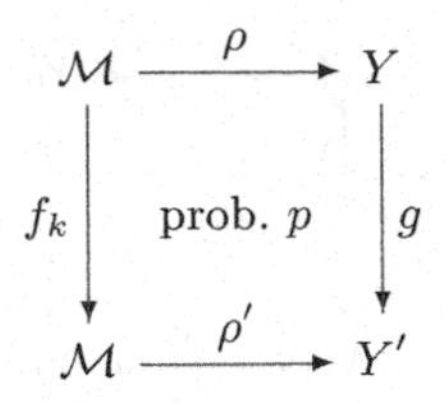

Fig. 6. A local property that holds with probability p

does not hold, we do have

$$\Pr_{X \leftarrow \mathcal{M}}[\rho'(f_k(X)) = g(\rho(X))] = p$$

for all $k \in \mathcal{K}$, where here the probability is taken with respect to the choice of X uniformly at random from $\mathcal{M}$. Written informally: $\rho' \circ f_k = g \circ \rho$ holds with probability p. (We could easily imagine many variants of this definition, for instance, by taking the probability over both the choice of X and k; however, we will not pursue such possibilities here.)

Probabilistic commutative diagrams share many useful properties with their deterministic cousins. First, probabilistic commutative diagrams can be composed. Suppose projections ρ, ρ' form a local property with prob. p for the first round, and ρ', ρ'' form a local property with prob. p' for the second round. Then ρ, ρ'' form a property for the composition of the first two rounds, and assuming our cipher is a Markov cipher [9], the composed property holds with prob. at least $p \cdot p'$. Second, probabilistic commutative diagrams for the whole cipher usually lead to distinguishing attacks. Suppose $\rho' \circ E_k = g \circ \rho$ holds with probability p, and $\rho' \circ \pi = g \circ \rho$ holds with probability q. Then there is a simple distinguishing attack that uses one known-plaintext/ciphertext pair (x, y) and has advantage $|p - q|$: we simply check whether $\rho'(y) \stackrel{?}{=} g(\rho(x))$.

Probabilistic commutative diagrams may appear fairly natural on first glance, but on further inspection, they seem to be lacking in some important respects. For our purposes, it will be useful to introduce a more general notion, which we term *stochastic commutative diagrams.* If we let the random variable X be uniformly distributed on $\mathcal{M}$, the maps ρ, ρ', E_k induce a Markov process on the pair of random variables $\rho(x)$, $\rho'(E_k(X))$. The associated transition matrix M is given by

$$M_{i,j} = \Pr_{X \leftarrow \mathcal{M}}[\rho'(E_k(X)) = j | \rho(X) = i].$$

Note that there is an implicit dependence on k, but for simplicity in this paper we will only consider the case where each key $k \in \mathcal{K}$ yields the same transition matrix M. (This corresponds to assuming that the Hypothesis of Stochastic Equivalence holds.) An example stochastic commutative diagram is shown pictorially in Fig. 7.

Stochastic commutative diagrams yield distinguishing attacks. Let M' be the transition matrix induced by the Markov process $\rho(X), \rho'(\pi(X))$, where π is

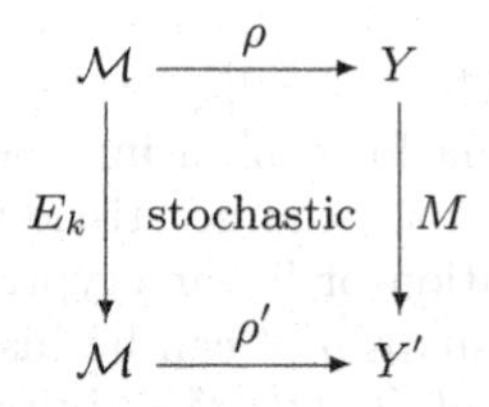

Fig. 7. A stochastic commutative diagram for E_k. Here M is the transition matrix of the Markov process $\rho(X) - \rho'(E_k(X))$

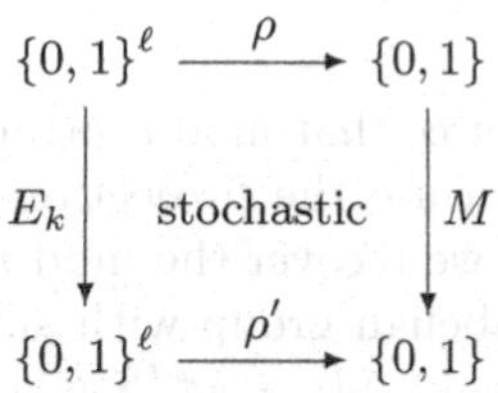

$\{0,1\}^\ell$ $\xrightarrow{\rho}$ $\{0,1\}$

E_k stochastic M

$\{0,1\}^\ell$ $\xrightarrow{\rho'}$ $\{0,1\}$

Fig. 8. A formulation of linear cryptanalysis as a stochastic commutative diagram attack. Here ρ and ρ' are linear maps

a random permutation. Then our stochastic commutative diagram yields a distinguishing attack that uses one chosen plaintext query and achieves advantage $\frac{1}{2}||M - M'||_\infty$.

Linear Cryptanalysis. Linear cryptanalysis may now be recognized as a special case of a stochastic commutative diagram attack. Suppose we have a ℓ-bit block cipher $E_k : \{0,1\}^\ell \to \{0,1\}^\ell$. In Matsui's linear cryptanalysis [10], the codebreaker somehow selects a pair of linear maps $\rho, \rho' : \{0,1\}^\ell \to \{0,1\}$, and then we use the stochastic commutative diagram shown in Fig. 8. For instance, the linear characteristic $\Gamma \to \Gamma'$ corresponds to the projections $\rho(x) = \Gamma \cdot x$, $\rho'(x) = \Gamma' \cdot x$.

In a linear attack, we obtain a 2×2 transition matrix M of the form

$$M = \begin{bmatrix} \frac{1}{2} + \frac{\epsilon}{2} & \frac{1}{2} - \frac{\epsilon}{2} \\ \frac{1}{2} - \frac{\epsilon}{2} & \frac{1}{2} + \frac{\epsilon}{2} \end{bmatrix}. \tag{1}$$

The transition matrix associated to a random permutation π is U, the 2×2 matrix where all entries are $1/2$. Therefore, the distinguishing advantage of a linear cryptanalysis attack using one known text is $\frac{1}{2}||M - U||_\infty = \epsilon/2$. It is not hard to verify that $\Theta(1/\epsilon^2)$ known texts suffice to obtain an attack with distinguishing advantage $1/2$ (say). Compare to Matsui's rule of thumb, which says that $8/(\epsilon/2)^2 = 32/\epsilon^2$ texts suffice.

Matsui's piling-up lemma can also be re-derived within this framework. Let M_1, M_2 be transition matrices for the first and second round, respectively, taking the form shown in Equation (1) albeit with ϵ replaced by ϵ_1, ϵ_2. It is not hard to verify that $M_1 M_2$ also takes the form shown in Equation (1), but with ϵ replaced by $\epsilon_1 \epsilon_2$. Hence $||M_1 M_2 - U||_\infty = ||M_1 - U||_\infty \times ||M_2 - U||_\infty$. This is exactly the piling-up lemma for computing the bias of a multi-round linear characteristic given the bias of the characteristic for each round.

After seeing this formulation of linear cryptanalysis, our use of the name "projection" to describe the maps ρ, ρ' can be justified as follows. Consider the vector subspace $\mathcal{V} = \{0, \Gamma\}$ of $\{0,1\}^\ell$. We obtain a canonical isomorphism of vector spaces $\mathcal{V} \cong \{0,1\}$. Then we can view $\rho : \{0,1\}^\ell \to \mathcal{V}$ as taking the form of a projection onto the subspace $\mathcal{V}$. In other words, we write $\{0,1\}^\ell$ as the direct sum $\{0,1\}^\ell = \mathcal{V} \oplus \mathcal{V}^T$, write each $x \in \{0,1\}^\ell$ as a sum $x = y \oplus z$ for $y \in \mathcal{V}$ and $z \in \mathcal{V}^T$, and then let $\rho(x) = y$ be the projection of x onto $\mathcal{V}$.

Mod n Cryptanalysis. Notice that mod n attacks also fall within this framework. If $\mathcal{M} = \mathbb{Z}/2^\ell\mathbb{Z}$, we can use the projection $\rho : \mathbb{Z}/2^\ell\mathbb{Z} \to \mathbb{Z}/n\mathbb{Z}$ given by $\rho(x) = x \bmod n$. In this way we recover the mod n attack.

In general, if $\mathcal{M}$ is any abelian group with subgroup $S \subseteq \mathcal{M}$, we may consider projections of the form $\rho : \mathcal{M} \to \mathcal{M}/S$ given by $\rho(x) = x \bmod S$. Linear cryptanalysis is simply the special case where $\mathcal{M} = (\{0,1\}^\ell, \oplus)$ and S has index 2, and mod n cryptanalysis is the special case where $\mathcal{M} = (\mathbb{Z}, +)$ and $S = n\mathbb{Z}$.

Linear Cryptanalysis with Multiple Approximations. Linear cryptanalysis with multiple approximations also falls naturally within this framework. Suppose we have a list of masks $\Gamma_1, \Gamma_2, \ldots, \Gamma_d \in \{0,1\}^\ell$, and assume that these masks are linearly independent as vectors in $\{0,1\}^\ell$. Let $\mathcal{V}$ be the vector subspace of dimension d spanned by $\Gamma_1, \ldots, \Gamma_d$, i.e., $\mathcal{V} = \{0, \Gamma_1, \Gamma_2, \Gamma_1 \oplus \Gamma_2, \ldots\}$. Choose the canonical isomorphism $\mathcal{V} \cong \{0,1\}^d$, i.e., $\sum_i c_i \Gamma_i \mapsto (c_1, \ldots, c_d)$. We can define the projection $\rho : \{0,1\}^\ell \to \{0,1\}^d$ by $\rho(x) = (\Gamma_1 \cdot x, \ldots, \Gamma_d \cdot x)$, or equivalently, as the projection $\rho : \{0,1\}^\ell \to \mathcal{V}$ from $\{0,1\}^\ell$ onto the subspace $\mathcal{V}$. We can build ρ' from $\Gamma'_1, \ldots, \Gamma'_d$ similarly. Then, we consider the Markov process M induced by E_k, ρ, ρ'. The distinguishing advantage of this attack is given by $\frac{1}{2}||M - M'||_\infty$, as before, except that now we are working with $2^d \times 2^d$ matrices. Notice that these d-bit projections ρ, ρ' simultaneously capture all 2^{2d} linear approximations of the form $\Gamma \to \Gamma'$ for some $\Gamma \in \mathcal{V}$, $\Gamma' \in \mathcal{V}'$, so this is a fairly powerful attack.

5 Higher-Order Attacks

The next idea is to examine plaintexts two (or more) at a time. If $f : \mathcal{M} \to \mathcal{M}$ is any function, let $\hat{f} : \mathcal{M}^2 \to \mathcal{M}^2$ be defined by $\hat{f}(x, x') = (f(x), f(x'))$. More generally, we can take $\hat{f} : \mathcal{M}^d \to \mathcal{M}^d$ and $\hat{f}(x_1, \ldots, x_d) = (f(x_1), \ldots, f(x_d))$ for any fixed d; this is known as a d-th order attack. Then, to distinguish E_k from π, the idea is to use stochastic commutative diagrams that separate $\hat{E}_k$ from $\hat{\pi}$.

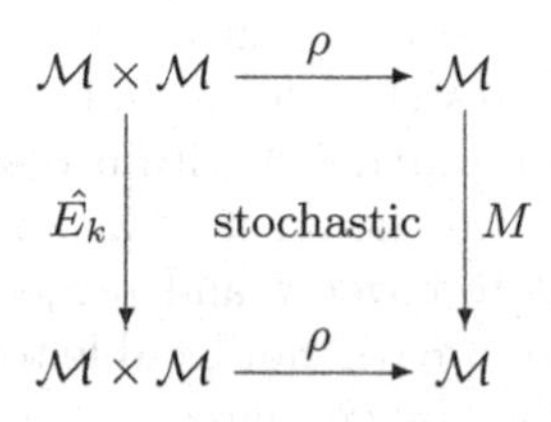

Fig. 9. The basis of differential-style attacks, as a commutative diagram. We use the projection $\rho(x, x') = x - x'$

Complementation Properties. Complementation attacks form a simple instance of a higher-order attack. Suppose there is some Δ so that $E_k(x \oplus \Delta) = E_k(x) \oplus \Delta$. Then we can define $\rho : \mathcal{M} \times \mathcal{M} \to \mathcal{M}$ by $\rho(x, x') = x - x'$, and we obtain the diagram shown in Fig. 9. Note that in this case the existence of the complementation property implies that $M_{\Delta,\Delta} = 1$. If M' denotes the transition matrix induced by an ideal cipher (namely, π), then $M'_{\Delta,j} = 1/(|\mathcal{M}| - 1)$ for each $j \neq 0$, and so $||M - M'||_\infty \geq 1 - 1/(|\mathcal{M}| - 1) + (|\mathcal{M}| - 2)/(|\mathcal{M}| - 1) = 2 - 2/(|\mathcal{M}| - 1)$. In other words, there is an attack using only 2 chosen plaintexts and achieving distinguishing advantage $1 - 1/(|\mathcal{M}| - 1)$.

Differential Cryptanalysis. A natural extension is to generalize the above attack by looking for some matrix element $M_{\Delta,\Delta'}$ with surprisingly large probability, rather than looking for a matrix element with probability 1. Indeed, such a modification yields exactly Biham & Shamir's differential cryptanalysis [2], and any large matrix element $M_{\Delta,\Delta'}$ gives us a differential $\Delta \to \Delta'$ with probability $p = M_{\Delta,\Delta'}$. Notice that when $p \gg 1/(|\mathcal{M}| - 1)$, we have $\frac{1}{2}||M - M'||_\infty \geq p - 1/(|\mathcal{M}| - 1)$, hence with 2 chosen plaintexts we obtain an attack with advantage $\approx p$. One can readily verify that with $2m$ chosen plaintexts, the iterated attack has advantage $1 - (1 - 1/p)^m - (1 - 1/(|\mathcal{M}| - 1))^m$, which is $\approx 1 - 1/e$ for $m = 1/p$. Thus, $2/p$ chosen plaintexts suffice to distinguish with good advantage.

Also, our framework easily models differential cryptanalysis with respect to other groups. For instance, when using additive differentials in the group $\mathcal{M} = (\mathbb{Z}/2^{64}\mathbb{Z}, \bullet\ )$, we can choose the projection $\rho : \mathcal{M} \times \mathcal{M} \to \mathcal{M}$ given by $\rho(x, x') = x - x' \bmod 2^{64}$.

Impossible Differential Cryptanalysis. Alternatively, we could look for matrix elements in M that are surprisingly small. If we look for entries that have probability 0, say $M_{\Delta,\Delta'} = 0$, then we obtain the impossible differential attack. In this case the differential $\Delta \to \Delta'$ will be impossible (it can never happen for the real cipher E_k). This yields an attack that can distinguish with good advantage once about $|\mathcal{M}|$ texts are available to the attacker.

Differential-Linear Cryptanalysis. Write $E_k = f' \circ f$. In a differential-linear attack [5], one covers the first half of the cipher (f) with a differential characteristic and approximates the second half of the cipher (f') with a linear characteristic. This can be modeled within our framework as follows. Our development so far suggests we should use the projections $\rho, \rho' : \mathcal{M}^2 \to \mathcal{M}$, $\rho(x, x') = \rho'(x, x') = x - x'$ to cover f and projections $\eta, \eta' : \mathcal{M} \to \{0, 1\}$ to cover f'. However, in this case, we will not be able to match η up with ρ', because neither their domains nor their ranges agree.

The solution is to introduce functions $\hat{\eta}, \hat{\eta}' : \mathcal{M}^2 \to \{0, 1\}$ given by $\hat{\eta}(x, x') = \eta(x - x')$ and $\hat{\eta}'(x, x') = \eta'(x - x')$. Suppose our differential characteristic has probability $p \gg 1/|\mathcal{M}|$, or in other words, ρ, ρ' commute with f with probability p. Then $\rho, \hat{\eta}$ will usually commute with $\hat{f}$ with non-trivial probability (heuristically, about $\frac{1}{2} + \frac{p}{2}$, though this is not guaranteed). Likewise, suppose our linear characteristic holds with probability $\frac{1}{2} \pm \frac{\epsilon}{2}$, or in other words, η, η' commute with f' with probability $\frac{1}{2} \pm \frac{\epsilon}{2}$. Then $\hat{\eta}, \hat{\eta}'$ will form a linear approximation for $\hat{f}'$ with probability $\frac{1}{2} \pm \frac{\epsilon^2}{2}$. These two properties can be composed to obtain a property $\rho, \hat{\eta}'$ for the whole cipher, typically with probability $\frac{1}{2} \pm \frac{p\epsilon^2}{2}$.

Hence a differential-linear attack with two chosen texts will typically have distinguishing advantage roughly $\frac{1}{2}p\epsilon^2$. Consequently, our framework predicts that such a cipher can be broken with $\Theta(1/p\epsilon^2)$ chosen texts. This corresponds closely to the classical estimate [1].

Higher-Order Differential Cryptanalysis. Higher-order differentials [7] can also be modeled within our framework. Let us give a simple example. If $f(X)$ is a polynomial of degree 2, then $f(X + \Delta_0 + \Delta_1) - f(X + \Delta_0) + f(X + \Delta_1) - f(X)$ is a constant polynomial, giving a way to distinguish f from random with 4 chosen plaintexts. This corresponds to choosing 4th order projections $\rho(w, x, y, z) = (x - w, y - w, z - x - y + w)$ and $\rho'(w, x, y, z) = z - x + y - w$, deriving a transition matrix M, and noticing that we have a matrix entry $M_{(\Delta_0, \Delta_1, 0), \Delta'}$ whose value is 1 for the real cipher but much smaller for a random permutation. More generally, if f is a polynomial of degree d, then the dth order differential of f is a constant, and the $d + 1$-th order differential is zero. Such higher order differential attacks can likewise be expressed be expressed as a higher-order commutative diagram attack.

Truncated Differential Cryptanalysis. Truncated differential attacks [8] also fit within our framework. Given a block cipher $E_k : \{0, 1\}^\ell \to \{0, 1\}^\ell$, we choose projections of the form $\rho : \{0, 1\}^\ell \times \{0, 1\}^\ell \to \{0, 1\}^m$, where $\rho(x, x') = \varphi(x - x')$ and $\varphi : \{0, 1\}^\ell \to \{0, 1\}^m$ is an appropriately chosen linear map.

Then we can look for an entry $M_{\Delta, \Delta'}$ in the transition matrix so obtained that has surprisingly large probability, and this will correspond to a truncated differential $\Delta \to \Delta'$ of the same probability. This truncated differential corresponds to the class of $2^{2\ell - 2m}$ conventional differentials $\delta \to \delta'$, where $\delta \in \varphi^{-1}(\Delta)$ and $\delta' \in \varphi^{-1}(\Delta')$. Alternately, we can look for an entry with surprisingly low probability, and this yields an impossible (or improbable) truncated differential that can be used in an attack.

In most truncated differential attacks, the linear map φ simply ignores part of the block. For instance, the truncated difference $(a, 0, 0, b)$ (where a, b are arbitrary differences) might correspond to the linear map $\varphi(w, x, y, z) = (x, y)$ and the projected value $\Delta = (0, 0)$. However, for maximum generality, we allow φ to be chosen as any linear map whatsoever.

The above account of truncated differentials is slightly naive. It leads to very large matrices, because a truncated difference of the form (say) $(a, 1, 2, b)$ is distinguished from the truncated difference $(a, 1, 3, b)$. However, in practice it is more common for cryptanalysts to care only about distinguishing between zero and non-zero words, with little reason to make any distinction between the different non-zero values.

Fortunately, our treatment can be amended to better incorporate typical cryptanalytic practice, as follows. Consider, as a concrete example, truncated differential attacks on Skipjack, where the block is $\mathcal{M} = \{0, 1\}^{64}$ and where attacks typically look at which of the four 16-bit words of the difference are zero or not. Consider the following 67 vector subspaces of $\{0, 1\}^{64}$: $\{0\}$, $\{(a, 0, 0, 0) : a \in \{0, 1\}^{16}\}$, $\{(0, a, 0, 0) : a \in \{0, 1\}^{16}\}$, ..., $\{(a, a, 0, 0) : a \in \{0, 1\}^{16}\}$, $\{(a, 0, a, 0) : a \in \{0, 1\}^{16}\}$, ..., $\{(a, b, 0, 0) : a, b \in \{0, 1\}^{16}\}$, $\{(a, 0, b, 0) : a, b \in \{0, 1\}^{16}\}$, ..., $\{(a, b, c, d) : a, b, c, d \in \{0, 1\}^{16}\}$. These can be put into one-to-one correspondence with the 67 vector subspaces of $\{0, 1\}^4$ in a natural way. Moreover, to any block $x \in \{0, 1\}^{64}$ we can associate its characteristic vector $(\chi_1, \ldots, \chi_{67})$, where χ_i is 1 if x is in the i-th subspace and 0 otherwise. This induces an equivalence relation $\sim$ on $\{0, 1\}^{64}$, where two blocks are considered equivalent if they have the same characteristic vector. We can now consider the projection $\rho : \{0, 1\}^{64} \to \{0, 1\}^{64}/\sim$ that maps x to its equivalence class under $\sim$. In this way we obtain a 67×67 transition matrix M that captures the probability of all 67^2 word-wise truncated differentials for Skipjack [11]. A similar construction can be used for ciphers of other word and block lengths. This leads to smaller transition matrices and a more satisfying theory of truncated differential cryptanalysis.

Generalized Truncated Differential Cryptanalysis. Armed with these ideas, we can now propose a new attack not previously seen in the literature. We retain the basic set-up from truncated differential attacks, but we replace the probabilistic commutative diagrams with stochastic commutative diagrams. In other words, instead of looking for a single entry in the transition matrix with unusually large (or small) probability, we use the matrix norm $\frac{1}{2}||M - M'||_\infty$. This approach allows us to exploit many small biases spread throughout the matrix M, rather than being confined to only taking advantage of one bias and ignoring the rest.

This may look like a very small tweak; however, it contributes considerable power to the attack. As we shall see, it subsumes all the previous attacks as special cases of generalized truncated differentials. The ability to unify many existing attacks, and to generate new attacks, using such a simple and natural-looking extension to prior work is one of the most striking features of our framework.

Generalized truncated differentials generalize conventional differential and impossible differential attacks. If there is a single entry in M with unusually large (or small) probability, then the matrix norm $\frac{1}{2}||M - M'||_\infty$ will also be large. Hence, the existence of a conventional differential or impossible differential attack with advantage ϵ implies the existence of a generalized truncated attack with the same advantage ϵ.

Likewise, linear cryptanalysis can also be viewed as a special case of generalized truncated differentials. As we argued before, if $\eta, \eta' : \{0,1\}^\ell \to \{0,1\}$ form a linear characteristic with probability $\frac{1}{2} \pm \frac{\epsilon}{2}$, then $\hat{\eta}, \hat{\eta}'$ given by $\hat{\eta}(x, x') = \eta(x - x')$, etc., has probability $\frac{1}{2} \pm \frac{\epsilon^2}{2}$. It may appear that we have diluted the power of the attack, because the distinguishing advantage has decreased from $\epsilon/2$ (for one known text in a linear attack) to $\epsilon^2/2$ (for one pair of texts in a generalized truncated attack). However, this is offset by an increase in the number of pairs of texts available in a generalized truncated attack: given a pool of n known texts, one can form n^2 pairs of texts. These two factors turn out to counterbalance each other. If k out of n texts follow η, η', then $k^2 + (n-k)^2$ out of n^2 pairs follow $\hat{\eta}, \hat{\eta}'$. Note that $h(k) = k^2 + (n-k)^2$ is a strictly increasing function of k, for $k \geq n/2$. Hence for any threshold T in a linear attack, there is a corresponding threshold $h(T)$ that makes the generalized truncated differential attack work with the same number of known texts and roughly the same distinguishing advantage. Consequently, the existence of a linear attack implies the existence of a generalized truncated attack with about the same advantage.

Differential-linear attacks are also subsumed by generalized truncated differentials. Because both a differential and a linear characteristic can be viewed as a generalized truncated differential, they can be concatenated. In a differential-linear attack, the transition is abrupt and binary, but generalized truncated attacks allow to consider other attacks, for instance with a gradual transition between differential- and linear-style analysis, or with hybrids partway between differential and linear attacks.

Intuitively, the power of generalized truncated differential attacks comes from the extra degrees of freedom available to the cryptanalyst. In a differential attack, the cryptanalyst can freely choose which matrix entry $M_{\Delta,\Delta'}$ to focus on, but has little control over ρ, ρ'. In a linear attack, the cryptanalyst can freely choose ρ, ρ' in some clever way, but has no choice over the matrix M. A generalized truncated differential attack allows the cryptanalyst to control both aspects of the attack at the same time.

6 Algebraic Attacks

One noticeable trend over the past few decades is that more and more block cipher designs have come to incorporate algebraic structure. For instance, the AES S-box is based on inversion in the field $GF(2^8)$. Yet this brings an opportunity

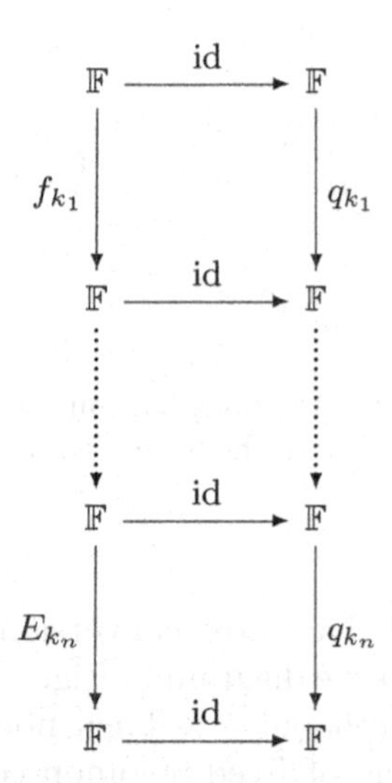

Fig. 10. The basis of an interpolation attack. Here each $q_{k_i}(X)$ is a polynomial over the field $\mathbb{F}$

for attacks that exploit this structure, and a rich variety of algebraic cryptanalytic methods have been devised: interpolation attacks, rational interpolation, probabilistic interpolation, and so on.

Interpolation Attacks. The basic interpolation attack [7] is easy to understand. We express each round f_{k_i} as a polynomial $q_{k_i}(X) \in \mathbb{F}[X]$ over some field $\mathbb{F}$, and in this way we obtain the commutative diagram shown in Fig. 10. Notice that composition of commutative diagrams allows to express the whole cipher as a polynomial $q_k(X) = q_{k_n}(\cdots(q_{k_2}(q_{k_1}(X)))\cdots)$. The polynomial $q_k(X)$ may depend on the key k in some possibly complex way, hence the attacker usually will not know $q_k(X)$ a priori. Consequently, a distinguishing attack based on this property must work a little differently.

The standard interpolation attack exploits the fact that $d+1$ points suffice to uniquely determine a polynomial of degree d. Given $\deg q_k(X) + 1$ known plaintext/ciphertext pairs for E_k, we can reconstruct the polynomial $q_k(X)$ using Lagrange interpolation (for instance), and then check one or two additional known texts for consistency with the recovered polynomial. This allows to distinguish any cipher with this property from a random permutation.

The idea can be generalized in many ways. We need not restrict ourself to univariate polynomials; we can generalize to multivariate polynomials $\mathbf{q}(\mathbf{X})$, where $\mathbf{X} = (X_1, \ldots, X_m)$ represents a vector of m unknowns, and where $\mathbf{q}(\mathbf{X}) = (q_1(\mathbf{X}), \ldots, q_m(\mathbf{X}))$ represents a vector of m multivariate polynomials. In this case, the number of texts needed corresponds to the number of coefficients of q_k not known to be zero.

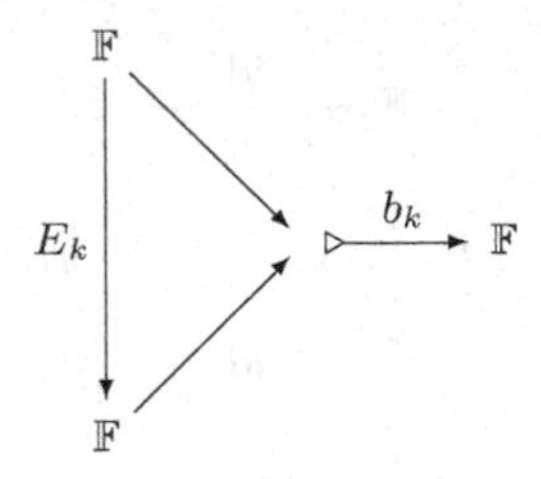

Fig. 11. The basis of a bivariate interpolation attack. Here $b_k(X, Y)$ represents a bivariate polynomial over $\mathbb{F}$, and the intended interpretation of the diagram is that $b_k(x, E_k(x)) = 0$ for all $x \in \mathbb{F}$

Also, one can naturally derive statistical versions of interpolation cryptanalysis by replacing the commutative diagram in Fig. 10 with a probabilistic commutative diagram with some probability p. Then noisy polynomial reconstruction techniques (e.g., list decoding of Reed-Solomon codes) will allow us to mount a distinguishing attack.

We can also use meet-in-the-middle techniques, using the polynomial q_k to cover the first half of the cipher and q'_k to cover the remaining rounds. Notice how these generalizations come naturally under the commutative diagram framework.

Rational Interpolation Attacks. Another generalization is the notion of rational interpolation attacks. If $q(X), q'(X)$ are any two polynomials with no common factor and where $q'(X)$ is not the zero polynomial, then $r(X) = q(X)/q'(X)$ is called a *rational polynomial.* It is then natural to consider the variant on the commutative diagram in Fig. 10 where each polynomial q_{k_i} is replaced by a rational polynomial r_{k_i}.

Note that rational polynomials are closed under composition, hence we can derive a global approximation for the whole cipher from rational approximations of the individual round functions. If we can express the cipher as a rational polynomial $E_k(x) = q_k(x)/q'_k(x)$, and if we have a supply of known plaintext/ciphertext pairs (x_i, y_i), then we obtain the equations $y_i \cdot q'_k(x_i) = q_k(x_i)$. Linear algebra reveals the rational polynomial $q_k(X)/q'_k(X)$, which gives us a distinguishing attack on the cipher.

Bivariate Interpolation. The notion of interpolation and rational interpolation can be generalized to obtain what might be called a *bivariate interpolation attack.* The idea is to seek a family of bivariate polynomials $b_k(X, Y) \in \mathbb{F}[X, Y]$ so that $b_k(x, f_k(x)) = 0$ for all $x \in \mathcal{M}$ and all $k \in \mathcal{K}$. This gives a local property of the round function f.

Local bivariate properties can be composed to obtain a global property for the whole cipher. Suppose the first round satisfies a bivariate relation $b(x, f(x)) = 0$

and the second round satisfies $b'(y, f'(y)) = 0$. Define

$$b''(X, Z) = \text{Res}_Y(b(X, Y), b'(Y, Z)).$$

Then the composition of the two rounds will satisfy the relation $b''(x, f'(f(x))) = 0$ for all $x \in \mathbb{F}$. The latter follows from a property of the resultant: given $f(Y), g(Y) \in \mathcal{R}[Y]$, the resultant $\text{Res}_Y(f(Y), g(Y))$ is a value in $\mathcal{R}$, and if f, g share a common root over $\mathcal{R}$, then the resultant will be zero. Letting $\mathcal{R} = \mathbb{F}[X, Z]$, $f(Y) = b(X, Y)$, and $g(Y) = b'(Y, Z)$ verifies the claimed result about $b''(X, Z)$.

In this way, we can compose bivariate relations for each round to obtain a bivariate relation for the whole cipher. Once we have a bivariate relation $b_k(x, E_k(x)) = 0$ for the whole cipher, we can use polynomial interpolation to reconstruct b_k given a sufficient quantity of known plaintext/ciphertext pairs. Unfortunately, in general the degree of the bivariate polynomial for the whole cipher can grow rapidly as the number of rounds increases.

Notice that interpolation attacks fall out as a special case of bivariate interpolation. If the first and second rounds of the cipher can be expressed as polynomials $q(X), q'(Y)$, this induces bivariate relations $b(X, Y) = q(X) - Y$ and $b'(Y, Z) = q'(Y) - Z$. Taking the resultant yields

$$b''(X, Z) = \text{Res}_Y(b(X, Y), b(Y, Z)) = \text{Res}_Y(q(X) - Y, q'(Y) - Z) = q'(q(X)) - Z,$$

which is nothing more than a round-about derivation of the obvious fact that the composition of first and second rounds may be expressed by the polynomial $q'(q(X))$.

Likewise, rational interpolation attacks are a special case of bivariate interpolation. Suppose the first and second rounds can be expressed as rational polynomials $p(X)/p'(X)$ and $q(Y)/q'(Y)$. We obtain the bivariate relations $b(X, Y) = p'(X) \cdot Y - p(X)$ and $b'(Y, Z) = q'(Y) \cdot Z - q(Y)$. Taking the resultant yields a bivariate relation for the composition of the first two rounds.

Probabilistic bivariate attacks have actually been suggested before by Jakobsen [6] and applied by others to DES [12], but it was not previously explained how to compose local approximations to obtain global approximations, nor was it noticed that bivariate attacks generalize and unify interpolation and rational interpolation.

7 Discussion

Closure Properties. The common theme here seems to be that closure properties enable cryptanalysis. For instance, differential and linear attacks exploit the fact that the set of linear functions is closed under composition: $g \circ f$ is linear if f, g are. Likewise for the set of polynomials, of rational polynomials, and so on. More generally, we may form a norm $\mathcal{N}(\cdot)$ on functions that grows slowly under composition and that corresponds somehow to the cost of an attack. Consider interpolation attacks: letting $\mathcal{N}(q) = \deg q$ for polynomials $q(X)$, we find

$\mathcal{N}(g \circ f) \leq \mathcal{N}(f) \times \mathcal{N}(g)$, hence if we can find low-degree properties for individual round functions, the corresponding global property for the whole cipher will have not-too-large degree. Perhaps other ways to place a metric space structure on the set of bijective functions $f : \mathcal{M} \to \mathcal{M}$ will lead to other cryptanalytic advances in the future.

Related Work. Commutative diagram cryptanalysis draws heavily on ideas found in previous frameworks, most notably Vaudenay's chi-squared cryptanalysis [13] and Harpes' partitioning cryptanalysis [4]. Vaudenay's work used linear projections $\rho, \rho' : \{0,1\}^\ell \to \{0,1\}^m$, and then applied the χ^2 statistical test to the pair $(\rho(X), \rho'(E_k(X))$. It turns out that the power of the χ^2 test is closely related to the ℓ_2 norm, $||M - M'||_2$, hence chi-squared cryptanalysis can be viewed as a variant of stochastic commutative diagrams where a different matrix norm is used. Partitioning cryptanalysis generalized this to allow arbitrary (not necessarily linear) projections ρ, ρ'.

We borrowed methods from Vaudenay's decorrelation theory [14] to calculate the distinguishing advantage of our statistical attacks. Also, Vaudenay shows how to build ciphers with provable resistance against all non-adaptive d-limited attacks, which corresponds to security against dth order commutative diagram attacks.

This work builds on an enormous quantity of work in the block cipher literature; it is a synthesis of many ideas that have previously appeared elsewhere. Due to space limitations, we have been forced to omit mention of a great deal of relevant prior work, and we apologize for all omissions.

8 Conclusion

We have introduced commutative diagram cryptanalysis and shown how it provides a new perspective on many prior attacks in the block cipher literature. We also described two new attack methods, generalized truncated differential cryptanalysis and bivariate interpolation, and demonstrated how they generalize and unify many previous attacks. It is an interesting open problem to extend this framework to incorporate more attacks, to discover more new attacks, or to build fast ciphers that are provably secure against commutative diagram cryptanalysis.

References

[1] E. Biham, O. Dunkelman, N. Keller, "Enhancing Differential-Linear Cryptanalysis," *ASIACRYPT 2002*, Springer-Verlag, LNCS 2501, pp.254–266. 26

[2] E. Biham, A. Shamir, *Differential Cryptanalysis of the Data Encryption Standard*, Springer-Verlag, 1993. 25

[3] D. Coppersmith, S. Halevi, C. Jutla, "Cryptanalysis of Stream Ciphers with Linear Masking," *CRYPTO 2002*, Springer-Verlag, LNCS 2442, pp.515–532. 18

[4] C. Harpes, J.L. Massey, "Partitioning Cryptanalysis," *FSE '97*, Springer-Verlag, LNCS 1267, pp.13–27. 32

[5] M. E. Hellman, S. K. Langford, "Differential-linear cryptanalysis," *CRYPTO '94*, Springer-Verlag, LNCS 839, pp.26–39. 26

[6] T. Jakobsen, "Cryptanalysis of Block Ciphers with Probabilistic Non-Linear Relations of Low Degree," *CRYPTO '98*, Springer-Verlag, LNCS 1462, pp.212–222. 31

[7] T. Jakobsen, L. R. Knudsen, "Attacks on Block Ciphers of Low Algebraic Degree," *J. Cryptology*, 14(3):197–210, 2001. 26, 29

[8] L. Knudsen, "Truncated and Higher Order Differentials," *FSE '94*, Springer-Verlag, LNCS 1008, pp.196–211. 26

[9] X. Lai, J. Massey, S. Murphy, "Markov Ciphers and Differential Cryptanalysis," *EUROCRYPT '91*, Springer-Verlag, LNCS 547, pp.17–38. 22

[10] M. Matsui, "Linear Cryptanalysis Method for DES Cipher," *EUROCRYPT '93*, Springer-Verlag, LNCS 765, pp.386–397. 23

[11] B. Reichardt, D. Wagner, "Markov truncated differential cryptanalysis of Skipjack," *SAC 2002*, Springer-Verlag, LNCS 2595, pp.110–128. 27

[12] T. Shimoyama, T. Kaneko, "Quadratic Relation of S-box and Its Application to the Linear Attack of Full Round DES," *CRYPTO '98*, Springer-Verlag, LNCS 1462, pp.200–211. 31

[13] S. Vaudenay, "An Experiment on DES: Statistical Cryptanalysis," *ACM CCS '96*, ACM Press, pp.139–147. 32

[14] S. Vaudenay, "Decorrelation: A Theory for Block Cipher Security," *J. Cryptology*, Springer-Verlag, 16(4):249–286, Sept. 2003. 18, 32

Algebraic Attacks on Summation Generators

Dong Hoon Lee, Jaeheon Kim, Jin Hong, Jae Woo Han, and Dukjae Moon

National Security Research Institute
161 Gajeong-dong, Yuseong-gu, Daejeon, 305-350, Korea
{dlee,jaeheon,jinhong,jwhan,djmoon}@etri.re.kr

Abstract. We apply the algebraic attacks on stream ciphers with memories to the summation generator. For a summation generator that uses n LFSRs, an algebraic equation relating the key stream bits and LFSR output bits can be made to be of degree less than or equal to $2^{\lceil \log_2 n \rceil}$, using $\lceil \log_2 n \rceil + 1$ consecutive key stream bits. This is much lower than the upper bound given by previous general results. We also show that the techniques of [6, 2] can be applied to summation generators using 2^k LFSRs to reduce the effective degree of the algebraic equation.

Keywords: stream ciphers, algebraic attacks, summation generators

1 Introduction

Among recent developments on stream ciphers, the algebraic attack has gathered much attention. In this attack, an algebraic equation relating the initial key bits and the output key stream bits is set up and solved through linearization techniques.

Algebraic attack was first applied to block ciphers and public key cryptosystems [8, 9, 3]. And its first successful application to stream cipher was done on Toyocrypt [4]. As the method was soon extended to LILI-128 [7], it gathered much attention. Stream ciphers that utilize memory were first thought to be much more resistant to these attacks, but soon it was shown that even these cases were subject to algebraic attacks [1, 5].

The summation generator proposed by Ruepel [12] is a nonlinear combiner with memory. It is known that the generator produces sequences whose period and correlation immunity are maximum, and whose linear complexity is conjectured to be close to the period. Hence it serves as a good building block for stream ciphers.

However, a correlation attack on the summation generator that uses two linear feedback shift registers (LFSR) was presented in [11], even though it is also stated in [11] that this attack is not plausible if there are more than two LFSRs in use. Another well known attack on summation generator is given by [10] and points in the opposite direction. It uses feedback carry shift registers (FCSR) to simulate the summation generator and indicates that for a fixed initial key size, breaking them into too many LFSRs will add to its weakness.

B. Roy and W. Meier (Eds.): FSE 2004, LNCS 3017, pp. 34–48, 2004.

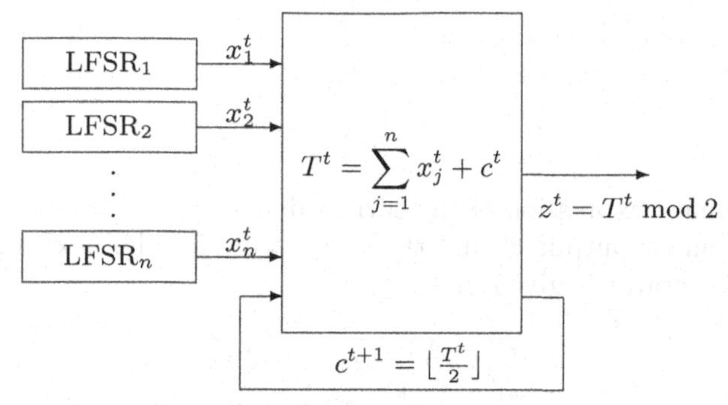

Fig. 1. Structure of summation generators

In this work, we study the summation generator with the algebraic attack in mind. We show that for a summation generator that uses n LFSRs, an algebraic equation relating the key stream bits and LFSR output bits can be made to be of degree less than or equal to $2^{\lceil \log_2 n \rceil}$, using $\lceil \log_2 n \rceil + 1$ consecutive key stream bits. This is much lower than the upper bound on the degree of algebraic equations that is guaranteed by the general works [1, 5]. We also show that the techniques of [6, 2] can be applied to summation generators using 2^k LFSRs to reduce the effective degree of the equation further.

The reader may confer to Tables 3 and 4, appearing in later sections, for a quick look on the improvements we have given to understanding the actual strength of algebraic attacks on summation generators. Actually, Table 3 should be a good reference in view of algebraic attacks for anyone considering the use of a summation generator.

The rest of this paper is organized as follows: We recall the definition of a summation generator and introduce elementary symmetric boolean functions in Section 2. We derive an algebraic equation that is satisfied by the summation generator on 4 LFSRs in Section 4. Induction is used in extending this to the summation generator on 2^k LFSRs in the following section. Section 5 deals with the general n case. We show that the degree of the equation can be reduced further in the case $n = 2^k$ using the technique of [6] in Section 6, and present our conclusion in Section 7.

2 Preliminaries

2.1 Summation Generators

The summation generator proposed by Rueppel [12] is a nonlinear combiner with memory. We consider a summation generator that uses n binary LFSRs (Figure 1). The output of the j-th LFSR at time t is denoted by $x_j^t \in \{0, 1\}$. Since we are dealing with binary values, we will not be using powers of these terms, hence the superscript t should not cause any confusion.

The current *carry* value from the previous stage is denoted by c^t. Note that the carry can be expressed in $k = \lceil \log_2 n \rceil$ bits. We shall let

$$c^t = (c^t_{k-1}, \ldots, c^t_1, c^t_0)$$

be the binary expression of the carry value.

The binary output z^t and the carry value for the next stage from the summation generator is given by

$$z^t = x^t_1 \oplus x^t_2 \oplus \cdots \oplus x^t_n \oplus c^t_0, \tag{1}$$

$$c^{t+1} = \lfloor (x^t_1 + x^t_2 + \cdots + x^t_n + c^t)/2 \rfloor. \tag{2}$$

2.2 Symmetric Polynomials

Let us denote by S^t_i, the i-th elementary symmetric polynomial in the variables $\{x^t_1, \ldots, x^t_n\}$. We view them as boolean functions rather than as polynomials. Explicitly, they are

$$\begin{aligned} S^t_0 &= 1, \\ S^t_1 &= \oplus^n_{j=1} x^t_j, \\ S^t_2 &= \oplus_{1 \le j_1 < j_2 \le n} x^t_{j_1} x^t_{j_2}, \\ &\vdots \\ S^t_n &= x^t_1 x^t_2 \cdots x^t_n. \end{aligned}$$

We shall also call these by the name *elementary symmetric boolean functions.*

For any fixed $0 \le b \le n$, consider the condition $\sum_j x^t_j = b$. This condition states that b of the n variable x^t_j are equal to 1 and that the others are equal to 0. Since S^t_a is symmetric for any $0 \le a \le n$, it makes sense to evaluate S^t_a under this condition. Let $m_{a,b}$ denote this value. It is also clear that the values $(m_{a,b})^n_{b=0}$ completely determines the value of S^t_a at an arbitrary input.

To calculate $m_{a,b}$, one has only to count how many of the monomials contained in S^t_a is nonzero. Hence we have

$$m_{a,b} \equiv \binom{b}{a} \pmod 2. \tag{3}$$

Now, consider the $(n+1) \times (n+1)$ matrix $M = (m_{a,b})$. The $n = 4$ case is given in Table 1, as an example. Denote by M', the $n \times n$ matrix obtained by removing the last row and column from M. When we want to make explicit the number of variable used in defining M and M', we shall write $M(n)$ and $M'(n)$.

Lemma 1. *For powers of* 2, *the matrix* M' *satisfies*

$$M'(2^{k+1}) = \begin{pmatrix} M'(2^k) & M'(2^k) \\ 0 & M'(2^k) \end{pmatrix}.$$

Table 1. The matrix M for $n = 4$

$\sum x_j^t$	0	1	2	3	4
S_0^t	1	1	1	1	1
S_1^t	0	1	0	1	0
S_2^t	0	0	1	1	0
S_3^t	0	0	0	1	0
S_4^t	0	0	0	0	1

Proof. Let us first divide M' into four parts and write

$$M' = \begin{pmatrix} \text{II} & \text{I} \\ \text{III} & \text{IV} \end{pmatrix}.$$

It is clear from (3) that the second quadrant, part II, is $M'(2^k)$, as claimed. The equation also shows that the lower triangular part of M' is zero. Hence the third quadrant is filled with zero, as claimed.

We now show that I, II, and IV are identical. Once more, referring to (3), it suffices to show

$$\binom{b}{a} \equiv \binom{2^k+b}{a} \equiv \binom{2^k+b}{2^k+a} \pmod 2 \tag{4}$$

for all $0 \le a, b < 2^k$. To this end, we may easily check that

$$\begin{aligned}(1+x)^{2^k+b} &= (1+x)^{2^k}(1+x)^b \\ &\equiv (1+x^{2^k})(1+x)^b \pmod 2 \\ &= (1+x)^b + x^{2^k}(1+x)^b.\end{aligned}$$

The coefficient of x^a that would appear in the expansion of the left hand side $(1+x)^{2^k+b}$ is the middle term of (4). Since $a < 2^k$, the x^a term in the right hand side may appear only in $(1+x)^b$ and is equal to the first term of (4). This shows the first equality. Similarly, comparison of the coefficients of x^{2^k+a} shows the equality between the first and the last terms of (4). This completes the proof.

This lemma allows one to write the matrix $M'(2^k)$ explicitly for any given k. Then, owing to (3), the matrix $M(n)$ for any n may be obtained as a submatrix of a big enough $M'(2^k)$. For example, Table 1 is a submatrix of $M'(8)$.

The basic theory on symmetric polynomials tells us that the set consisting of products of elementary symmetric *polynomials* forms a basis for the space of symmetric polynomials. In case of symmetric *boolean functions*, the following lemma can easily be seen to be true.

Lemma 2. *Any boolean function that is symmetric in its variables, may be written as a linear combination of the elementary symmetric boolean functions.*

Table 2. The next carry bits in relation to the current carry value

$c^t = (c_1^t, c_0^t)$	(0,0)	(0,1)	(1,0)	(1,1)
c_0^{t+1}	S_2^t	$S_1^t \oplus S_2^t$	$S_0^t \oplus S_2^t$	$S_0^t \oplus S_1^t \oplus S_2^t$
c_1^{t+1}	S_4^t	$S_3^t \oplus S_4^t$	$S_2^t \oplus S_4^t$	$S_1^t \oplus S_2^t \oplus S_3^t \oplus S_4^t$

3 Summation Generator on 4 LFSRs

We fix $n = 4$ throughout this section and derive an algebraic equation of degree 4 that is satisfied by the summation generator. The variables of the final equation will consist of the LFSR output bits and the key stream bits but will not contain any carry bits.

Since we are dealing with $4 = 2^2$ LFSRs, we have $k = 2$, and it suffices to use 2 bits in expressing the carry value. It is clear that the carry value should be symmetric with respect to the order of the LFSRs in use. Recalling Lemma 2, this implies that, when the current carry value c^t is fixed, the carry bits c_0^{t+1} and c_1^{t+1} may be expressed as linear combinations of the elementary symmetric boolean functions S_i^t. In the general case when the current carry bits are not fixed, they will be linear combinations of the S_i^t with boolean functions of the carry bits c_0^t and c_1^t used as coefficients.

For each value of c^t and LFSR inputs possible, we explicitly calculated c_0^{t+1} and c_1^{t+1}. We then used Table 1 to expressed them using the elementary symmetric boolean functions. The result is given in Table 2.

Lemma 3. *For a summation generator on* 4 *LFSRs, the following expresses the next stage carry bits as functions of the current LFSR output bits and current carry bits.*

$$c_0^{t+1} = S_2^t \oplus c_0^t S_1^t \oplus c_1^t \tag{5}$$

$$c_1^{t+1} = S_4^t \oplus c_0^t S_3^t \oplus c_1^t S_2^t \oplus c_0^t c_1^t S_1^t. \tag{6}$$

Proof. Using Table 2, we may write

$$\begin{aligned} c_0^{t+1} = &\{(1 \oplus c_0^t)(1 \oplus c_1^t)S_2^t\} \oplus \{c_0^t(1 \oplus c_1^t)(S_1^t \oplus S_2^t)\} \\ &\oplus \{(1 \oplus c_0^t)c_1^t(S_0^t \oplus S_2^t)\} \oplus \{c_0^t c_1^t(S_0^t \oplus S_1^t \oplus S_2^t)\} \end{aligned}$$

and

$$\begin{aligned} c_1^{t+1} = &\{(1 \oplus c_0^t)(1 \oplus c_1^t)S_4^t\} \oplus \{c_0^t(1 \oplus c_1^t)(S_3^t \oplus S_4^t)\} \\ &\oplus \{(1 \oplus c_0^t)c_1^t(S_2^t \oplus S_4^t)\} \oplus \{c_0^t c_1^t(S_1^t \oplus S_2^t \oplus S_3^t \oplus S_4^t)\}. \end{aligned}$$

Simplification of these equations gives the claimed statements.

Lemma 4. *For $n = 4$, the following expresses the carry bits of the summation generator as polynomials in the LFSR output bits and the key stream bits.*

$$c_0^t = S_1^t \oplus z^t. \tag{7}$$

$$c_1^t = S_2^t \oplus (1 \oplus z^t) S_1^t \oplus S_1^{t+1} \oplus z^{t+1}. \tag{8}$$

The first carry bit c_0^t is linear and the second carry bit c_1^t is of degree 2 in the LFSR output bits.

Proof. The first equation follows immediately from (1). It is a linear function on the variables x_j^t. Substituting this and its shift c_0^{t+1} into (5) gives

$$c_1^t = S_2^t \oplus (S_1^t \oplus z^t) S_1^t \oplus S_1^{t+1} \oplus z^{t+1}.$$

Now, since we are dealing with boolean functions, we have $(S_1^t)^2 = S_1^t$ and the second equation follows.

Finally, with this lemma, we may remove all occurrence of the carry bits from (6) to obtain the following proposition.

Proposition 1. *The following algebraic equation holds true for a summation generator on 4 LFSRs.*

$$0 = \begin{array}{l} S_4^t \oplus (1 \oplus z^t) S_3^t \oplus S_2^t S_1^{t+1} \\ \oplus (1 \oplus z^{t+1}) S_2^t \oplus S_2^{t+1} \oplus (1 \oplus z^t) S_1^t S_1^{t+1} \\ \oplus (1 \oplus z^t)(1 \oplus z^{t+1}) S_1^t \oplus (1 \oplus z^{t+1}) S_1^{t+1} \oplus S_1^{t+2} \\ \oplus z^{t+2}. \end{array}$$

It is of degree 4 in the output bits of the LFSRs and uses 3 consecutive key stream bits.

While simplifying the substitution of (7), (8), and shift of (8) into (6), we have used the equalities

$$S_1^t S_2^t = S_3^t \quad \text{and} \quad S_1^t S_3^t = S_3^t$$

of boolean functions.

4 Summation Generator on $n = 2^k$ LFSRs

Let us denote by $F_i^{t+1}(n)$, the (symmetric) polynomial that expresses the next stage carry bit c_i^{t+1} in terms of LFSR outputs x_j^t and current carry bits c_j^t. Since we shall be dealing with polynomials on different number of variables, we shall write $S_j^t(n)$ to denote the elementary symmetric boolean function on n variables. As an example, we saw in the previous section that

$$F_0^{t+1}(2^2) = S_2^t(2^2) \oplus c_0^t S_1^t(2^2) \oplus c_1^t, \tag{9}$$

$$F_1^{t+1}(2^2) = S_4^t(2^2) \oplus c_0^t S_3^t(2^2) \oplus c_1^t S_2^t(2^2) \oplus c_0^t c_1^t S_1^t(2^2). \tag{10}$$

Let us suppose that for some $n = 2^k$, the polynomials $F_i^{t+1}(2^k)$ are given by

$$F_i^{t+1}(2^k) = \oplus_j f_{i,j} S_j^t(2^k). \tag{11}$$

Here, each coefficient $f_{i,j}$ is a boolean function defined on the current carry bits $c_0^t, c_1^t, \ldots, c_{k-1}^t$. For example, we see from (10) that

$$f_{1,4} = 1, \quad f_{1,3} = c_0^t, \quad f_{1,2} = c_1^t, \quad f_{1,1} = c_0^t c_1^t, \quad f_{1,0} = 0,$$

for $k = 2$. The following proposition will allow us to inductively calculate all $F_i^{t+1}(2^k)$ for any i and k.

Proposition 2. *Suppose equation* (11) *holds for some* k*. Then we have*

$$F_i^{t+1}(2^{k+1}) = \oplus_j f_{i,j} S_j^t(2^{k+1}), \quad \text{for } i < k-1, \tag{12}$$

$$F_{k-1}^{t+1}(2^{k+1}) = \big(\oplus_j f_{k-1,j} S_j^t(2^{k+1}) \big) \oplus c_k^t, \tag{13}$$

$$F_k^{t+1}(2^{k+1}) = \big(\oplus_j f_{k-1,j} S_{j+2^k}^t(2^{k+1}) \big) \oplus c_k^t \big(\oplus_j f_{k-1,j} S_j^t(2^{k+1}) \big). \tag{14}$$

Proof. See the appendices.

As an immediate application of this proposition to (9) and (10), we may write

$$F_0^{t+1}(2^3) = S_2^t(2^3) \oplus c_0^t S_1^t(2^3) \oplus c_1^t, \tag{15}$$

$$F_1^{t+1}(2^3) = S_4^t(2^3) \oplus c_0^t S_3^t(2^3) \oplus c_1^t S_2^t(2^3) \oplus c_0^t c_1^t S_1^t(2^3) \oplus c_2^t, \tag{16}$$

$$\begin{aligned} F_2^{t+1}(2^3) = {} & S_8^t(2^3) \oplus c_0^t S_7^t(2^3) \oplus c_1^t S_6^t(2^3) \oplus c_0^t c_1^t S_5^t(2^3) \\ & \oplus c_2^t S_4^t(2^3) \oplus c_0^t c_2^t S_3^t(2^3) \oplus c_1^t c_2^t S_2^t(2^3) \oplus c_0^t c_1^t c_2^t S_1^t(2^3). \end{aligned} \tag{17}$$

Now, let us briefly recall the process we went through in Section 3 in obtaining the degree 4 equation of Proposition 1. We started out with three equations.

$$z^t = S_1^t \oplus c_0^t. \tag{18}$$

$$c_0^{t+1} = S_2^t \oplus c_0^t S_1^t \oplus c_1^t. \tag{19}$$

$$c_1^{t+1} = S_4^t \oplus c_0^t S_3^t \oplus c_1^t (S_2^t \oplus c_0^t S_1^t). \tag{20}$$

We used (18) to write c_0^t as a degree 1 equation that involves just the key stream bit and the LFSR outputs. This was substituted in the next equation (19) to write c_1^t as a degree 2 equation of the same kind. Finally, these expressions for c_0^t and c_1^t were substituted in the last equation to obtain the degree 4 equation that involves only key stream bits and LFSR output bits.

What would happen if we wanted to do the same for $n = 2^3$. We would start with the following set of equations.

$$z^t = S_1^t \oplus c_0^t. \tag{21}$$

$$c_0^{t+1} = S_2^t \oplus c_0^t S_1^t \oplus c_1^t. \tag{22}$$

$$c_1^{t+1} = S_4^t \oplus c_0^t S_3^t \oplus c_1^t S_2^t \oplus c_0^t c_1^t S_1^t \oplus c_2^t. \tag{23}$$

$$\begin{aligned} c_2^{t+1} = {} & S_8^t \oplus c_0^t S_7^t \oplus c_1^t S_6^t \oplus c_0^t c_1^t S_5^t \\ & \oplus c_2^t (S_4^t \oplus c_0^t S_3^t \oplus c_1^t S_2^t \oplus c_0^t c_1^t S_1^t). \end{aligned} \tag{24}$$

Notice that the first two equations here are identical to (18) and (19), as stated by (12) of Proposition 2. Hence, as before, c_0^t and c_1^t will be written as degree 1 and 2 polynomials. The main part of (23) is identical to (20), as stated by (13). They only differ in that c_2^t appears at the end of (23). Since (20) is of degree 4, equation (23) gives a degree 4 expression for c_2^t. Now, as given by (14), the right hand side of (24) may be broken into two big terms of degree (less than or equal to) 8. The first term is a degree 4 equation shifted by degree 4 and the second term is a product of two degree 4 equations. Finally, the left hand side of (24) is of degree 4. Hence, substitution of c_0^t, c_1^t and c_2^t into (24) gives a degree 8 polynomial connecting various LFSR output bits and key stream bits.

One can easily see that the above argument is general enough to be seen as the induction step needed in proving the following theorem.

Theorem 1. *Consider a summation generator on $n = 2^k$ LFSRs. There exists an algebraic equation connecting LFSR output bits and $k+1$ consecutive key stream bits in such a way that it is of degree 2^k in the LFSR output bits.*

5 The General Case

The following is an easy corollary to Theorem 1.

Theorem 2. *Consider a summation generator of n LFSRs. We shall let $k = \lceil \log_2 n \rceil$. There exists an algebraic equation connecting LFSR output bits and $k+1$ consecutive key stream bits in such a way that it is of degree less than or equal to 2^k in the LFSR output bits.*

Proof. We may model a summation generator on n LFSRs as a summation generator on 2^k LFSRs with $(2^k - n)$ of the LFSRs set to zero. Hence, our claim follows from Theorem 1.

For small n's, we explicitly calculated the algebraic equations. Table 3 compares the upper bounds on the degree of the algebraic equation claimed by various methods.

We believe this table is big enough to cover any practically usable summation generator and should serve as a good reference for anyone implementing a summation generator and considering its immunity to algebraic attacks.

Table 3. Degree bounds on algebraic equations for summation generators

n	2	3	4	5	6	7	8	9	10	11	12	13	14	15	16
[1, 5]	2	5	6	10	12	14	16	23	25	28	30	33	35	38	40
Thm 2	2	4	4	8	8	8	8	16	16	16	16	16	16	16	16
explicit calc.	2	3	4	6	6	7	8	12	12	13	14	14	14	15	16

6 Reducing the Degree Further for the $n = 2^k$ Case

For the case when $n = 2^k$, we may reduce the degree of the algebraic equation a little bit further, assuming that we have access to consecutive key stream bits. We will apply the fast algebraic attack, which is introduced by Courtois [6] and is improved by Armknecht [2].

Following the notation of [3, 6], we classify multivariate equations that relate key bits k_i and output bits z_j into types given by their degrees of k_i and z_j. We say that a polynomial is of type $k^d z^f$ if all of its monomial terms are of the form $k_{i_1} \cdots k_{i_d} z_{j_1} \cdots z_{j_f}$. Capital letters will be used in a similar manner to denote types of equations that may contain lower degree monomials also. For example, $K^2 = k^2 \cup k \cup 1$ and $KZ = kz \cup k \cup z \cup 1$.

In [6], *double-decker equations* (DDE) of degree (d, e, f) are defined to be any multivariate equation of type $K^d \cup K^e Z^f$. Cases where $d > e$ are of interest from the attacker's point of view. The following proposition states that the equations describing the summation generator on 2^k input LFSRs, given by Theorem 1, are DDEs.

Theorem 3. *A double-decker equation of degree $(2^k, 2^k - 1, 2^{k-1})$ that relates initial key bits and output stream bits exists for the summation generator on 2^k LFSRs.*

Proof. Let us, once more, recall the $n = 2^2$ case. The following is a very simple illustration of the process we went through in obtaining the degree 4 equation.

$$\begin{aligned}
(18) \quad &\Rightarrow \quad Z^1 = K^1 \oplus c_0^t \\
&\Rightarrow \quad c_0^t \text{ is of type } K^1 \cup Z^1 && (25) \\
(19), (25) \quad &\Rightarrow \quad K^1 \cup Z^1 = K^2 \cup (K^1 \cup Z^1)K^1 \oplus c_1^t \\
&\Rightarrow \quad c_1^t \text{ is of type } K^2 \cup K^1 Z^1 && (26) \\
(20), (26) \quad &\Rightarrow \\
K^2 \cup K^1 Z^1 &= (K^{2+2} \cup K^{1+2} Z^1) \cup (K^2 \cup K^1 Z^1)(K^2 \cup K^1 Z^1) \\
&\Rightarrow \quad \text{relation of type } K^4 \cup K^3 Z^2 && (27)
\end{aligned}$$

Let us next consider the $n = 2^3$ case also. Since changing the number of variables, i.e., replacing $S_j^t(2^2)$ with $S_j^t(2^3)$ does not change the degree of these equations, the degree 2^3 equation is obtained as follows.

$$\begin{aligned}
(18) = (21), (25) \quad &\Rightarrow \quad c_0^t \text{ is of type } K^1 \cup Z^1 && (28) \\
(19) = (22), (26) \quad &\Rightarrow \quad c_1^t \text{ is of type } K^2 \cup K^1 Z^1 && (29) \\
(20) \sim (23), (27) \quad &\Rightarrow \quad c_2^t \text{ is of type } K^4 \cup K^3 Z^2 && (30) \\
(24), (30) \quad &\Rightarrow \\
K^4 \cup K^3 Z^2 &= (K^{4+4} \cup K^{3+4} Z^2) \cup (K^4 \cup K^3 Z^2)(K^4 \cup K^3 Z^2) \\
&\Rightarrow \quad \text{relation of type } K^8 \cup K^7 Z^4 && (31)
\end{aligned}$$

This shows that the general case may be proved by induction. To prove the induction step, it suffices to show that

$$K^{2^k} \cup K^{2^k-1}Z^{2^{k-1}} = (K^{2^k+2^k} \cup K^{2^k-1+2^k}Z^{2^{k-1}}) \\ \cup (K^{2^k} \cup K^{2^k-1}Z^{2^{k-1}})(K^{2^k} \cup K^{2^k-1}Z^{2^{k-1}})$$

gives a DDE of type $K^{2^{k+1}} \cup K^{2^{k+1}-1}Z^{2^k}$. Checking the validity of this statement is trivial. And this completes the proof.

We may assume that all periods of the LFSRs used in the summation generator are relatively prime. Then the summation generator satisfies the requirement of the attack described in [6], if we have access to consecutive key stream bits. The following steps may be taken to reduce the complexity of the algebraic attack.

1. Compute an algebraic equation explicitly using the formulae given in Proposition 2.
2. Write the resulting DDE in the following form.
$$L^t(k) = R^t(k,z).$$
Here, $L^t(k)$ is the sum of all monomials of type K^d appearing in the equation and $R^t(k,z)$ is the sum of all other monomials of type K^eZ^f.
3. Fix an arbitrary nontrivial initial key k' and compute the value $L^t(k')$ for a sequence of length $2\binom{m}{d}$.
4. Using the Berlekamp-Massey algorithm, find a linear relation $\alpha = (\alpha_t)_t$ such that
$$\sum_t \alpha_t L^t(k') = 0.$$
We note that steps 1,2,3,4 are independent of the initial key k, hence we can pre-compute the relation α.
5. We have obtained an algebraic equation of degree e. It is given by
$$\sum_t \alpha_t R^t(k,z) = 0.$$
6. Apply the general algebraic attack given in [7] to the above equation.

Example 1. The DDE for $n = 2^2$ is given as follows:

$$\begin{pmatrix} S_4^t \oplus S_3^t \oplus S_1^{t+1} S_2^t \\ \oplus S_2^{t+1} \oplus S_2^t \oplus S_1^t S_1^{t+1} \\ \oplus S_1^{t+2} \oplus S_1^{t+1} \oplus S_1^t \end{pmatrix} = \begin{pmatrix} z^t S_3^t \oplus z^{t+1} S_2^t \oplus z^t S_1^t S_1^{t+1} \\ \oplus z^{t+1} S_1^{t+1} \oplus z^t z^{t+1} S_1^t \\ \oplus z^{t+1} S_1^t \oplus z^t S_1^t \oplus z^{t+2} \end{pmatrix}.$$

We take 4 LFSRs defined by the following characteristic polynomials.

$$L_1 : x^3 + x + 1$$
$$L_2 : x^5 + x^2 + 1$$
$$L_3 : x^7 + x + 1$$
$$L_4 : x^{11} + x^2 + 1$$

Table 4. Complexity comparison of attacks on summation generators

generator size		$(m, 2^k)$	$(128, 2^2)$	$(256, 2^2)$	$(256, 2^3)$
[1, 5]	data	$\binom{m}{2^{k-1}(k+1)}$	$2^{32.3}$	$2^{38.4}$	$2^{83.1}$
	computation	$\binom{m}{2^{k-1}(k+1)}^w$	$2^{90.8}$	$2^{107.9}$	$2^{233.2}$
[10]	data	$T = 2^{\frac{m}{2^k}+k+1}$	2^{35}	2^{67}	2^{36}
	computation	$T^2 \log_2 T \log_2 \log_2 T$	$2^{77.5}$	$2^{142.7}$	$2^{79.5}$
Thm 1	data	$\binom{m}{2^k}$	$2^{23.3}$	$2^{27.4}$	$2^{48.5}$
	computation	$\binom{m}{2^k}^w$	$2^{65.5}$	$2^{76.9}$	$2^{136.3}$
Thm 3	data	$\binom{m}{2^k-1}$	$2^{18.4}$	$2^{21.4}$	$2^{43.6}$
	computation	$\binom{m}{2^k-1}^w$	$2^{51.6}$	$2^{60.1}$	$2^{122.3}$

Since $\binom{26}{4} \simeq 15,000$, we compute 30,000 bits from the left hand side of the above equation for some arbitrary nontrivial key bits k'. After applying the Berlekamp-Massey algorithm, we found that the sequence has a linear relation of length $3892 < \binom{26}{4}$. With a consecutive key stream of length of the order $6520 = 3892 + (3-1) + (\binom{26}{3} - 1)$, one will be able to find the 26 bit initial key k.

Table 4 gives a simple comparison of the data and computational complexities needed for attacks on summation generators. Let w be the Gaussian elimination exponent. We shall use $w = \log_2 7$, as given by the Strassen algorithm. Let the summation generator with 2^k input LFSRs use an m-bit initial key.

We remark that for [10] and Theorem 3, the key stream needs to be consecutive. For [1, 5] and Theorem 1, the key stream need only be partially consecutive, i.e., we need groups of $k+1$ consecutive bits, but these groups may be far apart from each other. Hence, a straightforward comparison of data complexity might not be fair. Also, the values for [10] have been calculated assuming that the LFSRs in use have been chosen well, so that their 2-adic span is maximal.

7 Conclusion

We have applied the general results of [1, 5] and [6] on stream ciphers with memories to the summation generator. Our results show that the degree of algebraic equation obtainable and the complexity of the attack applicable are much lower than given by the general results.

For a summation generator that uses n LFSRs, the algebraic equation relating the key stream bits and LFSR output bits can be made to be of degree less than or equal to $2^{\lceil \log_2 n \rceil}$, using $\lceil \log_2 n \rceil + 1$ consecutive key stream bits. Under certain conditions, for the $n = 2^k$ case, the effective degree may further be reduced by 1. And for small n's we have summarized the degrees of the explicit equations in Table 3. The table should be taken into account by anyone using a summation generator.

References

[1] F. Armknecht and M. Krause, Algebraic attacks on combiners with memory, *Advances in Cryptology - Crypto 2003*, LNCS 2729, Springer-Verlag, pp. 162–175, 2003. 34, 35, 41, 44

[2] F. Armknecht, Improving Fast Algegraic Attacks, this proceeding, 2004. 34, 35, 42

[3] N. Courtois, The security of Hidden Field Equations (HFE), *CT-RSA 2001*, LNCS 2020, Springer-Verlag, pp. 266–281, 2001. 34, 42

[4] N. Courtois, Higher order correlation attacks, XL algorithm and Cryptanalysis of Toyocrypt, *ICISC 2002*, LNCS 2587, Springer-Verlag, pp. 182–199, 2002. 34

[5] N. Courtois, Algebraic attacks on combiners with memory and several outputs, E-print archive, 2003/125. 34, 35, 41, 44

[6] N. Courtois, Fast algebraic attack on stream ciphers with linear feedback, *Advances in Cryptology - Crypto 2003*, LNCS 2729, Springer-Verlag, pp. 176–194, 2003. 34, 35, 42, 43, 44

[7] N. Courtois and W. Meier, Algebraic attacks on stream ciphers with linear feedback, *Advances in Cryptology - Eurocrypt 2003*, LNCS 2656, Springer-Verlag, pp. 345–359, 2003. 34, 43

[8] N. Courtois and J. Pieprzyk, Cryptanalysis of block ciphers with overdefined systems of equations, *Asiacrypt 2002*, LNCS 2501, Springer-Verlag, pp. 267–287, 2002. 34

[9] A. Kipnis and A. Shamir, Cryptanalysis of the HFE public key cryptosystem by relinearization, *Advances in Cryptoloy - Crypto'99*, LNCS 1666, Springer-Verlag, pp. 19–30, 1999. 34

[10] A. Klapper and M. Goresky, Cryptanalysis based on 2-adic rational approximation, *Advances in Cryptology - Crypto '95*, LNCS 963, Springer-Verlag, pp. 262–273, 1995. 34, 44

[11] W. Meier and O. Staffelbach, Correlation Properties of Combiners with Memory in Stream Cipher, *Journal of Cryptology*, vol.5, pp. 67–86, 1992. 34

[12] R. A. Rueppel, Correlation immunity and the summation generator, *Advances in Cryptology - Crypto'85*, LNCS 219, Springer-Verlag, pp. 260–272, 1985. 34, 35

A Proof of Proposition 2, Equation (12)

To prove (12), we shall evaluate its right hand side at some arbitrary input value and show that it equals the next carry bit c_i^{t+1}. But before we do this, let us make some observations.

Note that we may take

$$c_i^{t+1} \equiv \lfloor (\sum_j x_j^t + c^t)/2^{i+1} \rfloor \pmod 2 \qquad (32)$$

as the definition of the carry bits. We may evaluate the right hand side of (11) at $\sum_j x_j^t = 0$ and equate it with the evaluation of (32) at the same point to shows $c_{i+1}^t = f_{i,0}$. Now, from the discussions in Section 2 on the matrix M, we know that at $\sum_j x_j^t = 2^k$, we have $S_0^t(2^k) = S_{2^k}^t(2^k) = 1$ and all other $S_j^t(2^k) = 0$. Once more, evaluating (11) and (32) at $\sum_j x_j^t = 2^k$ with this in mind shows $c_{i+1}^t = f_{i,0} \oplus f_{i,2^k}$. Since we already know $c_{i+1}^t = f_{i,0}$, this implies $f_{i,2^k} = 0$, or equivalently, that the term $S_{2^k}^t$ does not appear in the linear sum (11).

We shall now evaluate the right hand side of (12) at some fixed LFSR output values $(x_1^t, \ldots, x_{2^{k+1}}^t)$ and carry value c^t. Set

$$\begin{aligned} x &= \textstyle\sum_j x_j^t, \\ x' &= \text{remainder of } x \text{ divided by } 2^k, \\ c' &= \text{remainder of } c^t \text{ divided by } 2^k. \end{aligned}$$

From the above observation, we know that the right hand side of (12) contains some of the terms $S_0^t(2^{k+1}), \ldots, S_{2^k-1}^t(2^{k+1})$, but does not contain any of the terms $S_j(2^{k+1})$ for $j \geq 2^k$. We also know from discussions of Section 2 that the evaluation of each $S_j^t(2^{k+1})$ at x for $j < 2^k$ is equal to its evaluation at x'. Note also that the term $c_{2^k}^t$ does not appear as input to any of the coefficients $f_{i,j}$ in the right hand side of (12). Hence,

$$\begin{aligned} &\text{RHS of (12) at } x \text{ and } c^t \\ &\quad = \text{RHS of (12) at } x' \text{ and } c' \\ &\quad = \text{RHS of (11) at } x' \text{ and } c' \\ &\quad = \lfloor (x' + c')/2^{i+1} \rfloor \qquad \pmod 2 \\ &\quad = \lfloor (x + c^t)/2^{i+1} \rfloor \qquad \pmod 2 \\ &\quad = c_i^{t+1} \text{ at } x \text{ and } c^t. \end{aligned}$$

The condition $i \leq k - 2$ has been used in the fourth equality. And the last equality is just (32). We have completed the proof that (12) is a valid expression for the next carry bits.

B Proof of Proposition 2, Equation (13)

The proof for (13) is very similar to that of (12) and hence we shall be very brief. We ask the readers to read Appendix A before reading this section. The carry bit is given by

$$c_{k-1}^{t+1} \equiv \lfloor (\textstyle\sum_j x_j^t + c^t)/2^k \rfloor \qquad \pmod 2.$$

Evaluation of (11) at $\sum_j x_j^t = 0$ and 2^k shows that $f_{k-1,0} = 0$ and $f_{k-1,2^k} = 1$, hence we now always have $S_{2^k}^t$ as a linear term, with coefficient equal to 1, in the sums (11) and (13). The temporary values x, x', and c' may be defined as before. From the discussions of Section 2, one may write

$$(S_{2^k}^t(2^{k+1}) \text{ at } x) = \lfloor x/2^k \rfloor = (S_{2^k}^t(2^{k+1}) \text{ at } x') \oplus \lfloor x/2^k \rfloor \qquad \pmod 2$$

and

$$(S_j^t(2^{k+1}) \text{ at } x) = (S_j^t(2^{k+1}) \text{ at } x')$$

for $j < 2^k$. Also note that none of the terms $S_j(2^{k+1})$, for $j > 2^k$, appears in the sum (13) and that the term c_k^t visible in (13) is its only use in (13). Hence,

$$\begin{aligned}
&\text{RHS of (13) at } x \text{ and } c^t \\
&\quad = (\text{RHS of (13) at } x \text{ and } c') \oplus c_k^t \\
&\quad = (\text{RHS of (13) at } x' \text{ and } c') \oplus c_k^t \oplus \lfloor x/2^k \rfloor \pmod 2 \\
&\quad = (\text{RHS of (11) at } x' \text{ and } c') \oplus c_k^t \oplus \lfloor x/2^k \rfloor \pmod 2 \\
&\quad = \lfloor (x' + c')/2^k \rfloor \oplus c_k^t \oplus \lfloor x/2^k \rfloor \pmod 2 \\
&\quad = \lfloor (x + c^t)/2^k \rfloor \pmod 2 \\
&\quad = c_{k-1}^{t+1} \text{ at } x \text{ and } c^t.
\end{aligned}$$

This completes the proof.

C Proof of Proposition 2, Equation (14)

Define $x = \sum_j x_j^t$. Careful reading of Appendices A and B shows that the first term of (14) simplifies to

$$\oplus_j f_{k-1,j} S_{j+2^k}^t(2^{k+1}) = \begin{cases} 0 & \text{for } 0 \le x \le 2^k, \\ \lfloor (x - 2^k + c^t)/2^k \rfloor \oplus c_k^t & \text{for } 2^k < x \le 2^{k+1}, \end{cases} \tag{33}$$

and that the second term satisfies

$$c_k^t\big(\oplus_j f_{k-1,j} S_j^t(2^{k+1})\big) = c_k^t\big(\lfloor (x + c^t)/2^k \rfloor \oplus c_k^t\big). \tag{34}$$

It now suffices to compare the sum of these two values with

$$c_k^{t+1} \equiv \lfloor (x + c^t)/2^{k+1} \rfloor \pmod 2. \tag{35}$$

Define $x' = x - 2^k$ when $x > 2^k$ and define $c' = c^t - 2^k$ when $c_k^t = 1$.
Case 1) $x \le 2^k$, $c_k^t = 0$.
This is the most easy case.

$$(33) \oplus (34) = 0 = \lfloor (x + c^t)/2^{k+1} \rfloor.$$

Case 2) $x \le 2^k$, $c_k^t = 1$.

$$\begin{aligned}
(33) \oplus (34) &= \lfloor (x + c^t)/2^k \rfloor \oplus 1 \\
&= \lfloor (x + c')/2^k \rfloor \\
&= \lfloor \{(x + c') + 2^k\}/2^{k+1} \rfloor \\
&= \lfloor (x + c)/2^{k+1} \rfloor.
\end{aligned}$$

Case 3) $x > 2^k$, $c_k^t = 0$.

$$\begin{aligned}(33) \oplus (34) &= \lfloor (x' + c^t)/2^k \rfloor \\ &= \lfloor \{(x' + c^t) + 2^k\}/2^{k+1} \rfloor \\ &= \lfloor (x + c^t)/2^{k+1} \rfloor.\end{aligned}$$

Case 4) $x > 2^k$, $c_k^t = 1$.

$$\begin{aligned}(33) \oplus (34) &= \lfloor (x' + c^t)/2^k \rfloor \oplus \lfloor (x + c^t)/2^k \rfloor \\ &= \lfloor (x' + c^t)/2^k \rfloor \oplus (\lfloor (x' + c^t)/2^k \rfloor \oplus 1) \\ &= 1 \\ &= \lfloor (x + c^t)/2^{k+1} \rfloor.\end{aligned}$$

This completes the proof.

Algebraic Attacks on SOBER-t32 and SOBER-t16 without Stuttering

Joo Yeon Cho and Josef Pieprzyk*

Center for Advanced Computing – Algorithms and Cryptography
Department of Computing, Macquarie University, NSW, Australia, 2109
{jcho,josef}@ics.mq.edu.au

Abstract. This paper presents algebraic attacks on SOBER-t32 and SOBER-t16 without stuttering. For unstuttered SOBER-t32, two different attacks are implemented. In the first attack, we obtain multivariate equations of degree 10. Then, an algebraic attack is developed using a collection of output bits whose relation to the initial state of the LFSR can be described by low-degree equations. The resulting system of equations contains 2^{69} equations and monomials, which can be solved using the Gaussian elimination with the complexity of $2^{196.5}$. For the second attack, we build a multivariate equation of degree 14. We focus on the property of the equation that the monomials which are combined with output bit are linear. By applying the Berlekamp-Massey algorithm, we can obtain a system of linear equations and the initial states of the LFSR can be recovered. The complexity of attack is around $O(2^{100})$ with 2^{92} keystream observations. The second algebraic attack is applicable to SOBER-t16 without stuttering. The attack takes around $O(2^{85})$ CPU clocks with 2^{78} keystream observations.

Keywords: Algebraic attack, stream ciphers, linearization, NESSIE, SOBER-t32, SOBER-t16, modular addition, multivariate equations

1 Introduction

Stream ciphers are an important class of encryption algorithms. They encrypt individual characters of a plaintext message one at a time, using a stream of pseudorandom bits. Stream ciphers generally offer a better performance compared with block ciphers. They are also more suitable for implementations where computing resources are limited (mobile phones) or when characters must be individually processed (reducing the delay) [4].

Recently, there were two international calls for cryptographic primitives. NESSIE is an European initiative [2] and CRYPTREC [1] is driven by Japan. Many stream ciphers have been submitted and evaluated by the international cryptographic community. NESSIE announced the final decision at Feb. 2003 and none of candidates of stream ciphers was selected in the final report.

* Supported by ARC Discovery grant DP0451484.

B. Roy and W. Meier (Eds.): FSE 2004, LNCS 3017, pp. 49–64, 2004.

According to the final NESSIE security report [3], there were four stream cipher primitives which were considered during the phase II : BMGL [15], SNOW [12], SOBER-t16 and SOBER-t32 [17]. The security analysis of these stream ciphers is mainly focused on the distinguishing and guess-determine attacks. Note that ciphers for which such attacks exist are excluded from the contest (even if those attack do not allow to recover the secret elements of the cipher).

For SOBER-t16 and SOBER-t32, there are distinguishing attacks that are faster than the key exhaustive attack. SOBER-t16 is distinguishable from the truly random keystream with the work factor of 2^{92} for the version without stuttering and with the work factor of 2^{111} for the version with stuttering [13]. The same technique is used to construct a distinguisher for SOBER-t32 with complexity of $2^{86.5}$ for the non-stuttering version [13]. For the version with stuttering [14], the distinguisher has the complexity 2^{153}.

The distinguishing attacks are the weakest form of attack and normally identify a potential weakness that may lead to the full attack that allows to determine the secret elements (such as the initial state) of the cipher. Recent development of algebraic attacks on stream ciphers already resulted in a dramatic cull of potential candidates for the stream cipher standards. The casualties of the algebraic attacks include Toyocrypt submitted to CRYPTREC [7] and LILI-128 submitted to NESSIE [10].

In this paper, we present algebraic attacks on SOBER-t32 and SOBER-t16 without stuttering. For unstuttered SOBER-t32, two different attacks are implemented. In the first attack, we apply indirectly the algebraic attack on combiner with memory, even though SOBER-t32 does not include an internal memory state. We extract a part of most significant bits of the addition modulo 2^{32} and the carry generated from the part of less significant bits is regarded as the internal memory state (which is unknown). Our attack can recover the initial state of LFSR with the workload of $2^{196.5}$ by 2^{69} keystream observations, which is faster than the exhaustive search of the 256-bit key.

For the second attack, we build a multivariate equation of degree 14. This equation has a property that the monomials which are combined with output bit are linear. By applying the Berlekamp-Massey algorithm, we can obtain a system of simple linear equations and recover the initial states of the LFSR. The attack takes $O(2^{100})$ CPU clocks with around 2^{92} keystream observations.

We apply the second algebraic attack to SOBER-t16 without stuttering. Our attack can recover the initial state of LFSR with the workload of $O(2^{85})$ using 2^{78} keystream observations for unstuttered SOBER-t16.

This paper is organized as follows. In Section 2, an algebraic attack method is briefly described. The structure of SOBER-t32 is given in Section 3. In Section 4, the first algebraic attack on SOBER-t32 is presented. the second algebraic attack on SOBER-t32 is presented in Section 5. In Section 6, an algebraic attack on SOBER-t16 is presented. Section 7 concludes the paper.

2 Algebraic Attacks

2.1 Previous Works

The first application of algebraic approach for analysis of stream ciphers can be found in the Courtois' work [7]. The idea behind it is to find a relation between the initial state of LFSR and the output bits that is expressible by a polynomial of a low degree. By accumulating enough observations (and corresponding equations) the attacker is able to create a system of equations of a low algebraic degree. The number of monomials in these equations is relatively small (as the degree of each equation is low). We treat monomials as independent variables and solve the system by the Gaussian elimination. For Toyocrypt, the algebraic attack shown in [7] is probabilistic and requires the workload of 2^{92} with 2^{19} keystream observations. In [10], the authors showed that Toyocrypt is breakable in 2^{49} CPU clocks with 20K bytes of keystream. They also analyzed the NESSIE submission of LILI-128 showing that it is breakable within 2^{57} CPU clocks with 762 GBytes memory. Recently, the algebraic approach has been extended to analyze combiners with memory [5, 6, 8]. Very recently, the method that allows a substantial reduction of the complexity of all these attacks is presented in [9].

2.2 General Description of Algebraic Attacks

Let $S_0 = (s_0, \cdots, s_{n-1})$ be an initial state of the linear shift register at the clock 0. The state variables are next input to the nonlinear block producing the output $v_0 = NB(S_0)$ where NB is a nonlinear function transforming the state of the LFSR into the output. At each clock t, the state is updated according to the following relation $S_t = L(s_t, \cdots, s_{n-1+t})$ with L being a multivariate linear transformation. The output is $v_t = NB(S_t)$.

The general algebraic attack on such stream ciphers works as follows. For details see [7] or [10].

- Find a multivariate relation Q of a low degree d between the state bits and the bits of the output. Assume that the relation is $Q(S_0, v_0) = 0$ for the clock 0.
- The same relation holds for all consecutive clocks t so

$$Q(S_t, v_t) = Q(L^t(S_0), v_t) = 0$$

  where the state S_t at the clock t is a linear transformation of the initial state S_0 or $S_t = L^t(S_0)$. Note that all relations are of the same degree d.
- Given consecutive keystream bits $v_0, \cdots v_{M-1}$, we obtain a system of M equations with monomials of degree at most d. If we collect enough observations so the number of linearly independent equations is at least as large as the number T of monomials of degree at most d, then the system has a unique solution revealing the initial state S_0. We can apply the Gaussian reduction algorithm that requires $7 \cdot T^{\log_2 7}$ operations [18].

Finding relations amongst the input and output variables is a crucial task in each algebraic attack. Assume that we have a nonlinear block with n binary inputs and m binary outputs or simply the $n \times m$ S-box over $GF(2)$. The truth table of the box consists of 2^n rows. The columns point out all input and output variables (monomials of degree 1). We can add columns for all terms (monomials) of degree 2. There are $\binom{n+m}{2}$ such terms. We can continue adding columns for higher degree monomials until

$$2^n < \sum_{i=1}^{d} \binom{n+m}{i}$$

where d is the highest degree of the monomials. Informally, the extended truth table can be seen as a matrix having more columns than rows so there are some columns (monomials) that can be expressed as a linear combination of other columns establishing a required relations amongst monomials of the S-box.

3 Brief Description of SOBER-t32

3.1 Notation

All variables operates on 32-bit words. Refer to the Figures 1.

- $\oplus$: addition in $GF(2^{32})$, $\bullet$: addition modulo 2^{32}.
- $s_{i,j}$: the j-th bit of the state register s_i.
- $s_{i,j \to k}$: a consecutive bit stream from the j-th bit to k-th bit of s_i.
- $x = s_0 \bullet s_{16}$. x_i is the i-th bit of x.
- α : the output of the first S-box. α_i is the i-th bit of α.

3.2 SOBER-t32

SOBER-t32 is a word-oriented synchronous stream cipher. It operates on 32-bit words and has a secret key of 256 bits (or 8 words). SOBER-t32 consists of a linear feedback shift register (LFSR) having 17 words (or 544 bits), a nonlinear filter (NLF) and a form of irregular decimation called stuttering. The LFSR produces a stream S_t of words using operations over $GF(2^{32})$. The vector $S_t = (s_t, \cdots, s_{t+16})$ is known as the *state* of the LFSR at time t, and the state $S_0 = (s_0, \cdots, s_{16})$ is called the *initial state*. The initial state and a 32-bit, key-dependent constant called K are initialized from the secret key by the key loading procedure.

A Nonlinear Filter (NLF) takes some of the states, at time t, as inputs and produces a sequence v_t. Each output stream v_t is obtained as $v_t = NLF(S_t) = F(s_t, s_{t+1}, s_{t+6}, s_{t+13}, s_{t+16}, K)$. The function F is described in the following subsection. The stuttering decimates the stream that is produced by NLF and outputs the key stream. The detailed description is given in [17].

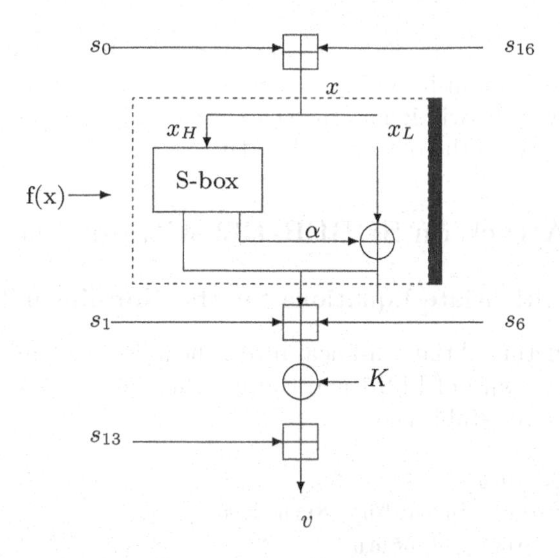

Fig. 1. The non-linear filter of SOBER-32 without stuttering

The Linear Feedback Shift Register. SOBER-t32 uses an LFSR of length 17 over $GF(2^{32})$. Each register element contains one 32-bit word. The contents of the LFSR at time t is denoted by $s_t, \cdots, s_{t+16}$. The new state of the LFSR is generated by shifting the previous state one step (operations are performed on words) and updating the state of the most significant word according to the following linear equation

$$s_{t+17} = s_{t+15} \oplus s_{t+4} \oplus \beta \cdot s_t$$

where $\beta = $ 0xc2db2aa3.

The Nonlinear Filter. At time t, the nonlinear filter takes five words from the LFSR states, $s_t, s_{t+1}, s_{t+6}, s_{t+13}, s_{t+16}$ and the constant K as the input and produces the output v_t. The nonlinear filter consists of function f, three adders modulo 2^{32} and the XOR addition. The value K is a 32-bit key-dependent constant that is determined during the initialization of LFSR. K is kept constant throughout the entire session. The function f translates a single word input into a word output. The output of the nonlinear filter, v_t, is equal to

$$v_t = ((f(s_t \bullet s_{t+16}) \bullet s_{t+1} \bullet s_{t+6}) \oplus K) \bullet s_{t+13}$$

where the function $f(x)$ is illustrated in Figure 1.

The function uses a S-box with 8-bit input and 32-bit output. The input x to the function $f(x)$ is split into two strings, x_H and x_L. The first string x_H is transformed by the S-box that gives α.

The Stuttering. The output of the stream cipher is obtained by decimating the output of the NLF in an irregular way. It makes correlation attack harder. In this paper, however, we will ignore the stuttering phase and assume that the attacker is able to observe the output v_t directly.

4 The First Attack on SOBER-t32 without Stuttering

4.1 Building Multivariate Equations for the Non-linear Filter

If we look at the structure of the non-linear filter, the following equations relating the state $S_0 = (s_0, \cdots, s_{16})$ of LFSR and the intermediate values x, α with the output keystream v are established.

$$\begin{cases} x_0 = s_{0,0} \oplus s_{16,0} \\ \alpha_0 = x_0 \oplus s_{1,0} \oplus s_{6,0} \oplus K_0 \oplus s_{13,0} \oplus v_0 \\ x_1 = s_{0,1} \oplus s_{16,1} \oplus s_{0,0}s_{16,0} \\ \alpha_1 = x_1 \oplus s_{1,1} \oplus s_{6,1} \oplus K_1 \oplus s_{13,1} \oplus v_1 \oplus K_0 s_{13,0} \oplus \\ \quad (x_0 \oplus \alpha_0)(s_{1,0} \oplus s_{6,0} \oplus s_{13,0}) \oplus s_{1,0}(s_{6,0} \oplus s_{13,0}) \oplus s_{6,0}s_{13,0} \\ \vdots \end{cases} \tag{1}$$

Lemma 1. *Variables x_i and α_i can be expressed as an equation of degree $d \geq i+1$ over the bits of the state variables.*

Let's consider the first modular addition $x = s_0 \bullet s_{16}$. If we denote *the carry* in the 24-th bit by ca, the addition modulo 2^{32} can be divided into two independent additions. One is the addition modulo 2^8 for the most significant 8 bits and another is the addition modulo 2^{24} for the remaining 24 bits. Then,

$$\begin{aligned} x_{24\to31} &= s_{0,24\to31} \bullet s_{16,24\to31} \bullet ca \quad (\text{modulo } 2^8) \\ x_{0\to23} &= s_{0,0\to23} \bullet s_{16,0\to23} \quad (\text{modulo } 2^{24}) \end{aligned}$$

Lemma 2. *If the carry in the 24-th bit position is regarded as an unknown, the degrees of equations which are related to each x_i $(24 \leq i \leq 31)$ are reduced to $(i-23)$.*

Now, we reconstruct a partial block of non-linear filter : the addition modulo 2^{32} and the S-box. These two blocks can be considered as a single S-box. This is going to open up a possibility of reduction of degree of relations derived for the complex S-box. Furthermore, we consider the carry in the 24-th bit of addition modulo 2^{32} as an another input variable that is unknown (so we will avoid using it in the relation and treat it as an unknown state [6, 8]).

The structure of the new block is shown in Figure 2. The addition is modified to add two 8-bit strings ($s_{0,24\to31}$ and $s_{16,24\to31}$) with the carry c_{24} that is unknown. Thus this part is an addition modulo 2^8. The output is put to the

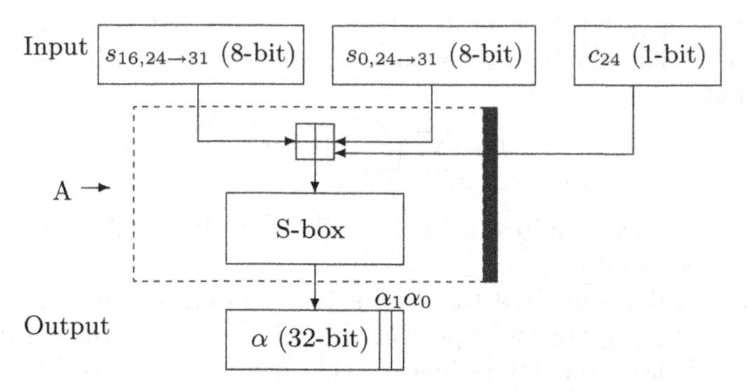

Fig. 2. A combined structure of partial addition modulo 2^8 and S-box

S-box that has 32-bit output and amongst the output bits the least significant two bits are α_1, α_0. If we regard the carry in addition modulo 2^n as an internal memory state, an algebraic attack on combiner with memory is able to be applied to the new block since the structure of block is very similar to the model which is analyzed in [6, 8].

Lemma 3. *Let A be the combined block of modular addition and S-box with input s_0, s_{16} and output α. If the carry in the 24-th bit c_{24} is considered as an unknown, there is a multivariate equation that relates $s_{0,24\to31}, s_{16,24\to31}$ and output bits α_0, α_1 without the carry bit c_{24}.*

Proof. Let's create the following matrix.

- Rows are all the possibilities for $s_{16,24\to31}, s_{0,24\to31}$ and carry bit c_{24}. There are 2^{17} rows enumerating all possible values for the input bits.
- The columns are all the monomials of degree up to 9 which are coming from the input bits $s_{16,24\to31}, s_{0,24\to31}$ and the output bits α_1, α_0. The number of columns is

$$\sum_{i=0}^{9}\binom{18}{i} = 2^{17} + \frac{1}{2}\binom{18}{9} \cong 2^{17} + 2^{14.5}$$

If we apply the Gaussian elimination to this matrix, definitely we can obtain equations of degree up to 9 because the number of columns is greater than that of rows.

Let's denote the equation which is derived above as $F(s_{0,24\to31}, s_{16,24\to31}, \alpha_0, \alpha_1)$ of degree 9. If α_0 and α_1 are replaced by Equation (1) for the multivariate equation F, the degree of F becomes at most 10 which consists of only state bits of the LFSR, K and output bits of the NLF. In Appendix A, we present a toy example for illustrating the idea of the attack.

4.2 Complexity

We can apply the general algebraic attack to unstuttered SOBER-t32 by using the equation F. The number of monomials T of degree up to 10 are chosen from 544 unknowns.

$$T = \sum_{i=0}^{10} \binom{544}{i} \cong 2^{69}$$

From [10], the attack requires roughly $7 \cdot T^{\log_2 7} \cong 2^{196.5}$ CPU clocks with 2^{69} keystream observations.

In CRYPTO'03, a method that can solve the multivariate equations more efficiently by pre-computation is presented [9]. Even though the constant factor of complexity is not precisely estimated, this method seems to allow a further reduction of the complexity of our attack. According to [9], the main workload of algebraic attack is distributed into pre-computation stage and Gaussian reduction stage. The pre-computation is operated by LFSR synthesis method, say the Berlekamp-Massey algorithm. This step needs to take $\mathcal{O}(S \log S + Sn)$ steps if the asymptotically fast versions of the Berlekamp-Massey algorithm is used, where S is the size of the smallest linear dependency and n is the number of the state bits. Then, Gaussian elimination is applied to the monomials of reduced degree.

For our attack, the pre-computation operation needs about $\mathcal{O}(2^{78.3})$ steps. Then, Gaussian reduction can be applied to the monomials of degree up to 9. It takes about $\mathcal{O}(2^{180})$ steps with $2^{126.4}$ bits of memory.

5 The Second Attack on SOBER-32 without Stuttering

5.1 An Observation of Modular Addition

Let us consider the function of modular addition $c = a \bullet b$ where $a = (a_{31}, \cdots, a_0)$, $b = (b_{31}, \cdots, b_0)$ and $c = (c_{31}, \cdots, c_0)$.

Lemma 4. *Let c_i be the i-th output bit of the modular addition. Then, $c_0 = a_0 \oplus b_0$, $c_1 = a_1 \oplus b_1 \oplus a_0 b_0$ and for $2 \leq i \leq 31$,*

$$c_i = a_i \oplus b_i \oplus a_{i-1} b_{i-1} \oplus \sum_{t=0}^{i-2} a_t b_t \{ \prod_{r=t+1}^{i-1} (a_r \oplus b_r) \}$$

Each c_i is expressed as a function of input bits of degree $i+1$.

Theorem 1. *Let c_i, $24 \leq i \leq 31$ be the i-th output bit of modular addition $c = a \bullet b$. If c_i is multiplied by $(1 \oplus a_{23} \oplus b_{23})$, then, the degree of $c_i \cdot (1 \oplus a_{23} \oplus b_{23})$ is reduced to $(i - 22)$.*

Proof. For $24 \leq i \leq 31$, c_i can be separated into two parts : one part includes $(a_{23} \oplus b_{23})$ and the remaining part does not. Therefore,

$$c_i = p(a_{23 \to i}, b_{23 \to i}) \oplus (a_{23} \oplus b_{23}) \cdot q(a_{0 \to i}, b_{0 \to i})$$

where p and q are functions of the input bits. Then,

$$c_i \cdot (1 \oplus a_{23} \oplus b_{23}) = p(a_{23\to i}, b_{23\to i}) \cdot (1 \oplus a_{23} \oplus b_{23})$$

From Lemma 4, the degree of $c_i \cdot (1 \oplus a_{23} \oplus b_{23})$ is $(i - 22)$.

5.2 Building a System Equation for the Non-linear Filter

If we look at the structure of the non-linear filter, we can easily see that the following equation holds. (see Figure 1)

$$\alpha_0 = s_{0,0} \oplus s_{16,0} \oplus s_{1,0} \oplus s_{6,0} \oplus s_{13,0} \oplus v_0 \oplus K_0 \quad (2)$$

An Equation for α_0. The bit α_0 is the least significant output bit of the S-box. α_0 can be represented by a non-linear equation where variables consist of only the input bits of the S-box. Let's construct the following matrix.

- Rows generate all the possibilities for $(x_{31}, \cdots, x_{24})$, so there are 2^8 rows enumerating all possible values for the input bits.
- The columns are all the monomials of degree up to 8 which are coming from the input bits $(x_{31}, \cdots, x_{24})$ and the least significant output bit α_0. The number of columns becomes $2^8 + 1$

If we apply the Gaussian elimination to this matrix, we can obtain a non-linear equation because the number of columns is greater than that of rows. Simulation shows that the degree of the equation for α_0 is 6. (See Appendix B)

In the next step, let's take a look at the first modular addition of the non-linear filter, which is $x = s_0 \bullet s_{16}$. By Theorem 1, for $24 \leq i \leq 31$, $x_i \cdot (1 \oplus s_{0,23} \oplus s_{16,23})$ becomes

$$x_i \cdot (1 \oplus s_{0,23} \oplus s_{16,23}) = g(s_{0,23\to i}, s_{16,23\to i})$$

where g is a multivariate equation of degree up to $(i-22)$. Let A_i be a monomial which is built over variables from the set $\{x_{24}, \cdots, x_{31}\}$. Then,

$$A_i \cdot (1 \oplus s_{0,23} \oplus s_{16,23}) = G_i(s_{0,23\to 31}, s_{16,23\to 31})$$

where G_i is a non-linear equation of degree d_{G_i}.

Lemma 5. *The degree of $A_i \cdot (1 \oplus s_{0,23} \oplus s_{16,23})$ is at most 16.*

We can see that the number of variables which give effect on the degree d_{G_i} is at most 18, which is the set of variable $\{s_{0,23\to 31}, s_{16,23\to 31}\}$. However, not all the monomials are available. For example, a monomial $s_{0,31} \cdot s_{0,30} \cdots s_{0,23} \cdot s_{16,31} \cdots s_{16,23}$, which is of degree 18, cannot happen. By careful inspection, we see that the degree of monomials is at most 16.

As shown in Appendix B, $\alpha_0 = \sum_i A_i$. If α_0 is multiplied by $(1 \oplus s_{0,23} \oplus s_{16,23})$, the monomials which include $(s_{0,23} \oplus s_{16,23})$ vanish.

Lemma 6. *The degree of $\alpha_0 \cdot (1 \oplus s_{0,23} \oplus s_{16,23})$ is at most 14.*

The degree of remaining monomials is expected to be not bigger than 16. However, computer simulation shows that the degree of monomials is not bigger than 14.

The Degree of Equation (2). If we multiply $(1 \oplus s_{0,23} \oplus s_{16,23})$ by Equation (2), then we get

$$\begin{aligned} &\alpha_0 \cdot (1 \oplus s_{0,23} \oplus s_{16,23}) \\ &= (s_{0,0} \oplus s_{16,0} \oplus s_{1,0} \oplus s_{6,0} \oplus s_{13,0} \oplus v_0 \oplus K_0) \cdot (1 \oplus s_{0,23} \oplus s_{16,23}) \end{aligned} \tag{3}$$

Let's consider the left part of the equation. The bit α_0 plays a major role in determining the degree of the equation. We know that $\alpha_0 \cdot (1 \oplus s_{0,23} \oplus s_{16,23})$ has the monomials of maximum degree 14. The right part equation becomes quadratic by multiplication. Therefore, we can obtain a multivariate equation of degree 14.

5.3 Applying an Algebraic Attack

Let's recall the recent algebraic attack introduced in [9]. Let S_t^d denote the monomials of state variables of the degree up to d and V_t^d denote the monomials of output variables of the degree up to d at clock t. Then, the final equation of degree 14 can be described as a following way.

$$S_t^{14} \oplus S_t V_t = 0 \tag{4}$$

If we put all the monomials on the left side which do not include the output variables,

$$S_t^{14} = S_t V_t \quad \text{or} \quad \begin{cases} \text{Left}(S_t) = S_t^{14} \\ \text{Right}(S_t, V_t) = S_t V_t \end{cases}$$

We can see that $L^t(S_0) = S_t$ where S_0 is the initial state of the state variables and L is a connection function which is linear over $GF(2)$. If we collect $N > \sum_i^{14} \binom{544}{i}$ consecutive equations, a linear dependency $\gamma = (\gamma_0, \ldots, \gamma_{N-1})$ for left side equations must exist and

$$\sum_{t=0}^{N-1} \gamma_t \cdot \text{Left}(L^t(S_0)) = 0, \quad \gamma_i \in GF(2)$$

Let's recover γ from the given sequence. (see [9]) We choose a non-zero random key S_0' and compute $2T$ outputs bits c_t of the left side equations.

$$c_t = \text{Left}(L^t(S_0')), \quad \text{for } t = 0, \ldots, 2T - 1$$

where $c_t \in GF(2)$. Then we apply the well-known Berlekamp-Massey algorithm to find the smallest connection polynomial that generates the sequence $c = (c_0, \ldots, c_{2T-1})$.

If we find γ successfully, the same linear dependency holds for the right hand side. So,

$$0 = \sum_{t=i}^{N+i-1} \gamma_{t-i} \cdot \text{Right}(L^t(S_0), V_t), \quad i = 0, 1, \ldots \tag{5}$$

We can see that Equation (5) is linear. If we collect and solve these equations for consecutive keystreams, we can recover the initial state bits of the LFSR with small complexity.

5.4 The Complexity of the Algebraic Attack

If we denote T as the number of monomials of degree up to 14 that are chosen from $n = 544$ unknowns, then

$$T = \sum_{i=0}^{14} \binom{544}{i} \cong 2^{91}$$

We see that recovering the linear dependency γ dominates the complexity of computation. It is estimated to take $O(T\log(T)+Tn)$ by using improved versions of the Berlekamp-Massey algorithm. [9, 11] Therefore, our attack is estimated to take around $O(2^{100})$ CPU clocks with around 2^{92} keystream observations.

For memory requirement, we need to store at most T bits of memory for the linear dependency γ. We need also some memory for Equation (5) but it is much smaller than T. Therefore, we expect that our attack needs around 2^{91} bits of memory.

6 Algebraic Attack on SOBER-t16 without Stuttering

The structure of SOBER-t16 is a very similar to that of SOBER-t32. Major differences from SOBER-t32 are

- operation based on 16-bit word
- the linear recurrence equation
- a S-box with 8-bit input and 16-bit output

For detail description of SOBER-t16, see [16].

We can apply a very similar algebraic attack presented in Section 5 for the unstuttered SOBER-t16. Let's look at Equation (2), which holds in SOBER-t16 as well. If we multiply Equation (2) by $(1 \oplus s_{0,7} \oplus s_{16,7})$, then we get

$$\begin{aligned} &\alpha_0 \cdot (1 \oplus s_{0,7} \oplus s_{16,7}) \\ &= (s_{0,0} \oplus s_{16,0} \oplus s_{1,0} \oplus s_{6,0} \oplus s_{13,0} \oplus v_0 \oplus K_0) \cdot (1 \oplus s_{0,7} \oplus s_{16,7}) \end{aligned} \tag{6}$$

Let's consider the left part of the equation. The bit α_0 plays a major role in determining the degree of the equation. Computer simulation shows that $\alpha_0 \cdot (1 \oplus s_{0,7} \oplus s_{16,7})$ has the monomials of maximum degree 14. The right part equation becomes quadratic by multiplication. Therefore, we can obtain a multivariate equation of degree 14. The remaining process for attack follows Section 5.3.

The Complexity. If we denote T as the number of monomials of degree up to 14 that are chosen from $n = 272$ unknowns, then

$$T = \sum_{i=0}^{14} \binom{272}{i} \cong 2^{76.5}$$

Therefore, our attack is estimated to take around $O(2^{85})$ CPU clocks with 2^{78} keystream observations. For memory requirement, we expect that our attack needs around $2^{76.5}$ bits of memory.

7 Conclusion

In this paper we present two algebraic attacks on SOBER-t32 without stuttering. For the first attack, we have built multivariate equations of degree 10. The carry at a specific bit position in addition modulo 2^n is regarded as an internal memory state. From some subset of the keystream observation, we can derive sufficient multivariate equations which are utilized in the algebraic attack. By solving these equations, we are able to recover all the initial states of LFSR and constant value K with roughly $2^{196.5}$ CPU clocks and 2^{69} keystream observations. Furthermore, fast algebraic attack with pre-computation allows more reduction of the complexity of our attack.

For the second attack, we derive a multivariate equation of degree 14 over the non-linear filter. The equation is obtained by multiplying the initial equation by a carefully chosen polynomial. Then, a new algebraic attack method is presented to recover the initial state bits of the LFSR. The attack is estimated to take $O(T\log(T)+Tn)\cong O(2^{100})$ CPU clocks with 2^{92} keystream observations.

By the similarity of the structure, we can apply the second algebraic attack to SOBER-t16 without stuttering. Our attack takes around $O(2^{85})$ CPU clocks using 2^{78} keystream observations for unstuttered SOBER-t16.

8 Acknowledgement

We are grateful to Greg Rose, Frederik Armknecht and unknown referees for their very helpful comments.

References

[1] Cryptrec. http://www.ipa.go.jp/security/enc/CRYPTREC/index-e.html. 49
[2] Nessie : New european schemes for signatures, integrity, and encryption. https://www.cryptonessie.org. 49
[3] Nessie security report. Technical Report V2.0, Feb. 2003. 50
[4] S. Vanstone A. Menezes, P. Oorschot. Handbook of Applied Cryptography. CRC Press, fifth edition, October 1996. 49
[5] F. Armknecht. A linearization attack on the bluetooth key stream generator. Cryptology ePrint Archive, Report 2002/191, 2002. http://eprint.iacr.org/. 51
[6] F. Armknecht and M. Krause. Algebraic attacks on combiners with memory. In Advances in Cryptology - CRYPTO 2003, volume 2729 / 2003, pages 162 - 175. Springer-Verlag, October 2003. 51, 54, 55
[7] N. Courtois. Higher order correlation attacks, xl algorithm and cryptanalysis of toyocrypt. Cryptology ePrint Archive, Report 2002/087, 2002. http://eprint.iacr.org/. 50, 51
[8] N. Courtois. Algebraic attacks on combiners with memory and several outputs. Cryptology ePrint Archive, Report 2003/125, 2003. http://eprint.iacr.org/. 51, 54, 55
[9] N. Courtois. Fast algebraic attacks on stream ciphers with linear feedback. In Advances in Cryptology - CRYPTO 2003, volume LNCS 2729, pages 176 - 194. Springer-Verlag, October 2003. 51, 56, 58, 59

[10] N. Courtois and W.Meier. Algebraic attacks on stream ciphers with linear feedback. In E. Biham, editor, Advances in Cryptology - EUROCRPYT 2003, LNCS 2656, pages 345 - 359. Springer-Verlag, January 2003. 50, 51, 56
[11] J. Dornstetter. On the equivalence between belekamp's and euclid's algorithms. IEEE Trans. on Information Theory, IT-33(3):428-431, May 1987. 59
[12] P. Ekdahl and T.Johansson. Snow. Primitive submitted to NESSIE, Sep. 2000. 50
[13] P. Ekdahl and T.Johansson. Distinguishing attacks on sober-t16 and t32. In V. Rijmen J. Daemen, editor, Fast Software Encryption, volume LNCS 2365, pages 210-224. Springer-Verlag, 2002. 50
[14] P. Ekdahl and T.Johansson. Distinguishing attacks on sober-t16 and t32. In Proceedings of the Third NESSIE Workshop, 2002. 50
[15] J. Hastad and M. Naslund. Bmgl: Synchronous key-stream generator with provable security. Primitive submitted to NESSIE, Sep. 2000. 50
[16] P.Hawkes and G.Rose. Primitive specification and supporting documentation for sober-t16 submission to nessie. In Proceedings of the first NESSIE Workshop, Belgium, Sep. 2000. 59
[17] P.Hawkes and G.Rose. Primitive specification and supporting documentation for sober-t32 submission to nessie. In Proceedings of the first NESSIE Workshop, Belgium, Sep. 2000. 50, 52
[18] V. Strassen. Gaussian elimination is not optimal. Numerische Mathematik, 13:354-356, 1969. 51

A A Toy Example to Build Equations by Reconstructing Block

This appendix illustrates a small example for building an low degree equation by reconstructing the non-linear block of SOBER-t32. Figure 3 represents the structure of new built block. The input and output operate on 4-bit. The most significant two bits of each input are added modulo 2^2 with carry and become the input of S-box. The S-box is 2×4 substitution defined by the following look-up table : $\{7,10,9,4\}$.

Let's denote as following

- $s_{0,H} = \{s_{0,3}, s_{0,2}\}$ and $s_{0,L} = \{s_{0,1}, s_{0,0}\}$
- $s_{16,H} = \{s_{16,3}, s_{16,2}\}$ and $s_{16,L} = \{s_{16,1}, s_{16,0}\}$
- $b_H = \{b_3, b_2\}$ and $b_L = \{b_1, b_0\}$

At first, we construct a matrix containing the input of modular addition and the S-box output.

- Rows are the possibilities for $\{carry, s_{0,3}, s_{0,2}, s_{16,3}, s_{16,2}\}$
- Columns are all monomials of degree to 3 which are coming from $\{s_{0,3}, s_{0,2}, s_{16,3}, s_{16,2}, b_1, b_0\}$.

Applying the Gaussian elimination, we can build at least 10 multivariate equations. The matrix below shows a part of the combination of monomials. Note

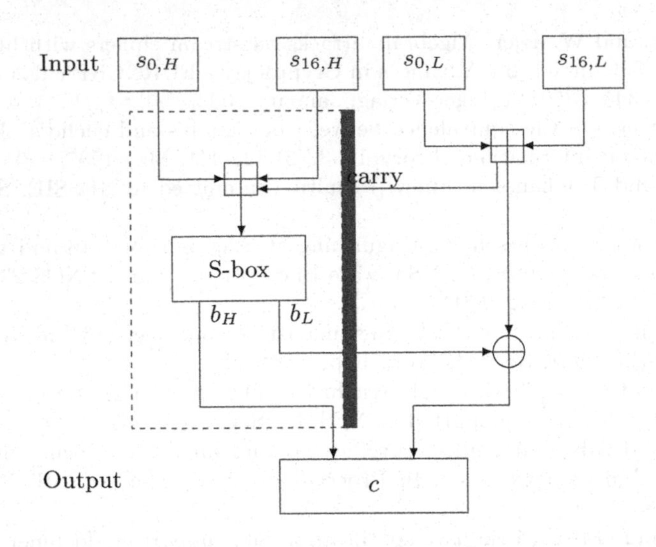

Fig. 3. A reconstructed block for the modular addition and S-box

that the number is represented by hexadecimal format. One of the equations that are built by the Gaussian elimination is

$$s_{0,2}s_{16,2}b_1 \oplus s_{0,2}s_{0,3}s_{16,2} \oplus s_{0,2}s_{16,2}s_{16,3} = 0$$

This equation is verified in the table below. Let us denote the equation as Q. We see that the Q is always zero for all possibilities of input value.

$$\left[\begin{array}{c|cccccccc}
1 & F & F & F & F & F & F & F & F \\
s_{16,2} & 5 & 5 & 5 & 5 & 5 & 5 & 5 & 5 \\
s_{16,3} & 3 & 3 & 3 & 3 & 3 & 3 & 3 & 3 \\
s_{0,2} & 0 & F & 0 & F & 0 & F & 0 & F \\
s_{0,3} & 0 & 0 & F & F & 0 & 0 & F & F \\
b_0 & A & 5 & A & 5 & 5 & A & 5 & A \\
b_1 & C & 9 & 3 & 6 & 9 & 3 & 6 & C \\
\vdots & & & & \vdots & & & & \\
s_{0,2}s_{16,2}b_1 & 0 & 1 & 0 & 4 & 0 & 1 & 0 & 4 \\
\vdots & & & & \vdots & & & & \\
s_{0,2}s_{0,3}s_{16,2} & 0 & 0 & 0 & 5 & 0 & 0 & 0 & 5 \\
\vdots & & & & \vdots & & & & \\
s_{0,2}s_{16,2}s_{16,3} & 0 & 1 & 0 & 1 & 0 & 1 & 0 & 1 \\
\vdots & & & & \vdots & & & &
\end{array}\right]$$

s_0	s_{16}	$s_0 \bullet s_{16}$	$(s_0 \bullet s_{16})_H$	b_1b_0	Q
0	0	0	0	3	0
0	1	1	0	3	0
0	2	2	0	3	0
⋮	⋮	⋮	⋮	⋮	⋮
0	F	F	3	0	0
1	0	1	0	3	0
⋮	⋮	⋮	⋮	⋮	⋮
F	F	E	3	0	0

B Algebraic Equations of S-Box in SOBER-t32 and SOBER-t16

For the S-box of SOBER-t32, we denote the 8-bit input stream of S-box as $(a_{31}, a_{30}, \ldots, a_{24})$ and the 32-bit output stream as $(b_{31}, b_{30}, \ldots, b_0)$. Each b_i can be also represented by the combination of monomials which are composed of only the input stream. In particular, the least significant output bit, b_0 is described as the combination of monomials of degree up to 6 as follows.

$$b_0 = 1 \oplus a_{24} \oplus a_{25} \oplus a_{24}a_{25} \oplus$$
$$a_{26} \oplus a_{25}a_{26} \oplus a_{25}a_{27} \oplus a_{24}a_{25}a_{27} \oplus a_{26}a_{27} \oplus$$
$$a_{24}a_{26}a_{27} \oplus a_{25}a_{26}a_{27} \oplus a_{25}a_{28} \oplus a_{24}a_{25}a_{28} \oplus a_{25}a_{26}a_{28} \oplus$$
$$a_{27}a_{28} \oplus a_{24}a_{27}a_{28} \oplus a_{24}a_{25}a_{27}a_{28} \oplus a_{26}a_{27}a_{28} \oplus a_{25}a_{26}a_{27}a_{28} \oplus$$
$$a_{29} \oplus a_{24}a_{29} \oplus a_{24}a_{25}a_{29} \oplus a_{26}a_{29} \oplus a_{24}a_{25}a_{26}a_{29} \oplus$$
$$a_{27}a_{29} \oplus a_{24}a_{25}a_{27}a_{29} \oplus a_{24}a_{26}a_{27}a_{29} \oplus a_{25}a_{26}a_{27}a_{29} \oplus a_{24}a_{28}a_{29} \oplus$$
$$a_{25}a_{28}a_{29} \oplus a_{24}a_{27}a_{28}a_{29} \oplus a_{25}a_{27}a_{28}a_{29} \oplus a_{24}a_{25}a_{27}a_{28}a_{29} \oplus a_{25}a_{26}a_{27}a_{28}a_{29} \oplus$$
$$a_{24}a_{25}a_{26}a_{27}a_{28}a_{29} \oplus a_{25}a_{26}a_{30} \oplus a_{24}a_{25}a_{26}a_{30} \oplus a_{24}a_{27}a_{30} \oplus a_{25}a_{27}a_{30} \oplus$$
$$a_{26}a_{27}a_{30} \oplus a_{24}a_{26}a_{27}a_{30} \oplus a_{28}a_{30} \oplus a_{24}a_{28}a_{30} \oplus a_{26}a_{28}a_{30} \oplus$$
$$a_{24}a_{25}a_{26}a_{28}a_{30} \oplus a_{27}a_{28}a_{30} \oplus a_{24}a_{27}a_{28}a_{30} \oplus a_{25}a_{26}a_{27}a_{28}a_{30} \oplus a_{29}a_{30} \oplus$$
$$a_{24}a_{29}a_{30} \oplus a_{25}a_{29}a_{30} \oplus a_{24}a_{25}a_{26}a_{29}a_{30} \oplus a_{27}a_{29}a_{30} \oplus a_{24}a_{27}a_{29}a_{30} \oplus$$
$$a_{24}a_{25}a_{27}a_{29}a_{30} \oplus a_{25}a_{26}a_{27}a_{29}a_{30} \oplus a_{24}a_{28}a_{29}a_{30} \oplus a_{25}a_{28}a_{29}a_{30} \oplus a_{24}a_{25}a_{28}a_{29}a_{30} \oplus$$
$$a_{26}a_{28}a_{29}a_{30} \oplus a_{25}a_{26}a_{28}a_{29}a_{30} \oplus a_{27}a_{28}a_{29}a_{30} \oplus a_{24}a_{27}a_{28}a_{29}a_{30} \oplus a_{25}a_{27}a_{28}a_{29}a_{30} \oplus$$
$$a_{24}a_{25}a_{27}a_{28}a_{29}a_{30} \oplus a_{26}a_{27}a_{28}a_{29}a_{30} \oplus a_{24}a_{26}a_{27}a_{28}a_{29}a_{30} \oplus a_{25}a_{26}a_{27}a_{28}a_{29}a_{30} \oplus a_{31} \oplus$$
$$a_{26}a_{31} \oplus a_{25}a_{26}a_{31} \oplus a_{24}a_{25}a_{26}a_{31} \oplus a_{24}a_{27}a_{31} \oplus a_{25}a_{26}a_{27}a_{31} \oplus$$
$$a_{24}a_{25}a_{26}a_{27}a_{31} \oplus a_{25}a_{28}a_{31} \oplus a_{24}a_{25}a_{28}a_{31} \oplus a_{26}a_{28}a_{31} \oplus a_{24}a_{26}a_{28}a_{31} \oplus$$
$$a_{24}a_{25}a_{26}a_{28}a_{31} \oplus a_{24}a_{27}a_{28}a_{31} \oplus a_{25}a_{27}a_{28}a_{31} \oplus a_{24}a_{25}a_{27}a_{28}a_{31} \oplus a_{24}a_{25}a_{26}a_{27}a_{28}a_{31} \oplus$$
$$a_{25}a_{29}a_{31} \oplus a_{24}a_{25}a_{29}a_{31} \oplus a_{24}a_{26}a_{29}a_{31} \oplus a_{25}a_{26}a_{29}a_{31} \oplus a_{24}a_{25}a_{26}a_{29}a_{31} \oplus$$
$$a_{27}a_{29}a_{31} \oplus a_{24}a_{27}a_{29}a_{31} \oplus a_{25}a_{27}a_{29}a_{31} \oplus a_{24}a_{25}a_{27}a_{29}a_{31} \oplus a_{26}a_{27}a_{29}a_{31} \oplus$$
$$a_{25}a_{26}a_{27}a_{29}a_{31} \oplus a_{28}a_{29}a_{31} \oplus a_{24}a_{28}a_{29}a_{31} \oplus a_{25}a_{28}a_{29}a_{31} \oplus a_{26}a_{28}a_{29}a_{31} \oplus$$
$$a_{25}a_{27}a_{28}a_{29}a_{31} \oplus a_{26}a_{27}a_{28}a_{29}a_{31} \oplus a_{25}a_{26}a_{27}a_{28}a_{29}a_{31} \oplus a_{30}a_{31} \oplus a_{24}a_{30}a_{31} \oplus$$
$$a_{24}a_{25}a_{30}a_{31} \oplus a_{26}a_{30}a_{31} \oplus a_{24}a_{26}a_{30}a_{31} \oplus a_{24}a_{25}a_{26}a_{30}a_{31} \oplus a_{24}a_{27}a_{30}a_{31} \oplus$$
$$a_{25}a_{26}a_{27}a_{30}a_{31} \oplus a_{24}a_{28}a_{30}a_{31} \oplus a_{24}a_{25}a_{28}a_{30}a_{31} \oplus a_{25}a_{26}a_{28}a_{30}a_{31} \oplus a_{27}a_{28}a_{30}a_{31} \oplus$$
$$a_{25}a_{27}a_{28}a_{30}a_{31} \oplus a_{24}a_{25}a_{27}a_{28}a_{30}a_{31} \oplus a_{24}a_{26}a_{27}a_{28}a_{30}a_{31} \oplus a_{29}a_{30}a_{31} \oplus a_{25}a_{26}a_{29}a_{30}a_{31} \oplus$$
$$a_{27}a_{29}a_{30}a_{31} \oplus a_{24}a_{27}a_{29}a_{30}a_{31} \oplus a_{25}a_{27}a_{29}a_{30}a_{31} \oplus a_{26}a_{27}a_{29}a_{30}a_{31} \oplus a_{28}a_{29}a_{30}a_{31} \oplus$$
$$a_{24}a_{28}a_{29}a_{30}a_{31} \oplus a_{24}a_{25}a_{28}a_{29}a_{30}a_{31} \oplus a_{26}a_{28}a_{29}a_{30}a_{31} \oplus a_{25}a_{27}a_{28}a_{29}a_{30}a_{31}$$

For S-box of SOBER-t16, we denote the 8-bit input stream of S-box as $(a_{15}, a_{14}, \ldots, a_8)$ and the 16-bit output stream as $(b_{15}, b_{14}, \ldots, b_0)$. Then, the least significant output bit, b_0 is described as the combination of monomials of degree up to 6 as follows.

$$b_0 = 1 \oplus a_8 \oplus a_9 \oplus a_8a_9 \oplus$$
$$a_{10} \oplus a_8a_{10} \oplus a_8a_9a_{10} \oplus a_{11} \oplus a_9a_{11} \oplus$$
$$a_8a_9a_{11} \oplus a_8a_{10}a_{11} \oplus a_8a_9a_{12} \oplus a_8a_{10}a_{12} \oplus a_{11}a_{12} \oplus$$
$$a_9a_{11}a_{12} \oplus a_8a_9a_{11}a_{12} \oplus a_9a_{10}a_{11}a_{12} \oplus a_9a_{10}a_{13} \oplus a_{11}a_{13} \oplus$$
$$a_8a_{11}a_{13} \oplus a_9a_{11}a_{13} \oplus a_8a_9a_{11}a_{13} \oplus a_{10}a_{11}a_{13} \oplus a_9a_{10}a_{11}a_{13} \oplus$$
$$a_{12}a_{13} \oplus a_9a_{12}a_{13} \oplus a_8a_9a_{12}a_{13} \oplus a_9a_{10}a_{12}a_{13} \oplus a_{11}a_{12}a_{13} \oplus$$
$$a_8a_{11}a_{12}a_{13} \oplus a_9a_{11}a_{12}a_{13} \oplus a_{10}a_{11}a_{12}a_{13} \oplus a_8a_{10}a_{11}a_{12}a_{13} \oplus a_9a_{10}a_{11}a_{12}a_{13} \oplus$$
$$a_9a_{14} \oplus a_{10}a_{14} \oplus a_8a_{10}a_{14} \oplus a_9a_{10}a_{14} \oplus a_8a_9a_{10}a_{14} \oplus$$
$$a_{11}a_{14} \oplus a_8a_{11}a_{14} \oplus a_9a_{11}a_{14} \oplus a_8a_9a_{11}a_{14} \oplus a_9a_{10}a_{11}a_{14} \oplus$$
$$a_8a_9a_{10}a_{11}a_{14} \oplus a_8a_{12}a_{14} \oplus a_9a_{12}a_{14} \oplus a_8a_9a_{12}a_{14} \oplus a_{10}a_{12}a_{14} \oplus$$

$a_8a_{10}a_{12}a_{14} \oplus a_9a_{10}a_{12}a_{14} \oplus a_8a_9a_{10}a_{12}a_{14} \oplus a_9a_{11}a_{12}a_{14} \oplus a_8a_{10}a_{11}a_{12}a_{14} \oplus$
$a_8a_9a_{13}a_{14} \oplus a_{10}a_{13}a_{14} \oplus a_8a_{10}a_{13}a_{14} \oplus a_8a_9a_{10}a_{13}a_{14} \oplus a_{11}a_{13}a_{14} \oplus$
$a_8a_9a_{11}a_{13}a_{14} \oplus a_8a_{10}a_{11}a_{13}a_{14} \oplus a_9a_{10}a_{11}a_{13}a_{14} \oplus a_8a_9a_{10}a_{11}a_{13}a_{14} \oplus a_{12}a_{13}a_{14} \oplus$
$a_8a_{12}a_{13}a_{14} \oplus a_9a_{12}a_{13}a_{14} \oplus a_8a_9a_{12}a_{13}a_{14} \oplus a_{10}a_{12}a_{13}a_{14} \oplus a_8a_{10}a_{12}a_{13}a_{14} \oplus$
$a_9a_{10}a_{12}a_{13}a_{14} \oplus a_8a_9a_{10}a_{12}a_{13}a_{14} \oplus a_9a_{11}a_{12}a_{13}a_{14} \oplus a_8a_{10}a_{11}a_{12}a_{13}a_{14} \oplus$
$a_9a_{10}a_{11}a_{12}a_{13}a_{14} \oplus$
$a_8a_{15} \oplus a_9a_{15} \oplus a_8a_9a_{10}a_{15} \oplus a_8a_{10}a_{11}a_{15} \oplus a_8a_9a_{10}a_{11}a_{15} \oplus$
$a_8a_9a_{12}a_{15} \oplus a_8a_{10}a_{12}a_{15} \oplus a_9a_{10}a_{12}a_{15} \oplus a_8a_9a_{10}a_{12}a_{15} \oplus a_{11}a_{12}a_{15} \oplus$
$a_8a_{11}a_{12}a_{15} \oplus a_8a_9a_{11}a_{12}a_{15} \oplus a_8a_9a_{10}a_{11}a_{12}a_{15} \oplus a_{13}a_{15} \oplus a_8a_9a_{13}a_{15} \oplus$
$a_8a_{10}a_{13}a_{15} \oplus a_{11}a_{13}a_{15} \oplus a_{10}a_{11}a_{13}a_{15} \oplus a_9a_{10}a_{11}a_{13}a_{15} \oplus a_{12}a_{13}a_{15} \oplus$
$a_9a_{12}a_{13}a_{15} \oplus a_8a_9a_{12}a_{13}a_{15} \oplus a_{10}a_{12}a_{13}a_{15} \oplus a_8a_{10}a_{12}a_{13}a_{15} \oplus a_9a_{10}a_{12}a_{13}a_{15} \oplus$
$a_{10}a_{11}a_{12}a_{13}a_{15} \oplus a_8a_{10}a_{11}a_{12}a_{13}a_{15} \oplus a_9a_{10}a_{11}a_{12}a_{13}a_{15} \oplus a_{14}a_{15} \oplus a_8a_{14}a_{15} \oplus$
$a_8a_{10}a_{14}a_{15} \oplus a_8a_9a_{10}a_{14}a_{15} \oplus a_{11}a_{14}a_{15} \oplus a_8a_{10}a_{11}a_{14}a_{15} \oplus a_9a_{10}a_{11}a_{14}a_{15} \oplus$
$a_8a_9a_{10}a_{11}a_{14}a_{15} \oplus a_{12}a_{14}a_{15} \oplus a_9a_{12}a_{14}a_{15} \oplus a_{10}a_{12}a_{14}a_{15} \oplus a_8a_{10}a_{12}a_{14}a_{15} \oplus$
$a_8a_9a_{10}a_{12}a_{14}a_{15} \oplus a_{11}a_{12}a_{14}a_{15} \oplus a_9a_{11}a_{12}a_{14}a_{15} \oplus a_{10}a_{11}a_{12}a_{14}a_{15} \oplus a_8a_{10}a_{11}a_{12}a_{14}a_{15} \oplus$
$a_9a_{10}a_{11}a_{12}a_{14}a_{15} \oplus a_{13}a_{14}a_{15} \oplus a_8a_{13}a_{14}a_{15} \oplus a_9a_{10}a_{13}a_{14}a_{15} \oplus a_8a_9a_{10}a_{13}a_{14}a_{15} \oplus$
$a_{11}a_{13}a_{14}a_{15} \oplus a_9a_{11}a_{13}a_{14}a_{15} \oplus a_8a_{10}a_{11}a_{13}a_{14}a_{15} \oplus a_9a_{10}a_{11}a_{13}a_{14}a_{15} \oplus a_8a_{12}a_{13}a_{14}a_{15} \oplus$
$a_9a_{12}a_{13}a_{14}a_{15} \oplus a_9a_{10}a_{12}a_{13}a_{14}a_{15} \oplus a_9a_{11}a_{12}a_{13}a_{14}a_{15}$

Improving Fast Algebraic Attacks

Frederik Armknecht*

Theoretische Informatik
Universität Mannheim, 68131 Mannheim, Germany
armknecht@th.informatik.uni-mannheim.de

Abstract. An algebraic attack is a method for cryptanalysis which is based on finding and solving a system of nonlinear equations. Recently, algebraic attacks where found helpful in cryptanalysing LFSR-based stream ciphers. The efficiency of these attacks greatly depends on the degree of the nonlinear equations. At Crypto 2003, Courtois [8] proposed Fast Algebraic Attacks. His main idea is to decrease the degree of the equations using a precomputation algorithm. Unfortunately, the correctness of the precomputation step was neither proven, nor was it obvious.

The three main results of this paper are the following: First, we prove that Courtois' precomputation step is applicable for cryptographically reasonable LFSR-based stream ciphers. Second, we present an improved precomputation algorithm. Our new precomputation algorithm is parallelisable, in contrast to Courtois' algorithm, and it is more efficient even when running sequentially. Third, we demonstrate the improved efficiency of our new algorithm by applying it to the key stream generator E_0 from the Bluetooth standard. In this case, we get a theoretical speed-up by a factor of about 8, even without any parallelism. This improves the fastest attack known. Practical tests confirm the advantage of our new precomputation algorithm for the test cases considered.

Keywords: Algebraic attacks, stream ciphers, linear feedback shift registers, Bluetooth

1 Introduction

Stream ciphers are designed for online encryption of secret plaintext bit streams $M = (m_1, m_2, \cdots)$, $m_i \in \mathbb{F}_2$, which have to pass an insecure channel. Depending on a secret key $\mathcal{K}$, the stream cipher produces a regularly clocked key stream $\mathcal{Z} = (z_1, z_2, \ldots)$, $z_i \in \mathbb{F}_2$, and encrypts M by adding both streams termwise over $\mathbb{F}_2$. The legal receiver who uses the same stream cipher and the same key $\mathcal{K}$, decrypts the received message by applying the same procedure.

Many popular stream ciphers are LFSR-based. They consist of some linear feedback shift registers (LFSRs) and an additional device, called the nonlinear

* This work has been supported by grant 620307 of the DFG (German Research Foundation).

B. Roy and W. Meier (Eds.): FSE 2004, LNCS 3017, pp. 65–82, 2004.

combiner. An LFSR produces a sequence over $\mathbb{F}_2$ depending on its initial state. They can be constructed very efficiently in hardware and can be chosen such that the produced sequence has a high period and good statistical properties. For these reasons, LFSR-based stream ciphers are widely used in cryptography.

A lot of different nontrivial approaches to the cryptanalysis of LFSR-based stream ciphers were discussed in the literature, e.g. fast correlation attacks (Meier, Staffelbach [18], Chepyzhov, Smeets [5], Johansson, Joensson [12, 13]), backtracking attacks (Golic [10], Zenner, Krause, Lucks [24], Fluhrer, Lucks [9], Zenner [23]), time-space tradeoffs (Biryukov, Shamir [4]), BDD-based attacks (Krause [15]) etc. For such stream ciphers many corresponding design criteria (correlation immunity, large period and linear complexity, good local statistics etc.) were developed, e.g. Rueppel [20]. Recently, a new kind of attack was proposed: algebraic attack. For some ciphers, algebraic attacks outmatched all previously known attacks (e.g. Courtois, Meier [7], Armknecht, Krause [1], Courtois [6]).

An algebraic attack consists of two steps: First find a system of equations in the bits of the secret key $\mathcal{K}$ and the output bits z_t. If the combiner is memoryless, the methods of [7] can be used. A general method, which applies to combiners with memory too, was presented in [1]. If enough low degree equations and known key stream bits are given, the secret key $\mathcal{K}$ can be recovered by solving this system of equations in a second step. For this purpose, several methods (Linearization, XL, XSL, Groebner bases, ...) exist. Most of them run faster if the degree of the equations is low. Hence, the search for systems of low degree equations is a desirable goal in algebraic attacks.

Having this in mind, fast algebraic attacks were introduced by Courtois at Crypto 2003 [8], using a very clever idea. Before solving the system of equations, the degree of the system of equations is decreased in a precomputation step. The trick is to eliminate all high degree monomials independent of the key stream bits which has to be done only once. For this purpose, an algorithm A was proposed, using the well known Berlekamp-Massey algorithm [17, 19]. Unfortunately, the correctness of A was neither proven, nor is it obvious. Therefore, the applicability of fast algebraic attacks on LFSR-based stream ciphers in general was still an open question.

In this paper, we finally give a positive answer to this question. We present a non-trivial proof that fast algebraic attacks work under rather weak assumptions which are satisfied by all LFSR-based stream ciphers we know. In particular, it is not necessary that the periods are pairwise co-prime which is not true in general (e.g., see the cipher used in the Bluetooth standard). To get the result, we prove some statements about minimal polynomials of linear recurring sequences. To the best of our knowledge, these statements have not been published elsewhere in open literature. Thus, our results may be of independent interest.

In general, an attacker has knowledge of the whole cipher except the secret key. Hence, it seems to be logical to exploit these information to perform the precomputation step more directly, reducing the number of operations. This was the motivation for developing a new algorithm B. Contrary to A, which runs strictly sequential, B can be performed partly in parallel. Thus, our new method improves the efficiency of fast algebraic attacks significantly.

We demonstrate this by applying B to the E_0 key stream generator used in the Bluetooth standard for wireless communication. A theoretical examination shows that our algorithm has a speed-up factor of about 8, even without any parallelism. This improves the fastest known attack against the E_0 cipher. Practical tests on reduced versions of E_0 confirm the advantage of B for the test cases considered.

The paper is organized as follows: in Section 2 we describe fast algebraic attacks. In Section 3 we give the missing correctness proof for fast algebraic attacks. In Section 4 we improve fast algebraic attacks by introducing a new precomputation step. In Section 5 we demonstrate the higher efficiency of our new method in theory and practice. Section 6 concludes the paper.

2 Fast Algebraic Attacks

Let $\mathcal{Z} := (z_t) := (z_t)_{t=0}^{\infty}$ be the key stream produced by an LFSR-based key stream generator, using a secret key $\mathcal{K} \in \{0,1\}^n$. $\mathcal{K}$ is defined as the initial states of the used LFSRs. At each clock t, the internal state of the cipher consists of $L^t(\mathcal{K})$ and some additional memory bits[1], where $L : \{0,1\}^n \to \{0,1\}^n$ is a linear boolean function known to the attacker. An algebraic attack works as follows:

The first step is to find a boolean function $\hat{F} \neq 0$ such that for an integer $\delta \geq 0$ the equation

$$\hat{F}(L^t(\mathcal{K}), \ldots, L^{t+\delta}(\mathcal{K}), z_t, \ldots, z_{t+\delta}) = 0 \tag{1}$$

is true for all clocks t. If the cipher is memoryless, the methods described in [7] can be used. Unfortunately, these methods neither work with ciphers using memory, nor are they guaranteed to find equations with the lowest degree. Therefore, in general the method proposed in [1] is the better choice. It can be applied to every cipher and finds for certain the equations with the lowest degree.

The second step is to use (1) to get a system of equations describing $\mathcal{K}$ in dependence of the observed key stream $\mathcal{Z}$. For each row $z_t, \ldots, z_{t+\delta}$ of output bits known to the attacker, replace these values in (1). The result is a valid equation in the bits of the secret key $\mathcal{K}$.

The third and final step is to recover the secret key $\mathcal{K}$ by solving this system of equations. One possibility to do so is the linearization method. Due to the linearity of L, all equations (1) have a degree $\leq \deg \hat{F}$. Therefore, the number m of different monomials occurring is limited. If the attacker has enough known key stream bits at his disposal the number of linearly independent equations equals m. By substituting each monomial by a new variable, the attacker gets

[1] Where zero memory bits are possible.

a linear system of equations in m unknowns which can be solved by Gaussian elimination or more refined methods like the one by Strassen [22].

In general, m will be about $\binom{n}{d}$. Hence, the lower the degree d the more efficient the attack. Therefore, an attacker using an algebraic attack will always try to find a system of low degree equations. A very clever approach, called "fast algebraic attack", to decrease the degree of a given system of equations was presented in [8]. We will discuss this method now. Suppose that (1) can be rewritten as

$$\begin{aligned} 0 &= \hat{F}(L^t(\mathcal{K}), \ldots, L^{t+\delta}(\mathcal{K}), z_t, \ldots, z_{t+\delta}) \\ &= F(L^t(\mathcal{K}), \ldots, L^{t+\delta}(\mathcal{K})) + G(L^t(\mathcal{K}), \ldots, L^{t+\delta}(\mathcal{K}), z_t, \ldots, z_{t+\delta}) \\ &=: F_t(\mathcal{K}) + G_t(\mathcal{K}, \mathcal{Z}) \end{aligned} \tag{2}$$

where the degree e of G in $\mathcal{K}$ is lower than the degree d of $\hat{F}$.[2] Furtheron, we assume that the attacker knows coefficients $c_0, \ldots, c_{T-1} \in \{0, 1\}$ such that

$$\sum_{i=0}^{T-1} c_i \cdot F_{t+i}(\mathcal{K}) = 0 \quad \forall t, \mathcal{K}. \tag{3}$$

Using (2) and (3), we get by $\sum_{i=0}^{T-1} c_i \cdot G_{t+i}(\mathcal{K}, \mathcal{Z}) = 0$ an equation in $\mathcal{K}$ and $\mathcal{Z}$ with a lower degree $e < d$. Therefore, the attacker can reduce the system of equations given by (1) with degree d into a new system of equations of degree $e < d$ where all equations are of the type $\sum c_i G_{t+i}$. Note, that this improves the third step enormously, but requires more known key stream bits z_t to perform step two.

Of course, it is vital for the whole approach that such coefficients $c_0, \ldots, c_{T-1}$ can be found efficiently. In [8], the following algorithm A was proposed:

1. Chose a reasonable[3] key $\hat{\mathcal{K}}$ and compute $\hat{z}_t := F_t(\hat{\mathcal{K}})$ for $t = 1, \ldots, 2T$.
2. Apply the Berlekamp-Massey algorithm to find $c_0, \ldots, c_{T-1}$ with

$$\sum_{i=0}^{T-1} c_i \cdot F_{t+i}(\hat{\mathcal{K}}) = 0 \quad \forall t. \tag{4}$$

It is known that the Berlekamp-Massey algorithm finds coefficients with the smallest value of T fulfilling (4). This needs about $\mathcal{O}(T^2)$ basic operations. Together with the first step, algorithm A has to perform about $\mathcal{O}(T^2 + 2T|\mathcal{K}|)$ steps. In general, the exact value of T is unknown but an upper bound is the maximum number of different monomials occurring.

The result of algorithm A is correct if (4) implies (3) which has not been proven in [8]. The only correctness argument indicated there was based on the

[2] For example, this assumption is true for the three ciphers E_0, Toyocrypt and LILI-128.

[3] In [8], $\hat{\mathcal{K}}$ can be any value in $\{0, 1\}^n$. But if one of the LFSRs is initialised with the all-zero state, the algorithm returns a wrong result in most cases. Therefore, $\hat{\mathcal{K}}$ has to be chosen such that the initial states are all non-zero.

assumption that the sequences produced by the LFSRs have pairwise co-prime periods. But this is not true in general. A counter-example is the key stream generator E_0 used in the Bluetooth standard for wireless communication: the two periods $2^{33} - 1$ and $2^{39} - 1$ share the common factor 7.

On the other hand algorithm A does not work correctly in general without any preconditions. An example is given in appendix D. This raises the question which preconditions are necessary for the correctness of algorithm A and if they are fulfilled by E_0.

One of the major achievements from this paper is to show the correctness of algorithm A under weaker assumptions. As we are not aware of any LFSR-based stream cipher for which these conditions are not true (including E_0), we assume that fast algebraic attacks as described in [8] can be mounted against most LFSR-based stream ciphers discussed in public.

3 Proof of Correctness

In this section, we prove the correctness of algorithm A under cryptographically reasonable assumptions. First, we repeat some known facts about linear recurring sequences.

Theorem 1. *(Lidl, Niederreiter [16]) A sequence $\mathcal{Z} = (z_t)$ over $\mathbb{F}_2$ is called a linear recurring sequence if coefficients $c_0, \ldots, c_{T-1} \in \{0,1\}$ (not all zero) exist such that $\sum c_i z_{t+i} = 0$ is true for all values $t \geq 1$. In this case, $\sum c_i x^i \in \mathbb{F}_2[x]$ is called a characteristic polynomial of the sequence $\mathcal{Z}$. Amongst all characteristic polynomials of $\mathcal{Z}$ exists one unique polynomial $min(\mathcal{Z})$ which has the lowest degree. We will call it the minimal polynomial of $\mathcal{Z}$. A polynomial $f(x) \in \mathbb{F}_2[x]$ is a characteristic polynomial of $\mathcal{Z}$ if and only if $min(\mathcal{Z})$ divides $f(x)$.*

From now on, a sequence $\mathcal{Z}$ will always stand for a linear recurring sequence. Furtheron, we will denote by $\overline{\mathbb{F}_2}$ the algebraic closure of the field $\mathbb{F}_2$.[4] By the roots of $f(x)\mathbb{F}_2[x]$, we will always mean the roots in $\overline{\mathbb{F}_2}$.

Definition 1. *Let $R_1, \ldots, R_\kappa \subseteq \overline{\mathbb{F}_2}$ be pairwise disjunct, $R := R_1 \dot{\cup} \ldots \dot{\cup} R_\kappa$. We say that a pair of vectors $(\alpha_1, \ldots, \alpha_n) \in R^n$, $(\beta_1, \ldots, \beta_m) \in R^m$ factorizes uniquely over $R_1, \ldots, R_\kappa$ if the following holds*

$$\alpha_1 \cdot \ldots \cdot \alpha_n = \beta_1 \cdot \ldots \cdot \beta_m \Rightarrow \prod_{\alpha_i \in R_l} \alpha_i \cdot \prod_{\beta_j \in R_l} (\beta_j)^{-1} = 1, \quad 1 \leq l \leq \kappa$$

For a monomial $\mu = \prod_{j=1}^{k} x_{i_j} \in \mathbb{F}_2[x_1, \ldots, x_n]$ with $\{i_1, \ldots, i_k\} \subseteq \{1, \ldots, n\}$ and $\alpha = (\alpha_1, \ldots, \alpha_n) \in R^n$, we define the vector $\overrightarrow{\mu(\alpha)} := (\alpha_{i_1}, \ldots, \alpha_{i_k}) \in R^k$.

Example 1. Set $R_1 := \{\alpha, \alpha\beta\}$ and $R_2 := \{\beta\}$ with $\beta \neq 1$. The pair of vectors (α, β) and $(\alpha\beta)$ does not factorize uniquely over R_1, R_2 because of $\alpha \cdot \beta = \alpha\beta$ but $\alpha \cdot (\alpha\beta)^{-1} = \beta^{-1} \neq 1$.

[4] I.e., $\overline{\mathbb{F}_2}$ is the smallest field such that $\mathbb{F}_2 \subset \overline{\mathbb{F}_2}$ and each polynomial $f(x) \in \mathbb{F}_2[x]$ has at least one root in $\overline{\mathbb{F}_2}$.

The motivation for this definition is that we need in our main theorem that certain products of roots of minimal polynomials are unique in the sense above (see appendix A. The main theorem of our paper is:

Theorem 2. *Let* $\mathcal{Z}_1 = (z_t^{(1)}), \ldots, \mathcal{Z}_\kappa = (z_t^{(\kappa)})$ *be sequences with pairwise co-prime minimal polynomials which have only non-zero roots. Let* R_i *denote the set of roots of* $min(\mathcal{Z}_i)$ *in* $\overline{\mathbb{F}_2}$, $F : \mathbb{F}_2^n \to \mathbb{F}_2$ *be an arbitrary boolean function and* $I := (i_1, \ldots, i_n) \in \{1, \ldots, \kappa\}^n$ *and* $\delta := (\delta_1, \ldots, \delta_\kappa) \in \mathbb{N}^\kappa$ *be two vectors.*

We set $R := R_{i_1} \times \ldots \times R_{i_n}$ *and divide* $F = \sum \mu_i$ *into a sum of monomials. Furtheron, for arbitrary* $d := (d_1, \ldots, d_\kappa) \in \mathbb{N}^\kappa$ *the sequences* $\mathcal{Z} := (z_t)$ *and* $\mathcal{Z}^{(d)} := (z_t^{(d)})$ *are defined by*

$$z_t := F(z_{t+\delta_1}^{(i_1)}, \ldots, z_{t+\delta_n}^{(i_n)}),\ z_t^{(d)} := F(z_{t+\delta_1+d_{i_1}}^{(i_1)}, \ldots, z_{t+\delta_n+d_{i_n}}^{(i_n)})$$

If all pairs of vectors $\overrightarrow{\mu_i(\alpha)}, \overrightarrow{\mu_j(\alpha')}$ *with* $\alpha, \alpha' \in R$ *factorize uniquely over* $R_1, \ldots, R_\kappa$*, then* $min(\mathcal{Z}) = min(\mathcal{Z}^{(d)})$.

What is the connection to algorithm A? From the theory of LFSRs, it can be easily argued that the sequence $\hat{\mathcal{Z}} = (\hat{z}_t)$ from algorithm A is a linear recurring sequence and that $\hat{c}_0, \ldots, \hat{c}_{T-1}$ correspond to its minimal polynomial m. $\hat{\mathcal{Z}}$ is produced in the way sequence $\mathcal{Z}$ is described in theorem 2 which assures that the minimal polynomial m remains unchanged if we shift each of the sequences produced by the LFSRs individually. As in the general the produced sequences have maximal period, the minimal polynomial found by algorithm A is the same for each possible key $\mathcal{K}$. In appendix B we show that the conditions of theorem 2 are satisfied for a large class of LFSR-based ciphers automatically, independent of F,I and δ. Before we can prove theorem 2, we need some statements about minimal polynomials.

Theorem 3. *([16], Theorem 6.21) Let* $\mathcal{Z} = (z_t)$ *be a sequence with characteristic polynomial* $f(x) = \prod_{i=1}^n (x - \alpha_i)$ *where the roots lie in* $\overline{\mathbb{F}_2}$*. If the roots* $\alpha_1, \ldots, \alpha_n$ *are all distinct, i.e. each root has multiplicity one, then for each t,* z_t *can be expressed in the following way:*

$$z_t = \sum_{i=1}^n A_i \alpha_i^t$$

where $A_1, \ldots, A_t \in \overline{\mathbb{F}_2}$ *are uniquely determined by the initial values of the sequence* $\mathcal{Z}$.

Theorem 4. *Let* $\mathcal{Z} = (z_t)$ *be a sequence with* $z_t = \sum_{i=1}^n A_i \alpha_i^t$ *with pairwise distinct elements* $\alpha_i \in \overline{\mathbb{F}_2}$ *and non-zero coefficients* A_i*. Let* $m(x) \in \mathbb{F}_2[x]$ *be the polynomial with the lowest degree such that* $m(\alpha_i) = 0$ *for* $1 \le i \le n$*. Then* $m(x)$ *is the minimal polynomial* $min(\mathcal{Z})$*. In particular, each root of* $min(\mathcal{Z})$ *has multiplicity one.*

Proof. We show that $f(x) \in \mathbb{F}_2[x]$ is a characteristic polynomial of $\mathcal{Z}$ if and only if $f(\alpha_i) = 0$ for all i. Thus, $m(x)$ is the characteristic polynomial with the lowest degree what is the definition of $min(\mathcal{Z})$. Let $f(x) = \sum_{k=0}^{r} c_k x^k$. Then for each t, we have

$$\sum_{k=0}^{r} c_k z_{t+k} = \sum_{k=0}^{r} c_k \left(\sum_{i=1}^{n} A_i \alpha_i^{t+k} \right) = \sum_{i=1}^{n} \left(A_i \sum_{k=0}^{r} c_k \alpha_i^k \right) \alpha_i^t = \sum_{i=1}^{n} (A_i f(\alpha_i))\, \alpha_i^t$$

For $1 \le i \le n$ and $0 \le t \le n-1$, let $M := (\alpha_i^t)$ be a Vandermonde-matrix of size $n \times n$. As the elements α_i are pairwise distinct, M is regular. Thus, the expression above equals to zero for each t if and only if $(A_1 f(\alpha_1), \ldots, A_n f(\alpha_n)) \in \{0,1\}^n$ is an element of the kernel of M, i.e. $A_i f(\alpha_i) = 0$. As the coefficients A_i were assumed to be non-zero, this is equivalent to $f(\alpha_i) = 0$ for all i. □

Proof of theorem 2. By theorem 4 all roots in R_i have multiplicity one. Therefore, by theorem 3, each sequence $\mathcal{Z}_i$ can be expressed by $z_t^{(i)} = \sum_{\alpha \in R_i} A_\alpha \alpha^t$ with unique coefficients A_α. For each i it holds

$$z_{t+\delta_i}^{(i)} = \sum_{\alpha \in R_i} A_\alpha \alpha^{t+\delta_i} = \sum_{\alpha \in R_i} \left(A_\alpha \alpha^{\delta_i} \right) \alpha^t$$

and therefore

$$z_t = F\Big(\sum_{\alpha \in R_{i_1}} (A_\alpha \alpha^{\delta_{i_1}}) \alpha^t, \ldots, \sum_{\alpha \in R_{i_n}} (A_\alpha \alpha^{\delta_{i_n}}) \alpha^t \Big)$$

We set $\mathcal{P} := \{\mu_i(\alpha) | \alpha \in R, 1 \le i \le l\}$. The sequences $\mathcal{Z}$ and $\mathcal{Z}^{(d)}$ can be expressed by

$$z_t = \sum_{\pi \in \mathcal{P}} A_\pi \pi^t, \quad z_t^{(d)} = \sum_{\pi \in \mathcal{P}} A_\pi^{(d)} \pi^t$$

with unique coefficients A_π and $A_\pi^{(d)}$. We show that A_π is non-zero if and only if $A_\pi^{(d)}$ is non-zero. Then the equality of $min(\mathcal{Z})$ and $min(\mathcal{Z}^{(d)})$ follows by theorem 4.

We express the coefficients A_π and $A_\pi^{(d)}$ in dependence of the coefficients A_α. For $\pi = \alpha_1 \cdot \ldots \cdot \alpha_m \in \mathcal{P}$, $\alpha_i \in \dot{\bigcup} R_i$, we define by π^d the product $\left(\prod_{\alpha_i \in R_1} \alpha_i^{d_1} \right) \cdot \ldots \cdot \left(\prod_{\alpha_i \in R_\kappa} \alpha_i^{d_\kappa} \right)$. As all pairs of vectors $\vec{\pi_1}, \vec{\pi_2}$ factorize uniquely over $R_1, \ldots, R_\kappa$, this expression is independent of the factorization $\alpha_1 \cdot \ldots \cdot \alpha_m$. Analogously, for $\alpha = (\alpha_1, \ldots, \alpha_n) \in R$ we set $\alpha^d := (\alpha_1^{d_{i_1}}, \ldots, \alpha_n^{d_{i_n}})$. With these definitions, we get

$$z_t = \sum_{\pi \in \mathcal{P}} \underbrace{\sum_{\mu_i} \sum_{\substack{\alpha \in \mathcal{R} \\ \mu_i(\alpha) = \pi}} \mu_i(A_\alpha)\, \pi^t}_{= A_\pi}$$

where $A_\alpha = (A_{\alpha_{i_1}}, \ldots, A_{\alpha_{i_n}})$. Therefore, the coefficients $A_\pi^{(d)}$ can be expressed by

$$A_\pi^{(d)} = \sum_{\mu_i} \sum_{\substack{\alpha \in \mathcal{R} \\ \mu_i(\alpha)=\pi}} \mu_i(A_\alpha \alpha^d) = \sum_{\mu_i} \sum_{\substack{\alpha \in \mathcal{R} \\ \mu_i(\alpha)=\pi}} \mu_i(A_\alpha)\mu_i(\alpha^d) = \sum_{\mu_i} \sum_{\substack{\alpha \in \mathcal{R} \\ \mu_i(\alpha)=\pi}} \mu_i(A_\alpha)\pi^d$$
$$= A_\pi \cdot \pi^d$$

As the roots of $m_i(x)$ are all non-zero by assumption, it is $\pi^d \neq 0$. Therefore, $A_\pi \neq 0$ iff $A_\pi^{(d)} \neq 0$. □

The proof shows why a precondition is necessary for the correctness of the pre-computation step. Otherwise, it could happen that for some π it is $A_\pi \neq 0$ but $A_\pi^{(d)} = 0$ (or vice versa). In this cases, the corresponding minimal polynomials could be different (see appendix D).

4 Improvements

In this section, we show how algorithm A can be improved. Let $min(F)$ denote the (unique) minimal polynomial found by algorithm A. Until now, we made no use of the knowledge of the minimal polynomials $m_1(x), \ldots, m_\kappa(x)$. The idea is to compute $min(F)$ and/or the parameter T more or less directly from the known minimal polynomials and F. For this purpose, we cite some statements about minimal polynomials.

Definition 2. *Consider two co-prime polynomials* $f(x) = \prod_{i=1}^n (x - \alpha_i)$ *and* $g(x) = \prod_{j=1}^m (x - \beta_j)$ *with no multiple roots. Then we define*

$$f(x) \otimes g(x) := \prod_{i,j} (x - \alpha_i \beta_j), \quad f(x) \otimes f(x) := \prod_{1 \le i < j \le n} (x - \alpha_i \alpha_j) \cdot f(x).$$

Theorem 5. *([16], Th. 6.57 + 6.67) Let* $\mathcal{Z}_1 = (z_t^{(1)}), \ldots, \mathcal{Z}_\kappa = (z_t^{(\kappa)})$ *be sequences with pairwise co-prime* $min(\mathcal{Z}_i)$. *Then*

$$min(\mathcal{Z}_1 + \ldots + \mathcal{Z}_\kappa) = min(\mathcal{Z}_1) \cdot \ldots \cdot min(\mathcal{Z}_\kappa)$$
$$min(\mathcal{Z}_i \cdot \mathcal{Z}_j) = min(\mathcal{Z}_i) \otimes min(\mathcal{Z}_j), \quad \forall i \neq j$$

where $\mathcal{Z}_1 + \ldots + \mathcal{Z}_\kappa := (z_t^{(1)} + \ldots + z_t^{(\kappa)})$ *and* $\mathcal{Z}_i \cdot \mathcal{Z}_j := (z_t^{(i)} \cdot z_t^{(j)})$.

Theorem 6. *(Key [14], Th. 1) Let* $\mathcal{Z} = (z_t)$ *be a sequence and* $l := \deg(min(\mathcal{Z}))$. *If* d *is an integer with* $1 \le d < l$, *then the sequence* $(z_t \cdot z_{t+d})$ *has the minimal polynomial* $min(\mathcal{Z}) \otimes min(\mathcal{Z})$ *of degree* $\frac{l(l+1)}{2}$.

Before we proceed further, we will take a closer look on the complexities of the operators "$\cdot$" and "$\otimes$".

Theorem 7. *(Schoenhage [21]) Let two polynomials* $f(x)$ *resp.* $g(x)$ *of* $\mathbb{F}_2[x]$ *be given of degrees* $\le m$. *Then, the product* $f(x) \cdot g(x)$ *can be computed with an effort of* $\mathcal{O}(m \log m \log \log m)$.

Theorem 8. *(Bostan, Flajolet, Salvy, Schost [3], Theorem 1) Let $f(x)$ resp. $g(x)$ be two co-prime polynomials of degree n resp. m with no multiple roots. Then the polynomial $f(x) \otimes g(x)$ can be computed directly within*

$$\mathcal{O}(\underbrace{nm\log^2(nm/2)\log\log(nm/2) + nm\log(nm)\log\log(nm)}_{=:\mathcal{T}(nm)})$$

operations in $\mathbb{F}_2$ without knowing the roots of $f(x)$ or $g(x)$.

This implies a kind of divide-and-conquer approach for computing the minimal polynomial from F. The trick is to split the function F into two or more functions $F_1, \ldots, F_l$ such that the corresponding minimal polynomials $min(F_i)$ are pairwise co-prime. Then by theorem 5 it holds $min(F) = min(F_1) \cdot \ldots \cdot min(F_l)$. In some cases, such a partition can be hard to find or may not even exist. If the minimal polynomials $min(F_i)$ are not pairwise co-prime the product $p(x) := min(F_1) \cdot \ldots \cdot min(F_l)$ is a characteristic polynomial of each possible sequence $\hat{\mathcal{Z}}$, i.e. the coefficients of $p(x)$ fulfill equation (3) also. Therefore, using $p(x)$ makes a fast algebraic attack possible though it may require more known key stream bits than really necessary.

We compare the effort of this approach to that of algorithm A. For simplicity we assume $l = 2$, i.e. $min(F) = min(F_1) \cdot min(F_2)$. Let $T_1 := \deg(min(F_1)) \leq \deg(min(F_2)) =: T_2$. Then $\deg(min(F)) = T_1 + T_2$. As said before, algorithm A needs about

$$\mathcal{O}(T_1^2 + T_2^2 + 2(T_1 + T_2)|\mathcal{K}| + 2T_1T_2))$$

basic operations. Instead of using algorithm A, we can do the following: First compute $min(F_1)$ and $min(F_2)$. In general, this can be done with algorithm A or in some cases by using the $\otimes$-product (see details later). If we use algorithm A, the complexity of these operations are $\mathcal{O}(T_1^2+2T_1)$ resp. $\mathcal{O}(T_2^2+2T_2)$. Notice that both operations can be performed in parallel. Having computed $min(F_1)$ and $min(F_2)$, the second and final step consists of computing the product $min(F_1) \cdot min(F_2) = min(F)$. By theorem 7, this has an effort of $\mathcal{O}(T_2 \log T_2 \log\log T_2)$ which implies an overall effort of

$$\mathcal{O}(T_1^2 + T_2^2 + 2(T_1 + T_2)|\mathcal{K}| + T_2 \log T_2 \log\log T_2)$$

In general, it is $\log T_2 \log\log T_2 \ll 2T_1$. Thus, our new approach has a lower runtime than algorithm A. The advantage increases if F can be divided in more than two parts.

In some cases, the precomputation step can be improved even a bit further. Assume that at least one of the F_i mentioned above can be written as a product $F_i = G_1 \cdot G_2$ such that $min(G_1)$ and $min(G_2)$ are co-prime. This is for example almost always the case when F_i is a monomial. Then, by theorem 5 it holds $min(F_i) = min(G_1) \otimes min(G_2)$ which implies a similar strategy. Let again be $T_1 := \deg(min(G_1)) \leq \deg(min(G_2)) =: T_2$. Using algorithm A would need about

$$\mathcal{O}(T_1^2T_2^2 + 2T_1T_2|\mathcal{K}|)$$

operations. Instead, we can we compute $min(G_1)$ and $min(G_2)$ with algorithm A in a first step. This takes $\mathcal{O}(T_1^2 + T_2^2 + 2(T_1 + T_2)|\mathcal{K}|)$ operations. If $min(G_1)$ and/or $min(G_2)$ are already known, than this step can be omitted. This is for example the case if F_i is the product of the output of two or several distinct LFSRs (see the E_0 example in appendix C). In the second step, we use the algorithm described in [3] to compute $min(G_1) \otimes min(G_2)$. The effort is $\mathcal{O}(\mathcal{T}(T_1T_2))$ which is in $\mathcal{O}(T_1T_2 \log^2(T_1T_2) \log\log(T_1T_2))$. Altogether, this approach needs

$$\mathcal{O}(T_1T_2 \log^2(T_1T_2) \log\log(T_1T_2) + T_1^2 + T_2^2 + 2(T_1 + T_2)|\mathcal{K}|)$$

operations. This shows the improvement. If we perform the operations of the first step in parallel, the time needed to get the result can be decreased further. In the following, we summarize our approaches by proposing the following new algorithm B:

Algorithm B

Given: Pairwise co-prime primitive polynomials $m_1(x), \ldots, m_k(x)$, an arbitrary boolean function F as described in section 2 and a partition $F = F_1 + \ldots + F_l$ such that the minimal polynomials $min(F_i)$ are pairwise co-prime

Task: Find $min(F)$

Algorithm:

- Compute the minimal polynomials $min(F_i)$ by using algorithm A. This can be done in parallel.
  If $F_i = G_1 \cdot G_2$ with co-prime $min(G_1)$ and $min(G_2)$ (e.g., F_i is a monomial), then the $\otimes$-product algorithm described in [3] can be used.
- Compute $min(F) = min(F_1) \cdot \ldots \cdot min(F_l)$ using the algorithm in [21].

5 Application to the E_0 Key Stream Generator

In this section, we demonstrate the efficiency of algorithm B on the E_0 key stream generator which is part of the Bluetooth standard for wireless communication [2]. It uses four different LFSRs with pairwise co-prime primitive minimal polynomials m_1, m_2, m_3, m_4 of degrees $T_1 = 25$, $T_2 = 31$, $T_3 = 33$ and $T_4 = 39$. The sequences produced by the LFSRs have the periods $2^{T_1} - 1, \ldots, 2^{T_4} - 1$. As the values $2^{T_3} - 1$ and $2^{T_4} - 1$ share the common factor 7, the assumption made in [8] is not satisfied.

In [1], a boolean function $\hat{F}$ was developed fulfilling equation (1). $\hat{F}$ can be divided as shown in (2) into $F + G$ with $\deg(F) = 4 > 3 = \deg(G)$. The function F is

$$F = \sum_{\substack{1 \le i < j \le 4 \\ 1 \le k < l \le 4}} z_t^{(i)} z_t^{(j)} z_{t+1}^{(k)} z_{t+1}^{(l)} + z_t^{(1)} z_t^{(2)} z_t^{(3)} z_t^{(4)}$$

Table 1. Fastest previous attacks against E_0

Attack	Data	Memory	Pre-computation	Attack Complexity
[1]	2^{24}	2^{48}		2^{68}
[8]	2^{24}	2^{37}	2^{46}	2^{49}
new (sequential)	2^{24}	2^{37}	2^{43}	2^{49}
new (parallel)	2^{24}	2^{37}	2^{42}	2^{49}

where $\mathcal{Z}_i = (z_t^{(i)})$ is the sequence produced by LFSR i. Let R_i be the set of roots of m_i. Inspired by the definition in theorem 2, we define

$$\mathcal{P} := \{\alpha_i\alpha_j\alpha_k\alpha_l | \ \alpha_s \in R_s, 1 \leq i < j \leq 4, 1 \leq k < l \leq 4\} \cup \{\alpha_1\alpha_2\alpha_3\alpha_4 \mid \alpha_i \in R_i\}.$$

We have checked with Maple that for all pairs $\pi_1, \pi_2 \in \mathcal{P}$ the pair of vectors $\vec{\pi_1}$ and $\vec{\pi_2}$ factorize uniquely over the union $R_1 \dot{\cup} R_2 \dot{\cup} R_3 \dot{\cup} R_4$. Hence, the weaker assumptions from theorem 2 are fulfilled here and a unique minimal polynomial $min(F)$ exists. Furtheron, it can be showed that F can be written as $F = F_1 + \ldots + F_{11}$ such that the minimal polynomials $min(F_i)$ are pairwise co-prime (see appendix C). By theorem 5, $min(F)$ can be expressed by

$$min(F) = \prod_{i=1}^{11} min(F_i) \tag{5}$$

The degree of $min(F)$ and thus the parameter T is upper bounded by

$$\sum_{1 \leq i < j \leq 4} \frac{T_i(T_i+1)T_j(T_j+1)}{4} + \sum_{1 \leq i < j < k \leq 4} T_iT_jT_k\frac{T_i+T_j+T_k-1}{2} + T_1T_2T_3T_4$$

In [2], the degrees T_1, T_2, T_3, T_4 are defined as 25, 31, 33, 39 respectively. Thus, the degree of $min(F)$ is $\leq 8.822.188 \approx 2^{23.07}$. The computation of $min(F)$ using algorithm A would need most about $2^{46.15}$ basic operations.

Equation (5) implies the usage of algorithm B. In the first step we apply algorithm A to compute the minimal polynomials $min(F_i)$, $i = 1, \ldots, 11$. This takes an overall number of basic operations of $\approx 2^{43.37}$. Exploiting parallelism, we have only to wait the time needed to perform $\approx 2^{41.91}$ basic operations.[5]

The second step is the computation of the product of these minimal polynomials. Here, this takes altogether about $\approx 2^{28.25}$ basic operations. Performed in sequential, algorithm B needs about $2^{43.37}$ basic operations which is almost 8 times faster than algorithm A. If we exploit the parallelism mentioned above, the number of basic operations we have to wait is about $2^{41.91}$ which is more than 16 times faster than in algorithm A. This improves the fastest known attack against the E_0 cipher. Table 1 sums up the fastest previous attacks known:

[5] Of course, the number of operations remains unchanged.

Table 2. Comparison of algorithm A and B on reduced versions of E_0

C_1	C_2	C_3	C_4	Algorithm A	Algorithm B	A/B
110	1100	10100	1000100	10 h 41 m 43 s	12 m 3 s	53.32
101	1001	10010	1000001	11 h 2 m 49 s	12 m 7 s	54.75
				10 h 50 m 0 s	11 m 59 s	54.30
				10 h 52 m 59 s	11 m 55 s	54.86
				10 h 53 m 31 s	11 m 58 s	54.65
110	1100	10100	10100010100	78 h 30 m 16 s	1 h 43 m 25 s	45.55
101	10100	1100000	11100010000	18 d 18 h 26 m 0 s	13 h 50 m 7 s	32.56

An ad-hoc implementation in Maple (without using parallelism and the algorithm of [3]), applied to reduced versions of E_0 with shorter LFSRs, confirmed the improved efficiency of our new algorithm. The results can be found in table 2. In the first four columns, the coefficients of the four minimal polynomials are given. The next two columns show the time consumptions of algorithm A and B respectively which are compared in the last column. In all cases, our new algorithm B was significantly faster than algorithm A, even without using parallelism. The speed-up factor was much higher than predicted theoretically and depended on the chosen minimal polynomials and the initial states.

In all cases, the degree of $min(F)$ was equal to the upper bound estimated on page 75. Hence, we expect that the upper bound is tight for the real E_0 key stream generator also.

6 Conclusion

In this paper, we discussed the fast algebraic attacks introduced by Courtois at Crypto 2003. Using a very clever idea, "traditional" algebraic attacks can be improved significantly in many cases by performing an efficient precomputation step. For this purpose, an algorithm A was proposed. Unfortunately, neither a correctness proof was given, nor was the correctness obvious.

In this paper, we gave the missing proof based on a cryptographically reasonable assumptions. To do so, it was necessary to prove some non-trivial statements about minimal polynomials of linear recurring sequences. To the best of our knowledge, these have not been published elsewhere in open literature.

In addition, we showed that the knowledge of the minimal polynomials of the LFSRs used in the cipher can help to improve fast algebraic attacks. For this reason, we developed an algorithm B which is based on a deeper understanding of minimal polynomials of linear recurring sequences.

Finally, we demonstrated the higher efficiency of algorithm B by applying it to the E_0 key stream generator used in the Bluetooth standard for wireless communication. In this case, theoretical analysis yielded that algorithm B runs almost 8 times faster than algorithm A. This improves the fastest known attack against E_0. An ad-hoc implementation applied to reduced versions of E_0 confirmed the advantage of our new algorithm for the test cases considered, even without using parallelism.

Acknowledgement

The author would like to thank Erik Zenner, Stefan Lucks, Matthias Krause and some unknown referees for helpful comments and discussions.

References

[1] Frederik Armknecht, Matthias Krause: *Algebraic attacks on Combiners with Memory*, Proceedings of Crypto 2003, LNCS 2729, pp. 162-176, Springer, 2003. 66, 67, 74, 75

[2] Bluetooth SIG, *Specification of the Bluetooth system*, Version 1.1, February 22, 2001. Available at http://www.bluetooth.com/. 74, 75

[3] Alin Bostan, Philippe Flajolet, Bruno Salvy, Eric Schost: *Fast Computation With Two Algebraic Numbers*, submitted, 2003. 73, 74, 76

[4] Alex Biryukov, Adi Shamir: *Cryptanalytic Time/Memory/Data tradeoffs for Stream Ciphers*, Proceedings of Asiacrypt 2000, LNCS 1976, pp. 1-13, Springer, 2000. 66

[5] Vladimor V. Chepyzhov, Ben Smeets: *On A Fast Correlation Attack on Certain Stream Ciphers*, Proceedings of Eurocrypt 1991, LNCS 547 pp. 176-185, Springer, 1991. 66

[6] Nicolas Courtois: *Higher Order Correlation Attacks, XL Algorithm and Cryptanalysis of Toyocrypt*, ICISC 2002, LNCS 2587. An updated version (2002) is available at http://eprint.iacr.org/2002/087/. 66

[7] Nicolas Courtois, Willi Meier: *Algebraic attacks on Stream Ciphers with Linear Feedback*, Eurocrypt 2003, Warsaw, Poland, LNCS 2656, pp. 345-359, Springer, 2003. An extended version is available at http://www.minrnak.org/toyolili.pdf. 66, 67

[8] Nicolas Courtois: *Fast Algebraic Attacks on Stream Ciphers with Linear Feedback*, Proceedings of Crypto '03, LNCS 2729, pp. 177-194, Springer, 2003. 65, 66, 68, 69, 74, 75

[9] Scott R. Fluhrer, Stefan Lucks: *Analysis of the E_0 Encryption System*, Proceedings of Selected Areas of Cryptography '01, LNCS 2259, pp. 38-48, Springer, 2001. 66

[10] Jovan Dj. Golic: *Cryptanalysis of Alleged A5 Stream Cipher*, Proceedings of Eurocrypt 1997, LNCS 1233, pp. 239-255, Springer, 1997. 66

[11] Rainer Goettfert, Harald Niederreiter: *On the Linear Complexity of Products of Shift-Register Sequences*, Proceedings of Eurocrypt '93, pp. 151-158, LNCS 765, Springer, 1994.

[12] Thomas Johansson, Fredrik Joensson: *Fast Correlation Attacks Based on Turbo Code Techniques*, Proceedings of Crypto 1999, LNCS 1666, pp. 181-197, Springer, 1999. 66

[13] Thomas Johansson, Fredrik Joensson: *Improved Fast Correlation Attacks on Stream Ciphers via Convolutional Codes*, Proceedings of Eurocrypt 1999, pp. 347-362, Springer, 1999. 66

[14] Edwin L. Key: *An Analysis of the Structure and Complexity of Nonlinear Binary Sequence Generators*, IEEE Transactions on Information Theory, Vol. IT-22, No. 6, November 1976. 72
[15] Matthias Krause: *BDD-Based Cryptanalysis of Key stream Generators*; Proceedings of Eurocrypt '02, pp. 222-237, LNCS 2332, Springer, 2002. 66
[16] Rudolf Lidl, Harald Niederreiter: *Introduction to finite fields an their applications*, Cambridge University Press, 1994. 69, 70, 72, 82
[17] J. L. Massey: *Shift-register synthesis and BCH decoding*, IEEE Trans. Information Theory, IT-15 (1969), pp. 122-127, 1969. 66
[18] Willi Meier, Othmar Staffelbach: *Fast Correlation Attacks on certain Stream Ciphers*, Journal of Cryptology, pp. 159-176, 1989. 66
[19] Alfred J. Menezes, Paul C. Oorschot, Scott A. Vanstone: *Handbook of Applied Cryptography*, Chapter 6, CRC Press. 66
[20] Rainer A. Rueppel: *Stream Ciphers*; Contemporary Cryptology: The Science of Information Integrity. G. Simmons ed., IEEE Press New York, 1991. 66
[21] A. Schoenhage: *Schnelle Multiplikation von Polynomen ueber Koerpern der Charakteristik 2*, Acta Informatica 7 (1977), pp. 395-398, 1977. 72, 74
[22] Volker Strassen: *Gaussian Elimination is Not Optimal*; Numerische Mathematik, vol 13, pp 354-356, 1969. 68
[23] Erik Zenner: *On the Efficiency of the Clock Control Guessing Attack*, Proceedings of ICISC 2002, LNCS 2587, Springer, 2002. 66
[24] Erik Zenner, Matthias Krause, Stefan Lucks: *Improved Cryptanalysis of the Self-Shrinking Generator ACISP 2001*, LNCS 2119, Springer, 2001. 66

A Motivation

In this section, we give a motivation for the somewhat technical definition 1. For this purpose we discuss some kind of "abstract example". Let $A = (a_t)$ resp. $B = (b_t)$ be two sequences produced by the co-prime minimal polynomials $m_a(x) = \prod(x - \alpha_i)$ and $m_b(x) = \prod(x - \beta_i)$. Then by theorem 3, a_t resp. b_t can be expressed by

$$a_t = \sum_i c_{\alpha_i}\alpha_i^t, \quad b_t = \sum_i c_{\beta_i}\beta_i^t$$

Consequently, the shifted sequences can be expressed by

$$a_{t+d_a} = \sum_i c_{\alpha_i}\alpha_i^{t+d_a} = \sum_i c_{\alpha_i}\alpha_i^{d_a}\alpha_i^t =: \sum_i \tilde{c}_{\alpha_i}\alpha_i^t$$

$$b_{t+d_b} = \sum_i c_{\beta_i}\beta_i^{t+d_b} = \sum_i c_{\beta_i}\beta_i^{d_b}\beta_i^t =: \sum_i \tilde{c}_{\beta_i}\beta_i^t$$

Furtheron, we define the sequences

$$\begin{aligned} z_t = a_t \cdot a_t + b_t \quad &= \textstyle\sum_{i,j} c_{\alpha_i}c_{\alpha_j}\alpha_i^t\alpha_j^t + \sum_i c_{\beta_i}\beta_i^t = \sum_\pi c_\pi \pi^t \\ \tilde{z}_t = a_{t+d_a} \cdot a_{t+d_a} + b_{t+d_b} \quad &= \textstyle\sum_\pi \tilde{c}_\pi \pi^t \end{aligned}$$

where $\pi \in P := \{\alpha_i \cdot \alpha_j\} \cup \{\beta_i\}$. If we assume that all roots α_i and β_j are non-zero[6] this holds for the elements in P too. The goal is to find a necessary

[6] This is for example true if $m_a(x)$ and $m_b(x)$ are irreducible and have a degree > 1.

precondition which guarantees that min(z_t)= min($\tilde{z}_t$) (regardless of the values of d_a and d_b). By theorem 4 we know that such a precondition is

$$c_\pi \neq 0 \Leftrightarrow \tilde{c}_\pi \neq 0 \tag{6}$$

How can we be sure that this is true in our case? We show what can go wrong on an example. Let $\pi \in P$ fixed. We distinguish now between two different cases:

1. π has the following two different representations in P: $\pi = \alpha_1 \cdot \alpha_2 = \alpha_3$
2. π has the following two different representations in P: $\pi = \alpha_1 \cdot \alpha_2 = \alpha_3 \cdot \beta_1$ with $\beta_1 \neq 1$

In the first case, we can express c_π and $\tilde{c}_\pi$ by

$$\begin{aligned}
c_\pi &= (c_{\alpha_1} \cdot c_{\alpha_2} + c_{\alpha_3}) \\
\tilde{c}_\pi &= (\tilde{c}_{\alpha_1} \cdot \tilde{c}_{\alpha_2} + \tilde{c}_{\alpha_3}) \\
&= (c_{\alpha_1} \cdot \alpha_1^{d_a} \cdot c_{\alpha_2} \cdot \alpha_2^{d_a} + c_{\alpha_3} \cdot \alpha_3^{d_a}) \\
&= (c_{\alpha_1} \cdot c_{\alpha_2} \cdot \underbrace{(\alpha_1 \cdot \alpha_2)}_{=\pi}{}^{d_a} + c_{\alpha_3} \cdot \underbrace{(\alpha_3)}_{=\pi}{}^{d_a}) \\
&= (c_{\alpha_1} \cdot c_{\alpha_2} + c_{\alpha_3}) \cdot \pi^{d_a} \\
&= c_\pi \pi^{d_a}
\end{aligned}$$

As we said before $\pi \in P$ is non-zero. Hence, it is $c_\pi \neq 0 \Leftrightarrow \tilde{c}_\pi \neq 0$ and condition 6 is fulfilled. Therefore, we can be sure that $min(\tilde{Z})$ is the same for all choices of d_a and d_b.

Now let us have a look at the second case. W.l.o.g., we assume $d_a < d_b$. Then

$$\begin{aligned}
c_\pi &= (c_{\alpha_1} \cdot c_{\alpha_2} + c_{\alpha_3} \cdot c_{\beta_1}) \\
\tilde{c}_\pi &= (\tilde{c}_{\alpha_1} \cdot \tilde{c}_{\alpha_2} + \tilde{c}_{\alpha_3} \cdot \tilde{c}_{\beta_1}) \\
&= (c_{\alpha_1} \cdot c_{\alpha_2} \cdot \pi^{d_a} + c_{\alpha_3} \cdot c_{\beta_1} \cdot \pi^{d_a} \cdot \beta_1^{d_b - d_a}) \\
&= (c_{\alpha_1} \cdot c_{\alpha_2} + c_{\alpha_3} \cdot c_{\beta_1} \cdot \beta_1^{d_b - d_a}) \cdot \pi^{d_a}
\end{aligned}$$

In this case (depending on c_{α_1}, c_{α_2}, c_{α_3}, c_{β_1}, β_1, d_a and d_b) it could happen that $c_\pi \neq 0$ but $\tilde{c}_\pi = 0$ (or vice versa).

The motivation for definition 1 was to avoid cases like case 2. To match the case considered here on definition 1, we set $R_1 := \{\alpha_i | i = \ldots\}$ (the roots of $m_a(x)$), $R_2 := \{\beta_j | j = \ldots\}$ and $R = R_1 \dot{\cup} R_2$. Let $V = (v_1, \ldots, v_n) \in R^n$ and $W := (w_1, \ldots, w_m) \in R^m$. Definition 1 was that V and W factorize uniquely over R_1, R_2 if

$$v_1 \cdot \ldots \cdot v_n = w_1 \cdot \ldots \cdot w_m \Rightarrow \prod_{v_i \in R_1} v_i \cdot \prod_{w_j \in R_1} w_j^{-1} = 1 \text{ and } \prod_{v_i \in R_2} v_i \cdot \prod_{w_j \in R_2} w_j^{-1} = 1$$

Now we check if the definitions are fulfilled for case 1 and 2 for the representations of π:

Case 1: It is $V = (v_1, v_2) = (\alpha_1, \alpha_2) \in R_1 \times R_1$ and $W = (w_1) = (\alpha_3) \in R_1$. As all of them are elements of R_1 we only have to check

$$v_1 \cdot v_2 \cdot w_1^{-1} = \pi \cdot \pi^{-1} = 1$$

Case 2: It is $V = (v_1, v_2) = (\alpha_1, \alpha_2) \in R_1 \times R_1$ and $W = (w_1, w_2) = (\alpha_3, \beta_1) \in R_1 \times R_2$. We have to check if $v_1 \cdot v_2 \cdot w_1^{-1} = 1$ and $w_2^{-1} = 1$. But this is not true as by assumption $w_2 = \beta_1 \neq 1$.

In the proof of theorem 2 it is shown that cases like the second one can be avoided if we require that the vectors $\vec{\pi}$ with $\pi \in P$ do all factorize uniquely over R_1, R_2.

B On the Practical Relevance of Theorem 2

In this section we show that theorem 2 applies to a large class of LFSR-based ciphers automatically. Let $m_1(x), \ldots, m_\kappa(x) \in \mathbb{F}_2[x]$ be the primitive minimal polynomials of the used LFSRs such that the roots are all pairwise distinct and non-zero. For the following classes of ciphers, the assumptions of theorem 2 are always satisfied:

1. The cipher is a filter generator, i.e. $\kappa = 1$.
2. The degrees of the minimal polynomials are pairwise co-prime.

Set $R := R_1 \dot\cup \ldots \dot\cup R_\kappa$. We define by $\mathcal{P} := \{\alpha_1 \cdot \ldots \cdot \alpha_n | \alpha_i \in R, n \in \mathbb{N}\}$ the set of all possible multiple products of elements in R. We show now that in both cases, all pairs of vectors $\vec{\alpha}, \vec{\beta}$ with $\alpha, \beta \in \mathcal{P}$ factorize uniquely over $R_1, \ldots, R_\kappa$. As $\mathcal{P}$ is a superset for all possible sets $\{\mu(\alpha)| \ldots\}$ this proves that the conditions of theorem 2 are satisfied. Let from now on $\vec{\alpha} = (\alpha_1, \ldots, \alpha_n)$ and $\vec{\beta} = (\beta_1, \ldots, \beta_m)$ denote two vectors such that $\alpha_1 \cdot \ldots \cdot \alpha_n$ and $\beta_1 \cdot \ldots \cdot \beta_m$ are elements in $\mathcal{P}$.

Let us start with the first case. Here, we have

$$\alpha_1 \cdot \ldots \cdot \alpha_n = \beta_1 \cdot \ldots \cdot \beta_m \Leftrightarrow \prod \alpha_i \cdot \prod \beta_j^{-1} = 1 \Leftrightarrow \prod_{\alpha_i \in R_1} \alpha_i \cdot \prod_{\beta_j \in R_1} \beta_j^{-1} = 1$$

This concludes the first case.

For the second case we remember the fact that $\mathbb{F}_{2^n} \subseteq \mathbb{F}_{2^m}$ iff n divides m. In particular, $\mathbb{F}_{2^n} \cap \mathbb{F}_{2^m} = \mathbb{F}_{2^c}$ with $c := gcd(n, m)$. We denote by T_i the degree of the minimal polynomial $m_i(x)$. Then the elements $\alpha, \alpha^{-1}, \alpha \in R_i$, and all multiple products are elements of $\mathbb{F}_{2^{T_i}}$. Let l be arbitrary with $1 \leq l \leq \kappa$ and set $S_l := T_1 \cdot \ldots \cdot T_{l-1} \cdot T_{l+1} \cdot \ldots \cdot T_\kappa$. Then $\alpha_1 \cdot \ldots \cdot \alpha_n = \beta_1 \cdot \ldots \cdot \beta_m$ implies

$$\gamma_l := \underbrace{\prod_{\alpha_i \in R_l} \alpha_i \prod_{\beta_j \in R_l} \beta_j^{-1}}_{\in \mathbb{F}_{2^{T_l}}} = \underbrace{\prod_{\alpha_i \notin R_l} \alpha_i \prod_{\beta_j \notin R_l} \beta_j^{-1}}_{\in \mathbb{F}_{2^{S_l}}}$$

Therefore, $\gamma_l \in \mathbb{F}_{2^{T_l}} \cap \mathbb{F}_{2^{S_l}}$. By assumption the values T_i are pairwise co-prime. Hence, it is $gcd(T_l, S_l) = 1$ and $\gamma_l \in \mathbb{F}_{2^1} = \mathbb{F}_2$. As the roots are all non-zero, γ_l equals to 1 for each choice of l. This concludes the second case.

C The E_0 Key Stream Generator

In this section, we show that theorem 2 and algorithm B are applicable to the E_0 cipher together with the following boolean function:

$$F = \sum_{\substack{1 \le i < j \le 4 \\ 1 \le k < l \le 4}} z_t^{(i)} z_t^{(j)} z_{t+1}^{(k)} z_{t+1}^{(l)} + z_t^{(1)} z_t^{(2)} z_t^{(3)} z_t^{(4)} \tag{7}$$

Let from now on denote i, j, k, l integers from the set $\{1, 2, 3, 4\}$. We define the following three sets of indices

$$\begin{aligned} I_2 &:= \{(i,j) \mid i < j\} \\ I_3 &:= \{(i,j,k) \mid i < j < k\} \\ I_4 &:= \{(i,j;k,l) \mid i < j, k < l, \{i,j\} \cup \{k,l\} = \{1,2,3,4\}\} \end{aligned}$$

Then, F can be rewritten as

$$F = \sum_{(i,j) \in I_2} \underbrace{z_t^{(i)} z_{t+1}^{(i)} z_t^{(j)} z_{t+1}^{(j)}}_{=:F_{(i,j)}} \tag{8}$$

$$+ \sum_{(i,j,k) \in I_3} \underbrace{\left(f_{ij} \cdot z_t^{(k)} z_{t+1}^{(k)} + f_{ik} \cdot z_t^{(j)} z_{t+1}^{(j)} + f_{jk} \cdot z_t^{(i)} z_{t+1}^{(i)}\right)}_{=:F_{(i,j,k)}} \tag{9}$$

$$\underbrace{+ \sum_{(i,j;k,l) \in I_4} z_t^{(i)} z_{t+1}^{(j)} z_t^{(k)} z_{t+1}^{(l)} + z_t^{(1)} z_t^{(2)} z_t^{(3)} z_t^{(4)}}_{\tilde{F}} \tag{10}$$

where $f_{ij} = z_t^{(i)} z_{t+1}^{(j)} + z_t^{(j)} z_{t+1}^{(i)}$. Let R_i be the set of roots of m_i, the minimal polynomial of the ith LFSR, and $R_i^2 := \{\alpha\alpha' | \alpha, \alpha' \in R_i, \alpha \neq \alpha'\}$. We define the sets :

$$\begin{aligned} R_{(i,j)} &:= \{\alpha_i \alpha_j \mid (i,j) \in I_2, \alpha_s \in R_s \cup R_s^2\} \\ R_{(i,j,k)} &:= \{\alpha_i \alpha_j \alpha_k \mid (i,j,k) \in I_3, \alpha_s \in R_s \cup R_s^2\} \\ \tilde{R} &:= \{\alpha_i \alpha_j \alpha_k \alpha_l \mid (i,j;k,l) \in I_4, \alpha_s \in R_s\} \end{aligned}$$

The first thing we observe is that the sets defined above are all pairwise disjunct. Furthermore, it is easy to see that the roots of $min(F_{(i,j)})$ are a subset of $R_{(i,j)}$ and so on. Thus, the minimal polynomials are pairwise co-prime and we can write by theorem 5 $min(F)$ as the product of 11 different minimal polynomials:

$$min(F) = \prod_{(i,j) \in I_2} min(F_{(i,j)}) \cdot \prod_{(i,j,k) \in I_3} min(F_{(i,j,k)}) \cdot min(\tilde{F})$$

Actually, even more can be said. Using theorem 5 and the fact that the polynomials m_i are pairwise co-prime, we have $min(F_{(i,j)}) = (m_i \otimes m_i) \otimes (m_j \otimes m_j)$ of degree $T_i(T_i+1)/2 \cdot T_j(T_j+1)/2$. Before we can estimate $min(F_{(i,j,k)})$ and $min(\tilde{F})$, we need the following theorem:

Theorem 9. *([16], Th. 6.55) Given sequences $\mathcal{Z}_1, \ldots, \mathcal{Z}_\kappa$, the minimal polynomial $min(\mathcal{Z}_1 + \ldots + \mathcal{Z}_\kappa)$ divides $lcm(min(\mathcal{Z}_1), \ldots, min(\mathcal{Z}_\kappa))$.*

First, we consider $min(F_{(i,j,k)})$. By theorem 5, it can be easy seen that

$$min(z_t^{(i)} z_{t+1}^{(j)}) = min(z_t^{(j)} z_{t+1}^{(i)}) = m_i \otimes m_j,$$

and by theorem 6, that $min(z_t^{(k)} z_{t+1}^{(k)}) = m_k \otimes m_k$. By theorem 9, the minimal polynomial $min(F_{(i,j,k)})$ divides the least common multiple of the polynomials $(m_i \otimes m_i) \otimes (m_j \otimes m_k)$, $(m_j \otimes m_j) \otimes (m_i \otimes m_k)$ and $(m_k \otimes m_k) \otimes (m_i \otimes m_j)$. Notice that all three polynomials share the common factor $m_i \otimes m_j \otimes m_k$.[7] Thus, the degree of $min(F_{(i,j,k)})$ is upper bounded by

$$\begin{aligned} & T_iT_jT_k + \frac{T_i(T_i-1)}{2}T_jT_k + \frac{T_j(T_j-1)}{2}T_iT_k + \frac{T_k(T_k-1)}{2}T_iT_j \\ = & T_iT_jT_K \frac{T_i + T_j + T_k - 1}{2} \end{aligned}$$

The minimal polynomial $min(\tilde{F})$ can be found exactly. We observe that

$$min(z_t^{(i)} z_{t+1}^{(j)} z_t^{(k)} z_{t+1}^{(l)}) = min(z_t^{(1)} z_t^{(2)} z_t^{(3)} z_t^{(4)}) = m_1 \otimes m_2 \otimes m_3 \otimes m_4 =: m$$

Thus, $min(\tilde{F})$ divides m. As $m_1, \ldots, m_4$ are irreducible, this holds for m too. Therefore, $min(\tilde{F})$ is equal to 1 or equal to m. But the first case would imply that the expression in (10) is the all-zero sequence which is obviously wrong. Ergo, $min(\tilde{F})$ is equal to m of degree $T_1T_2T_3T_4$. Summing up, for the degree T of $min(F)$ it holds

$$T \leq \sum_{(i,j) \in I_2} \frac{T_i(T_i+1)T_j(T_j+1)}{4} + \sum_{(i,j,k) \in I_3} T_iT_jT_k \frac{T_i + T_j + T_k - 1}{2} + T_1T_2T_3T_4$$

D Why Preconditions are Necessary

In this section, we show that algorithm A (and B) do not work properly without some preconditions. To do so we give an example. We consider the case of two LFSRs $\mathcal{L}_a$ and $\mathcal{L}_b$ with the primitive minimal polynomials $m_a(x) = 1 + x + x^2$ and $m_b(x) = 1 + x + x^4$. Notice that the polynomials are co-prime but the corresponding periods are not. Let a_t resp. b_t denote the outputs of LFSR $\mathcal{A}$ and $\mathcal{B}$ and define the sequence $\mathcal{Z} := (z_t)$ by

$$z_t := a_tb_t + a_t + b_t + a_ta_{t+1} + b_tb_{t+1}.$$

Let $\mathcal{K}_a = (a_1, a_2)$ be the initial state of LFSR $\mathcal{L}_a$ and $\mathcal{K}_b = (b_1, b_2, b_3, b_4)$ be the initial state of LFSR $\mathcal{L}_b$. For the correctness of the pre-computation step, $min(\mathcal{Z})$ should be the same for all non-zero choices of $\mathcal{K}_a$ and $\mathcal{K}_b$. This is not the case. For $\mathcal{K}_a = (1, 0)$ and $\mathcal{K}_b = (1, 1, 1, 0)$ it is $min(\mathcal{Z}) = 1 + x^2 + x^3 + x^6 + x^7 + x^9$. But for $\mathcal{K}_a = (0, 1)$ and $\mathcal{K}_b = (1, 1, 1, 1)$ it is $min(\mathcal{Z}) = 1 + x^{15}$.

---

[7] The reason is that f divides $f \otimes f$ and that $(f \cdot g) \otimes h = (f \otimes h) \cdot (g \otimes h)$.

Resistance of S-Boxes against Algebraic Attacks

Jung Hee Cheon[1] and Dong Hoon Lee[2]

[1] Department of Mathematics, Seoul National University
jhcheon@math.snu.ac.kr
[2] National Security Research Institute (NSRI)
dlee@etri.re.kr

Abstract. We develop several tools to derive linear independent multivariate equations from algebraic S-boxes. By applying them to maximally nonlinear power functions with the inverse exponents, Gold exponents, or Kasami exponents, we estimate their resistance against algebraic attacks. As a result, we show that S-boxes with Gold exponents have very weak resistance and S-boxes with Kasami exponents have slightly better resistance against algebraic attacks than those with the inverse exponents.

Keywords: Algebraic Attack, S-boxes, Boolean Functions, Nonlinearity, Differential Uniformity

1 Introduction

Recently, Courtois and Pieprzyk proposed an algebraic attack for block ciphers [4]. Their attack on AES [11] exploits algebraic properties of S-boxes: If we can obtain many equations of small number of monomials from S-boxes, a block cipher with the S-boxes can be represented by many equations of small number of variables. By solving these multivariate equations by so called the *XSL* algorithm, we may find the key of the block cipher.

In the AES case, they introduce another viewpoint of the S-box as a quadratic equation $xy = 1$ in x and y rather than as a higher degree equation $y = 1/x$ in x, and obtain additional quadratic equations by multiplying appropriate monomials. More precisely, they obtain 23 quadratic equations with a total of 81 distinct terms from the S-box of AES and show that the equations are linearly independent by simulation.

In this paper, we give a theoretical approach to obtain *linearly independent* multivariate equations from algebraic S-boxes. Multivariate equations are said to be linearly independent if they are linearly independent when every distinct monomial is considered as a new variable. We develop three tools to prove *linear independence*. The first tool is that if a vector Boolean function is nonlinear, their component functions should be linearly independent as multivariate equations. We apply this to $n \times n$ S-boxes x^{2^k+1} and $n \times 2n$ S-boxes $(x^{2^k+1}, x^{2^{k+1}+1})$ over $\mathbb{F}_{2^n}$ which are known to be nonlinear when $\gcd(n, 2k) = 1$ and $|k - n/2| > 1$, respectively [5]. The second one is that if for a vector Boolean function $F(x, y)$:

B. Roy and W. Meier (Eds.): FSE 2004, LNCS 3017, pp. 83–94, 2004.

$\mathbb{F}_{2^n} \times \mathbb{F}_{2^n} \to \mathbb{F}_{2^m}$ and $g : \mathbb{F}_{2^n} \to \mathbb{F}_{2^n}$, $F(x, g(x))$ has m linearly independent component functions, so has $F(x, y)$. The third one is that linear independence of multivariate functions is *invariant* under affine transformation of inputs and linear transformation of outputs.

By applying these tools, we can prove that $5n$ equations obtained from the inverse function $xy = 1$ in $\mathbb{F}_{2^n}$ (or its affine transformation) are linearly independent for any positive integer n. Further we apply them to estimate the resistance of power functions with well-known Gold exponents and Kasami exponents against algebraic attacks [7, 8]. Those S-boxes are the only power functions which are known to be maximally nonlinear (MN) and almost perfect nonlinear (APN) [6]. Note that 'MN' and 'APN' imply the best resistance against linear cryptanalysis and differential cryptanalysis, respectively [1, 2, 9]. Our analysis shows that the S-boxes with Gold exponents have very weak resistance and the S-boxes with Kasami exponents have better resistance against algebraic attacks while all of them have similar resistance against differential and linear cryptanalysis. It would be an interesting problem to apply algebraic attacks to the ciphers using Gold power functions as S-boxes such as MISTY [10] which is selected as standard block algorithms in NESSIE [12].

In Section 2, we introduce some preliminaries on nonlinearity, APN, and resistance against algebraic attacks. In Section 3, we propose some auxiliary lemmas used to show the linear independence of multivariate equations. In Section 4, we deal with the resistance of the above three families of S-boxes and compare them. We conclude in Section 5.

2 Preliminaries

In this section, we introduce the definitions of nonlinearity, APN, and resistance against algebraic attacks, and remind some useful results for algebraic S-boxes.

Definition 1. *A function $F : \mathbb{F}_{2^n} \to \mathbb{F}_{2^n}$ is called a* almost perfect nonlinear *(APN) if each equation*

$$F(x + a) - F(x) = b \quad \text{for } a \in \mathbb{F}_{2^n}^*, b \in \mathbb{F}_{2^n}$$

has at most two solutions $x \in \mathbb{F}_{2^n}$.

Note that APN functions have the best resistance against differential cryptanalysis. When n is odd, we have many classes of APN power functions. But when n is even, we have only two classes of APN power functions, that is, Gold exponents and Kasami exponents [7, 8, 6].

The *Hamming distance* between two Boolean functions $f : \mathbb{F}_{2^n} \to \mathbb{F}_2$ and $g : \mathbb{F}_{2^n} \to \mathbb{F}_2$ is the weight of $f + g$. The minimal distance between f and any affine function from $\mathbb{F}_{2^n}$ into $\mathbb{F}_2$ is the *nonlinearity* of f. Given a *vector Boolean function* $F = (f_1, \dots, f_m) : \mathbb{F}_{2^n} \to \mathbb{F}_{2^m}$, $b \cdot F$ denotes the Boolean function $b_1 f_1 + b_2 f_2 + \dots + b_m f_m$ for each $b = (b_1, b_2, \cdots, b_m) \in \mathbb{F}_{2^m}$. Then the nonlinearity of F is defined as minimal nonlinearity of component functions as follows:

Definition 2. *The nonlinearity of F, $\mathcal{N}(F)$, is defined as*

$$\mathcal{N}(F) = \min_{b \in \mathbb{F}_{2^m}^*} \mathcal{N}(b \cdot F) = \min_{b \neq 0, \phi \in \mathcal{A}} wt(b \cdot F + \phi)$$

where $\mathcal{A}$ is the set of all affine functions over $\mathbb{F}_{2^n}$.

It is known that $\mathcal{N}(F) \leq 2^{n-1} - 2^{\frac{n-1}{2}}$. If n is odd, $\mathcal{N}(F)$ can be maximal, we call such functions *maximally nonlinear (MN)* functions. For even n, it is an open question to determine the maximal value. It is known that if n is odd and F is maximally nonlinear then F is almost perfect nonlinear [6].

Now we define the resistance against algebraic attacks as in [4].

Definition 3. *Given r equations of t monomials in $\mathbb{F}_2^n$, we define $\Gamma = ((t-r)/n)^{\lceil (t-r)/n \rceil}$ as the resistance of algebraic attacks (RAA).*

This quantity was introduced by Courtois and Pieprzyk [4]. They showed that the S-box of AES and the S-boxes of Serpent have $\Gamma \approx 2^{22.9}$ and $\Gamma \approx 2^{8.0}$, respectively. They claimed it can be a serious weakness of these ciphers and Γ should be greater than 2^{32} for secure ciphers.

Note that this measure is not an exact measure of XSL algorithm and an improvement of algorithm on solving multivariate equations may result in different measures. However, it is true that this quantity reflects a difficulty of solving multivariate equations in some sense. Thus we will use this quantity to measure the resistance of algebraic attacks in this paper.

3 Auxiliary Lemmas

Definition 4. *Given Boolean functions $f_1, \dots, f_m$ from $\mathbb{F}_2^n$ to $\mathbb{F}_2$, they are said to be linearly independent over $\mathbb{F}_2$ if they are linearly independent as multivariate polynomials, or equivalently if $\sum_{i=1}^m a_i f_i(x) = 0$ for all $x \in \mathbb{F}_2^n$ with $a_1, \dots, a_m \in \mathbb{F}_2$ implies $a_1 = \cdots = a_m = 0$.*

Lemma 1. *Consider two vector Boolean functions $F(x, y) : \mathbb{F}_{2^n} \times \mathbb{F}_{2^n} \to \mathbb{F}_{2^m}$ and $g : \mathbb{F}_{2^n} \to \mathbb{F}_{2^n}$. If $F(x, g(x))$ has m linearly independent component functions, so does $F(x, y)$ in $\mathbb{F}_2[x_1, \dots, x_n, y_1, \dots, y_n]$.*

Proof. Suppose that $F(x, y) = (f_1(x, y), \dots, f_m(x, y))$ has m linearly dependent component functions, *i.e.* there are not-all-zero $a_1, \dots, a_m \in \mathbb{F}_2$ such that $\sum_{i=1}^m a_i f_i(x, y) = 0$. Then we have $\sum_{i=1}^m a_i f_i(x, g(x)) = 0$, which implies that $f_i(x, g(x))$'s are linearly dependent. It contradicts that $F(x, g(x))$ has m linearly independent components. Therefore $F(x, y)$ should have m linearly independent component functions.

Lemma 2. *Any permutation $F : \mathbb{F}_{2^n} \to \mathbb{F}_{2^n}$ has n linearly independent component functions.*

Proof. Suppose that there exist not-all-zero $a_1, \ldots, a_n \in \mathbb{F}_2$ such that $\sum_{i=1}^{n} a_i f_i(x) = 0$ for $F = (f_1, \ldots, f_n)$. Then the image of F is a subset of the hyperplane given by $\sum_{i=1}^{n} a_i f_i(x) = 0$. Since the hyperplane has dimension less than n, F can not be a permutation. Therefore if F is a permutation, its n component functions should be linearly independent.

Lemma 3. *Consider a vector Boolean function $F : \mathbb{F}_{2^n} \rightarrow \mathbb{F}_{2^m}$. If the nonlinearity of F is non-zero, F has m linearly independent component functions.*

Proof. Suppose that there exist not-all-zero $a_1, \ldots, a_m \in \mathbb{F}_2$ such that $\sum_{i=1}^{m} a_i f_i(x) = 0$ for $F = (f_1, \ldots, f_m)$. If we take $b = (a_1, \ldots, a_m)$, we can see that $b \cdot F$ is a zero function and so has zero nonlinearity. Thus the nonlinearity of F, the minimum of nonlinearity of the component functions, is also zero. Therefore any nonlinear function should have m linearly independent component functions.

For the nonlinearity of S-boxes, we have the following results [5]:

$$\mathcal{N}(x^{2^k+1}) \geq 2^{n-1} - 2^{\frac{n+\gcd(n,2k)}{2}-1}, \tag{1}$$

$$\mathcal{N}(x^3, x^5, \cdots, x^{2k+1}) \geq 2^{n-1} - k \cdot 2^{\frac{n}{2}}. \tag{2}$$

By applying these results to Lemma 3, we obtain the following corollary:

Corollary 1. *Let k be a positive integer.*
(1) If n does not divide $2k$, x^{2^k+1} has n linearly independent component functions.
(2) If $k \leq 2^{n/2-1}$, $F = (x^3, x^5, \cdots, x^{2k+1}) : \mathbb{F}_{2^n} \rightarrow \mathbb{F}_{2^{kn}}$ have kn linearly independent component functions.

3.1 Invariants under Transformations

Now we show that linear independence is invariant under invertible transformations of inputs and invertible linear transformations of outputs.

Lemma 4. *Let $T : \mathbb{F}_2^n \rightarrow \mathbb{F}_2^n$ be an invertible transformation and $S : \mathbb{F}_2^m \rightarrow \mathbb{F}_2^m$ an invertible linear transformation. A vector Boolean function $F : \mathbb{F}_2^n \rightarrow \mathbb{F}_2^m$ has m linearly independent component functions over $\mathbb{F}_2$ if and only if so does $S \circ F \circ T$.*

Proof. Since we consider invertible transformations T and S, we are enough to show that F has m linearly independent component functions when either $F \circ T$ or $S \circ F$ does.

Let $F(x) = (f_1(x), \ldots, f_m(x))$ for $x \in \mathbb{F}_2^n$. Assume $a_1, \ldots, a_m \in \mathbb{F}_2$ satisfies $\sum_{i=1}^{m} a_i f_i(x) = 0$ for all $x \in \mathbb{F}_2^n$. Since T is invertible, we have $\sum_{i=1}^{m} a_i f_i(Ty) = 0$ for all $y \in F_2^n$. Since $F \circ T$ has m linearly independent component functions, we have $a_1 = \cdots = a_m = 0$, which implies the independence of m component functions of F.

If we let $S^{-1} = (p_{ij})$ for p_{ij}'s $\in \mathbb{F}_2$ and $S \circ F = (g_1, \ldots, g_m)$, we have $f_i = \sum_{j=1}^{m} p_{ij} g_j$. If there are not-all-zero $a_1, \ldots, a_m \in \mathbb{F}_2$ satisfying $\sum_{i=1}^{m} a_i f_i(x) = 0$, we have

$$\sum_{i=1}^{m} \{\sum_{j=1}^{m} a_i p_{ij} g_j(x)\} = \sum_{j=1}^{m} \{\sum_{i=1}^{m} a_i p_{ij}\} g_j(x) = 0. \tag{3}$$

Since $g_1, \ldots, g_m$ are linearly independent, $\sum_{i=1}^{m} a_i p_{ij} = 0$ for all j. We can see $a_1 = \cdots = a_m = 0$ from the invertibility of $S^{-1} = (p_{ij})$. Hence m component functions of F should be linearly independent.

Remark that if S is an *affine* transformation, Lemma 4 does not hold. For example, $F : \mathbb{F}_2^2 \to \mathbb{F}_2^3 : (x_1, x_2) \mapsto (x_1 + 1, x_2 + 1, x_1 + x_2 + 1)$ has 3 linearly independent components, but after the affine transformation $S : \mathbb{F}_2^3 \to \mathbb{F}_2^3 : (x, y, z) \mapsto (x + 1, y + 1, z + 1)$ is taken to F, $S \circ F = (x_1, x_2, x_1 + x_2)$ is not linearly independent anymore. However, if we consider a constant term as one of variables, we can have this invariant property. That is, if $1, f_1, \ldots, f_m$ are linearly independent, $S(f_1, \ldots, f_m)$ are linearly independent. Also if all of f_i's do not have constant terms, independence property is preserved under an affine transformation S.

4 Independent Equations

From now on, we consider a polynomial over a finite field. If we fix a basis, this polynomial can be regarded as multivariate equations. Unless confused, we will consider a polynomial as multivariate equations without specifying a basis.

Because equations of higher degree than two do not help in the point of algebraic attacks to S-box, our purpose is to get linearly independent equations whose degree are at most two as many as possible. When we are given m quadratic equations from $F(x) = 0$, we can consider the following methods to get more quadratic equations:

1. Multiplication by linear or quadratic equations.
2. Composition with quadratic equations.

Note that composition of a monomial with affine equations gives only dependent equations and composition with equations of higher degree usually gives equations of higher degree.

The first case is restricted by the following lemma.

Lemma 5. *Suppose that $n > 2$ and $k \geq 1$. Assume that the Hamming weight of d is at most 2. The product x^m of two monomials x^{2^k+1} and x^d is linear or quadratic only in the following cases:*

1. *If $d = 1$, then* $m = \begin{cases} 4 & \text{if } k = 1, \text{(Linear)} \\ 2^k + 2 & \text{if } k \neq 1. \text{(Quadratic)} \end{cases}$
2. *If $d = 2^k$, then $m = 1 + 2^{k+1}$. (Quadratic)*

3. If $d = 3$, *then* $m = \begin{cases} 2^3 & \text{if } k = 2, \text{ (Linear)} \\ 2^k + 2^2 & \text{if } k \neq 2. \text{ (Quadratic)} \end{cases}$

4. If $d = 2^k + 1$, *then* $m = 2^{k+1} + 2$. *(Quadratic)*

5. If $d = 2^{k+1} + 2^k$, *then* $m = 2^{k+2} + 1$. *(Quadratic)*

Proof. It is sufficient to check the Hamming weight of $m = 2^k + 1 + d \mod (2^n - 1)$, since $x^{2^n-1} = 1$. Assume that $w(d) = 1$, i.e $d = 2^l$ for some $l < n$. Then m becomes $1 + 2^k + 2^l < (2^n - 1)$. Unless two of $\{0, k, l\}$ are equal, x^m is cubic. This covers first two cases of the lemma.

Assume that $w(d) = 2$, i.e $d = 2^l + 2^s$ for some $l < s < n$. Then m becomes $1 + 2^k + 2^l + 2^s < (2^n - 1)$. If all of $\{0, k, l, s\}$ are distinct, then x^m is quartic. Hence at least two of them are equal, especially $l = 0$ or $l = k$ since $0 < k$. If $l = 0$ then s should be 1 or k (Case 3 and 4). If $l = k$ then s should be $k + 1$ (Case 5). This completes the proof.

4.1 Inverse Exponents

First we count the number of linearly independent equations from $xy - 1 = 0$. A composition of $xy - 1 = 0$ with any quadratic equation gives a equation of degree larger than two. In order to get another quadratic equations, we must multiply linear or quadratic equations:

1. The original equation: $F(x, y) = xy - 1$
2. Multiplied by x: $G_0(x, y) = x^2y - x$
3. Multiplied by y: $H_0(x, y) = xy^2 - y$
4. Multiplied by x^3: $G_1(x, y) = x^4y - x^3$
5. Multiplied by y^3: $H_1(x, y) = xy^4 - y^3$

First, we must show that each of equations has n linearly independent component functions. Using Lemma 1 and Lemma 2, we can easily see that $F(x, y)$ has n linearly independent component functions since $F(x, y) = xy - 1$ is permutation for any nonzero y. Each component of G_0 and H_0 has a unique variable x_i and y_i respectively, hence they are linearly independent. Both G_1 and H_1 have n linearly independent components by Lemma 1, Lemma 4, and Corollary 1 using the following equations:

$$G_1(x, ax^{2^n-2}) = (a - 1)x^3$$
$$H_1(ay^{2^n-2}, y) = (a - 1)y^3$$

since any non-zero $(a - 1)$ is an invertible linear transformation.

In order to show that all components produced by the above polynomials are linearly independent, it is better to look at the matrix form. Each row corresponds to the equations from $G = (G_0, G_1)$, $H = (H_0, H_1)$, and F.

$$\begin{pmatrix} M_1 & 0 & M_2 & 0 \\ 0 & M_3 & M_4 & 0 \\ 0 & 0 & M_5 & M_6 \end{pmatrix} \begin{pmatrix} x_i x_j \\ y_i y_j \\ x_i y_j \\ 1 \end{pmatrix} = 0,$$

Table 1. The type and the number of distinct monomials

Eq.	Type	#
F	x_iy_j, 1	n^2+1
G_0	x_iy_j, x_i	n^2+n
H_0	x_iy_j, y_i	n^2+n
G_1	x_iy_j, x_ix_j, x_i	$\frac{3n(n+1)}{2}$
H_1	x_iy_j, y_iy_j, y_i	$\frac{3n(n+1)}{2}$

where each M_i represents a nonzero matrix and each monomial in the column vector represents all monomials of similar forms (For example, x_ix_j represents all x_ix_j for $1 \leq i, j \leq n$.).

It is sufficient to show that the rank of the coefficient matrix is $5n$. If we consider the coefficient matrix as a 3×3 block matrix, we can see that the rank is the sum of the ranks of M_1, M_3, and $(M_5\ M_6)$. Since F has n linearly independent components, we know that the rank of $(M_5\ M_6)$ is n.

Lemma 6. *Each of the ranks of M_1 and M_3 is $2n$.*

Proof. We refine the monomials x_ix_j for $1 \leq i, j \leq n$ as x_i and x_ix_k for $1 \leq i < k \leq n$. Then $M_1(x_ix_j)$ is expressed as the following:

$$M_1(x_ix_j) = \begin{pmatrix} A & B \\ C & D \end{pmatrix} \begin{pmatrix} x_i \\ x_ix_k \end{pmatrix}.$$

Since $(A\ B)$ represents the term x in G_0, A is the identity matrix of size n and $B = 0$. Since $(C\ D)$ represents the term $-x^3$ in G_1, we can write $-x^3 = C(x_i) + D(x_ix_k)$. Since $C(x_i)$ is a linear function over $\mathbb{F}_2^n$, the nonlinearity of $D(x_ix_k)$ is equal to that of x^3. Therefore $D(x_ix_k)$ has n linearly independent components by Lemma 3, hence the rank of D is n. This implies that the rank of M_1 is $2n$.

We can show that the rank of M_3 is also $2n$ by the similar argument.

Now we are ready to measure the resistance of S-boxes with inverse exponents by Γ value. The type and the number of distinct monomials in the equations from F, G, and H is as the following table.

From Table 1, we have the following theorem.

Theorem 1. *Consider $xy = 1$ in $\mathbb{F}_{2^n}$. Let t be the number of monomials and r the number of linearly independent equations. Then we can have the following parameters (r, t, Γ) for $xy = 1$:*

1. $\left(n, n^2+1, \left(\frac{n^2-n+1}{n}\right)^{\lceil \frac{n^2-n+1}{n} \rceil}\right)$ *for F*
2. $\left(2n, n^2+n+1, \left(\frac{n^2-n+1}{n}\right)^{\lceil \frac{n^2-n+1}{n} \rceil}\right)$ *for F and* $\{G_0$ *or* $H_0\}$
3. $\left(3n, n^2+2n+1, \left(\frac{n^2-n+1}{n}\right)^{\lceil \frac{n^2-n+1}{n} \rceil}\right)$ *for* F, G_0, *and* H_0
4. $\left(4n, \frac{(3n+2)(n+1)}{2}, \left(\frac{3n^2-3n+2}{2n}\right)^{\lceil \frac{3n^2-3n+2}{2n} \rceil}\right)$ *for* F, G_0, H_0 *and* $\{G_1$ *or* $H_1\}$
5. $\left(5n, 2n^2+n+1, \left(\frac{2n^2-4n+1}{n}\right)^{\lceil \frac{2n^2-4n+1}{n} \rceil}\right)$ *for all 5 polynomials*

4.2 Gold Exponents

When $\gcd(k,n)=1$, 2^k+1 is called a Gold exponent [7]. Note that any quadratic monomial can be changed into a monomial with a Gold exponent by an affine transformation. By multiplying monomials, we obtain

1. The original equation: $F_1(x,y) = x^{2^k+1} - y$
2. Multiplied by linear equations: $F_2(x,y) = x^{2^k+2} - xy$ and $F_3(x,y) = x^{2^{k+1}+1} - x^{2^k}y$
3. Multiplied by $x^{d_1}y^{d_2}$: $F_4(x,y) = x^4y - xy^2$ only for $k=1$
4. Composition with x^d: $F_5(x,y) = x^9 - y^3$ only for $k=1$.

Since the original equation consists of x^{2^k+1} and y, we should multiply monomials of type x^d or $x^{d_1}y^{d_2}$. In the first case, x^d should be linear so that we have $d=1$ or $d=2^k$ by Lemma 5. In the second case, $x^{2^k+1+d_1}$, y^{d_2}, x^{d_1}, and y^{1+d_2} should be linear so that $(d_1,d_2)=(1,1)$.

For composition case, if d is 2^s, the product produces only dependent equations on the original equations. Thus the Hamming weight of d should be two. Then $m = (2^k+1)(1+2^l) = 1+2^l+2^k+2^{k+l}$. Only when $l=k=1$, x^m can be quadratic.

F_1 has n independent component functions since each component contains distinct y_i. We can see that $F_2(x, ax^{2^k+1}) = (1-a)x^{2^k+2}$ and $F_3(x, ax^{2^k+1}) = (1-a)x^{2^{k+1}+1}$. When $k=1$, $F_2(x, ax^{2^k+1}) = (1-a)x^4$ and $F_3(x, ax^{2^k+1}) = (1-a)x^5$ are permutations unless $n \neq 2,4$. Also each of $F_4(x, ax^3) = (1-a)x^7$ and $F_5(x, a^{1/3}x^3) = (1-a)x^9$ has n linearly independent components if $\gcd(n,3)=1$ and $n \neq 2,4$ respectively. Thus F_4 and F_5 have by Lemma 1.

We show that all components produced by the above equations are linearly independent by the matrix argument similar to the inverse exponents case. At first, assume that $k=1$. Each row corresponds to the equations from F_1, F_2, F_3, F_4

and F_5.

$$\begin{pmatrix} M_1 & M_2 & M_3 & 0 & 0 \\ M_4 & 0 & 0 & 0 & M_5 \\ M_6 & 0 & M_7 & 0 & M_8 \\ 0 & 0 & 0 & 0 & M_9 \\ M_{10} & M_{11} & M_{12} & M_{13} & 0 \end{pmatrix} \begin{pmatrix} x_i \\ y_i \\ x_i x_k \\ y_i y_k \\ x_i y_j \end{pmatrix} = 0.$$

Each of M_2, M_4, M_7, M_9, and M_{13} represents $-y$, x^4, x^5, $x^4y - xy^2$, and y^3, respectively. Since all of them has n linearly independent component functions, each of the matrices has rank n. Further, if we consider the coefficient matrix by a 5×5 block matrix, we can easily convert it to a upper triangular matrix with diagonal M_2, M_4, M_7, M_9, and M_{13} by elementary row operations. Thus it has rank $5n$ and all components of the equations are linearly independent.

Next, assume that $k > 1$. Each row corresponds to the equations from F_1, F_2, and F_3.

$$\begin{pmatrix} M_1 & M_2 & M_3 & 0 \\ M_4 & 0 & M_5 & M_6 \\ M_7 & 0 & M_8 & M_9 \end{pmatrix} \begin{pmatrix} x_i \\ y_i \\ x_i x_k \\ x_i y_j \end{pmatrix} = 0.$$

Since M_2 represents $-y$, it is invertible. Thus we are enough to show that all components of F_2 and F_3 are linearly independent. Let $F(x, y) = (F_2(x, y), F_3(x, y))$. We have $F(x, ax^{2^k+1}) = ((1-a)x^{2^k+2}, (1-a)x^{2^{k+1}+1})$. By Corollary 1 we can see $(x^{2^{k-1}+1}, x^{2^{k+1}+1})$ and $(x^{2^{n-k+1}+1}, x^{2^{n-k-1}+1})$ are nonlinear if $k < n/2 - 1$ and $k > n/2 + 1$, respectively. Note that both of them are affine transformations of $F(x, ax^{2^k+1})$. Thus unless $|k - n/2| \leq 1$, $F(x, y)$ has $2n$ linearly independent component functions.

Theorem 2. *Consider $y = x^{2^k+1}$ with $\gcd(k, n) = 1$ in $\mathbb{F}_{2^n}$. Let t be the number of monomials and r the number of linearly independent equations. Then we can have the following parameters (r, t, Γ):*
(1) If $k = 1$, we can obtain 5 linearly independent polynomials. Thus we get the followings:

1. $\left(n, \frac{n(n+3)}{2}, \left(\frac{n+1}{2}\right)^{\lceil \frac{n+1}{2} \rceil}\right)$ *for F_1*
2. $\left(3n, \frac{n(3n+1)}{2}, \left(\frac{3n-5}{2}\right)^{\lceil \frac{3n-5}{2} \rceil}\right)$ *for F_2, F_3, and F_4 if $n \neq 2, 4$ and $\gcd(n, 3) = 1$*
3. $\left(4n, \frac{3n(n+1)}{2}, \left(\frac{3n-5}{2}\right)^{\lceil \frac{3n-5}{2} \rceil}\right)$ *for F_1, F_2, F_3, and F_4 if $n \neq 2, 4$ and $\gcd(n, 3) = 1$*
4. $\left(5n, n(2n+1), (2n-4)^{\lceil 2n-4 \rceil}\right)$ *for all polynomials if $n \neq 2, 4$ and $\gcd(n, 3) = 1$.*

(2) Otherwise, we can obtain 3 linearly independent polynomials. Thus we get the followings:

1. $\left(n, \frac{n(n+3)}{2}, \left(\frac{n+1}{2}\right)^{\lceil \frac{n+1}{2} \rceil}\right)$ *for F_1*
2. $\left(3n, \frac{3n(n+1)}{2}, \left(\frac{3n-3}{2}\right)^{\lceil \frac{3n-3}{2} \rceil}\right)$ *for F_1, F_2, and F_3 if $|k - n/2| \leq 1$.*

Table 2. Comparison of RAA for Almost Perfect Nonlinear Functions

Exponent	Alg. Deg.	# of Eqns	# of Monomials	RAA Γ	When $n=8$
Inverse	$n-1$	$3n$	n^2+2n+1	$\left(\frac{n^2-n+1}{n}\right)^{\lceil\frac{n^2-n+1}{n}\rceil}$	$\Gamma=2^{22.7}$
		$5n$	$2n^2+n+1$	$\left(\frac{2n^2-4n+1}{n}\right)^{\lceil\frac{2n^2-4+1}{n}\rceil}$	$\Gamma=2^{46.8}$
Gold	2	n	$\frac{n(n+3)}{2}$	$\left(\frac{n+1}{2}\right)^{\lceil\frac{n+1}{2}\rceil}$	$\Gamma=2^{10.8}$
$(k=1)$		$3n$	$\frac{n(3n+1)}{2}$	$\left(\frac{3n-5}{2}\right)^{\lceil\frac{3n-5}{2}\rceil}$	$\Gamma=2^{32.5}$
		$4n$	$\frac{3n(n+1)}{2}$	$\left(\frac{3n-5}{2}\right)^{\lceil\frac{3n-5}{2}\rceil}$	$\Gamma=2^{32.5}$
		$5n$	$n(2n+1)$	$(2n-4)^{\lceil 2n-4\rceil}$	$\Gamma=2^{43.0}$
Gold $(k>1)$	2	n	$\frac{n(n+3)}{2}$	$\left(\frac{n+1}{2}\right)^{\lceil\frac{n+1}{2}\rceil}$	$\Gamma=2^{10.8}$
$\lvert k-n/2\rvert>1$		$3n$	$\frac{3n(n+1)}{2}$	$\left(\frac{3n-3}{2}\right)^{\lceil\frac{3n-3}{2}\rceil}$	$\Gamma=2^{37.3}$
Kasami	$k+1$	n	n^2+n	n^n	$\Gamma=2^{24}$

4.3 Kasami Exponents

When $\gcd(n,k)=1$ and $k>1$, $2^{2k}-2^k+1$ is called a Kasami exponent [8]. A Kasami exponent has the Hamming weight $k+1$, but by applying composition by 2^k+1, we obtain a quadratic equation $F_1 : y^{2^k+1} - x^{2^{3k}+1}$.

By multiplying $x^{d_1}y^{d_2}$ to F_1, we have $x^{d_1+1+2^{3k}}y^{d_2} - x^{d_1}y^{d_2+1+2^k}$. Hence all x^{d_1}, y^{d_2}, $x^{d_1+1+2^{3k}}$, and $y^{d_2+1+2^k}$ should be linear monomials. It contradicts Lemma 5. Thus F_1 is the only quadratic equation we can obtain. F_1 has monomials of the type x_ix_j and y_iy_j. The number of monomials is n^2+n.

Theorem 3.[1] *Consider $y = x^{2^{2k}-2^k+1}$ with $\gcd(k,n)=1$ in $\mathbb{F}_{2^n}$. We can obtain n linearly independent equations in n^2+n variables. Then RAA is $\Gamma = n^n$.*

4.4 Comparison

Table 1 shows the comparison of the resistance of algebraic attacks. Surprisingly, more equations give larger RAA in each exponent. It is because RAA increases as $t-r$ increases and additional equations requires new variables more than new

[1] By substituting $x = z^{2^k+1}$, we can obtain two independent quadratic equations $x = z^{2^k+1}$ and $y = z^{2^{3k}+1}$ with $n(n+5)/2$ variables, which reduces its RAA significantly. It will be introduced in the full version of this paper [3].

equations. From the table, we can see that the power functions with Kasami exponents have slightly better resistance against algebraic attacks, and the power functions with Gold exponents have very weak resistant against algebraic attacks.

5 Conclusion

In this paper, we developed several tools to prove linear independence of multivariate equations from algebraic S-boxes. By applying these tools to APN power functions, we learned that a power function with a Gold exponent is very weak against algebraic attacks and a power function with a Kasami exponent has slightly stronger resistance against algebraic attacks. An open problem is to find S-boxes with $\Gamma > 2^{32}$ as indicated in [4]. Also, it is an interesting topic to apply algebraic attacks to block ciphers using a power function with a Gold exponent such as MISTY which is selected as standard block algorithms in NESSIE [12].

Acknowledgement

We are thankful to Hyun Soo Nam and Dae Sung Kwon for helpful discussions.

References

[1] E. Biham and A. Shamir, "Differential Cryptanalysis of DES-like Cryptosystems," *Journal of Cryptology*, vol. 4, pp. 3–72, 1991. 84

[2] E. Biham and A. Shamir, *Differential Cryptanalysis of the Data Encryption Standard*, Springer-Verlag, 1993. 84

[3] J. Cheon and D. Lee, "Almost Perfect Nonlinear Power Functions and Algebraic Attacks," Manuscript, 2004. 92

[4] N. Courtois and J. Pieprzyk, "Cryptanalysis of Block Ciphers with Overdefined Systems of Equations," *Proc. of Asiacrypt 2002*, LNCS 2501, Springer-Verlag, pp. 267–287, 2002. 83, 85, 93

[5] J. Cheon, S. Chee and C. Park, "S-boxes with Controllable Nonlinearity," *Advances in Cryptology - Eurocrypt'99*, Springer-Verlag, pp. 286–294, 1999. 83, 86

[6] H. Dobbertin, "Almost Perfect Nonlinear Power Functions on $GF(2^n)$: The Welch Case," *IEEE Trans. Inform. Theory*, Vol. 45, No. 4, pp. 1271-1275, 1999. 84, 85

[7] R. Gold, "Maximal Recursive Sequences with 3-valued Recursive Cross-correlation Functions," *IEEE Trans. Inform. Theory*, vol. IT-14, pp. 154–156, 1968. 84, 90

[8] T. Kasami, "The Weight Enumerators for Several Classes of Subcodes of the Second Order Binary Reed-Muller Codes," *Infor. Contr.*, Vol. 18, pp. 369–394, 1971. 84, 92

[9] M. Matsui, "Linear Cryptanalysis Method for DES cipher," *Advances in Cryptology - Eurocrypt'93*, Springer-Verlag, pp. 386–397, 1993. 84

[10] M. Matsui, "New Block Encryption Algorithm MISTY," *Proc. of FSE'97*, LNCS 1267, Springer-Verlag, pp. 54–68, 1997. 84

[11] Advances Encryption Standards. `http://csrc.nist.gov/CryptoToolkit/aes/`. 83

[12] New European Schemes for Signatures, Integrity, and Encryption. `https://www.cosic.esat.kuleuven.ac.be/nessie/`. 84, 93

Differential Attacks against the Helix Stream Cipher

Frédéric Muller

DCSSI Crypto Lab
51 boulevard de la Tour-Maubourg, 75700 Paris - 07 SP, France
frederic.muller@m4x.org

Abstract. In this paper, we analyze the security of the stream cipher Helix, recently proposed at FSE'03. Helix is a high-speed asynchronous stream cipher, with a built-in MAC functionality. We analyze the differential properties of its keystream generator and describe two new attacks. The first attack requires 2^{88} basic operations and processes only 2^{12} words of chosen plaintext in order to recover the secret key for any length up to 256 bits. However, it assumes the attacker can force nonces to be used twice. Our second attack relies on weaker assumptions. It is a distinguishing attack that detects internal state collisions after 2^{114} words of chosen plaintext.

1 Introduction

A stream cipher is a secret key cryptosystem that transforms a short random secret key K into a long pseudo-random sequence also called keystream, which is XORed to the plaintext to produce the ciphertext. Although it is possible to obtain a similar primitive with a block cipher in a "pseudo-random number generator" mode (like OFB or CFB [6]), it is generally not considered to offer optimal speed performances. To respond efficiency considerations, fast stream ciphers reveal useful in real-life applications, especially those using live data transmission. Many recent stream ciphers proposals have been made in that direction including SEAL [16], SNOW [2], Scream [10] or Sober-t32 [11].

However, the security of stream ciphers is still an issue (see [1, 3, 7]), especially when compared to the level of confidence in block ciphers security. For instance, all stream ciphers candidates for the NESSIE project [14] revealed various degrees of weakness allowing at least distinguishing attacks faster than exhaustive search, while no second round block cipher was successfully attacked. As a consequence, NESSIE did not select any stream cipher in its final portfolio. Thus the actual challenge is to design fast stream ciphers and provide a better confidence in their security level. Several new ciphers aim at reaching these expectations.

Helix was recently proposed at FSE'03 [5]. It is an asynchronous stream cipher based on a fast keystream generator. Its advantage over other new ciphers is to offer both confidentiality and integrity. Indeed, after encryption, Helix can

B. Roy and W. Meier (Eds.): FSE 2004, LNCS 3017, pp. 94–108, 2004.

produce a tag that guarantees the integrity of the message for very little additional computation and without requiring a second pass. This functionality is very useful in many applications where encryption and authentication must function together on streaming data. Recently, several block cipher modes of operation also providing integrity "almost for free" (see [9, 12, 15]) have been proposed, but some of them appear to be patented, which is supposedly not the case of Helix.

Moreover, the analysis of Helix is an interesting topic since new mechanisms that will be included in the new 802.11i standard for wireless networks are apparently fairly close to Helix [4, 18]. The new standard will have to repair some cryptologic flaws from the previous 802.11b standard, which resulted from weaknesses in RC4 key scheduling and from an improper use of initialization vectors [8].

In this paper, we analyze the security of Helix against chosen plaintext and chosen nonce attacks. We present two attacks which are both faster than exhaustive search. Our first attack recovers the secret key (for any length up to 256 bits) with time complexity of 2^{88} basic operations and using 2^{12} words of chosen plaintext. It assumes an attacker could force encryption of several messages using the same pair (key,nonce). Our second attack is based on internal state collisions and distinguishes Helix from random with data complexity of 2^{114} blocks. This attack uses chosen nonces and chosen plaintext but never re-uses a pair (key,nonce). Our paper is organized as follows : first, we briefly describe Helix. Then, in Section 3, we show two weaknesses of the cipher which are further developed in Section 4. In Section 5, we describe two attacks based on the previous observations.

2 Description of Helix

Helix offers two main features : encryption of a plain message and production of a Message Authentication Code (MAC) to ensure integrity. Several modes of operation for Helix are proposed by its authors - encryption only, MAC only, PRNG, ... Here, we describe briefly the mechanisms of Helix that are important in our attacks. More details about this design can be obtained in [5].

We mostly handle 32 bits values that we denote as *words*. Besides, $\oplus$ denotes bitwise addition on these values and $+$ addition modulo 2^{32}. $ROTL_n(x)$ is the circular rotation of the word x by n bits to the left. We also use the notations LSB and MSB to refer to the least and most significant bit of a word.

2.1 General Structure of the Cipher

Helix is an asynchronous stream cipher, based on an iterated block function applied to an internal state of 160 bits. The input consists in a secret key K of varying length, up to 256 bits, and a nonce N of 128 bits. The internal state before encryption of the i-th word of plaintext is represented as 5 words

$$(Z_0^{(i)}, \dots, Z_4^{(i)})$$

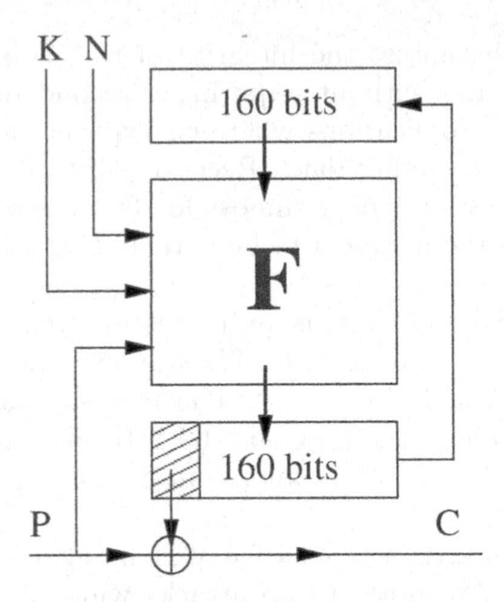

Fig. 1. The general structure of Helix

which are initialized for $i = 0$ using K and N. Details of this initialization mechanism are irrelevant here. The general structure of the encryption algorithm is described in Figure 1. It basically uses a block function F to update the internal state in function of the plaintext P, the key K and the nonce N.

More precisely, during the i-th round, the internal state is updated with F, using the i-th word of plaintext P_i and two words derived from K, N and i, denoted as $X_{i,0}$ and $X_{i,1}$. We refer to them as the "round key words". Hence,

$$(Z_0^{(i+1)}, \ldots, Z_4^{(i+1)}) = F(Z_0^{(i)}, \ldots, Z_4^{(i)}, P_i, X_{i,0}, X_{i,1})$$

The i-th keystream word, also denoted as S_i, is equal to $Z_0^{(i)}$. It is added to P_i to produce the i-th ciphertext word C_i. Thus,

$$S_i = Z_0^{(i)}$$
$$C_i = S_i \oplus P_i$$

This process is repeated until all words of the plaintext have been encrypted. Finally, a last step (described in [5]) can generate a tag of 128 bits that constitutes the MAC. More details on this general framework are given in the following sections.

2.2 The Block Function

The round function F of Helix mixes three types of basic operations on words: bitwise addition represented as $\oplus$, addition modulo 2^{32} represented as $\bullet$, and cyclic shifts represented as $<<<$. F relies on two consecutive applications of

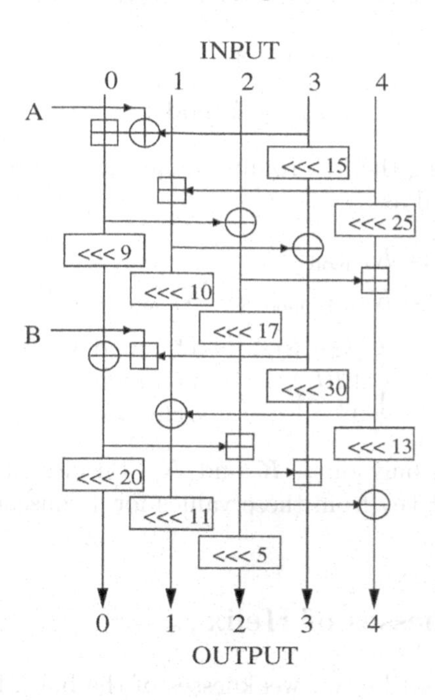

Fig. 2. The half-round "helix" function G

a single "helix" function, which constitutes half of the round function. This "helix" function is denoted as G and is represented in Figure 2.

G uses two auxiliary inputs (A, B). In the first half of the round function, $(A, B) = (0, X_{i,0})$ and in the second half, $(A, B) = (P_i, X_{i,1})$. Thus, the block function can be described by the following relations

$$(Y_0^{(i)}, \ldots, Y_4^{(i)}) = G(Z_0^{(i)}, \ldots, Z_4^{(i)}, 0, X_{i,0})$$
$$(Z_0^{(i+1)}, \ldots, Z_4^{(i+1)}) = G(Y_0^{(i)}, \ldots, Y_4^{(i)}, P_i, X_{i,1})$$

where $(Y_0^{(i)}, \ldots, Y_4^{(i)})$ is the internal state in the middle of the computation.

2.3 Role of K and N

To protect the cipher against related-key attacks, a first step is applied that computes a working key K from the actual secret key U. Independently of its length $l(U)$, K is always 256 bits long and is used in all subsequent operations instead of U. The derivation of K is based on 8 rounds of a Feistel network. The result is also represented as 8 words: $K_0, \ldots, K_7$.

Besides, Helix uses a nonce N to obtain different keystream sequences with the same secret key. N is always 128 bits long and is generally represented as 4 words: $N_0, \ldots, N_3$. An expansion phase turns it into a 256 bits value by creating 4 additional words $N_4, \ldots, N_7$ defined as

$$N_{k+4} := (k \bmod 4) - N_k$$

for $k = 0, \ldots, 3$. During the i-th round of encryption, the round key words $X_{i,0}$ and $X_{i,1}$ are computed as

$$\begin{aligned} X_{i,0} &:= K_{i \bmod 8} \\ X_{i,1} &:= K_{(i+4) \bmod 8} + N_{i \bmod 8} + X'_i + i + 8 \\ X'_i &:= \begin{cases} \lfloor (i+8)/2^{31} \rfloor & \texttt{if } i \bmod 4 = 3 \\ 4\, l(U) & \texttt{if } i \bmod 4 = 1 \\ 0 & \texttt{otherwise} \end{cases} \end{aligned}$$

These values depend only on i, K and N_i. Besides, it is straightforward to reconstruct the secret key from these values for 4 consecutive rounds when the nonce is known.

3 Some Weaknesses of Helix

In this section, we describe two weaknesses of the block function. They respectively concern the role of the plaintext words and the nonce words at each round.

3.1 Influence of Each Plaintext Word

Since Helix requires a plaintext-dependent keystream, it is reasonable to analyze the round function assuming an attacker can control the plaintext introduced. In general, an attacker should not be able to recover any information about the secret key or the internal state of the cipher, by observing the keystream corresponding to chosen plaintext.

Using the notations of Section 2, P_i denotes the i-th word of plaintext. It is introduced inside Helix internal state at the i-th advance. Then, at the beginning of the $(i+1)$-th advance, a new keystream word S_{i+1} is produced. From the description of Helix, one sees that P_i is introduced only in the second half of the block function (as the input A of Figure 2). It is XORed to $Y_3^{(i)}$, then added to $Y_0^{(i)}$. The result is then modified only once before the end of the round - excepting cyclic shifts - through a XOR with some intermediate value (referred to as a). However, it is easy to verify that a is actually independent of P_i. Thus S_{i+1} can be computed as

$$S_{i+1} = Z_0^{(i+1)} = ROTL_{20}(a \oplus ROTL_9(Y_0^{(i)} + (Y_3^{(i)} \oplus P_i)))$$

If the plaintext word $P'_i = P_i \oplus \Delta$ was introduced instead of P_i, then the next keystream word would be S'_{i+1}, such that

$$\begin{aligned} \delta &= S_{i+1} \oplus S'_{i+1} \\ &= ROTL_{29}((x + (y \oplus P_i)) \oplus (x + (y \oplus P_i \oplus \Delta))) \end{aligned}$$

where x and y respectively denote the intermediate words $Y_0^{(i)}$ and $Y_3^{(i)}$. Suppose that $P_i = 0$, then for any difference Δ on the plaintext,

$$\Delta' = ROTL_3(\delta) = (x + y) \oplus (x + (y \oplus \Delta)) \tag{1}$$

is the corresponding difference on the keystream. In Section 4, we will discuss how an attacker can take advantage of this differential property.

3.2 Influence of Each Nonce Word

Similar differential properties hold regarding each nonce word. Indeed, the nonce N serves two purposes in Helix :

- Fill the initial 160 bits of internal state.
- Derive two words $X_{i,0}$ and $X_{i,1}$ introduced at round i.

Concerning this second task, it appears from Section 2.3 that the two "key words" introduced at each round do not depend on the full nonce. Actually, the round key words at round i depend only on $N_{i \bmod 4}$. Therefore, if we consider two distinct nonces N and N' where only one word changes, the round function will essentially apply the same mapping on the internal state, for 3 rounds out of 4. This property has consequences on the propagation of state collisions.

Moreover, if only one nonce word N_i is modified to $N_i + \Delta$ then, for rounds j such that $j \bmod 4 \neq i$, both round key words remain unchanged. For other positions, $X_{j,1}$ is changed to $(X_{j,1} \pm \Delta)$ while $X_{j,0}$ is unchanged. Since $X_{j,1}$ is introduced at the very end of the block function, we have a differential property, like in Section 3.1. When all other inputs are unchanged, the difference on the keystream words resulting from this difference Δ on the nonce word N_i is

$$\Delta' = a \oplus (a \pm \Delta) \tag{2}$$

for some unknown internal value a (see Figure 2).

4 Differential Properties of Addition Modulo 2^{32}

We have seen that differential patterns on the plaintext or the nonce propagate to simple differential patterns on the keystream. More precisely, the differential property on the plaintext is related to a general problem concerning linear approximations of addition modulo 2^{32} that can be summarized by relation (1). In this section, we will describe various ways to take advantage of this observation.

4.1 Related Problems

A well known problem (see [13]) is, given two fixed words x and y, to find a pair (Δ, Δ') such that

$$\Delta' = (x + y) \oplus (x + (y \oplus \Delta)) \tag{3}$$

and that is observed with high probability. This problem has been studied from a theoretical point of view in [17]. However, in the present situation, we are looking things the other way around since x and y are unknown to us but we might be able to choose Δ and observe Δ'. More precisely, we want to

1. find statistical properties that can be easily detected in order to distinguish Helix from a random source.
2. recover some secret information about the internal state of Helix (the values of x and y for instance).

4.2 A "Dummy" Distinguisher

Suppose an attacker encrypts two messages that begin similarly, but, at some point, differ on one word by

$$\Delta = \texttt{0x80000000}$$

Then, the difference on the next keystream word (called Δ') is such that $\Delta' = \Delta$, since there is no propagation from MSBs to LSBs during an addition. Using this relation, the block function of Helix can be distinguished from a random source with two chosen messages, but this requires to use twice the same key and the same nonce. This attack scenario is discussed in Section 5. In the next section, we go further by trying to actually recover the two internal values x and y using relation (3).

4.3 Recovering x and y

In this section, we are interested in recovering the two intermediate values x and y involved in relation (3). Thus, we have to consider the following problem

Problem 1. Let x and y be two given constants of 32 bits. For any Δ,

$$\Delta' = (x + y) \oplus (x + (y \oplus \Delta)) \quad (4)$$

is given. How many (x, y) are possible solutions ? Give an efficient algorithm to recover these solutions.

First, it is easy to see that the solution is not always unique. Indeed, if $x = 0$, then Δ' does not depend on y. However, in average, the number of candidates is small. In this section, we propose an efficient algorithm to recover the two unknown values x and y with a limited number of observations. The following notations are used : w_j denotes the j-th bit of a word w. Besides, let c_j denote the carry bit at position j in the addition of x and $y \oplus \Delta$. For all j, $0 \leq j \leq 31$,

$$(x + (y \oplus \Delta))_j = x_j \oplus y_j \oplus \Delta_j \oplus c_j$$

and initially $c_0 = 0$. We also suppose that $x \neq 0$.

Claim. Let t, $0 \leq t \leq 30$, denote the position of the least significant bit '1' of x. Then, there are exactly 2^{t+3} valid pairs (x, y), solutions of the previous problem. Recovering these solutions can be done by testing at most 93 chosen values of Δ.

We use the following induction

- Assume all bits of x and y are known up to position $(i-1)$.
- If any $x_j = 1$ with $0 \leq j < i$, then
  - By choosing an appropriate value of Δ_k for $j \leq k < i$, it is possible to obtain any value of c_i (0 or 1), since everything is known up to position i.
  - In both cases, pick both values of Δ_i (0 and 1) and set all other bits of Δ to 0. The resulting value of Δ'_{i+1} depends only on the carry bit c_{i+1}.
  - Recover x_i and y_i by comparing the different distributions (see Table 1)
- Otherwise
  - Necessarily, $c_i = 0$
  - Using Table 1, it is still possible to recover x_i.
  - No information on y_i is obtained.

Therefore, by induction, all bits of x can be recovered from position 0 to 30 (it is impossible to recover x_{31} because no observation can be made about position 32 of Δ'). Similarly, all bits of y from position $(t+1)$ to 30 can be recovered. The other $t+3$ bits of x and y need to be guessed. When $x = 0$, our analysis remains valid by taking $t = 30$.

In fact, 3 queries are enough to distinguish the distributions in Table 1. Thus, at most $3 \times 31 = 93$ queries are sufficient to recover a valid solution (x, y). Besides, it is easy to verify that flipping the bit y_t will imply to flip all bits x_j and y_j for $t < j < 31$ in order to obtain an other valid solution, since all carry bits also get flipped. Therefore all solutions of the system can be expressed directly from a single solution, without any extra query.

We performed some experiments using various values of x and y and always identified with success the expected number of 2^{t+3} solutions.

Table 1. Distribution of Δ'_{i+1} depending on x_i and y_i

x_i	y_i	c_i	Δ_i	Δ'_{i+1}	x_i	y_i	c_i	Δ_i	Δ'_{i+1}
1	1	0	0	δ	0	1	0	0	δ
1	1	0	1	$\delta \oplus 1$	0	1	0	1	δ
1	1	1	0	δ	0	1	1	0	$\delta \oplus 1$
1	1	1	1	δ	0	1	1	1	δ
1	0	0	0	δ	0	0	0	0	δ
1	0	0	1	$\delta \oplus 1$	0	0	0	1	δ
1	0	1	0	$\delta \oplus 1$	0	0	1	0	δ
1	0	1	1	$\delta \oplus 1$	0	0	1	1	$\delta \oplus 1$

5 Attacks against Helix

In this section, two attacks against Helix are developed. The first one is a distinguishing attack using chosen plaintext, which is extended to a key recovery attack requiring 2^{88} basic operations and about 2^{12} block encryptions. A second attack takes advantage of choosing similar nonces to detect internal state collisions.

5.1 A Distinguishing Attack

In Section 3.1, we have shown that the introduction of a chosen difference on the plaintext from a fixed internal state results in predictable patterns on the keystream. However, to turn these observations into an attack, it is necessary to consider the following scenario

- The attacker requests encryption of some random message $P = (P_1, \dots, P_n)$ under some pair (key,nonce) $= (K, N)$. The resulting ciphertext is $C = (C_1, \dots, C_n)$.
- He requests encryption with (K, N) of an other message where P_{n-1} is replaced by $P'_{n-1} = P_{n-1} \oplus \Delta$. This yields the ciphertext $C' = (C'_1, \dots, C'_n)$.
- The attacker observes $\Delta' = C_n \oplus C'_n$.

In this case, we have seen that a real Helix output can be distinguished from a random output, by picking $\Delta = \texttt{0x80000000}$ (then, necessarily, $\Delta' = \Delta$).

5.2 A Simple Key Recovery Attack

Now, we wish to extend the observations of Section 4.3. This technique allowed an attacker to retrieve up to 64 bits of intermediate values by observing the keystream corresponding to well chosen plaintexts. Actually, this information leakage is an important weakness, since it reduces the entropy of the internal state. Using an appropriate guessing technique, one may hope to turn it into a key recovery attack. Such an attack is generally called a *guess-then-determine attack*, since an attacker will first *guess* some internal state bits and then *determine* the correct guess using available information.

First, let us consider the round number i of Helix encryption. We suppose an attacker has access to the keystream word $Z_0^{(i)}$ and to a few candidates for $Y_0^{(i)}$ and $Y_3^{(i)}$ as described in Section 4.3. These two intermediate words depend on the internal state at input of round i : $(Z_0^{(i)}, \dots, Z_4^{(i)})$ and on the first round key word $X_{i,0}$. This is represented in Figure 3 where each box is a 32 bits value and dashed boxes represent known values. An attacker may hope to use these conditions to reduce the number of possible internal states to

$$2^{128} \times 2^{32} \times 2^{-64} = 2^{96}$$

Actually, this number can be reached by guessing $Z_2^{(i)}$, $Z_3^{(i)}$ and $X_{i,0}$. Then the attacker can retrieve $Z_1^{(i)}$ and $Z_4^{(i)}$ by looking precisely at the function G

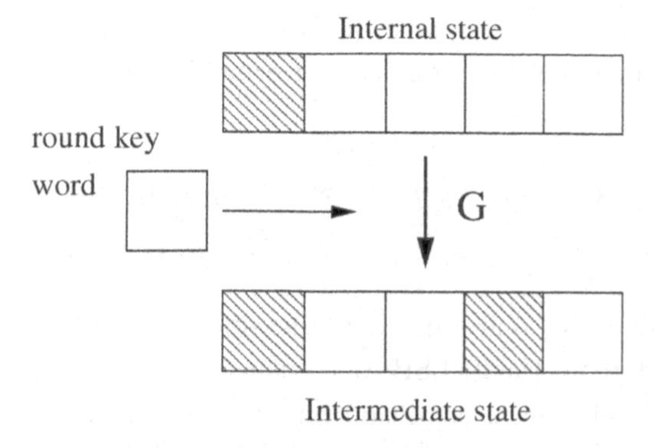

Fig. 3. The framework of the simple attack

(see Figure 2). Thus, the attacker can indeed find 2^{96} candidates for the internal state at the beginning of round i. To tell which candidate is correct, some of the previous rounds (say $\tau = 5$ rounds) need to be inverted. This can be done without increasing the number of candidates, provided $Y_0^{(i-j)}$ and $Y_3^{(i-j)}$ are known, for $0 \leq j < \tau$. For this purpose, the recovery technique of Section 4.3 needs to be applied τ times here. As long as it returns few solutions, an appropriate round inversion reduces the number of candidates - roughly by a factor 2^{32}. Thus, for $\tau = 5$ we eventually obtain a unique candidate, and enough "round key words" to directly retrieve the complete secret key.

To summarize, this simple attack requires to guess 96 bits of internal state and to apply τ times the technique described in Section 4.3 to recover intermediate values. However, this technique does not provide a unique solution, which increases the time complexity of the attack. Actually, only the round i is in the critical path and with probability $\frac{1}{2}$, the number of solutions here is only 8. In this "good" case, the complexity of the attack is $2^{96} \times 8 = 2^{99}$ basic instructions. In "bad" cases, there are more than 8 solutions at position i, but the attacker may easily find another position i' where there are only 8 solutions.

The data complexity corresponds to the encryption of $\tau \times 93$ pairs of messages of length at most $\tau = 5$ words. Thus, the number of plaintext blocks encrypted is

$$2 \times 5 \times 5 \times 93 \simeq 2^{12}$$

5.3 An Improved Attack

A more subtle guessing technique can be applied using bitwise analysis of the block function. The "subtle" attack consists in guessing only 2 words, $Z_3^{(i)}$ and $(Z_1^{(i)} + Z_4^{(i)})$, plus 17 LSBs of $Z_2^{(i)}$. Then, like in the "simple" attack, the attacker can obtain the 17 LSBs of $ROTL_{25}(Z_4^{(i)})$ and thus the 17 LSBs of $ROTL_{25}(Z_1^{(i)})$. Looking at the block function of round $i-1$, the attacker knows two output words, and has partial knowledge of the three other output words. Two relations can be written, involving one unknown intermediate word a

$$\begin{aligned} Z_3^{(i)} &= ROTL_{21}(Z_1^{(i)}) + a \\ Y_3^{(i-1)} &= ROTL_{28}(Z_1^{(i)}) \oplus ROTL_{21}(Z_2^{(i)}) \\ &\oplus ROTL_{26}(Z_4^{(i)}) \oplus ROTL_{19}(a) \end{aligned}$$

From the first relation, one sees that guessing the 4 LSBs of a will give the attacker a candidate for the 21 LSBs of a (using partial knowledge of $Z_1^{(i)}$). Then, using the second relation, a condition on bit number 13 of $Y_3^{(i-1)}$ is obtained. This condition eliminates half of the candidates. Then, each additional guessed bit of $Z_2^{(i)}$ provides one extra condition, that is immediately used to discard half of the guesses. This "early abort" technique results in a guessing complexity of

$$2^{32} \times 2^{32} \times 2^{17} \times 2^4 = 2^{85}$$

The backtracking can be performed here exactly as before to complete the attack. The resulting time complexity is reduced to $8 \times 2^{85} = 2^{88}$ guesses (each requiring a few boolean operations on 32 bit words). Furthermore, the existence of even better guessing techniques should be investigated.

5.4 Practical Impact

Previously, we have proposed a differential attack on Helix, using chosen plaintext. It requires to obtain twice the same internal state as input of the block function. Thus, the attacker needs to encrypt twice with the same key and the same nonce, and to introduce a difference in the plaintext at some point. However in [5], it is specified that "the sender must ensure that each (K, N) is used at most once to encrypt a message", otherwise Helix "loses its security properties". According to the authors, this requirement is not restrictive since it is underlying many similar situations in cryptography. For instance, when using a synchronous stream cipher, if secret key and nonce are unchanged, the same pseudo-random sequence is generated twice, which breaks the confidentiality. Similar problems may also be encountered when using a block cipher in OFB mode for instance. In general, a distinguishing attack is always possible when nonces are re-used. We believe the situation is more preoccupying in the case of Helix since we obtain key recovery attacks and not only distinguishing attacks.

On the one hand, there are situations where the previous scenario is not realistic. Indeed, the secret key may be used to communicate only in one direction. In this case, it is straightforward for the sender never to re-use the same nonce (he can use counters for instance). Apparently, this is true for wireless networks, where each pair of users have two separate secret keys, one for each direction. A differential attack cannot be applied there, unless the attacker gains physically access to the encryption machine and can force nonce repetition. This may be possible in some particular occasions, but in general it is a strong assumption.

On the other hand, in most situations, our differential attack scenario seems realistic. For instance, several users often need to share a secret key. Even if they split properly the nonce space, what happens if the same message is sent to multiple receivers ? An attacker can sit in the middle, and modify the ciphertext on one of the communication channels. Then, by comparing a "faulty" decryption with a correct decryption, he may obtain the kind of differential information he needs.

To conclude, we think the security impact of our attacks will highly depend on the context, but in general, one should expect the block function of Helix to resist better against differential attacks. Overall, the secrecy of the key cannot reasonably rely on the absence of nonce repetition.

5.5 A Chosen Nonce Attack

A weakness regarding the influence of each nonce word has been identified in Section 3.2. Here, we propose an extension to a distinguishing attack against Helix. Its complexity is much bigger than the previous attack. However it has the advantage of being based on weaker assumptions. Indeed, in this case, the attacker does not need to encrypt several messages with the same pair (key,nonce). Instead, we suppose that the same plaintext P is encrypted twice with the same secret key, but two distinct nonces N and N' such that

$$N = (N_0, N_1, N_2, N_3)$$
$$N' = (N_0, N_1, N_2, N_3 + \Delta)$$

Then, as argued in 3.2, the block function is essentially the same for any round i such that $i \bmod 4 \neq 3$. If a state collision occurs on the input of such a round, it will also propagate to a state collision for the input of the next round. Thus state collisions on inputs of rounds i such that $i \bmod 4 = 0$ imply collisions on 4 consecutive blocks of keystream. Moreover, the difference on the 5-th block can be predicted exactly (by picking $\Delta = 10\dots0_x$ for instance). Thus, we obtain a detectable condition on 160 bits of keystream. This is sufficient to detect state collisions with good probability.

Therefore, contrarily to what is claimed in [5], state collisions in Helix can be detected. However, the length of messages is not allowed to exceed 2^{62} blocks, so collisions are unlikely to be observed for practice purpose.

5.6 Forcing the Collisions

In this section, we show that the previous attack can be extended into a distinguisher against Helix with only 2^{114} encrypted blocks. This is an important result, since it constitutes a break of the cipher, according to the definition given by the authors [5].

The general idea is to work on a large set of nonces that will preserve collisions during a few rounds. Then these collisions can be detected by observing the corresponding keystream blocks. More precisely, we build a message P of the maximal authorized length 2^{62} words by repeating 2^{62} times the same word P_0. Then, P is encrypted under a fixed unknown secret key K using different nonces of the form

$$N^{(\delta,\Delta)} = (N_0 + \delta, N_1 + \delta, N_2 + \delta, N_3 + \Delta)$$

with four fixed constants $(N_0, \ldots, N_3)$. δ is of the form $8 \times x$ where x spans all values from 0 to 2^{20} and Δ spans all 2^{32} possible words. Therefore the number of blocks encrypted is

$$2^{62} \times 2^{32} \times 2^{20} = 2^{114}$$

As before, we consider any state collisions that occurs between two different nonces $N^{(\delta_1,\Delta_1)}$ and $N^{(\delta_2,\Delta_2)}$, at two different positions in the encryption, respectively i_1 and i_2. We would like this state collision to be preserved for several rounds, in order to detect some properties on the keystream, as in the previous Section. We are sure that the plaintext word introduced is always P_0, by construction. Furthermore we would like to have the same round key words for both encryptions. Hence, these positions should satisfy

$$i_1 \bmod 8 = i_2 \bmod 8 = 0$$

in order to have $X_{i_1+j,0} = X_{i_2+j,0}$ for all j. Besides, if

$$\delta_1 + i_1 = \delta_2 + i_2 \bmod 2^{32} \tag{5}$$

then $X_{i_1+j,1} = X_{i_2+j,1}$ when $j \bmod 4 \neq 3$. In this case, the state collision is preserved during at least 3 rounds. Concerning rounds $i_1 + 3$ and $i_2 + 3$, we would like to also preserve the collision, thus we need $X_{i_1+3,1} = X_{i_2+3,1}$ or

$$\Delta_1 + i_1 + X'_{i_1+3} = \Delta_2 + i_2 + X'_{i_2+3} \bmod 2^{32} \tag{6}$$

With these three assumptions, the state collision is preserved at least until the rounds $i_1 + 7$ and $i_2 + 7$ which results in collisions on 8 consecutive words of keystream.

To mount an attack, we first store sequences of 8 consecutive keystream words, for each message and for each position i such that $i \bmod 8 = 0$. Then, we look for a collision among the $\frac{2^{114}}{8} = 2^{111}$ entries in this table. This can be achieved by sorting the table, with complexity of $2^{111} \times 111 \simeq 2^{118}$ basic instructions. Then, since we consider objects of 256 bits, the number of "fortuitous" collisions in the table is

$$\frac{2^{111} \times 2^{111}}{2} \times 2^{-256} \simeq 0$$

Besides, when a "true" state collision occurs, a collision is also observed on the entries of the table, provided the additional assumptions (5) and (6) hold. (5) holds with probability 2^{-29}, since all terms are multiples of 8, and (6) holds with probability 2^{-32}. Therefore, the number of "true" collision observed in the table is in average

$$\frac{2^{111} \times 2^{111}}{2} \times 2^{-160} \times 2^{-29} \times 2^{-32} = 1$$

Thus we have considered enough encrypted data to detect some particular state collisions that are preserved during a few rounds. We achieve it by observing patterns of 8 consecutive words of keystream. For a true Helix output we expect to find a collision in the previous table, while it will not be the case for a random output. Actually this distinguishing attack can be slightly improved if we take into account the case $i \bmod 8 = 4$.

To conclude, we have proposed a distinguishing attack against Helix requiring the encryption of 2^{114} words of plaintext under chosen nonces. This attack is faster than exhaustive search, processes less than 2^{128} blocks of plaintext and respects the security requirements proposed in [5], since no pair (key,nonce) is ever re-used to encrypt different messages. Therefore, this attack constitutes a theoretical break of Helix.

6 Conclusion

This paper describes two attacks against the new stream cipher Helix. The first one recovers the secret key with a reasonably low complexity in time and data, so we think it should be considered as an important threat. The assumptions we use are quite usual (chosen plaintext, chosen nonce), but they are outside the security model proposed by the authors of the cipher.

However, we also propose a second attack, less efficient but which relies on weaker assumptions. This distinguishing attack constitutes a break of Helix according to the definition given by the authors. Both attacks result from weak differential properties of the encryption function regarding the plaintext and the nonce. In general, our attack illustrates the fact that one should be careful to protect new stream ciphers against differential-like attacks.

References

[1] D. Coppersmith, S. Halevi, and C. Jutla. Cryptanalysis of Stream Ciphers with Linear Masking. In M. Yung, editor, *Advances in Cryptology – Crypto'02*, volume 2442 of *Lectures Notes in Computer Science*, pages 515–532. Springer, 2002. 94

[2] P. Ekdahl and T. Johansson. SNOW - a New Stream Cipher. In *First Open NESSIE Workshop, KU-Leuven*, 2000. Submission to NESSIE. Available at `http://www.it.lth.se/cryptology/snow/`. 94

[3] P. Ekdahl and T. Johansson. Distinguishing Attacks on SOBER-t16 and t32. In J. Daemen and V. Rijmen, editors, *Fast Software Encryption – 2002*, volume 2365 of *Lectures Notes in Computer Science*, pages 210–224. Springer, 2002. 94

[4] N. Ferguson. Michael: an improved MIC for 802.11 WEP. Document 2-020. Available at `http://grouper.ieee.org/groups/802/11/`. 95

[5] N. Ferguson, D. Whiting, B. Schneier, J. Kelsey, S. Lucks, and T. Kohno. Helix, Fast Encryption and Authentication in a Single Cryptographic Primitive. In T. Johansson, editor, *Fast Software Encryption – 2003*, 2003. To appear. 94, 95, 96, 104, 105, 107

[6] FIPS PUB 81. *DES Modes of Operation*, 1980. 94

[7] S. Fluhrer. Cryptanalysis of the SEAL 3.0 Pseudorandom Function Family. In M. Matsui, editor, *Fast Software Encryption – 2001*, volume 2355 of *Lectures Notes in Computer Science*, pages 135–143. Springer, 2001. 94

[8] S. Fluhrer, I. Mantin, and A. Shamir. Weaknesses in the Key Scheduling Algorithm of RC4. In S. Vaudenay and A. M. Youssef, editors, *Selected Areas in Cryptography – 2001*, volume 2259 of *Lectures Notes in Computer Science*, pages 1–24. Springer, 2001. 95

[9] V. D. Gligor and P. Donescu. Fast Encryption and Authentication: XCBC Encryption and XECB Authentication Modes. In M. Matsui, editor, *Fast Software Encryption – 2001*, volume 2355 of *Lectures Notes in Computer Science*, pages 192–108. Springer, 2001. 95

[10] S. Halevi, D. Coppersmith, and C. Jutla. Scream : a Software-efficient Stream Cipher. In L. Knudsen, editor, *Fast Software Encryption – 2002*, volume 2332 of *Lectures Notes in Computer Science*, pages 195–209. Springer, 2002. 94

[11] P. Hawkes and G. Rose. Primitive Specification and Supporting Documentation for SOBER-t32. In *First Open NESSIE Workshop*, 2000. Submission to NESSIE. 94

[12] C. Jutla. Encryption Modes with Almost Free Message Integrity. In B. Pfitzmann, editor, *Advances in Cryptology – Eurocrypt'01*, volume 2045 of *Lectures Notes in Computer Science*, pages 529–544. Springer, 2001. 95

[13] H. Lipmaa and S. Moriai. Efficient Algorithms for Computing Differential Properties of Addition. In M. Matsui, editor, *Fast Software Encryption – 2001*, volume 2355 of *Lectures Notes in Computer Science*, pages 336–350. Springer, 2001. 99

[14] NESSIE - New European Schemes for Signature, Integrity and Encryption. `http://www.cryptonessie.org`. 94

[15] P. Rogaway, M. Bellare, J. Black, and T. Krovetz. OCB/ A Block-cipher Mode of Operation for Efficient Authenticated Encryption. In *Eight ACM Conference on Computer and Communications Security (CCS-8)*, pages 196–205. ACM Press, 2001. 95

[16] P. Rogaway and D. Coppersmith. A Software-optimized Encryption Algorithm. In R. Anderson, editor, *Fast Software Encryption – 1994*, volume 809 of *Lectures Notes in Computer Science*, pages 56–63. Springer-Verlag, 1994. 94

[17] J. Wallen. Linear Approximations of Addition Modulo 2^n. In T. Johansson, editor, *Fast Software Encryption – 2003*, 2003. To appear. 100

[18] IEEE P802.11, The Working Group for Wireless LANs. `http://grouper.ieee.org/groups/802/11/`. 95

Improved Linear Consistency Attack on Irregular Clocked Keystream Generators

Håvard Molland

The Selmer Center*, Dept. of Informatics
University of Bergen, P.B. 7800 N-5020 Bergen, Norway

Abstract. In this paper we propose a new attack on a general model for irregular clocked keystream generators. The model consists of two feedback shift registers of lengths l_1 and l_2, where the first shift register produces a clock control sequence for the second. This model can be used to describe among others the shrinking generator, the step-1/step-2 generator and the stop and go generator. We prove that the maximum complexity for attacking such a model is only $O(2^{l_1})$.

Keywords: Stream ciphers, irregular clocked generators, linear consistency test

1 Introduction

The goal in stream ciphers is to expand a short key into a long keystream $\mathbf{z}$ that is difficult to distinguish from a truly random bit stream. It should not be possible to reconstruct the short key from $\mathbf{z}$. The message is then encrypted by mod-2 additions with the keystream.

In this paper we analyze additive stream ciphers where the keystream is produced by an irregular clocked linear feedback shift register ($LFSR$). This model produces bit streams with high linear complexity, which is a important criteria for pseudo random sequences.

The cipher model we attack is composed of two $LFSR$s, $LFSR_{\mathrm{s}}$ of length l_{s} and $LFSR_{\mathrm{u}}$ of length l_{u}. $LFSR_{\mathrm{s}}$ produces a bit stream $\mathbf{s}$ and $LFSR_{\mathrm{u}}$ produces a bit stream $\mathbf{u}$. The bit stream $\mathbf{s}$ is sent through a function $D()$. Finally $D()$ outputs the clock control sequence of integers, $\mathbf{c}$, which is used to clock $LFSR_{\mathrm{u}}$. See Fig. 1 for an illustration, and Sec. 2.1 for a full description of the model. The effect of the irregular clocking is that $\mathbf{u}$ is irregularly decimated. The result from the decimation is the keystream $\mathbf{z}$. Thus, the positions of the bits in the original stream $\mathbf{u}$ are altered and the linearity of the stream are destroyed. This gives the keystream $\mathbf{z}$ high linear complexity.

There have been several previous attacks on this scheme. One popular method is to use the constrained Levenshtein distance (CLD) (also called edit distance),

* This work was supported by the Norwegian Research Council under Grant 146874/420.

B. Roy and W. Meier (Eds.): FSE 2004, LNCS 3017, pp. 109–126, 2004.

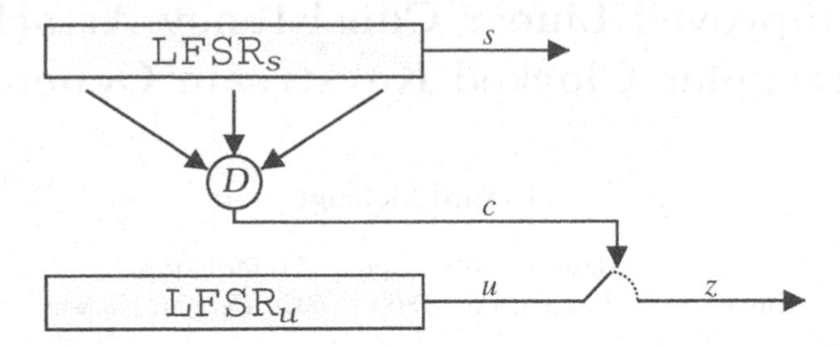

Fig. 1. The general model for irregular clocked keystream generators

which is the number of deletions, insertions, or substitutions required to transform one sequence into another. In [10, 9] they find the optimal edit distance and present efficient algorithms for its computation.

Another technique is to use the linear consistency test (LCT), see Handbook of Cryptography (HAC) [1] and [3]. Here the l_s clock control initialization bits are guessed and used to restore the positions the keystream bits had in $\mathbf{u}$. This gives the guess $\mathbf{u}^* = (.., *, z_i, ..., z_j, ...*, ..., z_k, ..., *, ...)$, where z_i, z_j, z_k are some keystream bits and the stars are the deleted bits. They now perform the LCT on $\mathbf{u}^*$, using the Gaussian algorithm on an equation set with l_u unknowns derived from $LFSR_u$ and $\mathbf{u}^*$. If the equation set is consistent the guess is outputted as the correct initialization bits for $LFSR_s$. The Gaussian algorithm will use about $\frac{1}{3}l_u^3$ calculations and the total complexity for the attack is $O(2^{l_s} \cdot l_u^3)$.

In [7, 8] and the recent paper [11] they guess only a few of the clock control bits before they reject/accept the guess, using the Gaussian algorithm. If the guessed bits pass the test, they do a exhaustive search on the remaining key space.

It is hard to estimate the running time for the attacks in [7, 8, 9, 10]. The attack in [11] is estimated to have a upper bound complexity $O(L^3 \cdot 2^{L\lambda})$, where $\lambda = \log A/(1 + \log A)$, $L = l_1 + l_2$ and A is the number of different clocking numbers from the $D()$ function.

Most of the previous LCT attacks have in common that they try to find the initialization bits for both $LFSR_s$ and $LFSR_u$ at once. We have a much more simple and algorithmic approach to the problem. The resulting algorithm is deterministic and has a lower and easily estimated running time which is independent from the number of clock control behaviors A, and the length l_u of $LFSR_u$. We will show that our attack has lower computational complexity than the previous LCT attacks.

We also do a test similar to the LCT, but our test is much more efficient since we are not using the Gaussian algorithm to reject or accept the initialization bits for $LFSR_s$. Our rejection test has constant complexity $O(K)$, where K is only 2 parity check operations in average. Thus, the total complexity for the attack becomes $O(2^{l_s})$.

The basic idea for the test is simple. From the generator polynomial $g_u(x)$ for $LFSR_u$ we derive a low weight cyclic equation that will hold for all bitstreams generated by $LFSR_u$. In Appendix C.3 we describe a modified version of Wagners General birthday algorithm [4] that finds the low weight cyclic equation. For each guess of $\mathbf{c}$ we generate the guess $\mathbf{u}^*$ for $\mathbf{u}$. Then we try the cyclic equation at a given number of entries in the $\mathbf{u}^*$ stream. If the equation hold every time, we can conclude that the bits are generated by $LFSR_u$, and it is most likely that we have the correct guess for $\mathbf{c}$. If the guess is wrong we have to test the equation at in average 2 entries before the guess is rejected. A naive implementation of this algorithm will, as shown in Section 3.2, have complexity $O(2^{l_s} \cdot N)$ where N is the length of the guess $\mathbf{u}^*$. The reason for this is that we have to calculate a new $\mathbf{u}^*$ for each guess for $\mathbf{c}$.

The real advantage in this paper is the new algorithm we present in Section 3.4. The algorithm is iterative and except for the first iteration it calculates each guess $\mathbf{u}^*$ using just a few operations. The idea is to go through the guesses for $\mathbf{c}$ cyclically. This way we can reuse most of $\mathbf{u}^*$ from one guess to another. In worst case our attack needs 2^{l_s} iterations to succeed, and we have the complexity $O(2^{l_s})$. Thus, by using the cyclic properties of feedback shift registers, we have got rid of the l_u^3 factor they have in the LCT attacks in [1, 3, 7, 8, 11]. In Section 4 we present some simulations of the algorithm.

2 A General Model for Irregular Clocked Generators

2.1 Description

We will first give a general description of irregular clocked generators.

Let $g_u(x)$ be the feedback polynomial for the shift register $LFSR_u$ of length l_u, and let $g_s(x)$ be the feedback polynomial for a shift register $LFSR_s$ of length l_s. $LFSR_u$ is called the *data generator*, and $LFSR_s$ is called the *clock control generator.*

From $g_s(x)$ we can calculate a clock control sequence $\mathbf{c}$ in the following way. Let $c_t = D(s_v, s_{v+1}, ..., s_{v+l_s-1}) \in \{a_1, a_2, ..., a_A\}, a_j \geq 0$ be a function where the input $(s_v, s_{v+1}, ..., s_{v+l_s-1})$ is the inner state of $LFSR_s$ after v feedback shifts and A is the number of values that c_t can take. Let p_j be the probability $p_j = \text{Prob}(c_t = a_j)$. The way $LFSR_s$ is clocked is defined by the specific generator. Often $LFSR_s$ and c_t are synchronized, which means that $v = t$.

$LFSR_u$ produces the stream $\mathbf{u} = (u_0, u_1, ...)$ The clock c_t decides how many times $LFSR_u$ is clocked before the output bit from $LFSR_u$ is taken as keystream bit z_t. Thus the keystream z_t is produced by $z_t = u_{k(t)}$, where $k(t)$ is the total sum of the clock at time t, that is $k(t) \leftarrow k(t-1) + c_t$.

Let $\mathbf{u} = (u_0, u_1, ..., u_{N-1})$ be the bit stream produced by the shift register $LFSR_u$. The resulting sequence will then be $z_t = u_{k(t)}, 1 < t < M$. This gives the following definition for the clocking of $LFSR_u$.

Definition 1. *Given bit stream* $\mathbf{u}$ *and clock control sequence* $\mathbf{c}$*, let* $\mathbf{z} = Q(\mathbf{c}, \mathbf{u})$ *be the function that generates* $\mathbf{z}$ *of length* M *by*

$$Q(\mathbf{c}, \mathbf{u}) : z_t \leftarrow u_{k(t)}, 0 \leq t < M$$

where $k(t) = \sum_{j=0}^{t} c_j - S,\ S \in \{0,1\}$

The parameter S only is for synchronization, and most often $S = 1$. Finally we let $\mathbf{s}^{\mathrm{I}} = (s_0, s_1, ..., s_{l_s-1})$ and $\mathbf{u}^{\mathrm{I}} = (u_0, u_1, ..., u_{l_s-1})$ be the initialization states for $LFSR_{\mathrm{s}}$ and $LFSR_{\mathrm{u}}$. Together, $\mathbf{s}^{\mathrm{I}}$ and $\mathbf{u}^{\mathrm{I}}$ defines *the secret key* for the given cipher system.

If $a_j \geq 1,\ \ 1 \leq j \leq A$, the function $Q(\mathbf{c}, \mathbf{u})$ can be looked on as a deletion channel with input $\mathbf{u}$ and output $\mathbf{z}$. The deletion rate is

$$P_{\mathrm{d}} = 1 - \frac{1}{\sum_{j=1}^{A} p_j a_j}. \tag{1}$$

Thus, given a stream $\mathbf{z}$ of length M, the expected length N of the stream $\mathbf{u}$ is

$$\mathrm{E}(N) = \frac{M}{(1 - P_{\mathrm{d}})} = M \sum_{j=1}^{A} p_j a_j. \tag{2}$$

2.2 Some Examples for Clock Control Generators

The Step-1/Step-2 Generator. The clocking function is defined by $Q(\mathbf{c}, \mathbf{u}) : z_t \leftarrow u_{k(t)}, 0 \leq t < M$, and $D(s_t) = 1 + s_t$. We see that the number of outputs is $A = 2$, with probabilities $p_j = 1/2,\ 1 \leq j \leq 2$. This gives $P_{\mathrm{d}} = 1 - \frac{1}{\frac{1}{2} + 2\frac{1}{2}} = \frac{1}{3}$, and $\mathrm{E}(N) = \frac{3}{2}M$. Since this generator is simple, we will use it in the examples in this paper.

Example 1. Assume we have a irregular clock control stream cipher as defined in Section 2.1, with $g_{\mathrm{s}}(x) = x^3 + x^2 + 1$. We let $\mathbf{s}^{\mathrm{I}} = (s_0, s_1, s_2) = (1, 0, 1)$ and we get $\mathbf{c}$ by $c_t = D(s_t)$:

$$\mathbf{c} = (2, 1, 2, 1, 1, 2, 2, 2, 1, 2, 1, 1, 2, 2, 2, 1, 2, 1, 1, 2, ...).$$

Let $g_{\mathrm{u}}(x) = x^4 + x^3 + 1$ and $LFSR_{\mathrm{u}}$ be initialized with $\mathbf{u}^{\mathrm{I}} = (1, 1, 0, 0)$. We get the following bit stream

$$\mathbf{u} = (1, 1, 0, 0, 1, 0, 0, 0, 1, 1, 1, 1, 0, 1, 0, 1, 1, 0, 0, 1, 0, 0, 0, 1, 1, 1, 1, 0, 1). \tag{3}$$

Using $\mathbf{c}$ on $\mathbf{u}$, the bits are discarded in this way,

$$\begin{aligned} \mathbf{u}^* = (&*, 1, 0, *, 1, 0, 0, *, 1, *, 1, *, 0, 1, *, \\ &1, 1, 0, *, 1, *, 0, *, 1, 1, *, 1, 0, 1) \end{aligned} \tag{4}$$

Finally the output bit from the cipher will be

$$\mathbf{z} = Q(\mathbf{c}, \mathbf{u}) = (1, 0, 1, 0, 0, 1, 1, 0, 1, 1, 1, 0, 1, 0, 1, 1, 1, 0, 1). \tag{5}$$

The LILI-128 Clock Control Generator. The clock control generator which is one of the building blocks in the LILI-128 cipher [12] is similar to the *step-1/step-2* generator but $\mathbf{c}$ has a larger range. The generator is defined by $Q(\mathbf{c}, \mathbf{u}) : z_t \leftarrow u_{k(t)}, 0 \leq t < M$, and $c_t = D(s_{t+i_1}, s_{t+i_2}) = 1 + s_{t+i_1} + 2s_{t+i_2}$. This gives $A = 4$, $p_j = \frac{1}{4}$, $1 \leq j \leq 4$, and $P_\mathrm{d} = 1 - \frac{1}{\sum_{j=1}^{4} \frac{1}{4} j} = \frac{3}{5}$, and the length of $\mathbf{u}$ is expected to be $N = \frac{5}{2}M$.

The Shrinking Generator. In the shrinking generator, the output bit u_k from $LFSR_\mathrm{u}$ is outputted as keystream bit z_t if the output s_k from $LFSR_\mathrm{s}$ equals one. If $s_k = 0$ then u_k is discarded.

To be able to attack the generator with our algorithm we must have the clock control sequence $\mathbf{c}$. The clock control sequence for the shrinking generator can be generated as follows. Let $y - 1$ be the number of consecutive zeros from $s_v = 1$, that is $(s_v, s_{v+1}, ..., s_{v+l_\mathrm{s}-1}) = (\underbrace{1, 0, ..., 0}_{y}, *, ..., *)$. Then the clocking function is defined as $D(s_v, s_{v+1}, ..., s_{v+l_\mathrm{s}-1}) = y$. It follows from the definition of the shrinking generator that $LFSR_\mathrm{s}$ and $LFSR_\mathrm{u}$ are synchronized, so $LFSR_\mathrm{s}$ must be clocked c_t times before the next bit is outputted. Thus the clock control sequence is $c_t = D(s_{k(t)}, s_{k(t)+1}, ..., s_{k(t)+l_\mathrm{s}-1})$,where $k(t) \leftarrow k(t-1) + c_t$ for each iteration. $Q(\mathbf{c}, \mathbf{u})$ is the same as for the generators above. If we analyze the clock control sequence, $c_t \in \{1, 2, ..., ..., l_\mathrm{s} - 1\}$, where $p_j = 1/2^j$, when l_s is a large number ($l_\mathrm{s} > 10$). This gives $A = l_\mathrm{s} - 1$, $P_\mathrm{d} = 1 - \frac{1}{\sum_{j=1}^{l_\mathrm{s}} \frac{1}{2^j} j} \approx 0.5$ and $\mathrm{E}(N) = 2M$, as intuitively expected.

3 A New Attack on Irregular Clocked Generators

The idea behind the attack is to guess the clock control sequence $\mathbf{c}$, and reconstruct the original positions the key stream bits in $\mathbf{z}$ had in $\mathbf{u}$ using the reversed function $Q^*(\mathbf{c}, \mathbf{z})$ defined below. From this we get a sequence $\hat{\mathbf{u}}$ looking similar to (4). When this is done, we test if $\hat{\mathbf{u}}$ is a sequence that could have be generated by $LFSR_\mathrm{u}$ using some linear equations we know hold over any sequences generated by $LFSR_\mathrm{u}$. If the test holds, we assume we have made the correct guess for $\mathbf{c}$. Knowing the correct $\mathbf{c}$, we can use the Gaussian algorithm as described in [3] to find the initialization bits for $\mathbf{u}$.

3.1 The Basics

First we state a definition.

Definition 2. *Given the clock control sequence* $\mathbf{c}$ *and keystream* $\mathbf{z}$*, let the function* $\mathbf{u}^* = Q^*(\mathbf{c}, \mathbf{z})$ *be the (not complete) reverse of* Q*, defined as*

$$Q^*(\mathbf{c}, \mathbf{z}) : u^*_{k(t)} \leftarrow z_t,\ 0 \leq t < M,$$

where $k(t) = \sum_{j=0}^{t} c_j - S$*, and* $u^*_k = *$ *for the entries* k *in* $\mathbf{u}^*$ *where* u^*_k *is deleted. When this occurs we say that* u^*_k *is not defined.*

The length of $\mathbf{u}^*$ will be $N^* = \sum_{j=0}^{M-1} c_j$. Note that the only difference between this definition and Definition 1, is that $\mathbf{u}$ and $\mathbf{z}$ have changed sides. Thus $Q^*(\mathbf{c}, \mathbf{z})$ is a reverse of the $Q(\mathbf{c}, \mathbf{u})$. But since some bits are deleted, the reverse is not complete and we get the stream $\mathbf{u}^*$. As seen in Example 1 we can reverse the keystream (5) back to (4) but not completely back to the original stream (3), since the deleted bits are not known.

The probability for a bit u_k^* being defined is $\text{Prob}(u_k^*) = 1 - P_\text{d}$. This happens when $k = k(t)$ holds for for some t, $0 \leq t < M$. It follows that the sum $\delta = u_k^* + u_{k+j_1}^* + ... + u_{k+j_{w-1}}^*$ will be defined if and only if all of the bits in the sum are defined. Thus the sum δ will be defined for given k in $\mathbf{u}^*$ with probability

$$P_\text{def} = \text{Prob}(u_k^*, u_{k+j_1}^*, ..., u_{k+j_{w-1}}^*) = (1 - P_\text{d})^w = \left(\frac{1}{\sum_{j=1}^{A} p_j a_j}\right)^w . \quad (6)$$

3.2 Naive Attack

Using definition 2, we first present a naive high complexity attack. In the next section we present a more advanced and low complexity version of the attack.

Let $\mathbf{s}^\text{I}$ be the initial state for $LFSR_\text{s}$, and let $L^v(\mathbf{s}^\text{I})$ be the inner state after v feedback shifts. Without loss of generality we assume $S = 1$ and that $LFSR_\text{s}$ is clocked once for each output c_t. Thus, $v = t$ and $c_t = D(L^t(\mathbf{s}^\text{I}))$ is the output from the clock generator after t feedback shifts.

We are given a keystream $\mathbf{z}$ of length M which is generated with $\mathbf{z} = Q(\mathbf{c}, \mathbf{u})$. Assume we have found an equation $u_k + u_{k+j_1} + ... + u_{k+j_{w-1}} = 0$ that holds over $\mathbf{u}$. First we guess the initial state $LFSR_\text{s}$ and generates the corresponding guess $\hat{\mathbf{c}}$ for $\mathbf{c}$ using the $D()$ function. Using definition 2 we can calculate $\mathbf{u}^* = Q^*(\hat{\mathbf{c}}, \mathbf{z})$. Then we try to find m (typically $m = l_\text{s} + 10$, we add 10 to prevent false alarms) entries in $\mathbf{u}^*$ where the equation is defined. If the equation holds for every entry it is defined, we assume we have found the correct guess for $\mathbf{s}^\text{I}$. If not, we make a new guess and do the test again. The pseudo code for this algorithm is given below.

Input The keystream $\mathbf{z}$ of length M

1. Preprocessing: Find an equation of low weight that holds over the stream $\mathbf{u}$ of length N.
2. For all possible guesses $\hat{\mathbf{s}}^\text{I}$ do the following:
3. Generate the clock control sequence $\hat{\mathbf{c}}$ of length M by $c_t = D(L^t(\hat{\mathbf{s}}^\text{I}))$.
4. Generate $\hat{\mathbf{u}}^*$ of average length $N = \frac{M}{(1-P_\text{d})}$ using $\hat{\mathbf{u}}^* = Q^*(\hat{\mathbf{c}}, \mathbf{z})$.
5. Find m entries $(k_1, k_2, ..., k_m)$ in the stream $\hat{\mathbf{u}}^*$ where the equation is defined.
6. If the equation holds for all the m entries over $\hat{\mathbf{u}}^*$, then stop the search and output the guess $\hat{\mathbf{s}}^\text{I}$ as the key for $LFSR_\text{s}$.

The problem with this algorithm is that for each guess for $\hat{\mathbf{s}}^\text{I}$, we have to generate a new clock control stream of length M and generate $\hat{\mathbf{u}}^* = Q^*(\hat{\mathbf{c}}, \mathbf{z})$ of length N. In larger examples, N and M will be large numbers, say around 10^6. Since the complexity is $O(N \cdot 2^{l_s})$, the run time for this algorithm will in many cases be worse than the algorithm in [3]. In the next session we present an idea that fixes this problem.

3.3 Final Idea

The problem in the previous section was that we had to generate M bits of the clock control stream for each guess for $\mathbf{s}^{\mathrm{I}}$. This can be avoided if we go through the guesses in a more natural way. We start by a initial guess $\hat{\mathbf{s}}^{\mathrm{I}} = (0, 0, ..., 1)$, and let the i'th guess be the internal state of the $LFSR_s$ after i feedback shifts.

Let $\mathbf{c}^i = (c_0^i, c_1^i, ..., c_{M-1}^i)$ be the i'th guess for the clock control sequence defined by $c_t^i = D(L^{i+t}(1, 0, ..., 0))$, $0 \leq t < M$. Let $\mathbf{u}^i = Q^*(\mathbf{c}^i, \mathbf{z})$ be the corresponding guess for $\mathbf{u}^*$ of length $N_i = \sum_{t=0}^{M-1} c_t^i$. We can now give a iterative method for generating $\mathbf{u}^{i+1}$ from $\mathbf{u}^i$.

Lemma 1. *We can transform* $\mathbf{u}^i$ *into* $\mathbf{u}^{i+1} = Q^*(\mathbf{c}^{i+1}, \mathbf{z})$ *using the following method: Delete the first* c_0^i *entries* $(*, ..., *, z_0)$ *in* $\mathbf{u}^i$*, append the* $c_{M-1}^{i+1} = c_M^i$ *entries* $(*, ..., *, z_M)$ *at the end, and replace* z_t *with* z_{t-1} *for* $1 \leq t \leq M$*.*

Proof. See Appendix B.

Lemma 1 gives us a fast method for generating all possible guesses for $\mathbf{u}$ given a keystream $\mathbf{z}$. See Table 1 for an intuitive example of how the lemma works. Next we prove a theorem that allows us to reuse the equation set defined for $\mathbf{u}^i$.

Theorem 1. *If the sum*

$$\beta_{\mathbf{u}^i,k} = u_k + u_{k+k_1} + ... + u_{k+k_{w-1}} = z_t + z_{t+j_1} + ... + z_{t+j_{w-1}} = \gamma_{\mathbf{z},t}$$

is defined over $\mathbf{u}^i$*, then the sum*

$$\beta_{\mathbf{u}^{i+1},k-c_0^i} = z_{t-1} + z_{t+j_1-1} + ... + z_{t+j_{w-1}-1} = \gamma_{\mathbf{z},t-1}$$

is defined over $\mathbf{u}^{i+1}$*.*

Proof. See Appendix B.

The main result from this theorem is that the equation set that is defined over $\mathbf{u}^i$ will still be defined over $\mathbf{u}^{i+1}$ if we shift the equations c_0^i entries to the left over $\mathbf{u}^{i+1}$. This means that we can just shift the equations 1 entry to the left over $\mathbf{z}$, and we will have an sum that is defined for the guess $\hat{\mathbf{s}}^{\mathrm{I}} = D(L^{i+1}(1, 0, ..., 0))$. Thus, the theorem indicates that we can go around a lot of computations if we let the i'th guess for the inner state of $LFSR_s$ be $L^i(1, 0, ..., 0)$.

Table 1. Example of a walk through of the key. The bits in bold font show how the pattern of defined bits in $\mathbf{u}^i$ shifts to the left, while the actual key bits stay relatively put. Also notice how the entries z_i in the patterns are replaced with z_{i-1} after one iteration. For example the sub stream $z_7, z_8, z_9, *, z_{10} \rightarrow z_6, z_7, z_8, *, z_9$. This means that if the sum $z_7 + z_8 + z_9 + z_{10}$ is defined for $\mathbf{c}^i$, then $z_6 + z_7 + z_8 + z_9$ will be defined for $\mathbf{c}^{i+1}$

Guessed clock sequence $\mathbf{c}^i$	Resulting 'known' bits of $\mathbf{u}^i = Q^*(\mathbf{c}^i, \mathbf{z})$.
$(2,1,1,2,2,2,1,2,1,1,2)$	$(*, z_0, z_1, z_2, *, z_3, *, z_4, *, z_5, z_6, *, \mathbf{z_7}, \mathbf{z_8}, \mathbf{z_9}, *, \mathbf{z_{10}})$
$(1,1,2,2,2,1,2,1,1,2,2)$	$(z_0, z_1, *, z_2, *, z_3, *, z_4, z_5, *, \mathbf{z_6}, \mathbf{z_7}, \mathbf{z_8}, *, \mathbf{z_9}, *, z_{10})$
$(1,2,2,2,1,2,1,1,2,2,2)$	$(z_0, *, z_1, *z_2, *, z_3, z_4, *, \mathbf{z_5}, \mathbf{z_6}, \mathbf{z_7}, *, \mathbf{z_8}, *, z_9, *, z_{10})$
$(2,2,2,1,2,1,1,2,2,2,1)$	$(*, z_0, *, z_1, *z_2, z_3, *, \mathbf{z_4}, \mathbf{z_5}, \mathbf{z_6}, *, \mathbf{z_7}, *, z_8, *, z_9, z_{10})$
$(2,2,1,2,1,1,2,2,2,1,2)$	$(*, z_0, *z_1, z_2, *, \mathbf{z_3}, \mathbf{z_4}, \mathbf{z_5}, *, \mathbf{z_6}, *, z_7, *, z_8, z_9, *, z_{10})$

3.4 The Complete Attack

We will now present a new algorithm that make use of the observations above. We start by analyzing $LFSR_{\mathrm{u}}$ (See Appendix C.3) to find an equation λ

$$\lambda : u_k + u_{k+j_1} + ... + u_{k+j_{w-1}} = 0$$

that holds over all $\mathbf{u}$ generated by $LFSR_{\mathrm{u}}$ for any $k \geq 0$. Let the first guess for the initialization state for $\mathbf{s}$ be $\hat{\mathbf{s}}^{\mathrm{I}} = (1, 0, 0, ..., 0)$, generate $\mathbf{c}^0$ by $c_t^0 = D(L^t(1, 0, ...0))$, $t < M$, and $\mathbf{u}^0 = Q^*(\mathbf{c}, \mathbf{z})$. Next we try to find m places $(k_1, k_2, ..., k_m)$ in $\mathbf{u}^0$ where the equation λ is defined. From this we get the equation set

$$\begin{array}{llll} u^0_{k_1} & + u^0_{k_1+j_1} & + ... + u^0_{k_1+j_{w-1}} & = 0 \\ u^0_{k_2} & + u^0_{k_2+j_1} & + ... + u^0_{k_2+j_{w-1}} & = 0 \\ & \vdots & & \vdots \\ u^0_{k_m} & + u^0_{k_m+j_1} & + ... + u^0_{k_m+j_{w-1}} & = 0 \end{array}$$

Since every $u_{k_x+j_y}$ in this equation set is defined in $\mathbf{u}^0$, we can replace $u_{k_x+j_y}$ with the corresponding bit z_t in the keystream $\mathbf{z}$. Thus, $\mathbf{u}^0$ is a sequence of pointers to $\mathbf{z}$ and we can write the equations over $\mathbf{z}$ as the equation set Ω :

$$\begin{array}{ll} z_{t_{1,1}} + z_{t_{1,2}} + ... + z_{t_{1,w}} & = 0 \\ z_{t_{2,1}} + z_{t_{2,2}} + ... + z_{t_{2,w}} & = 0 \\ \vdots & \vdots \\ z_{t_{m,1}} + z_{t_{m,2}} + ... + z_{t_{m,w}} & = 0 \end{array} \tag{7}$$

We are now finished with the precomputation.

Next, we test the equation set to see if all the equations hold. If not, we iterate using the algorithm below which outputs the correct $\mathbf{s}^{\mathrm{I}}$. Knowing $\mathbf{s}^{\mathrm{I}}$ it is easy to calculate $\mathbf{u}^{\mathrm{I}} = (u_0, u_1, ..., u_{l_{\mathrm{u}}-1})$ using the Gaussian algorithm once on an equation set derived from $\mathbf{s}^{\mathrm{I}}$ and $LFSR_{\mathrm{u}}$.

Input The keystream $\mathbf{z}$ of length M, the equation λ, the equation set Ω, the pointer sequence $\mathbf{u}^0$, the states $L^0(1,0,...0)$ and $L^M(1,0...,0)$, Set $i \leftarrow 0$

1. Calculate $c_0^i = D(L^i(1,0,...,0))$, and $c_{M-1}^{i+1} = c_M^i = D(L^{M+i}(1,0,...,0))$.
2. Use lemma 1 to generate $\mathbf{u}^{i+1} = Q^*(\mathbf{c}^{i+1}, \mathbf{z})$ and lower all indexes in the equation set Ω by one. Theorem 1 guarantees that the equations are defined over $\mathbf{u}^{i+1}$.
3. If the first equation in Ω gets a negative index, then remove the equation from Ω. Find a new index at the end of $\mathbf{u}^{i+1}$ where λ is defined, and add the new equation over $\mathbf{z}$ to Ω.
4. If the current equation set Ω holds, stop the algorithm and output $\mathbf{s}^\mathrm{I} = L^{i+1}(1,0,...,0)$ as the initialization state for $LFSR_\mathrm{s}$.
5. If δ does not hold, we set $i \leftarrow i+1$ and go to step 1.

Note 1. To reach the desired complexity (2^{l_s}) a few details on the implementation of the algorithm are needed. These details are given in Appendix A.

All changes during the iterations are done on $\mathbf{u}^i$ and the equation set Ω. Thus, each guess $L^i(1,0,...,0)$ for $\mathbf{s}^\mathrm{I}$ result in an unique equation set Ω. The $\mathbf{z}$ stream is never altered.

Example 2. We continue on the generator in Example 1. We have found the equation $u_k + u_{k+6} + u_{k+8} = 0$, which corresponds to the multiple $h(x) = 1 + x^6 + x^8$. We have $\mathbf{z}$ of length 19, and want to find $\mathbf{s}^\mathrm{I}$. The length of $\mathbf{u}^i$ will be $N \approx \frac{3}{2}19 = 28.5$. We set the first guess to $\mathbf{s}_0^\mathrm{I} = (1,0,0)$. From this we generate the clock control sequence using the function $c_t^0 = D(L^t(1,0,0))$, $0 \le t \le M-1$, and we get

$$\mathbf{c}^0 = (2,1,1,2,2,2,1,2,1,1,2,2,2,1,2,1,1,2,2,2).$$

Then we spread out the $\mathbf{z}$ stream corresponding to $\mathbf{c}$, that is $\mathbf{u}^0 = Q^*(\mathbf{c}^0, \mathbf{z})$. From this we get the sequence

$$\mathbf{u}^0 = (*, z_0, z_1, z_2, *, z_3, *, z_4, *, z_5, z_6, *, z_7, z_8, z_9, *, z_{10}, \\ *, z_{11}, *, z_{12}, z_{13}, *, z_{14}, z_{15}, z_{16}, *, z_{17}, *, z_{18}).$$

We search through $\mathbf{u}^0$ to find 4 entries where the equation $u_k + u_{k+6} + u_{k+8} = 0$ is defined. Since all the defined entries in $\mathbf{u}^0$ points to bits in the $\mathbf{z}$ stream, we get the following set of equations Ω over $\mathbf{z}$:

$$\begin{aligned} z_0 + z_4 + z_5 &= 0 \\ z_6 + z_{10} + z_{11} &= 0 \\ z_7 + z_{11} + z_{12} &= 0 \\ z_{13} + z_{17} + z_{18} &= 0 \end{aligned}$$

We test the equations to see if all the equations hold. If the set does not hold, we continue as follows. We shift the $LFSR_s$ once, and it will have $\mathbf{s}_1^{\mathrm{I}} = L^1(1,0,0) = (0,0,1)$ as inner state. We calculate $c_{M-1}^1 = c_M^0 = D(L^M(1,0,0))$. Then we use Lemma 1 to calculate $\mathbf{u}^1$ from $\mathbf{u}^0$. That is, we delete the $c_0^0 = 2$ entries $(*, z_0)$, append the $(c_{18}^1 = 2)$ entries $(*, z_{19})$ at the end, and at last replace the pointer z_t with z_{t-1} for $1 \leq t \leq M$. We get this guess for $\mathbf{u}$:

$$\mathbf{u}^1 = (z_0, z_1, *, z_2, *, z_3, *, z_4, z_5, *, z_6, z_7, z_8, *, z_9,$$
$$*, z_{10}, *, z_{11}, z_{12}, *, z_{13}, z_{14}, z_{15}, *, z_{16}, *, z_{17}, *, z_{18}).$$

If an equation is defined for the entry t in $\mathbf{z}$ for the guess $\mathbf{s}_0^{\mathrm{I}}$, it will now be defined for the entry $t-1$ in $\mathbf{z}$ for the guess $\mathbf{s}_1^{\mathrm{I}}$ as guaranteed by Theorem 1. From this Ω becomes:

$$z_{-1} + z_3 + z_4 = 0$$
$$z_5 + z_9 + z_{10} = 0$$
$$z_6 + z_{10} + z_{11} = 0$$
$$z_{12} + z_{16} + z_{17} = 0$$

We remove the first equation from Ω since it has a negative index, and find a new index at the end of $\mathbf{u}^1$ where λ is defined. We find the equation $z_{13}+z_{17}+z_{18} = 0$ and add it to Ω. We test the equations to see if all the equations hold. If the set does not hold, we continue the algorithm.

3.5 Complexity and Properties

Precomputation. If the generator polynomial $g_{\mathrm{u}}(x)$ for $LFSR_{\mathrm{u}}$ has sufficient low weight, say ≤ 10, we can use it directly in our algorithm with $w = \mathrm{weight}(g_{\mathrm{u}})$ and $h(x) = g_{\mathrm{u}}(x)$. In such a case we do not need much precomputation. The only precomputation is to generate $\mathbf{u}^0$ of length N, where the length of N is calculated below.

If $g_{\mathrm{u}}(x)$ has too high weight we use a modified version of Wagners algorithm for the generalized birthday problem [4] to find a multiple $h(x)=a(x)g(x)$ of weight $w = 2^r$ and degree l_{h}. The multiple $h(x)$ gives a new recursion of low weight. The fast search algorithm is described in Appendix C.3. See Table 2 for some multiples found by the algorithm.

When we have found a polynomial $h(x) = 1+x^{j_1}+...+x^{j_{w-1}}$ with $j_{w-1} = l_{\mathrm{h}}$, the corresponding equation λ over $\mathbf{u}$ is $u_k + u_{k+j_1} + ... + u_{k+j_{w-1}} = 0$. We want to find m places in the stream $\mathbf{u}$ where λ is defined. From equation (6) we have that an equation of weight w is defined at an random entry in $\mathbf{u}$ with a probability $P_{\mathrm{def}} = (1-P_{\mathrm{d}})^w$. Thus we must test around $m/(1-P_{\mathrm{d}})^w$ entries to find m equations over $\mathbf{z}$. To be able to do this $\mathbf{u}$ must have length

$$N > l_{\mathrm{h}} + \frac{m}{(1-P_{\mathrm{d}})^w}. \tag{8}$$

Table 2. The table shows some weight 4 multiples of different polynomials found using the algorithm in Appendix C.3. The algorithm used 1 hour and 15 minutes to find the multiple of the degree 80 polynomial, mostly due to heavy use of hard disc memory. The search for the multiple of the degree 60 polynomial took 14 seconds

$g(x)$	$h(x) = a(x)g(x)$
$x^{40} + x^{38} + x^{35} + x^{32} + x^{28} + x^{26} + x^{22} + x^{20} + x^{17} + x^{16} +$ $x^{14} + x^{13} + x^{11} + x^{10} + x^{9} + x^{8} + x^{6} + x^{5} + x^{4} + x^{3} + 1$	$x^{24275} + x^{6116}$ $+ x^{1752} + 1$
$x^{60} + x^{58} + x^{56} + x^{52} + x^{51} + x^{50} + x^{49} + x^{48} + x^{47} + x^{46} + x^{44} +$ $x^{41} + x^{40} + x^{39} + x^{36} + x^{29} + x^{28} + x^{27} + x^{26} + x^{25} + x^{23} + x^{21} +$ $x^{20} + x^{19} + x^{16} + x^{15} + x^{11} + x^{10} + x^{9} + x^{4} + x^{2} + x + 1$	$x^{2464041} + x^{1580916}$ $+ x^{131400} + 1$
$x^{80} + x^{79} + x^{78} + x^{76} + x^{75} + x^{69} + x^{68} + x^{57} + x^{56} + x^{55} + x^{54} + x^{52} + x^{49} +$ $x^{46} + x^{45} + x^{44} + x^{42} + x^{37} + x^{36} + x^{35} + x^{32} + x^{31} + x^{30} + x^{28} + x^{27} + x^{26} +$ $x^{24} + x^{23} + x^{21} + x^{20} + x^{19} + x^{13} + x^{12} + x^{10} + x^{8} + x^{6} + x^{4} + x^{3} + 1$	$x^{312578783} + x^{309946371}$ $+ x^{210261449} + 1$

To avoid false keys, we choose $m > l_s$. From the expectation (2) of N we have $E(M) = N(1 - P_\mathrm{d}) = (1 - P_\mathrm{d})l_\mathrm{h} + \frac{m}{(1-P_\mathrm{d})^{w-1}}$, and we have proved the following proposition:

Proposition 1. *Let an equation over* $\mathbf{u}$ *be defined by* $h(x)$ *of weight* w *and degree* l_h*. To get a equation set* Ω *of* $m > l_s$ *equations over* $\mathbf{z}$*, the length of the* $\mathbf{z}$ *stream must be*

$$M > (1 - P_\mathrm{d})l_\mathrm{h} + \frac{m}{(1 - P_\mathrm{d})^{w-1}}. \tag{9}$$

where $m \approx l_\mathrm{s} + 10$.

We see that the keystream length M is dependent of the degree l_h of $h(x)$ of weight $w = 2^r$. The degree l_h is then again highly dependent on the search algorithm we use to find $h(x)$. When we use the search algorithm in Appendix C.3 with the proposed parameters we show in the appendix that l_h will be in order of $l_\mathrm{u} \approx T_\mathrm{mem}(l_\mathrm{u}, r) = 2^{\frac{r+l}{r+1}}$.

Decoding. If this algorithm is implemented properly (Appendix A) it will have worst case complexity $O(2^{l_\mathrm{s}})$ with a very little constant factor. In average the number of iterations will be in the order of $2^{l_\mathrm{s}-1}$.

At each iteration i we shift the sliding window c_0^i to the right over $\mathbf{u}^i$. Then we shift the equation set 1 to the left over $\mathbf{z}$, and test it. If we have the wrong guess for $\mathbf{s}^\mathrm{I}$, each equation in the set will hold with a probability $\frac{1}{2}$. When we reach an equation that does not hold we know that the guess for $\mathbf{s}^\mathrm{I}$ is wrong and we break off the test. Thus the average number of equations we have to evaluate per guess is $\frac{\lim_{m\to\infty}\sum_{j=1}^{m} j\cdot 2^{l_\mathrm{s}}/2^j}{2^{l_\mathrm{s}}} = \lim_{m\to\infty}\sum_{i=1}^{m} i/2^i = 2$. This gives an average constant factor of 2 parity check tests for each of the 2^{l_s} guesses. Thus the complexity is $O(2 \cdot 2^{l_\mathrm{s}}) = O(2^{l_\mathrm{s}})$

Table 3. The attacks are done in C code on a 2.2 GHz Pentium IV running under Linux. Note how the running time is exactly the same for $l_u = 40$ and $l_u = 60$. We have set the number of equations to $m = 35$. The polynomials $g(x)$ and $h(x)$ are from Table 2

Degree l_s of $g_s(x)$	Degree l_u of $g_u(x)$	Degree l_h of $h(x)$	Number of Iterations	Decoding time	Length M of $\mathbf{z}$
25	40	24275	2^{25}	9 sec.	10000
26	40	24275	2^{26}	18 sec	10000
25	60	2464041	2^{25}	9 sec	1000000
26	60	2464041	2^{26}	18 sec	1000000

Each time an equation gets a negative index, we must delete it and search for a new equation at the end of $\mathbf{u}^i$. We expect to search through $1/(1-p_d)^{w-1}$ entries in $\mathbf{u}^i$ to find a new equation. This is done every $\frac{M-l_u(1-p_d)}{m}$ iteration in average, and will have little impact on the decoding complexity.

When we after i iterations have found the initialization bits for $LFSR_s$, we use the Gaussian algorithm on the linear equation set derived from $LFSR_u$ and $\mathbf{u}^i$to find the initialization bits for $LFSR_u$. This has complexity $O(l_u^3)$ and will have little effect on the everall complexity of the algorithm.

4 Simulations

We have done the attack on 4 small cipher systems, defined with clock control generator polynomials of degree 25 and 26, and the data generator polynomials of degree 40 and 60 from Table 2. The clock function $D()$ is the LILI-clock function as described in Section 2.2. Note that we only attack the irregular clocking building block in LILI and not the complete LILI-128 cipher. In LILI-128 the stream is filtered through a boolean function, and this is beyond the scope of this paper.

We have used Proposition 1 and Equation 8 to calculate the length M of $\mathbf{z}$ and length N of $\mathbf{u}$ (rounded up to nearest thousand and hundred thousand). The number of parity check equations over $\mathbf{z}$ is set to $m = 35 \approx l_s + 10$. Recall that the number of paritycheck equations does not effect the complexity. Table 3 shows how the running time of the attack is unchanged when the degree of $g_u(x)$ gets larger. The impact from a larger l_u is that we need longer keystream.

Normally we would stop the search when we have found the correct key. But then the running time would be highly dependent on where the key is in the keyspace. To avoid this we have gone trough the whole key space to be able to compare the different attacks in the table. In a real attack the average running time would be half the running times in Table 3. To compare with previous LCT attacks, the Gaussian factor $\frac{1}{3}l_{\mathrm{u}}^3$ would be around 72000 for $l_{\mathrm{u}} = 60$, and around 21333 for $l_{\mathrm{s}} = 40$. In our attack the constant factor is only 2 in average. Thus the same attacks presented in Table 3 would take several hours or even days using the previous LCT algorithm.

5 Conclusion

We have presented a new linear consistency attack with lower complexity than previous on a general model for irregular clocked stream ciphers. We have tested the attack in software and confirmed that the attack has a very low running time that follows the expected complexity $O(2^{l_{\mathrm{s}}})$. Thus the run time complexity is independent of the degree l_{u} of $LFSR_u$.

Further on, if we modify the algorithm, it will work on systems where noise is added on keystream $\mathbf{z}$. Using much higher m and giving each guess $\mathbf{s}^{\mathrm{I}}$ a metric, we can perform an correlation attack with complexity $O(m \cdot 2^{l_{\mathrm{s}}})$ on such systems. Initial tests seem very promising and we will come back to this matter in future work.

Acknowledgement

I would like to thank my supervisor prof. Tor Helleseth and Dr. Matthew Parker for helpful discussions and for reading and helping me in improving this paper.

References

[1] A. Menezes, P. van Oorschot, S. Vanstone, "Handbook of Applied Cryptography", CRC Press, 211-212, 1997. 110, 111

[2] D. Gollmann and W.G Chambers, "Clock-controlled shift registers: a review", *IEEE Journal on Selected Areas in Communications*, 7 (1989), 525-533.

[3] K. Zeng, C. Yang and Y. Rao, "On the linear consistency test (LCT) in cryptanalysis with applications", *Advances in Cryptology-CRYPTO '89* (LNCS 435), 164-174, 1990. 110, 111, 113, 115

[4] D. Wagner, "A Generalized Birthday problem", Advances in cryptology-*CRYPTO* '02, (LNCS 2442), 288-303, 2002. 111, 118, 125

[5] T. Johansson, and F. Jönsson, "Fast Correlation Attacks on Stream Ciphers via Convolutional Codes", *Advances in Cryptology*-EUROCRYPT'99, Lecture Notes in Computer Science, Vol. 1592, Springer-Verlag, 1999, pp. 347-362. 125

[6] Patrik Ekdahl, Willie Meier, Thomas Johannson, "Predicting the Shrinking Generator with Fixed Connections", EUROCRYPT 2003, Lecture Notes in Computer Science, Vol. , Springer-Verlag, 1999, 330-334.

[7] J. D. Golic, "Cryptanalysis of three mutually clock-controlled stop/go shift registers." *IEEE Trans. Inf Theory,* 46(3),525-533, 2000. 110, 111

[8] E. Zenner, M. Krause, and S. Lucks,"Improved cryptanalysis of the self-shrinking generator", *Proc ACISP '01,* (LNCS 2119), 21-35, 2001. 110, 111
[9] Jovan Dj. Golic, Miodrag J. Mihaljevic,"A Generalized Correlation Attack on a Class of Stream Ciphers Based on the Levenshtein Distance" *(Journal of Cryptology 3)*, 201-212, 1991. 110
[10] Jovan Dj. Golic and Slobodan V. Petrovic,"A Generalized Correlation Attack with a Probabilistic Constrained Edit Distance", (LNCS 658), 472, 1993. 110
[11] Erik Zenner, "On the Efficiency of the Clock Control Guessing Attack", *ICISC 2002*, (LNCS 2587). 110, 111
[12] L. Simpson, E. Dawson, J. Dj Golic, and W. Millan, "LILI keystream generator", *SAC'2000,* (LNCS 2012),248-261,2000. 113

Appendix

A Implementation Details

To reach the desired complexity $O(2^{l_s})$, the implementation of the algorithm needs some tricky details:

1. In Lemma 1 we get $\mathbf{u}^{i+1}$ by among other things deleting the c_0^i first bits of u^i. This is done using the sliding window technique, which means that we move the viewing to the right instead of shifting the whole sequence to the left. This way the shifting can be done in just one operation. To avoid heavy use of memory, we slide the window over an array of fixed length N, so that the entries that become free at the beginning of the array are reused. Thus, the left and right of the sliding window after i iterations will be
$$(left, right) = (i \bmod N, i + N_i \bmod N),$$
where $N¿N_i$, for all i, $0 \leq i < 2^{l_s}$
2. In lemma 1 every reference z_{t+1} in $\mathbf{u}$ is replaced with z_t for every $0 \leq t \leq M$, which would take M operations. If we skip the replacements we note that after i iterations the entry z_t in $\mathbf{u}$ will become z_{t+i}. It is also important to notice that when we write $\mathbf{u} = (..., z_0..., z_t, ..., z_M, ...)$, the entries $z_0, ..., z_t, ..., z_M$ are pointers from $\mathbf{u}$ to $\mathbf{z}$. They are not the actual key bits. Thus, in the implementation we do not replace z_t with z_{t-1}. But when we after i iterations in the search for equations find an equation $u_k^i + u_{k+j_1}^i + ... + u_{k+j_{w-1}}^i = 0$ that is defined, we replace the corresponding $z_{t_1} + z_{t_2} + ... + z_{t_w}$ with $z_{t_1-i} + z_{t_2-i} + ... + z_{t_w-i}$, to compensate.
3. We do not have to keep the whole clock control sequence $\mathbf{c}^i$ in memory. We only need the two clocks, c_0^i and c_{M-1}^{i+1}, since they are used by lemma 1 to generate $\mathbf{u}^{i+1}$.

B Proofs of Lemma 1 and Theorem 1

B.1 Proof of Lemma 1

Proof. Let $\mathbf{c}^i$ be the clocking integer sequence for a given i, $0 \leq i < 2^{l_s}$. We see that $c_t^{i+1} = c_{t+1}^i$, $0 \leq t < M - 1$, which means that pattern of the defined bits in $\mathbf{u}^{i+1}$ are the same as the pattern in $\mathbf{u}^i$ shifted c_0^i to the left. From this we deduce the following for given $1 \leq t \leq M$ and $k = \sum_{j=0}^{t-1} c_j^i - 1$: If $u_k^i = z_t$ for given k then $u_{k-c_0^i}^{i+1} = z_{t-1}$. If we delete the first c_0^i bits of $\mathbf{u}^i$ and get $\mathbf{u}'$ we will have that if $u'_k = z_t$ for given k, then $u_k^{i+1} = z_{t-1}$ for $k = \sum_{j=0}^{t-1} c_j^{i+1} - 1$. If we now replace every z_t in $\mathbf{u}'$ with z_{t-1} for $0 < t < M$ and get $\mathbf{u}''$ we see that $u''_k = u_k^{i+1}$, $0 \leq k < N_i - c_0^i$. To finally transform $\mathbf{u}''$ into $\mathbf{u}^{i+1}$ we just have to append the c_{M-1}^{i+1} entries $(*, ..., z_{M-1})$ at the end of $\mathbf{u}''$.

B.2 Proof of Theorem 1

Proof. Let

$$\mathbf{u}^i = (..., \underbrace{z_0}_{c_0^i - 1}, ..., *, ..., \underbrace{z_t}_{k}, ..., \underbrace{z_{t+j_1}}_{k+k_1}, ..., *, ..., \underbrace{z_{t+j_w}}_{k+k_{w-1}}, ..., \underbrace{z_{M-1}}_{N_i - 1})$$

be the stream of length N_i we get using $\mathbf{u}^i = Q^*(\mathbf{c}^i, \mathbf{z})$. The notation means that $u_{c_0^i-1}^i = z_0$, $u_k^i = z_t$, and $u_{N_i-1}^i = z_{M-1}$. We see that the sum $\beta_{\mathbf{u}_i,k} = u_k^i + u_{k+k_1}^i + ... + u_{k+k_{w-1}}^i$ is defined over $\mathbf{u}^i$. The corresponding sum over $\mathbf{z}$ will be $\gamma_{\mathbf{z},t} = z_t + z_{t+j_1} + ... + z_{t+j_{w-1}}$. Then the clock control sequence we get from $c_t^{i+1} = L^{i+1+t}(1, 0, ..., 0)$ will be

$$\mathbf{c}^{i+1} = (c_1^i, ..., c_{M-1}^i, c_{M-1}^{i+1}) = (c_0^{i+1}, ..., c_{M-1}^{i+1}).$$

The main observation here is the following: We transform $\mathbf{u}^i$ into $\mathbf{u}^{i+1}$ by deleting the first c_0^i entries $\underbrace{(*, ..., z_0)}_{c_0^i}$ in $\mathbf{u}^i$, appending $\underbrace{(*, ..., *, z_M)}_{c_M^{i+1}}$ at the end, and then replacing z_t with z_{t-1} for $1 \leq t \leq M$, as explained in lemma 1. From this we get the sequence

$$\mathbf{u}^{i+1} = (..., \underbrace{z_0}_{c_0^{i+1}-1}, ..., *, ..., \underbrace{z_{t-1}}_{k-c_0^i}, ..., \underbrace{z_{t+j_1-1}}_{k+k_1-c_0^i}, ..., *, ..., \underbrace{z_{t+j_w-1}}_{k+k_{w-1}-c_0^i}, ..., \underbrace{z_{M-1}}_{N_{i+1}-1}). \quad (10)$$

We can easily see from (10) that the sum $\beta_{\mathbf{u}^{i+1},k-c_0^i} = u_{k-c_0^i} + u_{k+k_1-c_0^i} + ... + c_{k+k_{w-1}-c_0^i}$ is defined since every entry in the sum is defined. The corresponding sum over $\mathbf{z}$ is $\gamma_{\mathbf{z},t-1} = z_{t-1} + z_{t+j_1-1} + ... + z_{t+j_{w-1}-1}$.

C Searching for Parity Check Equations

C.1 The Generator Matrix

Let $g(x) = 1 + g_{l-1}x + g_{l-2}x^2 + ... + g_1 x^{l-1} + x^l$, $g_i \in \mathcal{F}_2$, $g_l = g_0 = 1$ be the primitive feedback polynomial of degree l for a shift register that generates the sequence $\mathbf{u} = (u_0, u_1, ..., u_{N-1})$. The corresponding recurrence is $u_{t+l} = g_1 u_{t+l-1} + g_2 u_{t+l-2} + ... + g_l u_t$. Let α be defined by $g(\alpha) = 0$. From this we get the reduction rule $\alpha^l = g_1\alpha^{l-1} + g_2\alpha^{l-2} + ... + g_{l-1}\alpha + 1$. Then we can define the generator matrix for sequence u_t,$0 < t < N$ by the $l \times N$ matrix

$$G = [\alpha^0\alpha^1\alpha^2...\alpha^{N-1}]. \tag{11}$$

For each $i > l$, using the reduction rule α^i can be written as $\alpha^i = h^i_{l-1}\alpha^{l-1} + ... + h^i_2\alpha^2 + h^i_1\alpha + h^i_0$. We see that every column $i \geq l$ is a combination of the first l columns, and any column i in G can be represented by

$$\mathbf{g}_i = [h^i_0, h^i_1, ..., h^i_{l-1}]^{\mathbf{T}}. \tag{12}$$

Now the sequence $\mathbf{u}$ with length N and initialization state $\mathbf{u}^{\mathrm{I}} = (u_0, u_1, ..., u_{l-1})$, can be generated by

$$\mathbf{u} = \mathbf{u}^{\mathrm{I}}G.$$

The shift register is now turned in to a (N, l) block code.

C.2 Equations

Let $\mathbf{u}$ be a sequence generated by the generator polynomial $g(x)$ with degree l. It is well known that if we can find $w > 2$ columns in the generator matrix G, that sum to zero,

$$(\mathbf{g}_{j_0} + \mathbf{g}_{j_1} + \ldots + \mathbf{g}_{j_{w-1}})^{\mathbf{T}} = (0, 0, ..., 0), \tag{13}$$

for $l \leq j_0, j_1, ..., j_{w-1} < N$, we get an equation of the form

$$u_{j_0} + u_{j_1} + ... + u_{j_{w-1}} = 0. \tag{14}$$

The equation (13) can be formulated as $\alpha^{j_0} + \alpha^{j_1} + ... + \alpha^{j_{w-1}} = 0$. Thus, if (13) holds, the equation $\alpha^t(\alpha^{j_0} + \alpha^{j_1} + ... + \alpha^{j_{w-1}}) = \alpha^{j_0+t} + \alpha^{j_1+t} + ... + \alpha^{j_{w-1}+t} = 0$ also holds for $0 \leq t < N - j_{w-1}$. From this we can conclude that the equation is cyclic and can be written as

$$u_{t+j_0} + u_{t+j_1} + ... + u_{t+j_{w-1}} = 0, \tag{15}$$

for $0 \leq t < N - j_w$.

We can also use the indexes $j_1, j_2, ...j_w$ to formulate the polynomial $h(x) = x^{j_0} + x^{j_1} + ... + x^{j_{w-1}}$. If $j_0, j_1...$ is found using the method above, we will have the relationship $h(x) = g(x)a(x)$, for a polynomial $a(x)$. Thus $h(x)$ is a multiple of $g(x)$.

C.3 Fast Method for Finding an Multiple Weight $w = 2^r$

A previous and naive search algorithm for finding multiple $h(x)$ of weight w and degree ¡n is as follows. It corresponds to searching for w columns in G that sum to zero mod 2.

First sort the generator matrix G. Then for every choice of the columns $\mathbf{g}_{j_0}, \mathbf{g}_{j_1}, ..., \mathbf{g}_{j_{w-2}}$ in G search for the w'th column $\mathbf{g}_{j_{w-1}}$that gives $\mathbf{g}_{j_{w-1}} = \mathbf{g}_{j_0} +$

$\mathbf{g}_{j_1} + ... + \mathbf{g}_{j_{w-2}}$. This algorithm is not very effective and has the complexity $O(n^{w-1}\log n)$. By using hashing techniques we can get down to $O(n^{w-1})$. We can do better if we use the iterative method explained next. The algorithm is a modification of the Generalized birthday algorithm in [4].

First we sort the $n \times l$ generator matrix $G_1 = G$ in respect to the $l - B_1$ lowest entries in the columns, for a proper B_1. The columns that are equal in the lowest $l - B_1$ bits, will now be beside each other. If we sum them, the sum will be zero in the lowest $l - B_1$ bits. Next we go through the matrix and sum all the columns that are equal in the $l - B_1$ lowest entries and store the sums in a new matrix G_2. If we find m_1 sums, the matrix G_2 will have size $m_1 \times B_1$, since the m_1 sums we find will have 0 in the $l - B_1$ lowest entries. For each column i in G_2 we also store the indexes of the two columns from G that where summed to column i. Next we sort G_2 in respect to the $B_1 - B_2$ lowest bits, and do the same procedure over again and get a new matrix G_3 of size $m_2 \times B_2$.

We repeat the procedure until we in round r set B_r to be zero. After the r'th round we will hopefully have found 2^r columns in G that sum to zero. According to Section C.2 we will now have found an multiple of $g(x)$. This algorithm is much faster than the naive algorithm, but it "misses" a lot possible multiples and needs bigger matrix G.

Now we will present some new properties for this algorithm. The first round of the algorithm is similar to the well known search algorithm in [5] for finding equations of the type $c_0u_0 + c_1u_1 + ... + c_{B-1}u_{B-1} = u_i + u_j$. From this paper we have that the expected number of equations m_1 is given by $m_1 = n(n-1)/2^{l-B_1}$. When n is large we can approximate m_1 by

$$E(m_1) = \frac{n^2}{2^{l-B_1+1}}. \tag{16}$$

Since the algorithm is iterative we can use (16) again for the next round and we have $E(m_2) = \frac{m_1^2}{2^{B_1-B_2+1}} = \frac{N^4}{2^{2l-B_1-B_2+3}}$. Generally for each round i we will have

$$E(m_i) = \frac{m_{r-1}^2}{2^{B_{i-1}-B_i+1}} \tag{17}$$

for $B_0 = l$ and $m_0 = N$.

The iterative search algorithm has complexity $O(\sum_{i=0}^{r-1} m_i \log m_i)$ since we have to sort the matrices $G_1, G_2 ..., G_r$. Thus, it is not the complexity that limits the algorithm, but the memory. Given an polynomial $g(x)$ of degree l, we will now present a bound for needed memory for finding a multiple $h(x)$ of weight $w = 2^r$.

Assume that we have a computer with T_{mem} memory units and that one column in G_1 takes up one memory unit. Then it will be natural to use a column G of the maximum size $T_{\text{mem}} \times l$. To use the memory most efficiently, we will try find around $m_i = T_{\text{mem}}$ sums in each round i, that is $G_i = T_{\text{mem}} * B_{i-1}$. Thus we can set $N = m_1 = ... = m_{r-1} = T_{\text{mem}}$. We just need find one multiple, so $m_r = 1$. Setting these restriction we can now give an easy expression for how much memory that is needed to find a multiple of weight $w = 2^r$ of $g(x)$ of degree l.

Theorem 2. *Given a primitive polynomial $g(x)$ of degree l, and $r+1$ divides $r+l$, the expected amount of memory needed to find a weight $w = 2^r$ multiple $h(x)$ of $g(x)$ using the iterative search algorithm is*

$$T_{\text{mem}}(l_{\text{u}}, r) = 2^{\frac{r+l}{r+1}}, \tag{18}$$

with $B_i = i + l - i\frac{r+l}{r+1}$, $1 \leq i \leq r-1$, $B_r = 0$.

Proof. From equation (17) we have these formulas for $m_1, ..., m_r$:

$$\begin{aligned} m_1 &= \frac{n^2}{2^{l-B_1+1}}, \\ m_2 &= \frac{m_1^2}{2^{B_1-B_2+1}}, \\ &\vdots \\ m_r &= \frac{m_{r-1}^2}{2^{B_{r-1}-B_r+1}}. \end{aligned}$$

We require that $m_1 = m_2 = ... = m_{r-1} = n = T_{\text{mem}}$, and $m_r = 1$. We solve $n = m_1 = \frac{n^2}{2^{l-B_1+1}}$, in respect to B_1 and get

$$B_1 = 1 + l - \log n. \tag{19}$$

We use equation (17) and solve $n = \frac{n^2}{2^{B_{i-1}+B_i+1}}$ in respect to B_i and get

$$B_i = B_{i-1} + 1 - \log n. \tag{20}$$

Using (20) together with B_1, we get this expression for B_i :

$$B_i = i + l - i\log n. \tag{21}$$

Next we solve $m_r = 1$, that is $\frac{n^2}{2^{B_{r-1}-B_r+1}} = 1$. Solving in respect to n and putting in (21) for $i = r-1$ and setting $B_r = 0$, we get $n = 2^{\frac{r+l}{r+1}}$. The algorithm require that all the B_i's are integers. This will only be the case then we can set $n = 2^x$, for some x. If we want the expression to be exact, we get the requirement that $x = \frac{r+l}{r+1}$ must be an integer. Thus $r+1$ must divide $r+l$.

The theorem does not give a guarantee for finding a equations, it just say that we are expected to find one equation. Thus, in practical searches we may use around twice as many bits to assure success.

Correlation Attacks Using a New Class of Weak Feedback Polynomials

Håkan Englund, Martin Hell, and Thomas Johansson

Dept. of Information Techonolgy, Lund University
P.O. Box 118, 221 00 Lund, Sweden

Abstract. In 1985 Siegenthaler introduced the concept of correlation attacks on LFSR based stream ciphers. A few years later Meier and Staffelbach demonstrated a special technique, usually referred to as fast correlation attacks, that is very effective if the feedback polynomial has a special form, namely, if its weight is very low. Due to this seminal result, it is a well known fact that one avoids low weight feedback polynomials in the design of LFSR based stream ciphers.
This paper identifies a new class of such weak feedback polynomials, polynomials of the form $f(x) = g_1(x) + g_2(x)x^{M_1} + \ldots + g_t(x)x^{M_{t-1}}$, where $g_1, g_2, \ldots, g_t$ are all polynomials of low degree. For such feedback polynomials, we identify an efficient correlation attack in the form of a distinguishing attack.

1 Introduction

Stream cipher design and cryptanalysis are topics that have received lots of attention recently, due to new interesting designs that are very fast in software, see e.g. [3, 8, 20, 10, 13] and many others. One way to design a stream cipher is to use a Linear Feedback Shift Register (LFSR) sequence as input to a nonlinear function. The shift register is initialized using a short random string and the output from the cipher is a much longer string that has many random like properties. But the LFSR output is linear and in some way the linearity of the sequence must be destroyed through some nonlinear process.

Different attacks can be used to attack the nonlinear part of a cipher. A popular topic lately has been algebraic attacks [5, 6]. These attacks can be mounted if the nonlinear function gives rise to a large system of equations containing equations of low degree. A different attack is to use the correlation between the input and the output of the nonlinear function.

In 1985 Siegenthaler introduced the concept of correlation attacks on LFSR based stream ciphers. His basic idea was to perform a divide-and-conquer attack by exploring the correlation between the output sequence of the generator and the output sequence of one individual LFSR (assuming a nonlinear combination generator).

A few years later Meier and Staffelbach demonstrated a special technique, usually referred to as fast correlation attacks [17]. This attack is very effective if the feedback polynomial has a special form, namely, if its weight is very low.

B. Roy and W. Meier (Eds.): FSE 2004, LNCS 3017, pp. 127–142, 2004.

The ideas resemble a lot so-called iterative decoding of error correcting codes, for example low density parity check codes. Since then, many ideas concerning fast correlation attacks have been presented, see e.g. [1, 2, 14, 15, 19, 11]. Due to this seminal result, it is a well known fact that one avoids low weight feedback polynomials in the design of LFSR based stream ciphers.

This paper identifies a new class of such weak feedback polynomials, namely, polynomials of the form

$$f(x) = g_1(x) + g_2(x)x^{M_1} + \ldots + g_t(x)x^{M_{t-1}},$$

where $g_1, g_2, \ldots, g_l$ are all polynomials of low degree. For such feedback polynomials, we identify an efficient correlation attack in the form of a distinguishing attack.

In a distinguishing attack the key is not recovered, instead one tries to distinguish an observed keystream from a truly random stream. This attack is not as powerful as the key recovery attack, in which one finds the key that corresponds to the plaintext and the ciphertext. Whereas most previous work on correlation attacks have been focused on key recovery attacks, more recent work in cryptanalysis of stream ciphers have, to a large extent, been concerned with distinguishing attacks, see e.g. [9, 4].

It should also be noted that a distinguishing attack can sometimes be turned into a key recovery attack, in a similar way as for block ciphers.

The results of the paper are as follows. For the new class of such weak feedback polynomials, given above, we present an algorithm for launching an efficient fast correlation attack. The applicability of the algorithm is twofold.

Firstly, if the feedback polynomial is of the above form with a moderate number of polynomials ,g_i, the new algorithm will be much more powerful than applying any previously known (like Meier-Staffelbach) algorithm. This could be interpreted as feedback polynomials of the above kind should be avoided in designing new LFSR based stream ciphers.

Secondly, for an arbitrary feedback polynomial, a standard approach is to search for low weight multiples of the feedback polynomial and then to apply the Meier-Staffelbach approach to fast correlation attacks using a low weight multiple. We can do the same and search for multiples of the feedback polynomial that have the above form. It turns out that this approach is in general less efficient than searching for low weight polynomial, but we can always find specific instances of feedback polynomials where we do get an improvement.

The remaining parts of the paper are presented as follows. In Section 2 we give the basic preliminaries for the attack. In Section 3 we discuss how a basic distinguishing attack is mounted and in Section 4 we expand this attack by using vectors with noise variables. In Section 5 we give the consequences when tweaking the different parameters of the attack. Section 6 discusses the problem of finding a multiple of the characteristic polynomial. In Section 7 we compare our attack to the basic attack and in Section 8 we give our conclusions.

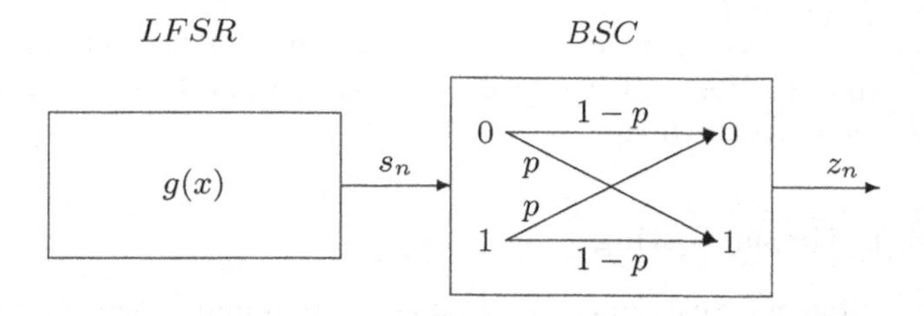

Fig. 1. Model for a correlation attack

2 Preliminaries

The model used for the attack is the standard model for a correlation attack, illustrated in Figure 1. For a more detailed description of this model we refer to, for example, [17]. The target stream cipher uses two different components, one linear and one nonlinear. The linear part is a LFSR and the nonlinear part can be modeled as a black box. One often illustrates the correlation attack scenario through the class of nonlinear combining generators, although it is applicable to the more general case described above. See for example [18].

It is common to view the problem of cryptanalysis as a decoding problem. The nonlinear function can then, through a linear approximation, be seen as a binary symmetric channel (BSC) with crossover probability p (correlation probability $1-p$) with $p \neq 0.5$.

The output bits from the LFSR are denoted s_n, $n = 1, 2, \ldots$, and the keystream bits are denoted z_n, $n = 1, 2, \ldots$. From the BSC it follows immediately that $P(s_n = z_n) = 1 - p \neq 0.5$. Assuming a known plaintext attack, the problem of ours is the following. Given the observed keystream sequence $\mathbf{z} = z_1, z_2 \ldots z_n$ we want to distinguish the keystream from a truly random process. If P_0 is the distribution induced by the cipher and P_1 is the uniform distribution, we try to determine if the underlying distribution D for the samples that we observe (z_n, $n = 1, 2, \ldots$) is more likely to be P_0 or P_1.

Example: We will here give an example in which such distinguishing attacks can be useful. Assume that the people in a small country are going to vote in a referendum. Alice is going to send her vote as a ciphertext $\mathbf{c} = \mathbf{m} + \mathbf{z}$, where $\mathbf{m}$ is the vote and $\mathbf{z}$ is the keystream. Now assume that there are two possible ways to vote, either $\mathbf{m} = \mathbf{m}^{(1)}$ or $\mathbf{m} = \mathbf{m}^{(2)}$, and that both of them include some large amount of data (e.g they are both pictures).

The attacker Eve, who can listen to the channel, receives the ciphertext $\mathbf{c}$. He makes the guess that Alice voted $\mathbf{m}^{(1)}$. Eve then adds $\mathbf{m}^{(1)}$ to the received ciphertext and depending on whether he made the right or wrong guess he gets

$$\hat{\mathbf{z}} = \mathbf{c} + \mathbf{m}^{(1)} = \begin{cases} \mathbf{z} & \text{correct guess} \\ \mathbf{m}^{(1)} + \mathbf{m}^{(2)} + \mathbf{z} & \text{wrong guess} \end{cases} .$$

If the guess was incorrect the result is a random sequence, assuming that $\mathbf{m}^{(1)} + \mathbf{m}^{(2)}$ is random. Hence if we apply a distinguisher to the vector $\hat{\mathbf{z}}$, Eve can determine how Alice voted.

2.1 Hypothesis Testing

In this section we give a brief introduction to binary hypothesis testing. The task of a binary hypothesis test is to decide which of two hypotheses H_0 or H_1 is the explanation for an observed measurement. Statistics provides methods to determine how many output symbols that are needed to make a correct decision and also how to carry out the actual hypothesis test. These two parts will be explained in the following.

Assume that we have a sequence of m independent and identically distributed (i.i.d.) random variables $X_1, X_2, \ldots, X_m$ over an alphabet $\mathcal{X}$. Its distribution is denoted $D(x) = Pr(X_i = x)$, $1 \leq i \leq m$ and the sample values obtained in an experiment are denoted $\mathbf{x} = x_1, x_2, \ldots, x_m$. We have the two hypotheses $H_0 : D = P_0$ and $H_1 : D = P_1$, where P_0 and P_1 are two different distributions. To distinguish between the two hypotheses, one defines a *decision function*, $\phi : \mathcal{X}^m \rightarrow \{0, 1\}$. $\phi(\mathbf{x}) = 0$ implies that H_0 is accepted and $\phi(\mathbf{x}) = 1$ implies that H_1 is accepted.

Two probabilities of error are associated with the decision function,

$$\begin{aligned} \alpha &= P(\phi(\mathbf{x}) = 1 | H_0 \textit{ is true}) \\ \beta &= P(\phi(\mathbf{x}) = 0 | H_1 \textit{ is true}). \end{aligned} \tag{1}$$

Let H_0 be the hypothesis that the distribution D is induced by the cipher and let H_1 be the hypothesis that D is uniform.

The overall probability of error, P_e, can be written as a weighted sum over α and β, i.e., $P_e = \pi_0 \alpha + \pi_1 \beta$, where π_0 and π_1 are the *a priori* probabilities of the two hypotheses. An important asymptotic result is the following,

$$P_e \approx 2^{-mC(P_0, P_1)}, \tag{2}$$

when m is large. The variable $C(P_0, P_1)$ is the *Chernoff information* between distributions P_0 and P_1. The Chernoff information is obtained through

$$C(P_0, P_1) = - \min_{0 \leq \lambda \leq 1} \log_2 \left(\sum_{x \in \mathcal{X}} (P_0(x))^{\lambda} (P_1(x))^{1-\lambda} \right). \tag{3}$$

It can be difficult to determine the exact value of λ but by picking just any value, e.g. $\lambda = 0.5$, it is possible to obtain a lower bound of the *Chernoff information* and hence, an upper bound of P_e. By using (2), the number of samples needed to distinguish P_0 and P_1 can be calculated for any error probability.

We also need to know how to perform the hypothesis test. The Neyman-Pearson lemma tells us how to carry out the actual test when we have a sequence of samples.

Lemma 1. *(Neyman-Pearson lemma) Let $X_1, X_2, \ldots, X_m$ be drawn i.i.d. according to mass function D. Consider the decision problem corresponding to the hypotheses $D = P_0$ vs. $D = P_1$. For $T \geq 0$ define a region*

$$\mathcal{A}_m(T) = \left\{ \frac{P_0(x_1, x_2, \ldots, x_m)}{P_1(x_1, x_2, \ldots, x_m)} > T \right\}.$$

Let $\alpha_m = P_0^m(\mathcal{A}_m^c(T))$ and $\beta_m = P_1^m(\mathcal{A}_m(T))$ be the error probabilities corresponding to the decision region $\mathcal{A}_m$. Let $\mathcal{B}_m$ be any other decision region with associated error probabilities α^* and β^*. If $\alpha^* \leq \alpha$, then $\beta^* \geq \beta$.

This tells us that the region $\mathcal{A}_m(T)$ that is determined by $\frac{P_0(\mathbf{x})}{P_1(\mathbf{x})} > T$, is the one that jointly minimizes α and β. In the hypothesis test we want α and β to be equal and hence $T = 1$. With the assumption that all x_n are independent we can rewrite the Newman-Pearson test using 2-logarithms. This gives us the test

$$\frac{P_0(x_1, x_2, \ldots, x_m)}{P_1(x_1, x_2, \ldots, x_m)} > 1 \quad \Rightarrow \quad \sum_{n=1}^{m} \left(\log_2 \frac{P_0(x_n)}{P_1(x_n)} \right) > 0\,. \tag{4}$$

The ratio in (4) is called a log-likelihood ratio, and the test is thus called a log-likelihood test.

This was a very brief overview of some tools that will be useful for us. For a more thorough treatment of hypothesis testing, we refer to any textbook on the subject, e.g. [7].

3 A Basic Distinguishing Attack from a Low Weight Feedback Polynomial

We start our investigation by simplifying the Meier-Staffelbach approach and turn their original ideas into a distinguishing attack.

Referring to the assumed model (Figure 1), the observed keystream output is considered as a noisy version of the sequence from the LFSR,

$$z_n = s_n + e_n, \tag{5}$$

where e_n, $n = 1, 2, \ldots$, are variables representing the noise introduced by the approximation. The noise has a biased distribution

$$P(e_n = 0) = p = 1/2 + \varepsilon,$$

where ε is usually rather small. The recursive computation of s_n is linear, and for a LFSR the computation of s_n will depend on the characteristic polynomial of the LFSR. The recurrence can be written as $s_n = \sum_{j=1}^{L} c_j s_{n-j}$ where L is the LFSR length and c_j, $j = 1, 2, \ldots$, are some known constants. By introducing $c_0 = 1$ the recurrence above can be put into the form

$$\sum_{j=0}^{L} c_j s_{n-j} = 0, \quad n \geq j.$$

By adding the corresponding positions in $\mathbf{z}$ we can get all s_n canceling out. What remains is just a sum of independent noise variables. In more detail, let us introduce

$$x_n = \sum_{j=0}^{L} c_j z_{n-j}.$$

Then $x_n = \sum_{j=0}^{L} c_j z_{n-j} = \sum_{j=0}^{L} c_j s_{n-j} + \sum_{j=0}^{L} c_j e_{n-j} = \sum_{j=0}^{L} c_j e_{n-j}$. Since the distribution of e_n is nonuniform it is possible to distinguish the sample sequence x_n, $n = 1, 2, \ldots$, from a truly random sequence. If we assume the binary case (all variables are binary), the sum of the noise will have a bias which according to the piling-up lemma [16] can be expressed as follows,

$$P(\sum_{j=0}^{L} c_j e_{n-j} = 0) = 1/2 + 2^{w-1}\varepsilon^w, \quad (6)$$

where w is the weight of $(c_0, c_1, \ldots, c_L)$. We also know that the required number of samples is in the order of $1/(2^{w-1}\varepsilon^w)^2$.

In this paper we choose to use, however, the Chernoff information as a measure of the distance between two distributions, as described in Section 2.1. If we consider the binary case above, the expression for the Chernoff information in equation (3) becomes

$$C(P_0, P_1) \geq -\log_2 \left(\sqrt{\left(\frac{1}{2} + 2^{w-1}\varepsilon^w\right)\frac{1}{2}} + \sqrt{\left(\frac{1}{2} - 2^{w-1}\varepsilon^w\right)\frac{1}{2}} \right).$$

In Table 1 in appendix we give the number of samples needed for some different w values. In the table we use $\varepsilon = 0.1$ and $\varepsilon = 0.01$.

Note that the ideas behind this simple attack has appeared in many attack scenarios before, even if it might not have been described exactly in this context before. We see that the weight of the characteristic polynomial is directly connected to the success of the attack.

4 A More General Distinguisher with Correlated Vectors

As we have seen in the previous section, low weight polynomials are easily attacked, but when the weight grows so does the required length and complexity (exponentially). At some point we argue that the attack is no longer realistic, or it might require more than an exhaustive key search. In this section we now describe a similar but more general approach that can be applied to another set of characteristic polynomials.

Consider a length L LFSR with characteristic polynomial

$$f(x) = f_0 + f_1 x + f_2 x^2 + f_3 x^3 + \ldots f_L x^L,$$

where $f_i \in \mathbb{F}_2$. We try to find a multiple, $a(x)$, of the characteristic polynomial so that this polynomial can be written as

$$a(x) = f(x)h(x) = g_1(x) + x^M g_2(x), \tag{7}$$

where $g_1(x)$ and $g_2(x)$ are polynomials of some small degree $\leq k$. It is possible that $f(x)$ is already on the form (7), then $h(x) = 1$. In the sequel we assume that such a polynomial $a(x)$ of the above form is given.

This will correspond to a shift register for which the taps are concentrated to two regions far away from each other. The linear recurrence relation can then be written as the two sums

$$\sum_{i=0}^{k} s_{n+i}a_i + \sum_{i=0}^{k} s_{n+M+i}a_{M+i} = 0, \tag{8}$$

where s_n is the nth output bit from the LFSR and a_i, $i = 0, 1, \ldots$, are the coefficients in the characteristic polynomial $a(x)$. We now consider the standard model for a correlation attack where the output of the cipher is considered as a noisy version of the LFSR sequence $z_n = s_n + e_n$. The noise variables e_n is introduced by the approximation of the nonlinear part of the cipher. Furthermore, the biased noise has distribution $P(e_n = 0) = 1/2 + \varepsilon$, and the variables are pairwise independent, i.e., $P(e_i, e_j) = P(e_i)P(e_j)$, $\forall i \neq j$.

Let us introduce the notation Q_n to be the sum

$$Q_n = \sum_{i=0}^{k} z_{n+i}a_i + \sum_{i=0}^{k} z_{n+M+i}a_{M+i} = \sum_{i=0}^{k} e_{n+i}a_i + \sum_{i=0}^{k} e_{n+M+i}a_{M+i}. \tag{9}$$

This can also be written as

$$\begin{aligned}
Q_0 &= e[0, k] \cdot \underline{g_1} + e[M, M+k] \cdot \underline{g_2}, \\
Q_1 &= e[1, k+1] \cdot \underline{g_1} + e[M+1, M+k+1] \cdot \underline{g_2}, \\
&\vdots \\
Q_{N-1} &= e[N-1, N+k-1] \cdot \underline{g_1} + e[M+N-1, M+N+k-1] \cdot \underline{g_2},
\end{aligned}$$

if we introduce $e[i, j] = (e_i, \ldots, e_j)$ for $i \leq j$ and $\underline{g_1} = (g_{1,0}, g_{1,1}, \ldots, g_{1,k})^T$ where $g_{1,j}$, $j = 0, 1, \ldots, k$ are the coefficients of the $g_1(x)$ polynomial. A corresponding notation is assumed for $\underline{g_2}$.

The noise variables (e_n, $n = 1, 2, \ldots$) are independent but Q_i values that are close to each other will not be independent in general. This is because of the fact that several Q_i will contain common noise variables. We can take advantage of this fact by moving to a vector representing the noise as follows.

Introduce the vectorial noise vector E_n of length N as

$$E_n = (Q_{N \cdot n}, \ldots, Q_{N(n+1)-1}). \tag{10}$$

Alternatively, E_n can be expressed as

$$E_n = (e_{N \cdot n}, \ldots, e_{N(n+1)+k-1}) \cdot G_1 + (e_{N \cdot n+M}, \ldots, e_{N(n+1)+M+k-1}) \cdot G_2, \quad (11)$$

where G_1 and G_2 are the $(N+k) \times N$ matrices

$$G_1 = \begin{pmatrix} g_{1,0} & & & \\ g_{1,1} & g_{1,0} & & \\ \vdots & \vdots & & g_{1,0} \\ g_{1,k} & g_{1,k-1} & \cdots & g_{1,1} \\ & g_{1,k} & & \vdots \\ & & & g_{1,k} \end{pmatrix} \text{ and } G_2 = \begin{pmatrix} g_{2,0} & & & \\ g_{2,1} & g_{2,0} & & \\ \vdots & \vdots & & g_{2,0} \\ g_{2,k} & g_{2,k-1} & \cdots & g_{2,1} \\ & g_{2,k} & & \vdots \\ & & & g_{2,k} \end{pmatrix}.$$

The remaining pieces to complete the distinguishing attack for the assumed polynomial are obvious. To prepare the attack, we derive the distribution of the E_n noise vector given in (11). This can be easily done for small polynomial degrees, since we know the distribution of the noice. This calculation results in a distribution which we denote by P_0. Our aim is to distinguish this distribution from the truly random case, so by P_1 we denote the uniform distribution.

When performing the attack, we collect a sample sequence

$$Q_0, Q_1, Q_2, \ldots, Q_B$$

by $Q_n = \sum_{i=0}^{k} z_{n+i} a_i + \sum_{i=0}^{k} z_{n+M+i} a_{M+i}$. This sample sequence is then transformed into a vectorial sample sequence $E_0, E_1, \ldots, E_{B'}$ by $E_n = (Q_{N \cdot n}, \ldots, Q_{N(n+1)-1})$.

The final step is to use optimal (Neyman-Pearson) hypothesis testing to decide whether $E_0, E_1, \ldots, E_{B'}$ is most likely to come from distribution P_0 or P_1. This proposed algorithm is summarized in Figure 2.

The performance of the algorithm depends on the polynomials, g_i, that are used. Figure 4 shows an example of how the number of vectors required for a successful attack depends on the vector length for a certain combination of

1. Find multiple.
2. For $t = 0 \ldots B$
   $Q_t = \sum_{i=0}^{k} z_{t+i} a_i + \sum_{i=0}^{k} z_{t+M+i} a_{M+i}$
   end for.
3. For $n = 0 \ldots B'$
   $E_n = (Q_{N \cdot n}, \ldots, Q_{N(n+1)-1})$
   end for.
4. Calculate $I = \sum_{n=0}^{B'} \left(\log_2 \frac{P_0(E_n)}{P_1(E_n)} \right)$.
If $I > 0$ then output **cipher**, else output **random**.

Fig. 2. Summary of proposed algorithm

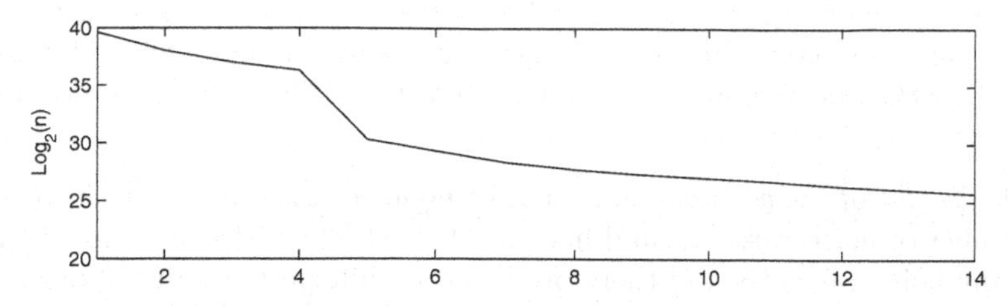

Fig. 3. The number of vectors needed as a function of the vector length N. In this example $g_1(x) = 1 + x + x^5 + x^6$ and $g_2(x) = 1 + x + x^7 + x^8$

two polynomials. $N = 1$ corresponds to the basic approach and we see that increasing the vector length will decrease the number of vectors needed. Note that $g_1(x)$ and $g_2(x)$ are just two examples of what the polynomials might look like, they do not represent a multiple of any specific primitive polynomial.

Finally, we note that we may generalize our reasoning with two groups to allow finding a multiple of arbitrarily many groups. Expression (7) for the multiple then becomes

$$a(x) = h(x)f(x) = g_1(x) + x^{M_1}g_2(x) + \ldots + x^{M_{t-1}}g_t(x), \tag{12}$$

where $g_i(x)$ is a polynomial of some small degree $\leq k$, and $M_1 < M_2 < \ldots < M_{t-1}$. It is clear that when t grows it is easier to find multiples of the characteristic polynomial with the desired properties. But it is also clear that when t grows, the attack becomes weaker.

This distinguishing attack that we propose may be mounted on ciphers using a shift register where "good" multiples easily can be found. Ciphers using a polynomial where many taps are close together might be attacked directly without finding any multiple. This attack may be viewed as a new design criteria, one should avoid LFSRs where multiples of the form (7) are easily found.

5 Tweaking the Parameters in the Attack

We have described the new attack in the previous section. We now discuss how various algorithmic parameters effect the results.

5.1 How $g_i(x)$ Effects the Results

Since the vectors E_n are correlated it is intuitive that the forms of $g_i(x)$ will effect the strength of the attack. Some combinations of polynomials will turn out to be much better than others.

We have tested a number of different polynomials $g_i(x)$ in order to find a set of rules describing how the form of the polynomials effects the result. A definite rule to decide which polynomials are best suited for the attack is hard to find. Some basic properties that characterizes a good polynomial can be found. The parameters that determine how a polynomial will effect the distribution of the noise vectors are the following.

- *The weight of the polynomial.* A small weight means that the we have a small number of noise variables and hence, a large bias. And a large weight means many noise variables and therefore a more uniform noise distribution.
- *Arrangement of the terms in a polynomial.* This is the same as the arrangement of the taps in the LFSR. If there are many taps close together, the corresponding noise variables will occur more frequently in the noise vector. This will significantly effect the distribution of the noise vectors.

Since M_i is typically large, all polynomials $g_i(x)$ can be considered independent. Their properties will have the same influence on the total result. However, the polynomials are combined to form the distribution. This combination is just a variant of the piling-up lemma so it is obvious that the total distribution is more uniform than the distribution of the individual polynomials. Depending on the form of the individual polynomials, the resulting distribution becomes more or less uniform. Some combinations are "better" than others. An example of this can be found in Figure 4 in the appendix.

5.2 Vector Length

The length of the vectors, denoted N impacts the effectiveness of the algorithm. The idea of using N larger than one is that we can get correlation between the vectors. Every time we increase N by 1 we will also increase the Chernoff information. Recalling (2), we see that increasing the Chernoff information means that we will decrease the number of vectors needed for our hypothesis test. The Chernoff information is however not a linear function of N. Depending on the form of the polynomials the increase can be much higher for some N than for others. The computational complexity is also higher when N gets higher since each vector have 2^N different values. The downside is that the complexity of the calculations increase with increasing N. So there is trade off between the number of samples needed and the complexity of the calculations. The largest gain in Chernoff information is usually achieved when going from $N = i$ to $N = i + 1$ for small i. This phenomenon can be seen in Figure 5 in the appendix.

5.3 Increase the Number of Groups

As we will show in Section 6.2, it is much easier to find a multiple if we allow the multiple to have more groups than just two. The drawback is that the distribution will become more uniform if more groups are used. This is similar to the binary case in which more taps in the multiple will cause a less biased distribution. The difference is that there is no equivalence to equation (6) for how much more uniform the distribution will be when having many groups, since this depends a lot on the polynomials used.

6 Finding Multiples of the Characteristic Polynomial of a Desired Form

We have many times assumed that a multiple of the feedback polynomial has a certain form. Here we briefly look at the problem of finding multiples of a certain form.

6.1 Finding Low Weight Multiples

According to the piling up lemma (6), the distribution becomes more uniform if the polynomial is of a high weight. Therefore, the first step in a correlation attack is to find a multiple that has a low weight. The multiple will produce the same sequence so the linear relation that describes the multiple will also satisfy the original LFSR sequence. There exist easy and efficient ways of finding a multiple of a given weight. The number of bits needed to actually start the attack depends on the degree of the multiple, which in turn depends the weight of the same. If we want to find a multiple of weight w of a polynomial that has degree L it can be shown [12] that the degree M of the multiple will be approximately:

$$M \geq 2^{\frac{L}{w-1}}. \tag{13}$$

In Table 2 in the appendix the result of this equation is listed for a LFSR of length 100 and a LFSR of length 1000.

6.2 Finding Multiples with Groups

Say that we want to find a multiple of the form

$$a(x) = h(x)f(x) = g_1(x) + x^M g_2(x),$$

where $f(x)$ is the characteristic polynomial of the cipher. The degree of $f(x)$ is L. This can be found by polynomial division. Assume that we have a $g_2(x)$ of degree smaller than k. We then multiply this polynomial with x^i. The result is divided by the original LFSR-polynomial. This gives us a quotient $q(x)$ and a remainder $r(x)$.

$$x^i g_2(x) = f(x) \cdot q(x) + r(x).$$

We have 2^k different $g_2(x)$-polynomials and i can be chosen in M different ways. The remainder $r(x)$ from the division is a polynomial with $0 \leq \deg(r(x)) < L$. If $r(x)$ has degree $\leq k$ we have found an acceptable $g_1(x)$. The probability of finding a polynomial of maximum degree k is $P(\deg(g_2(x)) \leq k) = 2^{k-L}$. If we would like it to be probable that we find at least one such polynomial, we need

$$2^k \cdot M \cdot \frac{2^k}{2^L} \geq 1 \;\; \Rightarrow \;\; M \geq 2^{L-2k}. \tag{14}$$

Examining the result in (14) we see that for modest values of k the length of the multiple will become quite large. Therefore we extend our reasoning to the case with arbitrarily many groups, then we have a multiple of the form

$$a(x) = h(x)f(x) = g_1(x) + x^{M_1}g_2(x) + \ldots + x^{M_{t-1}}g_t(x).$$

If we use the same reasoning as above we receive a new expression

$$2^k \cdot M_1 \cdot 2^k \cdot M_2 \cdot \ldots \cdot 2^k \cdot M_{t-1} \cdot \frac{2^k}{2^L} \geq 1 \Rightarrow M \geq 2^{\frac{L-tk}{t-1}}, \tag{15}$$

where it is assumed that $M_1, M_2, \ldots, M_{t-1} \leq M$. This gives us an upper bound on all M_i.

In (15) we see that by using larger values of t, i.e., more groups, we can lower the length of the multiple. One has to bear in mind though that a larger t, as stated in Section 5.3, will usually effect the Chernoff information in a negative way (from a cryptanalyst point of view). In Table 3 in the appendix we list some values on M needed to find a multiple, for some values of k and t.

7 Comparing the Proposed Attack with a Basic Distinguishing Attack

The applicability of our algorithm is twofold. Firstly, if the characteristic polynomial is of the form $f(x) = g_1(x) + g_2(x)x^{M_1} + \ldots + g_t(x)x^{M_{t-1}}$. Applying the basic algorithm to those LFSRs without finding any multiple first will be equivalent to applying our algorithm with $N = 1$. Since our algorithm has the ability to have vectors with noise variables ($N > 1$), it will be a significant improvement over the basic algorithm. Using the basic algorithm without first finding a multiple is naive, but if the length L of the LFSR is large the degree of the low weight multiple will also be large, see (13). So if $f(x)$ is of high degree then our algorithm can be more effective.

Our algorithm can also be applied to arbitrary characteristic polynomials. Then the approach is to first find a multiple of the polynomial that is of the form $f(x) = g_1(x) + g_2(x)x^{M_1} + \ldots + g_t(x)x^{M_{t-1}}$ and then apply the algorithm. By comparing the two equations (13) and (15) we see that it is not much harder to find a polynomial of some weight w than it is to find a polynomial with the same number of groups. Tables showing the corresponding M can be found in the appendix. Although our algorithm takes advantage of the fact that the taps are close together, it is still not enough to compensate for the larger amount of noise variables. In this case the proposed attack will give improvements only for certain specific instances of characteristic polynomials, e.g., those having a surprisingly weak multiple of the form $f(x) = g_1(x) + g_2(x)x^{M_1}$ but no low weight multiples where the weight is surprisingly low.

8 Conclusion and Future Work

Through a new correlation attack, we have identified a new class of weak feedback polynomials, namely, polynomials of the form $f(x) = g_1(x) + g_2(x)x^{M_1} + \ldots +$

$g_t(x)x^{M_{t-1}}$, where $g_1, g_2, \ldots, g_t$ are all polynomials of low degree. The correlation attack has been described in the form of a distinguishing attack. This was done mainly for simplicity, since the theoretical performance is easily calculated and we can compare with the basic attack based on low weight polynomials.

The next step in this direction would be to examine the possibility of turning these ideas into a key recovery attack. This could be done in a similar manner as the Meier Staffelbach approach. For example, we could try to derive many different relations (multiples) and apply some iterative decoding approach in vector form. The theoretical part of such an approach will probably be much more complicated.

References

[1] A. Canteaut, M. Trabbia. Improved fast correlation attacks using parity-check equations of weight 4 and 5. *Advances in Cryptology-Eurocrypt'2000*, volume 1807 of *Lecture Notes in Computer Science*, pages 573-588. Springer-Verlag, 2000. 128

[2] V. Chepyzhov, T. Johansson, B. Smeets. A simple algorithm for fast correlation attacks on stream ciphers. *Advances in Cryptology-FSE'2000*, volume 1978 of *Lecture Notes in Computer Science*, pages 181-195. Springer-Verlag, 2001. 128

[3] D. Coppersmith, S. Halevi and C. S. Jutla. SCREAM: a software efficient stream cipher. In J. Daemen and V. Rijmen, editors, *Advances in Cryptology-FSE'2002*, volume 2365 of *Lecture Notes in Computer Science*, pages 195-210. Springer-Verlag, 2002. 127

[4] D. Coppersmith, S. Halevi and C. S. Jutla. Cryptanalysis of stream ciphers with linear masking. In M. Young, editor. *Advances in Cryptology-Crypto'2002*, volume 2442 of *Lecture Notes in Computer Science*, pages 515-532. Springer-Verlag, 2002. 128

[5] N. Courtois and Josef Pieprzyk. Cryptanalysis of block ciphers with overdefined systems of equations. *Advances in Cryptology-Asiacrypt'2002*, volume 2501 of *Lecture Notes in Computer Science*, pages 267-287. Springer-Verlag, 2002. 127

[6] N. Courtois and W. Meier. Algebraic attacks on stream ciphers with linear feedback. *Advances in Cryptology-Eurocrypt'2003*, volume 2656 of *Lecture Notes in Computer Science*, pages 345-359. Springer-Verlag, 2003. 127

[7] T. Cover and J. A. Thomas. *Elements of information theory.* Wiley series in telecommunication. Wiley, 1991. 131

[8] P. Ekdahl and T. Johansson. A new version of the stream cipher SNOW. In K. Nyberg and H. Heys, editors. *Selected Areas in Cryptography-SAC 2002*, volume 2595 of *Lecture Notes in Computer Science*, pages 47-61. Springer-Verlag, 2003. 127

[9] P. Ekdahl and T. Johansson. Distinguishing attack on SOBER-t16 and SOBER-t32. In J. Daemen and V. Rijmen, editors. *Advances in Cryptology-FSE'2002*, volume 2365 of *Lecture Notes in Computer Science*, pages 210-224. Springer-Verlag, 2002. 128

[10] N. Ferguson, D. Whiting, B. Schneider, J. Kelsey, S. Lucks and T. Kohno. Helix: Fast Encryption and Authentication in a Single Cryptographic Primitive. In T. Johansson, editor, *Advances in Cryptology-FSE'2003*, volume 2887 of *Lecture Notes in Computer Science*, pages 330-346. Springer-Verlag, 2003. 127

[11] J. D. Golić. Intrinsic statistical weakness of keystream generators. *Advances in Cryptology-Asiacrypt'94*, volume 917 of *Lecture Notes in Computer Science*, pages 91-103. Springer-Verlag, 1995. 128

[12] J. D. Golić. Computation of low-weight parity-check polynomials. *Electronic Letters*, 32(21):1981-1982, October 1996. 137

[13] P. Hawkes and G. Rose. Primitive specification and supporting documentation for SOBER-t32 submission to NESSIE. In *Proceedings of First Open NESSIE Workshop, 2000.* 127

[14] T. Johansson, F. Jönsson. Improved fast correlation attacks on stream ciphers via convolutional codes. *Advances in Cryptology-Eurocrypt'99*, volume 1592 of *Lecture Notes in Computer Science*, pages 347-362. Springer-Verlag, 1999. 128

[15] T. Johansson, F. Jönsson. Fast correlation attacks based on turbo code techniques. *Advances in Cryptology-Crypto'99*, volume 1666 of *Lecture Notes in Computer Science*, pages 181-197. Springer-Verlag, 1999. 128

[16] M. Matsui. Linear cryptanalysis method for DES cipher. In T. Helleseth, editor, *Advances in Cryptology-Eurocrypt'93*, volume 765 of *Lecture Notes in Computer Science*, pages 386-397. Springer-Verlag, 1994. 132

[17] W. Meier and O. Staffelbach. Fast correlation attacks on stream ciphers.*Advances in Cryptology-Eurocrypt'88*,volume 330 of *Journal of Cryptology*,pages 310-314. Springer-Verlag, 1988. 127, 129

[18] A. Menezes, P. van Oorschot and S. Vanstone. *Handbook of Applied cryptography*, CRC Press, 1997. 129

[19] M. Mihaljevic, M. Fossorier and H. Imai. A low-complexity and high-performance algorithm for the fast correlation attack. *Advances in Cryptology-FSE'2000*, volume 1978 of *Lecture Notes in Computer Science*, pages 196-212. Springer-Verlag, 2001. 128

[20] D. Watanabe, S. Furuya, H. Yoshida, K. Takaragi and B. Preneel. A new keystream generator MUGI. In J. Daemen and V. Rijmen, editors, *Fast Software Encryption 2002*, volume 2365 of *Lecture Notes in Computer Science*, pages 179-194. Springer-Verlag, 2002. 127

A Tables and Figures

The appendix contains some tables and figures that can be used to compare some different parameters discussed in the paper. Table 1 shows the number of samples needed to distinguish a sequence from random in the basic binary case. The result is given for two different ε. Table 2 shows the expected degree of the multiple of $f(x)$ in the basic distinguishing attack, where w is the weight of the polynomial to be found and L is the length of the LFSR. Table 3 shows the corresponding values for our attack, where t denotes the number of groups and k the maximum number of taps allowed in each polynomial. The degree of the multiple is the same as the amount of plaintext needed before the actual attack can start.

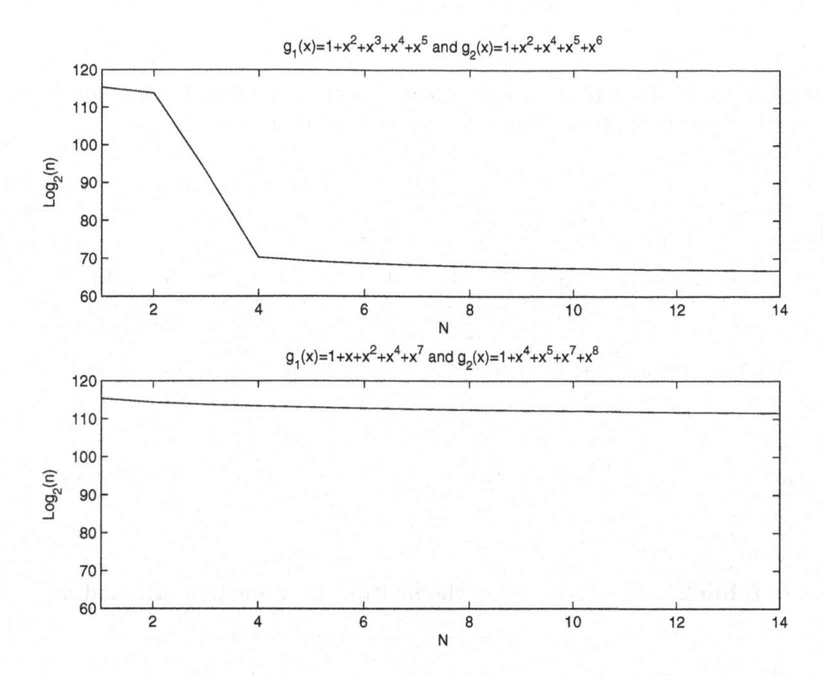

Fig. 4. The sequence length needed as a function of the vector length N (logarithmic)

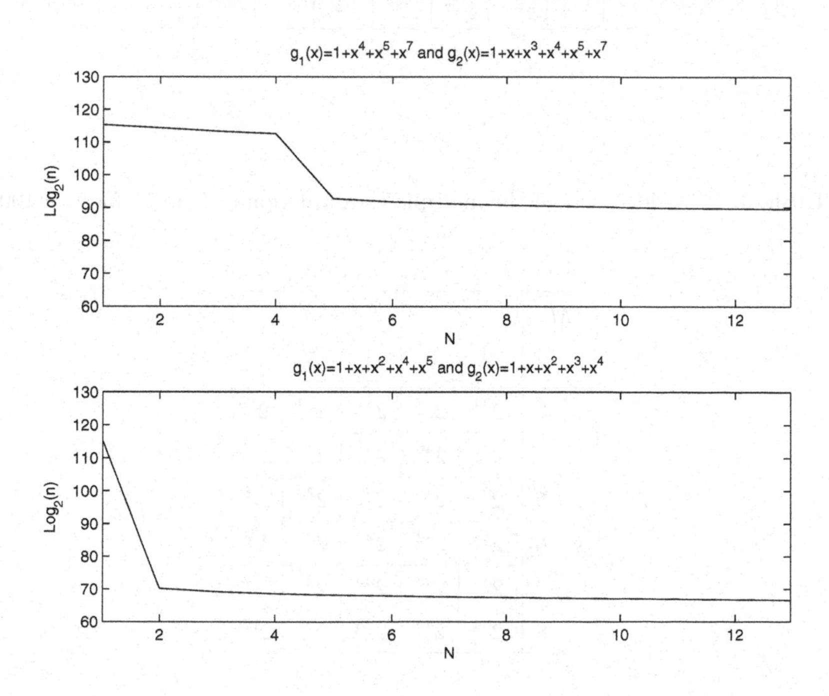

Fig. 5. The sequence length needed as a function of the vector length N (logarithmic)

Table 1. Number of samples needed for some different w in the binary case for $\varepsilon = 0.1$ and $\varepsilon = 0.01$. The number of vectors needed is calculated as $\frac{1}{C(P_0,P_1)}$

M	w													
	3	4	5	6	7	8	9	10	11	12	13	14	15	16
$\varepsilon = 0.1$	$2^{16.4}$	2^{21}	$2^{25.7}$	$2^{30.3}$	2^{35}	$2^{39.6}$	$2^{44.3}$	$2^{48.9}$	$2^{53.6}$	$2^{58.2}$	$2^{62.8}$	$2^{67.5}$	$2^{72.1}$	$2^{76.8}$
$\varepsilon = 0.01$	$2^{36.3}$	$2^{47.6}$	$2^{58.9}$	$2^{70.2}$	$2^{81.5}$	$2^{92.8}$	2^{104}	2^{115}	2^{127}	2^{138}	2^{149}	2^{160}	2^{172}	2^{183}

Table 2. The degree M of the multiple as a function of L and w

M	w														
	2	3	4	5	6	7	8	9	10	11	12	13	14	15	16
$L = 100$	2^{100}	2^{50}	2^{33}	2^{25}	2^{20}	2^{17}	2^{14}	2^{123}	2^{11}	2^{10}	2^{9}	2^{8}	2^{8}	2^{7}	2^{7}
$L = 1000$	2^{1000}	2^{500}	2^{333}	2^{250}	2^{200}	2^{167}	2^{143}	2^{125}	2^{111}	2^{100}	2^{91}	2^{83}	2^{77}	2^{71}	2^{67}

Table 3. The degree M of the multiple as a function of k and t for $L = 100$

M		t				
		2	3	4	5	6
	3	2^{94}	2^{47}	2^{31}	2^{24}	2^{19}
	4	2^{92}	2^{46}	2^{31}	2^{23}	2^{18}
	5	2^{90}	2^{45}	2^{30}	2^{23}	2^{18}
k	6	2^{88}	2^{44}	2^{29}	2^{22}	2^{18}
	7	2^{86}	2^{43}	2^{29}	2^{22}	2^{17}
	8	2^{84}	2^{42}	2^{28}	2^{21}	2^{17}

Minimum Distance between Bent and 1-Resilient Boolean Functions

Soumen Maity[1] and Subhamoy Maitra[2]

[1] Department of Mathematics, Indian Institute of Technology Guwahati
Guwahati 781 039, Assam, INDIA
soumen@iitg.ernet.in
[2] Applied Statistics Unit, Indian Statistical Institute
203, B T Road, Kolkata 700 108, INDIA
subho@isical.ac.in

Abstract. In this paper we study the minimum distance between the set of bent functions and the set of 1-resilient Boolean functions and present a lower bound on that. The bound is proved to be tight for functions up to 10 input variables. As a consequence, we present a strategy to modify the bent functions, by toggling some of its outputs, in getting a large class of 1-resilient functions with very good nonlinearity and autocorrelation. In particular, the technique is applied upto 12-variable functions and we show that the construction provides a large class of 1-resilient functions reaching currently best known nonlinearity and achieving very low autocorrelation values which were not known earlier. The technique is sound enough to theoretically solve some of the mysteries of 8-variable, 1-resilient functions with maximum possible nonlinearity. However, the situation becomes complicated from 10 variables and above, where we need to go for complicated combinatorial analysis with trial and error using computational facility.

Keywords: Autocorrelation, Bent Function, Boolean Function, Nonlinearity, Resiliency

1 Introduction

Construction of resilient Boolean functions with very good parameters in terms of nonlinearity, algebraic degree and other cryptographic parameters has received lot of attention in literature [15, 16, 18, 19, 8, 21, 2, 3]. In [17, 7], it had been shown how bent functions can be modified to construct highly nonlinear balanced Boolean functions. A recent construction method [12] presents modification of some output points of a bent function to construct highly nonlinear 1-resilient function. A natural question that arises in this context is "at least how many bits in the output of a bent function need to be changed to construct an 1-resilient Boolean function". The answer of this question gives the minimum distance between the set of bent functions and the set of 1-resilient functions. We here try to answer this question and show that the minimum distance for n-variable

B. Roy and W. Meier (Eds.): FSE 2004, LNCS 3017, pp. 143–160, 2004.

functions is

$$dBR_n(1) \geq 2^{\frac{n}{2}-1} + 2\left\lceil \frac{(r+1)(2^{\frac{n}{2}-1} - \sum_{i=0}^{r} \binom{n}{i}) + \sum_{i=1}^{r} i\binom{n}{i}}{n-r-1} \right\rceil,$$

where r is the integer such that $\sum_{i=0}^{r} \binom{n}{i} \leq 2^{\frac{n}{2}-1} + 1 < \sum_{i=0}^{r+1} \binom{n}{i}$ is satisfied. We also show that this result is tight for $n \leq 10$. The immediate corollary is the construction of 1-resilient Boolean functions with nonlinearity $\geq 2^{n-1} - 2^{\frac{n}{2}-1} - dBR_n(1)$ and maximum absolute value of autocorrelation spectra $\leq 4\, dBR_n(1)$. Interestingly, it is possible to get 1-resilient functions with better nonlinearity and autocorrelation than these bounds. In particular, we concentrate on construction of 1-resilient Boolean functions up to 12-variables with best known nonlinearity and autocorrelation. *Throughout the paper we consider the number of input variables* (n) *is even.*

The bent functions chosen in [12, Section 3] use the concept of perfect nonlinear functions and one example function each for 8, 10 and 12 variables were presented. However, it is not clear how a generalized construction of such bent functions can be achieved in that manner. We here identify a large subclass of Maiorana-McFarland type bent functions which can be modified to get 1-resilient functions with currently best known parameters. Further our construction is superior to [12] in terms of number of points that need to be toggled (we need less in case of 10, 12 variables), the nonlinearity (we get better nonlinearity for 12 variables) and autocorrelation (we get 1-resilient functions with autocorrelation values that were not known earlier for 10, 12 variables).

1.1 Preliminaries

A Boolean function on n variables may be viewed as a mapping from $\{0,1\}^n$ into $\{0,1\}$. A Boolean function $f(x_1, \ldots, x_n)$ is also interpreted as the output column of its *truth table* f, i.e., a binary string of length 2^n,
$f = [f(0,0,\cdots,0), f(1,0,\cdots,0), f(0,1,\cdots,0), \ldots, f(1,1,\cdots,1)]$.

The *Hamming distance* between two binary strings S_1, S_2 is denoted by $d(S_1, S_2)$, i.e., $d(S_1, S_2) = \#(S_1 \neq S_2)$. Also the *Hamming weight* or simply the weight of a binary string S is the number of ones in S. This is denoted by $wt(S)$. An n-variable function f is said to be *balanced* if its output column in the truth table contains equal number of 0's and 1's (i.e., $wt(f) = 2^{n-1}$).

Denote addition operator over $GF(2)$ by $\oplus$. An n-variable Boolean function $f(x_1, \ldots, x_n)$ can be considered to be a multivariate polynomial over $GF(2)$. This polynomial can be expressed as a sum of products representation of all distinct k-th order products ($0 \leq k \leq n$) of the variables. More precisely, $f(x_1, \ldots, x_n)$ can be written as

$$a_0 \oplus \bigoplus_{1\leq i\leq n} a_i x_i \oplus \bigoplus_{1\leq i<j\leq n} a_{ij} x_i x_j \oplus \ldots \oplus a_{12\ldots n} x_1 x_2 \ldots x_n,$$

where the coefficients $a_0, a_{ij}, \ldots, a_{12\ldots n} \in \{0,1\}$. This representation of f is called the *algebraic normal form* (ANF) of f. The number of variables in the highest order product term with nonzero coefficient is called the *algebraic degree*, or simply the degree of f and denoted by $deg(f)$.

Functions of degree at most one are called *affine* functions. An affine function with constant term equal to zero is called a *linear* function. The set of all n-variable affine (respectively linear) functions is denoted by $A(n)$ (respectively $L(n)$). The nonlinearity of an n-variable function f is

$$nl(f) = \min_{g \in A(n)} d(f, g),$$

i.e., the distance from the set of all n-variable affine functions.

Let $x = (x_1, \ldots, x_n)$ and $\omega = (\omega_1, \ldots, \omega_n)$ both belong to $\{0,1\}^n$ and

$$x \cdot \omega = x_1\omega_1 \oplus \ldots \oplus x_n\omega_n.$$

Let $f(x)$ be a Boolean function on n variables. Then the *Walsh transform* of $f(x)$ is a real valued function over $\{0,1\}^n$ which is defined as

$$W_f(\omega) = \sum_{x \in \{0,1\}^n} (-1)^{f(x) \oplus x \cdot \omega}.$$

In terms of Walsh spectra, the nonlinearity of f is given by

$$nl(f) = 2^{n-1} - \frac{1}{2} \max_{\omega \in \{0,1\}^n} |W_f(\omega)|.$$

For n-even, the maximum nonlinearity of a Boolean function can be $2^{n-1} - 2^{\frac{n}{2}-1}$ and the functions possessing this nonlinearity are called bent functions [14]. Further, for a bent function f on n variables, $W_f(\omega) = \pm 2^{\frac{n}{2}}$ for all ω.

In [9], an important characterization of correlation immune and resilient functions has been presented, which we use as the definition here. A function $f(x_1, \ldots, x_n)$ is m-resilient (respectively m-th order correlation immune) iff its Walsh transform satisfies

$$W_f(\omega) = 0, \text{ for } 0 \le wt(\omega) \le m \text{ (respectively } W_f(\omega) = 0, \text{ for } 1 \le wt(\omega) \le m).$$

As the notation used in [15, 16], by an (n, m, d, σ) function we denote an n-variable, m-resilient function with degree d and nonlinearity σ.

We will now define *restricted Walsh transform* which will be frequently used in this text. The *restricted Walsh transform* of $f(x)$ on a subset S of $\{0,1\}^n$ is a real valued function over $\{0,1\}^n$ which is defined as

$$W_f(\omega)|_S = \sum_{x \in S} (-1)^{f(x) \oplus x \cdot \omega}.$$

Now we present the following technical result.

Proposition 1. *Let $S \subset \{0,1\}^n$ and $b(x), f(x)$ be two n-variable Boolean functions such that $f(x) = 1 \oplus b(x)$ when $x \in S$ and $f(x) = b(x)$ otherwise. Then $W_f(\omega) = W_b(\omega) - 2W_b(\omega)|_S$.*

Proof. Take $\omega \in \{0,1\}^n$. Now

$$\begin{aligned}
W_f(\omega) &= \textstyle\sum_{x\in\{0,1\}^n}(-1)^{f(x)\oplus\omega\cdot x}\\
&= \textstyle\sum_{x\in\{0,1\}^n-S}(-1)^{f(x)\oplus\omega\cdot x} + \sum_{x\in S}(-1)^{f(x)\oplus\omega\cdot x}\\
&= \textstyle\sum_{x\in\{0,1\}^n-S}(-1)^{b(x)\oplus\omega\cdot x} - \sum_{x\in S}(-1)^{b(x)\oplus\omega\cdot x}\\
&\qquad \text{(since } f, b \text{ are same for the inputs } \notin S\\
&\qquad \text{and complement when the inputs } \in S)\\
&= \textstyle\sum_{x\in\{0,1\}^n-S}(-1)^{b(x)\oplus\omega\cdot x} + \sum_{x\in S}(-1)^{b(x)\oplus\omega\cdot x} - 2\sum_{x\in S}(-1)^{b(x)\oplus\omega\cdot x}\\
&= \textstyle\sum_{x\in\{0,1\}^n}(-1)^{b(x)\oplus\omega\cdot x} - 2\sum_{x\in S}(-1)^{b(x)\oplus\omega\cdot x}\\
&= W_b(\omega) - 2W_b(\omega)|_S.
\end{aligned}$$

□

Propagation Characteristics (PC) and Strict Avalanche Criteria (SAC) [13] are important properties of Boolean functions to be used in S-boxes. Further, Zhang and Zheng [22] identified related cryptographic measures called Global Avalanche Characteristics (GAC).

Let $\alpha \in \{0,1\}^n$ and f be an n-variable Boolean function. Define the autocorrelation value of f with respect to the vector α as

$$\Delta_f(\alpha) = \sum_{x\in\{0,1\}^n}(-1)^{f(x)\oplus f(x\oplus\alpha)},$$

and the absolute indicator

$$\Delta_f = \max_{\alpha\in\{0,1\}^n,\alpha\neq\bar{0}} |\Delta_f(\alpha)|.$$

A function is said to satisfy PC(k), if $\Delta_f(\alpha) = 0$ for $1 \le wt(\alpha) \le k$. Note that, for a bent function f on n variables, $\Delta_f(\alpha) = 0$ for all nonzero α, i.e., $\Delta_f = 0$.

Analysis of autocorrelation properties of correlation immune and resilient Boolean functions has gained substantial interest recently as evident from [20, 23, 11, 4]. In [11, 4], it has been identified that some well known construction of resilient Boolean functions are not good in terms of autocorrelation properties. Since the present construction is modification of bent functions which possess the best possible autocorrelation properties, we get very good autocorrelation properties of the 1-resilient functions. We present a bound on the Δ_f value of the 1-resilient functions and further achieve best known autocorrelation values for the cases $n = 8, 10, 12$.

2 The Distance

Initially we start with a simple technical result.

Proposition 2. $dBR_n(1) \ge 2^{\frac{n}{2}-1}$.

Proof. For a bent function b on n variables, $W_b(\omega) = \pm 2^{\frac{n}{2}}$. Hence the minimum distance from a bent function to balanced functions equals $2^{\frac{n}{2}-1}$. The 1-resilient functions are balanced by definition and hence the result. □

Now we present a restricted result. Let $b(x)$ be an n-variable bent function with $W_b(\omega) = +2^{\frac{n}{2}}$ for $wt(\omega) \leq 1$. We denote by $M_b(n,1)$ the minimum number of bits to be modified in the output column of $b(x)$ to construct an n variable 1-resilient function from $b(x)$.

Theorem 1. *Let $b(x)$ be an n-variable bent function with $W_b(\omega) = 2^{\frac{n}{2}}$ for $0 \leq wt(\omega) \leq 1$. Then*

$$M_b(n,1) \geq 2^{\frac{n}{2}-1} + 2\left\lceil \frac{(r+1)(2^{\frac{n}{2}-1} - \sum_{i=0}^{r}\binom{n}{i}) + \sum_{i=1}^{r} i\binom{n}{i}}{n-r-1} \right\rceil,$$

where r is the integer such that $\sum_{i=0}^{r}\binom{n}{i} \leq 2^{\frac{n}{2}-1} + 1 < \sum_{i=0}^{r+1}\binom{n}{i}$ is satisfied.

Proof. Let $S \subset \{0,1\}^n$ and $f(x)$ be an n-variable Boolean function obtained by modifying the $b(x)$ values for $x \in S$ and keeping the other bits unchanged. Then from Proposition 1, $W_f(\omega) = W_b(\omega) - 2W_b(\omega)|_S \; \forall \omega$, and in particular, $W_f(\omega) = 2^{\frac{n}{2}} - 2W_b(\omega)|_S$ for $0 \leq wt(\omega) \leq 1$.

It is known that, f is 1-resilient iff $W_f(\omega) = 0$ for $0 \leq wt(\omega) \leq 1$, i.e., iff $W_b(\omega)|_S = 2^{\frac{n}{2}-1}$ for $0 \leq wt(\omega) \leq 1$. Thus, our problem is to find a lower bound on $|S| = k$ with the constraint $W_b(\omega)|_S = 2^{\frac{n}{2}-1}$ for $0 \leq wt(\omega) \leq 1$.

Given $S = \{x^{i_1}, x^{i_2}, \ldots, x^{i_k}\} \subset \{0,1\}^n$, consider the matrices

$$\mathbf{S}^{k\times n} = (x^{i_1}, x^{i_2}, \ldots, x^{i_k})^T,\; b(\mathbf{S})^{k\times 1} = (b(x^{i_1}), b(x^{i_2}), \ldots, b(x^{i_k}))^T,$$

$$\text{and } (\mathbf{S} \oplus b(\mathbf{S}))^{k\times n} = (x^{i_1} \oplus b(x^{i_1}), x^{i_2} \oplus b(x^{i_2}), \ldots, x^{i_k} \oplus b(x^{i_k}))^T.$$

By A^T we mean transpose of a matrix A. Also by abuse of notation, $x^{i_j} \oplus b(x^{i_j})$ means the GF(2) addition (XOR) of the bit $b(x^{i_j})$ with each of the bits of x^{i_j}.

Now $W_b(\omega)|_S = 2^{\frac{n}{2}-1}$ for $0 \leq wt(\omega) \leq 1$ implies that there are exactly $\frac{k}{2} - 2^{\frac{n}{2}-2}$ many 1's in $b(\mathbf{S})$ and in each column of $\mathbf{S} \oplus b(\mathbf{S})$. Since all the rows of $\mathbf{S}$ are distinct and further $b(\mathbf{S})$ contains $\frac{k}{2} + 2^{\frac{n}{2}-2}$ many 0's, $\mathbf{S} \oplus b(\mathbf{S})$ should contain at least $\frac{k}{2} + 2^{\frac{n}{2}-2}$ distinct rows.

Consider that one such matrix $\mathbf{S} \oplus b(\mathbf{S})$ is formed. The number of 1's in the matrix is exactly $n \times (\frac{k}{2} - 2^{\frac{n}{2}-2})$ as each column contains exactly $\frac{k}{2} - 2^{\frac{n}{2}-2}$ many 1's and there are n columns. We know that there must be at least $\frac{k}{2} + 2^{\frac{n}{2}-2}$ many distinct rows. Thus the total number of 1's in these distinct rows must be $\leq n \times (\frac{k}{2} - 2^{\frac{n}{2}-2})$. Note that the minimum number of 1's in $\frac{k}{2} + 2^{\frac{n}{2}-2}$ many distinct rows is at least

$$\sum_{i=1}^{r} i\binom{n}{i} + (r+1)(\frac{k}{2} + 2^{\frac{n}{2}-2} - \sum_{i=0}^{r}\binom{n}{i})$$

(all the rows upto weight r and some of the rows with weight $r+1$). Hence,

$$\sum_{i=1}^{r} i\binom{n}{i} + (r+1)(\frac{k}{2} + 2^{\frac{n}{2}-2} - \sum_{i=0}^{r}\binom{n}{i}) \leq n \times (\frac{k}{2} - 2^{\frac{n}{2}-2}).$$

This gives,

$$k \geq 2\left\lceil \frac{(n+r+1)2^{\frac{n}{2}-2} + \sum_{i=1}^{r} i\binom{n}{i} - (r+1)\sum_{i=0}^{r}\binom{n}{i}}{n-r-1} \right\rceil.$$

Now we discuss how to choose this r. For this we need a easier lower bound on k which does not depend on r itself.

From Proposition 2, $k \geq 2^{\frac{n}{2}-1}$. We now show that $k \geq 2^{\frac{n}{2}-1} + 2$. This is because, to construct an 1-resilient function form bent function, the number of 1's in each column must be ≥ 1 (it cannot be 0 since then we will not be able to get distinct rows). As number of 1's in each column is $\frac{k}{2} - 2^{\frac{n}{2}-2}$, we get $\frac{k}{2} - 2^{\frac{n}{2}-2} \geq 1$, and hence $k \geq 2^{\frac{n}{2}-1} + 2$.

Since, $\frac{k}{2} + 2^{\frac{n}{2}-2}$ number of distinct rows has to be filled, we need to find the r such that $\sum_{i=0}^{r}\binom{n}{i} \leq \frac{k}{2} + 2^{\frac{n}{2}-2} < \sum_{i=0}^{r+1}\binom{n}{i}$. Putting the minimum value of k, i.e., $2^{\frac{n}{2}-1}+2$, we get r such that $\sum_{i=0}^{r}\binom{n}{i} \leq 2^{\frac{n}{2}-1} + 1 < \sum_{i=0}^{r+1}\binom{n}{i}$. □

As example, for $n = 8$, take $r = 1$ and $9 = \sum_{i=0}^{r}\binom{n}{i} \leq$ (the $=$ of $\leq$ is satisfied here) $2^{\frac{n}{2}-1} + 1 = 9 < \sum_{i=0}^{r+1}\binom{n}{i}$ is satisfied. For $n = 10$, take $r = 1$ and $11 = \sum_{i=0}^{r}\binom{n}{i} \leq$ (the $<$ of $\leq$ is satisfied here) $2^{\frac{n}{2}-1} + 1 = 17 < \sum_{i=0}^{r+1}\binom{n}{i}$ is satisfied.

Theorem 2. *Let $b(x)$ be any n-variable bent function. Then*

$$dBR_n(1) \geq 2^{\frac{n}{2}-1} + 2\left\lceil \frac{(r+1)(2^{\frac{n}{2}-1} - \sum_{i=0}^{r}\binom{n}{i}) + \sum_{i=1}^{r} i\binom{n}{i}}{n-r-1} \right\rceil,$$

where r is the integer such that $\sum_{i=0}^{r}\binom{n}{i} \leq 2^{\frac{n}{2}-1} + 1 < \sum_{i=0}^{r+1}\binom{n}{i}$ is satisfied.

Proof. Without loss of generality, assume that $W_b(\omega) = +2^{\frac{n}{2}}$ for $wt(\omega) = 0$. Let $G_1 = \{\omega | wt(\omega) = 1, W_b(\omega) = +2^{\frac{n}{2}}\}$ and $G_2 = \{\omega | wt(\omega) = 1, W_b(\omega) = -2^{\frac{n}{2}}\}$. Let $S \subset \{0,1\}^n$ and $f(x)$ be an n-variable Boolean function obtained by modifying the $b(x)$ values for $x \in S$ and keeping the other bits unchanged. Then from Proposition 1, $W_f(\omega) = W_b(\omega) - 2W_b(\omega)|_S \ \forall \omega$, and in particular, $W_f(\omega) = 2^{\frac{n}{2}} - 2W_b(\omega)|_S$ for $wt(\omega) = 0, \omega \in G_1$ and $W_f(\omega) = -2^{\frac{n}{2}} - 2W_b(\omega)|_S$ for $\omega \in G_2$.

Given f is 1-resilient, we need to find a lower bound on $|S| = k$ with the constraints $W_b(\omega)|_S = 2^{\frac{n}{2}-1}$ for $wt(\omega) = 0$ and $\omega \in G_1$ and $W_b(\omega)|_S = -2^{\frac{n}{2}-1}$ for $\omega \in G_2$.

Let $|G_1| = \lambda$. Using the same argument as in the proof of Theorem 1, our problem is to find a $k \times n$ binary matrix $\mathbf{S} \oplus b(\mathbf{S})$ with minimum number of rows k such that there are λ columns with exactly $\frac{k}{2} - 2^{\frac{n}{2}-2}$ many 1's in each column and exactly $\frac{k}{2} + 2^{\frac{n}{2}-2}$ many 1's in each of the remaining $n - \lambda$ columns. Further, there are at least $\frac{k}{2} + 2^{\frac{n}{2}-2}$ distinct rows.

Let $M_1^{k\times\lambda}$ (respectively $M_2^{k\times(n-\lambda)}$) be a binary matrix with exactly $\frac{k}{2} - 2^{\frac{n}{2}-2}$ (respectively $\frac{k}{2} + 2^{\frac{n}{2}-2}$) many 1's in each column. Let J be the $k \times (n-\lambda)$ matrix with all elements 1. Then the problem of "finding a binary matrix $(M_1 : M_2)$ with minimum number of rows k such that there are at least $\frac{k}{2} + 2^{\frac{n}{2}-2}$ distinct rows" is equivalent to "finding a binary matrix $(M_1 : J - M_2)$ with minimum number of rows k such that there are at least $\frac{k}{2} + 2^{\frac{n}{2}-2}$ distinct rows". Note that each column of $(M_1 : J - M_2)$ contains exactly $\frac{k}{2} - 2^{\frac{n}{2}-2}$ many 1's. Thus, the proof follows with the similar argument presented in Theorem 1. □

For $8 \leq n \leq 16$, it can be checked that $\sum_{i=0}^{1} \binom{n}{i} \leq 2^{\frac{n}{2}-1} + 1 < \sum_{i=0}^{1+1} \binom{n}{i}$ is satisfied. In these cases, the lower bound on k is attained for $r = 1$ itself. Thus we have the following result.

Corollary 1. *For even n, $8 \leq n \leq 16$, $dBR_n(1) \geq 2^{\frac{n}{2}-1} + 2\left\lceil \frac{2^{\frac{n}{2}}-n-2}{n-2} \right\rceil$.*

Assume that one can construct a bent function b on n variables such that $dBR_n(1)$ bits at the output column of b are changed to get an n-variable 1-resilient function f. It is clear that toggling of a single bit can reduce the nonlinearity at most by 1 and increase the maximum absolute value of the autocorrelation spectra (absolute indicator) by at most 4. Thus we have the following result.

Theorem 3. *$nl(f) \geq 2^{n-1} - 2^{\frac{n}{2}-1} - dBR_n(1)$ and $\Delta_f \leq 4\,dBR_n(1)$.*

Proof. This follows from $nl(f) \geq nl(b) - dBR_n(1)$ and $\Delta_f \leq \Delta_b + 4\,dBR_n(1)$, where b is a bent function. □

However, for the actual constructions of functions on 8, 10 and 12 variables, we will show that we get better nonlinearity and autocorrelation values than these bounds. For $n = 4, 6$, we refer the readers to Appendix A.

3 The 8-Variable 1-Resilient Functions

In the previous section we have presented a lower bound of the minimum distance between the bent and 1-resilient functions. However, it has not been discussed in Section 2 how exactly a construction is possible. Further to achieve the currently best known parameters (or even better than that, if possible) we may need to consider some other issues. In this section we consider the construction of an $(8, 1, 6, 116)$ function. Construction of this function was an important open question and the function has been first reported in [10] by interlinking combinatorial technique and computer search. Later this function has also been found by meta heuristic search (simulated annealing) in [5]. Further the function found in [5] has $\Delta_f = 24$, which is currently the best known value. We here follow the similar kind of technique used in [12]. In the course of discussion it will be clear that how our technique is an improvement over [12]. We present a generalized construction method of $(8, 1, 6, 116)$ functions by modifying Maiorana-McFarland type bent functions and in specific cases, these functions have the Δ_f value as low as 24, the best known one [5].

Construction 1. *Take a bent function $b(x)$ on 8 variables with the following properties :* **(1)** *$b(x) = 0$ for $wt(x) \leq 1$ and $b(x) = 1$ for $wt(x) = 8$,* **(2)** *$W_b(\omega) = 16$ for $wt(\omega) \leq 1$ and $W_b(\omega) = -16$ for $wt(\omega) = 8$. Define a set $S = \{x \in \{0,1\}^8 | wt(x) = 0, 1, 8\}$. Construct a function $f(x)$ as :*

$$f(x) = 1 \oplus b(x), \; if \; x \in S$$
$$= b(x), \quad otherwise.$$

From Corollary 1, we get that $dBR_8(1) \geq 10$ and we here choose exactly 10 positions and modify them. It is important to point out that we here start with bent functions with some specific properties. The reason for choosing such bent functions is to get an actual construction of 1-resilient function with very high nonlinearity.

Before presenting the theorem regarding the properties of f, let us enumerate the issues we improve here over the work presented in [12].

1. There is a gap in the proof of [12, Theorem 3]. Note the conditions imposed on the bent function b above. In the statement of [12, Theorem 3], only the conditions of item 1 has been considered and the conditions of the item 2 has not been considered as given in Construction 1. The conditions of item 2 has been implicitly assumed in the proof of [12, Theorem 3]. Fortunately, the bent function considered in [12, Section 3] satisfies the conditions of item 2. However, it should be noted that there exist bent functions which satisfy the conditions of item 1 and not all the conditions of item 2 and in that case the proof of [12, Theorem 3] does not go through.
2. The bent function chosen in [12, Section 3] uses the concept of perfect nonlinear functions and they presented one example function which satisfies the conditions of item 1 (and also conditions of item 2). However, it is not clear how a generalized construction of such bent functions can be achieved in that manner. It should also be noted that the example functions presented in [12] are basically Maiorana-McFarland type, even though they are designed in a different manner by using the concept of perfect nonlinear functions. We here identify a subclass of Maiorana-McFarland type bent functions which satisfy the conditions of both item 1 and 2. This gives a large class of $(8, 1, 6, 116)$ functions. In fact we show that there are more than $2^{46.297}$ many distinct (upto complementation) $(8, 1, 6, 116)$ functions f with $\Delta_f \leq 40$.
3. The proof of Theorem 4 below is much simpler than the proof of [12, Theorem 3] and it presents a clear picture of the Walsh spectra of the function f with respect to the spectra of the function b.

Theorem 4. *The function $f(x)$ as described in Construction 1 is an $(8, 1, 6, 116)$ function.*

Proof. Take $\omega \in \{0,1\}^8$ with $wt(\omega) = i$. Now

$$W_f(\omega) = \textstyle\sum_{x \in \{0,1\}^8} (-1)^{b(x) \oplus \omega \cdot x} - 2 \sum_{x \in S} (-1)^{b(x) \oplus \omega \cdot x} \text{ (from Proposition 1)}$$
$$= W_b(\omega) - 2\;(\;8 - 2wt(\omega) + 2(wt(\omega) \bmod 2)\;).$$

Now we explain how the last step is deduced. Note that $b(x) = 0$ when $wt(x) = 0$ and $b(x) = 1$, when $wt(x) = 8$. Thus,

$$\textstyle\sum_{x \in \{0,1\}^8 | wt(x) = 0,8} (-1)^{b(x) \oplus \omega \cdot x} = 0, \text{ when } wt(\omega) \text{ is even},$$
$$= 2, \text{ when } wt(\omega) \text{ is odd}.$$

Table 1. Relationship between Walsh spectra of f, g as described in Construction 1

$wt(\omega)$	0	1	2	3	4	5	6	7	8
$W_f(\omega) = W_b(\omega) +$	-16	-16	-8	-8	0	0	8	8	16

Moreover, $\sum_{x \in \{0,1\}^8 | wt(x)=1} (-1)^{b(x) \oplus \omega \cdot x} = 8 - 2wt(\omega)$, as

(i) $b(x) = 0$ when $wt(x) = 1$ and
(ii) $\omega \cdot x = 1$ at $wt(w)$ input points when $wt(x) = 1$.

Since $\sum_{x \in S} (-1)^{b(x) \oplus \omega \cdot x} = \sum_{x \in \{0,1\}^8 | wt(x)=0,8} (-1)^{b(x) \oplus \omega \cdot x}$ $+ \sum_{x \in \{0,1\}^8 | wt(x)=1} (-1)^{b(x) \oplus \omega \cdot x}$, we get,

$$W_f(\omega) = W_b(\omega) - 2\ (\ 8 - 2wt(\omega) + 2(wt(\omega) \bmod 2)\ ).$$

When $wt(\omega) \leq 1$, $W_f(\omega) = W_b(\omega) - 16 = 16 - 16 = 0$. Thus the function is 1-resilient.

Further, if $wt(\omega) = 8$, $W_f(\omega) = W_b(\omega) + 16 = -16 + 16 = 0$. For any other choice, i.e., for $2 \leq wt(\omega) \leq 7$, we have $|8 - 2wt(\omega) + 2(wt(\omega) \bmod 2)| \leq 4$ and hence, $|W_f(\omega)| \leq |W_b(\omega)| + 8 = 16 + 8 = 24$. Hence, $nl(f) = 2^{8-1} - \frac{24}{2} = 116$.

Since the function attains the maximum possible nonlinearity, the algebraic degree [1, 3] of the function must be $8 - 2 = 6$. □

Based on Table 1 and the previous discussion, we get related results with respect to (i) nonexistence of some 8-variable bent functions and (ii) some relationship between 8-variable bent functions and balanced Boolean functions with nonlinearity 118 (whose existence is not known till date). These results are placed in Appendix B.

3.1 A Subclass of Maiorana-McFarland Bent Functions

The original Maiorana-McFarland class of bent function is as follows [6]. Consider n-variable Boolean functions on (X, Y), where $X, Y \in \{0,1\}^{\frac{n}{2}}$ of the form $f(X, Y) = X \cdot \pi(Y) + g(Y)$ where π is a permutation on $\{0,1\}^{\frac{n}{2}}$ and g is any Boolean function on $\frac{n}{2}$ variables. The function f can be seen as concatenation of $2^{\frac{n}{2}}$ distinct (upto complementation) affine function on $\frac{n}{2}$ variables.

Once again we write what kind of bent function $b(x)$ on 8 variables we require.

1. $b(x) = 0$ for $wt(x) \leq 1$ and $b(x) = 1$ for $wt(x) = 8$,
2. $W_b(\omega) = 16$ for $wt(\omega) \leq 1$ and $W_b(\omega) = -16$ for $wt(\omega) = 8$.

In this case, $n = 8$, i.e., $\frac{n}{2} = 4$. We have to decide what permutations π on $\{0,1\}^4$ and what kind of functions g on $\{0,1\}^4$ we can take such that the conditions on b are satisfied. We present a set of conditions below, which taken all together, provides sufficient condition for construction of such functions. Before going into the conditions, let us fix the notation and ordering of input variables as $x = (x_1, x_2, x_3, x_4, x_5, x_6, x_7, x_8)$, $X = (X_1, X_2, X_3, X_4)$, and $Y = (Y_1, Y_2, Y_3, Y_4)$. Further we identify $X_1 = x_1, X_2 = x_2, X_3 = x_3, X_4 = x_4, Y_1 = x_5, Y_2 = x_6, Y_3 = x_7, Y_4 = x_8$.

1. First of all, the function b has the value 0 at the points $(0,0,0,0,0,0,0,0)$, $(1,0,0,0,0,0,0,0)$, $(0,1,0,0,0,0,0,0)$, $(0,0,1,0,0,0,0,0)$, $(0,0,0,1,0,0,0,0)$ and this condition is satisfied if we choose $\pi(0,0,0,0) = (0,0,0,0)$ and $g(0,0,0,0) = 0$.
2. Next we need function b should have value 0 at points $(0,0,0,0,1,0,0,0)$, $(0,0,0,0,0,1,0,0)$, $(0,0,0,0,0,0,1,0)$, $(0,0,0,0,0,0,0,1)$, and this condition is satisfied if we choose $g(Y) = 0$ for $wt(Y) = 1$.
3. We need b to be 1 when the input is $(1,1,1,1,1,1,1,1)$. Thus if $\pi(1,1,1,1)$ is a vector of odd weight then $g(1,1,1,1)$ need to be 0. otherwise if $\pi(1,1,1,1)$ is a vector of even weight then $g(1,1,1,1)$ has to be 1.
4. Since we have already decided that $\pi(0,0,0,0) = (0,0,0,0)$ and $g(0,0,0,0) = 0$, the $W_f(\omega)$ values for $\omega \in \{(0,0,0,0,1,0,0,0), (0,0,0,0,0,1,0,0), (0,0,0,0,0,0,1,0), (0,0,0,0,0,0,0,1)\}$ becomes $+2^{\frac{n}{2}} = 16$.
5. Further if $\pi(Y) \in \{(1,0,0,0), (0,1,0,0), (0,0,1,0), (0,0,0,1)\}$, then we take $g(Y) = 0$. This guarantees that $W_f(\omega)$ values for $\omega \in \{(1,0,0,0,0,0,0,0), (0,1,0,0,0,0,0,0), (0,0,1,0,0,0,0,0), (0,0,0,1,0,0,0,0)\}$ becomes $+2^{\frac{n}{2}} = 16$.
6. Lastly, if $\pi(Y) = (1,1,1,1)$, we have to fix $g(Y) = (wt(Y) + 1) \bmod 2$. This guarantees that $W_f(1,1,1,1,1,1,1,1) = -2^{\frac{n}{2}} = -16$.

Given a bent function from the Maiorana-McFarland class $f(X,Y) = X \cdot \pi(Y) + g(Y)$, the dual of such function f is $Y \cdot \pi^{-1}(X) + g(\pi^{-1}(X))$. It is interesting to check whether the above points can be replaced by more precise arguments using this idea.

Theorem 5. *Let $n = 8$, $x \in \{0,1\}^n$ and $X, Y \in \{0,1\}^{\frac{n}{2}}$. Let $b(x)$ be a Maiorana-McFarland type bent function $b(x) = b(X,Y) = X \cdot \pi(Y) + g(Y)$ where π is a permutation on $\{0,1\}^{\frac{n}{2}}$ and g is a Boolean function on $\frac{n}{2}$ variables with the following conditions.*

(1) *if $Y = (0,0,0,0)$, $\pi(Y) = Y$;*
(2) *if $wt(\pi(Y)) \leq 1$, or $wt(Y) \leq 1$, then $g(Y) = 0$;*
(3) *if $Y = (1,1,1,1)$, $g(Y) = (wt(\pi(Y)) + 1) \bmod 2$;*
(4) *if $wt(\pi(Y)) = 4$, $g(Y) = (wt(Y) + 1) \bmod 2$.*

Then **(1)** *$b(x) = 0$ for $wt(x) \leq 1$ and $b(x) = 1$ for $wt(x) = 8$,* **(2)** *$W_b(\omega) = 16$ for $wt(\omega) \leq 1$ and $W_b(\omega) = -16$ for $wt(\omega) = 8$.*

Further there are $\geq 2^{46.297}$ many distinct b's (upto complementation) satisfying these conditions and in turn there are $\geq 2^{46.297}$ many distinct (upto complementation) $(8,1,6,116)$ functions.

Proof. The proof of the properties of b is discussed above in detail. The count of such functions is arrived as follows. Note that there are $2^{\frac{n}{2}} = 16$ places for the permutation π.

Let there are i many Y's, $0 \leq i \leq 4$ such that $wt(\pi(Y)) = 1$ for $wt(Y) = 1$. There are 4 elements of weight 1 and 10 elements of weight 2 or 3. Thus the $\pi(Y)$'s for $wt(Y) = 1$ may be chosen in $\binom{4}{i}\binom{10}{4-i}$ ways. Note that $\pi(Y)$ can not be $(1,1,1,1)$ for $wt(Y) = 1$. Now there are two cases.

1. Consider that $\pi(1,1,1,1) = (1,1,1,1)$. Then the number of options is $\binom{4}{i} \cdot \binom{10}{4-i} \cdot 4! \cdot 10! \cdot 2^{6+i}$. This is because the 4 elements where $wt(Y) = 1$ can be permuted in 4! ways. The 4 elements where $wt(Y) = 2, 3$ can be permuted in 10! ways. The function $g(Y)$ is fixed when Y is $(0,0,0,0)$ (1 place, $g(Y) = 0$) or $wt(Y) = 1$ (4 places, $g(Y) = 0$) or $wt(\pi(Y)) = 1$ ($4 - i$ places, $g(Y) = 0$) or $wt(Y) = wt(\pi(Y)) = 4$ (1 place, $g(Y) = 1$). Thus $g(Y)$ is fixed in $10 - i$ places and we can put any choice from $\{0,1\}$ for $16 - (10 - i) = 6 + i$ places.
2. Consider that $\pi(1,1,1,1) \neq (1,1,1,1)$. Then the number of options is $\binom{4}{i} \cdot \binom{10}{4-i} \cdot 10 \cdot 4! \cdot 10! \cdot 2^{5+i}$. Choose one element of $wt(Y) \neq 4$ as $\pi(1,1,1,1)$. This can be done in 10 ways. The 4 elements where $wt(Y) = 1$ can be permuted in 4! ways. The 4 elements where $wt(Y) = 2, 3$ can be permuted in 10! ways. The function $g(Y)$ is fixed when Y is $(0,0,0,0)$ (1 place, $g(Y) = 0$) or $wt(Y) = 1$ (4 places, $g(Y) = 0$) or $wt(\pi(Y)) = 1$ ($4 - i$ places, $g(Y) = 0$) or $wt(Y) = 4$ (1 place, $g(Y) = 1$ if $wt(\pi(Y)) = 0$, else $g(Y) = 1$) or $wt(\pi(Y)) = 4$ (1 place, $g(Y) = (wt(Y) + 1) \bmod 2$). Thus $g(Y)$ is fixed in $11 - i$ places and we can put any choice from $\{0,1\}$ for $16 - (11 - i) = 5 + i$ places.

So the total number of options is $6 \sum_{i=0}^{4} \binom{4}{i} \cdot \binom{10}{4-i} \cdot 4! \cdot 10! \cdot 2^{6+i} = 6 \cdot 4! \cdot 10! \cdot 2^6 \sum_{i=0}^{4} \binom{4}{i} \cdot \binom{10}{4-i} \cdot 2^i \approx 2^{46.297492}$. □

Remark 1. Following Theorem 3, it is clear that for the function f as discussed in Theorem 4, $\Delta_f \leq 40$. Now we present the following specific case.

Consider $\pi(Y) = Y$ for all $Y \in \{0,1\}^4$, $g(Y) = 0$ for all $Y \in \{0,1\}^4 \setminus \{(1,1,1,1)\}$ and $g(Y) = 1$ for $Y = (1,1,1,1)$. Let $b(x) = b(X,Y) = X \cdot \pi(Y) + g(Y)$ and $f(x)$ is as given in Construction 1. Then f is an $(8,1,6,116)$ function with $\Delta_f = 24$.

Note that we get an $(8,1,6,116)$ function f with $\Delta_f = 24$ in this method which has earlier been found by simulated annealing and linear transformation in [5].

4 The 10-Variable 1-Resilient Functions

We here start with 10-variable bent functions. Theorem 1 and Theorem 2 do not directly provide the idea how the exact construction of an 1-resilient function from a bent function is possible. Let us now describe a method where we will be able to identify a subclass of 10-variable Maiorana-McFarland type bent functions for this purpose.

As described in Section 2, we need to modify at least $k = 22$ points (see Corollary 1). Now following Theorem 1 and Theorem 2, it is clear that we first need to select $\frac{k}{2} + 2^{\frac{n}{2}-2} = 19$ distinct points. Note that we can have 1 point of weight 0 and 10 points of weight 1. Thus we need to find out 8 more points from weight 2. Once these 19 points are selected, further there are 3 more points to be chosen.

$$
\mathbf{S} \oplus b(\mathbf{S}) = \begin{pmatrix}
x_{10} & x_9 & x_8 & x_7 & x_6 & x_5 & x_4 & x_3 & x_2 & x_1 \\
\hline
0 & 0 & 0 & 0 & 0 & 0 & 0 & 0 & 0 & 0 \\
0 & 0 & 0 & 0 & 0 & 0 & 0 & 0 & 0 & 1 \\
0 & 0 & 0 & 0 & 0 & 0 & 0 & 0 & 1 & 0 \\
0 & 0 & 0 & 0 & 0 & 0 & 0 & 1 & 0 & 0 \\
0 & 0 & 0 & 0 & 0 & 0 & 1 & 0 & 0 & 0 \\
0 & 0 & 0 & 0 & 0 & 1 & 0 & 0 & 0 & 0 \\
0 & 0 & 0 & 0 & 1 & 0 & 0 & 0 & 0 & 0 \\
0 & 0 & 0 & 1 & 0 & 0 & 0 & 0 & 0 & 0 \\
0 & 0 & 1 & 0 & 0 & 0 & 0 & 0 & 0 & 0 \\
0 & 1 & 0 & 0 & 0 & 0 & 0 & 0 & 0 & 0 \\
1 & 0 & 0 & 0 & 0 & 0 & 0 & 0 & 0 & 0 \\
0 & 0 & 0 & 0 & 0 & 0 & 0 & 1 & 1 & 0 \\
0 & 0 & 0 & 0 & 0 & 1 & 1 & 0 & 0 & 0 \\
0 & 0 & 0 & 0 & 0 & 1 & 0 & 0 & 0 & 1 \\
0 & 0 & 0 & 1 & 1 & 0 & 0 & 0 & 0 & 0 \\
0 & 0 & 1 & 1 & 0 & 0 & 0 & 0 & 0 & 0 \\
0 & 1 & 1 & 0 & 0 & 0 & 0 & 0 & 0 & 0 \\
1 & 1 & 0 & 0 & 0 & 0 & 0 & 0 & 0 & 0 \\
1 & 0 & 0 & 0 & 1 & 0 & 0 & 0 & 0 & 0 \\
\hline
0 & 0 & 0 & 0 & 0 & 0 & 0 & 0 & 0 & 0 \\
0 & 0 & 0 & 0 & 0 & 0 & 0 & 0 & 1 & 1 \\
0 & 0 & 0 & 0 & 0 & 0 & 1 & 1 & 0 & 0
\end{pmatrix}
$$

Now we refer to the $\mathbf{S} \oplus b(\mathbf{S})$ matrix given here. We present the first 19 points and after the horizontal line we show the next 3 points. Note that the choice of the all zero point and the points of weight 1 are clear from the discussion in Theorem 1. However, it is still to be sorted out how exactly the 8 points of weight 2 are chosen. We here do that by observation and choose the 8 points of weight 2 out of total $\binom{10}{2} = 45$ weight 2 points. The rest 3 points (one of weight 0 and other two of weight 2) are chosen properly to satisfy that weight of each column should be $\frac{k}{2} - 2^{\frac{n}{2}-2} = 3$. Now we need a bent function b on 10 variables with the property that $b(x) = 0$ when x is any of the first 19 points and $b(x) = 1$ when x is complement of any of the last 3 points. This means that the last three rows need to be complemented when they will be considered as input points in the function. Thus, we construct two sets S_1, S_2 as follows and then denote $S = S_1 \cup S_2$.
$S_1 = \{(0,0,0,0,0,0,0,0,0,0), (0,0,0,0,0,0,0,0,0,1), (0,0,0,0,0,0,0,0,1,0),$
$(0,0,0,0,0,0,0,1,0,0), (0,0,0,0,0,0,1,0,0,0), (0,0,0,0,0,1,0,0,0,0),$
$(0,0,0,0,1,0,0,0,0,0), (0,0,0,1,0,0,0,0,0,0), (0,0,1,0,0,0,0,0,0,0),$
$(0,1,0,0,0,0,0,0,0,0), (1,0,0,0,0,0,0,0,0,0), (0,0,0,0,0,0,0,1,1,0),$
$(0,0,0,0,0,1,1,0,0,0), (0,0,0,0,0,1,0,0,0,1), (0,0,0,1,1,0,0,0,0,0),$
$(0,0,1,1,0,0,0,0,0,0), (0,1,1,0,0,0,0,0,0,0), (1,1,0,0,0,0,0,0,0,0),$
$(1,0,0,0,1,0,0,0,0,0)\}$ and
$S_2 = \{(1,1,1,1,1,1,1,1,1,1), (1,1,1,1,1,1,1,1,0,0), (1,1,1,1,1,1,0,0,1,1)\}$.

Also consider
$S'_1 = \{(0,0,0,0,0,0,0,0,0,0), (0,0,0,0,0,0,0,0,0,1), (0,0,0,0,0,0,0,0,1,0),$
$(0,0,0,0,0,0,0,1,0,0), (0,0,0,0,0,0,1,0,0,0), (0,0,0,0,0,1,0,0,0,0),$
$(0,0,0,0,1,0,0,0,0,0), (0,0,0,1,0,0,0,0,0,0), (0,0,1,0,0,0,0,0,0,0),$
$(0,1,0,0,0,0,0,0,0,0), (1,0,0,0,0,0,0,0,0,0), (0,0,0,0,0,0,0,1,1,0),$
$(0,0,0,0,0,1,1,0,0,0), (0,0,0,0,0,1,0,0,0,1)\}$,
$S_3 = \{(0,0,0,0,0,0,0,1,0,1), (0,0,0,0,0,0,0,1,1,1), (0,0,0,0,0,0,1,0,0,1),$
$(0,0,0,0,0,0,1,0,1,0), (0,0,0,0,0,0,1,1,1,0), (0,0,0,0,0,1,0,0,1,1),$
$(0,0,0,0,1,1,1,0,0,1), (0,0,0,0,0,1,1,1,0,0), (0,0,0,0,0,1,1,1,1,1)\}$ and
$S_4 = \{(0,0,1,1,1,0,0,0,0,0), (0,1,1,1,0,0,0,0,0,0), (1,0,0,1,1,0,0,0,0,0),$
$(0,0,0,0,0,1,1,0,0,0), (1,1,0,0,1,0,0,0,0,0), (1,1,1,0,0,0,0,0,0,0),$
$(1,1,1,1,1,0,0,0,0,0)\}$. We will talk about these sets S'_1, S_3 and S_4 little later.

We now write the exact construction.

Construction 2. *We need a 10-variable bent function $b(x)$ with the following properties:*

1. *$b(x) = 0$ when $x \in S_1$ and $b(x) = 1$ when $x \in S_2$,*
2. *$W_b(\omega) = +32$ when $\omega \in S'_1 \cup S_3 \cup S_4$.*

The function $f(x)$ is as follows.

$$f(x) = 1 \oplus b(x), \text{ if } x \in S$$
$$= b(x), \quad \text{otherwise.}$$

From Theorem 1, it is clear that the function $f(x)$ is 1-resilient. Now we need to calculate the nonlinearity of f. In fact, we will prove that $nl(f) = 488$, the currently best known nonlinearity for 10-variable 1-resilient functions. By Proposition 1, $W_f(\omega) = W_b(\omega) - 2W_b(\omega)|_S$. Thus, it is important to analyse the values of $W_b(\omega)|_S$ for all $\omega \in \{0,1\}^{10}$. However, this can not be done in a nice way as it has been done in the 8-variable case in Theorem 4. So we use a computer program to calculate $W_b(\omega)|_S$ for all $\omega \in \{0,1\}^{10}$. Note that when $|W_b(\omega)|_S| \leq 8$, then at those points $|W_f(\omega)| \leq 48$. Thus, we have no restriction on the Walsh spectra of the bent function b at these points to get the nonlinearity 488 for f. However, we need to concentrate on the cases when $|W_b(\omega)|_S| \geq 12$. We have checked that this happens when $\omega \in S'_1 \cup S_3 \cup S_4$ and all these values are either $+12$ or $+16$. Thus as given in Construction 2, the Walsh spectra of the function b should be $+32$ at these points. Hence Construction 2 provides 10-variable 1-resilient functions having nonlinearity 488. Using similar technique as in Theorem 5, it is possible to get the count of such functions.

Note that we have not yet discussed the algebraic degree and autocorrelation properties of the functions. We now consider a specific case and check the algebraic degree and autocorrelation property.

Take $x = (x_1, x_2, x_3, x_4, x_5, x_6, x_7, x_8, x_9, x_{10})$, $X = (X_1, X_2, X_3, X_4, X_5)$, and $Y = (Y_1, Y_2, Y_3, Y_4, Y_5)$. Further we identify $X_1 = x_1, X_2 = x_2, X_3 = x_3, X_4 = x_4, X_5 = x_5, Y_1 = x_6, Y_2 = x_7, Y_3 = x_8, Y_4 = x_9, Y_5 = x_{10}$.

Consider a 10-variable Maiorana-McFarland type bent function

$$b(x) = b(X, Y) = X \cdot \pi(Y) + g(Y),$$

where π is a permutation on $\{0,1\}^5$ with $\pi(Y) = Y$ and g is a Boolean function on 5 variables which is a constant 0 function. It can be checked that this bent function satisfies the conditions required in Construction 2. Then we prepare f as given in Construction 2. We checked that nonlinearity of f is 488, algebraic degree is 8 and $\Delta_f = 48$. Now it is important to note the following two points.

1. The construction in [12, Theorem 4] required 26 points to be modified to get 1-resilient function from a bent function. We here need only 22 points to modify. Further, we have checked that the Δ_f value of the function constructed in [12] is 64. The function we construct here has $\Delta_f = 48$ and this is the best known value which is achieved for the first time here.
2. The $(10, 1, 8, 488)$ function was first constructed in [10] and we have checked that Δ_f value is 320 for that function. Thus our construction provides better parameter.

5 The 12-Variable Case

From Corollary 1, we find that $dBR_{12}(1) \geq 42$. However, it seems that it is not possible to construct an 1-resilient function by toggling 42 bits of a bent function. Instead we succeeded to construct a $(12, 1, 10, 2000)$ function f, with $\Delta_f = 120$ by toggling 44 points of a bent function. Thus taking $k = 44$, we have to first find $\frac{k}{2} + 2^{\frac{n}{2}-2} = 38$ distinct points. We select the all zero input point and the twelve input points each of weight one. Now there are $\binom{12}{2} = 66$ input points of weight two. Out of them we choose $38 - 13 = 25$ points by trial and error. These points are 2560, 2304, 2176, 2112, 1280, 1152, 1088, 640, 576, 320, 1536, 384, 40, 36, 34, 33, 20, 18, 17, 10, 9, 5, 24, 6, 2080 when written as decimal integers corresponding to 12-bit binary numbers. We need a bent function such that it will have output zero at these 38 input points. Next we take the six input points 4095, 3055, 3575, 3835, 3965, 4030. We need a bent function which provides output one at these six points. Now we present the bent function.

Take $x = (x_1, x_2, x_3, x_4, x_5, x_6, x_7, x_8, x_9, x_{10}, x_{11}, x_{12})$, $X = (X_1, X_2, X_3, X_4, X_5, X_6)$, and $Y = (Y_1, Y_2, Y_3, Y_4, Y_5, Y_6)$. Further we identify $X_1 = x_1, X_2 = x_2, X_3 = x_3, X_4 = x_4, X_5 = x_5, X_6 = x_6, Y_1 = x_7, Y_2 = x_8, Y_3 = x_9, Y_4 = x_{10}, Y_5 = x_{11}, Y_6 = x_{12}$. Consider a 12-variable Maiorana-McFarland type bent function $b(x) = b(X, Y) = X \cdot \pi(Y) + g(Y)$ where π is a permutation on $\{0,1\}^6$ with $\pi(Y) = Y$, except the cases $\pi(1,1,1,1,1,0) = (1,1,1,1,1,1)$ and $\pi(1,1,1,1,1,1) = (1,1,1,1,1,0)$. Here g is a Boolean function on 6 variables which is a constant 0 function.

The construction presented in [12] requires 54 points to be toggled and they could achieve a nonlinearity 1996. Thus our construction is clearly better. Further we get $\Delta_f = 120$ for the $(12, 1, 10, 2000)$ function that we construct here. This is the best known autocorrelation parameter which was not known earlier.

6 Conclusion

In this paper we present a lower bound on the minimum distance $dBR_n(1)$ between bent and 1-resilient functions on n variables, where n is even. We have also shown that it is possible to get 1-resilient functions by modifying exactly $dBR_n(1)$ many bits for $n = 4, 6, 8, 10$ which shows that the minimum distance is tight in these cases. For the case $n = 12$, we could not prove the bound is tight as we need to toggle at least 44 points of a bent function to get an 1-resilient function. The tightness of the bound for $n \geq 12$ remains an open question and to the best of our understanding, the bound is really not tight. The case for $n = 8$ could be nicely handled, but it starts to become complicated from $n = 10$ and requires some computer simulation.

A lot of open questions are still to be solved. First of all, a relatively hard question is to find out the minimum distance between bent and m-resilient functions on n variables, which we may denote as $dBR_n(m)$. It seems natural that $dBR_n(n-2) > dBR_n(n-3) > \ldots > dBR_n(1)$, though it needs a proof. Note that $(n-2)$-resilient functions on n variables are basically the affine functions, which are known to be at maximum distance from the bent function [14].

The functions we provide here possess currently best known parameters. The upper bound on nonlinearity of 1-resilient functions is $2^{n-1} - 2^{\frac{n}{2}-1} - 4$ for n even as described in [16]. The tightness of this bound [16] has been shown upto $n = 8$. For $n \geq 10$, there is no evidence of an 1-resilient function attaining that bound [16]. Our construction modifies $dBR_n(1) > 2^{\frac{n}{2}-1}$ many bits and it seems unlikely that modifying these many bits will result in a fall of nonlinearity only 4 for $n \geq 10$.

Acknowledgement

The authors like to thank the anonymous reviewer for important comments that improved the technical quality of the paper. The reviewer has kindly pointed out some technical problems in the submitted version of this draft. The authors also like to acknowledge Mr. Kishan Chand Gupta of Indian Statistical Institute, Kolkata, for detailed discussion on the proof of Theorem 1.

References

[1] C. Carlet. On the coset weight divisibility and nonlinearity of resilient and correlation immune functions. In *Sequences and Their Applications - SETA 2001*, Discrete Mathematics and Theoretical Computer Science, pages 131–144. Springer Verlag, 2001. 151

[2] C. Carlet. A larger Class of Cryptographic Boolean Functions via a Study of the Maiorana-McFarland Constructions. In *Advances in Cryptology - CRYPTO 2002*, number 2442 in Lecture Notes in Computer Science, pages 549–564. Springer Verlag, 2002. 143

[3] C. Carlet and P. Sarkar. Spectral domain analysis of correlation immune and resilient Boolean functions. *Finite Fields and Its Applications*, 8(1):120–130, January 2002. 143, 151

[4] P. Charpin and E. Pasalic. On Propagation Characteristics of Resilient Functions. In *SAC 2002*, number 2595 in Lecture Notes in Computer Science, pages 175–195. Springer-Verlag, 2003. 146

[5] J. Clark, J. Jacob, S. Stepney, S. Maitra and W. Millan. Evolving Boolean Functions Satisfying Multiple Criteria. In *INDOCRYPT 2002*, Volume 2551 in Lecture Notes in Computer Science, pages 246–259, Springer Verlag, 2002. 149, 153

[6] J. F. Dillon. Elementary Hadamard Difference sets. PhD Thesis, University of Maryland, 1974. 151

[7] H. Dobbertin. Construction of bent functions and balanced Boolean functions with high nonlinearity. In *Fast Software Encryption*, number 1008 in Lecture Notes in Computer Science, pages 61–74. Springer-Verlag, 1994. 143

[8] M. Fedorova and Y. V. Tarannikov. On the constructing of highly nonlinear resilient Boolean functions by means of special matrices. In *Progress in Cryptology - INDOCRYPT 2001*, number 2247 in Lecture Notes in Computer Science, pages 254–266. Springer Verlag, 2001. 143

[9] X. Guo-Zhen and J. Massey. A spectral characterization of correlation immune combining functions. *IEEE Transactions on Information Theory*, 34(3):569–571, May 1988. 145

[10] S. Maitra and E. Pasalic. Further constructions of resilient Boolean functions with very high nonlinearity. *IEEE Transactions on Information Theory*, 48(7):1825–1834, July 2002. 149, 156

[11] S. Maitra. Autocorrelation Properties of correlation immune Boolean functions. INDOCRYPT 2001, number 2247 Lecture Notes in Computer Science. Pages 242–253. Springer Verlag, December 2001. 146

[12] S. Maity and T. Johansson. Construction of Cryptographically Important Boolean Functions. In *INDOCRYPT 2002*, Volume 2551 in Lecture Notes in Computer Science, pages 234–245, Springer Verlag, 2002. 143, 144, 149, 150, 156

[13] B. Preneel, W. Van Leekwijck, L. Van Linden, R. Govaerts, and J. Vandewalle. Propagation characteristics of Boolean functions. In *Advances in Cryptology - EUROCRYPT'90*, Lecture Notes in Computer Science, pages 161–173. Springer-Verlag, 1991. 146

[14] O. S. Rothaus. On bent functions. *Journal of Combinatorial Theory, Series A*, 20:300–305, 1976. 145, 157

[15] P. Sarkar and S. Maitra. Construction of nonlinear Boolean functions with important cryptographic properties. In *Advances in Cryptology - EUROCRYPT 2000*, number 1807 in Lecture Notes in Computer Science, pages 485–506. Springer Verlag, 2000. 143, 145

[16] P. Sarkar and S. Maitra. Nonlinearity bounds and constructions of resilient Boolean functions. In *Advances in Cryptology - CRYPTO 2000*, number 1880 in Lecture Notes in Computer Science, pages 515–532. Springer Verlag, 2000. 143, 145, 157

[17] J. Seberry, X. M. Zhang, and Y. Zheng. Nonlinearly balanced Boolean functions and their propagation characteristics. In *Advances in Cryptology - CRYPTO'93*, pages 49–60. Springer-Verlag, 1994. 143

[18] Y. V. Tarannikov. On resilient Boolean functions with maximum possible nonlinearity. In *Progress in Cryptology - INDOCRYPT 2000*, number 1977 in Lecture Notes in Computer Science, pages 19–30. Springer Verlag, 2000. 143

[19] Y. V. Tarannikov. New constructions of resilient Boolean functions with maximal nonlinearity. In *Fast Software Encryption - FSE 2001*, to be published in Lecture Notes in Computer Science, pages 70–81 (in preproceedings). Springer Verlag, 2001. 143

[20] Y. V. Tarannikov, P. Korolev and A. Botev. Autocorrelation coefficients and correlation immunity of Boolean functions. In *ASIACRYPT 2001*, Lecture Notes in Computer Science. Springer Verlag, 2001. 146

[21] Y. Zheng and X. M. Zhang. Improved upper bound on the nonlinearity of high order correlation immune functions. In *Selected Areas in Cryptography - SAC 2000*, number 2012 in Lecture Notes in Computer Science, pages 264–274. Springer Verlag, 2000. 143

[22] X. M. Zhang and Y. Zheng. GAC - the criterion for global avalanche characteristics of cryptographic functions. *Journal of Universal Computer Science*, 1(5):316–333, 1995. 146

[23] Y. Zheng and X. M. Zhang. On relationships among propagation degree, nonlinearity and correlation immunity. In *Advances in Cryptology - ASIACRYPT'00*, Lecture Notes in Computer Science. Springer Verlag, 2000. 146

Appendix A

We are not interested in the case $n = 2$, since there is no nonlinear 2-variable 1-resilient functions.

We now consider the cases for $n = 4, 6$. Note that $r = 0$ for these two cases and then we arrive at $dBR_4(1) \geq 4$ and $dBR_6(1) \geq 6$. We have also checked that this bound is tight since we can construct 4-variable (respectively 6-variable) 1-resilient function by changing 4 (respectively 6) output points of 4-variable (respectively 6-variable) bent function.

For the 4-variable case, we have to take the rows of $\mathbf{S} \oplus b(\mathbf{S})$ as $\{0001, 0010, 0100, 1000\}$ due to the constraint that the number of 1's in each column has to be 1 and there are at least 3 distinct rows. Thus, take a bent function with truth table 0000011001010011 and toggle the function at the inputs

$$\{(0,0,0,1),(0,0,1,0),(1,0,0,0),(1,0,1,1)\}.$$

Then we get a $(4, 1, 2, 4)$ function with the truth table 0110011011000011.

For the 6-variable case, take a bent function with truth table 0000000001011010001111000110011001101001001100110101010100001111 and toggle the outputs at the input points $\{(0,0,0,0,0,1),(0,0,0,0,1,0),$ $(0,0,0,1,0,0),(0,0,1,0,0,0),(1,0,0,0,0,0),(1,0,1,1,1,1)\}$.
Then we get a $(6, 1, 4, 24)$ function with the truth table 0110100011011010001111000110011011101001001100100101010100001111.

Appendix B

Note that, in the Walsh spectra of a bent function on 8 variables, there are 120 values of +16 and 136 values of -16 or vice versa. It is known that even if that condition is satisfied for some Walsh spectra, the inverse Walsh transform may not produce a Boolean function. We here discuss that issue.

Lemma 1. *Consider a function $b(x)$ on 8 variables with the properties :*

1. $b(x) = 0$ *for* $wt(x) \leq 1$ *and* $b(x) = 1$ *for* $wt(x) = 8$,
2. $W_b(\omega) = 16$ *for* $wt(\omega) \leq 3$ *and* $W_b(\omega) = -16$ *for* $wt(\omega) \geq 6$.

This function can not be bent.

Proof. If such a function b is bent, then Table 1, we will get an 1-resilient function with nonlinearity 120. This is a contradiction. □

Corollary 2. *Consider a function $b(x)$ on 8 variables with the properties :*

1. $b(x) = 0$ *for* $wt(x) \leq 3$ *and* $b(x) = 1$ *for* $wt(x) \geq 6$,
2. $W_b(\omega) = 16$ *for* $wt(\omega) \leq 1$ *and* $W_b(\omega) = -16$ *for* $wt(\omega) = 8$.

Proof. The result follows from Lemma 1 and the duality property of bent functions. □

Next we present an important result related to the existence of balanced 8-variable function with nonlinearity 118.

Theorem 6. *Take a bent function $h(x)$ on 8 variables with the following properties :*

1. $h(x) = 0$ *for* $wt(x) \leq 1$ *and* $h(x) = 1$ *for* $wt(x) = 8$,
2. $W_h(\omega) = 16$ *for* $wt(\omega) \leq 2$ *and* $W_h(\omega) = -16$ *for* $wt(\omega) \geq 6$.

Define a set $T = \{x \in \{0,1\}^8 | wt(x) = 1\}$. *Construct a function $g(x)$ as :*

$$f(x) = 1 \oplus h(x), \text{ if } x \in T$$
$$= h(x), \quad \text{otherwise.}$$

Then g is a balanced 8-variable function with nonlinearity 118.

Proof. The proof is similar to the proof of Theorem 4. □

We have tried some heuristic search to find a bent function as mentioned in Theorem 6, but could not get any. Getting such a bent function or proving its nonexistence is an interesting open question.

Results on Rotation Symmetric Bent and Correlation Immune Boolean Functions

Pantelimon Stănică[1]*, Subhamoy Maitra[2], and John A. Clark[3]

[1] Mathematics Department, Auburn University Montgomery
Montgomery, AL 36124-4023, USA
pstanica@mail.aum.edu

[2] Applied Statistics Unit, Indian Statistical Institute
203, B T Road, Kolkata 700 108, INDIA
subho@isical.ac.in

[3] Department of Computer Science, University of York
York YO10 3EE, England
jac@cs.york.ac.uk

Abstract. Recent research shows that the class of Rotation Symmetric Boolean Functions (RSBFs), i.e., the class of Boolean functions that are invariant under circular translation of indices, is potentially rich in functions of cryptographic significance. Here we present new results regarding the Rotation Symmetric (rots) correlation immune (CI) and bent functions. We present important data structures for efficient search strategy of rots bent and CI functions. Further, we prove the nonexistence of homogeneous rots bent functions of degree ≥ 3 on a single cycle.

Keywords: Rotation Symmetric Boolean Function, Bent Functions, Balancedness, Nonlinearity, Autocorrelation, Correlation Immunity, Resiliency

1 Introduction

A variety of criteria for choosing Boolean functions with cryptographic applications (for secret key cryptosystems) have been identified. These are balancedness, nonlinearity, autocorrelation, correlation immunity, algebraic degree etc. The trade-offs among these criteria have received a lot of attention in Boolean function literature for a long time (see [7] and the references in this paper). The more criteria that have to be taken into account, the more difficult the problem is to obtain a Boolean function satisfying these properties.

It has been found recently that the class of RSBFs is extremely rich in terms of cryptographically significant Boolean functions. These functions have been analyzed in [4], where the authors studied the nonlinearity of these Boolean functions up to 9 variables and found encouraging results. This study has been

* This author is associated with the Institute of Mathematics "Simion Stoilow" of the Romanian Academy, Bucharest - Romania.

B. Roy and W. Meier (Eds.): FSE 2004, LNCS 3017, pp. 161–177, 2004.

extended in [15, 16] and important properties (further to [4]) of these functions up to 8 variables have been demonstrated. Also, the enumeration of RSBFs of specific degree has been discussed in [15, 16]. On the other hand, in [11], Pieprzyk and Qu studied these functions as components in the rounds of a hashing algorithm and research in this direction was later continued in [3].

The space of RSBFs is of size approximately $2^{\frac{2^n}{n}}$ for n-variable, which is of size n-th root of the total space 2^{2^n}. Thus any kind of search becomes comparatively easier and it has been shown in [15] that it is easy to get a 7-variable, 2-resilient RSBF with nonlinearity 56, which has earlier been considered as a function that is not easy to search for [10]. Moreover, these functions also possess the best known autocorrelation spectra. Thus it is important to present tools that can be used to efficiently search the space of RSBFs. We present important data structures, the matrices ${}_n\mathcal{A}$ and ${}_n\mathcal{B}$, that make this search and the study of bent functions more efficient.

Using these data structures, for the first time we could find 8-variable, 1-resilient, algebraic degree 6, nonlinearity 116, PC(1) functions with maximum absolute value in the autocorrelation spectra 32. Functions with such parameters have not been reported earlier. Moreover, interesting results are obtained for 9-variable correlation immune functions. The space for these functions in the Rotation Symmetric class is too large to execute exhaustive search. Hence we exploited simulated annealing technique to find these functions. The results found by simulated annealing are as follows. We could find 9-variable, 2-resilient, algebraic degree 6 and nonlinearity 240 functions and unbalanced 9-variable, 3rd order correlation immune, algebraic degree 5 and nonlinearity 240 functions. These functions have been posed as important open questions in [13, 14]. Note that the details of simulated annealing is not included in this paper and that has been published in [2].

In this paper, we also try to analyze the RSBFs class using combinatorial techniques in Section 3. We present enumerative results (based on constructive techniques) on balanced and correlation immune RSBFs. Further, we show that it is possible to transform a class of RSBFs to correlation immune functions depending on full rank of binary circulant matrices over $\mathbf{Z}_2$. In [15], it was observed that there is no homogeneous rots bent functions of degree ≥ 3 up to 10 variables. We here theoretically show the nonexistence of homogeneous rots bent functions of degree ≥ 3 on a single cycle for any (even) number of input variables ≥ 6.

2 Preliminaries

A Boolean function on n variables may be viewed as a mapping from $V_n = \{0,1\}^n$ into $\{0,1\}$. A Boolean function $f(x_1, \ldots, x_n)$ is also interpreted as the output column of its *truth table* f, i.e., a binary string of length 2^n, $f = [f(0,0,\cdots,0), f(1,0,\cdots,0), f(0,1,\cdots,0), \ldots, f(1,1,\cdots,1)]$.

The *Hamming distance* between S_1, S_2 is denoted by $d(S_1, S_2) = \#(S_1 \neq S_2)$. Also the *Hamming weight* or simply the weight of a binary string S is the number of ones in S. This is denoted by $wt(S)$. An n-variable function f is said to be *balanced* if its output column in the truth table contains equal number of 0's and 1's (i.e., $wt(f) = 2^{n-1}$).

Addition operator over $GF(2)$ is denoted by $\oplus$. An n-variable Boolean function $f(x_1, \ldots, x_n)$ can be considered to be a multivariate polynomial over $GF(2)$. This polynomial can be expressed as a sum of products representation of all distinct k-th order products ($0 \leq k \leq n$) of the variables. More precisely, $f(x_1, \ldots, x_n)$ can be written as

$$a_0 \oplus \bigoplus_{1 \leq i \leq n} a_i x_i \oplus \bigoplus_{1 \leq i < j \leq n} a_{ij} x_i x_j \oplus \ldots \oplus a_{12\ldots n} x_1 x_2 \ldots x_n,$$

where the coefficients $a_0, a_{ij}, \ldots, a_{12\ldots n} \in \{0, 1\}$. This representation of f is called the *algebraic normal form* (ANF) of f. The number of variables in the highest order product term with nonzero coefficient is called the *algebraic degree*, or simply the degree of f and denoted by $deg(f)$.

Take $0 \leq b \leq n$. An n-variable function is called *nondegenerate* on b variables if its ANF contains exactly b distinct input variables. A Boolean function is said to be *homogeneous* if its ANF contains terms of the same degree only.

Functions of degree at most one are called *affine* functions. An affine function with constant term equal to zero is called a *linear* function. The set of all n-variable affine (respectively linear) functions is denoted by $A(n)$ (respectively $L(n)$). The nonlinearity of an n-variable function f is

$$nl(f) = min_{g \in A(n)}(d(f, g)),$$

i.e., the distance from the set of all n-variable affine functions.

Let $x = (x_1, \ldots, x_n)$ and $\omega = (\omega_1, \ldots, \omega_n)$ both belonging to $\{0, 1\}^n$ and $x \cdot \omega = x_1\omega_1 \oplus \ldots \oplus x_n\omega_n$. Let $f(x)$ be a Boolean function on n variables. Then the *Walsh transform* of $f(x)$ is a real valued function over $\{0, 1\}^n$ which is defined as

$$W_f(\omega) = \sum_{x \in \{0,1\}^n} (-1)^{f(x) \oplus x \cdot \omega}.$$

In terms of Walsh spectra, the nonlinearity of f is given by

$$nl(f) = 2^{n-1} - \frac{1}{2} \max_{\omega \in \{0,1\}^n} |W_f(\omega)|.$$

In [5], an important characterization of correlation immune functions has been presented, which we use as the definition here. A function $f(x_1, \ldots, x_n)$ is m-th order correlation immune (respectively m-resilient) iff its Walsh transform satisfies

$$W_f(\omega) = 0, \text{ for } 1 \leq wt(\omega) \leq m \text{ (respectively } 0 \leq wt(\omega) \leq m).$$

As the notation used in [13, 14], by an (n, m, d, σ) function we denote an n-variable, m-resilient function with degree d and nonlinearity σ. Further by an

$[n, m, d, \sigma]$ function we denote an unbalanced n-variable, mth order correlation immune function with degree d and nonlinearity σ.

Propagation Characteristics (PC) and Strict Avalanche Criteria (SAC) [12] are important properties of Boolean functions to be used in S-boxes. Further, Zhang and Zheng [18] identified related cryptographic measures called Global Avalanche Characteristics (GAC).

Let $\alpha \in \{0,1\}^n$ and f be an n-variable Boolean function. Let us denote the autocorrelation value of the Boolean function f with respect to the vector α as

$$\Delta_f(\alpha) = \sum_{x \in \{0,1\}^n} (-1)^{f(x) \oplus f(x \oplus \alpha)},$$

and the absolute indicator

$$\Delta_f = \max_{\alpha \in \{0,1\}^n, \alpha \neq \bar{0}} |\Delta_f(\alpha)|.$$

A function is said to satisfy PC(k), if

$$\Delta_f(\alpha) = 0 \ for \ 1 \leq wt(\alpha) \leq k.$$

2.1 Rotation Symmetric Boolean Functions

Let $x_i \in \{0,1\}$ for $1 \leq i \leq n$. For $1 \leq k \leq n$, we define

$$\begin{aligned} \rho_n^k(x_i) &= x_{i+k}, \quad \text{if } i+k \leq n, \text{ and} \\ &= x_{i+k-n}, \text{ if } i+k > n. \end{aligned}$$

Let $(x_1, x_2, \ldots, x_{n-1}, x_n) \in V_n$. We can extend the definition of ρ_n^k on tuples and monomials as $\rho_n^k(x_1, x_2, \ldots, x_n) = (\rho_n^k(x_1), \rho_n^k(x_2), \ldots, \rho_n^k(x_n))$ and $\rho_n^k(x_{i_1} x_{i_2} \cdots) = \rho_n^k(x_{i_1}) \rho_n^k(x_{i_2}) \cdots$.

Definition 1. *A Boolean function f is called* Rotation Symmetric *if for each input $(x_1, \ldots, x_n) \in \{0,1\}^n$, $f(\rho_n^k(x_1, \ldots, x_n)) = f(x_1, \ldots, x_n)$ for $1 \leq k \leq n$.*

Following [15], let us denote

$$G_n(x_1, \ldots, x_n) = \{\rho_n^k(x_1, \ldots, x_n), \text{ for } 1 \leq k \leq n\}.$$

Note that $G_n(x_1, \ldots, x_n)$ generates a partition in the set V_n. Let g_n be the number of such partitions. Using Burnside's lemma, it can be shown (see also [15]) that the number of n-variable RSBFs is

$$2^{g_n}, \text{ where } g_n = \frac{1}{n} \sum_{k|n} \phi(k)\, 2^{\frac{n}{k}},$$

ϕ being Euler's *phi*–function. Further the following result has been proved regarding n-variable RSBFs of some specific degree. The number of

(i) degree w homogeneous functions is $2^{g_{n,w}} - 1$,

(ii) the number of degree w functions is $(2^{g_{n,w}} - 1)2^{\sum_{i=0}^{w-1} g_{n,i}}$ and

(iii) the number of functions with degree at most w is $2^{\sum_{i=0}^{w} g_{n,i}}$, where $g_{n,w}$ is defined as follows (see also [15]).

Consider $G_n(x_1, \ldots, x_n)$, where $wt(x_1, \ldots, x_n)$ is exactly w, and define $g_{n,w}$ as the number of partitions over the n bit binary strings of weight w (total number $\binom{n}{w}$), determined by G_n. Further, denote by $h_{n,w}$ the number of distinct sets $G_n(x_1, \ldots, x_n)$, where $wt(x_1, \ldots, x_n) = w$ and $|G_n(x_1, \ldots, x_n)| = n$, that is, the number of long cycles of weight w. It is easy to see that $h_{n,w} < g_{n,w}$. Write $k|m$, if k $(1 < k \leq m)$ is a proper divisor of m. The following results were obtained in [15].

(i) $g_{n,w} = \frac{1}{n}\binom{n}{w}$, if $gcd(n, w) = 1$. Also, $g_{n,0} = g_{n,n} = 1$.

(ii) $g_{n,w} = \frac{1}{n}\left(\binom{n}{w} - \sum_{k|gcd(n,w)} \frac{n}{k} \cdot h_{\frac{n}{k},\frac{w}{k}}\right) + \sum_{k|gcd(n,w)} h_{\frac{n}{k},\frac{w}{k}}$, if $w < n$.

Filiol and Fontaine [4] discussed the set of idempotent Boolean functions in an experimental setting. Let $\mathcal{B} = (b_1, \ldots, b_n)$ be a basis of F_2^n (which is identified with F_{2^n}). An *idempotent* f is a Boolean function on F_{2^n} that satisfies $f^2 = f$. Define the *Mattson-Solomon (MS) polynomial* by

$$MS_f(Z) = \sum_{j=0}^{2^n-2} A_j Z^{2^n-j-1}, \ where \ A_j = \sum_{i=0}^{2^n-1} f(\alpha^i)\alpha^{ij},$$

where α is a primitive element of F_{2^n}. Using the representation

$$f = \sum_{g \in F_{2^n}^*} f(g)(g)$$

(in the multiplicative algebra $F_2[F_{2^n}, \times]$), one gets that f is an idempotent iff $f(g) = f(g^2)$, $\forall\, g$; the coefficients of the MS polynomial belong to F_2; $A_j = A_k$ for all k in the 2-cyclotomic class of j $(\{j, 2j, \ldots, 2^{n-1}j\})$; the ANF of f á(using a normal basis $(\gamma, \gamma^2, \ldots, \gamma^{2^{n-1}})$ remains invariant under circular shift. This gives that the corpus of idempotents is the same as the class of Rotation Symmetric Boolean functions. For $n = 5, 7$, they found idempotents of highest nonlinearity (12, respectively 56) of degrees 2, 3 (for $n = 5$), and degrees $2, 3, 4, 5, 6$ (for $n = 7$). For $n = 6, 8$ they found all idempotents of highest nonlinearity (28, respectively 120), of degrees $2, 3$, respectively, $2, 3, 4$. They were not able to find all idempotent functions for $n = 8$, though. Finally, for $n = 9$, they found 1142395 functions (up to equivalence) with nonlinearity 240, some of which are balanced, of degrees $2, 3, 4, 5, 6, 7$.

3 Study on RSBFs

Motivated by [4, 15], in this section we will investigate the richness of the RSBFs class in terms of cryptographic properties and present some important data structures. The data structures will help in running the search algorithms very fast. In this direction we start with a few technical results. In the preliminaries, we have defined $G_n(x_1,\ldots,x_n) = \{\rho_n^k(x_1,\ldots,x_n),\ \text{for } 1 \le k \le n\}$. As example, for $n = 4$ we get the following partition of $\{0,1\}^n$:

$$\begin{aligned}
G_4(0,0,0,0) &= \{(0,0,0,0)\};\\
G_4(0,0,0,1) &= \{(0,0,0,1),(0,0,1,0),(0,1,0,0),(1,0,0,0)\};\\
G_4(0,0,1,1) &= \{(0,0,1,1),(0,1,1,0),(1,0,0,1),(1,1,0,0)\};\\
G_4(0,1,0,1) &= \{(0,1,0,1),\ (1,0,1,0)\};\\
G_4(0,1,1,1) &= \{(0,1,1,1),(1,0,1,1),(1,1,0,1),(1,1,1,0)\};\\
G_4(1,1,1,1) &= \{(1,1,1,1)\}.
\end{aligned}$$

Note that there are g_n such partitions, and the lexicographically first element of each part is considered as the representative element. We denote these representative elements by $\Lambda_{n,i}$ where i varies from 0 to $g_n - 1$ and representative elements are again arranged lexicographically. That is, in the above example, $\Lambda_{4,0} = (0,0,0,0), \Lambda_{4,1} = (1,0,0,0), \Lambda_{4,2} = (1,1,0,0), \Lambda_{4,3} = (1,0,1,0), \Lambda_{4,4} = (1,1,1,0), \Lambda_{4,5} = (1,1,1,1)$.

By RSTT (rotation symmetric truth table) we mean the g_n-bit long binary string

$$[f(\Lambda_{n,0}), f(\Lambda_{n,1}), \ldots, f(\Lambda_{n,g_n-1})],$$

which gives the complete information of the function f when it is rots.

Lemma 1. *Let $u, v \in \{0,1\}^n$ and $u \neq v$ with $u \in G_n(v)$. Let f be an n-variable RSBF. Then $W_f(u) = W_f(v)$, which implies that the Walsh spectra of f can be at most g_n valued.*

Proof. First we show that for $a \in \{0,1\}$,

$$\sum_{x \in G_n(\Lambda_{n,i})} (-1)^{a \oplus x \cdot u} = \sum_{x \in G_n(\Lambda_{n,i})} (-1)^{a \oplus x \cdot v}.$$

Since $u \in G_n(v)$, $u = \rho_n^k(v)$ for some k. Now $\sum_{x \in G_n(\Lambda_{n,i})} (-1)^{a \oplus x \cdot u}$ $= \sum_{x \in G_n(\Lambda_{n,i})} (-1)^{a \oplus \rho_n^k(x) \cdot \rho_n^k(u)} = \sum_{y \in G_n(\Lambda_{n,i})} (-1)^{a \oplus y \cdot v}$ (take $y = \rho_n^k(x)$) $=$ $\sum_{x \in G_n(\Lambda_{n,i})} (-1)^{a \oplus x \cdot v}$.

$W_f(u) = \sum_{x \in \{0,1\}^n} (-1)^{f(x) \oplus x \cdot u} = \sum_{i=0}^{g_n - 1} \sum_{x \in G_n(\Lambda_{n,i})} (-1)^{f(x) \oplus x \cdot u}$
$=$ (using the above result) $\sum_{i=0}^{g_n - 1} \sum_{x \in G_n(\Lambda_{n,i})} (-1)^{f(x) \oplus x \cdot v} = W_f(v)$. □

Note that, Lemma 1 helps to run any heuristic in a much smaller space. Now we define an important matrix called ${}_n\mathcal{A}$ with respect to the set of n-variable RSBFs as:

$${}_n\mathcal{A}_{i,j} = \sum_{x \in G_n(\Lambda_{n,i})} (-1)^{x \cdot \Lambda_{n,j}}.$$

See the following example corresponding to 6-variable case.

i	0	1	2	3	4	5	6
$\Lambda_{6,i}$	000000	000001	000011	000101	000111	001001	001011
i	7	8	9	10	11	12	13
$\Lambda_{6,i}$	001101	001111	010101	010111	011011	011111	111111

$$
{}_6\mathcal{A} = \begin{bmatrix}
1 & 1 & 1 & 1 & 1 & 1 & 1 & 1 & 1 & 1 & 1 & 1 & 1 & 1 \\
6 & 4 & 2 & 2 & 0 & 2 & 0 & 0 & -2 & 0 & -2 & -2 & -4 & -6 \\
6 & 2 & 2 & -2 & 2 & -2 & -2 & -2 & 2 & -6 & -2 & -2 & 2 & 6 \\
6 & 2 & -2 & 2 & -2 & -2 & -2 & -2 & -2 & 6 & 2 & -2 & 2 & 6 \\
6 & 0 & 2 & -2 & 0 & -6 & 0 & 0 & -2 & 0 & 2 & 6 & 0 & -6 \\
3 & 1 & -1 & -1 & -3 & 3 & 1 & 1 & -1 & -3 & -1 & 3 & 1 & 3 \\
6 & 0 & -2 & -2 & 0 & 2 & -4 & 4 & 2 & 0 & 2 & -2 & 0 & -6 \\
6 & 0 & -2 & -2 & 0 & 2 & 4 & -4 & 2 & 0 & 2 & -2 & 0 & -6 \\
6 & -2 & 2 & -2 & -2 & -2 & 2 & 2 & 2 & 6 & -2 & -2 & -2 & 6 \\
2 & 0 & -2 & 2 & 0 & -2 & 0 & 0 & 2 & 0 & -2 & 2 & 0 & -2 \\
6 & -2 & -2 & 2 & 2 & -2 & 2 & 2 & -2 & -6 & 2 & -2 & -2 & 6 \\
3 & -1 & -1 & -1 & 3 & 3 & -1 & -1 & -1 & 3 & -1 & 3 & -1 & 3 \\
6 & -4 & 2 & 2 & 0 & 2 & 0 & 0 & -2 & 0 & -2 & -2 & 4 & -6 \\
1 & -1 & 1 & 1 & -1 & 1 & -1 & -1 & 1 & -1 & 1 & 1 & -1 & 1
\end{bmatrix}
$$

This matrix is of size $g_n \times g_n$. Now for an n-variable RSBF f, we have $W_f(\omega) = \sum_{x \in \{0,1\}^n} (-1)^{f(x) \oplus x \cdot \omega} = \sum_{i=0}^{g_n - 1} \sum_{x \in G_n(\Lambda_{n,i})} (-1)^{f(x) \oplus x \cdot \omega}$
$= \sum_{i=0}^{g_n-1} (-1)^{f(\Lambda_{n,i})} \sum_{x \in G_n(\Lambda_{n,i})} (-1)^{x \cdot \Lambda_{n,j}}$, if $\omega \in G_n(\Lambda_{n,j})$. Thus, $W_f(\Lambda_{n,j}) = \sum_{i=0}^{g_n-1} (-1)^{f(\Lambda_{n,i})} {}_n\mathcal{A}_{i,j}$. To summarize, we have the following result.

Proposition 1. $W_f(\Lambda_{n,j}) = \sum_{i=0}^{g_n-1} (-1)^{f(\Lambda_{n,i})} {}_n\mathcal{A}_{i,j}$.

In terms of Proposition 1, we can list the following.

Lemma 2. *Let f be an n-variable RSBF.*

1. $nl(f) = 2^{n-1} - \frac{1}{2} \max_{\Lambda_{n,j}, 0 \le j < g_n} |\sum_{i=0}^{g_n-1} (-1)^{f(\Lambda_{n,i})} {}_n\mathcal{A}_{i,j}|$.
2. *f is balanced iff*
$$\sum_{i=0}^{g_n-1} (-1)^{f(\Lambda_{n,i})} {}_n\mathcal{A}_{i,0} = 0.$$
3. *f is m-th order CI (respectively m-resilient) iff*
$$\sum_{i=0}^{g_n-1} (-1)^{f(\Lambda_{n,i})} {}_n\mathcal{A}_{i,j} = 0, \ \textit{for } 1 \ (\textit{respectively } 0) \le wt(\Lambda_{n,j}) \le m.$$
4. *f is bent iff*
$$\sum_{i=0}^{g_n-1} (-1)^{f(\Lambda_{n,i})} {}_n\mathcal{A}_{i,j} = \pm 2^{\frac{n}{2}} \ \textit{for } 0 \le j \le g_n - 1.$$

Theorem 1. *The number of balanced RSBFs is exactly* $2\pi_n$*, where* π_n *is the number of partitions of the space* V_n *as* $V_n = A_n \cup B_n$*, where* A_n *and* B_n *have the same cardinal, and both include complete cycles of any length. Further, if* $n = p$ *is an odd prime, then the number of balanced RSBFs is* $2 \cdot \binom{(2^p-2)/p}{(2^{p-1}-1)/p}$*; if* $n = p^a$ $(a > 1)$ *and* p *is an odd prime, then the number of balanced RSBFs is* $2 \cdot \pi_n$*, with* $\pi_n \geq \binom{x}{x/2} \cdot \prod_{i=1}^{a} \binom{x_i}{x_i/2}$*, where* $x_i = \dfrac{2^{p^i} - 2^{p^{i-1}}}{p^i}$*, and*

$$x = p^{-a}\left(2^{p^a} + \sum_{j=1}^{a} \phi(p^j) \cdot 2^{p^{a-j}}\right) - \sum_{i=1}^{a} x_i - 2.$$

Proof. Using item 2 of Lemma 2, to determine balanced RSBFs, it suffices to find the RSBFs satisfying $\sum_{i=0}^{g_n-1}(-1)^{f(\Lambda_{n,i})}{}_n\mathcal{A}_{i,0} = 0$. According to the definition ${}_n\mathcal{A}_{i,0} = \sum_{i=0}^{g_n-1}(-1)^{x\cdot\Lambda_{n,0}} = \#G_n(\Lambda_{n,i})$. Since the values of $(-1)^{f(\Lambda_{n,i})}$ are either ± 1, and f is constant on $G_n(v)$ for any v, we get the first claim.

If $n = p$ is prime, the number of long cycles is $h_p = \frac{2^p-2}{p}$ and the number of short cycles is 2 (the trivial ones) (see Subsection 2.1). Therefore, to partition $V_n = A_n \cup B_n$ (with A_n, B_n having the same number of elements), we need to place a short cycle in each of A_n, B_n, and the rest of $p \cdot h_p$ elements must be placed half in A_n and half in B_n (keeping together cycles). That can be done in $\binom{(2^p-2)/p}{(2^{p-1}-1)/p}$ ways. The second claim is proved.

If $n = p^a$ $(a > 1)$, the number of short cycles of length p^i (for any $i = 1, \ldots, a-1$) is $x_i = (2^{p^i} - 2^{p^{i-1}})/p^i$ (see Subsection 2.1). For each i, we can put half of the cycles in A_n, and half in B_n. The same can be done with the long cycles. Since the number of long cycles is x, the result is proved. □

For example, consider the case for 4-variable balanced RSBFs. We have

$$V_4 = G_4(\Lambda_{4,0}) \cup G_4(\Lambda_{4,1}) \cup G_4(\Lambda_{4,2}) \cup G_4(\Lambda_{4,3}) \cup G_4(\Lambda_{4,4}) \cup G_4(\Lambda_{4,5}).$$

Now consider

$$W_4 = G_4(\Lambda_{4,0}) \cup G_4(\Lambda_{4,3}) \cup G_4(\Lambda_{4,5}).$$

Hence

$$V_4 = W_4 \cup G_4(\Lambda_{4,1}) \cup G_4(\Lambda_{4,2}) \cup G_4(\Lambda_{4,4}).$$

Therefore, a balanced RSBF must be 1 at the output corresponding to any two of $W_4, G_4(\Lambda_{4,1}), G_4(\Lambda_{4,2}), G_4(\Lambda_{4,4})$. Hence $\pi_4 = 3$ and there are 6 balanced RSBFs on 4-variables. The reason we do not exhaust all possibilities in the second part of the previous theorem is because we can get a different partition of V_n, satisfying the requirements, by placing more short cycles in A_n (or B_n) as long as one ends up with the same number of elements in A_n, B_n.

Note that we have defined $\rho_n^k(x_{i_1}x_{i_2}\cdots) = \rho_n^k(x_{i_1})\rho_n^k(x_{i_2})\cdots$ in Subsection 2.1. By abuse of notation let us denote

$$G_n(x_{i_1}x_{i_2}\ldots x_{i_l}) = \{\rho_n^k(x_{i_1}x_{i_2}\ldots x_{i_l}), \text{ for } 1 \leq k \leq n\}.$$

We select the representative element of $G_n(x_{i_1}x_{i_2}\ldots x_{i_l})$ as the lexicographically first element. As example, the representative element of $\{x_1x_2x_3,\ x_2x_3x_4,\ x_3x_4x_1,\ x_4x_1x_2\}$ is $x_1x_2x_3$. Note that it is also clear that the term x_1 will always exist in the lexicographically first element (the representative element).

We now define the *short algebraic normal form* (SANF) of an RSBF. An RSBF $f(x_1,\ldots,x_n)$ can be written as

$$a_0 + a_1x_1 + \sum a_{1j}x_1x_j + \ldots + a_{12\ldots n}x_1x_2\ldots x_n,$$

where the coefficients $a_0, a_1, a_{1j}, \ldots, a_{12\ldots n} \in \{0,1\}$, and the existence of a representative term $x_1x_{i_2}\ldots x_{i_l}$ implies the existence of all the terms from the set $G_n(x_1x_{i_2}\ldots x_{i_l})$ in the ANF. This representation of f is called the *short algebraic normal form* (SANF) of f. Note that the number of terms in each summation ($\sum$) corresponding to same degree terms depends on the number of short and long cycles.

As example consider the ANF of a 4-variable RSBF $x_1 + x_2 + x_3 + x_4 + x_1x_2x_3 + x_2x_3x_4 + x_3x_4x_1 + x_4x_1x_2$. Its SANF is $x_1 + x_1x_2x_3$.

We can easily identify a monomial $x_{i_1}x_{i_2}\cdots x_{i_k}$ as a binary string of length n where the positions $i_1, i_2, \cdots, i_k$ contain '1' and the rest of the positions contain '0'. By abuse of notation we associate the n-bit patterns with monomials. It is clear that all the monomials in $G_n(\Lambda_{n,i})$ will either be present in the ANF or all of them will be absent if the Boolean function is rotation symmetric. Let us define another matrix ${}_n\mathcal{B}$ as

$${}_n\mathcal{B}_{i,j} = \bigoplus_{e \in G_n(\Lambda_{n,j})} e|_{\Lambda_{n,i}}.$$

That is, we take an RSBF (say h) with all the monomials coming from a single Rotation Symmetric group (say represented by $\Lambda_{n,j}$). Then we check what is the value of h at the representative input points $\Lambda_{n,i}$ and put that in the location ${}_n\mathcal{B}_{i,j}$ which contains either 0 or 1. Given ${}_n\mathcal{B}_{i,j}$ and the SANF of an RSBF, one can directly get the RSTT of the RSBF. The example for ${}_6\mathcal{B}$ is as follows.

$${}_6\mathcal{B} = \begin{bmatrix}
1&0&0&0&0&0&0&0&0&0&0&0&0&0\\
1&1&0&0&0&0&0&0&0&0&0&0&0&0\\
1&0&1&0&0&0&0&0&0&0&0&0&0&0\\
1&0&0&1&0&0&0&0&0&0&0&0&0&0\\
1&1&0&1&1&0&0&0&0&0&0&0&0&0\\
1&0&0&0&0&1&0&0&0&0&0&0&0&0\\
1&1&1&1&0&1&1&0&0&0&0&0&0&0\\
1&1&1&1&0&1&0&1&0&0&0&0&0&0\\
1&0&1&0&0&1&1&1&1&0&0&0&0&0\\
1&1&0&1&0&0&0&0&0&1&0&0&0&0\\
1&0&0&1&1&1&1&1&0&1&1&0&0&0\\
1&0&0&0&0&0&0&0&0&0&0&1&0&0\\
1&1&0&0&1&0&1&1&0&1&0&1&1&0\\
1&0&0&0&0&1&0&0&0&0&0&1&0&1
\end{bmatrix}$$

Note that the matrices ${}_n\mathcal{A}$, ${}_n\mathcal{B}$ help to perform the search much faster than the naive Boolean function implementation.

3.1 Correlation Immune (CI) and Resilient RSBFs

We start our discussion with construction of 1st order CI RSBFs. Note that the second column of the matrix ${}_n\mathcal{A}$ is instrumental in the analysis of first order CI functions. From item 3 of Lemma 2 we get that f is 1st order CI if $\sum_{i=0}^{g_n-1}(-1)^{f(\Lambda_{n,i})}\,{}_n\mathcal{A}_{i,j} = 0$ for $wt(\Lambda_{n,j}) = 1$, i.e., when $j = 1$, i.e., $\Lambda_{n,j} = \Lambda_{n,1}$. Note that ${}_n\mathcal{A}_{i,1} = \frac{(n-2wt(\Lambda_{n,i}))}{k}$ for cycles of length $\frac{n}{k}$, where $k\,(1 \le k \le n)$ is a divisor of n. See the second column of ${}_6\mathcal{A}$ as example. Thus we have the following result.

Theorem 2. *An n-variable Rotation Symmetric Boolean function f is 1st order CI iff $\sum_{i=0}^{g_n-1}(-1)^{f(\Lambda_{n,i})}\frac{(n-2wt(\Lambda_{n,i}))}{k_i} = 0$, where $|G_n(\Lambda_{n,i})| = \frac{n}{k_i}$.*

Based on this we present the following enumerative result when n is prime.

Corollary 1. *There are at least $2\prod_{w=1}^{\frac{n-1}{2}}\sum_{k=0}^{g_{n,w}}\binom{g_{n,w}}{k}^2$ many 1st order CI RSBFs on n variables, where n is an odd prime. In this case, $g_{n,w} = \frac{\binom{n}{w}}{n}$.*

Proof. For n prime we know that $g_n = \frac{2^n-2}{n} + 2$. There are $\frac{2^n-2}{n}$ full cycles and two trivial short cycles (all zero and all one). Thus it is clear that ${}_n\mathcal{A}_{i,1} = n - 2wt(\Lambda_{n,i})$ for $1 \le i \le g_n - 2$ and ${}_n\mathcal{A}_{0,1} = 1$, ${}_n\mathcal{A}_{g_n-1,1} = -1$. Note that, ${}_n\mathcal{A}_{i_1,1} = -{}_n\mathcal{A}_{i_2,1}$, when $wt(\Lambda_{n,i_1}) = n - wt(\Lambda_{n,i_2})$. Now consider an assignment of 0 or 1 value at output corresponding to the $g_{n,w}$ classes where $wt(\Lambda_{n,i_1}) = w$. We have to put the same number of 0's and 1's corresponding to the $g_{n,n-w}$ classes where $wt(\Lambda_{n,i_2}) = n - w$. The two trivial cycles should also have the same value at the output, either both zero or both 1. This satisfies the condition that $\sum_{i=0}^{g_n-1}(-1)^{f(\Lambda_{n,i})}\,{}_n\mathcal{A}_{i,1} = 0$, i.e., $W_f(\Lambda_{n,1}) = 0$, i.e., f is 1st order CI. Hence the number of possible options is $2 \times \prod_{w=1}^{\frac{n-1}{2}}(\sum_{k=0}^{g_{n,w}}\binom{g_{n,w}}{k}\cdot\binom{g_{n,w}}{k})$. □

Note that similar strategy can be exploited for higher order correlation immune or resilient RSBFs. However, in those cases, the analysis will be more involved.

3.2 A Large Subclass of RSBFs that Are Transformable to 1st Order CI Functions

We first investigate the independence of the vectors of a full cycle, i.e., the vectors in $G_n(\Lambda_{n,i})$ when $|G_n(\Lambda_{n,i})| = n$.

Lemma 3. *Consider the elements of $G_n(\Lambda_{n,i})$ for some i, where $|G_n(\Lambda_{n,i})| = n$. Let $\Lambda_{n,i} = (a_1, a_2, \ldots, a_n)$ of weight w and the positions of 1's in $\Lambda_{n,i}$ be $s_1 = 1, s_2, \ldots, s_w$. The vectors in $G_n(\Lambda_{n,i})$ are linearly dependent (over $\mathbf{Z}_2$) iff there is an n-th root of unity μ such that $1 + \mu^{s_2} \cdots + \mu^{s_w} = 0$, over $\mathbf{Z}_2$.*

Proof. The set $\{(a_1, a_2, \ldots, a_n), (a_n, a_1, \ldots, a_{n-1}), \ldots\}$ is linear dependent over $\mathbf{Z}_2$ if and only if the matrix

$$\mathrm{circ}(a_1, a_2, \ldots, a_n) = \begin{bmatrix} a_1 & a_2 \; a_3 \ldots & a_n \\ a_n & a_1 \; a_2 \ldots & a_{n-1} \\ a_{n-1} & a_n \; a_1 \ldots & a_{n-2} \\ \vdots & & \vdots \\ a_2 & a_3 \; a_4 \ldots & a_1 \end{bmatrix}$$

has zero determinant over $\mathbf{Z}_2$. We observe that the matrix is circular and it is known that the determinant of a circular matrix is given by

$$\det(\mathrm{circ}(a_1, a_2, \ldots, a_n)) = \prod_{\mu} \left(a_1 + a_2\mu + a_3\mu^2 + \cdots + a_n\mu^{n-1}\right),$$

where the product runs over all the n number of n-th roots of unity. Since a_i's are 1 in the positions described by s_j's and 0 elsewhere, we get that

$$\det(\mathrm{circ}(a_1, a_2, \ldots, a_n)) = \prod_{\mu} (1 + \mu^{s_2} + \cdots + \mu^{s_w}),$$

which is zero if and only if one of the factors is zero, that is, iff there exists an n-th root of unity such that $1 + \mu^{s_2} \cdots + \mu^{s_w} = 0$ (over $\mathbf{Z}_2$). □

Corollary 2. *Take n arbitrary. If $wt(\Lambda_{n,i})$ is even, then the full cycle generated by $\Lambda_{n,i}$ is dependent.*

Proof. We have $\det(\mathrm{circ}(\Lambda_{n,i})) = \prod_{\mu} (1 + \mu^{s_2} + \cdots + \mu^{s_w}) = 0$, since $1 + \mu^{s_2} + \cdots + \mu^{s_w} = 0$ (in $\mathbf{Z}_2$), for $\mu = 1$ (which is an n-th root of unity, for any n). □

Now we present some examples. Take the cycle generated by $(1, 1, 0, 0)$ in V_4. The circular determinant is $\det(\mathrm{circ}(1, 1, 0, 0)) = \prod_{\mu} (1 + \mu) = 0$, since $\mu = -1$ is one of the 4-roots of unity. Another example is the cycle generated over V_6 by $(1, 1, 1, 0, 1, 0)$. We have $\det(\mathrm{circ}(1, 1, 1, 0, 1, 0)) = \prod_{\mu} (1 + \mu + \mu^2 + \mu^4) = 0$, since $\mu = 1$ (a 6-root of unity) satisfies $1 + \mu + \mu^2 + \mu^4 = 0$ over $\mathbf{Z}_2$. On the other hand, the full cycle generated in V_6 by $(1, 1, 0, 0, 1, 0)$ is linearly independent.

Corollary 3. *Let n be a positive integer, and p be the least odd prime occurring in the factorization of n. Take $\Lambda_{n,i}$ (a generator of a full cycle), of odd weight w and $s_w \leq p - 2$. Then the full cycle generated by $\Lambda_{n,i}$ is independent.*

Proof. As before, under the above conditions, if we have dependence, then there is an n-th root of unity μ, such that $P(\mu) = 0$, where $P(x) = x^{s_w} + \cdots + x^{s_2} + 1$. Since w is odd, $\mu \neq \pm 1$. There exists $k \mid n$ such that μ is a primitive k-th root of unity. Therefore, the cyclotomic polynomial $\Phi_k(x)$ divides $P(x)$ over $\mathbf{Z}_2$ (see [6], Ch. 2 & 3). If $k < p$, then it must be that k is a power of 2, say 2^l (since k is a divisor of n, and p is the least odd prime dividing n). But that is impossible, since then $\Phi_k(x)$ will divide $x^{2^l} - 1$, so (over $\mathbf{Z}_2$) $1 = \mu^{2^l} = \mu$. Therefore, $k \geq p$.

Assume $k = 2^l p_1^{\alpha_1} \cdots p_r^{\alpha_r}$. If $r \geq 1$, then $p_i \geq p$, so $\phi(k) \geq \phi(p_i) = p_i - 1 \geq p - 1$. But the degree of $P(x)$ is at most $p-2$ and that of $\Phi_k(x)$ is greater than or equal to $p-1$. That is a contradiction. If $r = 0$, then $k = 2^l > p$, and the previous case's argument applies. □

Corollary 3 is the best we can get in that direction, as we see taking the cycles in V_{14} generated by $(1,1,1,1,1,0,0,0,\ldots)$ and $(1,1,1,1,1,1,1,0,\ldots)$. Now, the prime 7 is the least odd prime dividing $n = 14$. The weight w and s_w of the first generator is 5 and the cycle is independent; the weight w and s_w of the second generator is 7 and the cycle is dependent.

With the background of Lemma 3, Corollary 2 and Corollary 3 we present the following result.

Theorem 3. *Let f be an n-variable RSBF with $W_f(\Lambda_{n,j}) = 0$ for some j such that $G_n(\Lambda_{n,j})$ contains n independent vectors. Then the function f can be transformed to a 1st order correlation immune function g which may or may not be RSBF. Further if f is balanced, i.e., $W_f(\overline{0}) = 0$, then g is 1-resilient.*

Proof. Given the set of n independent vectors, at which the values of the Walsh spectra are 0, it is possible to apply linear transformation on the function f to get a function g which is 1st order correlation immune (using the methods of [8]). Note that, after the linear transformation, the Rotation Symmetric property of g is not guaranteed. □

Theorem 3 presents a simple method to get 1st order CI or 1-resilient functions easily from RSBFs satisfying some conditions. Moreover, the combinatorially interesting point is that the conditions are related to full rank of binary circulant matrices over $\mathbf{Z_2}$ and n-th roots of unity as described in Lemma 3.

3.3 Search for Important Functions

Recall the notation $g_{n,w}$ in Subsection 2.1. It is clear that for an RSBF, the $\binom{n}{w}$ many monomials of degree w are partitioned into $g_{n,w}$ many groups and the monomials of each group are either present or absent together. Now the search technique works as follows.

1. Choose a candidate RSBF (say f) represented by its SANF.
2. Use ${}_n\mathcal{B}$ to get the RSTT of f from the SANF.
3. Use ${}_n\mathcal{A}$ and the RSTT of f to analyze the Walsh Spectra of f.

Let us now consider the (8, 1, 6, 116) functions. These functions are of lot of interest as evident from [7, 1, 9]. Note that so far there was no evidence of $(8,1,6,116)$ functions with PC(1) property. We here show that there are such functions in the RSBFs class. We consider $f(\overline{0}) = 0$, and there can not be any term of degree 7, 8 in the ANF. Thus we need to take any combination from $\sum_{i=1}^{5} g_{8,i}$ groups and at least one group from $g_{8,6}$ groups. This search space is of size $2^{\sum_{i=1}^{5} g_{8,i}}(2^{g_{8,6}} - 1)$. Note that $g_{8,1} = 1$, $g_{8,2} = g_{8,6} = 4$, $g_{8,3} = g_{8,5} =$

$7, g_{8,4} = 10$. Thus we need to search a space of size $2^{29}(2^4 - 1) \approx 2^{33}$ and the search needed little more than a day on a Pentium 1.6 GHz computer with 256 MB RAM using Linux 7.2 operating system. We searched the complete space and found 10272 such functions. The Δ_f (autocorrelation values) of the functions are 32 (2176 many), 40 (1024 many), 48 (128 many), 64 (6688 many) and 128 (256 many). Next we searched the set of these 10272 functions for the propagation property. There are 2672 such functions. The Δ_f (autocorrelation values) of the functions are 32 (384 many), 40 (256 many), 64 (1936 many) and 128 (96 many). Thus we have the following theorem.

Theorem 4. *There are* 10272 *many* $(8, 1, 6, 116)$ *RSBFs* f *with* $f(\overline{0}) = 0$. *Among them we have* 2672 *many* $(8, 1, 6, 116)$ *RSBFs which are also* PC(1) *and out of them* 384 *many functions have* Δ_f *value as low as* 32.

The following one is the truth table (in Hex) of an $(8, 1, 6, 116)$, PC(1) RSBF with $\Delta_f = 32$.

```
0055 6267 7d59 2d7a 3be6 32c3 4da2 3bcc
0f8b fd3c 5a49 b05a 31f6 c94c 5e9a e4a0
```

Next we concentrate on 9-variable functions. As we discuss, it will be clear that even if the search space is reduced, it is not possible to go for an exhaustive search. Thus we attempted heuristic search using simulated annealing. Note that the details of simulated annealing is not included in this paper and that has been published in [2].

Let us consider the (9, 2, 6, 240) functions with $f(\overline{0}) = 0$. There can not be any term of degree 7, 8, 9. Thus we need to take any combination from $\sum_{i=1}^{5} g_{9,i}$ groups and at least one group from $g_{9,6}$ groups. Now $g_{9,1} = 1$, $g_{9,2} = 4$, $g_{9,3} = g_{9,6} = 10$, $g_{9,4} = g_{9,5} = 14$. Thus the search space is of size $2^{\sum_{i=1}^{5} g_{9,i}}(2^{g_{9,6}} - 1) = 2^{43}(2^{10} - 1) \approx 2^{53}$. With the current computational facility this search would be extremely time consuming. Hence we attempted heuristic search in this case and succeeded to get such functions. Note that this function was posed as an important open question in [13, 14]. The best possible functions that have been achieved earlier [13] are $(9, 2, 6, 232)$ and $(9, 2, 5, 240)$, i.e., the first one has smaller nonlinearity (than the upper bound 240) when the algebraic degree was maximum and the second one has smaller algebraic degree (maximum upper bound 6) when the nonlinearity was maximum.

Next we consider the (9, 3, 5, 240) functions with $f(\overline{0}) = 0$. There can not be any term of degree 6, 7, 8, 9. Thus we need to take any combination from $\sum_{i=1}^{4} g_{9,i}$ groups and at least one group from $g_{9,5}$ groups. Thus the search space is of size $2^{\sum_{i=1}^{4} g_{9,i}}(2^{g_{9,5}} - 1) = 2^{29}(2^{14} - 1) \approx 2^{43}$. Though this search space is not extremely large, with our current implementation it is expected to take almost 3 years to complete the search on a single Pentium 1.6 GHz computer with 256 MB RAM using Linux 7.2 operating system. Hence we attempted heuristic search, but could not succeed. Instead we could achieve unbalanced $[9, 3, 5, 240]$ functions, which were also not known earlier.

4 Rotation Symmetric Bent Functions

Let us now discuss a sieving strategy for rots bent functions. Given the matrix ${}_n\mathcal{A}$, a rots bent function needs to satisfy item 5 of Lemma 2. Thus the idea is to get the RSTT of the function which can be seen as a column of g_n elements. Now one needs to calculate $\sum_{i=0}^{g_n-1}(-1)^{f(\Lambda_{n,i})}\,{}_n\mathcal{A}_{i,j}$ and check whether this is equal to $\pm 2^{\frac{n}{2}}$ for $0 \leq j \leq g_n - 1$. The first time it fails for some j, we terminate checking that function and go for the next. This gives a very good performance for search strategies.

At the time of the search we can consider that $b(\overline{0}) = 0$ and the function is free from linear terms. Moreover, for a bent function, the maximum possible algebraic degree is $\frac{n}{2}$. Here the matrix ${}_n\mathcal{B}$ comes into play. We need to consider only those columns of ${}_n\mathcal{B}$ where $2 \leq wt(\Lambda_{n,j}) \leq \frac{n}{2}$. Then we choose all the linear combinations of those columns and then search for the bent functions. Thus the algorithm needs to check $2^{\sum_{i=2}^{\frac{n}{2}} g_{n,i}} - 1$ combinations as we ignore the all zero combination. Note that in this case once we get any $\sum_{i=0}^{g_n-1}(-1)^{f(\Lambda_{n,i})}\,{}_n\mathcal{A}_{i,j}$ not equal to $\pm 2^{\frac{n}{2}}$ for $0 \leq j \leq g_n - 1$, then we need not check the function further for bentness and check the next function. Thus the process of sieving is much faster.

Filiol and Fontaine [4] counted all the bent functions b on 8-variables where $b(\overline{0}) = 0$ and b is free from linear terms. There are 3776 such functions and in total $3776 \times 4 = 15104$ many. With the matrices ${}_6\mathcal{A}, {}_6\mathcal{B}$, and using our sieving method we need just one minute on a Pentium 1.6 GHz computer with 256 MB RAM using Linux 7.2 operating system. The number of functions to be checked is $2^{\sum_{i=2}^{4} g_{n,i}} - 1 = 2^{21} - 1$ for $n = 8$.

Note that, $g_{10} = 108$ and $g_{n,2} = 5$, $g_{n,3} = 12$, $g_{n,4} = 22$, $g_{n,5} = 26$. Thus the search required is $2^{65} - 1$ and with the current computational facility, it is not possible to exhaust this set easily. That is the reason some kind of heuristic search is required in this case and we found enough number of bent functions in each attempt using simulated annealing. We can also increase the speed of the algorithm by noting that there can not be any single cycle rots bent function of degree ≥ 3. In [15] it has been observed that up to 10-variables, there is no rotation symmetric homogeneous bent function with degree > 2 and it has been conjectured that it is true for any even n. Our result on single cycle rots bent functions provides a partial answer to that.

We have already denoted $V_n = \{0,1\}^n$. For a Boolean function $f : V_{2n} \to V_1$, let k_i $(i = 1, \ldots, 4)$ be the number of input bits 1 (i.e., x with $f(x) = 1$) in each of the quarters of f. If S is a bit string, by $(S)_u$ or S_u we shall mean the string obtained by concatenation of u copies of S. The concatenation of two strings u, v will be denoted by uv or $u|v$. Further, $\bar{h}$ is the complement of h, and for fixed integer d, $\hat{h}$ is equal to h (bit string in V_s) with the last 2^{s-d} bits of its truth table complemented. Let $A = 0,0,1,1$; $B = 0,1,0,1$; $C = 0,1,1,0$; $D = 0,0,0,0$; $U = 1,0,0,0$; $V = 0,0,0,1$; $X = 0,1,0,0$; $Y = 0,0,1,0$. The following result was a central proposition in [17].

Proposition 2. *Let $f : V_{2n} \to V_1$ be a bent Boolean function (not necessarily homogeneous) and the corresponding k_i $(i = 1, 2, 3, 4)$. Then (i) three of k_i's are equal and one is different, and (ii) $\min(k_1, k_2, k_3, k_4) \geq 2^{2n-3} - 2^{n-1}$.*

The following lemma (Lemma 11 of [3]) turns out to be quite useful. It gives the truth table of every monomial of arbitrary degree.

Lemma 4 ([3]). *The truth table of any monomial $x_{i_1} \cdots x_{i_s}$ of degree s is*

$$\begin{aligned}
&\left(D_{2^{n-i_1-2}} \cdots \left(D_{2^{n-i_s-2}} \bar{D}_{2^{n-i_s-2}}\right)_{2^{i_s-i_{s-1}-1}}\right)_{2^{i_1-1}},\\
&\textit{if } 1 \leq i_1 < \cdots < i_s \leq n-2,\\
&\left(D_{2^{n-i_1-2}} \cdots \left(D_{2^{n-i_{s-1}-2}} M_{2^{n-i_{s-1}-2}}\right)_{2^{i_{s-1}-i_{s-2}-1}}\right)_{2^{i_1-1}}, \qquad (1)\\
&\textit{where } M = A \textit{ or } B \textit{ if } i_s = n-1, \textit{ respectively } i_{s-1} < n-1 \textit{ and } i_s = n,\\
&\left(D_{2^{n-i_1-2}} \cdots \left(D_{2^{n-i_{s-2}-2}} V_{2^{n-i_{s-2}-2}}\right)_{2^{i_{s-2}-i_{s-3}-1}}\right)_{2^{i_1-1}},\\
&\textit{if } i_{s-1} = n-1 \textit{ and } i_s = n.
\end{aligned}$$

Theorem 5. *There are no homogeneous RSBFs with a single full cycle of degree $d \geq 3$ on V_n $(n \geq 6$ even) that are bent.*

Proof. Any full one-cycle RSBF is affinely equivalent to an RSBF f generated by $x_1 x_2 \ldots x_d$. We show now that the first quarter in the truth table of f has weight strictly less than $2^{2n-3} - 2^{n-1}$, thus contradicting Proposition 2. Therefore, f it is not bent.

An immediate application of Lemma 4 gives that, for $i \leq n-d-2$, the truth table of $x_i x_{i+1} \ldots x_{i+d} = \left(D_{2^{n-i-2}} \cdots (D_{2^{n-i-d-2}} \bar{D}_{2^{n-i-d-2}})\right)_{2^{i-1}}$, $x_{n-d} \cdots x_{n-2} x_{n-1} = (D_{2^{d-2}} \cdots (DA))_{2^{n-d-1}}$, and $x_{n-d+1} \cdots x_{n-1} x_n = (D_{2^{d-3}} \cdots (DV))_{2^{n-d}}$, therefore the first quarter of the truth table of f is given by the first quarter of

$$\begin{aligned}
&\sum_{i=1}^{n-d-2} \left(D_{2^{n-i-2}} \cdots (D_{2^{n-i-d-2}} \bar{D}_{2^{n-i-d-2}})\right)_{2^{i-1}} + (D_{2^{d-1}-1} A)_{2^{n-d-1}}\\
&+ (D_{2^{d-2}-1} V)_{2^{n-d}}\\
&= \sum_{i=1}^{n-d-2} \left(D_{2^{n-i-1}-2^{n-i-d-2}} \bar{D}_{2^{n-i-d-2}}\right)_{2^{i-1}} + (D_{2^{d-2}-1} V D_{2^{d-2}-1} Y)_{2^{n-d-1}} \qquad (2)
\end{aligned}$$

To see that it is so, observe that the only terms missing are $x_1 x_{n-d+2} \cdots x_{n-1} x_n + \cdots$. But all these contain $x_1 \cdots x_{n-1} x_n$. Therefore, in all the missing terms, $i_1 = 1, i_{s-1} = n-1, i_s = n$, so the last case of Lemma 4 implies that they all have 0 in the first quarter of their truth table, so all these terms do not contribute anything to the weight of the first quarter of f.

For easy writing, denote the first quarter in the truth table of f (on V_n) by h_d^{n-2}. Let $n = d+2$ and consider h_d^d. Since the first quarter of the truth table of f (on V_n), that is h_d^d, is obtained by taking the last two variables $x_{d+1} = x_{d+2} = 0$, and since the degree is d, it follows easily that h_d^d is nonzero only for $x_1 = x_2 = \cdots = x_d = 1$, that is, $h_d^d = D_{2^{d-2}-1} V$. Inductively on s, by using the displayed relation (2), we obtain the recurrence $h_d^s = h_d^{s-1}\hat{h}_d^{s-1}$ (write the displayed relation (2) for $s-1$ and s, and look at how the first quarter of that expression for s changes from the expression for $s-1$; this is why we needed the definition for $\hat{h}$, to explain that change). As example, let $d=3$, and f be the RSBF generated by $x_1x_2x_3$. Write $fq(f)$ for the first quarter of f. If $n=5$, then the RSTT of $fq(f) = 00000001 = DV$; if $n=6$, then $fq(f) = DV(\widehat{DV}) = DVDY$; if $n=7$, then $fq(f) = DVDY(\widehat{DVDY}) = DVDY\ DVD\overline{Y}$.

When d is fixed we shall write h_d^s as h^s. Using the recurrence and Maple (a trademark of *Waterloo Maple*) we obtained easily that the sequence of weights of h^n for the first few values of n, say $d \le n \le d+10$ is

n	d	$d+1$	$d+2$	$d+3$	$d+4$	$d+5$	$d+6$	$d+7$	$d+8$	$d+9$	$d+10$
$wt(h_d^n)$	1	2	6	14	32	72	156	336	712	1496	3120

(3)

Fixing d, and using the recurrence $h^s = h^{s-1}\hat{h}^{s-1}$, we get

$$h^s = h^{s-1}h^{s-2}h^{s-3}\bar{\hat{h}}^{s-4}\hat{\hat{h}}^{s-4} \quad \text{and} \quad \hat{h}^s = h^{s-1}h^{s-2}\bar{\hat{h}}^{s-3}h^{s-4}\hat{h}^{s-4}.$$

Therefore, denoting by w^s the weight of h^s, and by $\hat{w}^s$ the weight of $\hat{h}^s$, we arrive at the identities $\hat{w}^s = 2w^{s-1} + 2w^{s-2} - w^s + 2^{s-2}$, and $w^s = w^{s-1} + \hat{w}^{s-1}$. We deduce ($s \ge 6$)

$$wt(h^s) = 2\left(wt(h^{s-2}) + wt(h^{s-3})\right) + 2^{s-3}. \tag{4}$$

Next we want to prove that $wt(h_d^s) < wt(h_3^s) < 2^{s-1} - 2^{\lfloor \frac{s+2}{2} \rfloor}$, $s \ge 5, d > 3$. From these inequalities we derive the theorem. The first inequality on weights follows easily from the recursive definition of h_d^s. The second inequality will be proved by induction. If $s=5$, then $wt(h_d^5) = 6 < 2^4 - 2^3 = 8$; if $s=6$, then $wt(h_d^6) = 14 < 2^5 - 2^4 = 16$; if $s=7$, then $wt(h_d^7) = 32 < 2^6 - 2^4 = 48$. They are certainly true. Assume the inequality true for all values from 5 to $n-1$. Now, for dimension n, $wt(h_d^{n-2}) = 2\left(wt(h^{n-4}) + wt(h^{n-5})\right) + 2^{n-5} \le 2\left(2^{n-5} - 2^{\lfloor \frac{n-2}{2} \rfloor} + 2^{n-6} - 2^{\lfloor \frac{n-3}{2} \rfloor}\right) + 2^{n-5} = 2^{n-3} - 2^{\lfloor \frac{n-2}{2} \rfloor + 1} - 2^{\lfloor \frac{n-3}{2} \rfloor + 1} < 2^{n-3} - 2^{\lfloor \frac{n}{2} \rfloor}$, since $2^{\lfloor \frac{n-2}{2} \rfloor + 1} + 2^{\lfloor \frac{n-3}{2} \rfloor + 1} > 2^{\lfloor \frac{n}{2} \rfloor}$. □

References

[1] J. Clark, J. Jacob, S. Stepney, S. Maitra and W. Millan. Evolving Boolean Functions Satisfying Multiple Criteria. In *INDOCRYPT 2002*, Volume 2551 in Lecture Notes in Computer Science, pages 246–259, Springer Verlag, 2002. 172

[2] J. Clark, J. Jacob, S. Maitra and P. Stanica. Almost Boolean Functions: The Design of Boolean Functions by Spectral Inversion. In *CEC 2003, the 2003 Congress on Evolutionary Computation*, Volume 3 in the proceedings, page 2173–2180, IEEE Press, December 8–12, 2003, Canberra, Australia. 162, 173

[3] T.W. Cusick and P. Stănică. Fast Evaluation, Weights and Nonlinearity of Rotation-Symmetric Functions. *Discrete Mathematics*, pages 289-301, vol 258, no 1-3, 2002. 162, 175

[4] E. Filiol and C. Fontaine. Highly nonlinear balanced Boolean functions with a good correlation-immunity. In *Advances in Cryptology - EUROCRYPT'98*. Springer-Verlag, 1998. 161, 162, 165, 174

[5] X. Guo-Zhen and J. Massey. A spectral characterization of correlation immune combining functions. *IEEE Transactions on Information Theory*, 34(3):569–571, May 1988. 163

[6] R. Lidl and H. Niederreiter. Introduction to finite fields and their applications. Cambridge University Press, 1994. 171

[7] S. Maitra and E. Pasalic. Further constructions of resilient Boolean functions with very high nonlinearity. *IEEE Transactions on Information Theory*, 48(7):1825–1834, July 2002. 161, 172

[8] S. Maitra and P. Sarkar. Cryptographically significant Boolean functions with five-valued walsh spectra. *Theoretical Computer Science*, Volume 276, Number 1–2, pages 133-146, 2002. 172

[9] S. Maity and T. Johansson. Construction of Cryptographically Important Boolean Functions. In *INDOCRYPT 2002*, Volume 2551 in Lecture Notes in Computer Science, pages 234–245, Springer Verlag, 2002. 172

[10] E. Pasalic, S. Maitra, T. Johansson and P. Sarkar. New constructions of resilient and correlation immune Boolean functions achieving upper bound on nonlinearity. In *Workshop on Coding and Cryptography - WCC 2001*, Paris, January 8–12, 2001. Electronic Notes in Discrete Mathematics, Volume 6, Elsevier Science, 2001. 162

[11] J. Pieprzyk and C. X. Qu. Fast Hashing and Rotation-Symmetric Functions. *Journal of Universal Computer Science*, pages 20-31, vol 5, no 1 (1999). 162

[12] B. Preneel, W. Van Leekwijck, L. Van Linden, R. Govaerts, and J. Vandewalle. Propagation characteristics of Boolean functions. In *Advances in Cryptology - EUROCRYPT'90*, Lecture Notes in Computer Science, pages 161–173. Springer-Verlag, 1991. 164

[13] P. Sarkar and S. Maitra. Construction of nonlinear Boolean functions with important cryptographic properties. In *Advances in Cryptology - EUROCRYPT 2000*, number 1807 in Lecture Notes in Computer Science, pages 485–506. Springer Verlag, May 2000. 162, 163, 173

[14] P. Sarkar and S. Maitra. Nonlinearity bounds and constuction of resilient Boolean functions. In Mihir Bellare, editor, *Advances in Cryptology - Crypto 2000*, pages 515–532, Berlin, 2000. Springer-Verlag. Lecture Notes in Computer Science Volume 1880. 162, 163, 173

[15] P. Stănică and S. Maitra. Rotation Symmetric Boolean Functions – Count and Cryptographic Properties. In *R. C. Bose Centenary Symposium on Discrete Mathematics and Applications*, December 2002. Electronic Notes in Discrete Mathematics, Elsevier, Volume 15. 162, 164, 165, 174

[16] P. Stănică and S. Maitra. A constructive count of Rotation Symmetric functions. *Information Processing Letters*, 88:299–304, 2003. 162

[17] T. Xia, J. Seberry, J. Pieprzyk, C. Charnes. Homogeneous bent functions of degree n in $2n$ variables do not exist for $n > 3$. *Discrete Mathematics*, (to appear). 174

[18] X-M. Zhang and Y. Zheng. GAC – the criterion for global avalanche characteristics of cryptographic functions. *Journal of Universal Computer Science*, 1(5):316–333, 1995. 164

A Weakness of the Linear Part of Stream Cipher MUGI

Jovan Dj. Golić

System on Chip, Telecom Italia Lab, Telecom Italia
Via Reiss Romoli 274, I-10148 Turin, Italy
jovan.golic@tilab.com

Abstract. The linearly updated component of the stream cipher MUGI, called the buffer, is analyzed theoretically by using the generating function method. In particular, it is proven that the intrinsic response of the buffer, without the feedback from the nonlinearly updated component, consists of binary linear recurring sequences with small linear complexity 32 and with extremely small period 48. It is then shown how this weakness can in principle be used to facilitate the linear cryptanalysis of MUGI with two main objectives: to reconstruct the secret key and to find linear statistical distinguishers.

Keywords: Stream ciphers, combiners with memory, linear finite-state machines, linear cryptanalysis

1 Introduction

MUGI is a specific keystream generator for stream cipher applications proposed in [8]. Due to its design rationale, it is suitable for software implementations. Efficient hardware implementations in terms of speed are also possible, but are more complex with respect to the gate count than the usual designs based on linear feedback shift registers (LFSRs), nonlinear combining functions, and irregular clocking. MUGI has been evaluated within the CRYPTREC project of the Information-technology Promotion Agency for possible electronic government applications in Japan.

It is interesting to note that in mathematical terms, the structure of MUGI as well as of PANAMA, which is a stream cipher of a similar type previously proposed in [1], is essentially one of a combiner with memory, which is a well-known type of keystream generators (see [5]). Specific features are the following:

- the nonlinear combining function has a large internal memory size and is based on a round function of the block cipher AES [2]
- the driving linear finite-state machine (LFSM) providing input to the combining function is not an LFSR with a primitive connection polynomial
- the LFSM receives feedback from a part of the internal memory of the combining function

B. Roy and W. Meier (Eds.): FSE 2004, LNCS 3017, pp. 178–192, 2004.

- the output at a given time is a binary word taken from the internal memory of the combining function and then bitwise added to the plaintext word to produce the ciphertext word.

A security analysis of MUGI is presented in [9] and [10]. The main claims are essentially that MUGI is not vulnerable to common attacks on block ciphers and also to some attacks on stream ciphers. In particular, the linear cryptanalysis method for block ciphers [7] is adapted to deal with MUGI and to show that particular linear approximations cannot be effective for MUGI. However, some general methods for analyzing stream ciphers based on combiners with memory, most notably the so-called linear cryptanalysis of stream ciphers [3], [4], [5], and [6], are not addressed. Since MUGI can essentially be regarded as a combiner with memory, such methods are in principle also applicable to MUGI. Also, the underlying LFSM of MUGI, the so-called buffer, is not analyzed in [9] and [10].

Recall that linear cryptanalysis of product block ciphers composed of an iterated round function is based on the fact that the input and output to the block cipher are known, in the known plaintext scenario, and that all the intermediate outputs are unknown. Linear cryptanalysis of stream ciphers is essentially different from the linear cryptanalysis of block ciphers because of the underlying iterative structure in which the initial state is unknown and the output sequence produced from a sequence of internal states is known, in the known plaintext scenario. It essentially consists in finding linear relations among the unknown internal variables, possibly conditioned on the known output sequence, that hold with probabilities different from one half. It has two main objectives:

- to reconstruct the initial state of the keystream generator as well as the secret key
- to derive a linear statistical distinguisher which can distinguish the output sequence from a purely random sequence, defined as a sequence of mutually independent uniformly distributed random variables.

The main objective of this paper is to show that the linear part of MUGI, that is, the buffer is analyzable and that it is surprisingly weak. The second objective is to investigate how this weakness can in principle be used to facilitate the cryptanalysis of MUGI, especially the linear cryptanalysis.

The paper is organized as follows. Section 2 contains a brief description of MUGI. Analysis of the LFSM of MUGI is presented in Section 3, with some elements shown in the Appendix, a related transformation that eliminates the LFSM from the underlying system of nonlinear recurrences is given in Section 4, and the framework for the linear cryptanalysis of MUGI is outlined in Section 5. Section 6 contains a summary of the established weaknesses of MUGI and problems for future investigation.

2 Description of MUGI

A concise description of MUGI is specified here in as much detail as needed for the analysis. More details can be found in [8].

The keystream generator is essentially a combiner with memory, with specific properties described in Section 1. It is a finite-state machine (FSM) whose internal state has two components:

- a linearly updated component, called buffer, $b = b_0 b_1 \cdots b_{15}$, where each b_i is a 64-bit word; the size of this component is 1024 bits
- a nonlinearly updated component, called state, $a = a_0 a_1 a_2$, where each a_i is a 64-bit word; the size of this component is 192 bits.

The next-state or update function is invertible and has two components, $\phi = (\rho, \lambda)$, where ρ updates a and λ updates b, that is, $(a^{(t+1)}, b^{(t+1)}) = (\rho(a^{(t)}, b^{(t)}), \lambda(a^{(t)}, b^{(t)}))$.

The ρ component is a nonlinear function defined in terms of an invertible $(64 \times 64)-$bit function F by a kind of Feistel structure. More precisely:

$$\begin{aligned} a_0^{(t+1)} &= a_1^{(t)} \\ a_1^{(t+1)} &= a_2^{(t)} \oplus F(a_1^{(t)} \oplus b_4^{(t)}) \oplus C_1 \\ a_2^{(t+1)} &= a_0^{(t)} \oplus F(a_1^{(t)} \oplus {}^{<17}b_{10}^{(t)}) \oplus C_2. \end{aligned} \tag{1}$$

The ρ function is invertible for any given $b_4^{(t)}$ and ${}^{<17}b_{10}^{(t)}$. Here, for a 64-bit word x, ${}^{<i}x$ and ${}^{>i}x$ denote the rotations of x by i bits to the left and right, respectively. The function F is derived from the round function of AES and is a composition of (G, G) and a permutation of 8 bytes, where a $(32 \times 32)-$bit function G is a composition of a parallel combination of 4 $(8 \times 8)-$bit S-boxes and a linear $(32 \times 32)-$bit function, MixColumn, of AES (see [2]). The byte-permutation is (4,5,2,3,0,1,6,7) and C_1 and C_2 are 64-bit constants.

The λ component is a linear function defined by the following equations:

$$\begin{aligned} b_i^{(t+1)} &= b_{i-1}^{(t)}, \quad i \neq 0, 4, 10 \\ b_0^{(t+1)} &= b_{15}^{(t)} \oplus a_0^{(t)} \\ b_4^{(t+1)} &= b_3^{(t)} \oplus b_7^{(t)} \\ b_{10}^{(t+1)} &= b_9^{(t)} \oplus {}^{<32}b_{13}^{(t)}. \end{aligned} \tag{2}$$

The λ function is invertible for any given $a_0^{(t)}$.

The 64-bit output of the keystream generator at time t is defined as $a_2^{(t)}$.

The initial internal state of the keystream generator is produced from a 128-bit secret key K and a 128-bit initialization vector IV in three stages, by using the keystream generator itself. In the first two stages only ρ is used and in the third stage both ρ and λ are used. At the beginning of the third stage, the initial value of a, $a^{(0)}$, depends on both K and IV, but the initial value of the buffer b, $b^{(0)}$, depends on K only. The third stage consists of iterating ϕ (i.e., both ρ and λ) 15 times, without producing the output $(a_2^{(t)})_{t=0}^{15}$, and the keystream generation starts from the 16-th iteration on.

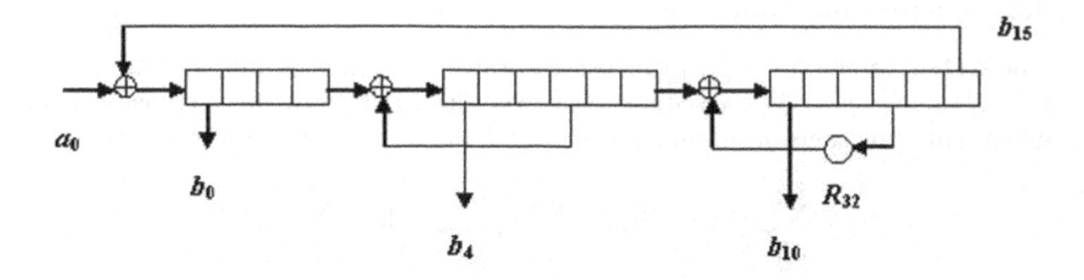

Fig. 1. The buffer as LFSM

3 Analysis of Buffer

In this section, the buffer is analyzed as a non-autonomous LFSM with one input sequence, namely, $a_0 = (a_0^{(t)})_{t=0}^{\infty}$. The input sequence and all the internal sequences in the buffer are 64-bit sequences. Our objective is to derive expressions for the internal sequences in the buffer in terms of the input sequence a_0 and the initial state of the buffer, $b^{(0)} = b_0^{(0)} b_1^{(0)} \cdots b_{15}^{(0)}$.

In view of the λ update function, the 16 internal sequences in the buffer can be divided in three groups, in each group the sequences being phase shifts of each other (see Fig. 1, where R_j denotes the rotation by j bits to the left, which is a linear transformation of a 64-bit word).

3.1 Linear Recurrences

From the λ update function, we directly obtain the following linear recurrences, all holding for $t \geq 1$:

$$\begin{aligned} b_4^{(t)} &= b_4^{(t-4)} \oplus b_0^{(t-4)} \\ b_{10}^{(t)} &= R_{32} b_{10}^{(t-4)} \oplus b_4^{(t-6)} \\ b_0^{(t)} &= a_0^{(t-1)} \oplus b_{10}^{(t-6)} . \end{aligned} \tag{3}$$

In vectorial notation where vectors are represented as one-column matrices, R_{32} is represented as a matrix. The initial state of the buffer can now be represented as $(b_0^{(t)})_{t=-3}^{0} (b_4^{(t)})_{t=-5}^{0} (b_{10}^{(t)})_{t=-5}^{0}$. Then, by eliminating b_0, we obtain

$$\begin{aligned} b_4^{(t)} &= b_4^{(t-4)} \oplus b_{10}^{(t-10)} \oplus a_0^{(t-5)}, \quad t \geq 5 \\ b_{10}^{(t)} &= b_4^{(t-6)} \oplus R_{32} b_{10}^{(t-4)}, \quad t \geq 1. \end{aligned} \tag{4}$$

This is a system of two 64-bit linear recurrences (that is, 128 binary linear recurrences) in terms of 64-bit sequences b_4 and b_{10}, where a_0 is regarded as a given 64-bit sequence.

3.2 Generating Functions

The system can be solved by using the generating function technique dealing with the z-transforms of 64-bit sequences. In vectorial notation, the z-transforms or generating functions of b_4, b_{10}, and a_0 are defined as formal power series

$$B_4 = \sum_{t=0}^{\infty} b_4^{(t)} z^t, \quad B_{10} = \sum_{t=0}^{\infty} b_{10}^{(t)} z^t, \quad A_0 = \sum_{t=0}^{\infty} a_0^{(t)} z^t, \tag{5}$$

respectively. It is shown in the Appendix how to convert the system (4) of linear recurrences from the time domain into the generating function domain. Namely, letting I denote the 64×64 identity matrix, we thus obtain the representation

$$(1 \oplus z^4)B_4 \oplus z^{10} B_{10} \;=\; z^5 A_0 \oplus \Delta_1 \tag{6}$$

$$z^6 B_4 \oplus (I \oplus z^4 R_{32}) B_{10} \;=\; \Delta_2 \tag{7}$$

where

$$\Delta_1 \;=\; b_4^{(0)} \oplus z^4 b_0^{(0)} \oplus \sum_{t=1}^{3} (b_4^{(t-4)} \oplus b_0^{(t-4)}) z^t \oplus \sum_{t=5}^{9} b_{10}^{(t-10)} z^t \tag{8}$$

$$\Delta_2 \;=\; \sum_{t=1}^{5} b_4^{(t-6)} z^t \oplus b_{10}^{(0)} \oplus R_{32} \sum_{t=1}^{3} b_{10}^{(t-4)} z^t \tag{9}$$

are 64-dimensional vectors (64×1 matrices) whose elements are polynomials in z defined by the initial state of the buffer and whose degrees are at most 9 and 5, respectively. Essentially, we obtain a system of 128 linear equations with coefficients being polynomials in z and with unknowns being 128 generating functions of 64 binary sequences in b_4 and 64 binary sequences in b_{10}.

3.3 Solution

The system has a unique solution which can be found in the following way. First, by elimination we obtain

$$F(z) B_4 \;=\; z^5 (I \oplus z^4 R_{32}) A_0 \oplus (I \oplus z^4 R_{32}) \Delta_1 \oplus z^{10} \Delta_2 \tag{10}$$

$$F(z) B_{10} \;=\; z^{11} A_0 \oplus z^6 \Delta_1 \oplus (1 \oplus z^4) \Delta_2 \tag{11}$$

where

$$F(z) \;=\; I \oplus z^4 (I \oplus R_{32}) \oplus z^8 R_{32} \oplus z^{16} I \tag{12}$$

denotes a 64×64 matrix whose coefficients are polynomials in z of degree at most 16.

When regarded over a field of rational functions in z, $F(z)$ is invertible as is seen from the following equation:

$$\begin{aligned} F(z)F(z) &= (I \oplus z^4(I \oplus R_{32}) \oplus z^8 R_{32} \oplus z^{16} I)^2 \\ &= I \oplus z^8(I \oplus R_{32}^2) \oplus z^{16} R_{32}^2 \oplus z^{32} I \\ &= I \oplus z^8(I \oplus I) \oplus z^{16} I \oplus z^{32} I \ = \ (1 \oplus z \oplus z^2)^{16} I \end{aligned} \tag{13}$$

because of $R_{32}^2 = I$. Thus we get that $F(z)/f(z)$ is the inverse of $F(z)$, where

$$f(z) \ = \ 1 \oplus z^{16} \oplus z^{32} \ = \ (1 \oplus z \oplus z^2)^{16} \ = \ \frac{1 \oplus z^{48}}{1 \oplus z^{16}}. \tag{14}$$

Accordingly, we obtain the solution for the generating functions B_4 and B_{10} in the form of

$$\begin{aligned} B_4 &= \frac{1}{f(z)} z^5 G(z) A_0 \oplus \frac{1}{f(z)} \Delta_1' \\ &= \frac{1}{1 \oplus z^{48}} z^5 (1 \oplus z^{16}) G(z) A_0 \oplus \frac{1}{1 \oplus z^{48}} (1 \oplus z^{16}) \Delta_1' \end{aligned} \tag{15}$$

$$\begin{aligned} B_{10} &= \frac{1}{f(z)} z^{11} F(z) A_0 \oplus \frac{1}{f(z)} \Delta_2' \\ &= \frac{1}{1 \oplus z^{48}} z^{11} (1 \oplus z^{16}) F(z) A_0 \oplus \frac{1}{1 \oplus z^{48}} (1 \oplus z^{16}) \Delta_2' \end{aligned} \tag{16}$$

where

$$G(z) \ = \ F(z)(I \oplus z^4 R_{32}) \ = \ (1 \oplus z \oplus z^2 \oplus z^3 \oplus z^4)^4 I \oplus z^{20} R_{32} \tag{17}$$

denotes a 64×64 matrix whose coefficients are polynomials in z of degree at most 20, and

$$\Delta_1' \ = \ G(z)\Delta_1 \oplus z^{10} F(z) \Delta_2 \tag{18}$$

$$\Delta_2' \ = \ F(z)(z^6 \Delta_1 \oplus (1 \oplus z^4)\Delta_2) \tag{19}$$

are 64-dimensional vectors (64×1 matrices) whose elements are polynomials in z defined by the initial state of the buffer and whose degrees are at most 31.

3.4 Discussion

Both b_4 and b_{10} have two components, one being a linear transform of the input sequence a_0 to the buffer and the other being a linear transform of the initial conditions contained in Δ_1 and Δ_2. For both b_4 and b_{10}, the other, intrinsic component consists of 64 binary linear recurring subsequences each produced by an LFSR with the feedback polynomial $f(z)$, or alternatively, by a cycling LFSR with the feedback polynomial $1 \oplus z^{48}$. The following properties should then be considered as serious weaknesses of the buffer design.

- The exponent (period) of $f(z)$ is only 48; the period of each of the intrinsic binary subsequences is thus equal to 48 or divides 48.
- The degree of $f(z)$ is only 32, and with an appropriate design it could have been as large as 1024, which is the bit-size of the internal state of the buffer.
- Due to their extremely small period, the statistical properties of the intrinsic binary subsequences are very bad.
- In common designs of keystream generators, the linear component, with the feedback from the nonlinear component disconnected, ensures a large period of the corresponding internal state sequence which itself very likely provides a lower bound on the period of the keystream sequence as well as good statistical properties. Consequently, the design of MUGI does not satisfy this criterion.
- The polynomial $f(z)$ defines a linear sequential transform of any buffer sequence, in particular b_4 or b_{10}, that is equal to a linear sequential transform of the input sequence, which is produced by the nonlinear component. Its low degree, 32, small number of nonzero coefficients, 3, and small period, 48, facilitate the initial state reconstruction and finding statistical distinguishers for the keystream sequence (see Sections 4 and 5).
- The intrinsic binary subsequences do not depend on the initialization vector, but only on the secret key. Namely, for each $1 \leq j \leq 64$, the j-th binary subsequence of both b_4 and b_{10} solely depends on $(b_{i,j}^{(0)})_{i=0}^{15}(b_{i,(j+32)_{\mathrm{mod}\ 64}}^{(0)})_{i=0}^{15}$, that is, on the j-th and the $(j+32)_{\mathrm{mod}\ 64}$-th binary subsequences of $b^{(0)}$, which are defined by 32 secret key bits only. This is because the mixing between different binary subsequences in the buffer, provided by the linear transform R_{32}, is not good. Divide-and-conquer secret key reconstruction attacks may be facilitated by this property.

4 Elimination of Buffer

The obtained expressions (15) and (16) for the generating functions of the 64-bit buffer sequences b_4 and b_{10}, respectively, can be transformed into the time domain and then appropriately substituted in the recurrences (1) for the update function ρ. In this way, we can derive the recurrences involving only the state sequences a_1 and a_2, where the output sequence a_2 is assumed to be known, in the known plaintext scenario, except for the first 16 outputs $(a_2^{(t)})_{t=0}^{15}$, which are discarded. Namely, from (1) we first eliminate a_0 and use the fact that F is invertible to get for $t \geq 0$

$$a_1^{(t)} \oplus b_4^{(t)} = F^{-1}(a_1^{(t+1)} \oplus a_2^{(t)} \oplus C_1) \tag{20}$$

$$a_1^{(t)} \oplus {}^{<17}b_{10}^{(t)} = F^{-1}(a_1^{(t-1)} \oplus a_2^{(t+1)} \oplus C_2) \tag{21}$$

where F^{-1} is the inverse of F and formally $a_1^{(-1)} = a_0^{(0)}$. Then, by converting (15) and (16) into the time domain we get the following linear recurrences holding for $t \geq 48$:

$$b_4^{(t)} \oplus b_4^{(t-48)} = a_0^{(t-5)} \oplus a_0^{(t-9)} \oplus a_0^{(t-13)} \oplus a_0^{(t-17)} \oplus a_0^{(t-25)} \oplus a_0^{(t-29)} \oplus a_0^{(t-33)} \oplus a_0^{(t-37)} \oplus {}^{<32}a_0^{(t-25)} \oplus {}^{<32}a_0^{(t-41)} \quad (22)$$

$$b_{10}^{(t)} \oplus b_{10}^{(t-48)} = a_0^{(t-11)} \oplus a_0^{(t-15)} \oplus a_0^{(t-31)} \oplus a_0^{(t-43)} \oplus {}^{<32}a_0^{(t-15)} \oplus {}^{<32}a_0^{(t-19)} \oplus {}^{<32}a_0^{(t-31)} \oplus {}^{<32}a_0^{(t-35)}. \quad (23)$$

Finally, by combining (20) with (22) and (21) with (23), we get the following recurrences involving a_1 and a_2 only, which hold for $t \geq 48$:

$$a_1^{(t)} \oplus a_1^{(t-48)} \oplus F^{-1}(a_1^{(t+1)} \oplus a_2^{(t)} \oplus C_1) \oplus F^{-1}(a_1^{(t-47)} \oplus a_2^{(t-48)} \oplus C_1) = a_1^{(t-6)} \oplus a_1^{(t-10)} \oplus a_1^{(t-14)} \oplus a_1^{(t-18)} \oplus a_1^{(t-26)} \oplus a_1^{(t-30)} \oplus a_1^{(t-34)} \oplus a_1^{(t-38)} \oplus {}^{<32}a_1^{(t-26)} \oplus {}^{<32}a_1^{(t-42)} \quad (24)$$

$$a_1^{(t)} \oplus a_1^{(t-48)} \oplus F^{-1}(a_1^{(t-1)} \oplus a_2^{(t+1)} \oplus C_2) \oplus F^{-1}(a_1^{(t-49)} \oplus a_2^{(t-47)} \oplus C_2) = {}^{<17}a_1^{(t-12)} \oplus {}^{<17}a_1^{(t-16)} \oplus {}^{<17}a_1^{(t-32)} \oplus {}^{<17}a_1^{(t-44)} \oplus {}^{<49}a_1^{(t-16)} \oplus {}^{<49}a_1^{(t-20)} \oplus {}^{<49}a_1^{(t-32)} \oplus {}^{<49}a_1^{(t-36)}. \quad (25)$$

The recurrences are nonlinear because of nonlinear F^{-1}. As the first 16 outputs are not known, it is interesting to consider (24) and (25) only for $t \geq 64$.

One can also obtain different recurrences by using the polynomial $f(z)$ of degree 32 instead of the polynomial $1 \oplus z^{48}$. They will hold for $t \geq 32$, but each will involve F^{-1} three times which makes them less useful. In principle, there are at least two general ways of using the recurrences (24) and (25).

- One way is to try to eliminate a_1, possibly with certain approximation probabilities, thus yielding a recurrence in a_2 holding with a certain probability which will represent a statistical distinguisher between the keystream sequence and a purely random sequence (see Section 5).
- The other way is to assume that a_2 is known, in the known plaintext scenario, and try to solve the corresponding nonlinear equations for a_1, possibly by an algebraic approach or a probabilistic approach using approximations to the nonlinear equations.

5 Linear Cryptanalysis of MUGI

A general way of conducting the linear cryptanalysis of stream ciphers is to linearize the (vectorial) next-state function and the output function, with certain approximation probabilities, and to analyze the LFSM resulting from these linear approximations (see [3], [4], [5], and [6]). In particular, this method is also applicable if only one bit of the internal state is known at a time. This is essentially different from the linear cryptanalysis of iterated block ciphers where one concatenates mutually correlated linear functions of the input and output to the round function. The obtained LFSM is in fact an LFSM approximation to the keystream generator, which itself is a nonlinear autonomous FSM. Our objective here is to present a framework for conducting the linear cryptanalysis of MUGI by using the results from Sections 3 and 4.

Linearizing the next-state function of MUGI reduces to linearizing the nonlinear function F or its inverse F^{-1}. More precisely, we linearize equations (20) and (21), where, for convenience, $t-1$ is substituted for t, and the sequences b_4 and b_{10} are determined by (15) and (16), respectively. In turn, linearizing F^{-1} reduces to linearizing the S-boxes of AES. The effectiveness of the linear cryptanalysis depends on the underlying linear approximations and the corresponding approximation probabilities.

A linear approximation to an S-box is a pair of input, α, and output, β, linear functions with a nonzero correlation coefficient $c(\alpha,\beta) = \Pr(\alpha = \beta) - \Pr(\alpha \neq \beta)$. It is well known that the maximal correlation coefficient magnitude is $1/8$. This value is relatively small, but allows a lot of freedom when choosing the linear approximations. In particular, for any given α, there are 5 values of β with $c(\alpha,\beta) = 1/8$, 16 values of β with $c(\alpha,\beta) = 7/64$, and 36 values of β with $c(\alpha,\beta) = 6/64$.

A linear approximation to F^{-1} consists of linear approximations to the 64 component Boolean functions and each of the 64 correlation coefficients is determined by the corresponding active S-box. More generally, one can also consider linear approximations to any 64 linearly independent linear combinations of the 64 component Boolean functions of F^{-1}. In this case more than just one S-box can be active for each linear combination considered, so that the resulting correlation coefficient is the product of the involved correlation coefficients of active S-boxes. In the more general scenario, let L_2 define a set of 64 linear approximations to 64 linearly independent linear combinations of the component Boolean functions of F^{-1} defined by L_1F^{-1}. More precisely, an invertible matrix L_1 is first applied to both (20) and (21) and then a matrix L_2 is substituted for L_1F^{-1} on the right-hand sides of the equations. We thus get

$$L_1a_1^{(t-1)} \oplus L_1b_4^{(t-1)} = L_2a_1^{(t)} \oplus L_2a_2^{(t-1)} \oplus L_2C_1 \oplus e_1^{(t)} \tag{26}$$

$$L_1a_1^{(t-1)} \oplus L_1{}^{<17}b_{10}^{(t-1)} = L_2a_1^{(t-2)} \oplus L_2a_2^{(t)} \oplus L_2C_2 \oplus e_2^{(t)} \tag{27}$$

where e_1 and e_2 are the 64-bit approximation-error sequences whose binary component subsequences are expressed as nonbalanced Boolean functions of the corresponding inputs to F^{-1}. The greater the imbalance, the better the underlying linear approximations.

Equations (26) and (27) in fact define an LFSM with input sequences e_1 and e_2. The LFSM can be solved for a_1 and a_2 by using the generating function method. Let A_1, A_2, E_1, E_2, and $A_0 = zA_1 \oplus a_0^{(0)}$ denote the generating functions of a_1, a_2, e_1, e_2, and a_0, respectively. Then after some algebraic manipulations, by using the expressions (15) and (16) for B_4 and B_{10}, in terms of $1 \oplus z^{48}$, respectively, we get

$$F_1(z)A_1 = z(1 \oplus z^{48})L_2A_2 \oplus (1 \oplus z^{48})E_1 \oplus \Delta_1'' \oplus (1 \oplus z^{48})C_1' \quad (28)$$

$$F_2(z)A_1 = (1 \oplus z^{48})L_2A_2 \oplus (1 \oplus z^{48})E_2 \oplus \Delta_2'' \oplus (1 \oplus z^{48})C_2' \quad (29)$$

where $C_1' = C_1/(1 \oplus z)$ and $C_2' = C_2/(1 \oplus z)$ denote the generating functions of the constant sequences, of period equal to 1, corresponding to the constants C_1 and C_2, respectively, and

$$F_1(z) = (1 \oplus z^{48})(zL_1 \oplus L_2) \oplus z^7(1 \oplus z^{16})L_1G(z) \quad (30)$$

$$F_2(z) = z\left((1 \oplus z^{48})(L_1 \oplus zL_2) \oplus z^{12}(1 \oplus z^{16})L_1R_{17}F(z)\right) \quad (31)$$

$$\Delta_1'' = (1 \oplus z^{16})(zL_1\Delta_1' \oplus z^6L_1G(z)a_0^{(0)}) \oplus (1 \oplus z^{48})L_2a_1^{(0)} \quad (32)$$

$$\Delta_2'' = z(1 \oplus z^{16})L_1R_{17}\Delta_2' \oplus z\left(z^{11}(1 \oplus z^{16})L_1R_{17}F(z) \oplus (1 \oplus z^{48})L_2\right)a_0^{(0)} \oplus (1 \oplus z^{48})L_2a_2^{(0)}. \quad (33)$$

Note that Δ_1'' and Δ_2'' are 64-dimensional vectors whose elements are polynomials in z defined by the initial state of the whole keystream generator ($b^{(0)}$ and $a_0^{(0)}a_1^{(0)}a_2^{(0)}$) and whose degrees are at most 49. Here E_1 and E_2 are generating functions of unknown, but nonbalanced sequences, so that (28) and (29) in fact constitute a system of binary linear recurrences each holding with a probability different from one half.

To eliminate the unknown A_1 from the system, assume for simplicity that $F_1(z)$ is invertible, where $F_1(z)^{-1} = F_1(z)^*/\det F_1(z)$, $F_1(z)^*$ being the adjunct matrix of $F_1(z)$. (As L_1 is invertible, it is likely that $F_1(z)$ or $F_2(z)$ are invertible.) We then obtain a correlation equation

$$\frac{F_2(z)F_1(z)^*\Delta_1'' \oplus \det F_1(z)\Delta_2''}{1 \oplus z^{48}} = (F_2(z)F_1(z)^*E_1 \oplus \det F_1(z)E_2) \oplus (zF_2(z)F_1(z)^* \oplus \det F_1(z)I)\,L_2A_2 \oplus (F_2(z)F_1(z)^*C_1' \oplus \det F_1(z)C_2')\,. \quad (34)$$

In the time domain, the left-hand side of (34) is a 64-bit sequence, x , in the generating function domain denoted as X, which depends on the initial conditions, and the last two terms on the right-hand side of (34) are linear sequential transforms of the (known) sequence a_2 and of the constant sequences corresponding to C_1 and C_2 , respectively. The first term on the right-hand side of (34), $E = F_2(z)F_1(z)^*E_1 \oplus \det F_1(z)E_2$, is the noise term depending on the performed linear approximations. Consequently, (34) means that the linear recurring sequence x depending on the initial conditions which is ultimately periodic with period of only 48 (or smaller) is termwise correlated to a linear sequential transform of a_2 and of the constant sequences corresponding to C_1 and C_2. Equivalently, this is the case for each of the 64 constituent binary subsequences.

5.1 Initial State Reconstruction

If $t = 0$ is taken as the initial time, then the first 16 elements of the output sequence, $(a_2^{(t)})_{t=0}^{15}$, are unknown, and the initial state of the buffer, $b^{(0)}$, represented through Δ_1 and Δ_2, depends on the secret key only. Alternatively, if $t = 16$ is taken as the initial time, then the output sequence is known, but the initial state of the buffer, $b^{(16)}$, depends on the initialization vector as well. Note that the initial state of MUGI, $b^{(0)}$ and $a_0^{(0)}a_1^{(0)}a_2^{(0)}$, can be obtained from any internal state of MUGI (e.g., $b^{(16)}$ and $a_0^{(16)}a_1^{(16)}a_2^{(16)}$) by reversing the next-state function. Furthermore, the 128-bit secret key can be obtained from $b_0^{(0)}b_1^{(0)}$ by reversing the update function ρ.

The effectiveness of the correlation equation (34) is determined by how much the probabilities for the 64 underlying binary noise subsequences of the resulting noise sequence e deviate from one half. Let $c_i = 1 - 2p_i$ denote the (positive or negative) correlation coefficient of the i-th binary noise subsequence of e, where p_i is the probability that the noise bit is equal to 1. The correlation coefficients of the involved approximation-error sequences e_1 and e_2 depend on the linearization of S-boxes and their magnitudes are equal to 2^{-3} or are close to this value if only one S-box is active per each linear approximation involved. If we assume that these subsequences are mutually independent sequences of mutually independent and identically distributed binary random variables, then the correlation coefficient magnitude is given as $|c_i| = 2^{-3m_i}$ where m_i is the total number of binary terms from e_1 and e_2 present in this subsequence, that is, the total number of nonzero binary coefficients of the polynomials in the i-th row of $F_2(z)F_1(z)^*$ and in $\det F_1(z)$. Here we used a well-known fact that the correlation coefficient of a binary sum of mutually independent binary random variables is equal to the product of their individual correlation coefficients (see also the piling-up lemma from [7]).

Accordingly, the periodic part of the 64-bit linear recurring sequence x, that is, any corresponding segment composed of 48 consecutive 64-bit words can in principle be statistically reconstructed by a sort of repetition attack based on the fact that any bit of the periodic part of x is repeated with period 48. If c denotes the underlying correlation coefficient for a considered bit of x, then the bit can be reconstructed by the mojority count from the given output segment of length $O(48c^{-2})$ with complexity $O(c^{-2})$. It is easy to see that the periodic part of x depends (linearly) only on $b^{(0)}$ and $a_0^{(0)}$ and not on $a_1^{(0)}$. Namely, this is due to

$$X = \frac{(1 \oplus z^{16})F_2(z)F_1(z)^* \left(zL_1\Delta_1' \oplus z^6 L_1 G(z) a_0^{(0)}\right)}{1 \oplus z^{48}}$$

$$\oplus \frac{\det F_1(z)(1 \oplus z^{16}) \left(zL_1 R_{17} \Delta_2' \oplus z^{12} L_1 R_{17} F(z) a_0^{(0)}\right)}{1 \oplus z^{48}}$$

$$\oplus F_2(z)F_1(z)^* L_2 a_1^{(0)} \oplus \det F_1(z) \left(zL_2 a_0^{(0)} \oplus L_2 a_2^{(0)}\right) \tag{35}$$

where Δ_1' and Δ_2' depend only on $b^{(0)}$. Accordingly, the initial time of the 48-word segment of x to be reconstructed should on one hand be sufficiently large so as to render the terms depending on $a_1^{(0)}$ and $a_2^{(0)}$ in (35) vanish and, on the other, should be at least 16 so that the involved terms of the output sequence a_2 in (34) are all known.

The recovered 48 64-bit words of the periodic part of x define a system of $48 \cdot 64$ binary linear equations in the unknown $17 \cdot 64$ initial state bits, $b^{(0)}$ and $a_0^{(0)}$, which can then be obtained by solving the system. The 128-bit secret key can be obtained from $b_0^{(0)} b_1^{(0)}$ by reversing the update function ρ.

Alternatively, if the objective is to recover the initial state of MUGI, regardless of the initialization algorithm and the secret key, but taking into account the fact that the first 16 elements of the output sequence, $(a_2^{(t)})_{t=0}^{15}$, are unknown, then $t = 16$ should be taken as the initial time, $b^{(16)}$ and $a_0^{(16)}$ should be reconstructed by the repetition attack described above, and the 64 bits of $a_1^{(16)}$ can be reconstructed by the exhaustive search in view of the fact that $a_2^{(16)}$ is known. Finally, the initial state, $b^{(0)}$ and $a_0^{(0)} a_1^{(0)} a_2^{(0)}$, is then obtained from the internal state $b^{(16)}$ and $a_0^{(16)} a_1^{(16)} a_2^{(16)}$ by reversing the next-state function.

The feasibility of the attack can be estimated by computing the noise correlation coefficients, as described above, for different vectorial linear transformations L_1 and L_2. Assuming for simplicity that the underlying correlation coefficient $|c| = 2^{-3m}$ is the same for all the bits of x to be reconstructed, the attack would in theory be effective if it is faster than the exhaustive search over the initial states. Note that the exhaustive search requires 18 64-bit output values to be produced for each guessed initial state or, more precisely, for each guessed internal state at time $t = 16$. Roughly speaking, the attack is effective if $48 \cdot 2^{6m} < 2^{18 \cdot 64}$, that is, if $m \leq 191$, where the underlying complexity units are assumed to be the same. Of course, it has to be noted that the required output sequence length would be of the same order of magnitude as the complexity.

5.2 Linear Statistical Weakness

The equation (34) can also be put into the form

$$L(z)A_2 = \frac{1 \oplus z^{48}}{1 \oplus z} (F_2(z)F_1(z)^* C_1 \oplus \det F_1(z) C_2)$$

$$\oplus \ (F_2(z)F_1(z)^* \Delta_1'' \oplus \det F_1(z) \Delta_2'') \oplus (1 \oplus z^{48})E \qquad (36)$$

where the matrix $L(z) = (1 \oplus z^{48})(zF_2(z)F_1(z)^* \oplus \det F_1(z)I)$ defines a linear sequential transform of the output sequence a_2. This equation specifies a linear statistical distinguisher between the output sequence and a purely random sequence.

Namely, all the terms except the noise term on the right-hand side of (36) are polynomials in z and as such vanish in the time domain after a sufficiently large t depending on the degrees of the involved polynomials. So, (36) means that a linear sequential transform of the output sequence is termwise correlated to the all-zero 64-bit sequence where the approximation/correlation noise is defined by $(1 \oplus z^{48})E$. Equivalently, the 64 constituent binary subsequences, obtained as linear sequential transforms of the output sequence, are bitwise correlated to the all-zero binary sequence, where the corresponding correlation coefficients can be approximated as squares of the correlation coefficients of the corresponding binary noise subsequences of e. If c is such a correlation coefficient, then the output sequence length required for detecting the weakness in the corresponding binary subsequence is $O(c^{-4})$. The output sequence length required to detect the weakness by using all the 64 subsequences is then $O(c^{-4}/64)$.

The feasibility of the attack can be estimated by computing the noise correlation coefficients for different vectorial linear transformations L_1 and L_2. Assuming for simplicity that the underlying correlation coefficient $|c| = 2^{-3m}$ is the same for all the bits of x to be reconstructed, the attack would in theory be effective if the total required output sequence length, proportional to the complexity, is smaller than the expected period for the size of the internal state, that is, if $2^{12m}/64 < 2^{18 \cdot 64}$, that is, if $m \leq 96$.

6 Conclusions

Our main finding is that the linearly updated component of MUGI, the so-called buffer, is not designed properly. It is proven that if the feedback from the nonlinearly updated component is disconnected, then the binary subsequences of the buffer, comprising its intrinsic response, are linear recurring sequences with the linear complexity of only 32 and with the period of only 48. Accordingly, the buffer neither provides a large lower bound on the period nor ensures good statistics of the output sequences of MUGI, unlike the usual designs of keystream generators. Furthermore, as each such subsequence depends on only 32 bits of the initial state of the buffer, the mixing between the 1024 bits of the initial state of the buffer is not good. In addition, it is pointed out that the 128-bit secret key can directly be recovered from the reconstructed internal state of MUGI at any time, which is not desirable.

As a consequence of this small period, it is shown that the buffer sequence can be eliminated from the update equations for the nonlinearly updated component of MUGI, the so-called state, thus yielding nonlinear recurrences involving only the output sequence and a part of the state sequence. It is then pointed out how the weakness of the buffer can be used to facilitate the linear cryptanalysis of MUGI. This is achieved by developing a framework for the linear cryptanalysis with two main objectives: to reconstruct the initial state or the secret key and to find linear statistical distinguishers for MUGI. This framework can be used for future experiments to investigate if the proposed attacks are feasible. More precisely, the feasibility depends on the magnitude of the corresponding correlation coefficients which can be estimated by examining the corresponding matrices with polynomial elements that result from the chosen linearizations of the S-boxes.

Appendix: Generating Function Representation

Firstly, by virtue of (5), the system of linear recurrences (4) results in

$$\sum_{t=0}^{\infty} b_4^{(t)} z^t = z^4 \sum_{t=5}^{\infty} b_4^{(t-4)} z^{t-4} \oplus z^{10} \sum_{t=5}^{\infty} b_{10}^{(t-10)} z^{t-10} \oplus z^5 \sum_{t=5}^{\infty} a_0^{(t-5)} z^{t-5} \tag{37}$$

$$\sum_{t=0}^{\infty} b_{10}^{(t)} z^t = z^6 \sum_{t=1}^{\infty} b_4^{(t-6)} z^{t-6} \oplus z^4 R_{32} \sum_{t=1}^{\infty} b_{10}^{(t-4)} z^{t-4}. \tag{38}$$

Then, we get

$$\begin{aligned}(1 \oplus z^4) B_4 \oplus z^{10} B_{10} &= z^5 A_0 \oplus \sum_{t=0}^{4} b_4^{(t)} z^t \oplus z^4 b_4^{(0)} \oplus \sum_{t=5}^{9} b_{10}^{(t-10)} z^t \\ &= z^5 A_0 \oplus (1 \oplus z^4) b_4^{(0)} \oplus \sum_{t=1}^{4} (b_4^{(t-4)} \oplus b_0^{(t-4)}) z^t \oplus \sum_{t=5}^{9} b_{10}^{(t-10)} z^t \\ &= z^5 A_0 \oplus b_4^{(0)} \oplus z^4 b_0^{(0)} \oplus \sum_{t=1}^{3} (b_4^{(t-4)} \oplus b_0^{(t-4)}) z^t \oplus \sum_{t=5}^{9} b_{10}^{(t-10)} z^t \end{aligned} \tag{39}$$

$$(I \oplus z^4 R_{32}) B_{10} \oplus z^6 B_4 = \sum_{t=1}^{5} b_4^{(t-6)} z^t \oplus b_{10}^{(0)} \oplus R_{32} \sum_{t=1}^{3} b_{10}^{(t-4)} z^t. \tag{40}$$

By using the simplified notation (8) and (9), we then directly obtain (6) and (7).

Acknowledgement

This work is based on a result of evaluation requested by the Japanese CRYPTREC project: http://www.ipa.go.jp/security/enc/CRYPTREC/index-e.html.

References

[1] J. Daemen and C. Claap, "Fast hashing and stream encryption with PANAMA," Fast Software Encryption - FSE '98, *Lecture Notes in Computer Science*, vol. 1372, pp. 60-74, 1998. 178

[2] J. Daemen and V. Rijmen, *The Design of Rijndael: AES - The Advanced Encryption Standard*. Berlin: Springer-Verlag, 2002. 178, 180

[3] J. Dj. Golić, "Correlation via linear sequential circuit approximation of combiners with memory," Advances in Cryptology - EUROCRYPT '92, *Lecture Notes in Computer Science*, vol. 658, pp. 124-137, 1993. 179, 186

[4] J. Dj. Golić, "Linear cryptanalysis of stream ciphers," Fast Software Encryption - FSE '94, *Lecture Notes in Computer Science*, vol. 1008, pp. 154-169, 1995. 179, 186

[5] J. Dj. Golić, "Correlation properties of a general combiner with memory," *Journal of Cryptology*, vol. 9, pp. 111-126, 1996. 178, 179, 186

[6] J. Dj. Golić, "Linear models for keystream generators," *IEEE Transactions on Computers*, vol. 45, pp. 41-49, 1996. 179, 186

[7] M. Matsui, "Linear cryptanalysis method for DES cipher," Advances in Cryptology - EUROCRYPT '93, *Lecture Notes in Computer Science*, vol. 765, pp. 159-169, 1994. 179, 188

[8] D. Watanabe, S. Furuya, H. Yoshida, and K. Takaragi, MUGI Pseudorandom number generator, Specification, Ver. 1.2, 2001, available at http://www.sdl.hitachi.co.jp/crypto/mugi/index-e.html. 178, 179

[9] D. Watanabe, S. Furuya, H. Yoshida, and K. Takaragi, MUGI Pseudorandom number generator, Self-evaluation report, Ver. 1.1, 2001, available at http://www.sdl.hitachi.co.jp/crypto/mugi/index-e.html. 179

[10] D. Watanabe, S. Furuya, H. Yoshida, K. Takaragi, and B. Preneel, "A new keystream generator MUGI," Fast Software Encryption 2002, *Lecture Notes in Computer Science*, vol. 2365, pp. 179-194, 2002. 179

Vulnerability of Nonlinear Filter Generators Based on Linear Finite State Machines

Jin Hong[1], Dong Hoon Lee[1], Seongtaek Chee[1], and Palash Sarkar[2]

[1] National Security Research Institute
161 Gajeong-dong, Yuseong-gu, Daejeon, 305-350, Korea
{jinhong,dlee,chee}@etri.re.kr
[2] Indian Statistical Institute
203, B.T. Road, Kolkata 700108, India
palash@isical.ac.in

Abstract. We present a realization of an LFSM that utilizes an LFSR. This is based on a well-known fact from linear algebra. This structure is used to show that a previous attempt at using a CA in place of an LFSR in constructing a stream cipher did not necessarily increase its security. We also give a general method for checking whether or not a nonlinear filter generator based on an LFSM allows reduction to one that is based on an LFSR and which is vulnerable to Anderson information leakage.

Keywords: Stream cipher, nonlinear filter model, LFSR, CA, Anderson information leakage

1 Introduction

Linear feedback shift registers (LFSR) are one of the most useful building blocks for constructing stream ciphers. There are classical models of memoryless synchronous stream ciphers that utilize LFSR's : the nonlinear filter model (NF) and the nonlinear combiner model (NC).

For the NF model, building on previous works[9, 3], Anderson[1] showed that much information about the state of the LFSR may be obtained from the output key stream, if the distribution of possible states of LFSR's in relation to output stream blocks is not uniform. And for a random NF model, this is usually quite irregular.

In the paper [10], presented at CRYPTO 2002, a model that combines the NF and NC models was introduced. This model utilized a cellular automaton (CA) instead of an LFSR to eliminate the above mentioned information leakage of NF models. In the paper, it is claimed that this non-uniform distribution stems from the fact that a particular state bit of an LFSR affects the output key stream several times. This would be unavoidable in an NF generator based on an LFSR. It is also claimed that this property can be avoided through the use of a CA, thus removing Anderson information leakage. In this paper, we show this claim to be incorrect.

B. Roy and W. Meier (Eds.): FSE 2004, LNCS 3017, pp. 193–209, 2004.

CA is a special case of linear finite state machines (LFSM) and can be viewed as a one-dimensional array of cells. The cells change state at each clock tick, and the new state of a cell is completely determined by its present state and those of its left and right neighbors. CA's have been applied to various fields such as biological system, fault-tolerant computation, VLSI design, and cryptography. (See [12] for a survey on the general theory of CA.) In the cryptographical field, CA's have been used in designing hash functions and stream ciphers[8, 13]. It was believed that from the security perspective, a CA would give properties better than those of an LFSR. However, we shall show that the use of a CA in place of an LFSR does not necessarily increase security.

Recalling a well-known fact from linear algebra, we give a way to realize an LFSM, utilizing an LFSR. In short, the realization is done by attaching a linear map to an LFSR. We understand that, due to its simplicity, this could have been known to experts of this field. But we could not find any references, and it seems that this fact was not looked at from the security perspective.

The realization could be of interest in its own. For example, it gives a natural way of running a CA in the reverse direction, something which was thought to be a complex procedure. But as will be shown through the examination of arguments in [10], this realization also has grave consequences in the use of LFSM's as cryptographic building blocks.

The paper is organized as follows. Section 2 shall present the simple mathematical fact that is the starting point of this paper. This is used in Section 3 in realizing an LFSM using an LFSR and a linear map. In Section 4, we review the notion of Anderson information leakage and examine the system given in [10]. Using the realization of a CA which utilizes an LFSR, we shall argue that the system did not achieve its design goal. The section that follows presents an explicit example confirming these arguments. Next, in Section 6 we give a general method for checking whether or not a given NF generator based on an LFSM admits a reduction to a NF generator based on an LFSR that is vulnerable to Anderson information leakage. The last section closes the paper with some concluding remarks. Some readers might want to read Appendix E, which contains remarks on what further developments the basic idea of this paper might bring.

2 Basic Facts and Definitions

In this section we shall recall some elementary facts from linear algebra and introduce two classes of linear finite state machines, CA and LFSR.

2.1 Linear Algebra

Let us denote by I the $n \times n$ identity matrix. The characteristic polynomial of a matrix M with entries in the binary field $\mathbf{F}_2$ is defined to be the polynomial

$$\mathrm{char}(M) = \det(xI - M) \in \mathbf{F}_2[x]. \tag{1}$$

We define the *companion matrix* of a monic polynomial

$$p(x) = a_0 + a_1 x + \cdots + a_{n-1}x^{n-1} + x^n \in \mathbf{F}_2[x] \tag{2}$$

to be the matrix

$$\begin{pmatrix} 0 & 1 & 0 & 0 & \cdots & 0 & 0 \\ 0 & 0 & 1 & 0 & \cdots & 0 & 0 \\ 0 & 0 & 0 & 1 & & 0 & 0 \\ \vdots & & & & & & \vdots \\ 0 & 0 & & & & & 0 \\ 0 & 0 & 0 & \cdots & \cdots & 0 & 1 \\ a_0 & a_1 & a_2 & \cdots & \cdots & a_{n-2} & a_{n-1} \end{pmatrix}. \tag{3}$$

We shall accept the following statement as a fact.

Fact 1 *Let $p(x)$ be the characteristic polynomial of a square matrix M. Denote by L, the companion matrix of the monic polynomial $p(x)$. If $p(x)$ is irreducible, then there exists an invertible (basis transition) matrix T satisfying*

$$TMT^{-1} = L.$$

This is the only fact from linear algebra we shall need in this paper. Readers familiar with linear algebra can look up Appendix A to see justification for this fact.

Remark 1. The matrix T appearing in this fact is not unique. If the size of the square matrix M is n, there can be up to $2^n - 1$ of them.

2.2 Linear Finite State Machine

An n-bit *linear finite state machine* (LFSM), denoted by $\mathcal{M}$, is a pair $(\mathbf{F}_2^n, M)$, where M is an $n \times n$ matrix. The internal state of $\mathcal{M}$ is described by an n-bit vector $\mathbf{v} = (v_0, \ldots, v_{n-1}) \in \mathbf{F}_2^n$. The evolution of $\mathcal{M}$ over discrete time $t \geq 0$ is described by a sequence of n-bit vectors $\mathbf{v}^{(0)}, \mathbf{v}^{(1)}, \ldots$, satisfying

$$\mathbf{v}^{(t+1)} = M\mathbf{v}^{(t)}. \tag{4}$$

Here, $\mathbf{v}^{(0)}$ is the initial state. For $t \geq 0$, we shall write

$$\mathbf{v}^{(t)} = (v_0^{(t)}, v_1^{(t)}, \ldots, v_{n-1}^{(t)}).$$

It is well known that if the characteristic polynomial of M is *primitive* over $\mathbf{F}_2$, then each of the sequences

$$\mathbf{v}_i = (v_i^{(t)})_{t \geq 0} \tag{5}$$

has period $2^n - 1$ [7]. This is the maximum possible period obtainable for the state sequence of an LFSM. The most popular subclasses of the LFSM's are CA's and LFSR's.

2.3 Cellular Automaton

A *cellular automaton* (CA) is an LFSM with a defining matrix M which is tri-diagonal. If the upper and lower subdiagonal entries of M are all equal to 1, then it is called a 90/150 CA. Visually, the general matrix defining a 90/150 CA will be of the form

$$\begin{pmatrix} c_0 & 1 & 0 & 0 \cdots & \cdots & 0 \\ 1 & c_1 & 1 & 0 \cdots & \cdots & 0 \\ 0 & 1 & c_2 & 1 & & 0 \\ \vdots & & & & & \vdots \\ & & & & 1 & 0 \\ 0 & & \cdots & 1 & c_{n-2} & 1 \\ 0 & & \cdots & 0 & 1 & c_{n-1} \end{pmatrix}, \tag{6}$$

where each c_i is either 0 or 1. We shall only consider 90/150 CA's in this paper. The sequences obtained from such a CA satisfies the following relation. For each $0 \le i \le n-1$ and $t \ge 0$,

$$v_i^{(t+1)} = v_{i-1}^{(t)} \oplus c_i v_i^{(t)} \oplus v_{i+1}^{(t)},$$

where we take $v_{-1}^{(t)} = v_n^{(t)} = 0$.

2.4 Linear Feedback Shift Register

A *linear feedback shift register* (LFSR) corresponding to a monic polynomial $p(x)$ given by (2) is an LFSM with the defining matrix set to the companion matrix of $p(x)$ given by (3). So if we write the internal state of the LFSR at time $t \ge 0$ as $\mathbf{v}^{(t)} = (v_0^{(t)}, v_1^{(t)}, \ldots, v_{n-1}^{(t)})$, we have

$$v_i^{(t+1)} = v_{i+1}^{(t)}$$

for each $0 \le i \le n-2$, and

$$v_{n-1}^{(t+1)} = a_0 v_0^{(t)} \oplus a_1 v_1^{(t)} \oplus \cdots \oplus a_{n-1} v_{n-1}^{(t)}.$$

Hence, register contents will be shifted to the *left* by one cell during the evolution process.

3 Reducing an LFSM to an LFSR

In this section, we shall see how the contents of the previous section relate to each other. We give a way to realize an LFSM, utilizing an LFSR and a linear map. It seems that the method we are going to give is known to the experts of this field. But we could not find any references, so it is explained here for completeness.

Let us be given an LFSM $\mathcal{M}$ defined by a matrix M. Denote the characteristic polynomial of M by $p(x)$ and the companion matrix of $p(x)$ by L. Notice that the matrix L defines an LFSR. We say that this LFSR is *associated* with the LFSM $\mathcal{M}$.

Suppose that $p(x)$ is irreducible. Then we know from Fact 1 that there exists some invertible matrix T such that

$$TMT^{-1} = L. \tag{7}$$

Now, recalling that the evolution of LFSM internal state is given by (4), if the initial state of the LFSM $\mathcal{M}$ was $\mathbf{v}$, the state $\mathbf{v}^{(t)}$ of the LFSM at time $t \geq 0$ will be given by

$$\mathbf{v}^{(t)} = M^t \mathbf{v}.$$

Here, M^t denotes M multiplied t times and not the transpose of M. Similarly, if the initial state of the LFSR defined by the matrix L was $\mathbf{w}$, the state $\mathbf{w}^{(t)}$ of the LFSR at time $t \geq 0$ will be given by

$$\mathbf{w}^{(t)} = L^t \mathbf{w}.$$

Now, if we had $\mathbf{w} = T\mathbf{v}$, using (7), we can easily check the following sequence of equalities.

$$\mathbf{v}^{(t)} = M^t \mathbf{v} = (T^{-1}LT)^t \mathbf{v} = T^{-1}L^t T\mathbf{v} = T^{-1}L^t \mathbf{w} = T^{-1}\mathbf{w}^{(t)}. \tag{8}$$

This shows that an LFSM is intimately related to the LFSR defined by its characteristic polynomial.

Proposition 1. *The current internal state $\mathbf{v}^{(t)}$ of an LFSM which starts at the initial state $\mathbf{v}^{(0)}$ may be calculated using the internal state $\mathbf{w}^{(t)}$ of the associated LFSR through the simple linear equation*

$$\mathbf{v}^{(t)} = T^{-1}\mathbf{w}^{(t)} \tag{9}$$

by initializing the LFSR with $\mathbf{w}^{(0)} = T\mathbf{v}^{(0)}$.

So, even though an LFSM seems much more complicated than an LFSR, the two are only apart by a simple linear transformation.

The only hypothesis on the LFSM we have used in this section is that its characteristic polynomial be irreducible. In most cryptographic applications of an LFSM, the characteristic polynomial will be taken to be primitive, in order to achieve maximal period, so this is not a very restricting assumption. Hence any cryptographic system that bases its safety on the complexity of an LFSM, compared to an LFSR, may not be as safe as it seems at first sight. Since CA's are just a special type of LFSM's the same can be said of systems using CA's.

4 Security of Nonlinear Filter Models Utilizing a CA

In this section, we shall present a system which has tried to use a CA in place of an LFSR in order to remove some unwanted property of a stream cipher. We shall apply the theory of Section 3 to show that the attempt did not succeed in achieving its goal.

4.1 The NF-CA Model

In the paper [10], a memoryless synchronous stream cipher called the *filter-combiner* (FC) model was introduced. We shall not present the whole FC model in this paper, but use only a small part of the model in explaining one of the main arguments of that work. The referenced paper contains more than what is presented here.

Let $\mathcal{M} = (\mathbf{F}_2^n, M)$ be a CA. We assume that the characteristic polynomial $p(x)$ of the CA is primitive. It is known that each of the n sequences given by (5) are all exactly the same periodic sequence with only the starting points different. Hence they are relative shifts of each other.

We apply a nonlinear filter f with good properties, for example, high resiliency and nonlinearity, on the cells of the CA to obtain a stream cipher. The system is to satisfy the following loosely stated constraints. We refer the reader to the original paper [10] for exact statements.

1. The number r of cells used as inputs to f is small relative to the size n of the CA. ($r \leq \log_2 n$)
2. The starting points of the sequences obtained from the cells used as inputs to f is (almost) evenly distributed within the common periodic sequence.
3. The number of bits encrypted using the system does not come close to $2^n/r$.

We shall call this reduced model by the name NF-CA, a nonlinear filter model utilizing a CA. The paper claims that under these constraints the NF-CA is resistant to Anderson information leakage [1].

Anderson information leakage is an observation on the nonlinear filter model (a stream cipher that applies a nonlinear filter on an LFSR) that allows one to gather information on the initial state of the LFSR from the key stream. More explanation is given in the next subsection.

The author of [10] believed that Anderson information leakage was fundamentally due to using the same bit more than once as input to the nonlinear filter in obtaining the key stream. This reasoning was also somewhat vaguely stated in the paper [1]. Hence the main objective behind the above constrains was to remove the possibility of any part of the periodic sequence being used more than once.

4.2 Anderson Information Leakage

Consider the filter model of stream ciphers. This is a stream cipher that uses cell states of an LFSR as inputs to a nonlinear filter in obtaining a key stream. We shall write this model as NF-LFSR for short.

Suppose we use k consecutive cells of the LFSR as inputs to the nonlinear filter f. Let us take the convention, as given in Section 2.4, that the contents of the register are being shifted to the left at each step. If we fix the contents of k cells used as inputs to f, we can calculate one bit of output from the NF-LFSR. Similarly, if we fix contents of the k cells and also $(k-1)$ more cells that lie immediately to their right, we can calculate k output bits from the NF-LFSR.

Now, suppose we classify all possible $(2k-1)$-bit states according to the k-bit output key stream it will give. In the ideal case, each class will contain exactly 2^{k-1} elements. That this distribution usually is not uniform was investigated by Anderson [1] to show that much information about the state of the LFSR may be obtained from the output key stream.

He gives an explicit example using a 2-resilient nonlinear filter that uses 5 variables. The above mentioned table is constructed to show that it is indeed irregular. To show that actually useful information may be found, he lists all possible 9-bit initial states that can give the 5-bit output stream `11010`.

001010101	001110010	100110010	101110001	110110001
001110001	100110001	101001011	101110010	110110010

If we look closely at these values, we see that there is only a single `0` among all the 5th bits, and a single `1` among both 6th and 7th. In other words, if the key stream we obtain is `11010`, then at the starting point of this key stream, the state of the 5th cell of the LFSR would have been `1` with probability 0.9. Likewise, state of the 6th and 7th bit would have been `0` with probability 0.9.

Irregularity in the distribution of initial states classified according to output stream blocks contains potential for the NF-LFSR giving out information on the initial LFSR state.

We remark that some further developments of Anderson's idea appear in [6, 2, 4].

4.3 Information Leakage of the NF-CA

Let the initial state of the NF-CA, or equivalently, that of the CA $\mathcal{M} = (\mathbf{F}_2^n, M)$ be denoted by $\mathbf{v}^{(0)}$. We shall add dummy variables to the nonlinear filter f and view it as defined on the whole CA. Then the t-th output key stream bit c_t of the NF-CA will be given by

$$c_t = f(\mathbf{v}^{(t)}). \tag{10}$$

We may follow through the arguments of Section 3 in constructing an associated LFSR and finding a linear transformation T satisfying (7). And by applying (9) to the above equation, we may write

$$c_t = f \circ T^{-1}(\mathbf{w}^{(t)}), \tag{11}$$

where we have taken the initial state of the associated LFSR to be $\mathbf{w}^{(0)} = T\mathbf{v}^{(0)}$. Notice that since T^{-1} is a simple linear map, we may view the map $g = f \circ T^{-1}$ as just another normal nonlinear filter. That is, we have

$$c_t = g(\mathbf{w}^{(t)}). \tag{12}$$

We see that the right hand side is now the output of a normal NF-LFSR.

Proposition 2. *The NF-CA which uses nonlinear filter f on a CA initialized to $\mathbf{v}^{(0)}$ may be realized as an NF-LFSR. This is done by applying the nonlinear filter $g = f \circ T^{-1}$ to the associated LFSR and initializing it to $\mathbf{w}^{(0)} = T\mathbf{v}^{(0)}$.*

Now, we do not yet have any criterion for measuring an NF-LFSR's resistance to Anderson information leakage. And, as stated in Anderson's paper [1], random NF-LFSR's tend to leak a lot of information. Hence there is a non-dismissible chance of (12) and hence (10) representing a stream cipher which is not resistant to Anderson information leakage.

Remark 2. Anderson information leakage does not seem to be applicable to the nonlinear combiner model. Hence Anderson information leakage is probably not applicable to the FC model of [10]. But, once again, this is due to the use of *combiner* part of the FC model rather than from the three constraints of Section 4.1.

5 Explicit Example of Leaking NF-CA

We have constructed a small but concrete example to verify that it is possible for an NF-CA to satisfy all three of the constraints introduced in Section 4.1 and still show Anderson information leakage.

5.1 CA and Its Relation to an LFSR

Consider the 90/150 CA represented by a matrix M of the form (6) with the diagonal entries given by

$$(c_0, c_1, \ldots, c_{22}) = (1,1,1,0,0,1,0,0,1,1,1,1,1,0,0,1,1,1,0,0,1,0,0).$$

For clarity, we have written down the explicit matrix M in Appendix B. Characteristic polynomial $p(x)$ of M is

$$1 + x^2 + x^4 + x^7 + x^9 + x^{10} + x^{11} + x^{13} + x^{14} + x^{15} + x^{17} + x^{19} + x^{20} + x^{22} + x^{23}.$$

This is a primitive polynomial so that each of the 23 cells of the CA gives a sequence of period $(2^{23} - 1)$. Let L be the companion matrix of $p(x)$. To define T, we let $T_1 = (1, 0, 0, \ldots, 0)$ be its top row and recursively fix the i-th row T_i by setting

$$T_i = MT_{i-1} \quad \text{for } 1 < i \leq 23. \tag{13}$$

The actual matrix T may be found in Appendix C.

Checking

$$TMT^{-1} = L \tag{14}$$

is easy. We remark that the such a T is not unique. Any invertible matrix obtained through the process (13) starting with an arbitrary nonzero vector will satisfy (14). For the above T, its inverse, T^{-1}, is given in Appendix D.

5.2 Shifts between CA Cells and the Nonlinear Filter

Let $m_i^{(t)}$ be the sequences generated by the i-th cell of the CA for $1 \leq i \leq 23$. Then the relative shifts between $m_1^{(t)}$ and $m_i^{(t)}$ are

$$0 \ \mathbf{1988170} \ \mathbf{8388605} \ 5964510 \ \mathbf{4125305} \ 3763873 \ \mathbf{6190462} \ 6778815 \ \cdots.$$

These have been calculated using a program implementing [11]. From this, one can check that the relative shifts between the four sequences obtained from the 2nd, 3rd, 5th, and 7th cells are quite close to 2^{21} or 2^{22}.

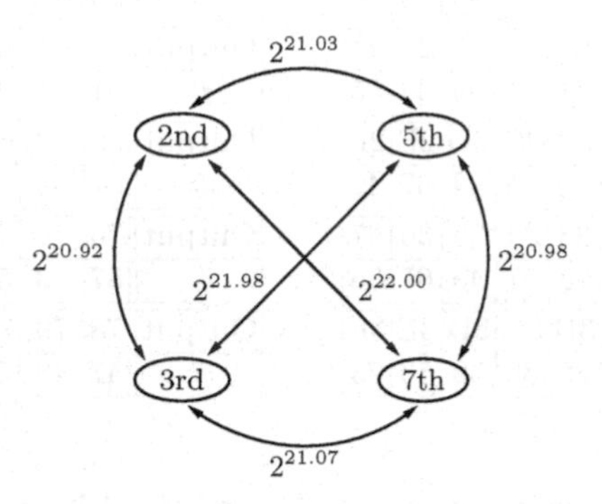

For example, the shift between $m_3^{(t)}$ and $m_2^{(t)}$ is

$$\begin{aligned} 1988170 - 8388605 &= -6400435 \\ &\equiv 1988172 \pmod{(2^{23}-1)} \\ &\bullet \ 2^{20.92301\cdots}. \end{aligned}$$

Hence, if we apply the nonlinear filter given by

$$f = m_2 \oplus m_3 \oplus (m_5 \cdot m_7) \tag{15}$$

on the CA, Rule 2 of Section 4.1 is satisfied. It is easily checked that f is a 1-resilient function, so we are not using a very bad filter.

That Rule 1 is also satisfied is easily checked by calculating

$$\log_2 23 = 4.52356\cdots \geq 4.$$

To satisfy Rule 3, we just need to use less than 2^{21} bits from the NF-CA we shall create. This should not be a problem since we shall be using less than 20 key stream bits.

5.3 Equivalent NF-LFSR

We may recall from equations (10) and (11) that

$$f(\mathbf{v}^{(t)}) = (f \circ T^{-1})(\mathbf{w}^{(t)}).$$

Table 1. The number of possible input states for each 7-bit output

Output	00	01	02	03	04	05	06	07
#	38	51	61	73	51	75	73	89
Output	08	**09**	0A	0B	0C	0D	0E	0F
#	65	**29**	67	63	69	69	87	63
Output	10	11	12	13	14	15	16	17
#	79	63	45	37	87	67	77	57
Output	18	19	1A	1B	1C	1D	1E	1F
#	49	73	43	59	73	85	59	71
Output	20	21	22	23	24	25	26	27
#	73	93	51	71	61	69	39	55
Output	28	29	2A	2B	2C	2D	2E	2F
#	79	83	77	49	75	43	57	49
Output	30	31	32	33	34	35	36	37
#	65	49	99	75	57	45	67	55
Output	38	39	3A	3B	3C	3D	3E	3F
#	63	71	69	85	39	59	53	73

Output	40	41	42	43	44	45	46	47
#	51	83	65	89	43	55	65	61
Output	48	49	4A	4B	4C	4D	4E	4F
#	77	53	87	71	53	33	71	67
Output	50	51	52	53	54	55	56	57
#	87	59	85	57	75	59	41	49
Output	58	59	5A	5B	5C	5D	5E	5F
#	65	101	59	63	61	69	39	55
Output	60	61	62	63	64	65	66	67
#	61	61	47	55	69	89	47	83
Output	68	69	6A	6B	6C	6D	6E	6F
#	67	59	57	41	91	79	73	45
Output	70	71	72	73	74	75	**76**	77
#	57	53	59	55	69	53	**103**	63
Output	78	79	7A	7B	7C	7D	7E	7F
#	47	43	53	81	51	75	73	89

And in terms of the state $(l_1, l_2, \ldots, l_{23})$ of the LFSR, the nonlinear filter f translates into the nonlinear filter

$$g = (f \circ T^{-1}) = (l_1 \oplus l_2 \oplus l_3) \oplus (l_2 \oplus l_4 \oplus l_5) \cdot (l_1 \oplus l_2 \oplus l_3 \oplus l_5 \oplus l_7).$$

We have used the 2nd, 3rd, 5th, and 7th rows of the explicitly calculated T^{-1} given in Appendix D. Notice that the 7th bit from the LFSR is the rightmost bit used to find states of the 4 CA cells we have chosen to use for the NF-CA.

5.4 Information Leakage

Since the 7th bit of LFSR remains in effect until we obtain 7 bits of the output stream, we trace the 7-bit output stream by running the obtained NF-LFSR on all possible 13-bit states of the leftmost part of the LFSR. Input count for each output is given in Table 1.

The 7-bit output stream is represented in hexadecimal notation and # denotes the corresponding possible input state count. Leftmost bit of the 7 bits in hexadecimal notation(exclude the leading 0 from the 8 bits) is the first output bit.

In the ideal case, all the counts should be equal to (or, at least near) $2^6 = 64$. But as we see in Table 1, this is not the case. It is quite irregular. Number as big as 103 appears and number as small as 29 also appears. So this shows the potential of this structure leaking information.

For example, let us look at the following list of all 13-bit input states that give the 7-bit output stream `0x09 = 0001001`. As given by the table, there are 29 such states.

1101000000100	0011000000110	0110101110001	0110010001011
0011000000100	1011000000110	0110101110101	1110010001011
1011000000100	0111000000110	0110100011101	0110100101011
0111000000100	0110101110110	0110101110011	0110101110111
0110101110100	0110100011110	1101000001011	0110100011111
0110100011100	0000011111110	0011000001011	
0110100101010	0000011100001	1011000001011	
1101000000110	0000000110001	0111000001011	

We find that, with probability 28/29, only one of the bits 4, 5, and 6 is equal to 1. In particular, sum of the three bits is equal to 1 with probability 28/29. Anderson information leakage theory is applicable to this structure. Therefore, applying a nonlinear filter having good cryptographic properties to a CA and using cells of large relative shifts, as suggested in [10], does not necessarily prevent Anderson information leakage.

6 Checking the Vulnerability of a Given NF-LFSM

In this section, we shall give a general method for checking whether or not a given NF-LFSM allows reduction to a vulnerable NF-LFSR. Since CA and LFSR are subclasses of LFSM, our method applies even to NF-LFSR. That is, we can check whether the nonlinear filter to an NF-LFSR may be rewritten in a form that shows information leakage.

Let us be given an NF-LFSM defined by a matrix M of size n and a nonlinear filter f. Denote by L the companion matrix of M and write

$$\mathcal{Z}(L) = \{Z \in \mathrm{GL}(n) \mid ZL = LZ\}$$

for the centralizer of L.

For a given companion matrix L, it is easy to write down $\mathcal{Z}(L)$ more explicitly. The following lemma may be proved through a straightforward application of $ZL = LZ$.

Lemma 1. *For the companion matrix L given by (3), the centralizer $\mathcal{Z}(L)$ consists of elements $Z = (z_{i,j})_{i,j=0}^{n-1} \in \mathrm{GL}(n)$ satisfying*

$$\begin{aligned}
z_{i+1,0} &= a_0 z_{i,n-1}, \\
z_{i+1,1} &= a_1 z_{i,n-1} \oplus z_{i,0}, \\
z_{i+1,2} &= a_2 z_{i,n-1} \oplus z_{i,1}, \\
&\vdots \\
z_{i+1,n-1} &= a_{n-1} z_{i,n-1} \oplus z_{i,n-2}
\end{aligned}$$

for all $0 \le i < n-1$.

The important implication of this lemma is that every entry of a $Z \in \mathcal{Z}(L)$ may be written as a linear sum of the terms belonging to its first row in a uniform way. For example,

$$\mathcal{Z}\left(\begin{pmatrix} 0 & 1 & 0 \\ 0 & 0 & 1 \\ a & b & c \end{pmatrix}\right) = \left\{ \begin{pmatrix} x & y & z \\ az & x \oplus bz & y \oplus cz \\ ay \oplus acz & az \oplus by \oplus bcz & x \oplus bz \oplus cy \oplus cz \end{pmatrix} \in \mathrm{GL}(3) \right\}.$$

Now, fix any matrix $\bar{T}$ satisfying $\bar{T}M\bar{T}^{-1} = L$. It is an easy exercise in linear algebra to show that the set of all T satisfying (7) is given by $\mathcal{Z}(L)\bar{T} := \{Z\bar{T} \mid Z \in \mathcal{Z}(L)\}$. And since $Z \in \mathcal{Z}(L)$ if and only if $Z^{-1} \in \mathcal{Z}(L)$, we have the following proposition.

Proposition 3. *The set of all T^{-1} satisfying (7) is given by*

$$\bar{T}^{-1}\mathcal{Z}(L) := \{\bar{T}^{-1}Z \mid Z \in \mathcal{Z}(L)\}.$$

We are now ready to give a general method for checking whether or not a given NF-LFSM allows reduction to a vulnerable NF-LFSR. If a nonlinear filter applied to an LFSR uses a small number of variables and if those variables correspond to LFSR cells that are close to each other, then such an NF-LFSR is vulnerable to Anderson information leakage. Otherwise, the NF-LFSR is highly immune to information leakage. Hence, for a given NF-LFSM, it suffices to check the possibility of finding a T, for which the filter $g = f \circ T^{-1}$ given by Proposition 2 uses variables from a small clustered set.

Decide on a (small) number $s < n$. If it is possible to choose T so that all variables used by g falls within some s consecutive LFSR cells, we shall conclude that there is a high probability that the NF-LFSM yields to Anderson information leakage. Otherwise we shall presume that the NF-LFSM does not leak information.

Procedure for Checking Vulnerability.

1. From the matrix M of size n, defining the LFSM, calculate its characteristic polynomial and the associated LFSR L.
2. Fix any matrix $\bar{T}$ satisfying $\bar{T}M\bar{T}^{-1} = L$. Using the idea of (13) is one way to do this.
3. Write $\mathcal{Z}(L)$ in the form given by Lemma 1, so that all entries of lower rows are expressed as linear combinations of the first row terms. We shall denote the first row terms by $x_0, \ldots, x_{n-1}$.
4. Recalling Proposition 3, multiply $\bar{T}^{-1}$ to the general element of $\mathcal{Z}(L)$ obtained in the previous step to express the general T^{-1}. All entries of the general T^{-1} will again be linear combinations of x_j.

$\star$. Let r be the number of rows used by f. Note that from the general expression of T^{-1}, which is an $n \times n$ array of linear sums over x_j, only the r rows that that correspond to variables used by the nonlinear filter f will have any significance.

5. Remove all rows not corresponding to variables used by f.

⋆. Now, suppose that for some explicit nontrivial values of the variables x_j, all the remaining entries evaluate to zero, except for those contained in the first s columns. Then $g = f \circ T^{-1}$ would used only s variables for the T^{-1} evaluated at the explicit values.

6. Temporarily remove the first s columns from the remaining array of T^{-1} entries.
7. Check whether setting all remaining entries to zero yields a nontrivial solution.
8. If a nontrivial solution is found, conclude that the NF-LFSM allows reduction to a vulnerable NF-LFSR.
9. Otherwise, bring back the array of linear sums obtained after Step 5.
10. Unless we've tried all possible consecutive s columns, (temporarily) remove the next set of s consecutive columns and go back to Step 7.
11. If no nontrivial solution is found, conclude that the NF-LFSM resists Anderson information leakage.

Notice that at Step 7, we have a set of $r \times (n-s)$ equations in n variables. For most interesting values of r and s, the number of equations would be larger than the number of variables. But our (small number of) testings show that nontrivial solutions do occur from time to time.

We close this section by adding that the complexity of this process can easily be seen to be of polynomial order in n.

7 Conclusion

We have seen that an LFSM (or a CA) is intimately connected to an LFSR by the simple relation (9). This structure allows one to realize an LFSM using an LFSR and a linear transformation. Since an LFSR is much simpler than an LFSM, this will have implications on the security of any system that (iteratively) uses an LFSM as one of its building blocks, and has used it assuming that it is more complex than an LFSR.

An example of such an attempt has been examined in this paper. Using a CA in place of an LFSR in an attempt to remove Anderson information leakage from a nonlinear filter model has failed.

We have also given a general method for checking whether or not a given NF-LFSM allows reduction to an NF-LFSR which is vulnerable to information leakage.

References

[1] Ross Anderson, Searching for the optimum correlation attack. *Proceedings of FSE*, LNCS 1008, pp. 137–143, Springer-Verlag, 1995. 193, 198, 199, 200

[2] Jovan Dj. Golić, On the security of nonlinear filter generators. *Fast Software Encryption*, LNCS 1039, pp. 173–188, Springer-Verlag, 1996. 199

[3] Jovan Dj. Golic, Correlation via linear sequential circuit approximation of combiners with memory. *Advances in Cryptology - EUROCRYPT'92,* LNCS 658, pp. 113-123, Springer-Verlag. 193
[4] Jovan Dj. Golic, Andrew Clark, and Ed Dawson, Generalized inversion attack on nonlinear filter generators. *IEEE Transations on Computers*, vol. 49 (10), pp. 1100–1109, October, 2000. 199
[5] Thomas W. Hungerford, *Algebra.* GTM 73, Springer-Verlag, 1980. 206
[6] Sangjin Lee, Seongtaek Chee, Sangjoon Park, and Sungmo Park, Conditional correlation attack on nonlinear filter generators. *Advances in Cryptology - ASIACRYPT'96,* LNCS 1163, pp. 360–367, Springer-Verlag, 1996. 199
[7] R. Lidl and H. Niederreiter. *Introduction to finite fields and their applications.* Cambridge University Press, 1994. 195
[8] Miodrag Mihaljević, Yuliang Zheng, and Hideki Imai, A Family of Fast Dedicated One-Way Hash Functions Based on Linear Cellular Automata over GF(q), *IEICE Trans. Fundamentals*, vol.E82-A(a), pp. 1–8, January 1999. 194
[9] W. Meier and O. Staffelbach, Fast correlation attacks on certain stream ciphers. *Journal of Cryptology,* vol.1, pp. 159–176, 1989. 193
[10] Palash Sarkar, The filter-combiner model for memoryless synchronous stream ciphers. *Advances in Cryptology - CRYPTO 2002,* LNCS 2442, pp. 533–548, Springer-Verlag, 2002. 193, 194, 198, 200, 203
[11] Palash Sarkar, Computing shifts in 90/150 cellular automata sequences. *Finite Fields and their Applications*, vol. 9 (2), pp. 175–186, April 2003. 201
[12] Palash Sarkar, Brief History of Cellular Automata, *ACM Computing Serveys*, vol. 32 (1), pp. 80–107, March 2000. 194
[13] Palash Sarkar, Hiji-bij-bij: A New Stream Cipher with a Self-Synchronizing Mode of Operation, ICAR e-print 2003-014, 2003. `http://eprint.iacr.org`. 194

A Explanation for Fact 1

Let us denote by E, a vector space over a field $\mathbf{K}$. We have gathered some well-known facts from linear algebra in the following theorem. Basic definitions and proofs may be found in standard textbooks (for example, [5]).

Theorem 1. *Let $\phi : E \to E$ be a linear transformation. Then the following statements hold.*

1. *There exists monic polynomials of positive degree $p_1, \ldots, p_t \in \mathbf{K}[x]$ and ϕ-cyclic subspaces $E_1, \ldots, E_t$ of E such that $E = E_1 \oplus \cdots \oplus E_t$ and $p_1 | p_2 | \cdots | p_t$.*
2. *The sequence $p_1, \ldots, p_t$ is uniquely determined by E and ϕ. (This is called the* invariant factors *of ϕ.)*
3. *If E is a ϕ-cyclic space and ϕ has minimal polynomial $p(x)$ of degree r, then $\dim_K E = r$ and there exists an ordered basis of E relative to which the matrix of ϕ is the* companion matrix *of $p(x)$.*
4. *The characteristic polynomial of ϕ is the product of its invariant factors.*

From these statements, we may easily obtain the following corollary, stated in terms of matrices.

Corollary 1. *Let A be an $n \times n$ matrix with entries in the field $\mathbf{K}$. If the characteristic polynomial of A is irreducible, then the matrix A is* similar *to the companion matrix of the characteristic polynomial.*

To make it easier for those without a mathematical background and to make everything explicit, we shall explain this corollary, giving out some basic definitions.

The companion matrix of a monic polynomial $p(x) = a_0 + a_1x + \cdots + a_{n-1}x^{n-1} + x^n \in \mathbf{K}[x]$ is usually defined to be the matrix

$$\begin{pmatrix} 0\,0 & 0 & \cdots & 0 & -a_0 \\ 1\,0 & 0 & \cdots & 0 & -a_1 \\ 0\,1 & 0 & \cdots & 0 & -a_2 \\ \vdots & & & \vdots & \vdots \\ \vdots\ \vdots & & & 0 & \vdots \\ 0\,0 & \cdots & 0 & 1 & -a_{n-1} \end{pmatrix}.$$

We can see that the form given by (3) is the transpose of this one, if we take into consideration the fact that we were dealing with the binary field there.

Now, let $p(x)$ be the characteristic polynomial of a square matrix A and let B be the companion matrix of $p(x)$. Corollary 1 states that if $p(x)$ is irreducible, then there exists some invertible matrix C satisfying

$$CA\,C^{-1} = B.$$

Notice that we may take the transpose of both sides to obtain

$$(C^T)^{-1}A^TC^T = B^T.$$

It is clear that the characteristic polynomial of A is equal to that of A^T. Hence, Fact 1 follows.

B Matrix M Defining the CA

```
1 1 0 0 0 0 0 0 0 0 0 0 0 0 0 0 0 0 0 0 0 0 0
1 1 1 0 0 0 0 0 0 0 0 0 0 0 0 0 0 0 0 0 0 0 0
0 1 1 1 0 0 0 0 0 0 0 0 0 0 0 0 0 0 0 0 0 0 0
0 0 1 0 1 0 0 0 0 0 0 0 0 0 0 0 0 0 0 0 0 0 0
0 0 0 1 0 1 0 0 0 0 0 0 0 0 0 0 0 0 0 0 0 0 0
0 0 0 0 1 1 1 0 0 0 0 0 0 0 0 0 0 0 0 0 0 0 0
0 0 0 0 0 1 0 1 0 0 0 0 0 0 0 0 0 0 0 0 0 0 0
0 0 0 0 0 0 1 0 1 0 0 0 0 0 0 0 0 0 0 0 0 0 0
0 0 0 0 0 0 0 1 1 1 0 0 0 0 0 0 0 0 0 0 0 0 0
0 0 0 0 0 0 0 0 1 1 1 0 0 0 0 0 0 0 0 0 0 0 0
0 0 0 0 0 0 0 0 0 1 1 1 0 0 0 0 0 0 0 0 0 0 0
0 0 0 0 0 0 0 0 0 0 1 1 1 0 0 0 0 0 0 0 0 0 0
0 0 0 0 0 0 0 0 0 0 0 1 1 1 0 0 0 0 0 0 0 0 0
0 0 0 0 0 0 0 0 0 0 0 0 1 0 1 0 0 0 0 0 0 0 0
0 0 0 0 0 0 0 0 0 0 0 0 0 1 0 1 0 0 0 0 0 0 0
0 0 0 0 0 0 0 0 0 0 0 0 0 0 1 1 1 0 0 0 0 0 0
0 0 0 0 0 0 0 0 0 0 0 0 0 0 0 1 1 1 0 0 0 0 0
0 0 0 0 0 0 0 0 0 0 0 0 0 0 0 0 1 1 1 0 0 0 0
0 0 0 0 0 0 0 0 0 0 0 0 0 0 0 0 0 1 0 1 0 0 0
0 0 0 0 0 0 0 0 0 0 0 0 0 0 0 0 0 0 1 0 1 0 0
0 0 0 0 0 0 0 0 0 0 0 0 0 0 0 0 0 0 0 1 1 1 0
0 0 0 0 0 0 0 0 0 0 0 0 0 0 0 0 0 0 0 0 1 0 1
0 0 0 0 0 0 0 0 0 0 0 0 0 0 0 0 0 0 0 0 0 1 0
```

C Matrix T

```
1 0 0 0 0 0 0 0 0 0 0 0 0 0 0 0 0 0 0 0 0 0 0
1 1 0 0 0 0 0 0 0 0 0 0 0 0 0 0 0 0 0 0 0 0 0
0 0 1 0 0 0 0 0 0 0 0 0 0 0 0 0 0 0 0 0 0 0 0
0 1 1 1 0 0 0 0 0 0 0 0 0 0 0 0 0 0 0 0 0 0 0
1 0 1 1 1 0 0 0 0 0 0 0 0 0 0 0 0 0 0 0 0 0 0
1 0 0 0 1 1 0 0 0 0 0 0 0 0 0 0 0 0 0 0 0 0 0
1 1 0 1 1 0 1 0 0 0 0 0 0 0 0 0 0 0 0 0 0 0 0
0 0 0 1 1 0 0 1 0 0 0 0 0 0 0 0 0 0 0 0 0 0 0
0 0 1 1 1 1 1 0 1 0 0 0 0 0 0 0 0 0 0 0 0 0 0
0 1 0 0 0 1 1 0 1 1 0 0 0 0 0 0 0 0 0 0 0 0 0
1 1 1 0 1 0 1 0 0 0 1 0 0 0 0 0 0 0 0 0 0 0 0
0 1 0 0 0 0 0 1 0 1 1 1 0 0 0 0 0 0 0 0 0 0 0
1 1 1 0 0 0 1 0 0 0 1 0 1 0 0 0 0 0 0 0 0 0 0
0 1 0 1 0 1 0 1 0 1 1 0 1 1 0 0 0 0 0 0 0 0 0
1 1 0 0 0 1 0 0 0 0 0 0 1 1 0 0 0 0 0 0 0 0 0
0 0 1 0 1 1 1 0 0 0 0 0 1 1 1 1 0 0 0 0 0 0 0
0 1 1 0 1 1 1 1 0 0 0 1 0 0 0 0 1 0 0 0 0 0 0
1 0 0 0 1 1 0 1 1 0 1 1 1 0 0 1 1 1 0 0 0 0 0
1 1 0 1 1 0 0 1 0 0 0 1 0 1 1 0 1 0 1 0 0 0 0
0 0 0 1 1 1 1 0 1 0 1 1 0 1 1 0 1 0 0 1 0 0 0
0 0 1 1 0 1 1 0 1 0 0 0 0 1 1 0 1 1 1 0 1 0 0
0 1 0 1 0 0 1 0 1 1 0 0 1 1 1 0 0 1 1 0 1 1 0
1 1 0 0 1 1 0 0 0 0 1 1 0 0 1 1 1 0 1 0 0 1 1
```

D Matrix T^{-1}

```
1 0 0 0 0 0 0 0 0 0 0 0 0 0 0 0 0 0 0 0 0 0 0
1 1 0 0 0 0 0 0 0 0 0 0 0 0 0 0 0 0 0 0 0 0 0
0 0 1 0 0 0 0 0 0 0 0 0 0 0 0 0 0 0 0 0 0 0 0
1 1 1 1 0 0 0 0 0 0 0 0 0 0 0 0 0 0 0 0 0 0 0
0 1 0 1 1 0 0 0 0 0 0 0 0 0 0 0 0 0 0 0 0 0 0
1 1 0 1 1 1 0 0 0 0 0 0 0 0 0 0 0 0 0 0 0 0 0
1 1 1 0 1 0 1 0 0 0 0 0 0 0 0 0 0 0 0 0 0 0 0
1 0 1 0 1 0 0 1 0 0 0 0 0 0 0 0 0 0 0 0 0 0 0
1 0 1 1 1 1 1 0 1 0 0 0 0 0 0 0 0 0 0 0 0 0 0
0 1 0 0 1 0 0 0 1 1 0 0 0 0 0 0 0 0 0 0 0 0 0
1 1 0 1 0 0 1 0 0 0 1 0 0 0 0 0 0 0 0 0 0 0 0
1 1 1 1 0 0 1 1 1 1 1 1 0 0 0 0 0 0 0 0 0 0 0
0 1 0 1 1 0 0 0 0 0 1 0 1 0 0 0 0 0 0 0 0 0 0
1 0 0 0 0 1 1 1 1 1 0 0 1 1 0 0 0 0 0 0 0 0 0
0 0 0 1 1 0 1 1 1 1 0 0 1 1 1 0 0 0 0 0 0 0 0
1 0 0 0 1 0 1 0 0 0 1 0 1 0 1 1 0 0 0 0 0 0 0
1 1 0 1 0 1 0 0 1 1 1 1 0 0 0 0 1 0 0 0 0 0 0
0 0 1 1 0 1 0 0 1 0 1 0 0 0 1 1 1 1 0 0 0 0 0
1 1 1 1 1 0 1 0 0 0 0 0 0 0 1 0 1 0 1 0 0 0 0
0 1 0 0 1 0 0 1 1 0 1 0 0 0 1 0 1 0 0 1 0 0 0
1 1 0 1 1 1 1 0 1 1 0 1 0 0 1 1 1 1 1 0 1 0 0
1 1 1 1 1 0 0 0 0 0 0 1 1 0 0 0 1 0 0 0 1 1 0
1 0 1 0 0 0 1 0 1 1 0 1 1 1 1 1 1 0 1 0 1 1 1
```

E Remarks on Further Developments

Here, we have gathered some remarks on what other implications the basic idea of this paper might have.

Remark 3. The arguments of this paper need not be constrained to the binary field. An LFSM which uses cells representing elements of any finite field can be realized using an LFSR over the same finite field.

Remark 4. One can deduce from Fact 1 that any two square matrices with a common irreducible characteristic polynomial are related by an invertible matrix. So even though we have focused on the relationship between an LFSM and an LFSR, the arguments of this paper can be applied in realizing an LFSM using a different (and maybe simpler) LFSM.

Remark 5. Those familiar with linear algebra will know that it is easy to extend the idea of this paper to the case when the characteristic polynomial of an LFSM is not irreducible. In such a case, the resulting realization will use several LFSR's whose sizes add up to the size of the original LFSM.

Remark 6. One can view the idea of this paper from a different direction and use it in realizing an LFSR with an LFSM. In this case, we can exploit our freedom over the choice of transition matrixes. So, for example, applying it to an NF-LFSR, one can turn any nonlinear filter of high resiliency into a filter having correlation of degree one.

Remark 7. In a way, Anderson information leakage is fundamentally due to the fact that inputs to different variables of the filter is supplied by sequences that are just small shifts of each other. This paper shows that as long as LFSM's are used, this is unavoidable. So it might be a good idea to look at *nonlinear* feedback shift registers now.

VMPC One-Way Function and Stream Cipher

Bartosz Zoltak

bzoltak@vmpcfunction.com
http://www.vmpcfunction.com

Abstract. A simple one-way function along with its proposed application in symmetric cryptography is described. The function is computable with three elementary operations on permutations per byte. Inverting the function, using the most efficient method known to the author, is estimated to require an average computational effort of about 2^{260} operations. The proposed stream cipher based on the function was designed to be efficient in software implementations and, in particular, to eliminate the known weaknesses of the alleged RC4 keystream generator while retaining most of its speed and simplicity.

Keywords: one-way function; stream cipher; cryptanalysis; RC4; lower bound

1 Introduction

A simple transformation of permutations appearing to be hard to invert is described together with its proposed practical application in a software-efficient symmetric encryption algorithm.

The transformation, here termed "VMPC" function as an abbreviation of "Variably Modified Permutation Composition", is a combination of elementary operations on permutations and integers. The simplest variant of the function can be coded with three basic "MOV" instructions from the Intel 80x86 processor instruction set per byte. When applied on 256-element permutations (a comfortable size in practical cryptographic applications), the function requires an estimated average of 2^{260} computational operations to be inverted using the most efficient method known to the author.

The very low computational cost required to obtain practical one-way property makes the function a plausible candidate for cryptographic applications. The simplicity of the function could also raise a question whether it might be possible to prove a lower bound on the complexity of inverting it. This currently is an open problem and a possible subject of future research.

A proposition of an encryption algorithm constructed as a stream cipher based on the VMPC function is described in sections 8-14. The cipher was designed to be both efficient in software implementations and to resist the known attacks against this kind of algorithms (like the alleged RC4 keystream generator) – in particular against attacks distinguishing the keystream from a truly random source and attacks against the cipher's Key Scheduling Algorithm (KSA).

B. Roy and W. Meier (Eds.): FSE 2004, LNCS 3017, pp. 210–225, 2004.

2 Definition of the VMPC Function

Notation:
n, P, Q : P, Q : n-element permutations. For simplicity of further references P and Q are assumed to be one-to-one mappings $A \to A$; $A = \{0, 1, \ldots, n-1\}$
k : Level of the function; $k < n$
$+$: addition modulo n

Definition 1. *A k-level VMPC function, referred to as $VMPC_k$, is such transformation of P into Q, where*

$$Q[x] = P[P_k[P_{k-1}[\ldots[P_1[P[x]]]\ldots]]],$$

$x \in \{0, 1, \ldots, n-1\}$,
P_i is an n-element permutation such that $P_i[x] = f_i(P[x])$, where f_i is any function such that $P_i[x] \neq P[x] \neq P_j[x]$ for $i \in \{1, 2, ..., k\}, j \in \{1, 2, ..., k\}, i \neq j$. For simplicity of further references f_i is assumed to be $f_i(x) = x + i$

Example: $Q = \text{VMPC}_1(P) : Q[x] = P[P[P[x]] + 1]$

3 Difficulty of Inverting the VMPC Function

n-element permutation P has to be recovered from the n-element permutation Q, where $Q = \text{VMPC}_k(P)$.

By definition each element of Q is formed by $k + 2$, usually different, elements of P. One element of Q can be formed by many possible configurations of elements of P (e.g. for $Q = \text{VMPC}_1(P)$: $Q[X] = Y$ can be formed by $P[X] = a, P[a] = b, P[b + 1] = Y$ for any reasonable combination of values of a and b).

All possible configurations are equally likely to be correct. If any of them is chosen, it needs to be verified with all of those elements of Q which use any of the elements of P included in the picked configuration.

Each element of P is usually used to form $k + 2$ different elements of Q thus $(k+2) \times (k+1)$ new elements of Q usually need to be inverted (all $k+2$ elements of P used to form each of these elements of Q need to be revealed) to verify the elements of P from the picked configuration.

Because the cycle structure of P is corrupted by the addition operation(s) it is usually impossible to find two different elements of Q, which share at least $k + 1$ elements of P.

Instead only such element of Q can usually be found, name it $Q[x]$, which shares only one of the $k + 2$ elements of P with another element of Q. This forces k elements of P used to form $Q[x]$ to be guessed to invert $Q[x]$.

However at each new guessed element of P there usually occur $k + 1$ new elements of Q which use this element of P and which need to be inverted to verify the guess.

The algorithm falls into a loop, where at every step usually k new elements of P need to be guessed to enable continuation of verification of the previously

Table 1. Assembler implementation of 1-level VMPC function

Instruction	Description
MOV AL, [Pm] + EAX	Store (EAX=AL)-th element of Pm in AL
MOV AL, [Pm] + EAX	Store (EAX=AL)-th element of Pm in AL
MOV AL, [Pm] + EAX+1	Store ((EAX=AL)+1)-th element of Pm in AL

guessed elements. In consequence the $k+2$ elements of P picked at the beginning of the process indirectly depend on all n elements of Q.

The described scenario is the case usually. In some circumstances the verification process can be simplified by benefiting from coincidences (where for example it is possible to find two elements of Q, which share more than one element of P (e.g. for $Q = \text{VMPC}_1(P) : Q[2] = 3 : P[2] = 4, P[4] = 8, P[9] = 3$ and $Q[1] = 8 : P[1] = 9, P[9] = 3, P[4] = 8$)).

A proposed algorithm for inverting the VMPC function (Section 6) was optimized to benefit from the possible coincidences. The average number of elements of P which need to be guessed for $n = 256$ was reduced to about 34 for 1-level VMPC function, to about 57 for 2-level VMPC, to about 77 for 3-level VMPC and to about 92 for 4-level VMPC function.

Searching through half of the possible states of these elements of P requires on average about 2^{260} steps for 1-level VMPC function, about 2^{420} for 2-level VMPC, about 2^{550} for 3-level VMPC and about 2^{660} steps for 4-level VMPC function.

4 A 3-instruction Implementation of the VMPC Function

Implementation of 1-level VMPC function, where $Q[x] = P[P[P[x]] + 1]$, for 256-element permutations P and Q in assembly language is described.

Assume that:

- Pm is a 257-byte array indexed by numbers from 0 to 256. The P permutation is stored in the Pm array at indexes from 0 to 255 ($Pm[0...255] = P$) and $Pm[256] = Pm[0]$
- the EAX 32-bit register specifies which element of the Q permutation to compute ("AL" always denotes 8 least significant bits of EAX, here EAX=AL)

The 3 MOV instructions in Table 1 store the EAX-th element of permutation Q, where Q=VMPC$_1(P)$, in the AL (and EAX) register.

5 Example Values of the VMPC Function

Values of 1,2,3 and 4-level VMPC function of an example 10-element permutation P are given in Table 2:

Table 2. Example values of the VMPC function

index	0	1	2	3	4	5	6	7	8	9
P	2	0	4	3	6	9	7	8	5	1
Q_1=VMPC$_1(P)$	9	3	8	6	5	4	1	7	2	0
Q_2=VMPC$_2(P)$	0	9	2	5	8	7	3	1	6	4
Q_3=VMPC$_3(P)$	3	4	9	5	0	2	7	6	1	8
Q_4=VMPC$_4(P)$	8	5	3	1	6	7	0	2	9	4

6 Proposed Algorithm for Inverting the VMPC Function

The algorithm outputs an n-element permutation P satisfying $Q = \text{VMPC}_k(P)$.

Notation:
P : n-element table the searched permutation will be stored in
X, Y : temporary variables
Argument, *Value*, *Base*, *Parameter* of an element of P:
For an element $P[x] = y$: x is termed the *Argument*; x can be either the *Base* or the *Parameter*. y is termed the *Value*; y is the *Parameter* or the *Base* respectively.

Example: For an element $P[3] = 5$: If *Argument* 3 is the *Parameter*, *Value* 5 is the *Base*.

1.1) Reveal one element of P by assuming $P[X] = Y$, where
  X and Y are any values from $\{0, 1, ..., n-1\}$
1.2) Choose at random whether X is the *Base* and Y the *Parameter*
  or Y the *Base* and X the *Parameter* of the element $P[X] = Y$.
  Denote $P[X] = Y$ as the *Current* element of P.
2) Reveal all possible elements of P by running the Deducing Process
  (see Sect. 6.1)
3) If n elements of P are revealed with no contradiction:
  Terminate the algorithm and output P
4) If fewer than n elements of P are revealed with no contradiction:
  4.1) Reveal a new element of P by running the Selecting Process
    (see Sect. 6.2).
    Denote the revealed element as the *Current* element of P.
  4.2) Save the *Parameter* of the *Current* element of P
  4.3) Go to step 2
5) If a contradiction occurred in step 2:
  5.1) Remove all elements of P revealed in step 2 when
    the *Current* element of P had been revealed
  5.2) Increment modulo n the *Parameter* of the *Current* element of P
  5.3) If the *Parameter* of the *Current* element of P returned
    to the value saved in step 4.2:

5.3.1) Remove the *Current* element of P

5.3.2) Denote the element, which had been the *Current* element of P directly before the element removed in step 5.3.1 became the *Current* one, as the *Current* element of P

5.3.3) Go to step 5.1

6) Go to step 2

6.1 The Deducing Process

The Deducing Process reveals all possible elements of P given Q and given the already revealed elements of P.

Notation as in Section 6, with:

C, A : temporary variables

Statement y : A sequence of $k+2$ elements of P used to compute $Q[y]$

Word x of *Statement* y : The x-th element of the sequence of $k+2$ elements of P used to compute $Q[y]$

Example: For $Q = \text{VMPC}_2(P) : Q[x] = P[P[P[P[x]] + 1] + 2]$:

Assume that $P[2] = 3, P[3] = 5, P[6] = 2, P[4] = 7$, which forms $Q[2] = 7$.
The elements $P[2] = 3, P[3] = 5, P[6] = 2, P[4] = 7$ form *Statement* 2.
The element $P[2] = 3$ is *Word* 1 of *Statement* 2; $P[3] = 5$ is *Word* 2 of *Statement* 2, etc.

1.1) Set C to 0

1.2) Set A to 0

2) If the element $P[A]$ is revealed:

2.1) If the element $P[A]$ and k other revealed elements of P fit a general pattern of $k+1$ *Words* of any *Statement*: Deduce the remaining *Word* of that *Statement* (see *Example 6.1.1*)

2.2) If the deduced *Word* is not a revealed element of P:

2.2.1) Reveal the deduced *Word* as an element of P

2.2.2) Set C to 1

2.3) If the deduced *Word* contradicts any already revealed element of P (see *Example 6.1.2*): Output a contradiction and terminate the Deducing Process

3.1) Increment A

3.2) If A is lower than n: Go to step 2

3.3) If C is equal 1: Go to step 1.1

Example 6.1.1. For $Q = \text{VMPC}_2(P) : Q[x] = P[P[P[P[x]] + 1] + 2]$:

Assume that $Q[0] = 9$ and that the following elements of P are revealed:
$P[0] = 1$, $P[1] = 3$, $P[8] = 9$.
Word 3 of *Statement* 0 can be deduced as $P'[4] = 6$ $(P'[3 + 1] = 8 - 2)$

Example 6.1.2. For $Q = \text{VMPC}_2(P) : Q[x] = P[P[P[P[x]] + 1] + 2]$:

Assume that $Q[7] = 2$ and that the following elements of P are revealed:
$P[1] = 8$, $P[9] = 3$, $P[5] = 2$ and $P[6] = 1$.
Word 1 of *Statement* 7, deduced as $P'[7] = 1$,
contradicts the already revealed element $P[6] = 1$

6.2 The Selecting Process (for k not higher than 4)

The Selecting Process selects such new element of P to be revealed which is expected to maximize the number of elements of P possible to deduce in further steps of the inverting algorithm. The Selecting Process outputs a selected *Base* and a randomly chosen *Parameter* of a new element of P.

Notation as in Section 6.1, with:

B, R : temporary variables
Ta, Tv : temporary tables
$Weight$: table of numbers: $Weight[0; 1; 2; 3; 4] = (0; 2; 5; 9; 14)$.
Example: $Weight[3] = 9$

1.1) Set Ta and Tv to 0
1.2) Set B to 0
1.3) Set R to 1
2) Count the number of revealed elements of P which fit a general pattern of *Words* of a *Statement* in which an unrevealed element of P with *Argument* B would be *Word* R. Increment $Ta[B]$ by *Weight* of this number (see *Example 6.2.1*)
3) Count the number of revealed elements of P which fit a general pattern of *Words* of a *Statement* in which an unrevealed element of P with *Value* B would be *Word* R. Increment $Tv[B]$ by *Weight* of this number
4.1) Increment R
4.2) If R is lower than $k + 3$: Go to step 2
4.3) Increment B
4.4) If B is lower than n: Go to step 1.3
5.1) Pick any index of Ta or Tv for which the number stored in tables Ta and Tv is maximal
5.2) If the index picked in step 5.1 is an index of Ta:
   5.2.1) Store this index in variable X
   5.2.2) Generate a random number $Y \in \{0, 1, \ldots, n - 1\}$, such that an element of P with *Value* Y is not revealed
   5.2.3) Output $P'[X] = Y$, where X is the *Base* and Y is the *Parameter*
5.3) If the index picked in step 5.1 is an index of Tv:
   5.3.1) Store this index in variable Y
   5.3.2) Generate a random number $X \in \{0, 1, \ldots, n - 1\}$,

such that an element of P with *Argument* X is not revealed
5.3.3) Output $P'[X] = Y$, where Y is the *Base* and X is the *Parameter*

Example 6.2.1. For $Q = \text{VMPC}_2(P) : Q[x] = P[P[P[P[x]] + 1] + 2]$:

Assume that $B = 8$, $R = 2$, $Q[6] = 9$ and that $P[6] = 8$ and $P[2] = 9$ are revealed.
There are two revealed elements of P which fit a general pattern of *Words* of a *Statement* in which $P[8]$ would be *Word* 2: $P[6] = 8$ and $P[2] = 9$:

Word 1	*Word* 2	*Word* 3	*Word* 4
$P[6] = 8$	$P[8] = y$	$P[y + 1] = 0$	$P[2] = 9$

$(Ta[8] = Ta[8] + Weight[2] = Ta[8] + 5)$

7 Example Complexities of Inverting the VMPC Function

Complexity of inverting the VMPC function (Table 3) was approximated as an average number of times the Deducing Process (step 2) needs to be run by the inverting algorithm described in Section 6 until permutation P satisfies $Q = \text{VMPC}_k(P)$.

Average numbers of elements of P which need to be assumed are given in Table 3 in brackets.

Complexities of inverting the VMPC function of the following levels were approximated:

$$Q = \text{VMPC}_1(P) : Q[x] = P[P[P[x]] + 1]$$
$$Q = \text{VMPC}_2(P) : Q[x] = P[P[P[P[x]] + 1] + 2]$$
$$Q = \text{VMPC}_3(P) : Q[x] = P[P[P[P[P[x]] + 1] + 2] + 3]$$
$$Q = \text{VMPC}_4(P) : Q[x] = P[P[P[P[P[P[x]] + 1] + 2] + 3] + 4]$$

Example: For 1-level VMPC function applied on 256-element permutations: on average about 34 elements of P need to be assumed. Searching through half of the possible states of these elements requires about 2^{260} calls to the Deducing Process.

8 Design Objectives for the VMPC Stream Cipher and its KSA

The Cipher should require no initial outputs to be discarded directly after running the KSA.

Probability that the Cipher's output will enter a short cycle should be negligibly low.

Output generated by the Cipher should be free from statistical biases.

Effort required to recover the internal state from the Cipher's output should be higher than a brute-force search of all possible 512-bit keys.

The KSA should resist related-key attacks and attacks against the scheme of using the Initialization Vector (IV).

The KSA should provide random-looking diffusion of changes of one byte of the key of size up to 512 bits onto the generated permutation and onto output generated by the Cipher.

9 Description of the VMPC Stream Cipher

The algorithm generates a stream of 8-bit values.

Variables:

P : 256-byte table storing a permutation initialized by the VMPC KSA
s : 8-bit variable initialized by the VMPC KSA
n : 8-bit variable
L : desired length of the keystream in bytes

10 Description of the VMPC Key Scheduling Algorithm

The VMPC Key Scheduling Algorithm (KSA) transforms a cryptographic key and (optionally) an Initialization Vector into a 256-element permutation P and initializes variable s.

Variables as in Section 9, with:

c : fixed length of the cryptographic key in bytes, $16 \leq c \leq 64$
K : c-element table storing the cryptographic key
z : fixed length of the Initialization Vector in bytes, $16 \leq z \leq 64$
V : z-element table storing the Initialization Vector
m : 16-bit variable

Table 3. Example complexities of inverting the VMPC function

Function n	$VMPC_1$	$VMPC_2$	$VMPC_3$	$VMPC_4$
6	$2^{4,1}$ (2,3)	$2^{5,5}$ (3,1)	$2^{6,1}$ (3,3)	$2^{6,9}$ (3,8)
8	$2^{5,5}$ (2,7)	$2^{7,5}$ (3,4)	$2^{8,8}$ (4,0)	$2^{9,8}$ (4,4)
10	$2^{7,1}$ (3,0)	$2^{9,7}$ (4,0)	$2^{11,5}$ (4,7)	$2^{13,0}$ (5,2)
16	$2^{11,5}$ (3,8)	$2^{16,6}$ (5,4)	$2^{20,4}$ (6,6)	$2^{23,3}$ (7,5)
32	2^{24} (6,0)	2^{37} (9,1)	2^{47} (11,5)	2^{54} (13,4)
64	2^{53} (10,2)	2^{84} (16,2)	2^{108} (21,0)	2^{127} (24,9)
128	2^{117} (18)	2^{190} (30)	2^{245} (40)	2^{292} (47)
256	2^{260} (34)	2^{420} (57)	2^{550} (77)	2^{660} (92)

Table 4. VMPC Stream Cipher

1. $n = 0$

2. Repeat steps 3-6 L times:
   3. $s = P[(s + P[n])$ modulo $256]$
   4. Output $P[(P[P[s]] + 1)$ modulo $256]$
   5. $Temp = P[n]$
      $P[n] = P[s]$
      $P[s] = Temp$
   6. $n = (n + 1)$ modulo 256

Table 5. VMPC Key Scheduling Algorithm

1. $s = 0$
2. for n from 0 to 255: $P[n] = n$

3. for m from 0 to 767: execute steps 4-6:
   4. $n = m$ modulo 256
   5. $s = P[(s + P[n] + K[m$ modulo $c])$ modulo $256]$
   6. $Temp = P[n]$
      $P[n] = P[s]$
      $P[s] = Temp$

7. If Initialization Vector is used: execute step 8:

8. for m from 0 to 767: execute steps 9-11:
   9. $n = m$ modulo 256
   10. $s = P[(s + P[n] + V[m$ modulo $z])$ modulo $256]$
   11. $Temp = P[n]$
       $P[n] = P[s]$
       $P[s] = Temp$

11 VMPC Stream Cipher Test Vectors

Table 6 gives 16 test output-bytes generated by the VMPC Stream Cipher for a given 16-byte key (K) and a given 16-byte Initialization Vector (V):

12 Performance of the VMPC Stream Cipher

Performance of a moderately optimized 32-bit assembler implementation of the VMPC Stream Cipher and its KSA, measured on an Intel Pentium 4, 2.66 GHz processor, is given in tables 7 and 8.

Table 6. Test-output of the VMPC Stream Cipher

K; $c = 16$ [hex]	96, 61, 41, 0A, B7, 97, D8, A9, EB, 76, 7C, 21, 17, 2D, F6, C7
V; $z = 16$ [hex]	4B, 5C, 2F, 00, 3E, 67, F3, 95, 57, A8, D2, 6F, 3D, A2, B1, 55

Output-byte position [dec]	0	1	2	3	252	253	254	255
Output-byte value [hex]	A8	24	79	F5	B8	FC	66	A4

Output-byte position [dec]	1020	1021	1022	1023	102396	102397	102398	102399
Output-byte value [hex]	E0	56	40	A5	81	CA	49	9A

Table 7. Perfomance of the VMPC Stream Cipher

MBytes/s	MBits/s	cycles/byte
210	1680	12.7

13 Analysis of the VMPC Stream Cipher

13.1 Theoretical Probability of Equal Consecutive Outputs

Probability of two consecutive outputs being equal appears to be an important parameter for a cipher based, as VMPC is, on an internal permutation (P). A sole permutation is obviously distinguishable from a truly random stream as its values never repeat. The construction of a cipher based on an internal permutation should corrupt the regular structure of the permutation in such way as to force the outputs to repeat with a random-looking probability.

This section explains theoretically why the probability of consecutive outputs generated by the VMPC Stream Cipher being equal is the same as we would expect from a random oracle, i.e. that $Prob(out[x] = out[x+1]) = 2^{-N}$, where N is the word-size of the Cipher in bits; the standard value of N is 8.

To compute the probability, two scenarios need to be considered: (1) - there is no swap in step 5 (Table 4) and (2) - there is a swap in step 5.

In (1): $Prob(\text{no-swap}) = Prob(s[x] = n[x]) = 2^{-N}$

As a result of (1) permutation P will have the same arrangement of elements in steps x and $x+1$. This implies a distinction into two sub-scenarios - (1a) where $s[x] = s[x+1]$ and (1b) where $s[x] \neq s[x+1]$, which directly affects whether $out[x] = out[x+1]$ or $out[x] \neq out[x+1]$.

Table 8. Perfomance of the VMPC KSA for 128-, 256- and 512-bit keys

keys/s	milliseconds/key
310 000	0.0032

In (1a): $Prob(s[x] = s[x + 1]) = 2^{-N}$
and (1aR): $Prob(out[x] = out[x + 1]) = 1$

In (1b): $Prob(s[x] \neq s[x + 1]) = 1 - 2^{-N}$
and (1bR): $Prob(out[x] = out[x + 1]) = 0$

In (2): $Prob(\text{swap}) = Prob(s[x] \neq n[x]) = 1 - 2^{-N}$
and (2R): $Prob(out[x] = out[x + 1]) = 2^{-N}$,
regardless of the relation between $s[x]$ and $s[x+1]$ because P (and VMPC$_1(P)$) in steps x and $x+1$ are different permutations. This probability was also confirmed experimentally.

By combining the probabilities in scenarios (1)(1a)(1aR), (1)(1b)(1bR) and (2)(2R) we get:

$$Prob(out[x] = out[x + 1]) =$$
$$= 2^{-N} \times 2^{-N} \times 1 + 2^{-N} \times (1 - 2^{-N}) \times 0 + (1 - 2^{-N}) \times 2^{-N} = 2^{-N}$$

which ends the proof.

13.2 Recovering the Cipher's Internal State

A method analogous to the Forward Tracking Algorithm proposed by Mister and Tavares in [3] was applied to break the VMPC Stream Cipher. Following this approach an average of over 2^{900} computational operations is estimated to be required to recover the Cipher's internal state from its output.

13.3 Digraph and Trigraph Probabilities

Frequencies of occurrence of each of the possible 2^{16} pairs of consecutive output values ($out[x]$, $out[x+1]$) were measured in a stream of $2^{40.1}$ output bytes. None of the measured frequencies showed a statistically significant deviation from its expected value of 1 / 65536.

Frequencies of occurrence of each of the possible 2^{24} triplets of two consecutive output values and the n variable ($out[x]$, $out[x + 1]$, n) were measured in a stream of $2^{41.85}$ output bytes. None of the measured frequencies showed a statistically significant deviation from its expected value of 1 / 16777216.

Frequencies of occurrence of each of the possible 2^{24} trigraphs of consecutive output values ($out[x]$, $out[x + 1]$, $out[x + 2]$) were measured in a stream of $2^{41.6}$ output bytes. None of the measured frequencies showed a statistically significant deviation from its expected value of 1 / 16777216.

13.4 Single-Output Probabilities

Frequencies of occurrence of each of the possible 2^8 output values ($out[x]$) were measured in a stream of $2^{41.85}$ output bytes. None of the measured frequencies showed a statistically significant deviation from its expected value of 1 / 256.

Table 9. VMPC Stream Cipher cycle lengths

M	Cycle lengths
4	200, 88, 40, 36, 12, 8
5	1 860, 640, 295, 110, 45, 25, 20, 5
6	15 510, 5 580, 2 508, 936, 516, 510, 252, 90, 12, 6
7	215 089, 23 821, 3 990, 2 485, 1 015, 392, 70, 56, 28, 14
8	2 401 728, 79 504, 53 512, 42 120, 2 136, 1 032, 288, 96, 24, 16 (2 different cycles of length 16 possible), 8
9	20 355 471, 2 908 098, 2 728 890, 1 359 855, 949 725, 609 174, 299 592, 125 091, 27 306, 13 068, 6 219, 5 067, 2 853, 2 538, 180, 90, 18 (3 different cycles of length 18 possible), 9
10	113 748 840, 99 425 590, 75 813 290, 37 178 940, 20 169 740, 9 955 030, 3 239 140, 2 349 150, 572 500, 363 830, 45 520, 8 730, 7 520, 700, 390, 370, 40 (17 different cycles of length 40 possible), 20, 10 (2 different cycles of length 10 possible)

Frequencies of occurrence of each of the possible 2^{16} configurations of an output value and the n variable ($out[x]$, n) were measured in a stream of $2^{39.4}$ output bytes. None of the measured frequencies showed a statistically significant deviation from its expected value of 1 / 65536.

13.5 First-Outputs Probabilities

Frequencies of occurrence of each of the possible 2^8 values on each of the first 256 byte-positions of the keystream generated directly after running the KSA were measured in a sample of $2^{40.3}$ bytes of the Cipher's output for $2^{32.3}$ different keys. None of the measured frequencies showed a statistically significant deviation from its expected value of 1 / 256. [In [6] Mantin and Shamir show that the second output of RC4 takes on value 0 with probability 1 / 128 rather than 1 / 256.]

13.6 Short Cycles

Probability of Entering a Cycle Not Longer than X. Following Knuth's [1], probability of entering a cycle not longer than X for an n-element random permutation is X/n.

To compare cycle lengths in the output of the VMPC Stream Cipher to cycle lengths in a random permutation, the Cipher was scaled down to use M-element permutations for $M \in \{4, 5, \ldots, 10\}$.

The total number of $M! \times M^2$ possible internal states of the Cipher is determined by all possible configurations of permutation P and variables s and n.

The observed cycle lengths, listed in the Appendix, do not show an appreciable difference from a model of cycles in a random $(M! \times M^2)$-element permutation.

Probability of entering a cycle not longer than X by the VMPC Stream Cipher is conjectured from this to be approximately $X/(256! \times 256^2) \cong X/2^{1700}$. An example estimate is that probability that the Cipher's output will enter a cycle not longer than 2^{850} is about 1 / 2^{850}.

Finney States. In [10] Finney defined a theoretical class of internal states of RC4 which produce a short cycle of length 65280 by swapping $P[n] = 1$ in each step (the KSA of RC4 prevents the cipher's internal state from entering this class). The class is diagnosed by $n + 1 = s$ and $P[n + 1] = 1$.

Such phenomenon is possible because step $s = s + P[n]$ of the state-transformation function of RC4 retains the linear structure of $P[n]$ in variable s ($P[n]$, after the increment of n, is always equal 1).
The VMPC Stream Cipher uses an additional table-lookup ($s = P[s + P[n]]$), which, assuming that P was properly initialized, corrupts a possible linear structure of $P[n]$ (or s) and prevents situations analogous to Finney states from occurring.

13.7 Binary Derivatives of Bit Output Sequences

This family of tests was inspired by Golic's [8], where the author describes a statistical bias in the second binary derivative of the least significant bit output sequence of RC4. The author finds that the bias allows the attacker to distinguish RC4 output from a truly random source using about $64^N/225$ outputs, where N is the cipher's word-size in bits (e.g. for $N = 8$ the required length is about 2^{40})[1].

Output generated by the VMPC Stream Cipher showed no bias in this family of tests. The following objectives were taken in testing VMPC here:

$N = 7$ word-size was chosen to make the tested algorithm as close to the real 8-bit one as possible while significantly decreasing the output-sequence length required to reveal the bias for RC4 (for 7-bit RC4 - about $2^{34.2}$ outputs according to the original estimates in [8]). First, second and third binary derivatives of all 7 bits output sequences were tested (21 frequencies of ($out_k[x] + out_k[x+A]$=1) were measured for $k \in \{0, 1, \ldots, 6\}$, $A \in \{1, 2, 3\}$, where $out_k[x]$ denotes k-th bit of x-th output word).

In a sequence of $2^{44.8}$ (about $10^{13.5}$) VMPC outputs tested according to this approach none of the measured frequencies showed a statistically significant deviation from its expected value of 0.5.

13.8 Equal Neighboring Outputs Probabilities

Frequencies of occurrence of situations where there occurs a given number (0, 1, 2, 3, 4, 5 and over 5) of direct (generated consecutively) and indirect (separated by one output byte) equal neighboring outputs in the consecutive 256-byte

[1] Authors of [5] consider this estimate somewhat optimistic and suggest that the required keystream length for $N = 8$ is about $2^{44.7}$.

sub-streams of the Cipher's output and the average total number of direct and indirect equal neighboring outputs – showed no statistically significant deviation from their expected values in a sample of $2^{43.1}$ bytes of the Cipher's output.

13.9 Statistical Tests on the Cipher's Output

Keystreams generated by the VMPC Stream Cipher were tested by two popular batteries of statistical tests – the DIEHARD battery [11] and the NIST statistical tests suite [12]. No bias was found by any of the 15 tests included in the DIEHARD battery or by any of the 16 tests from the NIST suite.

14 Analysis of the VMPC Key Scheduling Algorithm

The VMPC Key Scheduling Algorithm was tested for diffusion of changes of the cryptographic key onto the generated permutation and onto the Cipher's output. A change of one byte of the cryptographic key of size 128, 256 and 512 bits appears to cause a random-looking change in the generated permutation and in the Cipher's output.

The KSA was designed to provide the diffusion without the use of the Initialization Vector and the tests were run without the IV. The Initialization Vector would obviously mix the generated permutation further, which would improve the diffusion effect.

14.1 Given Numbers of Equal Permutation Elements Probabilities

Frequencies of occurrence of situations where in two permutations, generated from keys differing in one byte, there occurs a given number (0, 1, 2, 3, 4, 5) of equal elements in the corresponding positions and the average number of equal elements in the corresponding positions – showed no statistically significant deviation from their expected values in samples of $2^{33.2}$ pairs of 128-, 256- and 512-bit keys.

14.2 Given Numbers of Equal Cipher's Outputs Probabilities

Frequencies of occurrence of situations where in two 256-byte streams generated by the VMPC Stream Cipher directly after running the KSA for keys differing in one byte, there occurs a given number (0, 1, 2, 3, 4, 5) of equal values in the corresponding byte-positions and the average number of equal values in the corresponding byte-positions – showed no statistically significant deviation from their expected values in samples of $2^{33.2}$ pairs of 128-, 256- and 512-bit keys.

14.3 Equal Corresponding Permutation Elements Probabilities

Frequencies of occurrence of situations where the elements in the corresponding positions of permutations generated from keys differing in one byte are equal (for each of the 256 positions) – showed no statistically significant deviation from their expected value in samples of $2^{33.2}$ pairs of 128-, 256- and 512-bit keys.

15 Conclusions

A simple one-way function together with a description of the most efficient method of inverting it found have been presented. An open problem is whether the simplicity of the function helps make a hypothetical attempt to prove a lower bound on the complexity of inverting it worth undertaking.

A proposed stream cipher which employs the function was given together with some analysis of the cipher's cryptographic strength, statistical properties of the cipher's output and statistical properties of the cipher's Key Scheduling Algorithm.

The analyses performed so far did not reveal any weakness in the design and indicated that the cipher has a number of security advantages over the alleged RC4 keystream generator while retaining most of its speed and simplicity.

References

[1] Donald E. Knuth: The Art of Computer Programming, vol. 1. *Fundamental Algorithms* Third Edition. Addison Wesley Longman, 1997. 221

[2] Donald E. Knuth: The Art of Computer Programming, vol. 2. *Seminumerical Algorithms* Third Edition. Addison Wesley Longman, 1998.

[3] Serge Mister, Stafford E. Tavares: Cryptanalysis of RC4-like Ciphers. Proceedings of SAC 1998, LNCS, vol. 1556, Springer-Verlag, 1999. 220

[4] Lars R. Knudsen, Willi Meier, Bart Preneel, Vincent Rijmen, Sven Verdoolaege: Analysis Methods for (Alleged) RC4. Proceedings of ASIACRYPT 1998, LNCS, vol. 1514, Springer-Verlag, 1998.

[5] Scott R. Fluhrer, David A. McGrew: Statistical Analysis of the Alleged RC4 Keystream Generator. Proceedings of FSE 2000, LNCS, vol. 1978, Springer-Verlag, 2001. 222

[6] Itsik Mantin, Adi Shamir: A Practical Attack on Broadcast RC4. Proceedings of FSE 2001, LNCS, vol. 2355, Springer-Verlag, 2002. 221

[7] Scott Fluhrer, Itsik Mantin, Adi Shamir: Weaknesses in the Key Scheduling Algorithm of RC4. Proceedings of SAC 2001, LNCS, vol. 2259, Springer-Verlag 2001.

[8] Jovan Dj. Golic: Linear Statistical Weakness of Alleged RC4 Keystream Generator. Proceedings of EUROCRYPT 1997, LNCS, vol. 1233, Springer-Verlag 1997. 222

[9] Alexander L. Grosul, Dan S. Wallach: A Related-Key Cryptanalysis of RC4. Technical Report TR-00-358, Department of Computer Science, Rice University, 2000.

[10] Hal Finney: An RC4 Cycle That Can't Happen. Post in sci.crypt, 1994. 222

[11] George Marsaglia: DIEHARD battery of statistical tests with documentation. `http://stat.fsu.edu/~geo/diehard.html`. 223

[12] NIST statistical tests suite with documentation. `http://csrc.nist.gov/rng`. 223

Appendix: Cycle Lengths Observed in the Output of the VMPC Stream Cipher

The observed cycle lengths in the output of the scaled down variants of the Cipher for $M \in \{4, 5, \ldots, 10\}$ are listed in Table 9. M denotes the number of elements in the P permutation. All addition operations performed by the Cipher here are additions modulo M.

A New Stream Cipher HC-256

Hongjun Wu

Institute for Infocomm Research
21 Heng Mui Keng Terrace, Singapore 119613
hongjun@i2r.a-star.edu.sg

Abstract. Stream cipher HC-256 is proposed in this paper. It generates keystream from a 256-bit secret key and a 256-bit initialization vector. HC-256 consists of two secret tables, each one with 1024 32-bit elements. The two tables are used as S-Box alternatively. At each step one element of a table is updated and one 32-bit output is generated. The encryption speed of the C implementation of HC-256 is about 1.9 bit per clock cycle (4.2 clock cycle per byte) on the Intel Pentium 4 processor.

1 Introduction

Stream ciphers are used for shared-key encryption. The modern software efficient stream ciphers can run 4-to-5 times faster than block ciphers. However, very few efficient and secure stream ciphers have been published. Even the most widely used stream cipher RC4 [25] has several weaknesses [14, 16, 22, 9, 10, 17, 21]. In the recent NESSIE project all the six stream cipher submissions cannot meet the stringent security requirements [23]. In this paper we aim to design a very simple, secure, software-efficient and freely-available stream cipher.

HC-256 is the stream cipher we proposed in this paper. It consists of two secret tables, each one with 1024 32-bit elements. At each step we update one element of a table with non-linear feedback function. Every 2048 steps all the elements of the two tables are updated. At each step, HC-256 generates one 32-bit output using the 32-bit-to-32-bit mapping similar to that being used in Blowfish [28]. Then the linear masking is applied before the output is generated.

In the design of HC-256, we take into consideration the superscalar feature of modern (and future) microprocessors. Without compromising the security, we try to reduce the dependency between operations. The dependency between the steps is reduced so that three consecutive steps can be computed in parallel. At each step, three parallel additions are used in the feedback function and three additions are used to combine the four table lookup outputs instead of the addition-xor-addition being used in Blowfish (similar idea has been suggested by Schneier and Whiting to use three xors to combine those four terms [29]).

With the high degree of parallelism, HC-256 runs very efficiently on the modern processor. We implemented HC-256 in C and tested its performance on the Pentium 4 processor. The encryption speed of HC-256 reaches 1.93 bit/cycle.

This paper is organized as follows. We introduce HC-256 in Section 2. The security of HC-256 is analyzed in Section 3. Section 4 discusses the implementation and performance of HC-256. Section 5 concludes this paper.

B. Roy and W. Meier (Eds.): FSE 2004, LNCS 3017, pp. 226–244, 2004.

2 Stream Cipher HC-256

In this section, we describe the stream cipher HC-256. From a 256-bit key and a 256-bit initialization vector, it generates keystream with length up to 2^{128} bits.

2.1 Operations, Variables and Functions

The following operations are used in HC-256:

$+$: $x + y$ means $x + y \bmod 2^{32}$, where $0 \leq x < 2^{32}$ and $0 \leq y < 2^{32}$
$\boxminus$: $x \boxminus y$ means $x - y \bmod 1024$
$\oplus$: bit-wise exclusive OR
$||$: concatenation
$\gg$: right shift operator. $x \gg n$ means x being right shifted n bits.
$\ll$: left shift operator. $x \ll n$ means x being left shifted n bits.
$\ggg$: right rotation operator. $x \ggg n$ means $((x \gg n) \oplus (x \ll (32-n))$ where $0 \leq n < 32$, $0 \leq x < 2^{32}$.

Two tables P and Q are used in HC-256. The key and the initialization vector of HC-256 are denoted as K and IV. We denote the keystream being generated as s.

P : a table with 1024 32-bit elements. Each element is denoted as $P[i]$ with $0 \leq i \leq 1023$.
Q : a table with 1024 32-bit elements. Each element is denoted as $Q[i]$ with $0 \leq i \leq 1023$.
K : the 256-bit key of HC-256.
IV : the 256-bit initialization vector of HC-256.
s : the keystream being generated from HC-256. The 32-bit output of the ith step is denoted as s_i. Then $s = s_0||s_1||s_2||\cdots$

There are six functions being used in HC-256. $f_1(x)$ and $f_2(x)$ are the same as the $\sigma_0^{\{256\}}(x)$ and $\sigma_1^{\{256\}}(x)$ being used in the message schedule of SHA-256 [24]. For $g_1(x)$ and $h_1(x)$, the table Q is used as S-box. For $g_2(x)$ and $h_2(x)$, the table P is used as S-box.

$$\begin{aligned}
f_1(x) &= (x \ggg 7) \oplus (x \ggg 18) \oplus (x \gg 3)\\
f_2(x) &= (x \ggg 17) \oplus (x \ggg 19) \oplus (x \gg 10)\\
g_1(x,y) &= ((x \ggg 10) \oplus (y \ggg 23)) + Q[(x \oplus y) \bmod 1024]\\
g_2(x,y) &= ((x \ggg 10) \oplus (y \ggg 23)) + P[(x \oplus y) \bmod 1024]\\
h_1(x) &= Q[x_0] + Q[256 + x_1] + Q[512 + x_2] + Q[768 + x_3]\\
h_2(x) &= P[x_0] + P[256 + x_1] + P[512 + x_2] + P[768 + x_3]
\end{aligned}$$

where $x = x_3||x_2||x_1||x_0$, x is a 32-bit word, x_0, x_1, x_2 and x_3 are four bytes. x_3 and x_0 denote the most significant byte and the least significant byte of x, respectively.

2.2 Initialization Process (Key and IV Setup)

The initialization process of HC-256 consists of expanding the key and initialization vector into P and Q (similar to the message setup in SHA-256) and running the cipher 4096 steps without generating output.

1. Let $K = K_0||K_1||\cdots||K_7$ and $IV = IV_0||IV_1||\cdots||IV_7$, where each K_i and IV_i denotes a 32-bit number. The key and IV are expanded into an array W_i $(0 \leq i \leq 2559)$ as:

$$W_i = \begin{cases} K_i & 0 \leq i \leq 7 \\ IV_{i-8} & 8 \leq i \leq 15 \\ f_2(W_{i-2}) + W_{i-7} + f_1(W_{i-15}) + W_{i-16} + i & 16 \leq i \leq 2559 \end{cases}$$

2. Update the tables P and Q with the array W.

$$P[i] = W_{i+512} \quad \text{for } 0 \leq i \leq 1023$$
$$Q[i] = W_{i+1536} \quad \text{for } 0 \leq i \leq 1023$$

3. Run the cipher (the keystream generation algorithm in Subsection 2.3) 4096 steps without generating output.

The initialization process completes and the cipher is ready to generate keystream.

2.3 The Keystream Generation Algorithm

At each step, one element of a table is updated and one 32-bit output is generated. An S-box is used to generate only 1024 outputs, then it is updated in the next 1024 steps. The keystream generation process of HC-256 is given below ("$\boxminus$" denotes "−" modulo 1024, s_i denotes the output of the i-th step).

```
i = 0;
repeat until enough keystream bits are generated.
{
    j = i mod 1024;
    if (i mod 2048) < 1024
    {
        P[j] = P[j] + P[j ⊟ 10] + g1( P[j ⊟ 3], P[j ⊟ 1023] );
        s_i = h1( P[j ⊟ 12] ) ⊕ P[j];
    }
    else
    {
        Q[j] = Q[j] + Q[j ⊟ 10] + g2( Q[j ⊟ 3], Q[j ⊟ 1023] );
        s_i = h2( Q[j ⊟ 12] ) ⊕ Q[j];
    }
    end-if
    i = i + 1;
}
end-repeat
```

2.4 Encryption and Decryption

The keystream is XORed with the message for encryption. The decryption is to XOR the keystream with the ciphertext.

3 Security Analysis of HC-256

We start with a brief review of the attakcs on stream ciphers. Many stream ciphers are based on the linear feedback shift registers (LFSRs) and a number of correlation attacks, such as [30, 31, 19, 11, 20, 4, 15], were developed to analyze them. Later Golić [12] devised the linear cryptanalysis of stream ciphers. That technique could be applied to a wide range of stream ciphers. Recently Coppersmith, Halevi and Jutla [6] developed the distinguishing attacks (the linear attack and low diffusion attack) on stream ciphers with linear masking.

The correlation attacks cannot be applied to HC-256 because HC-256 uses non-linear feedback functions to update the two tables P and Q. The output function of HC-256 uses the 32-bit-to-32-bit mapping similar to that being used in Blowfish. The analysis on Blowfish shows that it is extremely difficult to apply linear cryptanalysis [18] to the large secret S-box. The large secret S-box of HC-256 is updated during the keystream generation process and it is almost impossible to develop linear relations linking the input and output bits of the S-box. Vaudenay has found some differential weakness of the randomly generated large S-box [32]. But it is very difficult to launch differential cryptanalysis [2] against HC-256 since it is a synchronous stream cipher for which the keystream generation is independent of the message.

In this section, we will analyze the security of the secret key, the randomness of the keystream, and the security of the initialization process.

3.1 Period

The 65547-bit state of HC-256 ensures that the period of the keystream is extremely large. But the exact period of HC-256 is difficult to predict. The average period of the keystream is estimated to be about 2^{65546} (if we assume that the invertible next-state function of HC-256 is random). The large number of states also completely eliminates the threat of time-memory tradeoff attack on stream ciphers [1, 13].

3.2 The Security of the Key

We begin with the study of a modified version of HC-256 (with no linear masking). Our analysis shows that even for this weak version of HC-256, it is impossible to recover the secret key faster than exhaustive key search. The reason is that the keystream is generated from a highly non-linear function ($h_1(x)$ or $h_2(x)$), so the keystream leaks very small amount of information at each step. Recovering P and Q requires the partial information leaked from a lot of steps. Because the tables are updated in a highly non-linear way, it is difficult to retrieve the informtion of P and Q from those leaked information.

HC-256 With No Linear Masking. For HC-256 with no linear masking, the output at the ith step is generated as $s_i = h_1(P[i \bullet 12])$ or $s_i = h_2(Q[i \bullet 12])$. If two outputs generated from the same S-box are equal, then very likely those two inputs to the S-box are equal. According to the analysis on the randomness of the outputs of $h_1(x)$ and $h_2(x)$ given in Subsection 3.3, $s_{2048\times\alpha+i} = s_{2048\times\alpha+j}$ $(0 \leq i < j < 1024)$ with probability about 2^{-31}. If $s_{2048\times\alpha+i} = s_{2048\times\alpha+j}$, then at the $(2048 \times \alpha + j)$-th step, $P[i \bullet 12] = P[j \bullet 12]$ with probability about 0.5 (31-bit information of the table P is leaked). We note that for every 1024 steps in the range $(2048 \times \alpha, 2048 \times \alpha + 1024)$, the same S-box is used in $h_1(x)$. The probability that there are two equal outputs is $\binom{1024}{2} \times 2^{-31} \approx 2^{-12}$. In average each output leaks $\frac{2^{-12}\times 31}{1024} \approx 2^{-17}$ bit information of the table P. To recover P, we need to analyze at least $\frac{1024\times 32}{2^{-17}} \approx 2^{32}$ outputs. Recovering P from those 2^{32} outputs involves very complicated non-linear equations and solving them is computationally infeasible. Recovering Q is as difficult as recovering P. We note that the table Q is used as S-box to update P, and vice versa. P and Q interact in such a complicated way and recovering them from the keystream cannot be faster than exhaustive key search.

HC-256. The analysis above shows that the secret key of HC-256 with no linear masking is secure. With the linear masking, the information leakage is greatly reduced and it would be even more difficult to recover the secret key from the keystream. We thus conclude that the key of HC-256 cannot be recovered faster than exhaustive key search.

3.3 Randomness of the Keystream

In this subsection, we investigate the randomness of the keystream of HC-256. Because the large, secret and frequently updated S-boxes are used in the cipher, the efficient attack is to analyze the randomness of the overall 32-bit words. Under this guideline, we developed some attacks against HC-256 with no linear masking. Then we show that the linear masking eliminates those threats.

3.3.1 Keystream of HC-256 with No Linear Masking

The attacks on HC-256 with no linear masking is to investigate the security weaknesses in the output and feedback functions. We developed two attacks against HC-256 with no linear masking.

Weakness of $h_1(x)$ and $h_2(x)$. For HC-256 with no linear masking, the output is generated as $s_i = h_1(P[i \bullet 12])$ or $s_i = h_2(Q[i \bullet 12])$. Because there is no difference between the analysis of $h_1(x)$ and $h_2(x)$, we use $h(x)$ to refer $h_1(x)$ and $h_2(x)$ here. Assume that $h(x)$ is a 32-bit-to-32-bit S-box $H(x)$ with randomly generated secret elements and the inputs to H are randomly generated. Because the elements of the $H(x)$ are randomly generated, the output of $H(x)$

is not uniformly distributed. If a lot of outputs are generated from $H(x)$, some values in the range $[0, 2^{32})$ never appear and some appear with probability larger than 2^{-32}. Then it is straightforward to distinguish the outputs from random signal. However each $H(x)$ in HC-256 is used to generate only 1024 outputs, then it gets updated. The direct computation of the distribution of the outputs of $H(x)$ from those 1024 outputs cannot be successful. Instead, we consider the collision between the outputs of $H(x)$. The following theorem gives the collision rate of the outputs of $H(x)$.

Theorem 1. *Let H be an m-bit-to-n-bit S-box and all those n-bit elements are randomly generated, where $m \geq n$ and n is a large integer. Let x_1 and x_2 be two m-bit random inputs to H. Then $H(x_1) = H(x_2)$ with probability about $2^{-n} + 2^{-m}$.*

Proof. If $x_1 = x_2$, then $H(x_1) = H(x_2)$. If $x_1 \neq x_2$, then $H(x_1) = H(x_2)$ with probability 2^{-n}. $x_1 = x_2$ with probability 2^{-m} and $x_1 \neq x_2$ with probability $1 - 2^{-m}$. The probability that $H(x_1) = H(x_2)$ is $2^{-m} + (1 - 2^{-m}) \times 2^{-n} \approx 2^{-n} + 2^{-m}$.

Attack 1. According to Theorem 1, for the 32-bit-to-32-bit S-box H, the collision rate of the outputs is $2^{-32} + 2^{-32} = 2^{-31}$. With 2^{35} pairs of $(H(x_1), H(x_2))$, we can distinguish the output from random signal with success rate 0.761. (The success rate can be improved to 0.996 with 2^{36} pairs.) Note that only 1024 outputs are generated from the same S-box H, so 2^{26} outputs are needed to distinguish the keystream of HC-256 with no linear masking.

Experiment. To compute the collision rate of the outputs of HC-256 (with no linear masking), we generated 2^{39} outputs (2^{48} pairs). The collision rate is $2^{-31} - 2^{-40.09}$. The experiment confirms that the collision rate of the outputs of $h(x)$ is very close to 2^{-31}, and approximating $h(x)$ with randomly generated S-box has negligible effect on the attack.

Remark 1. The distinguishing attack above can be slightly improved if we consider the differential attack on Blowfish. Vaudenay [32] has pointed out that the collision in a randomly generated S-box in Blowfish can be applied to distinguish the outputs of Blowfish with reduced round number (8 rounds). The basic idea of Vaudenay's differential attack is that if $Q[i] = Q[j]$ for $0 \leq i, j < 256$, $i \neq j$, then for $a_0 \oplus a_0' = i \oplus j$, $h_1(a_3||a_2||a_1||a_0) = h_1(a_3||a_2||a_1||a_0')$ with probability 2^{-7}, where each a_i denotes an 8-bit number. We can detect the collision in the S-box with success rate 0.5 since that S-box Q is used as inputs to $h_2(x)$ to produce 1024 outputs. If $Q[i] = Q[j]$ for $256\alpha \leq i, j < 256\alpha + 256$, $0 \leq \alpha < 4$, $i \neq j$, and x_1 and x_2 are two random inputs (note that we cannot introduce or identify inputs with particular difference to $h(x)$), then the probability that $h_1(x_1) = h_1(x_2)$ becomes $2^{-31} + 2^{-32}$. However the chance that there is one useful collision in the S-box is only $\frac{\binom{256}{2} \times 4}{2^{32}} = 2^{-15}$. The average collision rate becomes $2^{-15} \times (2^{-31} + 2^{-32}) + (1 - 2^{-15}) \times 2^{-31} = 2^{-31} + 2^{-47}$. The increase in collision rate is so small that the collision in the S-box has negligible effect on this attack.

Weakness of the Feedback Function. The table P is updated with the non-linear feedback function $P[i \bmod 1024] = P[i \bmod 1024] + P[i \bullet\ 10] + g_1(P[i \bullet\ 3], P[i \bullet\ 1023])$. The following attack is to distinguish the keystream by exploiting this relation.

Attack 2. Assume that the $h(x)$ is a one-to-one mapping function. Consider two groups of outputs (s_i, s_{i-3}, s_{i-10}, s_{i-2047}, s_{i-2048}) and (s_j, s_{j-3}, s_{j-10}, s_{j-2047}, s_{j-2048}). If $i \neq j$ and $1024 \times \alpha + 10 \leq i, j < 1024 \times \alpha + 1023$, they are equal with probability about 2^{-128}. The collision rate is 2^{-160} if the outputs are truely random. 2^{-128} is much larger than 2^{-160}, so the keystream can be distinguished from random signal with about 2^{128} pairs of such five-tuple groups of outputs. Note that the S-box is updated every 1024 steps, 2^{119} outputs are needed in the attack.

The two attacks given above show that the HC-256 with no linear masking does not generate secure keystream.

3.3.2 Keystream of HC-256

With the linear masking being applied, it is no longer possible to exploit those two weaknesses separately and the attacks given above cannot be applied directly. We need to remove the linear masking first. We recall that at the ith step, if $(i \bmod 2048) < 1024$, the table P is updated as

$$P[i \bmod 1024] = P[i \bmod 1024] + P[i \bullet\ 10] + g_1(P[i \bullet\ 3], P[i \bullet\ 1023])$$

We know that $s_i = h_1(P[i \bullet\ 12]) \oplus P[i \bmod 1024]$. For $10 \leq (i \bmod 2048) < 1023$, this feedback function can be written alternatively as

$$\begin{aligned} s_i \oplus h_1(z_i) = (s_{i-2048} \oplus h'_1(z_{i-2048})) + (s_{i-10} \oplus h_1(z_{i-10})) + \\ g_1(s_{i-3} \oplus h_1(z_{i-3}), s_{i-2047} \oplus h'_1(z_{i-2047})) \end{aligned} \tag{1}$$

where $h_1(x)$ and $h'_1(x)$ indicate two different functions since they are related to different S-boxes; z_j denotes the $P[j \bullet\ 12]$ at the j-th step. The linear masking is removed successfully in (1). However, it is very difficult to apply (1) directly to distinguish the keystream. To simplify the analysis, we attack a weak version of (1). We replace all the '+' in the feedback function with '$\oplus$' and write (1) as

$$\begin{aligned} & s_i \oplus s_{i-2048} \oplus s_{i-10} \oplus (s_{i-3} \ggg 10) \oplus (s_{i-2047} \ggg 23) \\ & = h_1(z_i) \oplus h'_1(z_{i-2048})) \oplus h_1(z_{i-10})) \oplus (h_1(z_{i-3}) \ggg 10) \oplus \\ & \quad \oplus (h'_1(z_{i-2047}) \ggg 23) \oplus Q[r_i], \end{aligned} \tag{2}$$

where $r_i = (s_{i-3} \oplus h_1(z_{i-3}) \oplus s_{i-2047} \oplus h'_1(z_{i-2047})) \bmod 1024$. Because of the random nature of $h_1(x)$ and Q, the right hand side of (2) is not uniformly distributed. But each S-box is used in only 1024 steps, these 1024 outputs are not sufficient to compute the distribution of $s_i \oplus s_{i-2048} \oplus s_{i-10} \oplus (s_{i-3} \ggg$

$10) \oplus (s_{i-2047} \ggg 23)$. Instead we need to study the collision rate. The effective way is to eliminate the term $h_1(z_i)$ before analyzing the collision rate.

Replace the i with $i+10$. For $10 \leq i \bmod 2048 < 1013$, (2) can be written as

$$\begin{aligned} & s_{i+10} \oplus s_{i-2038} \oplus s_i \oplus (s_{i+7} \ggg 10) \oplus (s_{i-2037} \ggg 23) \\ & = h_1(z_{i+10}) \oplus h_1'(z_{i-2038})) \oplus h_1(z_i) \oplus (h_1(z_{i+7}) \ggg 10) \oplus \\ & \quad \oplus (h_1'(z_{i-2037}) \ggg 23) \oplus Q[r_{i+10}] \end{aligned} \tag{3}$$

For the left-hand sides of (2) and (3) to be equal, i.e., for the following equation

$$\begin{aligned} & s_i \oplus s_{i-2048} \oplus s_{i-10} \oplus (s_{i-3} \ggg 10) \oplus (s_{i-2047} \ggg 23) = \\ & s_{i+10} \oplus s_{i-2038} \oplus s_i \oplus (s_{i+7} \ggg 10) \oplus (s_{i-2037} \ggg 23) \end{aligned} \tag{4}$$

to hold, we require that (after eliminating the term $h_1(z_i)$)

$$\begin{aligned} & h_1(z_{i-10}) \oplus h_1'(z_{i-2048}) \oplus (h_1(z_{i-3}) \ggg 10) \\ & \quad \oplus (h_1'(z_{i-2047}) \ggg 23) \oplus Q[r_i] \\ & = h_1(z_{i+10}) \oplus h_1'(z_{i-2038}) \oplus (h_1(z_{i+7}) \ggg 10) \\ & \quad \oplus (h_1'(z_{i-2037}) \ggg 23) \oplus Q[r_{i+10}] \end{aligned} \tag{5}$$

For $22 \leq i \bmod 2048 < 1013$, we note that $z_{i-10} = z_i \oplus z_{i-2048} \oplus (z_{i-3} \ggg 10) \oplus (z_{i-2047} \ggg 23)$, and $z_{i+10} = z_i \oplus z_{i-2038} \oplus (z_{i+7} \ggg 10) \oplus (z_{i-2037} \ggg 23)$. Approximate (5) as

$$H(x_1) = H(x_2) \tag{6}$$

where H denotes a random secret 106-bit-to-32-bit S-box, x_1 and x_2 are two 106-bit random inputs, $x_1 = z_{i-3} || z_{i-2047} || z_{i-2048} || r_i$ and $x_2 = z_{i+7} || z_{i-2037} || z_{i-2038} || r_{i+10}$. (The effect of z_i is included in H.) According to Theorem 1, (6) holds with probability $2^{-32} + 2^{-106}$. So (4) holds with probability $2^{-32} + 2^{-106}$. We approximate the binomial distribution with the normal distribution. The mean $\mu = Np$ and the standard deviation $\sigma = \sqrt{Np(1-p)}$, where N is the total number of equations (4), and $p = 2^{-32} + 2^{-106}$. For random signal, $p' = 2^{-32}$, the mean $\mu' = Np'$ and the standard deviation $\sigma' = \sqrt{Np'(1-p')}$. If $|u - u'| > 2(\sigma + \sigma')$, i.e. $N > 2^{184}$, the output of the cipher can be distinguished from random signal with success rate 0.9772.

After verifying the validity of 2^{184} equations (4), we can successfully distinguish the keystream from random signal. We note that the S-box is updated every 1024 steps, so only about 2^{10} equations (4) can be obtained from 1024 steps in the range $1024 \times \alpha \leq i < 1024 \times \alpha + 1024$. To distinguish the keystream from random signal, 2^{184} outputs are needed in the attack.

The attack above can be improved by exploiting the relation $r_i = (s_{i-3} \oplus h_1(z_{i-3}) \oplus s_{i-2047} \oplus h_1'(z_{i-2047})) \bmod 1024$. If $(s_{i-3} \oplus s_{i-2047}) \bmod 1024 = (s_{i+7} \oplus s_{i-2037}) \bmod 1024$, then (6) holds with probability $2^{-32} + 2^{-96}$ and 2^{164} equations (4) are needed in the attack. Note that only about one equation (4) can now be obtained from 1024 steps in the range $1024 \times \alpha \leq i < 1024 \times \alpha + 1024$.

To distinguish the keystream from random signal, 2^{174} outputs are needed in the attack.

We note that the attack above can only be applied to HC-256 with all the '+' in the feedback function being replaced with '$\oplus$'. To distinguish the keystream of HC-256, more than 2^{174} outputs are needed. So we conclude that it is impossible to distinguish a 2^{128}-bit keystream of HC-256 from random signal.

3.4 Security of the Initialization Process (Key/IV Setup)

The initialization process of the HC-256 consists of two stages, as given in Subsection 2.2. We expand the key and IV into P and Q. At this stage, every bit of the key/IV affects all the bits of the two tables and any difference in the related keys/IVs results in uncontrollable differences in P and Q. Then we run the cipher 4096 steps without generating output so that the P and Q become more random. After the initialization process, we expect that any difference in the keys/IVs would not result in biased keystream.

4 Implementation and Performance of HC-256

The direct C implementation of the encryption algorithm given in Subsection 2.3 runs at about 0.6 bit/cycle on the Pentium 4 processor. The program size is very small but the speed is only about 1.5 times that of AES [7]. At each step in the direct implementation, we need to compute (i mod 2048), $i \bullet 3$, $i \bullet 10$ and $i \bullet 1023$. And at each step there is a branch decision based on the value of (i mod 2048). These operations affect greatly the encryption speed. The optimization process is to reduce the amount of these operations.

4.1 The Optimized Implementation of HC-256

This subsection describes the optimized C implementation of HC-256 given in Appendix B ("hc256.h"). In the optimized code, loop unrolling is used and only one branch decision is made for every 16 steps. The experiment shows that the branch decision in the optimized code affects the encryption speed by less than one percent.

There are several fast implementations of the feedback functions of P and Q. We use the implementation given in Appendix B because it achieves the best consistency on different platforms. The details of that implementation are given below. The feedback function of P is given as

$$P[i \bmod 1024] = P[i \bmod 1024] + P[i \bullet 10] + g_1(P[i \bullet 3], P[i \bullet 1023])$$

A register X containing 16 elements is introduced for P. If (i mod 2048) < 1024 and i mod 16 $= 0$, then at the begining of the ith step, $X[j] = P[(i - 16 + j) \bmod 1024]$ for $j = 0, 1, \cdots 15$, i.e. the X contains the values of $P[i \bullet 16], P[i \bullet$

$15], \cdots, P[i \bullet\ 1]$. In the 16 steps starting from the ith step, the P and X are updated as

$$
\begin{aligned}
P[i] &= P[i] + X[6] + g_1(\,X[13], P[i+1]\,);\\
X[0] &= P[i];\\
P[i+1] &= P[i+1] + X[7] + g_1(\,X[14], P[i+2]\,);\\
X[1] &= P[i+1];\\
P[i+2] &= P[i+2] + X[8] + g_1(\,X[15], P[i+3]\,);\\
X[2] &= P[i+2];\\
P[i+3] &= P[i+3] + X[9] + g_1(\,X[0], P[i+4]\,);\\
X[3] &= P[i+3];\\
&\cdots\\
P[i+14] &= P[i+14] + X[4] + g_1(\,X[11], P[i+15]\,);\\
X[14] &= P[i+14];\\
P[i+15] &= P[i+15] + X[5] + g_1(\,X[12], P[(i+1) \bmod 1024]\,);\\
X[15] &= P[i+15];
\end{aligned}
$$

Note that at the ith step, two elements of $P[i \bullet\ 10]$ and $P[i \bullet\ 3]$ can be obtained directly from X. Also for the output function $s_i = h_1(P[i \bullet\ 12]) \oplus P[i \bmod 1024]$, the $P[i \bullet\ 12]$ can be obtained from X. In this implementation, there is no need to compute $i \bullet\ 3$, $i \bullet\ 10$ and $i \bullet\ 12$.

A register Y with 16 elements is used in the implementation of the feedback function of Q in the same way as that given above.

To reduce the memory requirement and the program size, the initialization process implemented in Appendix B is not as straightforward as that given in Subsection 2.2. To reduce the memory requirement, we do not implement the array W in the program. Instead we implement the key and IV expansion on P and Q directly. To reduce the program size, we implement the feedback functions of those 4096 steps without involving X and Y.

4.2 Performance of HC-256

Encryption Speed. We use the C codes given in Appendix B and C to measure the encryption speed. The processor used in the test is Pentium 4 (2.4 GHz, 8 KB Level 1 data cache, 512 KB Level 2 cache, no hyper-threading). The speed is measured by repeatedly encrypting the same 512-bit buffer for 2^{26} times (The buffer is defined as 'static unsigned long DATA[16]' in Appendix C). The encryption speed is given in Table 1.

The C implementation of HC-256 is faster than the C implementations of almost all the other stream ciphers. (However different designers may have made different efforts to optimize their codes. And the encryption speed may be measured in different ways. So the speed comparison is not absolutely accurate.) SEAL [26] is a software-efficient cipher and its C implementation runs at the

Table 1. The speed of the C implementation of HC-256 on Pentium 4

Operating System	Compiler	Optimization option	Speed (bit/cycle)
Windows XP (SP1)	Intel C++ Compiler 7.1	-O3	1.93
	Microsoft Visual C++ 6.0 Professional (SP5)	-Release	1.81
Red Hat Linux 9 (Linux 2.4.20-8)	Intel C++ Compiler 7.1	-O3	1.92
	gcc 3.2.2	-O3	1.83

speed of about 1.6 bit/cycle on Pentium III processor. Scream [5] runs at about the same speed as SEAL. The C implementation of SNOW2.0 [8] runs at about 1.67 bit/cycle on Pentium 4 processor. TURING [27] runs at about 1.3 bit/cycle on the Pentium III mobile processor. The C implementation of MUGI [33] runs at about 0.45 bit/cycle on the Pentium III processor. The encryption speed of Rabbit [3] is about 2.16 bit/cycle on Pentium III processor, but it is programmed in assembly language inline in C.

Remark 2. In HC-256, there is dependency between the feedback and output functions since the $P[i \bmod 1024]$ (or $Q[i \bmod 1024]$) being updated at the ith step is immediately used as linear masking. This dependency reduces the speed of HC-256 by about 3%. We do not remove this dependency from the design of HC-256 for security reason. Our analysis shows that each term being used as linear masking should not have been used in an S-box in the previous steps, otherwise the linear masking could be removed much easier. In our optimized implementation, we do not deal with this dependency because its effect on the encryption speed is very limited on the Pentium 4 processor.

Initialization Process. The key setup of HC-256 requires about 74,000 clock cycles (measured by repeating the setup process 2^{16} times on the Pentium 4 processor with Intel C++ compiler 7.1). This amount of computation is more than that required by most of the other stream ciphers (for example, the initialization process of Scream takes 27,500 clock cycles). The reason is that two large S-boxes are used in HC-256. To eliminate the threat of related key/IV attack, the tables should be updated with the key and IV thoroughly and this process requires a lot of computations. So it is undesirable to use HC-256 in the applications where key (or IV) is updated frequently.

5 Conclusion

In this paper, we proposed a software-efficient stream cipher HC-256. Our analysis shows that HC-256 is very secure. However, the extensive security analysis of any new cipher requires a lot of efforts from many researchers. We thus invite and encourage the readers to analyze the security of HC-256.

Finally we explicitly state that HC-256 is available royalty-free and HC-256 is not covered by any patent in the world.

References

[1] S. Babbage, "A Space/Time Tradeoff in Exhaustive Search Attacks on Stream Ciphers", European Convention on Security and Detection, IEE Conference publication, No. 408, May 1995. 229

[2] E. Biham, A. Shamir, "Differential Cryptanalysis of DES-like Cryptosystems", in *Advances in Cryptology - Crypto'90*, LNCS 537, pp. 2-21, Springer-Verlag, 1991. 229

[3] M. Boesgaard, M. Vesterager, T. Pedersen, J. Christiansen, and O. Scavenius, "Rabbit: A New High-Performance Stream Cipher", in *Fast Software Encryption (FSE'03)*, LNCS 2887, pp. 307-329, Springer-Verlag, 2003. 236

[4] V.V. Chepyzhov, T. Johansson, and B. Smeets. "A Simple Algorithm for Fast Correlation Attacks on Stream Ciphers", in *Fast Software Encryption (FSE'00)*, LNCS 1978, pp. 181-195, Springer-Verlag, 2000. 229

[5] D. Coppersmith, S. Halevi, and C. Jutla, "Scream: A Software-Efficient Stream Cipher", in *Fast Software Encryption (FSE'02)*, LNCS 2365, pp. 195-209, Springer-Verlag, 2002. 236

[6] D. Coppersmith, S. Halevi, and C. Jutla, "Cryptanalysis of Stream Ciphers with Linear Masking", in *Advances in Cryptology - Crypto 2002*, LNCS 2442, pp. 515-532, Springer-Verlag, 2002. 229

[7] J. Daeman and V. Rijmen, "AES Proposal: Rijndael", available on-line from NIST at `http://csrc.nist.gov/encryption/aes/rijndael/`. 234

[8] P. Ekdahl and T. Johansson, "A new version of the stream cipher SNOW", in *Selected Areas in Cryptology (SAC 2002)*, LNCS 2595, pp. 47–61, Springer-Verlag, 2002. 236

[9] S. Fluhrer and D. McGrew, "Statistical Analysis of the Alleged RC4 Keystream Generator", in *Fast Software Encryption (FSE'00)*, LNCS 1978, pp. 19-30, 2001. 226

[10] S. Fluhrer, I. Mantin, and A. Shamir. "Weaknesses in the Key Scheduling Algorithm of RC4", in *Selected Areas in Cryptography (SAC 2001)*, LNCS 2259, pp. 1-24, Springer-Verlag, 2001. 226

[11] J.D. Golić, "Towards Fast Correlation Attacks on Irregularly Clocked Shift Registers", in *Advances in Cryptography - Eurocrypt'95*, pages 248-262, Springer-Verlag, 1995. 229

[12] J.D. Golić, "Linear Models for Keystream Generator". *IEEE Trans. on Computers*, 45(1):41-49, Jan 1996. 229

[13] J.D. Golić, "Cryptanalysis of Alleged A5 Stream Cipher", in *Advances in Cryptology - Eurocrypt'97*, LNCS 1233, pp. 239 - 255, Springer-Verlag, 1997. 229

[14] J.D. Golić, "Linear Statistical Weakness of Alleged RC4 Keystream Generator", in *Advancesin Cryptology - Eurocrypt'97*, pp. 226 - 238, Springer-Verlag, 1997. 226

[15] T. Johansson and F. Jönsson. "Fast Correlation Attacks through Reconstruction of Linear Polynomials", in *Advances in Cryptology - CRYPTO 2000*, LNCS 1880, pp. 300-315, Springer-Verlag, 2000. 229

[16] L. Knudsen, W. Meier, B. Preneel, V. Rijmen and S. Verdoolaege, "Analysis Methods for (Alleged) RC4", in *Advances in Cryptology - Asiacrypt'98*, LNCS 1514, pp. 327-341, Springer-Verlag, 1998. 226

[17] I. Mantin and A. Shamir, "A Practical Attack on Broadcast RC4", in *Fast Software Encryption (FSE'01)*, LNCS 2355, pp. 152-164, Springer-Verlag, 2002. 226
[18] M. Matsui, "Linear Cryptanalysis Method for DES Cipher", in *Advances in Cryptology – Eurocrypt'93*, LNCS 765, pp. 386-397, Springer-Verlag, 1994. 229
[19] W. Meier and O. Staffelbach, "Fast Correlation Attacks on Certain Stream Ciphers". *Journal of Cryptography*, 1(3):159-176, 1989. 229
[20] M. Mihaljević, M. P. C. Fossorier, and H. Imai, "A Low-Complexity and High-Performance Algorithm for Fast Correlation Attack", in *Fast Software Encryption (FSE'00)*, pp. 196-212, Springer-Verlag, 2000. 229
[21] I. Mironov, "(Not So) Random Shuffles of RC4", in *Advances in Cryptology – Crypto 2002*, LNCS 2442, pp. 304-319, Springer-Verlag, 2002. 226
[22] S. Mister, and S. E. Tavares, "Cryptanalysis of RC4-like Ciphers", in *Selected Areas in Cryptography (SAC'98)*, LNCS 1556, pp. 131-143, Springer-Verlag, 1999. 226
[23] NESSIE, "NESSIE Project Announces Final Selection of Crypto Algorithms", available at
https://www.cosic.esat.kuleuven.ac.be/nessie/deliverables/press_release_feb27.pdf. 226
[24] National Institute of Standards and Technology, "Secure Hash Standard (SHS)", available at http://csrc.nist.gov/cryptval/shs.html. 227
[25] R. L. Rivest, "The RC4 Encryption Algorithm". RSA Data Security, Inc., March 12, 1992. 226
[26] P. Rogaway and D. Coppersmith, "A Software Optimized Encryption Algorithm". *Journal of Cryptography*, 11(4), pp. 273-287, 1998. 235
[27] G. G. Rose and P. Hawkes, "Turing: a Fast Stream Cipher". *Fast Software Encryption (FSE'03)*, LNCS 2887, pp. 290-306, Springer-Verlag, 2003. 236
[28] B. Schneier, "Description of a New Variable-Length Key, 64-bit Block Cipher (Blowfish)", in *Fast Software Encryption (FSE'93)*, LNCS 809, pp. 191-204, Springer-Verlag, 1994. 226
[29] B. Schneier and D. Whiting, "Fast Software Encryption: Designing Encryption Algorithms for Optimal Software Speed on the Intel Pentium Processor", in *Fast Software Encryption (FSE'97)*, LNCS 1267, pp. 242-259, Springer-Verlag, 1997. 226
[30] T. Seigenthaler. "Correlation-Immunity of Nonlinear Combining Functions for Cryptographic Applications". *IEEE Transactions on Information Theory*, IT-30:776-780,1984. 229
[31] T. Seigenthaler. "Decrypting a Class of Stream Ciphers Using Ciphertext Only". *IEEE Transactions on Computers*, C-34(1):81-85, Jan. 1985. 229
[32] S. Vaudenay, "On the Weak Keys of Blowfish", in *Fast Software Encryption (FSE'96)*, LNCS 1039, pp. 27-32, Springer-Verlag, 1996. 229, 231
[33] D. Watanabe, S. Furuya, H. Yoshida, K. Takaragi, and B. Preneel, "A New Keystream Generator MUGI", in *Fast Software Encryption (FSE'02)*, LNCS 2365, pp. 179-194, Springer-Verlag, 2002. 236

A Test Vectors of HC-256

Let $K = K_0||K_1||\cdots||K_7$ and $IV = IV_0||IV_1||\cdots||IV_7$. The first 512 bits of keystream are given for different values of key and IV.

1. The key and IV are set as 0.

```
8589075b 0df3f6d8 2fc0c542 5179b6a6
3465f053 f2891f80 8b24744e 18480b72
ec2792cd bf4dcfeb 7769bf8d fa14aee4
7b4c50e8 eaf3a9c8 f506016c 81697e32
```

2. The key is set as 0, the IV is set as 0 except that $IV_0 = 1$.

```
bfa2e2af e9ce174f 8b05c2fe b18bb1d1
ee42c05f 01312b71 c61f50dd 502a080b
edfec706 633d9241 a6dac448 af8561ff
5e04135a 9448c434 2de7e9f3 37520bdf
```

3. The IV is set as 0, the key is set as 0 except that $K_0 = 0x55$.

```
fe4a401c ed5fe24f d19a8f95 6fc036ae
3c5aa688 23e2abc0 2f90b3ae a8d30e42
59f03a6c 6e39eb44 8f7579fb 70137a5e
6d10b7d8 add0f7cd 723423da f575dde6
```

B The Optimized C Implementation of HC-256 ("hc256.h")

```
#include <stdlib.h>

typedef unsigned long uint32;
typedef unsigned char uint8;

uint32 P[1024],Q[1024];
uint32 X[16],Y[16];
uint32 counter2048; // counter2048 = i mod 2048;

#ifndef _MSC_VER
#define rotr(x,n)   (((x)>>(n))|((x)<<(32-(n))))
#else
#define rotr(x,n)   _lrotr(x,n)
#endif

#define h1(x,y) {                  \
     uint8 a,b,c,d;                \
     a = (uint8) (x);              \
```

```
     b = (uint8) ((x) >> 8);  \
     c = (uint8) ((x) >> 16); \
     d = (uint8) ((x) >> 24); \
     (y) = Q[a]+Q[256+b]+Q[512+c]+Q[768+d]; \
}

#define h2(x,y) {               \
     uint8 a,b,c,d;             \
     a = (uint8) (x);           \
     b = (uint8) ((x) >> 8);  \
     c = (uint8) ((x) >> 16); \
     d = (uint8) ((x) >> 24); \
     (y) = P[a]+P[256+b]+P[512+c]+P[768+d]; \
}

#define step_A(u,v,a,b,c,d,m){         \
     uint32 tem0,tem1,tem2,tem3;       \
     tem0 = rotr((v),23);              \
     tem1 = rotr((c),10);              \
     tem2 = ((v) ^ (c)) & 0x3ff;       \
     (u) += (b)+(tem0^tem1)+Q[tem2];   \
     (a) = (u);                        \
     h1((d),tem3);                     \
     (m) ^= tem3 ^ (u) ;               \
}

#define step_B(u,v,a,b,c,d,m){         \
     uint32 tem0,tem1,tem2,tem3;       \
     tem0 = rotr((v),23);              \
     tem1 = rotr((c),10);              \
     tem2 = ((v) ^ (c)) & 0x3ff;       \
     (u) += (b)+(tem0^tem1)+P[tem2];   \
     (a) = (u);                        \
     h2((d),tem3);                     \
     (m) ^= tem3 ^ (u) ;               \
}

void encrypt(uint32 data[])  //each time it encrypts 512-bit data
{
   uint32 cc,dd;
   cc = counter2048 & 0x3ff;
   dd = (cc+16)&0x3ff;

   if (counter2048 < 1024)
   {
```

```
        counter2048 = (counter2048 + 16) & 0x7ff;
        step_A(P[cc+0], P[cc+1], X[0], X[6], X[13],X[4], data[0]);
        step_A(P[cc+1], P[cc+2], X[1], X[7], X[14],X[5], data[1]);
        step_A(P[cc+2], P[cc+3], X[2], X[8], X[15],X[6], data[2]);
        step_A(P[cc+3], P[cc+4], X[3], X[9], X[0], X[7], data[3]);
        step_A(P[cc+4], P[cc+5], X[4], X[10],X[1], X[8], data[4]);
        step_A(P[cc+5], P[cc+6], X[5], X[11],X[2], X[9], data[5]);
        step_A(P[cc+6], P[cc+7], X[6], X[12],X[3], X[10],data[6]);
        step_A(P[cc+7], P[cc+8], X[7], X[13],X[4], X[11],data[7]);
        step_A(P[cc+8], P[cc+9], X[8], X[14],X[5], X[12],data[8]);
        step_A(P[cc+9], P[cc+10],X[9], X[15],X[6], X[13],data[9]);
        step_A(P[cc+10],P[cc+11],X[10],X[0], X[7], X[14],data[10]);
        step_A(P[cc+11],P[cc+12],X[11],X[1], X[8], X[15],data[11]);
        step_A(P[cc+12],P[cc+13],X[12],X[2], X[9], X[0], data[12]);
        step_A(P[cc+13],P[cc+14],X[13],X[3], X[10],X[1], data[13]);
        step_A(P[cc+14],P[cc+15],X[14],X[4], X[11],X[2], data[14]);
        step_A(P[cc+15],P[dd+0], X[15],X[5], X[12],X[3], data[15]);
    }
    else
    {
        counter2048 = (counter2048 + 16) & 0x7ff;
        step_B(Q[cc+0], Q[cc+1], Y[0], Y[6], Y[13],Y[4], data[0]);
        step_B(Q[cc+1], Q[cc+2], Y[1], Y[7], Y[14],Y[5], data[1]);
        step_B(Q[cc+2], Q[cc+3], Y[2], Y[8], Y[15],Y[6], data[2]);
        step_B(Q[cc+3], Q[cc+4], Y[3], Y[9], Y[0], Y[7], data[3]);
        step_B(Q[cc+4], Q[cc+5], Y[4], Y[10],Y[1], Y[8], data[4]);
        step_B(Q[cc+5], Q[cc+6], Y[5], Y[11],Y[2], Y[9], data[5]);
        step_B(Q[cc+6], Q[cc+7], Y[6], Y[12],Y[3], Y[10],data[6]);
        step_B(Q[cc+7], Q[cc+8], Y[7], Y[13],Y[4], Y[11],data[7]);
        step_B(Q[cc+8], Q[cc+9], Y[8], Y[14],Y[5], Y[12],data[8]);
        step_B(Q[cc+9], Q[cc+10],Y[9], Y[15],Y[6], Y[13],data[9]);
        step_B(Q[cc+10],Q[cc+11],Y[10],Y[0], Y[7], Y[14],data[10]);
        step_B(Q[cc+11],Q[cc+12],Y[11],Y[1], Y[8], Y[15],data[11]);
        step_B(Q[cc+12],Q[cc+13],Y[12],Y[2], Y[9], Y[0], data[12]);
        step_B(Q[cc+13],Q[cc+14],Y[13],Y[3], Y[10],Y[1], data[13]);
        step_B(Q[cc+14],Q[cc+15],Y[14],Y[4], Y[11],Y[2], data[14]);
        step_B(Q[cc+15],Q[dd+0], Y[15],Y[5], Y[12],Y[3], data[15]);
    }
}

//The following defines the initialization functions

#define f1(x)  (rotr((x),7) ^ rotr((x),18) ^ ((x) >> 3))
#define f2(x)  (rotr((x),17) ^ rotr((x),19) ^ ((x) >> 10))
#define f(a,b,c,d) (f2((a)) + (b) + f1((c)) + (d))
```

```
#define feedback_1(u,v,b,c) {        \
   uint32 tem0,tem1,tem2;            \
   tem0 = rotr((v),23); tem1 = rotr((c),10); \
   tem2 = ((v) ^ (c)) & 0x3ff;       \
   (u) += (b)+(tem0^tem1)+Q[tem2]; \
}

#define feedback_2(u,v,b,c) {        \
   uint32 tem0,tem1,tem2;            \
   tem0 = rotr((v),23); tem1 = rotr((c),10); \
   tem2 = ((v) ^ (c)) & 0x3ff;       \
   (u) += (b)+(tem0^tem1)+P[tem2]; \
}

void initialization(uint32 key[], uint32 iv[])
{
   uint32 i,j;

   //expand the key and iv into P and Q
   for (i = 0; i < 8; i++)    P[i] = key[i];
   for (i = 8; i < 16; i++)   P[i] = iv[i-8];

   for (i = 16; i < 528; i++)
       P[i] = f(P[i-2],P[i-7],P[i-15],P[i-16])+i;
   for (i = 0; i < 16; i++)
       P[i] = P[i+512];
   for (i = 16; i < 1024; i++)
       P[i] = f(P[i-2],P[i-7],P[i-15],P[i-16])+512+i;

   for (i = 0;  i < 16;  i++)
       Q[i] = P[1024-16+i];
   for (i = 16; i < 32;  i++)
       Q[i] = f(Q[i-2],Q[i-7],Q[i-15],Q[i-16])+1520+i;
   for (i = 0;  i < 16;  i++)
       Q[i] = Q[i+16];
   for (i = 16; i < 1024;i++)
       Q[i] = f(Q[i-2],Q[i-7],Q[i-15],Q[i-16])+1536+i;

   //run the cipher 4096 steps without generating output
   for (i = 0; i < 2; i++) {
       for (j = 0;  j < 10;   j++)
           feedback_1(P[j],P[j+1],P[(j-10)&0x3ff],P[(j-3)&0x3ff]);
       for (j = 10; j < 1023; j++)
           feedback_1(P[j],P[j+1],P[j-10],P[j-3]);
```

```
            feedback_1(P[1023],P[0],P[1013],P[1020]);
        for (j = 0;  j < 10;   j++)
            feedback_2(Q[j],Q[j+1],Q[(j-10)&0x3ff],Q[(j-3)&0x3ff]);
        for (j = 10; j < 1023; j++)
            feedback_2(Q[j],Q[j+1],Q[j-10],Q[j-3]);
            feedback_2(Q[1023],Q[0],Q[1013],Q[1020]);
    }

    //initialize counter2048, and tables X and Y
    counter2048 = 0;
    for (i = 0; i < 16; i++) X[i] = P[1008+i];
    for (i = 0; i < 16; i++) Y[i] = Q[1008+i];
}
```

C Test HC-256 ("test.c")

```
//This program prints the first 512-bit keystream
//then measure the average encryption speed

#include "hc256.h"
#include <stdio.h>
#include <time.h>

int main()
{
    uint32 key[8],iv[8];
    static uint32 DATA[16]; // the DATA is encrypted

    clock_t start, finish;
    double duration, speed;
    uint32 i;

    //initializes the key and IV
    for (i = 0; i < 8; i++) key[i]=0;
    for (i = 0; i < 8; i++) iv[i]=0;

    //key and iv setup
    initialization(key,iv);

    //generate and print the first 512-bit keystream
    for (i = 0; i < 16; i++) DATA[i]=0;
    encrypt(DATA);
    for (i = 0; i < 16; i++) printf(" %8x ", DATA[i]);

    //measure the encryption speed by encrypting
```

```
    //DATA repeatedly for 0x4000000 times
    start = clock();
    for (i = 0; i < 0x4000000; i++)  encrypt(DATA);
    finish = clock();

    duration = ((double)(finish - start))/ CLOCKS_PER_SEC;
    speed = ((double)i)*32*16/duration;

    printf("\n The encryption takes %4.4f seconds.\n\
            The encryption speed is %13.2f bit/second \n",\
            duration,speed);
    return (0);
}
```

A New Weakness in the RC4 Keystream Generator and an Approach to Improve the Security of the Cipher*

Souradyuti Paul and Bart Preneel

Katholieke Universiteit Leuven, Dept. ESAT/COSIC
Kasteelpark Arenberg 10, B-3001 Leuven-Heverlee, Belgium
{souradyuti.paul,bart.preneel}@esat.kuleuven.ac.be

Abstract. The paper presents a new statistical bias in the distribution of the first two output bytes of the RC4 keystream generator. The number of outputs required to reliably distinguish RC4 outputs from random strings using this bias is only 2^{25} bytes. Most importantly, the bias does not disappear even if the initial 256 bytes are dropped. This paper also proposes a new pseudorandom bit generator, named RC4A, which is based on RC4's exchange shuffle model. It is shown that the new cipher offers increased resistance against most attacks that apply to RC4. RC4A uses fewer operations per output byte and offers the prospect of implementations that can exploit its inherent parallelism to improve its performance further.

1 Introduction

RC4 is the most widely used software based stream cipher. The cipher has been integrated into TLS/SSL and WEP implementations. The cipher was designed by Ron Rivest in 1987 and kept as a trade secret until it was leaked out in 1994. RC4 is extremely fast and its design is simple.

The RC4 stream cipher is based on a secret internal state of $N = 256$ bytes and two index-pointers of size $n = 8$ bits. In this paper we present a bias in the distribution of the first two output bytes. We observe that the first two output words are equal with probability that is significantly less than expected. Based on this bias we construct a distinguisher with non-negligible advantage that distinguishes RC4 outputs from random strings with only 2^{24} pairs of output bytes when $N = 256$. More significantly, the bias remains detectable even after discarding the initial N output bytes. This fact helps us to create another practical distinguisher with only 2^{32} pairs of output bytes that works 256 rounds away from the beginning when $N = 256$.

A second contribution of the paper is a modified RC4 keystream generator, within the scope of the existing model of an exchange shuffle, in order to achieve

* This work was partially supported by the Concerted Research Action GOA-MEFISTO-666 of the Flemish government.

B. Roy and W. Meier (Eds.): FSE 2004, LNCS 3017, pp. 245–259, 2004.

KSA (K, S)	PRGA(S)
for $i = 0$ to $N-1$	$i = 0$
$S[i] = i$	$j = 0$
$j = 0$	Output Generation loop
	$i = (i+1) \bmod N$
for $i = 0$ to $N-1$	$j = (j + S[i]) \bmod N$
$j = (j+S[i]+K[i \bmod l]) \bmod N$	Swap$(S[i], S[j])$
Swap$(S[i], S[j])$	Output=$S[(S[i] + S[j]) \bmod N]$

Fig. 1. The Key Scheduling Algorithm (KSA) and the Pseudo-Random Generation Algorithm (PRGA)

better security. The new cipher is given the name RC4A. We compare its security to the original RC4. Most of the known attacks on RC4 are less effective on RC4A. The new cipher needs fewer instructions per byte and it is possible to exploit the inherent parallelism inherent to improve its performance.

1.1 Description of RC4

RC4 runs in two phases (description in Fig. 1). The first part is the key scheduling algorithm KSA which takes an array S or S-box to derive a permutation of $\{0, 1, 2, \ldots, N-1\}$ using a variable size key K. The second part is the output generation part PRGA which produces pseudo-random bytes using the permutation derived from KSA. Each iteration or loop or 'round' produces one output value. Plaintext bytes are *bit-wise* XORed with the output bytes to produce ciphertext. In most of the applications RC4 is used with word length $n = 8$ bits and $N = 256$. The symbol l denotes the *byte-length* of the secret key.

1.2 Previous Attacks on RC4

RC4 came under intensive scrutiny after it was made public in 1994. Finney [1] showed a class of internal states that RC4 never enters. The class contains all the states for which $j = i + 1$ and $S[j] = 1$. A fraction of N^{-2} of all possible states fall under Finney's forbidden states. It is simple to show that these states are connected by a cycle of length $N(N-1)$. We know that RC4 states are also connected in a cycle (because the *next state function* of RC4 is a bijective mapping on a finite set) and the initial state, where $i = 0$ and $j = 0$, is not one of Finney's forbidden states.

Jenkins [6] detected a probabilistic correlation between the secret information (S, j) and the public information (i, output). Golić [4] showed a positive correlation between the second binary derivative of the least significant bit output sequence and 1. Using this correlation RC4 outputs can be distinguished from a perfectly random stream of bits by observing only $2^{44.7}$ output bytes. Fluhrer and McGrew [3] observed stronger correlations between consecutive bytes. Their distinguisher works using $2^{30.6}$ output bytes. Properties of the state transition graph of RC4 were analyzed by Mister and Tavares [11]. Grosul and Wallach demonstrated a related key attack that works better on very long keys [5]. Andrew Roos also discovered classes of weak keys [15].

Knudsen *et al.* have attacked versions of RC4 with $n < 8$ by their backtracking algorithm in which the adversary guesses the internal state and checks if an anomaly occurs in later stage [7]. In the case of contradiction the algorithm backtracks through the internal states and re-guesses. So far this remains the only efficient algorithm which attempts to discover the secret internal state of this cipher.

The most serious weakness in RC4 was observed by Mantin and Shamir [9] who noted that the probability of a zero output byte at the second round is twice as large as expected. In broadcast applications a practical ciphertext only attack can exploit this weakness.

Fluhrer *et al.* [2] have recently shown that if some portion of the secret key is known then RC4 can be broken completely. This is of practical importance because in the Wired Equivalence Privacy Protocol (WEP in short) a fixed secret key is concatenated with IV modifiers to encrypt different messages. In [16] it is shown that the attack is feasible.

Pudovkina [14] has attempted to detect a bias, only analytically, in the distribution of the first, second output values of RC4 and digraphs under certain uniformity assumptions.

Mironov modeled RC4 as a Markov chain and recommended to dump the initial $12 \cdot N$ bytes of the output stream (at least $3 \cdot N$) in order to obtain uniform distribution of the initial permutation of elements [10].

More recently Paul and Preneel [12] have formally proved that only a known elements of the S-box along with two index-pointers cannot predict more than a output bytes in the next N rounds. They have also designed an efficient algorithm to deduce certain special RC4-states known as *Non-fortuitous Predictive States.*

1.3 Organization

The remainder of this paper is organized as follows. Section 2 introduces the new statistical bias in RC4. Section 3 reflects on the design principles of RC4. Our new variant RC4A is introduced in Sect. 4 and its security is analyzed in Sect. 5. Section 6 presents our concluding remarks.

2 The New Weakness

Our major observation is that the distribution of the first two output bytes of RC4 is not uniform. We noted that the probability that the first two output bytes are equal is $\frac{1}{N}(1 - \frac{1}{N})$. This fact is, in a sense, counter-intuitive from the results obtained by Fluhrer and McGrew [3] who showed that the first two outputs take the value (0, 0) with probability significantly larger than $1/N^2$. Pudovkina [14] analytically obtained, under certain assumptions, that the first two output bytes

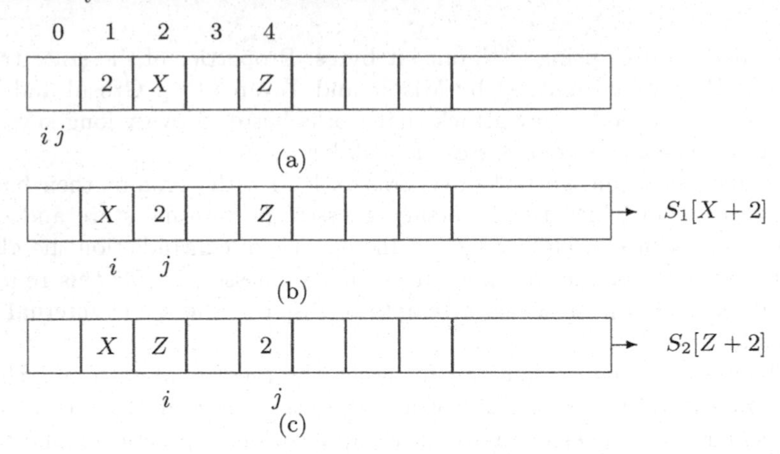

Fig. 2. (a) Just before the beginning of the output generation. (b) First output $S_1[X+2]$ is produced. (c) Second output $S_2[Z+2]$ is produced. $S_1[X+2] \neq S_2[Z+2]$

are equal with probability larger than $1/N$. However, experiments revealed that this result is incorrect. Our main objective is to find a reasonable explanation for this particular bias.

Throughout the paper $S_r[l]$ and O_r denote the lth element of the S-box after the swapping at round r and the output byte generated at that round respectively. Similarly, i_r and j_r should be understood. All arithmetic operations are computed modulo N.

2.1 Motivational Observation

Theorem 1. *If $S_0[1] = 2$ then the first two output bytes of RC4 are always different.*

Proof. Fig. 2 shows the execution of the first two rounds of the output generation. We note that, $O_1 = S_1[X+2]$ and $O_2 = S_2[Z+2]$. Clearly $X+2$ and $Z+2$ point to two different cells of the array. Therefore, the first two outputs are the same if ($X = 0$ and $Z = 2$) or if ($X = 2$ and $Z = 0$). But this is impossible because $X \neq Z \neq 2$ as they are permutation elements. □

2.2 Quantifying the Bias in the First Two Outputs

Theorem 1 immediately implies a bias in the first two outputs of the cipher that are captured in the following corollaries.

Corollary 1. *If the first two output bytes are equal then* $S_0[1] \neq 2$.

Proof. This is an obvious and important deduction from Theorem 1. This fact can be used to speed up exhaustive search by a factor of $\frac{N}{N-1}$. □

Corollary 2. *The probability that the first two output bytes are equal is* $(1 - 1/N)/N$ *(assuming that* $S_0[1] = 2$ *occurs with probability* $1/N$ *and that for the rest of the permutations for which* $S_0[1] \neq 2$ *the first two output bytes are equal with probability* $1/N$*).*[1]

Proof. If $S_0[1] = 2$ occurs with probability $1/N$ then the first two output bytes are different for a fraction of $1/N$ of the total keys (see Theorem 1). The output bytes are equal with probability $1/N$ for each of the other keys, i.e., a fraction of $(1 - 1/N)$ of the total keys. Combining these two,

$$\begin{aligned} P[O_1 = O_2] &= P[O_1 = O_2 | S_0[1] = 2] \cdot P[S_0[1] = 2] + \\ &\quad P[O_1 = O_2 | S_0[1] \neq 2] \cdot [S_0[1] \neq 2] \\ &= 0 \cdot \frac{1}{N} + \frac{1}{N} \cdot (1 - \frac{1}{N}) \\ &= \frac{1}{N} \cdot (1 - \frac{1}{N}). \end{aligned}$$

□

2.3 Distinguisher Based on the Weakness

A *distinguisher* is an efficient algorithm which distinguishes a stream of bits from a perfectly random stream of bits, that is, a stream of bits that has been chosen according to the uniform distribution. There are two ways an adversary can attempt to distinguish between strings, one generated by a pseudorandom generator and the other by a perfectly random source. In the first case the adversary selects *only* one key randomly and produces keystream, seeded by the chosen key, long enough to detect a bias. In this scenario the adversary is "weak" as she has a keystream produced by a single key and therefore the *distinguisher* is called a *weak distinguisher.* In the other case the adversary may use *any number of* randomly chosen keys and the respective keystreams generated by those keys. In this case the adversary is "strong" because she may collect outputs to her advantage from many keystreams to detect a bias. Therefore, the *distinguisher* so constructed is termed a *strong distinguisher.* A bias present in the output at time t in a single stream may hardly be detected by a *weak distinguisher* but a *strong distinguisher* can easily discover the anomaly with fewer bytes. This fact was wonderfully used by Mantin and Shamir [9] to identify a strong bias toward zero in the second output byte of RC4. We also construct a *strong* distinguisher for RC4 based on the non-uniformity of the first two output bytes. We use the following result (see [9] for a proof).

Theorem 2. *If an event* e *occurs in a distribution* X *with probability* p *and in* Y *with probability* $p(1+q)$ *then, for small* p *and* q, $O(\frac{1}{pq^2})$ *samples are required to distinguish* X *from* Y *with non-negligible probability of success.*

In our case, X, Y and e are the distribution of the first two output bytes collected from a perfectly random source, the distribution of these variables from RC4 and

[1] Note that this condition is more relaxed than assuming that the initial permutation is distributed uniformly.

the occurrence of equal successive bytes respectively. Therefore, the number of required samples equals $O(N^3)$ (by Corollary 2, $p = 1/N$ and $q = -1/N$).

Experimental observations agree well with the theoretical results. For $N = 256$, with 2^{24} pairs of the first two output bytes, generated from as many randomly chosen keys, our distinguisher separates RC4 outputs from random strings with an advantage of 40% when the threshold is set at 65408. Empirical results show that the expected number of pairs, for which the bytes are equal, trails the expected number from a random source by *1.21 standard deviations* of the binomial distribution with parameters $(2^{24}, 1/256)$.

2.4 Bias after Dropping the First N Output Bytes

A similar but a smaller bias is also expected in the output bytes O_{t+1} and O_{t+2}, where $t = 0 \bmod N$ and $t > 0$, if we assume that $P[S_t[1] = 2 \cap j_t = 0] = 1/N^2$ and the expected probability that $O_{t+1} = O_{t+2}$ for rest of the internal states is $1/N$. Almost in a similar manner, we can compute $P[O_{t+1} = O_{t+2}]$, where $t = 0 \bmod N$ and $t > 0$.

$$\begin{aligned}
P[O_{t+1} = O_{t+2}] &= P[O_{t+1} = O_{t+2} | S_t[1] = 2 \cap j_t = 0] \cdot P[S_t[1] = 2 \cap j_t = 0] \\
&\quad + P[O_{t+1} = O_{t+2} | S_t[1] \neq 2 \cup j_t \neq 0] \cdot P[S_t[1] \neq 2 \cup j_t \neq 0] \\
&= 0 \cdot \frac{1}{N^2} + \frac{1}{N} \cdot (1 - \frac{1}{N^2}) \\
&= \frac{1}{N} \cdot (1 - \frac{1}{N^2}). \qquad \square
\end{aligned}$$

Therefore, the required number of samples needed to establish the distinguisher is $O(N^5)$ according to Theorem 2 (note that in this case $p = 1/N$ and $q = -1/N^2$).

However, our experiments for $N = 256$ show that the bias can be detected much sooner: 2^{32} pairs of the output bytes (each pair is chosen from rounds $t = 257$ and $t = 258$) are sufficient for a distinguisher, where the theory predicts that this should be 2^{40}. In our experiments, the expected number of pairs, for which the bytes are equal, trails the expected number from a random source by *1.13 standard deviations* of the binomial distribution with parameters $(2^{32}, 1/256)$. A large number of experiments showed that the frequency of simultaneous occurrence of $j = 0$ and $S[1] = 2$ at the 256th round is much higher than expected. This phenomenon accounts for the optimistic behavior of our distinguisher. However, it is still unknown how to quantify the bias in $P[j_{256} = 0 \cap S_{256}[1] = 2]$ theoretically. It is worth noting at this point that Mironov, based on an idealized model of RC4, has suggested to drop the initial $3 \cdot N$ bytes (more conservatively the initial $12 \cdot N$ bytes) in order to obtain a uniform distribution of the initial permutation thereby ruling out any possibility of a strong attack [10].

2.5 Possibility of a Weak Distinguisher

The basic fact examined in the Theorem 1 can be used to characterize a general internal state which would produce unequal consecutive bytes. One can see that for an RC4 internal state, if $i_t = j_t$ and $S_t[i_t + 1] = 2$ then $O_{t+1} \neq O_{t+2}$. If we assume uniformity of RC4 internal states when the value of i is fixed, then the above observation allows for a weak distinguisher for RC4 (i.e., distinguishing RC4 using a single stream generated by a randomly chosen key). However, extensive experiments show that, in case of a single stream, the bias in the output bytes due to these special states is much weaker. With a sample size of 2^{32} pairs of bytes, the expected number of pairs for which the output bytes are equal trails the expected number from a random source by *0.21 standard deviations* of the binomial distribution with parameters $(2^{32}, 1/256)$ (compared to *1.13 standard deviations* for the strong distinguisher with the same sample size). The experimentally obtained standard deviation of the distribution of the number of pairs with equal members in a sample of 2^{32} pairs equals $0.9894 \cdot \sigma$ where σ is the standard deviation of the binomial distribution with parameters $(2^{32}, 1/256)$. The closeness of these two distributions shows that such a *weak distinguisher* is less effective than the strong distinguisher with respect to the required number of outputs. Further experimental work is required to determine the effectiveness of distinguishers which require outputs for fewer keys (say 2^{20} rather than 2^{32}) but longer output streams than just a pair of consecutive bytes for each key.

It is still unclear how this correlation between the internal and external states can be used to mount a full attack on RC4. However, the observation reveals a weakness in the working principle of the cipher, even if N output bytes are dropped.

3 Analyzing RC4 Design Principles

The pseudorandom bit generation in RC4 is divided into two stages (see Sect. 1.1). The key scheduling algorithm KSA intends to turn an identity permutation S into a pseudorandom permutation of elements and the pseudorandom byte generation algorithm PRGA issues one output byte from a pseudorandom location of S in every round. At every round the secret internal state S is changed by the swapping of elements, one in a known location and another pointed to by a 'random' index. The whole idea is inspired by the principle of an exchange shuffle to obtain a 'random' distribution of a deck of cards [8]. Therefore, the security of RC4 in this model of exchange shuffle depends mainly on the following three factors.

- Uniform distribution of the initial permutation of elements (that is, S).
- Uniform distribution of the value of the index-pointer of the element to be swapped with the element contained in a known index (that is, the index-pointer j).
- Uniform distribution of the value of the index-pointer from which the output is issued in a round (i.e., $S[i] + S[j]$).

Note that the above three conditions are necessary conditions of the security of the cipher but by no means they can be sufficient. This fact is wonderfully demonstrated by Golić in [4] using the *Linear Sequential Circuit Approximation* (LSCA for short) methods capturing the basic flaw in the model that the arrangement of the S-box elements in two successive rounds can be approximated to be identical because of 'negligible' changes of S in two successive rounds.

In this paper we take the key scheduling algorithm of RC4 for granted and assume that the distribution of the initial permutation of elements is uniform; we focus solely on the pseudorandom output generation process. As a consequence, the adversary concentrates on deriving the secret internal state S (not the secret key K) from the known outputs exploiting correlations between the internal state and the output bytes.

Most of the known attacks on RC4 to derive part of the secret internal state are based on fixing a few elements of the S-box along with the two index pointers i and j that give information about the outputs at certain rounds with probability 1 or very close to it. This at once results in *distinguishing attacks* on the cipher and helps to derive the secret internal state with probability that is significantly larger than expected. Paul and Preneel have proved that under reasonable assumptions the maximum probability with which a part of the internal state (i.e., certain S-box elements and the value of j) can be predicted by observing known outputs is $1/N$ which is too high [12]. Note that these correlations between the internal state and the external state immediately violate the 'randomness' criteria of an ideal cipher.

The only algorithm attempting to derive the entire internal state of RC4 from the sequence of outputs is by Knudsen *et al.* which is based on a "guess on demand" strategy [7]. The expected complexity of the algorithm is much smaller than a trivial exhaustive search because it implicitly uses the weakness of RC4 that a part of internal state can be guessed with non-trivial probability by observing certain outputs.

In the following discussion we modify RC4 in an attempt to achieve tighter security than the original cipher within the scope of the existing model of exchange shuffle without degrading its speed.

4 RC4A: An Attempt to Improve RC4

4.1 RC4A Design Principles

As most of the existing known plaintext attacks on RC4 harness the stronger correlations between the internal and external states (in generic term b-predictive a-state attack [9, 12]), in principle, making the output generation dependent on more random variables weakens the correlation between them, i.e., the probability to guess the internal state by observing output sequence can be reduced. The larger the number of the variables the weaker will be the correlation between them. On the other hand, intuitively, the large number of variables increases the time complexity as it involves more arithmetic operations.

1. Set $i = 0, j_1 = j_2 = 0$
2. $i{+}{+}$
3. $j_1 = j_1 + S_1[i]$
4. Swap($S_1[i], S_1[j_1]$)
5. $I_2 = S_1[i] + S_1[j_1]$
6. Output=$S_2[I_2]$
7. $j_2 = j_2 + S_2[i]$
8. Swap($S_2[i], S_2[j_2]$)
9. $I_1 = S_2[i] + S_2[j_2]$
10. Output=$S_1[I_1]$
11. Repeat from step 2.

Fig. 3. Pseudo-random Generation Algorithm of RC4A

4.2 RC4A Description

We take one randomly chosen key k_1. Another key k_2 is also generated from a pseudorandom bit generator (e.g. RC4) using k_1 as the seed. Applying the Key Scheduling Algorithm, as described in Fig. 1, we construct two S-boxes S_1 and S_2 using the keys k_1 and k_2 respectively. As mentioned before we assume that S_1 and S_2 are two random permutations of $\{0, 1, 2, \ldots, N-1\}$. In this new scheme the Key Scheduling Algorithm is assumed to produce a uniform distribution of permutation of $\{0, 1, 2, \ldots, N-1\}$. Therefore, our effort focuses on the security of the Pseudo-Random Generation Algorithm. In Fig. 3, we show the pseudo-code of the pseudorandom byte generation algorithm of RC4A. All the arithmetic operations are computed modulo N. The transition of the internal states of the two S-boxes are based on an exchange shuffle as before. Here we introduce two variables j_1 and j_2 corresponding to S_1 and S_2 instead of one. The only modification is that the index-pointer $S_1[i] + S_1[j]$ evaluated on S_1 produces output from S_2 and vice-versa (see steps 5, 6 and 9, 10 of Fig. 3). The next round starts after each output generation.

RC4A Uses Fewer Instructions per Output Byte than RC4. To produce two successive output bytes the i pointer is incremented once in case of RC4A where it is incremented twice to produce as many output words in RC4.

Parallelism in RC4A. The performance of RC4A can be further improved by extracting the parallelism latent in the algorithm. The parallel steps of the algorithm can be easily found by drawing a dependency graph of the steps shown in Fig. 3. In the following list the parallel steps of RC4A are shown within brackets.

1. (3, 7).
2. (4, 5, 9).
3. (6, 10).
4. (8,2).

The existence of many parallel steps in RC4A is certainly an important aspect of this new cipher and it offers the possibility of a faster stream cipher if RC4A is implemented efficiently.

5 Security Analysis of RC4A

The RC4A pseudorandom bit generator has passed all the statistical tests listed in [13]. RC4A achieves two major gains over RC4. By making every byte depend on at least two random values (e.g. O_1 depends on $S_1[1]$, $S_1[j_1]$ and $S_2[S_1[1] + S_1[j_1]]$) of S_1 and S_2 the secret internal state of RC4A becomes $N!^2 \times N^3$. So, for $N = 256$, the number of secret internal states for RC4A is approximately 2^{3392} when the number is only 2^{1700} for RC4.

Let the events E_A and E_B denote the occurrences of an *internal state* (i.e., a known elements in the S-boxes, i, j_1 and j_2) and the corresponding b outputs when the i value of an *internal state* is known. We assume uniformity of the *internal state* and the corresponding *external state* for any fixed i value of the *internal state*. Assuming that a is much smaller than N and disregarding the small bias induced in E_B due to E_A, we apply Bayes' Rule to get

$$P[E_A|E_B] = \frac{P[E_A]}{P[E_B]} P[E_B|E_A] \approx \frac{N^{-(a+2)}}{N^{-b}} \cdot 1 = N^{b-a-2} . \tag{1}$$

Paul and Preneel [12] have proved that $a \geq b$ for RC4 for small values of a. We omit the proof as it is quite rigorous and beyond the scope of this paper. Exactly the same technique can be applied here to prove that a known elements of the S-boxes along with i, j_1 and j_2 cannot predict more than a elements for small values of a. Therefore, the maximum probability with which any internal state of RC4A can be predicted from a known output sequence equals $1/N^2$ compared to $1/N$ for RC4. In the following sections we describe how RC4A resists the two major attacks on it: one attempts to derive the entire internal state deterministically and another to derive a part of the internal state probabilistically.

5.1 Precluding the Backtracking Algorithm by Knudsen et al.

As mentioned earlier that the "guess on demand" backtracking algorithm by Knudsen *et al.* is so far the best algorithm to deduce the internal state of RC4 from the known plaintext [7]. Now we briefly discuss the functionality of the variant of the algorithm to be applied for RC4A.

The algorithm simulates RC4A by observing only the output bytes in recursive function calls. The values of $S[i]$ and $S[j]$ in one S-box are guessed from the permutation elements to agree with the output and its possible location in the other S-box. If they match then the algorithm calls the round function for the next round. If an anomaly occurs then it backtracks through the previous rounds and re-guesses. The number of outputs m, needed to uniquely determine the entire internal state, is bounded below by the inequality, $2^{nm} > (2^n!)^2$. Therefore, $m \geq 2N$ (note, $N = 2^n$).

Theorem 3 (RC4 vs RC4A). *If the expected computational complexity to derive the secret internal state of RC4A from known $2N$ initial output bytes with the algorithm by Knudsen et al. is C_{rc4a} and if the corresponding complexity for RC4 using N known initial output bytes is C_{rc4} then C_{rc4a} is much higher than C_{rc4} and C_{rc4a} can be approximated to C_{rc4}^2 under certain assumptions.*[2]

Proof. According to the algorithm by Knudsen *et al.*, the internal state of RC4 is derived using only the first N output bytes, that is, simulating RC4 for the first N rounds. The variant of this algorithm which works on RC4A uses the initial $2N$ bytes, thereby runs for the first $2N$ rounds.

Let the algorithms A_1 and A_2 derive the secret internal states for RC4 and RC4A respectively. At every round the S-boxes are assigned either 0, 1, 2, or 3 elements and move to the next round.

Let, at the tth round, A_2 go to the next round after assigning k elements an expected number of $m_{k,t}$ times. So the number of value assignments in the tth round is $\sum_{k=0}^{3} k \cdot m_{k,t}$. Note, each of the $\sum_{k=0}^{3} m_{k,t}$ iterations gives rise to an S-box arrangement in the next round. It is possible that we reach some S-box arrangements from which no further transition to the next rounds is possible because of contradictions. In such case, we assume assignment of zero elements in the S-box till the Nth round is reached. Let the number of S-box arrangements at the tth round from which these $\sum_{k=0}^{3} m_{k,t}$ arrangements are generated is L_t. Consequently,

$$\sum_{k=0}^{3} m_{k,t} = L_{t+1} \,. \tag{2}$$

Now we set,

$$\sum_{k=0}^{3} k \cdot m_{k,t} = \tilde{k}_t \cdot \sum_{k=0}^{3} m_{k,t} = \tilde{k}_t \cdot L_{t+1} \,. \tag{3}$$

In Eqn. (3), $\tilde{k}_t$ is the expected number elements which are assigned to the S-boxes in each iteration in that particular round. If each of L_t is assumed to produce an expected $\tilde{L}_{t+1}$ number of S-box arrangements in the next round then Eqn. (3) becomes,

$$\sum_{k=0}^{3} k \cdot m_{k,t} = \tilde{k}_t \cdot (\tilde{L}_{t+1} \cdot L_t) \,. \tag{4}$$

[2] The complexity is measured in terms of the number of value assignments.

Denoting the total number of value assignments in the tth round by $C(t)$, it is easy to note from Eqn. (4),

$$C(t) = \tilde{k}_t \cdot (\tilde{L}_{t+1} \cdot L_t) . \tag{5}$$

Proceeding this way it can be shown that,

$$C(t+s) = \tilde{k}_{t+s} \cdot L_t \cdot \prod_{i=t}^{t+s} \tilde{L}_{i+1} . \tag{6}$$

If $t = 1$ then $L_t = 1$. Setting $t + s = n$ in Eqn. (6), we get,

$$C(n) = \tilde{k}_n \cdot \prod_{i=1}^{n} \tilde{L}_{i+1} . \tag{7}$$

From Eqn. (2) and Eqn. (3), $C(n)$ can be evaluated $\forall n \in \{1, 2, \ldots 2N\}$ when $m_{k,t}$ is known $\forall (k, t) \in \{0, 1, 2, 3\}\{1, 2, \ldots 2N\}$.

It is important to note that on a random output sequence $\tilde{k}_{2f-1} \approx \tilde{k}_{2f}$ and $\tilde{L}_{2f} \approx \tilde{L}_{2f+1}$ $\forall f \in \{1, 2, \ldots N\}$. The reason behind the approximation is that, with the algorithm by Knudsen *et al.*, the difference between the expected number of assignments in the S-boxes in the $(2f - 1)$th and the $2f$th rounds is very small. Therefore, the overall complexity C_{rc4a} becomes,

$$\begin{aligned}
C_{rc4a} &= \sum_{n=1}^{2N} C(n) \\
&= \sum_{n=1}^{2N} (\tilde{k}_n \cdot \prod_{i=1}^{n} \tilde{L}_{i+1}) \\
&= \tilde{k}_1 \cdot \tilde{L}_2 + \sum_{i=1}^{N} (\tilde{k}_{2i} \cdot \prod_{j=1}^{i} \tilde{L}_{2j+1}^2) + \sum_{i=2}^{N} (\tilde{k}_{2i-1} \cdot \tilde{L}_{2i} \cdot \prod_{j=1}^{i-1} \tilde{L}_{2j}^2) \\
&= \tilde{k}_1 \cdot \tilde{L}_2 + \sum_{i=1}^{N} (\tilde{k}_{2i-1} \cdot \prod_{j=1}^{i} \tilde{L}_{2j}^2) + \sum_{i=2}^{N} (\tilde{k}_{2i-1} \cdot \tilde{L}_{2i} \cdot \prod_{j=1}^{i-1} \tilde{L}_{2j}^2) .
\end{aligned}$$

Replacing $\tilde{k}_q$ and $\tilde{L}_q$ by $x_{\frac{q+1}{2}}$ and $g_{\frac{q}{2}}$ we get,

$$C_{rc4a} = \sum_{i=1}^{N} (x_i \cdot \prod_{j=1}^{i} g_j^2) + x_1 \cdot g_1 + \sum_{i=2}^{N} (x_i \cdot g_i \cdot \prod_{j=1}^{i-1} g_j^2) . \tag{8}$$

Applying a similar technique as above it is easy to see that,

$$C_{rc4} = \sum_{i=1}^{N} (x_i \cdot \prod_{j=1}^{i} g_j) . \tag{9}$$

Again we note that the difference between the expected number of elements that are already assigned in S_1 for RC4A at round $(2t-1)$ and the expected number of elements in S for RC4 at round t is negligible. Therefore, the corresponding $\tilde{k}_t$ and $\tilde{L}_{t+1}$ for RC4 can be approximated to $\tilde{k}_{2t-1}$ and $\tilde{L}_{2t}$ for RC4A.

As the g_i's are real numbers greater than 1 and the x_i's are non-negative real numbers, from Eqn. (8) and Eqn. (9) it is easy to see that $C_{rc4a} \gg C_{rc4}$.

We observe from the algorithm that $x_i \in \{y : 0 \leq y \leq 3, y \in \mathbb{R}\}$. It is clear from the algorithm that x_i decreases as i increases. Intuitively, x_i is less than one in the last rounds. Therefore, assuming $C_{rc4a} \approx \prod_{i=1}^{N} g_i^2$ and $C_{rc4} \approx \prod_{i=1}^{N} g_i$, we get $C_{rc4a} \approx C_{rc4}^2$. □

By Theorem 3, the expected complexity to deduce the secret internal state of RC4A ($N = 256$) with the algorithm by Knudsen *et al.* is 2^{1558} when the corresponding complexity is 2^{779} for RC4.

5.2 Resisting the Fortuitous States Attack

Fluhrer and McGrew discovered certain RC4 states in which only m known consecutive S-box elements participate in producing the next m successive outputs. Those states are defined to be *Fortuitous States* (see [3, 12] for a detailed analysis). *Fortuitous States* increase the probability to guess a part of internal state in a known plaintext attack (see Eqn. (1)). The larger the probability of the occurrence of a fortuitous state, the smaller will be the number of required rounds to obtain one of them.

RC4A also weakens the fortuitous state attack by a large degree. A moment's reflection shows that RC4A does not have any fortuitous state of length 1. Now we will compare the probability of the occurrence of a fortuitous state of length $2a$ in RC4A to that of length a in RC4. It is easy to note that a fortuitous state of length $2a$ of RC4A implies and is implied by two fortuitous states of length a of RC4 appearing simultaneously in S_1 and S_2. If C denotes the number of fortuitous states of length a of RC4 then the expected number of fortuitous states of length $2a$ in RC4A is C^2/N. Let P_a denote the probability of the occurrence of a fortuitous state of length a in RC4 and P_{2a} denote the probability of the occurrence of a fortuitous state of length $2a$ in RC4A. Then, for small values of a, $P_a = \frac{C}{N^{a+2}}$ and $P_{2a} = \frac{C^2}{N^{2a+4}}$ which immediately implies $P_{2a} < P_a$.

5.3 Resisting other Attacks

One can see that the strong positive bias of the second output byte of RC4 toward zero [9], and the bias described in the first part of this paper are also diminished in this new cipher as more random variables are required to be fixed for the biased state to occur.

5.4 Open Problems and Directions for Future Work

Although RC4A has an improved security over the original cipher against most of the known plaintext attacks, it is still as vulnerable as RC4 against the attack by Golić which uses the positive correlation between the second binary derivative of the least significant bit output sequence and 1. The weakness originates from the slow change of the S-box in successive rounds that seems to be inherent in any model based on exchange shuffle. Therefore, this still remains an open problem whether it is possible to remove this weakness from the output words of the stream cipher based on an exchange shuffle while retaining all of its speed and security.

Our work leaves room for more research. It is worthwhile to note that one output byte generation in this existing model of exchange shuffle involves two random pointers; j and $S[i]+S[j]$. In RC4 both the pointers fetch values from a single S-box. We obtained better results by making $S[i]+S[j]$ fetch value from a different S-box. What if we obtain $S[j]$ from another S-box and generate output using three S-boxes?

6 Conclusions

In this paper we have described a new statistical weakness in the first two output bytes of the RC4 keystream generator. The weakness does not disappear even after dropping the initial N bytes. Based on this observation, we recommend to drop at least the initial $2N$ bytes of RC4 in all future applications of it. In the second part of the paper we attempted to improve the security of RC4 by introducing more random variables in the output generation process thereby reducing the correlation between the internal and the external states.

As a final comment we would like to mention that the security of RC4A could be further improved. For example, one could introduce key-dependent values of i and j at the beginning of the first round, and one could address the weaknesses of the Key Scheduling Algorithm. In this paper, we have assumed that the original Key Scheduling Algorithm produces a uniform distribution of the initial permutation of elements, which is certainly not correct.

Acknowledgements

We are grateful to Adi Shamir for helpful comments. The authors also thank Kenneth G. Paterson for pointing out the possibility of a *weak* distinguisher using our observation. We would like to thank Joseph Lano for providing us with experimental results. Thanks are due to Alex Biryukov, Christophe De Cannière, Jorge Nakahara Jr. and Dai Watanabe for many useful discussions. The authors also acknowledge the constructive comments of the anonymous reviewers.

References

[1] H. Finney, "An RC4 cycle that can't happen," Post in `sci.crypt`, September 1994. 246

[2] S. Fluhrer, I. Mantin, A. Shamir, "Weaknesses in the Key Scheduling Algorithm of RC4," *SAC 2001* (S. Vaudenay, A. Youssef, eds.), vol. 2259 of *LNCS*, pp. 1-24, Springer-Verlag, 2001. 247

[3] S. Fluhrer, D. McGrew, "Statistical Analysis of the Alleged RC4 Keystream Generator," *Fast Software Encryption 2000* (B. Schneier, ed.), vol. 1978 of *LNCS*, pp. 19-30, Springer-Verlag, 2000. 246, 247, 257

[4] J. Golić, "Linear Statistical Weakness of Alleged RC4 Keystream Generator," *Eurocrypt '97* (W. Fumy, ed.), vol. 1233 of *LNCS*, pp. 226-238, Springer-Verlag, 1997. 246, 252

[5] A. Grosul, D. Wallach, "A related key cryptanalysis of RC4," *Department of Computer Science, Rice University, Technical Report TR-00-358,* June 2000. 247

[6] R. Jenkins, "Isaac and RC4," Published on the Internet at `http://burtleburtle.net/bob/rand/isaac.html`. 246

[7] L. Knudsen, W. Meier, B. Preneel, V. Rijmen, S. Verdoolaege, "Analysis Methods for (Alleged) RC4," *Asiacrypt '98* (K. Ohta, D. Pei, eds.), vol. 1514 of *LNCS*, pp. 327-341, Springer-Verlag, 1998. 247, 252, 254

[8] D.E. Knuth, *"The Art of Computer Programming," vol. 2, Seminumerical Algorithms,* Addison-Wesley Publishing Company, 1981. 251

[9] I. Mantin, A. Shamir, "A Practical Attack on Broadcast RC4," *Fast Software Encryption 2001* (M. Matsui, ed.), vol. 2355 of *LNCS*, pp. 152-164, Springer-Verlag, 2001. 247, 249, 252, 257

[10] I. Mironov, "Not (So) Random Shuffle of RC4," *Crypto 2002* (M. Yung, ed.), vol. 2442 of *LNCS*, pp. 304-319, Springer-Verlag, 2002. 247, 250

[11] S. Mister, S. Tavares, "Cryptanalysis of RC4-like Ciphers," *SAC '98* (S. Tavares, H. Meijer, eds.), vol. 1556 of *LNCS*, pp. 131-143, Springer-Verlag, 1999. 247

[12] S. Paul, B. Preneel, "Analysis of Non-fortuitous Predictive States of the RC4 Keystream Generator," *Indocrypt 2003* (T. Johansson, S. Maitra, eds.), vol. 2904 of *LNCS*, pp. 52-67, Springer-Verlag, 2003. 247, 252, 254, 257

[13] B. Preneel *et al.*, "NESSIE Security Report," Version 2.0, IST-1999-12324, February 19, 2003, `http://www.cryptonessie.org`. 254

[14] M. Pudovkina, "Statistical Weaknesses in the Alleged RC4 keystream generator," *Cryptology ePrint Archive 2002–171,* IACR, 2002. 247

[15] A. Roos, "Class of weak keys in the RC4 stream cipher," Post in `sci.crypt`, September 1995. 247

[16] A. Stubblefield, J. Ioannidis, A. Rubin, "Using the Fluhrer, Mantin and Shamir attack to break WEP," *Proceedings of the 2002 Network and Distributed Systems Security Symposium,* pp. 17–22, 2002. 247

Improving Immunity of Feistel Ciphers against Differential Cryptanalysis by Using Multiple MDS Matrices

Taizo Shirai* and Kyoji Shibutani

Ubiquitous Technology Laboratories, Sony Corporation
7-35 Kitashinagawa 6-chome, Shinagawa-ku, Tokyo, 141-0001 Japan
{taizo.shirai,kyoji.shibutani}@jp.sony.com

Abstract. A practical measure to estimate the immunity of block ciphers against differential and linear attacks consists of finding the minimum number of active S-Boxes, or a lower bound for this minimum number. The evaluation result of lower bounds of differentially active S-boxes of AES, Camellia (without FL/FL^{-1}) and Feistel ciphers with an MDS based matrix of branch number 9, showed that the percentage of active S-boxes in Feistel ciphers is lower than in AES. The cause is a *difference cancellation* property which can occur at the XOR operation in the Feistel structure. In this paper we propose a new design strategy to avoid such difference cancellation by employing multiple MDS based matrices in the diffusion layer of the F-function. The effectiveness of the proposed method is confirmed by an experimental result showing that the percentage of active S-boxes of the newly designed Feistel cipher becomes the same as for the AES.

Keywords: MDS, Feistel cipher, active S-boxes, multiple MDS design

1 Introduction

Throughout recent cryptographic primitive selection projects, such as AES, NESSIE and CRYPTREC projects, many types of symmetric key block ciphers have been selected for widely practical uses [16, 17, 18]. A highly regarded design strategy in a lot of well-known symmetric-key block ciphers consists in employing small non-linear functions (S-box), and designing a linear diffusion layer to achieve a high value of the minimum number of active S-boxes [2, 4, 17, 18].

If the diffusion layer guarantees a sufficient minimum number of differentially active S-boxes, and the S-boxes have low maximum differential probability, the resistance against differential attacks will be strong enough. Let a be the lower bound on the minimum number of active S-boxes, and DP_{max} be the maximum differential probability (MDP) of S-boxes. It is guaranteed that there is no differential path whose differential characteristic probability (DCP) is higher than

* The first author was a guest researcher at Katholieke Universiteit Leuven, Dept. ESAT/SCD-COSIC from 2003 to 2004.

B. Roy and W. Meier (Eds.): FSE 2004, LNCS 3017, pp. 260–278, 2004.

$(DP_{max})^a$. For instance, in the case of a 128-bit block cipher using 8-bit bijective S-boxes with $DP_{max} = 2^{-6}$, the necessary condition to rule out any path with $DCP > 2^{-128}$ is that a should be at least 22. In order to determine the appropriate number of rounds of a fast and secure cipher, it is thus essential to have an accurate estimation of the lower bound a [1, 4]. Regarding this problem, finding an optimal linear diffusion is one of the research topics included in the future research agenda of cryptology proposed by STORK project in EU [19].

Comparing the minimum number of active S-boxes of two well-known ciphers, AES and Camellia without FL/FL^{-1} (denoted by Camellia*), it is shown that the ratio of the minimum number of active S-boxes to the total number of S-boxes for Camellia* is lower than for AES. Even if the diffusion matrix of Camellia* is replaced by a 8×8 MDS based matrix with branch number 9 (which is called a MDS-Feistel cipher), the ratio won't increase significantly and there is an apparent gap between these Feistel ciphers (Camellia* and MDS-Feistel) and a SPN cipher AES.

We found that the low percentage of active S-boxes in a MDS-Feistel structure is due to a *difference cancellation* which always occurs in the differential path that realizes the minimum number of active S-boxes. In such a case, the output difference of the F-function in the i-th round will be canceled completely by the output difference of $i+2j$-th round ($j > 0$). It is obvious that one of the conditions for difference cancellations is employing an unique diffusion matrix for all F-functions.

In this work, we propose a new design strategy to avoid the difference cancellation in Feistel ciphers with SP-type F-function. We call this new strategy *multiple MDS Feistel structure design.* The basic principle of this design is as follows. Let $2r, m$ be the round number of the Feistel structure and the number of S-boxes in the F-function, respectively. We then employ $q(< r)$ $m \times m$ MDS matrices. Furthermore they are chosen that any m columns in these q MDS matrices also satisfy the MDS property. Then, at first these MDS matrices are allocated in the odd-round F-functions, then they are allocated in the even-round F-functions again keeping the involution property. This construction removes chances of difference cancellation within consecutive $2q+1$ rounds.

We will also show an evaluation result that confirms the effectiveness of the new design, which shows that our new design strategy makes the Feistel cipher achieve a high ratio for the minimum number of active S-boxes. The new design has a ratio that is at the same level as AES.

Our results open a way to design faster Feistel ciphers keeping its advantage that the same implementation can be used for encryption and decryption except the order of subkeys.

This paper is organized as follows: In Sect. 2, we describe some definitions used in this paper. In Sect. 3, we compare the minimum number of active S-boxes of various ciphers. In Sect. 4, we explain how difference cancellation occurs. In Sect. 5, we propose our new design strategy, *multiple MDS Feistel structure design.* In Sect. 6, we investigate the effect of the multiple MDS Feistel structure design. Finally in Sect. 7, we discuss the new design and future research.

2 Preliminaries

In this section, we state some definitions and notions that are used in the rest of this paper.

Definition 1. active S-box
An S-box which has non-zero input difference is called **active S-box**.

Definition 2. χ function
For any difference $\Delta X \in GF(2^n)$, a function $\chi : GF(2^n) \rightarrow \{0,1\}$ is defined as follows:

$$\chi(\Delta X) = \begin{cases} 0 \; if \;\; \Delta X = \mathbf{0} \\ 1 \; if \;\; \Delta X \neq \mathbf{0} \end{cases}$$

For any differential vector $\Delta X = (\Delta X[1], \Delta X[2], \ldots, \Delta X[m]) \in GF(2^n)^m$, the truncated difference $\delta X \in \{0,1\}^m$ is defined as

$$\delta X = \chi(\Delta X) = (\chi(\Delta X[1]), \chi(\Delta X[2]), \ldots, \chi(\Delta X[m]))$$

Definition 3. (truncated) Hamming weight of vector in $GF(2^n)^m$
Let $v = (v_1, v_2, \ldots, v_m) \in GF(2^n)^m$. the Hamming weight of a vector v is defined as follows:

$$w_h(v) = \sharp\{v_i | v_i \neq 0, 1 \leq i \leq m\}.$$

Theorem 1. *[7] A $[k+m, k, d]$ linear code with generator matrix $G = [I_{k\times k} M_{k\times m}]$, is MDS iff every square submatrix (formed from any i rows and any i columns, for any $i = 1, 2, ..., min\{k, m\}$) of $M_{k\times m}$ is nonsingular.*

From the above theorem, we call a matrix M is a MDS matrix if every square submatrix is nonsingular.

Definition 4. Branch Number
Let $v = (v_1, v_2, \ldots, v_m) \in GF(2^n)^m$. The branch number $\mathcal{B}$ of a linear mapping $\theta : GF(2^n)^m \rightarrow GF(2^n)^m$ is defined as:

$$\mathcal{B}(\theta) = \min_{v \neq 0}\{w_h(v) + w_h(\theta(v))\}.$$

If M is a $m \times m$ MDS matrix and $\theta : x \rightarrow Mx$, then $\mathcal{B}(\theta) = m + 1$.

Definition 5. Feistel structure using SP-type F-function
A SP-type F-function is defined as the following: Let n be a bit width of bijective S-boxes, and m be a number of S-boxes employed in a F-function. In the i-th round F-function, (1) mn bit round key $k_i \in GF(2^n)^m$ and data $x_i \in GF(2^n)^m$ are XORed: $w_i = x_i \oplus k_i$. (2)w_i is split into m pieces of n-bit data, then each n-bit data is input to a corresponding S-box. (3) The output values of S-boxes regarded as $z_i \in GF(2^n)^m$ are transformed by an $m \times m$ matrix M over $GF(2^n)$: $y_i = Mz_i$.

A Feistel structure using SP-type F-function is shown in Fig. 1.

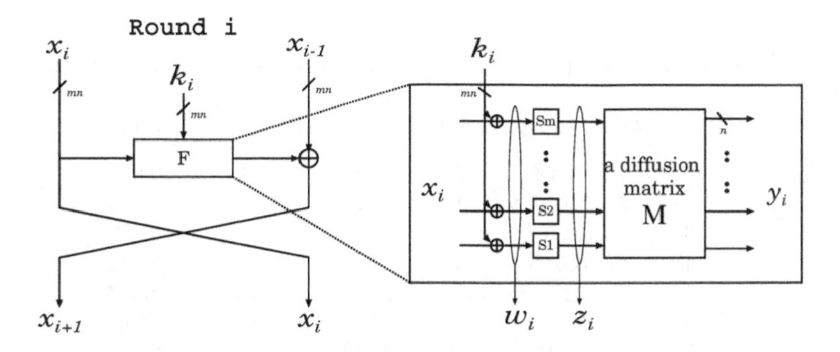

Fig. 1. The general model of a SP-type F-function

3 Comparison of the Minimum Number of Active S-Boxes

At first in this paper, we compare the lower bound of the minimum number of active S-boxes of 3 typical ciphers: AES, Camellia without FL/FL^{-1} (we call it Camellia*) and a Feistel cipher using a 8×8 MDS matrix with branch number 9 (we call it MDS-Feistel cipher). Note that we assumed that the MDS-Feistel uses eight 8-bit bijective S-boxes in the F-function like Camellia*, therefore the block sizes of these block ciphers are all 128-bit.

The lower bound estimation for these ciphers have been obtained as follows.

- AES: The wide trail strategy guarantees $\mathcal{B}^2 = 25$ active S-boxes in 4 consecutive rounds. The lower bound is obtained by using Matsui's truncated path search technique which is slightly modified to analyze AES [2, 8, 9, 12]. Let $a(r)$ be the minimum number of active S-boxes for r rounds, then the conjectured $a(r)$ from the estimation is $a(0) = 0, a(1) = 1, a(2) = 5, a(3) = 9$, and then $a(r) = a(r-4) + 25$ for $(r \geq 4)$.
- Camellia*: The lower bound is obtained from Shirai et al.'s result. They used an improved estimation method based on Matsui's technique which was also used by the designers' evaluation of Camellia [2, 8, 9, 14]. The improved method discards algebraic contradiction in difference paths [12, 13].
- MDS-Feistel: The lower bound is obtained by also using Matsui's truncated path search technique which is slightly modified to analyze the MDS-Feistel's round function [2, 8, 9, 12]. Shimizu has also shown a similar but limited result for the lower bound by using a method not based on Matsui's approach, and he has conjectured an equation $a(r) = \lfloor r/4 \rfloor (\mathcal{B}+1) + (r \mod 4) - 1$ [11]. We confirmed that our result matches the Shimizu's conjectured equation.

Table 1 shows the lower bound on the number of active S-boxes for r-round ciphers, and the ratio of active S-boxes to all S-boxes in the r-round cipher. Fig. 2 shows a graph of the ratios of active S-boxes to all S-boxes.

Table 1. the lowerbound of the minimum number of active S-boxes

Round	AES	(ratio)	Camellia*	(ratio)	MDS($\mathcal{B} = 9$)	(ratio)
1	1	6.3%	0	0.0%	0	0.0%
2	5	15.6%	1	6.3%	1	6.3%
3	9	18.8%	2	8.3%	2	8.3%
4	25	39.1%	7	21.9%	9	28.1%
5	26	32.5%	9	22.5%	10	25.0%
6	30	31.3%	12	25.0%	11	22.9%
7	34	30.1%	14	25.0%	12	21.4%
8	50	39.1%	16	25.0%	19	29.7%
9	51	35.4%	20	27.8%	20	27.7%
10	55	34.4%	22	23.8%	21	26.3%
11	59	33.5%	24	27.5%	22	25.0%
12	75	39.1%	-	-	29	30.2%
13	76	36.5%	-	-	30	28.8%
14	80	36.5%	-	-	31	27.7%
15	84	35.0%	-	-	32	26.7%
16	100	39.1%	-	-	39	30.5%
∞	-	39.1%	-	-	-	34.4%

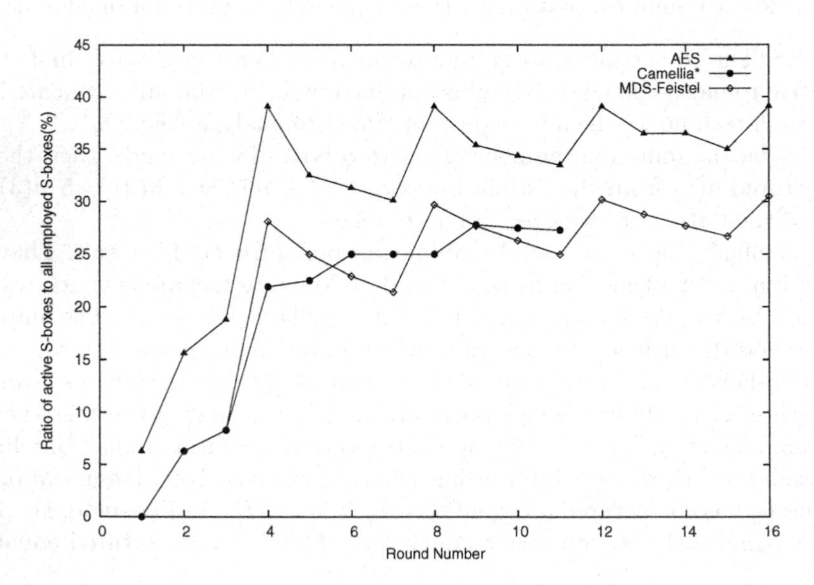

Fig. 2. The percentage of active S-boxes for AES, Camellia* and MDS-Feistel

The fact that the minimum numbers of active S-boxes are smaller for Feistel ciphers (Camellia* and MDS-Feistel) than for AES is not unexpected, because there are only half as many S-boxes in each round (8 in Feistel ciphers, 16 in AES). However, Fig. 2 shows a non-trivial fact that also the percentage of active S-boxes is lower in these Feistel ciphers than in AES. Also, we note that even with a MDS matrix of branch number 9, which is the best possible branch number for an 8×8 matrix, the construction doesn't gain significantly compared to Camellia*, which uses a non-MDS matrix of branch number 5 .

The percentage of active S-boxes indicates how many S-boxes are effectively used in all existing S-boxes for the first consecutive rounds, and can be considered as a reference of efficiency of the diffusion property for ciphers which have the same block length and the same S-box bit length.

As described in [1], if we choose 8-bit S-boxes with maximum differential probability (MDP) 2^{-6}, 22 active S-boxes is a necessary condition to rule out the existence of differential characteristics with a probability higher than 2^{-128}. In the case of AES, 22 active S-boxes are already achieved by only 4 rounds. However, Feistel ciphers require more than 11 rounds to guarantee 22 active S-boxes. Because the minimum number of active S-boxes is often taken into consideration when determining the round number of a block cipher, if more active S-boxes can be guaranteed in SP-type Feistel ciphers we may be able to design fewer rounds (it means fast) ciphers.

In the following sections, we analyze a mechanism that explains why the ratio for MDS-Feistel ciphers is low, and we propose a new design strategy that achieves more active S-boxes and thus enables us to construct Feistel structures with fewer rounds.

4 Difference Cancellation

By our analysis of the MDS-Feistel cipher, we found that every path which contains the minimum number of active S-boxes includes a particular phenomenon where differences generated at certain round are canceled after some rounds at an XOR operation. We will call this phenomenon *difference cancellation*.

The left half of Fig. 3 shows an example of the 3-round difference cancellation. Differences are represented in the truncated way in which 8-bit difference data is represented as 0 or 1, depending on whether each difference is 0 or not [6]. The 3-round difference cancellation starts from the difference $\delta x_{i-1} = (00000000)$,and ends with the difference $\delta x_{i+3} = (00000000)$ again. This means that a certain difference is generated and then canceled between these two 0-differences. In this case, the full hamming weight difference $\delta y_i = (11111111)$ is canceled by $\delta y_{i+2} = (11111111)$ at once. Consequently there's no active S-boxes in $i+3$-round.

Similarly, 5-round difference cancellation, which is shown in the right half of Fig. 3, have the form $\delta x_i = \delta x_{i+4} = (00000001), \delta x_{i+2} = (00000000)$, and output differences of both active S-boxes in the i-th and $i+4$-th round are equal.

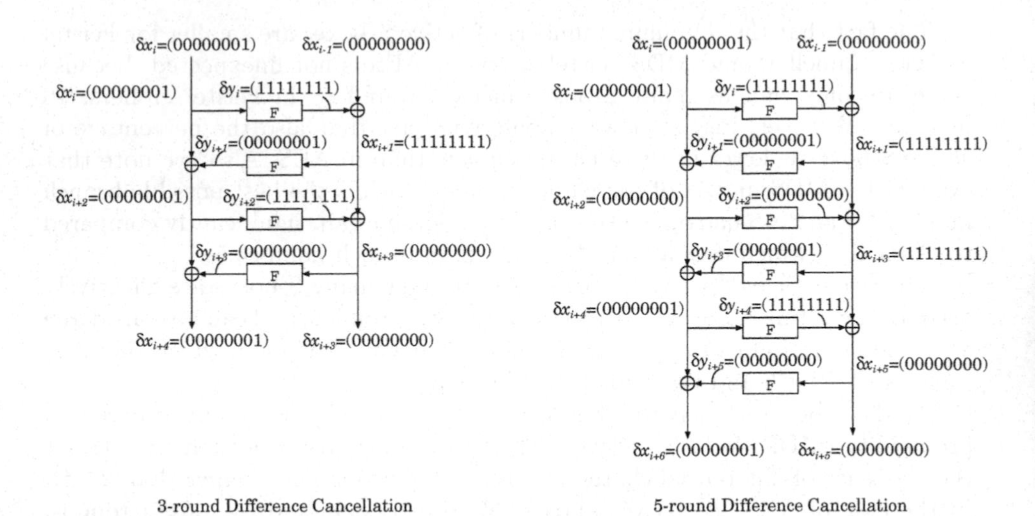

3-round Difference Cancellation

5-round Difference Cancellation

Fig. 3. Difference Cancellation

In both cases, a truncated difference (11111111) generated by one active S-box is canceled by a truncated differences (11111111) which is also generated by one active S-box. These difference cancellations are derived from 2 active S-boxes, and we call this type of difference cancellation *2-derived difference cancellation.*

An interesting fact is that at least one of these 3-round or 5-round 2-derived difference cancellations can be found in every differential path of more than 6-round that realizes the minimum number of active S-boxes in the MDS-Feistel cipher. Details are shown in Appendix A.

4.1 Observation on Difference Cancellation

Let $X, Y \in \{0,1\}^8$ and $X \overset{M}{\rightarrow} Y$ denotes that a truncated difference X can produce a truncated difference Y by a matrix M. Let M_{MDS} be a 8×8 MDS matrix, then the following property of the M_{MDS} contributes to occur the 2-derived difference cancellation.

$$(00000001) \overset{M_{MDS}}{\longrightarrow} (11111111) \quad and \quad (11111111) \overset{M_{MDS}}{\longrightarrow} (00000001)$$

These transitions are appeared in the above 3-round and 5-round difference cancellation several times.

More precisely, let $C_M(X,Y) = \{0,1\}$ be a function which shows the capability of connection between truncated difference X and Y defined as,

$$C_M(X,Y) = \begin{cases} 1 & if \quad X \overset{M}{\rightarrow} Y \\ 0 & else \end{cases}$$

We can observe that a 2-derived difference cancellation can occur if there exists at least one set of truncated differences X, Y where $w_h(X) = 1$ which satisfy $C_M(X,Y) \cdot C_M(Y,X) \neq 0$.

From MDS property, any $m \times m$ MDS matrix M_{MDS} holds the condition $C_{M_{MDS}}(X,Y) = 1$ for all $w_h(X) + w_h(Y) \geq m+1$ and $w_h(X) = w_h(Y) = 0$. Otherwise $C_{M_{MDS}}(X,Y) = 0$. It is obvious that at least one set X, Y where $w_h(X) = 1$ satisfy $C_{M_{MDS}}(X,Y) * C_{M_{MDS}}(Y,X) \neq 0$, thus 2-derived difference cancellation can occur in a MDS matrix construction.

This observation explains why Camellia*'s lower bounds are not too low even though it employs a non-MDS matrix M_{Ca} of branch number 5. For any choice of X, Y where $w_h(X) = 1$, $C_{M_{Ca}}(X,Y) \cdot C_{M_{Ca}}(Y,X) = 0$ always. Thus M_{Ca} never produces the 2-derived difference cancellation, and it keeps a moderate number of active S-boxes.

However, even though 2-derived difference cancellation is avoided by choosing Camellia type matrix, if certain X, Y where $w_h(X) = 2$ satisfying $C_M(X,Y) \cdot C_M(Y,X) \neq 0$ exists, then a 4-derived difference cancellation would be a building block for a small number of active S-boxes, and a significant gain of the number of active S-boxes may not be expected.

In the next section another approach to avoid m-derived difference cancellation will be introduced by using multiple MDS matrices in a Feistel structure.

5 Multiple MDS Feistel Structure Design

5.1 Basic Strategy

Suppose that some intermediate differential data $\Delta x_{i-1} = 0$, and that the output of F-function in every 2 rounds is added to the data, (ex. $\Delta y_i, \Delta y_{i+2}, ..\Delta y_{i+2j}$). Consider a situation where the differential data Δx_{i+2j+1} become 0 after XORing the output of the F-function in the $i+2j$-th round, caused by a difference cancellation as shown in Fig. 4.

In the difference cancellation, the following condition exists:

$$\sum_{k=0}^{j} \Delta y_{i+2k} = 0\,. \tag{1}$$

Therefore,

$$M \sum_{k=0}^{j} \Delta z_{i+2k} = 0\,. \tag{2}$$

When a nonsingular matrix M is employed, we obtain that

$$\sum_{k=0}^{j} \Delta z_{i+2k} = 0\,. \tag{3}$$

The above equation shows that a difference cancellation occurs by only 2 active S-boxes in $\Delta z_{i+2k}, (0 \leq k \leq j)$ in the minimum case, which is exactly the case of

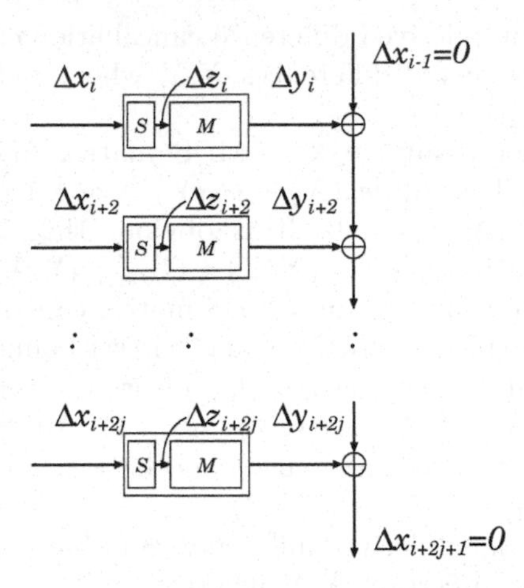

Fig. 4. Difference Cancellation

2-derived difference cancellations shown in the previous section. Now we consider a setting with multiple matrices in which a different matrix is used in each F-function. Let M_i be a diffusion matrix employed in the i-th round. Obviously, the transformation from (1) to (2) is not correct when the matrices M_i are different from each other. In such setting, we can rewrite (1) as

$$M_i \Delta z_i + M_{i+2} \Delta z_{i+2} + \ldots + M_{i+2j} \Delta z_{i+2j} = 0 \,. \tag{4}$$

The above condition can be written as the product of a large $m \times m(j+1)$ matrix and a vector with $m(j+1)$ elements:

$$[M_i M_{i+2} \cdots M_{i+2j}] \begin{bmatrix} \Delta z_i \\ \Delta z_{i+2} \\ \vdots \\ \Delta z_{i+2j} \end{bmatrix} = 0 \,. \tag{5}$$

If these matrices are chosen to satisfy that there is no combination of l column vectors that are dependent of each other $(2 \leq l \leq m)$ in the matrix, k-derived difference cancellation $(k \leq l)$ would never happen in the consecutive $2j$ rounds. From this observation, we introduce a strategy to choose matrices $M_i, .., M_{i+2j}$ for which any choice of m column vectors are independent of each other in the large matrix $[M_i, .., M_{i+2j}]$.

5.2 Construction Steps

We propose a new design strategy that employs multiple MDS matrices in the Feistel network, in order to avoid an occurrence of m-derived difference cancellation in any consecutive $2q$ rounds where q is the number of employed matrices.

The construction steps are as follows. Without loss of generality, we assume the round number is $2r$.

1. Choose $q(\leq r)$ MDS matrices: $M_0, M_1, \ldots, M_{q-1}$.
2. Check that any m of qm column vectors in all M_i matrices hold the MDS property.
3. Assign matrix $M_{(i \mod q)}$ to the $2i+1$-th round $(0 \leq i < r)$.
4. Assign matrix $M_{(i \mod q)}$ to the $2r-2i$-th round $(0 \leq i < r)$ (reverse order).

In this construction, since any m columns of the large matrix $[M_i M_{i+2} \cdots M_{i+2q-2}]$ have been chosen to generate MDS which has m independent column vectors, there is no chance to generate m-derived difference cancellation in any consecutive $2q-1$ rounds. Fig. 5 shows the construction of the example setting $r = 6, q = 3, 6$ respectively.

When m, n are small, we can randomly generate MDS matrices and check MDS conditions of column vectors in Step 2. However, if m, n are large, it might be difficult to search such a set of matrices. In such a case, we can use the algorithm to generate Reed-Solomon code's generation matrix. The algorithm can generate a large MDS matrix immediately, because its complexity is $O(N^3)$ where N is the dimension of the matrix [7]. Once the $qm \times qm$ MDS matrix M_L is made, any combination of m rows of M_L can be regarded as a matrix $[M_0, M_1, .., M_{q-1}]$ with proposed additional MDS property from Theorem 1. We can find a 128×128 MDS matrix on $GF(2^8)$ from [256, 128, 129] extended RS codes, thus 16 MDS matrices of dimension 8 satisfying the condition of Step 2. can be found in it. We show an example of a set of such matrices in Appendix B.

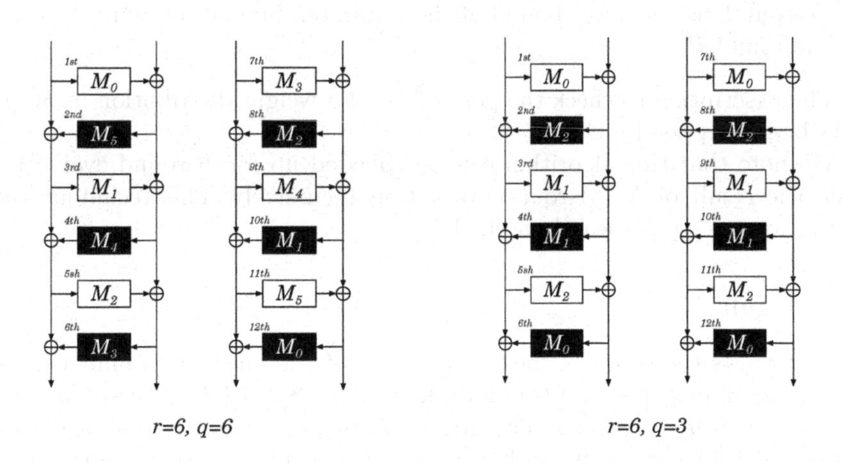

Fig. 5. Examples of the New Design ($r = 3, q = 3, 6$)

6 Evaluation of the Proposed Construction

We estimated a lower bound for the number of active S-boxes of the new construction with $m = 8, r = 6$ (12-round cipher) for the number of matrices $q = 1..6$. We adopted a weight based approach in the evaluation algorithm, because the known truncated path approach which is employed in the other cipher's evaluation requires too huge memory space and time consumption.

6.1 Algorithm

The following algorithm outputs a lower bound for the number of active S-boxes of our proposed construction based on the weight based approach . Let the round number be R.

1. Set $L = \infty$.
2. For each possible combination of the weight $0, 1, .., 8$ in $\delta x_0, \delta x_1, .., \delta x_{R+1}$ (There are 9^{R+2} candidates):
   (a) For $i = 2$ to $R + 1$ do the following,
      i. For $j = 2$ to $j \leq i$, $j \leftarrow j + 2$ do the following,
         A. Check whether the given weight combination of $w_h(\delta x_{i-j})$, $w_h(\delta x_i)$ and the list of given weight of active S-boxes in the $w_h(\delta x_{i-j+1})$, $w_h(\delta x_{i-j+3})$,.., $w_h(\delta x_{i-1})$ are possible or not in the weight context of the given MDS property.
         B. If the check passed then continue the loop, else exit the loop.
   (b) If all checks passed, count the total number of active S-boxes A in the path. If $A < L$, set $L = A$.
3. Output L as the lower bound of the minimum number of active S-boxes for the round R.

The description to check the possibility of a weight distribution in Step A is described in Appendix C.

We note that this algorithm can be speeded up for R-round evaluation by using the result of R – 1-round evaluation recursively. This technique can be seen in Matsui's path search method [8].

6.2 Result

Table 2 shows the result of the lower bound of the minimum number of active S-boxes for four types of 12-round multiple-MDS Feistel ciphers, the cases of $m = 8, r = 6, q = 1, 2, 3, 6$. The graph of the ratio of active S-boxes to total number of S-boxes is shown in Fig. 6. It can be confirmed that the result of the case $q = 1$ is the same as the result of the MDS-Feistel cipher shown in Table 1. Moreover, the results shows that the lower bounds for the cases $q = 3$ and $q = 6$ are always the same.

The numbers are significantly increased in the case of $q = 2$ compared to the case $q = 1$. However there is no gain in 8 and 9 rounds. In the case of $q \geq 3$,

Table 2. Result of Evaluation

Round	$q = 1$ (rate)		$q = 2$ (rate)		$q = 3$ (rate)		$q = 6$ (rate)	
1	0	0.0%	0	0.0%	0	0.0%	0	0.0%
2	1	6.3%	1	6.3%	1	6.3%	1	6.3%
3	2	8.3%	2	8.3%	2	8.3%	2	8.3%
4	9	28.1%	9	28.1%	9	28.1%	9	28.1%
5	10	25.0%	10	25.0%	10	25.0%	10	25.0%
6	11	22.9%	18	37.5%	18	37.5%	18	37.5%
7	12	21.4%	18	32.1%	18	32.1%	18	32.1%
8	19	29.7%	19	29.7%	20	31.3%	20	31.3%
9	20	27.7%	20	27.8%	27	37.5%	27	37.5%
10	21	26.3%	27	33.8%	28	35.0%	28	35.0%
11	22	25.0%	28	31.8%	32	36.4%	32	36.4%
12	29	30.2%	36	37.5%	36	37.5%	36	37.5%
avrg.(4-12)		26.3%		31.5%		33.4%		33.4%

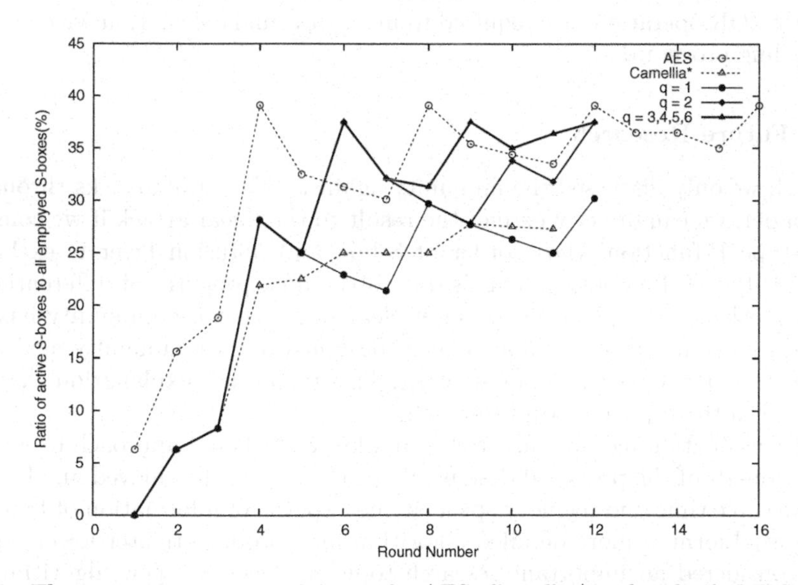

Fig. 6. The percentage of active S-boxes for AES, Camellia* and multiple-MDS Feistel

the lower bound is even higher than for the case $q = 2$ when we have more than 8 rounds, and more than 22 active S-boxes are guaranteed in 9 rounds. The ratio of the new design successfully came to the level of AES after more than 6 rounds. These results show that the design with triple MDS matrices has enough advantages over single MDS matrix design. Also, our experiment indicates that not so many MDS matrices seem to be required to get a benefit from the proposed design.

7 Discussion

7.1 Implementation Aspects

This new construction requires an additional implementation cost because it employs multiple matrices. Multiple diffusion matrices require additional gate size in hardware and lookup tables in memory in software implementation. However the speed impact in hardware is expected to be negligible because only switching circuits for matrices will be added. Detailed observations on hardware implementation of many types of SP-type Feistel networks can be found in [15].

If all lookup tables can be stored in the fastest cache memory, not much time cost would be expected. If b matrices of dimension 8 on $GF(2^8)$ are employed in the 128-bit block setting, the cipher requires $16b$ KB lookup tables at maximum in which the size of each entry is 64-bit. Since some recent 64-bit CPUs have 64KB first cache memory for data, 48 KB lookup table required by 3 matrices would be acceptable. In such a setting, it is estimated that only $8R$ table lookups and $9R$ XOR operations are required to finish R round calculation without a key scheduling procedure.

7.2 Future Research

Though we only discussed the immunity against differential attacks throughout this paper, we can directly extend the result to the linear attack if we construct a PS-type F-function whose order of S-box and diffusion layer is exchanged from SP-type F-function[5]. This is due to the dual property of differential and linear masks [5, 10]. However, it is not clear so far that the immunity has been gained for the linear attack if a cipher is designed to have immunity against the differential attack based on our strategy. This theoretical explanation should be included in the topic of future research.

Our evaluation method adopted a simple weight based approach to estimate lower bounds of the proposed designs. Since the approach achieved an algorithm with feasible time and memory space at the expense of information of truncated differential form, a more detailed algorithm may produce tighter lower bounds. It is considered an important research topic to develop a new algorithm that counts lower bounds of the minimum number of active S-boxes more strictly for the new design.

Acknowledgement

The authors would like to thank Bart Preneel, Alex Biryukov, Christophe De Cannière and Joseph Lano of COSIC, K.U.Leuven for their helpful discussions and suggestions. Furthermore, we thank the anonymous reviewers for helpful feedback.

References

[1] K. Aoki, T. Ichikawa, M. Kanda, M. Matsui, S. Moriai, J. Nakajima and T. Tokita, "Specification of Camellia - a 128-bit Block Cipher," Primitive submitted to NESSIE by NTT and Mitsubishi, 2000, See also http://info.isl.ntt.co.jp/camellia. 261, 265

[2] K. Aoki, T. Ichikawa, M. Kanda, M. Matsui, S. Moriai, J. Nakajima and T. Tokita, "Camellia: A 128-Bit Block Cipher Suitable for Multiple Platforms - Design and Analysis," Selected Area in Cryptography, SAC 2000, LNCS 2012, pp.39-56, 2000. 260, 263

[3] E. Biham and A. Shamir," Differential Cryptanalysis of DES-like Cryptosystems," CRYPTO '90, LNCS 537, pp.2-21, 1991.

[4] J. Daemen and V. Rijmen, "The Design of Rijndael: AES - The Advanced Encryption Standard," Springer-Verlag, 2002. 260, 261

[5] M. Kanda, "Practical Security Evaluation against Differential and Linear Cryptanalysis for Feistel Ciphers with SPN Round Function," Selected Areas in Cryptography, SAC 2000, LNCS 2012, pp. 324-338, 2000. 272

[6] L. R.Knudsen, "Truncated and Higher Order Differentials," Fast Software Encryption - Second International Workshop, LNCS 1008, pp.196-211, 1995. 265

[7] R. Lidl and H. Niederreiter, "Finite Fields," Encyclopedia of mathematics and its applications 20, Cambridge Univ. Press, 1997. 262, 269, 276

[8] M.Matsui, "Differential Path Search of the Block Cipher E2," Technical Report ISEC99-19, IEICE, 1999.(written in Japanese) 263, 270

[9] M. Matsui and T. Tokita,"Cryptanalysis of Reduced Version of the Block Cipher E2," Fast Software Encryption, FSE'99, LNCS 1636, 1999. 263

[10] M. Matsui, "Linear Cryptanalysis of the Data Encryption Standard," EUROCRYPT '93, LNCS 765, pp.386-397, 1994. 272

[11] H. Shimizu, "On the security of Feistel cipher with SP-type F function," 7A-3, SCIS 2001, 2001. (written in Japanese) 263

[12] T. Shirai, S. Kanamaru and G. Abe, "Improved Upper Bounds of Differential and Linear Characteristic Probability for Camellia", Fast Software Encryption, FSE2002, LNCS 2365, pp.128-142, 2002. 263

[13] T. Shirai, "Differential, Linear, Boomerang and Rectangle Cryptanalysis of Reduced-Round Camellia," preproceedings of Third NESSIE Workshop, 2002. 263

[14] S. Moriai, M. Sugita, K. Aoki and M. Kanda, "Security of E2 against Truncated Differential Cryptanalysis," Selected Areas in Cryptography, SAC'99, LNCS 1758, pp.106-117, 2000. 263

[15] L. Xiao and H. M. Heys,"Hardware Performance Characterization of Block Cipher Structures," Topics in Cryptology - CT-RSA 2003, The Cryptographers' Track at the RSA Conference 2003, Proceedings pp.176-192, 2003. 272

[16] National Institute of Standards and Technology, Advanced Encryption Standard, FIPS 197, 2001. 260
[17] NESSIE project - New European Schemes for Signatures, Integrity, and Encryption, http://www.cryptonessie.org. 260
[18] CRYPTREC project, http://www.ipa.go.jp/security/enc/CRYPTREC/. 260
[19] STORK project - Strategic Roadmap for Crypto, Public Document, D6: Open problems in Cryptology version 2.1, chapter 4, 2003, available at http://www.stork.eu.org/. 261

Appendix A

All the minimum differentially active S-boxes paths for more than 5 rounds of MDS-Feistel cipher using an 8×8 MDS matrix can be represented by only the following eight types of differential paths as building blocks in Fig. 7.

Pattern (A),(B) and (C) are prefix patterns which appear only at the beginning of differential paths, and pattern (X), (Y) and (Z) are suffix patterns which appear only at the end of differential paths. Pattern (P) and (Q) are middle iteration patterns which appear at the middle of differential paths and are sometimes iterated more than once depending on the total number of rounds. (P) and (Q) are respectively shown as 3-round and 5-round differential cancellations in Sect. 4.

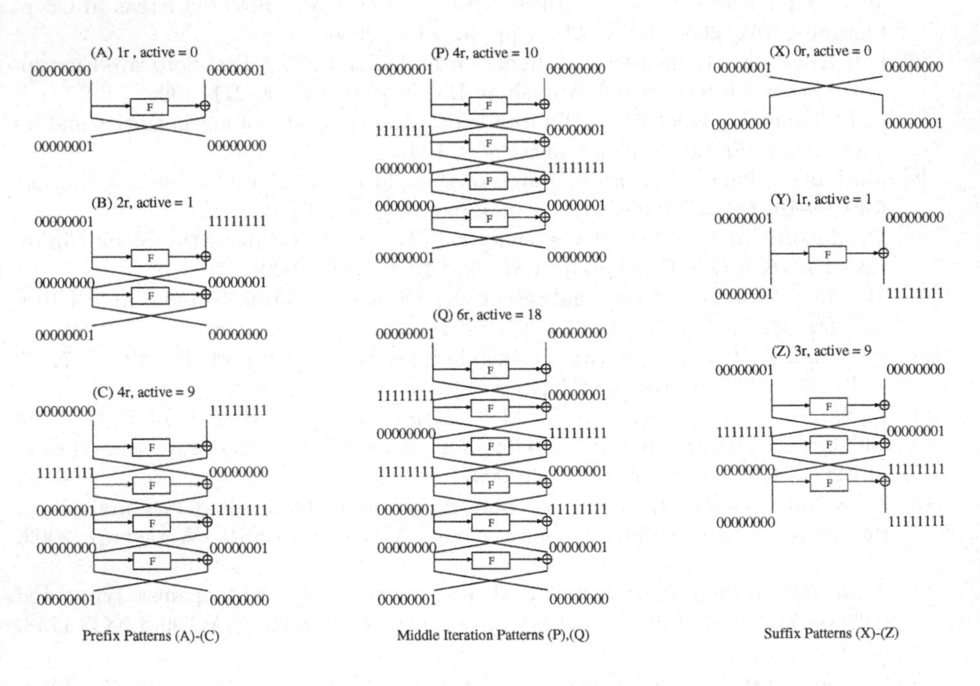

Fig. 7. Building Blocks based on $\delta A = (00000001)$

Table 3. All path patterns of the minimum number of active S-boxes for each round

R.	1	2	3	4
M. A.	0	1	2	9
R	5	6	7	8
M. A.	10	11	12	19
Pat.	APX BZ CY	APY BPX	BPY	APZ AQY BQX CPX
R.	9	10	11	12
M. A.	20	21	22	29
Pat.	APPX BPZ BQY CPY	APPY BPPX	BPPY	APPZ APQY, AQPY BPQX, BQPX CPPX
R.	13	14	15	16
M. A.	30	31	32	39
Pat.	APPPX BPPZ BPQY, BQPY CPPY	APPPY BPPPX	BPPPY	APPPZ APPQY, APQPY, AQPPY BPPQX, BPQPX, BQPPY CPPPX
R.	17	18	19	20
M. A.	40	31	32	39
Pat.	APPPPX BPPPZ BPPQY, BPQPY, BQPPY CPPPY	APPPPY BPPPPX	BPPPPY	APPPPZ APPPQY, APPQPY, APQPPY, AQPPPY BPPPQX, BPPQPX, BPQPPX, BQPPPX CPPPPX
R.	4n+1	4n+2	4n + 3	4n + 4
M. A.	10n	10n+1	10n+2	10n + 9
Pat. $(R. \geq 5)$	AP^nX $BP^{n-1}Z$ $B(P^{n-2}Q)Y (n > 1)$ $CP^{n-1}Y$	AP^nY BP^nX	BP^nY	AP^nZ $A(P^{n-1}Q)Y$ $B(P^{n-1}Q)X$ CP^nX

R. : Round Number
M. A. : the minimum numbers of active S-boxes
Pat. : path expressions constructed from basic patterns.
P^k: iteration of pattern P for k times
(P^kQ): all possible patterns generated from pattern P for k times and Q for once

Each pattern in Fig. 7 shows a representative path using truncated difference (00000001), and each pattern also contains 7 other different paths by replacing (00000001) with one of (00000010), (00000100), .., (10000000).

Table 3 shows the search results for 5-round to 20-round differential paths of MDS-Feistel. The *Patterns* field shows the path patterns expressed by their building blocks. It means that any resulting path can be expressed by one of the path patterns in the corresponding field. We can see that there is a 4-round regularity. The last three rows shows the regularity in a generalized form. From this experimental result, any differential path with a minimum number of active S-boxes for more than 6-rounds contains at least one (P) or (Q). This is the reason why difference cancellation should be avoided in order to gain more active S-boxes, as described in Sect. 4.

Appendix B

In this appendix, we show a set of example 8×8 MDS matrices in which any combination of any 8 columns form a MDS matrix, which was obtained from the right part of a $[256, 128, 129]$ Reed-Solomon code's generation matrix in standard form [7].

In the following example we employed a primitive polynomial $p(x) = x^8 + x^4 + x^3 + x^2 + 1$. Let α be a root of p(x), we set a parity check matrix H as:

$$H = \begin{bmatrix} 1 & 1 & \cdots & 1 & 1 & 1 \\ \alpha^{254} & \alpha^{253} & \cdots & \alpha & 1 & 0 \\ \vdots & \vdots & \ddots & \vdots & \vdots & \vdots \\ (\alpha^{254})^{126} & (\alpha^{253})^{126} & \cdots & \alpha^{126} & 1 & 0 \\ (\alpha^{254})^{127} & (\alpha^{253})^{127} & \cdots & \alpha^{127} & 1 & 0 \end{bmatrix}.$$

Then we calculated a generation matrix $G = [I_{128\times128} M_{128\times128}]$ The following 16 matrices are obtained from the first 8 rows of $M_{128\times128}$ simply by splitting every 8 columns as $[M_0 M_1 ... M_{15}]$. Each element is expressed in a hexadecimal value corresponding to a binary representation of elements in $GF(2^8)$.

$$M_0 = \begin{pmatrix} 9d & b4 & d3 & 5d & 84 & ae & ec & b9 \\ 29 & 34 & 39 & 60 & 5c & 81 & 25 & 13 \\ 67 & 6a & d2 & e3 & 4b & db & 9d & 4 \\ 8e & d7 & e6 & 1b & 8b & 9e & 3a & 91 \\ d9 & e5 & 4d & dd & c6 & 5 & f0 & ad \\ 2a & f7 & 67 & 72 & b1 & 7 & f2 & 27 \\ 42 & e6 & a0 & 4 & f1 & 4 & 7d & 8c \\ 55 & 63 & fa & 51 & c & d9 & 28 & d6 \end{pmatrix} M_1 = \begin{pmatrix} b8 & f1 & 65 & ef & d2 & c3 & 7b & f4 \\ 3a & f6 & 2d & 6a & 1e & cc & 5e & a4 \\ 4a & 97 & a3 & b9 & f4 & 2b & a0 & 76 \\ 82 & 5f & a2 & c1 & bf & 30 & 69 & 2d \\ 59 & 89 & 10 & 2d & 4 & bc & fb & 5c \\ 1d & 69 & eb & 4e & c8 & b8 & b0 & 2d \\ 31 & 1b & 22 & 29 & 71 & 51 & 37 & 63 \\ e0 & 7b & b5 & 5a & f2 & 81 & cd & 81 \end{pmatrix} M_2 = \begin{pmatrix} d0 & 46 & a6 & a7 & e1 & b7 & 16 & d2 \\ fd & b3 & 84 & 18 & 7a & cc & 31 & e7 \\ ca & 9b & d3 & 9c & 66 & b1 & 12 & af \\ 79 & ec & 6a & a8 & c1 & 55 & e2 & 14 \\ 56 & f8 & a0 & 79 & 3a & 4b & 13 & 27 \\ 77 & e1 & 26 & 19 & 77 & bd & 3a & f6 \\ c2 & 5 & 33 & 9c & d1 & 3 & 1e & 5 \\ 2b & fd & 5b & aa & 3a & a4 & 47 & c5 \end{pmatrix}$$

$$M_3 = \begin{pmatrix} ce & a2 & 8 & 3d & c2 & c4 & c0 & a6 \\ 9a & e1 & 65 & f6 & 5f & b5 & 5e & 2 \\ b0 & 7 & 6 & 6f & bb & 1f & 8 & 3e \\ d7 & 53 & 23 & 62 & 21 & ee & 58 & f9 \\ eb & fa & 91 & 69 & 3e & a2 & c & 14 \\ 62 & b4 & e5 & 2a & b2 & aa & b7 & d2 \\ dd & d & d3 & 18 & db & 2e & 8b & 65 \\ 7f & b8 & 7b & 70 & 2f & 44 & 8a & d5 \end{pmatrix} M_4 = \begin{pmatrix} 1c & 42 & 16 & 4d & f4 & b5 & f2 & 71 \\ 1f & 92 & c4 & 36 & 92 & 21 & 7f & 50 \\ cb & 84 & 4c & cd & 1e & a9 & 4 & 4 \\ 40 & ff & d7 & cd & 40 & 24 & 3f & 6b \\ c3 & a0 & b9 & 75 & f8 & 74 & ce & 88 \\ b4 & f5 & d1 & b8 & a3 & b1 & ed & 13 \\ c6 & 7b & 56 & 7c & 6b & 7 & f & d5 \\ c4 & c2 & 3f & 4b & ca & 8d & 19 & 76 \end{pmatrix} M_5 = \begin{pmatrix} 24 & 75 & 60 & ec & cf & de & 60 & 4f \\ a7 & 6b & 38 & eb & d8 & 14 & 93 & b8 \\ 16 & 21 & bb & 24 & c4 & 5c & c4 & 6a \\ 86 & 6b & 8c & 5a & d6 & f4 & f1 & 4c \\ 25 & 50 & 1a & e2 & fa & b0 & 85 & ed \\ 75 & 79 & d4 & ed & c0 & 81 & 34 & 4a \\ 2f & f0 & e7 & ae & ae & 25 & 1f & 49 \\ 2b & 6d & a2 & cb & 13 & 38 & 77 & 91 \end{pmatrix}$$

$$M_6 = \begin{pmatrix} 86 & 33 & 2e & b3 & 64 & c1 & 36 & 8a \\ b6 & aa & da & ee & 13 & 82 & 7c & e5 \\ 63 & 56 & fb & ec & c3 & d6 & e7 & bd \\ b & 38 & f9 & da & 53 & e & 15 & b0 \\ 2f & 5c & a & 12 & ca & ee & 67 & d5 \\ 4f & c4 & 70 & c & 17 & b8 & e3 & 5b \\ 19 & b8 & 4f & b9 & 2f & 5e & 9a & 6a \\ d2 & 64 & a3 & 1 & d0 & 4c & df & 4c \end{pmatrix} M_7 = \begin{pmatrix} 7f & d7 & 37 & 4a & 8b & ca & f & e9 \\ fa & 1f & 16 & 7e & e5 & 2e & 64 & 15 \\ 74 & e1 & 2 & 6a & c7 & 4f & bc & fd \\ ed & ea & a8 & 12 & b7 & ad & fa & c5 \\ 1a & 16 & dc & 8a & 8d & 29 & ef & aa \\ 58 & 19 & 7c & f5 & d1 & 49 & 1e & fe \\ 69 & b2 & 53 & d4 & 14 & 47 & 2b & b1 \\ cd & 4e & f4 & 21 & c5 & 55 & 5e & fb \end{pmatrix} M_8 = \begin{pmatrix} 50 & cd & 21 & ff & 88 & 97 & 8b & c \\ c & 9c & 8a & bd & d4 & 9e & 38 & a1 \\ ad & d3 & 5f & cd & 8c & a4 & 27 & 22 \\ 88 & 56 & c9 & 75 & 93 & 2f & 79 & 11 \\ 2 & 96 & 26 & df & b & 36 & b0 & da \\ 6c & ee & 8d & 46 & 2d & f0 & 6d & 2e \\ 2f & ff & 7 & 81 & 1f & d8 & 11 & b7 \\ ca & 1 & fd & 93 & c4 & af & c9 & 5c \end{pmatrix}$$

$$M_9 = \begin{pmatrix} 9c & ae & 3e & 9e & 58 & ca & c8 & 77 \\ 59 & 99 & d4 & 93 & bc & fd & 97 & 7f \\ be & de & 1b & 3b & cc & 16 & a7 & e1 \\ 26 & 6b & 39 & 16 & 6a & a6 & 75 & d3 \\ 97 & a8 & c0 & 1 & 13 & f4 & 86 & 1d \\ 97 & 56 & ba & 43 & d8 & d7 & fe & 14 \\ 4 & 22 & 13 & 40 & f5 & 4e & 91 & ab \\ 25 & 2c & d5 & 12 & 4d & b4 & 9c & 40 \end{pmatrix} M_{10} = \begin{pmatrix} 51 & f4 & a9 & 82 & c8 & f7 & d9 & f6 \\ ee & 8e & 98 & bd & df & 93 & 45 & fe \\ bf & c6 & 7c & be & e7 & 7f & 62 & 9d \\ 1e & 32 & 82 & f & dd & e9 & de & e6 \\ 4b & 2b & 3c & 80 & 2f & 9 & f & 55 \\ bc & 9d & c2 & 1f & 8a & 50 & 5a & 17 \\ ad & 6f & 2d & 2c & 59 & e1 & b0 & 59 \\ 17 & 1b & 28 & 8d & fe & bb & 18 & 95 \end{pmatrix} M_{11} = \begin{pmatrix} d7 & bf & 93 & 96 & 9 & ae & 2b & 49 \\ 96 & 3d & 44 & f9 & 2d & c & d6 & e6 \\ e5 & ba & b0 & 4c & 66 & aa & d8 & 22 \\ 3 & f2 & b & 99 & e2 & b3 & 9d & 4b \\ 1d & 4b & 36 & f1 & 4 & 6c & bf & 5e \\ 56 & 85 & 31 & aa & 89 & c5 & a6 & 3f \\ fd & 45 & cd & ac & a5 & 3c & 9b & b6 \\ 7e & fe & ce & ba & 1d & 8d & db & bd \end{pmatrix}$$

$$M_{12} = \begin{pmatrix} 9d & 16 & 3a & 75 & a0 & f2 & 4a & c2 \\ 60 & 1b & 81 & 75 & a2 & 6a & bb & 28 \\ 81 & be & 64 & 7 & 18 & 87 & 16 & f6 \\ ac & d2 & 4b & 19 & ed & 8e & 97 & 58 \\ 92 & ea & 18 & 9d & 8a & 7 & ac & cb \\ 74 & e3 & 79 & 44 & c & 13 & 2e & 77 \\ 7d & d7 & 1b & fc & fb & b2 & bb & df \\ e3 & 39 & 1b & 59 & a & e5 & c0 & 8c \end{pmatrix} M_{13} = \begin{pmatrix} 54 & c8 & 7d & 23 & 25 & 3f & a9 & 99 \\ ca & 9 & 40 & c2 & 89 & 23 & 53 & 6d \\ 6d & 90 & 68 & 73 & fc & 73 & d1 & c9 \\ 91 & e & c9 & 7b & 7b & 91 & 37 & 4d \\ fe & b1 & 1a & 7 & 6a & 8f & 9f & 8f \\ 8c & f4 & f2 & cf & 9d & 1f & 66 & 34 \\ 1b & ca & 16 & cb & 9b & b0 & af & 99 \\ 9f & c9 & fe & 87 & f2 & ab & f7 & c1 \end{pmatrix} M_{14} = \begin{pmatrix} a1 & 6d & 5 & 75 & 5 & 9c & 74 & d9 \\ 32 & ba & fc & 4a & e4 & 50 & af & c6 \\ 78 & 80 & 3f & 7a & cf & fb & ae & 5e \\ a9 & 69 & 18 & 42 & e2 & cb & c & f \\ 81 & 24 & 57 & ce & 7c & 64 & 42 & ea \\ 79 & fc & c2 & b & 28 & 31 & 5f & 11 \\ e3 & 55 & 29 & 47 & de & 2 & dc & bf \\ a8 & 3b & a9 & 6b & bb & 4c & 87 & 25 \end{pmatrix}$$

$$M_{15} = \begin{pmatrix} 49 & 7f & 18 & 34 & 55 & 8c & 7c & 4c \\ 14 & 39 & 2b & 5d & 40 & 93 & 78 & 10 \\ 32 & 85 & d & be & 60 & 8c & 8a & 61 \\ cd & 62 & e9 & c7 & 53 & 2c & 5f & 6a \\ cc & 67 & 39 & 3e & 3e & 4c & 67 & 97 \\ 6a & 41 & 46 & fa & 1e & 4d & 68 & f1 \\ 1 & 58 & e8 & b4 & fe & fa & c7 & 34 \\ 99 & a6 & c2 & fa & 33 & 80 & 9 & ee \end{pmatrix}$$

Appendix C

In our evaluation algorithm, a check procedure to judge whether a given weight distribution is possible or not was employed.

Let M_i $(1 \leq i \leq 2r)$ be the i-th round diffusion matrix, and let N_j $(1 \leq j \leq q)$ be q matrices designed by our proposed design strategy. In any combination of Δx_i and $\Delta x_{i+2j} \in GF(2^n)^m$, there is the following relation,

$$\Delta x_i + \Delta x_{i+2j} = \sum_{k=1}^{j} M_{i+2k-1} \Delta z_{i+2k-1} \tag{6}$$

In the case of $q = r$, all M_k in the above equation are guaranteed to be different because they are chosen from N_j without overlap.

Then we consider weight conditions on both sides of (6). Let W_1 be $w_h(\Delta x_i + \Delta x_{i+2j})$. We obtain the following inequality.

$$|w_h(\Delta x_i) - w_h(\Delta x_{i+2j})| \leq W_1 \leq min(m, w_h(\Delta x_i) + w_h(\Delta x_{i+2j})) \tag{7}$$

On the other hand, let W_2 be $w_h(\sum_{k=1}^{j} M_{i+2k-1} \Delta z_{i+2k-1})$. If there is at least one nonzero hamming weight in $w_h(\Delta z_j)$,

$$max(m + 1 - \sum_{k=1}^{j} w_h(\Delta z_{i+2k-1}), 0) \leq W_2 \leq m \tag{8}$$

If all hamming weight of $w_h(\Delta z_j)$ are 0, obviously $W_2 = 0$.

In the evaluation algorithm, we compare the above weight conditions W_1 and W_2 implied by a given weight distribution. In the checking procedure, it is judged false if these two weight range conditions have no overlap, because it means there is no path with such a weight distribution.

In $q < r$ and $j > q$ settings, some of the matrices N_j appear more than once in equation (6). In that case, we can rewrite (6) in partially bundle form,

$$\Delta x_i + \Delta x_{i+2j} = \sum_{k=1}^{q} N_k \Delta z'_k \,, \tag{9}$$

where

$$\Delta z'_k = \sum_{\{l|N_k=M_l\}} \Delta z_l \,. \tag{10}$$

In this case, we also have to consider the weight condition for XOR of multiple element to treat $\Delta z'_k = \sum_{\{l|N_k=M_l\}} \Delta z_l$. Let $j \geq 1$, W_3 be $w_h(\sum_{i=1}^{j} \Delta a_i)$, and let w_{max} be $max(w_h(\Delta a_1), w_h(\Delta a_2), .., w_h(\Delta a_j))$, we obtain the range of W_3 as,

$$max(0, 2w_{max} - \sum_{i=1}^{j} w_h(\Delta a_i)) \leq W_3 \leq min(m, \sum_{i=1}^{j} w_h(\Delta a_i)) \tag{11}$$

Then this weight condition W_3 can be used to determine the weight condition of W_2 for the equation (9).

ICEBERG : An Involutional Cipher Efficient for Block Encryption in Reconfigurable Hardware*

Francois-Xavier Standaert, Gilles Piret, Gael Rouvroy, Jean-Jacques Quisquater, and Jean-Didier Legat

UCL Crypto Group
Laboratoire de Microelectronique
Universite Catholique de Louvain
Place du Levant, 3, B-1348, Louvain-La-Neuve, Belgium
{standaert,piret,rouvroy,quisquater,legat}@dice.ucl.ac.be

Abstract. We present a fast involutional block cipher optimized for reconfigurable hardware implementations. ICEBERG uses 64-bit text blocks and 128-bit keys. All components are involutional and allow very efficient combinations of encryption/decryption. Hardware implementations of ICEBERG allow to change the key at every clock cycle without any performance loss and its round keys are derived "on-the-fly" in encryption and decryption modes (no storage of round keys is needed). The resulting design offers better hardware efficiency than other recent 128-key-bit block ciphers. Resistance against side-channel cryptanalysis was also considered as a design criteria for ICEBERG.

Keywords: block cipher design, efficient implementations, reconfigurable hardware, side-channel resistance.

1 Introduction

In October 2000, NIST (National Institute of Standards and Technology) selected Rijndael as the new Advanced Encryption Standard. The selection process included performance evaluation on both software and hardware platforms. However, as implementation versatility was a criteria for the selection of the AES, it appeared that Rijndael is not optimal for reconfigurable hardware implementations. Its highly expensive substitution boxes are a typical bottleneck but the combination of encryption and decryption in hardware is probably as critical.

In general, observing the AES candidates [1, 2], one may assess that the criteria selected for their evaluation led to highly conservative designs although the context of certain cryptanalysis may be considered as very unlikely (e.g. more than 2^{100} chosen plaintexts). More recent designs of the NESSIE[1] project

* This work has been funded by the Wallon region (Belgium) through the research project TACTILS http://www.dice.ucl.ac.be/crypto/TACTILS/T_-home.html

[1] NESSIE: New European Schemes for Signatures, Integrity, and Encryption. See http://www.cryptonessie.org.

B. Roy and W. Meier (Eds.): FSE 2004, LNCS 3017, pp. 279–299, 2004.

(e.g. Khazad [3], Misty [4]) provide an improved efficiency. They also allowed interesting comparisons between Feistel networks (e.g. Misty) and substitution-permutation networks, with respect to hardware efficiency. Although Khazad is not a Feistel network, its structure is designed so that by choosing all components to be involutions, the inverse operation of the cipher differs from the forward operation in the key scheduling only. ICEBERG is also based on an involutional structure but allows to derive the keys more efficiently than Khazad. Moreover, the combination of encryption and decryption is improved due to a simplified diffusion layer.

Reconfigurable hardware devices usually enable high performance encryption/ decryption solutions for real-time applications of multi-Gbps data streams. Video-processing is the typical context where high throughput has to be provided at low hardware cost. Although present encryption algorithms may provide very high encryption rates, it is often at the cost of expensive designs. The main feature of ICEBERG is that is has been defined in order to allow very efficient reconfigurable hardware implementations. An additional criteria was the simplicity of the design. ICEBERG is scalable for different architectures (loop, unrolled, pipeline) and FPGA[2] technologies. As a consequence, ASIC[3] implementations are also efficient. All its components easily fit into 4-input lookup tables and its key scheduling allows to derive the round keys "on-the-fly" in encryption and decryption modes. This involves no storage requirements for the round keys. The resulting design offers better hardware efficiency than other recent 128-key-bit block ciphers. As a consequence, very low-cost hardware crypto-processors and high throughput data encryption are potential applications of ICEBERG.

Finally, resistance against side-channel cryptanalysis was also considered as a design criteria for ICEBERG. Small substitution tables are used in order to allow efficient boolean masking. Moreover, the key agility offers the opportunity to consider new encryption modes where the key is changed frequently in order to make the averaging of side-channel traces unpractical.

This paper is structured as follows. Section 2 presents the design goals of ICEBERG and section 3 gives its specifications. The security analysis of ICEBERG is in section 4 and its performance analysis in section 5. Finally, conclusions are in section 6. Some tables and proofs are given in appendixes.

2 Design Goals

Present reconfigurable components like FPGAs are usually made of reconfigurable logic blocks combined with fast access memories (RAM blocks) and high speed arithmetic circuits [5, 6]. Basic logic blocks of FPGAs include a 4-input function generator (called lookup table, LUT), carry logic and a storage element.

As reconfigurable components are divided into logic elements and storage elements, an efficient implementation will be the result of a better compromise

[2] FPGA : Field Programmable Gate Array.

[3] ASIC : Application Specific Integrated Circuit.

between combinatorial logic used, sequential logic used and resulting performances. These observations lead to different definitions of implementation efficiency [7, 8, 9, 10, 11, 12]:

1. In terms of performances, let the efficiency of a block cipher be the ratio *Throughput (Mbits/s)/Area (LUTs, RAM blocks)*.
2. In terms of resources, the efficiency is easily tested by computing the ratio *Nbr of LUTs/Nbr of registers*: it should be close to one.

The general design goal of ICEBERG is to provide an efficient algorithm for reconfigurable hardware implementations meeting the usual security requirements of block ciphers. More precisely, our design goals were:

1. Good security properties: ICEBERG has a resistance against known attacks comparable to recently published block ciphers (AES, NESSIE).
2. Hardware implementation efficiency (as previously defined).
3. Hardware implementation versatility: ICEBERG is scalable for different architectures (loop, unrolled, pipeline) and FPGA technologies.
4. Resistance against side-channel cryptanalysis.

3 Specifications

3.1 Block and Key Size

Let n be the block bit-size and k be the key bit-size. The state X is represented as a n-bit vector where $X(i)$ $(0 \leq i < n)$ represents the ith bit from the right. Alternatively, X can be represented as an array of $\frac{n}{4}$ 4-bit blocks, where X_j is the jth block from the right. ICEBERG operates on 64-bit blocks and uses a 128-bit key. It is an involutional iterative block cipher based on the repetition of R identical key-dependent round functions.

3.2 The Non-Linear Layer γ

Function γ consists of the successive application of non-linear substitution boxes and bit permutations (i.e. wire crossings).

Substitution layers S_0, S_1: The substitution layers S_j consists of the parallel application of substitution boxes s_j to the blocks of the state.

$$S_j : \mathbb{Z}_{2^4}^{16} \rightarrow \mathbb{Z}_{2^4}^{16} : x \rightarrow y = S_j(x) \Leftrightarrow y_i = s_j(x_i) \quad 0 \leq i \leq 15 \qquad (1)$$

Tables of S-boxes s_0, s_1 are given in

Bit permutation layer $P8$: The permutation layer $P8$ consists of the parallel application of 8 permutations $p8$ to the state, where $p8$ consists of bit permutations on 8-bit blocks of data. Table of $p8$ is given in appendix B.

$$P8 : \mathbb{Z}_{2^8}^8 \rightarrow \mathbb{Z}_{2^8}^8 : x \rightarrow y = P8(x) \Leftrightarrow y(8i + j) = x(8i + p8(j)) \quad 0 \leq i \leq 7,\ 0 \leq j \leq 7 \tag{2}$$

Based on previous descriptions, the non-linear layer γ can be expressed as:

$$\gamma : \mathbb{Z}_2^{64} \rightarrow \mathbb{Z}_2^{64} : \gamma \equiv S_0 \circ P8 \circ S_1 \circ P8 \circ S_0 \tag{3}$$

For cryptanalytic and software implementation purposes, γ may also be viewed as a unique layer consisting of the application of 8 identical 8×8 S-boxes for which the table is given in appendix B.

3.3 The Key Addition Layer σ_K

The affine key addition σ_K consists of the bitwise exclusive or (XOR, $\oplus$) of a key vector K.

$$\sigma_K : \mathbb{Z}_2^{64} \rightarrow \mathbb{Z}_2^{64} : x \rightarrow y = \sigma_K(x) \Leftrightarrow y(i) = x(i) \oplus K(i) \quad 0 \leq i \leq 63 \tag{4}$$

3.4 The Linear Layer ϵ_K

Function ϵ_K consists of the successive application of binary matrix multiplications and wire crossing layers, combined with the key addition layer for efficiency purposes. We describe it as:

$$\epsilon_K : \mathbb{Z}_2^{64} \rightarrow \mathbb{Z}_2^{64} : \epsilon_K \equiv P64 \circ P4 \circ \sigma_K \circ M \circ P64 \tag{5}$$

where M, $P64$, and $P4$ are defined as follows:

Matrix multiplication layer M: The matrix multiplication layer M is based on the parallel application of a simple involutional matrix multiplication. Let $V \in \mathbb{Z}_2^{4\times4}$ be a binary involutional (i.e. such that $V^2 = I_n$) matrix:

$$V = \begin{bmatrix} 0\,1\,1\,1 \\ 1\,0\,1\,1 \\ 1\,1\,0\,1 \\ 1\,1\,1\,0 \end{bmatrix}$$

M is then defined as:

$$M : \mathbb{Z}_{2^4}^{16} \rightarrow \mathbb{Z}_{2^4}^{16} : x \rightarrow y = M(x) \Leftrightarrow y_i = V \cdot x_i \quad 0 \leq i \leq 15 \tag{6}$$

We define diffusion boxes D as performing multiplication by V. Table of D is given in appendix B.

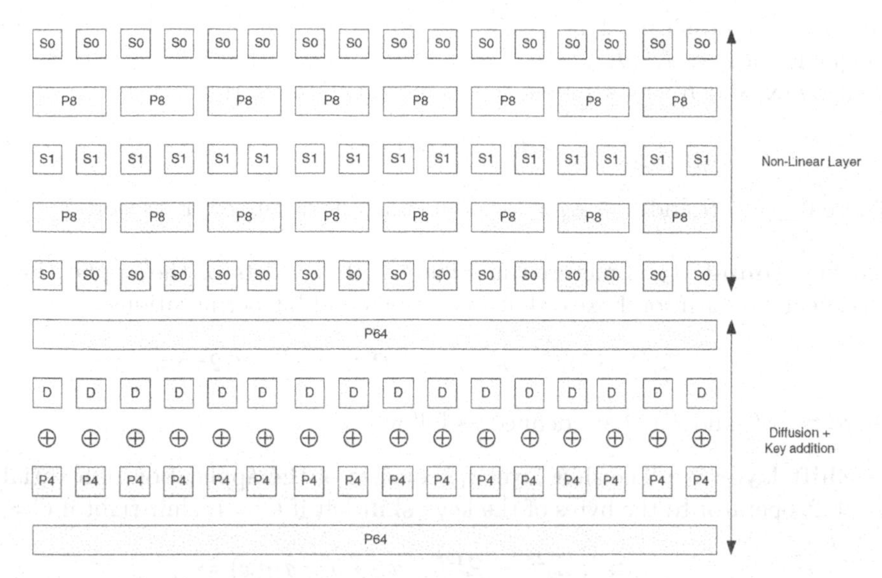

Fig. 1. The round function ρ_K.

Bit permutation layer *P64*: Permutation $P64$ performs bit permutations on 64-bit blocks of data.

$$P64 : \mathbb{Z}_2^{64} \rightarrow \mathbb{Z}_2^{64} : x \rightarrow y = P64(x) \Leftrightarrow y(i) = x(P64(i)) \quad 0 \leq i \leq 63 \tag{7}$$

Table of permutation $P64$ is given in appendix B.

Bit permutation layer *P4*: The permutation layer $P4$ consists of the parallel application of 16 permutations $p4$ to the state. $p4$ consists of bit permutations on 4-bit blocks of data. Table of $p4$ is given in appendix B.

$$P4 : \mathbb{Z}_{2^4}^{16} \rightarrow \mathbb{Z}_{2^4}^{16} : x \rightarrow y = P4(x) \Leftrightarrow y_i(j) = x_i(p4(j)) \\ 0 \leq i \leq 15,\ 0 \leq j \leq 3 \tag{8}$$

The purpose of permutation $P4$ is to efficiently distinguish encryption from decryption. It will become clearer in section 3.8 and appendix A.

3.5 The Round Function ρ_K

Finally, the whole round function can be expressed as:

$$\rho_K : \mathbb{Z}_2^{64} \rightarrow \mathbb{Z}_2^{64} : \rho_K \equiv \epsilon_K \circ \gamma \tag{9}$$

It is illustrated in Figure 1.

3.6 The Key Schedule

The key scheduling process consists of key expansion and key selection.

The key expansion: This process expands the cipher key $K \in \mathbb{Z}_2^{128}$ into a sequence of keys $K^0, K^1, ..., K^R$ also $\in \mathbb{Z}_2^{128}$. We set the initial key $K^0 = K$. Then we expand K^0 by a simple key round function β_C so that:

$$K^{i+1} = \beta_C(K^i) \tag{10}$$

Where $0 \le i \le R$ and $C \in \mathbb{Z}_2$ is a round constant discussed in section 3.9.

The **key round** β_C is pictured in Figure 2. It consists of the application of non-linear substitution boxes, shift operations and bit permutations:

$$\beta_C : \mathbb{Z}_2^{128} \rightarrow \mathbb{Z}_2^{128} : \beta_C \equiv \tau_C \circ P128 \circ S' \circ P128 \circ \tau_C \tag{11}$$

where τ_C, S', and $P128$ are defined as follows:

- **Shift layer τ_C:** The shift layer τ_C consists of the application of a variable shift operator to the bytes of the key : shift left if $C = 1$, shift right if $C = 0$.

$$\begin{aligned} &\tau_C : \mathbb{Z}_2^{128} \rightarrow \mathbb{Z}_2^{128} : x \rightarrow y = \tau_C(x) \Leftrightarrow \\ &\text{if } C = 0 : y(i) = x((i+8) \mod 128) \quad 0 \le i \le 127 \\ &\text{if } C = 1 : y(i) = x((i-8) \mod 128) \quad 0 \le i \le 127 \end{aligned} \tag{12}$$

- **Substitution layer S':** The substitution layer S' consists of the parallel application of substitution boxes s_0 to the blocks of the key.

$$S' : \mathbb{Z}_{2^4}^{32} \rightarrow \mathbb{Z}_{2^4}^{32} : x \rightarrow y = S'(x) \Leftrightarrow y_i = s_0'(x_i) \quad 0 \le i \le 31 \tag{13}$$

  Table of S-box s_0 is given in appendix B.
- **Bit permutations layer $P128$:** $P128$ performs bit permutation on 128-bit blocks of data.

$$P128 : \mathbb{Z}_2^{128} \rightarrow \mathbb{Z}_2^{128} : x \rightarrow y = P128(x) \Leftrightarrow y(i) = x(P128(i)) \quad 0 \le i \le 127 \tag{14}$$

  Table of $P128$ is given in appendix B.

The key selection: From every 128-bit vector K^i, we first apply a simple compression function E that selects 64 bits corresponding to key bytes of K^i having odd indices. We denote the resulting key as $K64^i$. Then we apply a **key selection layer** (ϕ) that consists of the parallel application of a selection function X to the blocks of the key.

$$\begin{aligned} \phi_{sel} : \mathbb{Z}_{2^4}^{16} \rightarrow \mathbb{Z}_{2^4}^{16} : K64^i \rightarrow RK_{sel}^i = \phi_{sel}(K64^i) \Leftrightarrow \\ RK_{sel,j}^i = X_{sel}(K64_j^i) \quad 0 \le j \le 15 \end{aligned} \tag{15}$$

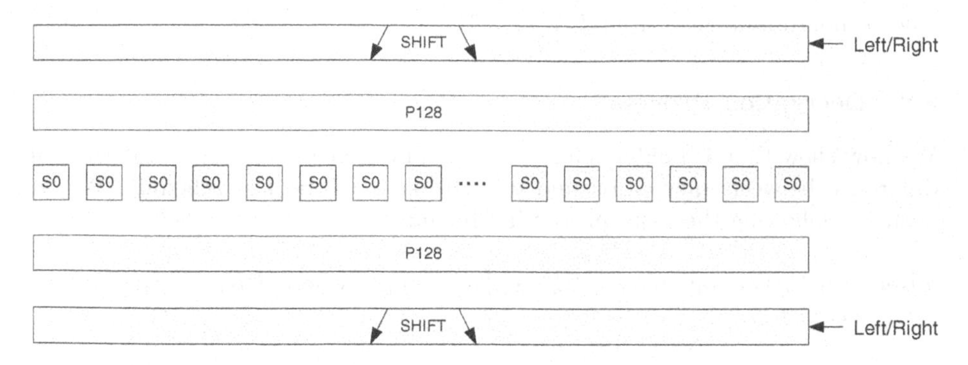

Fig. 2. The key round β_C.

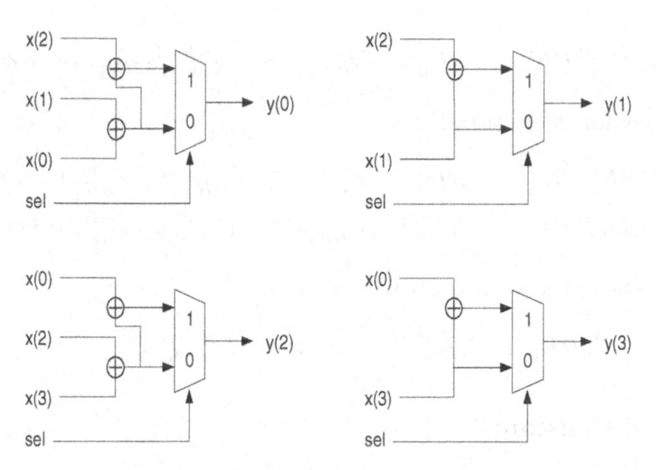

Fig. 3. The key selection function X_{sel}.

The **selection function** X takes 4-bit inputs and a selection bit *sel*:

$$X_{sel} : \mathbb{Z}_{2^4} \to \mathbb{Z}_{2^4} : x \to y = X_{sel}(x) \Leftrightarrow$$
$$\begin{cases} y(0) = (x(0) \oplus x(1) \oplus x(2)) \cdot sel \vee (x(0) \oplus x(1)) \cdot \overline{sel} \\ y(1) = (x(1) \oplus x(2)) \cdot sel \vee x(1) \cdot \overline{sel} \\ y(2) = (x(2) \oplus x(3) \oplus x(0)) \cdot sel \vee (x(2) \oplus x(3)) \cdot \overline{sel} \\ y(3) = (x(3) \oplus x(0)) \cdot sel \vee x(3) \cdot \overline{sel} \end{cases} \tag{16}$$

This selection process is represented in Figure 3. As a result, we obtain a 64-bit round key denoted by RK_1^i if $sel = 1$ and RK_0^i if $sel = 0$.

3.7 Encryption Process

ICEBERG is defined for the cipher key K, as the transformation ICEBERG$[K] = \alpha_R[RK_1^0, RK_1^1,.. .,RK_0^R]$ applied to the plaintext where:

$$\alpha_R[RK_1^0, RK_1^1, ..., RK_0^R] = \sigma_{RK_0^R} \circ \gamma \circ (\bigcirc_{r=1}^{R-1} \rho_{RK_1^r}) \circ \sigma_{RK_1^0} \tag{17}$$

The standard number of rounds is $R = 16$.

3.8 Decryption Process

We now show that ICEBERG is an involutional cipher in the sense that the only difference between encryption and decryption is in the key schedule. We will need the following theorem, proven in appendix A:

Theorem 1. For any $K64 \in \mathbb{Z}_{2^4}^{16}$: $\epsilon_{RK_0}^{-1} = \epsilon_{RK_1}$, where $RK_0 = \phi_0(K64)$ and $RK_1 = \phi_1(K64)$.

Then, the decryption process can be obtained as follows. We start from the encryption process:

$$\alpha_R[RK_1^0, RK_1^1, ..., RK_0^R] = \sigma_{RK_0^R} \circ \gamma \circ (\bigcirc_{r=1}^{R-1} \epsilon_{RK_1^r} \circ \gamma) \circ \sigma_{RK_1^0}$$

Then we have for decryption:

$$\alpha_R^{-1}[RK_1^0, RK_1^1, ..., RK_0^R] = \sigma_{RK_1^0} \circ (\bigcirc_{r=R-1}^{1} \gamma \circ \epsilon_{RK_1^r}^{-1}) \circ \gamma \circ \sigma_{RK_0^R}$$

$$\Leftrightarrow \alpha_R^{-1}[RK_1^0, RK_1^1, ..., RK_0^R] = \sigma_{RK_1^0} \circ \gamma \circ (\bigcirc_{r=R-1}^{1} \epsilon_{RK_1^r}^{-1} \circ \gamma) \circ \sigma_{RK_0^R} \quad (19)$$

Finally the above theorem leads to:

$$\alpha_R^{-1}[RK_1^0, RK_0^1, ..., RK_0^R] = \sigma_{RK_1^0} \circ \gamma \circ (\bigcirc_{r=R-1}^{1} \epsilon_{RK_0^r} \circ \gamma) \circ \sigma_{RK_0^R}$$

3.9 Round Constants

ICEBERG is an involutional cipher in the sense that the only difference between encryption and decryption is in the key schedule. Moreover, if properly chosen, the round constants allow to compute keys "on-the-fly" in encryption and decryption modes. Basically, we would like round keys to satisfy:

$$\begin{aligned} K^0 &= K^R \\ K^1 &= K^{R-1} \\ K^2 &= K^{R-2} \\ &\dots \end{aligned} \quad (21)$$

This involves that R is even. Then, if the first half of round constants (i.e. until round 8) is 0 (shift left) and the second half is 1 (shift right), the resulting round keys will satisfy Equation (21).

As a consequence, the only difference between encryption and decryption is the selection function ϕ of the key bits, as $\epsilon_{RK_1}^{-1} \equiv \epsilon_{RK_0}$.

4 Security Analysis

4.1 Design Properties of the Components

S-Boxes: The non-linear layer γ may be viewed as made out of the parallel application of 8 copies of the same 8×8 S-box. We designed this S-box such that it has the following properties:

- It is an involution.
- Its δ-parameter[4] is 2^{-5}.
- Its λ-parameter[5] is 2^{-2}.
- Its nonlinear order ν is maximum, namely 7.

For efficiency purposes, the s-box was generated from a fixed permutation $p8$ and small 4×4 s-boxes s_0 and s_1 that perfectly fits into 4-input LUTs. The generation of the s-boxes is detailed in Appendix C.

The bit permutations: $P64$ and $P128$ were designed such as to disturb as much as possible the bit alignment inside bytes, in order to provide resistance against some attacks. A remarkable property of $P64$ and $P128$ is that 2 bits from the same byte are always mapped to 2 bits belonging to different bytes. $p8$ is involutional and allows to generate good substitution boxes. Finally, $p4$ allows the selection function X_{sel} to be implemented in one LUT.

The Diffusion Layer: Due to the fact that we attached much importance to hardware implementation aspects in the design of the diffusion layer, it is not optimal. More precisely, it is easy to see that its byte branch number is 4, as the bit branch number of $p4 \circ \sigma_K \circ D$ is 4, and because of the remarkable property of $P64$ we have just mentioned. The diffusion boxes D were designed so that their combination with the key addition layer σ_K can be done inside one LUT.

The key round: The key round has been chosen for its efficiency properties, as well as in order to provide resistance against key schedule cryptanalysis and slide attacks:

1. Non periodicity is provided by the shift operation τ_C.
2. Non linearity is provided by non-linear S-boxes.
3. Good diffusion properties are provided by the combination of shifts, S-boxes and bit permutations.

Moreover, the shift layer τ_C is used in order to allow the property (19) to be respected. The selection function X_{sel} is necessary to prove the property of Appendix A and is designed such that it fits into a single LUT.

---

[4] δ equals the probability of the best differential approximation.

[5] We define the *bias* of a linear approximation that holds with probability p as $\epsilon = |p - 1/2|$. The *λ-parameter* of a S-box is equal to 2 times the bias of its best linear approximation.

4.2 Strength Against Known Attacks

Linear and Differential Cryptanalysis: From the properties of the S-box and the diffusion layer, we can compute that a differential characteristic [13] over 2 rounds of ICEBERG has probability at most $(2^{-5})^4 = 2^{-20}$. Also, a linear characteristic [14] over these 2 rounds has input-output correlation at most $(2^{-2})^4 = 2^{-8}$. Therefore loose bounds can be computed for the full cipher (16 rounds):

- The probability of the best differential characteristic is smaller than 2^{-160}.
- The input-output correlation of the best linear characteristic is smaller than 2^{-64}.

The security margin is very likely big enough to prevent variants of differential and linear attacks, such as boomerang [15] and rectangle [16] attacks, multiple linear cryptanalysis [17], non-linear approximations of outer rounds [18],... Note also that the security margin of ICEBERG against linear and differential cryptanalysis is comparable to the one of Khazad. This is probably more than it is necessary, as resistance against structural attacks [19, 20] was probably more determinant in the choice of the number of rounds of Khazad (8), than security margins against linear and differential cryptanalysis.

Truncated and Impossible Differentials: Truncated differentials were introduced in [21], and impossible differentials in [22, 23]. They typically apply to ciphers operating on well-aligned data blocks (often bytes), such as Khazad or the AES (and many others). However our cipher does not enter in this category because of the $P64$ layer, which makes it very difficult to attack this way. Therefore such an attack on the full 16-round cipher seems very unlikely.

Square Attacks: Like truncated differential attacks, square attacks [19] generally apply to ciphers operating on well-aligned data blocks. Therefore the $P64$ layer should prevent them efficiently, at least on more than a few rounds. More precisely, consider a batch of $256^m (m \in \{1...7\})$ plaintexts, such that $8 - m$ bytes remain constant for all of them (these bytes are said *passive*), while the concatenation of the m other bytes takes every possible value (it is said *active*). This property is preserved by the γ layer. On the contrary, the $P64$ layer makes all bytes garbled (i.e. not active nor passive), which prevents pushing a basic "square characteristic" further. The same type of argument could be applied to truncated differential attacks.

Interpolation Attacks: Interpolation attacks [24] are made possible when the S-box has a simple algebraic structure, allowing to express the cipher as a sufficiently simple polynomial or rational expression. The diffusion layer also has a role with this respect. As the S-box of ICEBERG has no simple algebraic expression, it prevents interpolation attacks for more than a few rounds of our cipher.

Higher Order Differential Cryptanalysis: It was introduced by Knudsen in [21], and relies on finding high order differentials being a constant for the whole cipher. But as the nonlinear order of the S-box we use is maximal, namely 7, we can expect that the maximal value of 63 for the non-linear order of the cipher is reached after a few rounds of ICEBERG.

Slide Attacks: Slide attacks [25, 26] work against ciphers using a periodic key schedule. Although the sequence of subkeys produced by the key schedule of ICEBERG is not periodic, it has a particular structure, namely:

$$(K^0, K^1, ..., K^7, K^8, K^7, ..., K^0)$$

The key schedule of the GOST cipher has some similarities with the one of ICEBERG. Vulnerability of some variants and reduced-round versions of GOST against slide attacks is examined in [26]. However none of the attacks presented there seems to be applicable to our cipher.

Related-Key Attacks: The first related-key attack has been described in [27], and is the related-key counterpart of the slide attack. Let us examine a slightly simplified version of ICEBERG, where the initial key addition $\sigma_{RK_1^0}$ is replaced by a normal round $\rho_{RK_1^0}$, and the final $\sigma_{RK_0^R} \circ \gamma$ is also replaced by a normal round $\rho_{RK_0^R}$. Then if 2 keys K and K^* are such that $K^1{=}K^{*0}$, and 2 plaintexts P and P^* are such that $P^* = \rho_{RK_1^0}(P)$, encryption of P under K and of P^* under K^* will process the same way (with a difference of 1 round) during 8 rounds. However then round keys, and hence computation, will differ; therefore such a related key attack does not work against our key schedule. Forgetting the simplification we made on the first and last round of ICEBERG, a related-key attack becomes even more difficult. Differential related-key attacks [28] are also very unlikely to be applicable to ICEBERG, due to the good diffusion and nonlinearity of its key schedule.

Weak keys: The design properties of the key round prevent ICEBERG from having weak keys. The only remarkable property of the key round is in the selection function X_{sel} where some symbols are independent of the selection bit. Namely, hexadecimal input symbols $0, 2, 8, A$ become $0, C, 3, F$ regardless of $sel = 0$ or $sel = 1$. However, this point is very unlikely to be an exploitable weakness.

Biryukov's observations on Involutional Ciphers: Observations of Biryukov on Khazad and Anubis [29] remain valid for ICEBERG. However this study could at best threaten 5 rounds of our cipher, while it is made out of 16 rounds.

Side-channel cryptanalysis: Although cryptosystem designers frequently assume that secret parameters will be manipulated in closed reliable computing environments, Kocher et *al.* stressed in 1998 [30] that actual computers and microchips leak information correlated to the data handled. Side-channel attacks based on time, power and electromagnetic measurements were successfully applied to smart card implementations of block ciphers. Protecting implementations against side-channel attacks is usually difficult and expensive. Masking all the data with random boolean values is suggested in several papers [31, 32] and the use of small substitution tables allows to implement this efficiently, although it is still an expensive solution.

The key agility provided by ICEBERG (changing the key at every plaintext block is for free) also offers interesting opportunities to prevent most side-channel attacks by defining new encryption modes where the key is changed sufficiently often. As most side-channel attacks need to collect several leakage traces to remove the noise from useful information, changing the key frequently, even in a well chosen deterministic way (e.g. LFSR-based), would make most attacks somewhat unpractical. Actually, only template attacks [33] allow to extract information from a single sample but the context is also more specific as they require that an adversary has access to an experimental device (identical to the device attacked) that he can program to his choosing.

5 Performance Analysis

ICEBERG has been designed in order to allow very efficient reconfigurable hardware implementations, as defined in section 2. For this purpose, we applied the following design rules:

1. All components easily fit into 4-input LUTs. Practically, ICEBERG is made of the parallel application of 4-input-bit transforms combined with bit permutations or shifts.
2. All components are involutional so that encryption and decryption can be made with the same hardware. The only difference between encryption and decryption is in the selection bit of ϕ_{sel}.
3. The key expansion allows to derive round keys "on-the-fly" in encryption and decryption modes. There is no need to store the round keys and the key can be changed in one clock cycle.
4. The scheduling of the algorithm is balanced so that the round and key round can be made in the same number of clock cycles.
5. The non-linear layer can be efficiently implemented into the RAM blocks available in most modern FPGAs.

5.1 Hardware Implementations

As all components easily fit into 4-input LUTs, we can directly evaluate the hardware cost of ICEBERG:

Component	HW cost (LUTs)	Component	HW cost (LUTs)
S_0, S_1	64	τ_C	128
γ	192	S'	128
ϵ_K	64	Key round β_C	384
Round ρ_K	256	ϕ_{sel}	64
Complete round + key round			704

As a comparison, Khazad needs 576 LUTs for its round and 768 LUTs for its key round [11], with a more expensive encryption/decryption structure. AES Rijndael is even more critical as its round needs 2608 LUTs and its key round 768 LUTs [12]. Although comparisons between hardware implementations are made difficult by their high dependency on the design methodology, we may reasonably expect to have ICEBERG encryption/decryption for the cost of Khazad encryption only and half the cost of Rijndael encryption, with comparable encryption rates.

Moreover, the parallel nature of ICEBERG allows to implement every possible throughput/area tradeoff as well as very efficient pipeline. The maximum pipeline can be obtained by inserting registers after every LUT which allows to reach an optimal ratio $Nbr\ of\ LUTs/Nbr\ of\ registers = 1$. Loop and unrolled architectures are easily implementable with LUT-based or RAM-based substitution boxes. As a consequence, ICEBERG offers various and efficient implementation opportunities. Its regular structure makes the classical design optimizations easily reachable. Software efficiency is not a design goal of ICEBERG. It is briefly discussed in Appendix D.

6 Conclusion

This paper presented the platform-specific encryption algorithm ICEBERG and the rationale behind its design. ICEBERG is based on a fast involutional structure in order to provide very efficient hardware implementation opportunities. We showed the specificity of this type of platform in block cipher design. We also underlined that the overall structure of a cipher is important for efficiency purposes (for example, in designing rounds and key rounds that can be made in the same number of clock cycles, or in allowing "on the fly" key derivation in both encryption and decryption modes). We believe ICEBERG to be as secure as AES and NESSIE candidates and much more efficient for reconfigurable hardware implementations. ICEBERG also offers free opportunities to defeat most side-channel attacks by using adequate encryption modes.

Acknowledgements

The authors are grateful to Paulo Barreto for providing valuable comments and help during the design of ICEBERG.

References

[1] NIST Home page, http://csrc.nist.gov/CryptoToolkit/aes/. 279

[2] J.Daemen, V.Rijmen, *The Block Cipher Rijndael*, Smart Card Research and Applications, pp 288-296, Springer-Verlag, LNCS 1820, 2000. 279

[3] P.Barreto, V.Rijmen, *The KHAZAD Legacy-Level Block Cipher*, Submission to NESSIE project, available from http://www.cosic.esat.kuleuven.ac.be/nessie/ 280, 298

[4] M.Matsui, *Supporting Document of MISTY1*, , Submission to NESSIE project, available from http://www.cosic.esat.kuleuven.ac.be/nessie/ 280

[5] Xilinx: *Virtex 2 FPGAs Data Sheet*, http://www.xilinx.com. 280

[6] Altera: *Stratix 1.5V FPGAs Data Sheet*, http://www.altera.com. 280

[7] M.McLoone and J. V.McCanny, *High Performance Single Ship FPGA Rijndael Algorithm Implementations*, in the proceedings of CHES 2001: The Third International CHES Workshop, Lecture Notes In Computer Science, LNCS2162, pp 65-76, Springer-Verlag. 281

[8] V.Fischer and M.Drutarovsky, *Two Methods of Rijndael Implementation in Reconfigurable Hardware*, in the proceedings of CHES 2001: The Third International CHES Workshop, Lecture Notes In Computer Science, LNCS2162, pp 65-76, Springer-Verlag. 281

[9] A.Satoh et al, *A Compact Rijndael Hardware Architecture with S-Box Optimization*, Advances in Cryptology - ASIACRYPT 2001, LNCS 2248, pp239-254, Springer-Verlag. 281

[10] Helion Technology, *High Performance AES (Rijndael) Cores for XILINX FPGA*, *http://www.heliontech.com*. 281

[11] F. X.Standaert,G.Rouvroy,J. J.Quisquater,J. D.Legat, *Efficient FPGA Implementations of Block Ciphers KHAZAD and MISTY1*, in the proceedings of the Third NESSIE Workshop, November 6-7 2002, Munich, Germany. 281, 291

[12] F. X.Standaert,G.Rouvroy,J. J.Quisquater,J. D.Legat, *A Methodology to Implement Block Ciphers in Reconfigurable Hardware and its Application to Fast and Compact AES Rijndael*, in the proceedings of FPGA 2003: the Field Programmable Logic Array Conference, February 23-25 2003, Monterey, California. 281, 291

[13] E.Biham, A.Shamir, *Differential cryptanalysis of DES-like cryptosystems (Extended abstract)*, Proceedings of Crypto 90, pp 2-21, Springer-Verlag, LNCS 537, 1990. 288

[14] M.Matsui, *Linear cryptanalysis method for DES cipher*, Proceedings of EuroCrypt 93, pp 386-397, Springer-Verlag, LNCS 765, 1993. 288

[15] D.Wagner, *The Boomerang Attack*, Proceedings of FSE 99, pp 156-170, Springer-Verlag, LNCS 1636, 1999. 288

[16] E.Biham, O.Dunkelman, N.Keller, *The rectangle Attack - Rectangling the Serpent*, Proceedings of Eurocrypt 2001, pp 340-357, Springer-Verlag, LNCS 2045, 2001. 288

[17] B. S.Kaliski, M. J. B.Robshaw, *Linear Cryptanalysis using Multiple Approximations*, Proceedings of Crypto 94, pp.26-39, Springer-Verlag, LNCS 0839, 1994. 288

[18] L.Knudsen, M. J. B.Robshaw, *Non-Linear Approximations in Linear Cryptanalysis*, Proceedings of Eurocrypt 96, pp 224-236, Springer-Verlag, LNCS 1070, 1996. 288

[19] J.Daemen, L.Knudsen, V.Rijmen, *The Block Cipher SQUARE*, Proceedings of FSE 1997, pp 149-165, Springer-Verlag, LNCS 1267, 1999. 288

[20] N.Ferguson, J.Kelsey, S.Lucks, and al., *Improved Cryptanalysis of Rijndael*, Proceedings of FSE 2000, pp 213-230, Springer-Verlag, LNCS 1978, 2000. 288

[21] L.Knudsen, *Truncated and Higher Order Differentials*, Proceedings of FSE 94, pp 196-211, Springer-Verlag, LNCS 1008, 1995. 288, 289

[22] E.Biham, A.Biryukov and A.Shamir, *Cryptanalysis of Skipjack Reduced to 31 Rounds Using Impossible Differentials*, Proceedings of Eurocrypt 99, pp 12-23, Springer-Verlag, LNCS 1592, 1999. 288
[23] E.Biham, A.Biryukov and A.Shamir, *Miss in the Middle Attacks on IDEA, Khufu, and Khafre*, Proceedings of FSE 99, pp 124-138, Springer-Verlag, LNCS 1636, 1999. 288
[24] T.Jakobsen and L.Knudsen, *The Interpolation Attack on Block Ciphers*, Proceedings of FSE 97, pp 28-40, Springer-Verlag, LNCS 1267, 1997. 288
[25] A.Biryukov, D.Wagner, *Slide Attacks*, Proceedings of FSE'99, pp 245-259, Springer Verlag, LNCS 1636, 1999. 289
[26] A.Biryukov, D.Wagner, *Advanced Slide Attacks*, Proceedings of Eurocrypt 00, pp 589-606, Springer Verlag, LNCS 1807, 2000. 289
[27] E.Biham, *New Type of Cryptanalytic Attacks Using Related Key*, Proceedings of Eurocrypt 93, pp 229-246, Springer-Verlag, LNCS 765, 1994. 289
[28] J.Kelsey, B.Schneier, D.Wagner, *Related-Key Cryptanalysis of 3-WAY, Biham-DES, CAST, DES-X, NewDES, RC2, and TEA*, Proceedings of AusCrypt'92, pp 196-208, Springer-Verlag, LNCS 718, 1993. 289
[29] A.Biryukov, *Analysis of Involutional Ciphers: Khazad and Anubis*, Proceedings of FSE 2003, Springer-Verlag, to appear. 289
[30] P.Kocher, J.Jaffe, B.Jun, *Differential Power Analysis*, in the proceedings of CRYPTO 99, Lecture Notes in Computer Science 1666, pp 398-412, Springer-Verlag. 290
[31] L.Goubin,J.Patarin, *DES and Differential Power Analysis: The Duplication Method*, in the proceedings of CHES 1999, Lecture Notes in Computer Science 1717, pp 158-172, Springer-Verlag. 290
[32] S.Chari et *al.*, *Towards Sound Approaches to Counteract Power-Analysis Attacks*, in the proceedings of CRYPTO 1999, Lecture Notes in Computer Science 1666, pp 398-412, Springer-Verlag. 290
[33] S.Chari, J.Rao, P.Rohatgi, *Template Attacks*, in the proceedings of CHES 2002, Lecture Notes in Computer Science 2523, pp 13-28, Springer-Verlag. 290
[34] A.Pfitzmann,R.Aßmann, *More Efficient Software Implementations of (Generalized) DES*, Institut fur Rechnerent und Fehlertoleranz, Univ.Karlsruhe, Interner Bericht 18/90. 299
[35] E.Biham, *A Fast New DES Implementation in Software*, Technion - Computer Science Department, Technical Report CS0891 - 1997. 299
[36] A. M.Youssef, S. E.Tavares, H.Heys, *A New Class of Substitution-Permutation Networks*, Proceedings of Selected Areas in Cryptography (SAC 96), pp 132-147, 1996.
[37] H. M. Heys, S. E. Tavares, *Known Plaintext Cryptanalysis of Tree-Structured Block Ciphers*, Electronics Letters, Vol. 31, pp 784-785, May 1995.
[38] L.Knudsen, *Block Ciphers - Analysis, Design and Applications*, Doctoral Dissertation, DAIMI PB 485, Aarhus University, Denmark, 1994.
[39] J.Daemen, *Cipher and Hash Function Design*, Doctoral Dissertation, March 1995, KULeuven.
[40] V.Rijmen, *Cryptanalysis and Design of Iterated Block Ciphers*, Doctoral Dissertation, October 1997, KULeuven.

A Proof of theorem 1.

We have to prove that $P4 \circ \sigma_{RK_1} \circ M \equiv M \circ \sigma_{RK_0} \circ P4$. Inputs and outputs of every transform are represented in Figure 4. We simply write down relations between them. First we encrypt with RK_1:

$$\begin{aligned} b_0 &= a_1 \oplus a_2 \oplus a_3 \\ b_1 &= a_0 \oplus a_2 \oplus a_3 \\ b_2 &= a_0 \oplus a_1 \oplus a_3 \\ b_3 &= a_0 \oplus a_1 \oplus a_2 \end{aligned}$$

$$\begin{aligned} c_0 &= a_1 \oplus a_2 \oplus a_3 \oplus k_0 \oplus k_1 \oplus k_2 \\ c_1 &= a_0 \oplus a_2 \oplus a_3 \oplus k_1 \oplus k_2 \\ c_2 &= a_0 \oplus a_1 \oplus a_3 \oplus k_2 \oplus k_3 \oplus k_0 \\ c_3 &= a_0 \oplus a_1 \oplus a_2 \oplus k_3 \oplus k_0 \end{aligned}$$

$$\begin{aligned} d_0 &= a_0 \oplus a_2 \oplus a_3 \oplus k_1 \oplus k_2 \\ d_1 &= a_1 \oplus a_2 \oplus a_3 \oplus k_0 \oplus k_1 \oplus k_2 \\ d_2 &= a_0 \oplus a_1 \oplus a_2 \oplus k_3 \oplus k_0 \\ d_3 &= a_0 \oplus a_1 \oplus a_3 \oplus k_2 \oplus k_3 \oplus k_0 \end{aligned}$$

Then, when we decrypt with RK_0:

$$\begin{aligned} e_0 &= d_1 \oplus d_2 \oplus d_3 = a_1 \oplus k_0 \oplus k_1 \\ e_1 &= d_0 \oplus d_2 \oplus d_3 = a_0 \oplus k_1 \\ e_2 &= d_0 \oplus d_1 \oplus d_3 = a_3 \oplus k_2 \oplus k_3 \\ e_3 &= d_0 \oplus d_1 \oplus d_2 = a_2 \oplus k_3 \end{aligned}$$

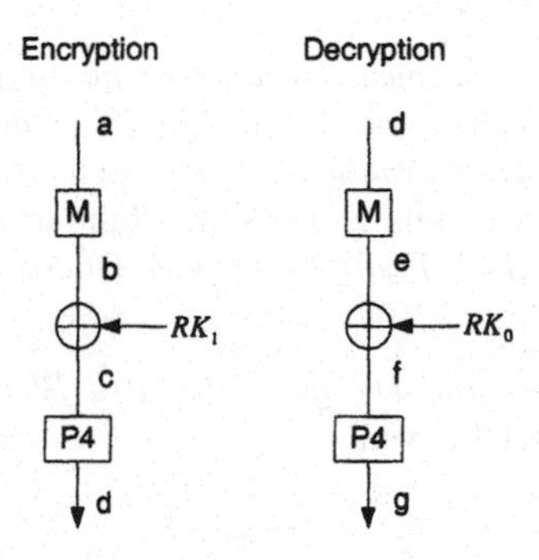

Fig. 4. Theorem 1.

From (16), we have:

$$f_0 = a_1$$
$$f_1 = a_0$$
$$f_2 = a_3$$
$$f_3 = a_2$$

And finally, permutation $P4$ finishes the decryption:

$$g_0 = a_0$$
$$g_1 = a_1$$
$$g_2 = a_2$$
$$g_3 = a_3$$

Remark that permutation $P4$ allows the selection function X_{sel} to be efficiently implemented in LUTs as it has at most 4 inputs: 3 key bits and a selection bit.

B Tables.

0	1	2	3
1	0	3	2

Table 1. p4.

0	1	2	3	4	5	6	7
0	1	4	5	2	3	6	7

Table 2. p8.

0	1	2	3	4	5	6	7	8	9	a	b	c	d	e	f
d	7	3	2	9	a	c	1	f	4	5	e	6	0	b	8

Table 3. s_0.

0	1	2	3	4	5	6	7	8	9	a	b	c	d	e	f
4	a	f	c	0	d	9	b	e	6	1	7	3	5	8	2

Table 4. s_1.

0	1	2	3	4	5	6	7	8	9	a	b	c	d	e	f
0	e	d	3	b	5	6	8	7	9	a	4	c	2	1	f

Table 5. D.

	00	01	02	03	04	05	06	07	08	09	0a	0b	0c	0d	0e	0f
00	24	c1	38	30	e7	57	df	20	3e	99	1a	34	ca	d6	52	fd
10	40	6c	d3	3d	4a	59	f8	77	fb	61	0a	56	b9	d2	fc	f1
20	07	f5	93	cd	00	b6	62	a7	63	fe	44	bd	5f	92	6b	68
30	03	4e	a2	97	0b	60	83	a3	02	e5	45	67	f4	13	08	8b
40	10	ce	be	b4	2a	3a	96	84	c8	9f	14	c0	c4	6f	31	d9
50	ab	ae	0e	64	7c	da	1b	05	a8	15	a5	90	94	85	71	2c
60	35	19	26	28	53	e2	7f	3b	2f	a9	cc	2e	11	76	ed	4d
70	87	5e	c2	c7	80	b0	6d	17	b2	ff	e4	b7	54	9d	b8	66
80	74	9c	db	36	47	5d	de	70	d5	91	aa	3f	c9	d8	f3	f2
90	5b	89	2d	22	5c	e1	46	33	e6	09	bc	e8	81	7d	e9	49
a0	e0	b1	32	37	ea	5a	f6	27	58	69	8a	50	ba	dd	51	f9
b0	75	a1	78	d0	43	f7	25	7b	7e	1c	ac	d4	9a	2b	42	e3
c0	4b	01	72	d7	4c	fa	eb	73	48	8c	0c	f0	6a	23	41	ec
d0	b3	ef	1d	12	bb	88	0d	c3	8d	4f	55	82	ee	ad	86	06
e0	a0	95	65	bf	7a	39	98	04	9b	9e	a4	c6	cf	6e	dc	d1
f0	cb	1f	8f	8e	3c	21	a6	b5	16	af	c5	18	1e	0f	29	79

Table 6. 8 x 8 substitution box.

0	1	2	3	4	5	6	7	8	9	10	11	12	13	14	15
0	12	23	25	38	42	53	59	22	9	26	32	1	47	51	61
16	17	18	19	20	21	22	23	24	25	26	27	28	29	30	31
24	37	18	41	55	58	8	2	16	3	10	27	33	46	48	62
32	33	34	35	36	37	38	39	40	41	42	43	44	45	46	47
11	28	60	49	36	17	4	43	50	19	5	39	56	45	29	13
48	49	50	51	52	53	54	55	56	57	58	59	60	61	62	63
30	35	40	14	57	6	54	20	44	52	21	7	34	15	31	63

Table 7. P64.

0	1	2	3	4	5	6	7	8	9	10	11	12	13	14	15
76	110	83	127	67	114	92	97	98	65	121	106	78	112	91	82
16	17	18	19	20	21	22	23	24	25	26	27	28	29	30	31
71	101	89	126	72	107	81	118	90	124	73	88	64	104	100	85
32	33	34	35	36	37	38	39	40	41	42	43	44	45	46	47
109	87	75	113	120	66	103	115	122	108	95	69	74	116	80	102
48	49	50	51	52	53	54	55	56	57	58	59	60	61	62	63
84	96	125	68	93	105	119	79	123	86	70	117	111	77	99	94
64	65	66	67	68	69	70	71	72	73	74	75	76	77	78	79
28	9	37	4	51	43	58	16	20	26	44	34	0	61	12	55
80	81	82	83	84	85	86	87	88	89	90	91	92	93	94	95
46	22	15	2	48	31	57	33	27	18	24	14	6	52	63	42
96	97	98	99	100	101	102	103	104	105	106	107	108	109	110	111
49	7	8	62	30	17	47	38	29	53	11	21	41	32	1	60
112	113	114	115	116	117	118	119	120	121	122	123	124	125	126	127
13	35	5	39	45	59	23	54	36	10	40	56	25	50	19	3

Table 8. P128.

C Generation of the ICEBERG S-box [3]

As $P8$ is fixed, the only part of the ICEBERG S-box structure still unspecified consists of the s_0 and s_1 involutions, which are generated pseudo-randomly in a verifiable way.

The searching algorithm starts with two copies of a simple involution without fixed points (namely, the negation mapping $u \mapsto \bar{u} = u \oplus \texttt{0xF}$), and pseudo-randomly derives from each of them a sequence of 4×4 substitution boxes ("mini-boxes") with the optimal values $\delta = 1/4$, $\lambda = 1/2$, and $\nu = 3$. At each step, in alternation, only one of the sequences is extended with a new mini-box. The most recently generated mini-box from each sequence is taken, and the pair is combined according to the ICEBERG S-box shuffle structure; finally, the resulting 8×8 S-box, if free of fixed points, is tested for the design criteria regarding δ, λ, and ν.

Given a mini-box at any point during the search, a new one is derived from it by choosing two pairs of mutually inverse values and swapping them, keeping the result an involution without fixed points; this is repeated until the running mini-box has optimal values of δ, λ, and ν.

The pseudo-random number generator is implemented using the AES cipher Rijndael in counter mode, with a fixed key consisting of 128 zero bits and an initial counter value consisting of 128 zero bits.

The following pseudo-code fragment illustrates the computation of the chains of mini-boxes and the resulting S-box:

```
procedure ShuffleStructure(s0, s1)
    for w ← 0 to 255 do
        u0 ← s0[w ≫ 4];   v0 ← s0[w & 0x0F];
        u1 ← (u0 & 0xC) | ((v0 & 0xC) ≫ 2);   v1 ← (v0 & 0x3) | ((u0 & 0x3) ≪ 2);
        u0 ← s1[u1];   v0 ← s1[v1];
        u1 ← (u0 & 0xC) | ((v0 & 0xC) ≫ 2);   v1 ← (v0 & 0x3) | ((u0 & 0x3) ≪ 2);
        S[w] ← (s0[u1] ≪ 4) | s0[v1];
    end for
    return S;
end procedure

procedure SearchRandomSBox()
    // initialize mini-boxes to the negation involution:
    for u ← 0 to 255 do
        s0[u] ← ū; s1[u] ← ū;
    end for
    // look for S-box conforming to the design criteria:
    repeat
        // swap mini-boxes (update the "older" one only)
        s0 ↔ s1;
        // randomly generate a "good" GF(2^4) involution free of fixed points:
```

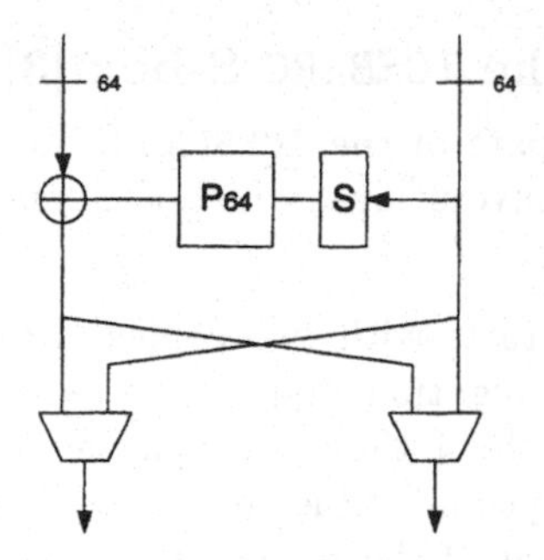

Fig. 5. A modified ρ_K.

repeat
  repeat
    // randomly select x and y such that
    // $x \neq y$ and $s_1[x] \neq y$ (this implies $s_1[y] \neq x$):
    $z \leftarrow$ RandomByte(); $x \leftarrow z \gg 4$; $y \leftarrow z \,\&\,$ 0x0F;
  until $x \neq y \wedge s_1[x] \neq y$;
  // swap entries:
  $u \leftarrow s_1[x]$; $v \leftarrow s_1[y]$;
  $s_1[x] \leftarrow v$; $s_1[u] \leftarrow y$;
  $s_1[y] \leftarrow u$; $s_1[v] \leftarrow x$;
until $\delta(s_1) = 1/4 \wedge \lambda(s_1) = 1/2 \wedge \nu(s_1) = 3$;
// build S-box from the mini-boxes:
$S \leftarrow$ ShuffleStructure(s_0, s_1);
// test the design criteria:
until #FixedPoints(S) $= 0 \vee \delta(S) \leqslant 2^{-5} \wedge \lambda(S) \leqslant 2^{-2} \wedge \nu(S) = 7$;
return S;
end procedure

D Software implementations

Software efficiency is not a design goal of ICEBERG. Nevertheless, its round function may be implemented using a table lookup approach as it is suggested in [34] and is therefore comparable to the one of Khazad. The key expansion of ICEBERG is actually its most critical part as a lookup table implementation requires separate tables for transforms $\tau_C \circ P128$ and transforms $S' \circ P128 \circ \tau_C$. Note that ICEBERG is also susceptible to be implemented in bitslice mode as suggested in [35].

If a software-efficient key schedule is wanted, an alternative key round based on a small Feistel structure can be used, illustrated in Figure 5. We just use a conditional switch of the two 64-bit vectors so that we can encrypt during half the rounds and decrypt afterwards in order to satisfy Equation (21). This will only slightly affect hardware performances (an additional multiplexor is necessary to select the round keys).

Related Key Differential Attacks on 27 Rounds of XTEA and Full-Round GOST*

Youngdai Ko[1], Seokhie Hong[1], Wonil Lee[1], Sangjin Lee[1], and Ju-Sung Kang[2]

[1] Center for Information Security Technologies (CIST)
Korea University, Anam Dong, Sungbuk Gu, Seoul, Korea
{koyd,hsh,wonil,sangjin}@cist.korea.ac.kr

[2] Section 0741, Information Security Technology Division, ETRI
161 Kajong-Dong, Yusong-Gu, Taejon, 305-350, Korea
jskang@etri.re.kr

Abstract. In this paper, we present a related key truncated differential attack on 27 rounds of XTEA which is the best known attack so far. With an expected success rate of 96.9%, we can attack 27 rounds of XTEA using $2^{20.5}$ chosen plaintexts and with a complexity of $2^{115.15}$ 27-round XTEA encryptions. We also propose several attacks on GOST. First, we present a distinguishing attack on full-round GOST, which can distinguish it from a random permutation with probability $1 - 2^{-64}$ using a related key differential characteristic. We also show that H. Seki et al.'s idea combined with our related key differential characteristic can be applied to attack 31 rounds of GOST . Lastly, we propose a related key differential attack on full-round GOST. In this attack, we can recover 12 bits of the master key with 2^{35} chosen plaintexts, 2^{36} encryption operations and an expected success rate of 91.7%.

Keywords: Related key differential attack, Distinguishing attack, XTEA, GOST, Differential characteristic

1 Introduction

XTEA [10] was proposed as a modified version of TEA [7] by R. Needham and D. Wheeler in order to resist related key attacks [5]. XTEA is a very simple block cipher using only exclusive-or operations, additions, and shifts. Until now, the best known result on XTEA is a truncated differential attack on 23 rounds of XTEA (8~30 or 30~52) proposed in [3]. In this paper, we present related key truncated differential attacks on 25 (1~25) and 27 (4~30) rounds of XTEA.

GOST was proposed in the former Soviet Union [2]. It has a very simple round function and key schedule. GOST uses key addition modulo 2^{32} in each round function. So, the probability of a differential characteristic depends not only on the value of input-output differences but also on the value of the round key. In order to reduce the effect of the round key addition, H. Seki et al. introduced a specific set of differential characteristics and proposed a differential attack on

* This work is supported by the MOST research fund(M1-0326-08-0001).

B. Roy and W. Meier (Eds.): FSE 2004, LNCS 3017, pp. 299–316, 2004.

Table 1. Various attacks on reduced-round XTEA

Attack method	paper	Rounds	# of Chosen Plaintexts	Total Complexity
Impossible Diff. attack	[6]	14	$2^{62.5}$	2^{85}
Diff. attack	[3]	15	2^{59}	2^{120}
Truncated Diff. attack	[3]	23	$2^{20.55}$	$2^{120.65}$
R·K Truncated Diff.	this paper	27	$2^{20.5}$	$2^{115.15}$

Table 2. Various attacks on GOST

Attack method	paper	Rounds	# of C·P	Total Complexity
R·K Diff. attack	[4]	24	theoretical	theoretical
A set of Diff. Char.	[8]	13	2^{51}	Not mentioned.
R·K Diff. attack	[8]	21	2^{56}	Not mentioned.
Distinguishing attack	this paper	full	2	2
R·K Diff. attack	this paper	31	2^{26}	2^{39}
R·K Diff. attack	this paper	full	2^{35}	2^{36}

13 rounds of GOST as well as a related key differential attack on 21 rounds of GOST [8].

Here, we present several attacks on GOST. First, we introduce a distinguishing attack on full-round GOST which can distinguish it from a random oracle with probability $1 - 2^{-64}$ using a related key differential characteristic. We also present a related key differential attack on 31 rounds of GOST using our related key differential characteristic combined with H. Seki et al.'s set of differential characteristics. Finally, we describe a related key differential attack on full-round GOST. In this attack, we can recover 12 bits of the master key with 2^{35} chosen plaintexts, 2^{36} encryption operations and an expected success rate of 91.7%.

Table. 1 and 2 depict recent results on XTEA and GOST, respectively.

The following is the outline of this paper. In Section 2, we present the notations used in this paper. In Section 3, we describe an 8-round related key truncated differential characteristic of XTEA and propose related key truncated differential attacks on 25 (1~25) and 27 (4~30) rounds of XTEA. In Section 4, we present a distinguishing attack and a related key differential attack on 31 rounds of GOST and full-round GOST. We conclude in Section 5.

2 Notations

Here, we describe several notations used in this paper. Let $\bullet$, $\oplus$, $\cdot$, $\ll$ and $\gg$ be addition modulo 2^{32}, exclusive-or, multiplication modulo 2^{32} and left and right shift operations, respectively. Let $\bullet$ be left rotation and $\|$ be concatenation of two binary strings. Let e_i be a 32-bit binary string in which the i-th bit is one and the others are zero. Let $A[i]$ be the i-th bit of a 32-bit block A. Let $A[i \sim j]$ denote $A[j] \| A[j-1] \| \cdots \| A[i]$.

3 Related Key Truncated Differential Attacks on XTEA

In this section, we first briefly describe the XTEA algorithm and introduce an 8-round related key truncated differential characteristic of XTEA, which is similar to that of [3] [1]. Then, we show that related key differential cryptanalysis can be applied to attack several reduced-round versions of XTEA using this 8-round related key truncated differential characteristic.

3.1 Description of XTEA

XTEA is a 64-round Feistel block cipher with 64-bit block size and 128-bit key size. Operations used in XTEA are just exclusive-or, additions and shifts. As shown in Fig. 1, XTEA has a very simple round function. Let δ be the constant value $9e3779b9_x$ and $P = (L_n, R_n)$ be the input to the n-th round, for $1 \leq n \leq 64$. Then the output of the n-th round is (L_{n+1}, R_{n+1}), where $L_{n+1} = R_n$ and R_{n+1} is computed as follows :

For each i $(1 \leq i \leq 32)$, if $n = 2i - 1$

$$R_{n+1} = L_n \bullet (((R_n \ll 4 \oplus R_n \gg 5) \bullet R_n) \oplus ((i-1) \cdot \delta \bullet K_{((i-1)\cdot\delta \gg 11) \& 3}),$$

and if $n = 2i$,

$$R_{n+1} = L_n \bullet (((R_n \ll 4 \oplus R_n \gg 5) \bullet R_n) \oplus (i \cdot \delta \bullet K_{(i\cdot\delta \gg 11) \& 3}).$$

XTEA has a very simple key schedule: the 128-bit master key K is split into four 32-bit blocks K_0, K_1, K_2, K_3. Then, for $r = 1, \cdots, 64$, the round keys K_r are derived from the following equation :

$$K_r = \begin{cases} K_{(\frac{r-1}{2}\cdot\delta \gg 11) \& 3} & \text{if } r \text{ is odd} \\ K_{(\frac{r}{2}\cdot\delta \gg 11) \& 3} & \text{if } r \text{ is even} \end{cases}$$

Table. 3 depicts the entire key schedule.

[1] There is an explicit separation between our truncated differential characteristic and that of [3]. We use the internal difference caused by the specific related key and plaintext pair, whereas S. Hong et al. [3] used the difference resulting from the plaintext pair only. So, we call this characteristic related key truncated differential characteristic in this paper.

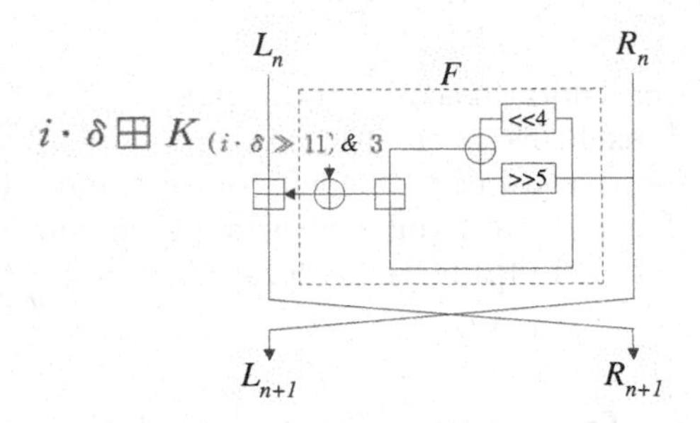

Fig. 1. $2i$-th round of XTEA

3.2 8-round Related Key Truncated Differential Characteristic

In [3], S. Hong et al. suggested an 8-round truncated differential characteristic in order to attack 23 rounds of XTEA (8~30 or 30~52). Here, we construct a similar 8-round related key truncated differential characteristic. See Fig. 2. Let Ψ be our 8-round related key truncated differential characteristic described in Fig. 2. Let γ be 0^{32} or e_{30} and RK_i be the round key of the i-th round. We consider identical input values (zero difference) to the i-th round and a related round key pair RK_i and $RK_i' = RK_i \oplus e_{30}$. Then, the value of the 30-th bit of the right output difference in the i-th round is always one, and the other bits are all zero (except for the 31-st bit). Note that the 31-st bit is unknown. (However, we do not need to consider this value). That is, the output difference of the i-th round is $(0, e_{30})$ or $(0, e_{31} \oplus e_{30})$ with probability 1. As shown in Fig. 2, there are three possible colors for every bit: white, black, and gray. Every white bit denotes a zero difference. The bit which we focus on is the black bit. Note that

Table 3. Key schedule of XTEA

Round	1	2	3	4	5	6	7	8	9	10	11	12	13	14	15	16
Key	K_0	K_3	K_1	K_2	K_2	K_1	K_3	K_0	K_0	K_0	K_1	K_3	K_2	K_2	K_3	K_1
Round	17	18	19	20	21	22	23	24	25	26	27	28	29	30	31	32
Key	K_0	K_0	K_1	K_0	K_2	K_3	K_3	K_2	K_0	K_1	K_1	K_1	K_2	K_0	K_3	K_3
Round	33	34	35	36	37	38	39	40	41	42	43	44	45	46	47	48
Key	K_0	K_2	K_1	K_1	K_2	K_1	K_3	K_0	K_0	K_3	K_1	K_2	K_2	K_1	K_3	K_1
Round	49	50	51	52	53	54	55	56	57	58	59	60	61	62	63	64
Key	K_0	K_0	K_1	K_3	K_2	K_2	K_3	K_2	K_0	K_1	K_1	K_0	K_2	K_3	K_3	K_2

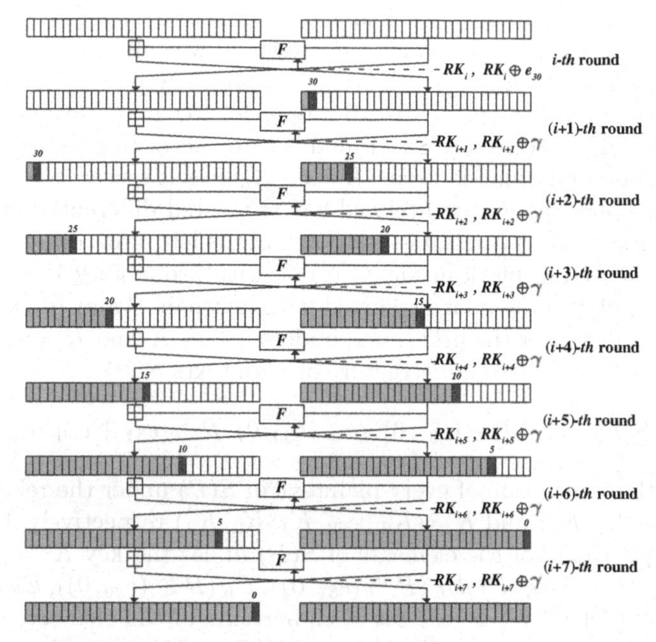

Fig. 2. 8-round related key truncated differential characteristic Ψ ($\gamma = 0^{32}$ or e_{30})

the value of the black bit does not change throughout Ψ, while its position is shifted up to 5 bits to the right per round. And the values of the gray bits are irrelevant. That is, if for each j ($i+1 \le j \le i+7$) the relation between RK_j and RK'_j is $RK'_j = RK_j \oplus \gamma$ ($\gamma = 0^{32}$ or e_{30}), then by the property of the round function F of XTEA, the black bit will be located at bit position 0 in the left output difference with probability 1, after $(i+7)$ rounds.

3.3 Related Key Truncated Differential Attacks on XTEA

Using the above 8-round related key truncated differential characteristic, we can attack 25 (1~25) and 27 (4~30) rounds of XTEA. Here, we apply conventional related key differential cryptanalysis as described in [4]. We exploit the property that if there exists a related key pair (K, K') such that a non-zero input difference of two plaintexts can be changed into a zero output difference, then we can bypass several rounds for free in our attack.

Attack on 25 (1~25) Rounds of XTEA. We consider the related key pair $K = (K_0, K_1, K_2, K_3)$ and $K' = (K_0 \oplus e_{30}, K_1, K_2, K_3)$ in order to attack 25 (1~25) rounds of XTEA. Then, according to the key schedule of XTEA, K_0 and $K_0 \oplus e_{30}$ are used in the first round. Assume that there exist plaintext-ciphertext pairs, (P, C) and (P', C') respectively encrypted under the master keys K and K', such that the first round output value of the two plaintexts P

and P' under the round keys K_0 and $K_0 \oplus e_{30}$ are the same, i.e. such that the output difference of the first round is zero. (Here, we call the first 32-bit blocks of K and K', (K_0 and $K_0 \oplus e_{30}$) 'the related round key pair'). Then, due to the key schedule of XTEA, we can bypass 6 rounds for free in our attack. This means that the input difference to the 8-th round is zero. According to the key schedule of XTEA, the related round key pair, K_0 and $K_0 \oplus e_{30}$ is reused in the 8-th round. Now, we can apply the 8-round related key truncated differential characteristic described in Fig. 2. As a result, bit position 0 of the left input difference to round 16, which is colored in black in Fig. 3, is one with probability 1.

In order to obtain the above assumed plaintext pair P and P', which has the same output value after the first round under the key K and K' respectively, we consider the following 1-round structure of plaintexts $S(P)$.

$$S(P) = \{P, P \oplus (e_{31}, 0), P \oplus (e_{30}, 0), P \oplus (e_{31} \oplus e_{30}, 0)\}$$

We request the encryption of every plaintext in $S(P)$ under the related key pair $K = (K_0, K_1, K_2, K_3)$ and $K' = (K_0 \oplus e_{30}, K_1, K_2, K_3)$ respectively. Let $C(P)$ be the set of ciphertexts of the elements of $S(P)$ under the key $K = (K_0, K_1, K_2, K_3)$, i.e. $C(P) = \{E_K(P), E_K(P \oplus (e_{31}, 0)), E_K(P \oplus (e_{30}, 0)), E_K(P \oplus (e_{31} \oplus e_{30}, 0)), \}$. And let $C'(P)$ be the set of ciphertexts of the elements of $S(P)$ under the key $K' = (K_0 \oplus e_{30}, K_1, K_2, K_3)$, i.e. $C'(P) = \{E_{K'}(P), E_{K'}(P \oplus (e_{31}, 0)), E_{K'}(P \oplus (e_{30}, 0)), E_{K'}(P \oplus (e_{31} \oplus e_{30}, 0)), \}$. Then, it is easy to see that there are exactly four plaintext pairs that have the same output value after the first round, i.e. we can obtain the required zero output difference after the first round. We denote these four plaintext pairs as (P_u, P'_u) where $1 \leq u \leq 4$. We also denote $(C_u = E_K(P_u), C'_u = E_{K'}(P'_u))$ as the ciphertexts corresponding to plaintext P_u under key K and plaintext P'_u under key K', respectively.

Now we use the above property of the black bit located in bit position 0 of the left input difference in round 16, in order to attack 25 (1~25) rounds of XTEA and recover 111 bits of the subkey derived from master key K.

Algorithm **??** describes how to recover 111 bits of subkey material from the ciphertexts. We compute the difference of every dotted bit position and the values of the bit pair of every gray bit position in order to get the black bit of the left half of the input difference in round 16. (See Fig. 3). In detail, in order to compute the black bit of the input difference of round 16, ($L_{16}[0]$), we need to know the differences of $R_{17}[0]$, $L_{17}[0]$, and $L_{17}[5]$, respectively. Also, in order to know the differences of $R_{17}[0]$, $L_{17}[0]$, and $L_{17}[5]$ we need to know the differences of $L_{18}[10]$ and $R_{18}[5]$. (For the knowledge of these differences, we need $L_{18}[0 \sim 10]$, $R_{18}[0 \sim 5]$, and $K_0[0 \sim 4]$.). Due to the structure of the round function of XTEA, the key and output bit positions related to the black bit increase by 5 bits per round. Consequently, if we guess all the bits of K_0, K_2, K_3, and 15 bits of K_1, and we also get the output pair after the 25-th round, we can compute the black bit of the input difference in round 16.

In *Algorithm* **??**, σ denotes the function which outputs the one-bit difference in $L_{16}[0]$ using a given ciphertext pair and the guessed key bits. Let $\mathbf{K}$ be the concatenation of $K_1[0 \sim 14]$, K_2, and K_3, i.e., $\mathbf{K} = K_1[0 \sim 14] || K_2 || K_3$. Note that

$\mathbf{K}$ is a 79-bit string. In *Algorithm* 1, we guess K_0 and $\mathbf{K}$. Using this algorithm, we are able to find 111 bits of $K = (K_0, K_1, K_2, K_3)$.

Input : m structures : $S(P^1), S(P^2), \cdots, S(P^m)$,
corresponding m pairs of ciphertexts :
$(C(P^1), C'(P^1)), \cdots, (C(P^m), C'(P^m))$
Output : 111-bit partial key value of $K = (K_0, K_1, K_2, K_3)$

---

1. For $K_0 = 0, 1, \ldots, 2^{32} - 1$
   1.1. For $i = 1, \ldots, m$
      1.1.1 Find the four plaintext pairs $(P_u^i, P_u^{i\prime})$, $1 \le u \le 4$, such that for each u, the following two conditions hold:
      (a) $P_u^i = (LP^i \oplus v)||RP^i$ and $P_u^{i\prime} = (LP^i \oplus w)||RP^i$ for some $v, w \in \{0, e_{31}, e_{30}, (e_{31} \oplus e_{30})\}$.
      (b) $[(LP^i \oplus v) \boxplus F(RP^i, K_0)] \oplus [(LP^i \oplus w) \boxplus F(RP^i, K_0 \oplus e_{30})] = 0$
      // Here, we use the notation LP^i for the left half and RP^i for the right half of the plaintext P^i.//
   1.2. For $\mathbf{K} = 0, 1, \ldots, 2^{79} - 1$
      1.2.1. For $i = 1, \ldots, m$
         1.2.1.1 For $u = 1, \ldots, 4$
            Compute $\sigma_u^i = \sigma(C_u^i, C_u^{i\prime}, K_0, \mathbf{K})$
            If $i = m$, $u = 4$, and $\sigma_u^i = 1$, then output K_0, $\mathbf{K}$ and stop.
            Else if $\sigma_u^i = 0$, goto 1.2.

Algorithm 1. Related key truncated differential attack on 25 rounds of XTEA

The output of the algorithm is the right value of some 111 bits of $K = (K_0, K_1, K_2, K_3)$ with high probability if m is sufficiently large. For each i ($0 \le i \le 2^{111} - 1$), the probability that the attack algorithm outputs the i-th key-candidate is $(1 - 2^{-4m})^i$. So, the average success rate of this attack is

$$2^{-111} \sum_{i=0}^{2^{111}-1} (1 - 2^{-4m})^i = 2^{4m-111}(1 - (1 - 2^{-4m})^{2^{111}}) \approx 2^{4m-111}(1 - e^{-2^{111-4m}}).$$

Let k be a key candidate ($0 \le k \le 2^k - 1$). For m structures of plaintexts, the expected number of trials, required until each k is determined as a wrong value, is $1 + 2^{-1} + 2^{-2} + \cdots + 2^{-4m+1} = 2 - 2^{-4m+1}$. If a key k is right, then the number of trials is exactly $4m$. Thus, the average number of trials in the attack is

$$2^{-k} \sum_{i=0}^{2^k-1} i \cdot (2 - 2^{-4m+1}) + 4m = 4m + (1 - 2^{-4m})(2^k - 1).$$

So, if we get ≈ 29 plaintext structures, the attack on 25 rounds of XTEA will succeed on average with probability 96.9%. This success rate implies that our

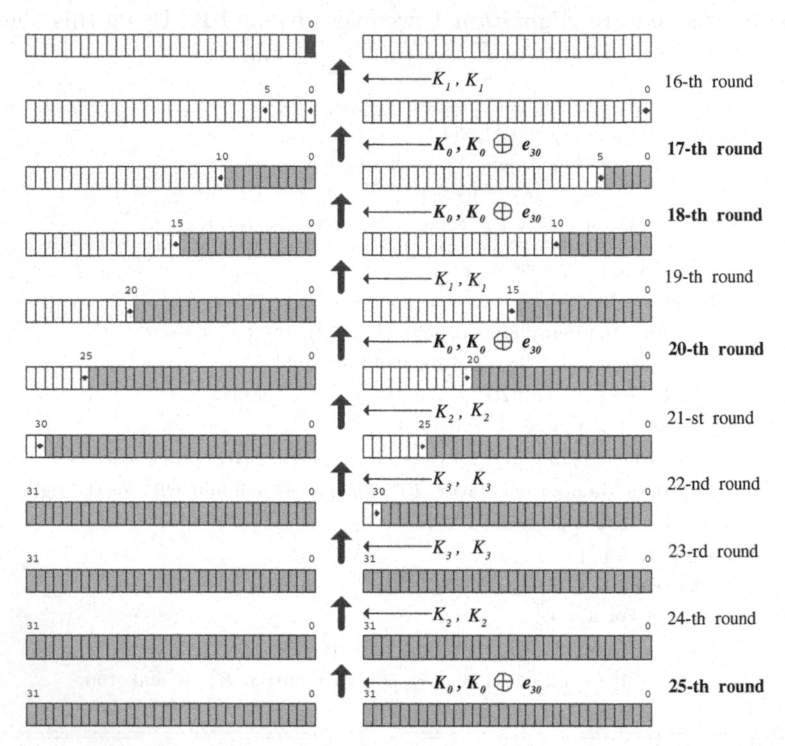

Fig. 3. Related key Truncated Differential Attack on 25 rounds of XTEA

attack reduces almost all key spaces efficiently. It has a data complexity of $29 \cdot 4 = 116$ chosen-plaintexts and time complexity of $(116+(1-2^{-116})(2^{111}-1)) \cdot \frac{6.5}{25} \cdot 2 \approx 2^{110.05}$ 25-round XTEA encryptions.

Attack on 27 (4~30) Rounds of XTEA. In the key schedule of XTEA, K_3 is not used from the 24-th round until the 30-th round. This means that we may expand more rounds for free. Using this observation, we can attack 27 (4~30) rounds of XTEA. This attack only differs from the attack on 25 (1~25) rounds of XTEA in two aspects. One is the use of the related key pair $K = (K_0, K_1, K_2, K_3)$ and $K' = (K_0, K_1 \oplus e_{30}, K_2, K_3)$. The other is the use of 2-round plaintext structures, $S'(P)$.

First, we describe what we mean by a 2-round plaintext structure. Let P be a plaintext and A be the set of all 32-bit values whose lower 22 bits are fixed to $10 \cdots 0$. We define the 2-round structure of plaintexts $S'(P)$ as follows :

$$S'(P) = \{P\} \cup \{P \oplus (w, v) | w \in A, v \in \triangle X\},$$

Table 4. Various attacks on 27 rounds of XTEA

variant rounds	Key Bits	relation of keys
13-th $\sim$ 39-th	K_1, K_2, K_3 : 32 bits, respectively, K_0 : 15 bits	$K \oplus K' = (0, 0, 0, e_{30})$
17-th $\sim$ 43-rd	K_0, K_1, K_3 : 32 bits, respectively, K_2 : 15 bits	$K \oplus K' = (0, e_{30}, 0, 0)$
22-nd $\sim$ 48-th	K_1, K_2, K_3 : 32 bits, respectively, K_0 : 20 bits	$K \oplus K' = (0, 0, e_{30}, 0)$
31-st $\sim$ 57-th	K_0, K_2, K_3 : 32 bits, respectively, K_1 : 15 bits	$K \oplus K' = (e_{30}, 0, 0, 0)$
35-th $\sim$ 61-st	K_0, K_1, K_2 : 32 bits, respectively, K_3 : 15 bits	$K \oplus K' = (0, 0, e_{30}, 0)$

where $\triangle X$ is the following set :

$$\begin{aligned}\triangle X = \{&01000010\cdots0, 01000110\cdots0, 01001110\cdots0,\\ &01011110\cdots0, 01111110\cdots0, 00111110\cdots0,\\ &11000010\cdots0, 11000110\cdots0, 11001110\cdots0,\\ &11011110\cdots0, 11111110\cdots0, 10111110\cdots0\}\end{aligned}$$

Note that $S'(P)$ contains 12,289 chosen-plaintexts and there are 12,288 plaintext pairs of the form $(P, P\oplus(w, v))$ where $w \in A$ and $v \in \triangle X$. We consider encryptions of these plaintexts under the keys $K = (K_0, K_1, K_2, K_3)$ and $K' = (K_0, K_1 \oplus e_{30}, K_2, K_3)$, respectively. Then, for every subkey K_2, there exist (w_1, v_1), (w_2, v_2) and (w_3, v_3) such that the second round output differences of $(P, P\oplus(w_1, v_1))$, $(P, P\oplus(w_2, v_2))$, $(P, P\oplus(w_3, v_3))$ are respectively $(e_{30}, 0)$, $(e_{31}, 0)$ and $(e_{30} \oplus e_{31}, 0)$. Note that this attack is starting from the 4-th round. That is, the 4-th and 5-th round correspond to the first and second round, respectively. This means that there exist four plaintexts in $S'(P)$ which have the same property as the elements of the 1-round structure $S(P)$ described in the attack on 25 rounds of XTEA. Furthermore, in the key schedule, the related round key pair K_1 and $K_1 \oplus e_{30}$ is first used in the 6-th round and then again in the 11-th round. So we can again get the same output values after the third round (6-th round) of encryption, i.e. the output difference after the third round is zero. Thus, we can bypass 5 rounds for free. Then, Ψ is applied from the eighth round (11-th round) through the fifteenth round (18-th round). Therefore, with similar methods as for the attack on 25 (1$\sim$25) rounds of XTEA, we can recover 116 bits of the master key K (K_0, K_1, K_2, and $K_3[0\sim19]$). Overall, we use 121 structures to attack 27 (4$\sim$30) rounds with an expected success rate of 96.9%. This requires $(121 * 12289 = 1486949) \approx 2^{20.5}$ chosen-plaintexts and $(121 + (1 - 2^{-121})(2^{116} - 1) \cdot \frac{7.5}{27} \cdot 2 \approx 2^{115.15}$ 27-round XTEA encryptions.

In addition, with similar methods, various attacks on 27 rounds of XTEA are possible. Table 4. depicts these attacks. 'Key Bits' denotes the total number of bits in K_0, K_1, K_2, and K_3 recovered by the attack.

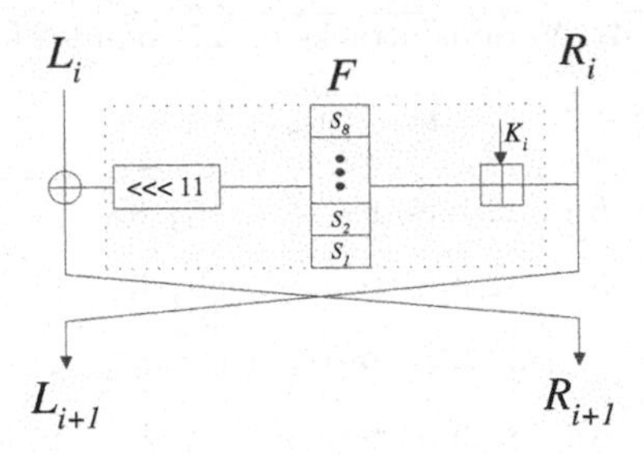

Fig. 4. i-th round of GOST

4 Related Key Differential Attacks on GOST

In this section, we describe the specification of GOST and briefly introduce H. Seki et al.'s differential cryptanalysis of a reduced-round version of it [8]. Next, we show that we can distinguish full-round GOST from a random oracle with probability $1 - 2^{-64}$ using a related key differential characteristic and also present a related key differential attack on 31 rounds of GOST. Finally, we propose a related key differential attack on full-round GOST.

4.1 Description of GOST and Previous Work

GOST is a 32-round Feistel block cipher with 64-bit block size and 256-bit key size. It iterates a simple round function F composed of key additions, eight different 4×4 S-boxes S_i $(1 \leq i \leq 8)$ and cyclic rotations. See Fig. 4.

The key schedule of GOST is very simple. The 256-bit master key K is split into eight 32-bit blocks $K_1, \cdots, K_8$, i.e. $K = (K_1, \cdots, K_8)$ and each round uses one of them as shown in Table 5.

Due to the subkey addition operation in the round function, the differential properties of GOST vary not only with the values of the input and output differences, but also with the value of the subkey itself. In order to minimize the dependence of the differential probability on the key, H. Seki et al. introduced the idea of using a set of differential characteristics [8]. They use two differential sets $\Delta = \{0abc\}$ and $\nabla = \{abc0\}$ where $a, b, c \in \{0, 1\}$, which respectively represent nonzero 4-bit input and output differences of an S-box. In addition,

Table 5. Key schedule of GOST

Round	$1 \ldots 8$	$9 \ldots 16$	$17 \ldots 24$	$25 \ldots 32$
Key	$K_1 \ldots K_8$	$K_1 \ldots K_8$	$K_1 \ldots K_8$	$K_8 \ldots K_1$

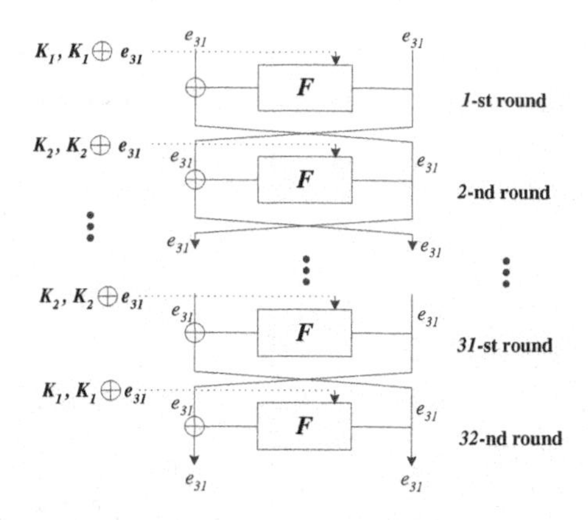

Fig. 5. 32-round related key differential characteristic of GOST

they computed the following average probability of differentials for each S-box represented by p_{S_i}.

$$p_{S_i} \ : \ Prob\{\Delta \stackrel{S_i}{\rightarrow} \nabla\}$$

The value of $Prob\{\Delta \stackrel{S_i}{\rightarrow} \nabla\}$ varies from 0.30 to 0.75 depending both on the S-box S_i and on the key value. See Table. 6 (for more details, refer to [8]). Using this set of characteristics, they attack 13 rounds of GOST and also present a combined related key attack on 21 rounds of GOST.

4.2 Related Key Differential Attacks on GOST

Now we present several attacks on GOST.

Distinguishing Attack. We can distinguish full-round GOST from a truly random permutation with probability $1-2^{-64}$ using a related key differential characteristic. Here, we consider an attacker that has two oracles $\mathcal{O}$ and $\mathcal{O}'$. $\mathcal{O}$ is the oracle which, given a plaintext P, outputs a ciphertext $E_K(P)$ under key $K = (K_1, \cdots, K_8)$. $\mathcal{O}'$ is the oracle which, given a plaintext P', outputs a ciphertext $E_{K'}(P')$ under key $K' = (K_1 \oplus e_{31}, K_2 \oplus e_{31}, \cdots, K_8 \oplus e_{31})$. Note that the key $K = (K_1, \cdots, K_8)$ is unknown to the attacker. However he knows the relation $K \oplus K' = (e_{31}, \cdots, e_{31})$.

Let us first consider the function E as GOST. In this case, if we query $\mathcal{O}$ for $P = (P_L, P_R)$, and $\mathcal{O}'$ for $P' = (P_L \oplus e_{31}, P_R \oplus e_{31})$ respectively, and obtain the corresponding ciphertexts C and C', then the output difference $C \oplus C'$ of full-round GOST is always (e_{31}, e_{31}). More specifically, it is easy to see that for every round, the input difference of each S-box after key addition is zero with probability 1. Therefore the difference between the plaintexts, (e_{31}, e_{31}) is maintained after every round. In other words, this is a 32-round related key differential characteristic with probability 1 (See Fig. 5).

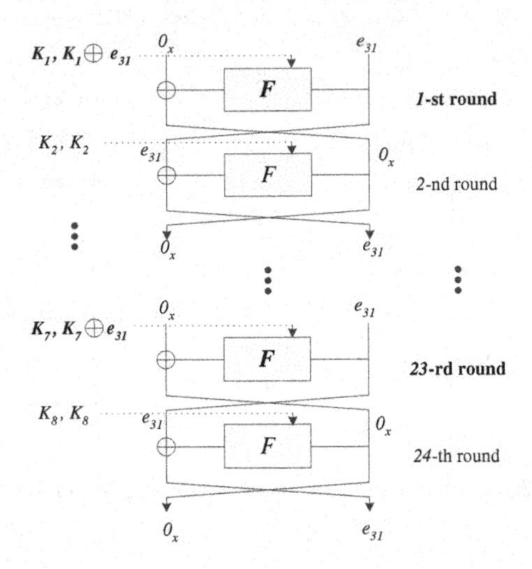

Fig. 6. 24-round related key differential characteristic (1∼24) with probability 1

If we consider the function E as a truly random permutation, then the output pair of the truly random permutation is unpredictable so that we can't obtain any information from it. So we can successfully distinguish full-round GOST from a truly random permutation with high probability, namely $1-2^{-64}$. Note that this distinguishing attack is possible with only two chosen plaintexts under the given key relation.

Related Key Differential Attack on 31 Rounds of GOST. We use the differential probability for each S-box presented in [8], as their differential characteristic enables us to mount a related key differential attack on 31 rounds of GOST. For this attack we consider the following two related keys K and K'.

$$K = (K_1, K_2, K_3, K_4, K_5, K_6, K_7, K_8)$$
$$K' = (K_1 \oplus e_{31}, K_2, K_3 \oplus e_{31}, K_4, K_5 \oplus e_{31}, K_6, K_7 \oplus e_{31}, K_8)$$

We request the encryption of $P = (P_L, P_R)$ under key K and of $P' = (P_L, P_R \oplus e_{31})$ under key K'. Then we obtain a 24-round related key differential characteristic with probability 1. See Fig. 6. With this 24-round related key differential characteristic, we can bypass 24 rounds for free in our attack. As shown in Fig. 6., the output difference of the 24-th round is $(0, e_{31})$, i.e. the input difference of the 25-th round is $(0, e_{31})$. We use the set of differential characteristics [8] mentioned in Section 4.1 in order to construct another 6-round related key differential characteristic from the 25-th round through the 30-th round. See Fig. 7. In this figure, # denotes an element of the differential set $\Delta = \{0abc\}$, where $a, b, c \in \{0, 1\}$. In the 25-th round, the input difference of

Table 6. Average differential probability of each S-box

ps_1	ps_2	ps_3	ps_4	ps_5	ps_6	ps_7	ps_8
0.43	0.38	0.37	0.37	0.37	0.35	0.47	0.45

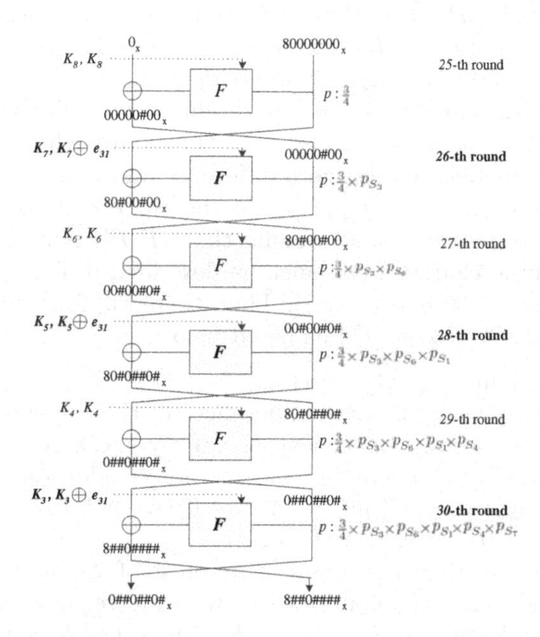

Fig. 7. 6-round related key differential characteristic (rounds 25∼30)

round function F, 80000000_x becomes $00000\#00_x$ with probability $\frac{3}{4}$. [2] After the 25-th round, each average related key differential probability is computed using Table. 6 [8]. Thus, combining these two related key differential characteristics, we construct a 30-round related key differential characteristic from the 1-st round to the 30th-round of GOST with probability about $2^{-23.33}$.

Now, we consider a $1R$ [1] related key differential attack on 31 rounds of GOST using the above constructed 30-round related key differential characteristic. Considering 2^{26} chosen plaintext pairs, there remain about 2^{12} ciphertext pairs after the filtering step. Among them, we expect that there exist at least 5 right pairs. A wrong key is counted with probability 2^{-17} by the above 30-round related key differential characteristic. The signal-to-noise ratio S_N [1] of this related key differential characteristic is about $2^{22.67}$. Thus, according to [9], we can

[2] We only need to compute the probability $Prob\{1000 \xrightarrow{S_8} \nabla = \{abc0\}\}$ because of the structure of the round function F. This probability is easily checked by simulation and also represented in [8].

recover the 32 bits of the 31-st round subkey with about 2^{26} chosen plaintexts and time complexity $(2^{32} \times 2^{26} \times 2^{-14} \times \frac{1}{31}) \approx 2^{39}$ with an expected success rate of 97.9%.

Full Rounds Attack on GOST. In this section, we suggest an algorithm to find 12 bits of K_1 with high probability of success.

Consider $P = (P_L, P_R)$ and $P' = (P_L \oplus e_{30}, P_R \oplus e_{30})$ encrypted under keys $K = (K_1, \cdots, K_8)$ and $K' = (K_1 \oplus e_{30}, K_2 \oplus e_{30}, \cdots, K_8 \oplus e_{30})$ respectively. Then, after key addition, in each round the input difference becomes 0 with probability 2^{-1}. Thus, we can construct a 30-round related key differential characteristic with probability 2^{-30} as shown in Fig. 8. In this figure, white bits denote a zero difference, black bits denote a nonzero difference and gray bits are unknown.

Let $C = (C_L, C_R)$, $C' = (C'_L, C'_R)$ be the ciphertexts of P and P' under keys K and K', respectively and assume that (P, P') is a right pair for a related key differential characteristic such as described in Fig. 8. (i.e. the output difference after round 30 is (e_{30}, e_{30})). Then there are four types of differential characteristics $C1, C2, C3$ and $C4$ as listed below.

C1. $C_R \oplus C'_R = e_{30}$ and $C_L \oplus C'_L = e_{30}$.
This case means that the input differences of the S-boxes in round 31 and round 32 are 0, so we can recover $K_1[30]$ by checking $C_R[30] + K_1[30] = C'_R[30] + (K_1[30] \oplus 1)$ where "+" means integer addition.

C2. $C_R \oplus C'_R = e_{30}$, $(C_L \oplus C'_L)[0 \sim 6] = 0, (C_L \oplus C'_L)[11 \sim 29] = 0$ and $(C_L \oplus C'_L)[31] = 0$. (Refer to Fig. 8*a*.)
This case means that the input difference of S_8 in round 31 is zero, but in round 32 it is nonzero, so we can recover $K_1[30]$ by checking $C_R[30] + K_1[30] \neq C'_R[30] + (K_1[30] \oplus 1)$. Also if such a pair is given, $K_1[28], K_1[29], K_1[31]$ can be recovered by checking

$S8(C_R[28 \sim 31] + K_1[28 \sim 31]) \oplus C_L[7 \sim 10] = S8(C'_R[28 \sim 31] + K'_1[28 \sim 31]) \oplus C'_L[7 \sim 10]$ or

$S8(C_R[28 \sim 31] + 1 + K_1[28 \sim 31]) \oplus C_L[7 \sim 10] = S8(C'_R[28 \sim 31] + 1 + K'_1[28 \sim 31]) \oplus C'_L[7 \sim 10]$.

If we add $C_R[0 \sim 27]$ to $K_1[0 \sim 27]$, a carry may occur at the 27-th bit position, so we need to check the above two equations. We denote $S8(C_R[28 \sim 31] + K_1[28 \sim 31]) \oplus C_L[7 \sim 10]$ by $F_{k_j}(C_R[28 \sim 31]) \oplus C_L[7 \sim 10]$ in *Algorithm* 2.

C3. $C_R \oplus C'_R \neq e_{30}$, $(C_R \oplus C'_R)[7] = 0$
This case means that the input difference of S_8 in round 31 is nonzero and $(C_R \oplus C'_R)[7] = 0$. In this case, if we know $K_1[0 \sim 11]$, we can compute the exact value of $(F_{K_1}(C_R))[11 \sim 22]$ and $(F_{K_1}(C'_R))[11 \sim 22]$. So $K_1[8 \sim 11]$ can be recovered by checking

$(F_{K_1[0\sim11]}(C_R))[19 \sim 22] \oplus C_L[19 \sim 22] = (F_{K'_1[0\sim11]}(C'_R))[19 \sim 22] \oplus C'_L[19 \sim 22]$.

Note that $(F_{K_1[0\sim7]}(C_R))[11 \sim 18] \oplus C_L[11 \sim 18] = (F_{K'_1[0\sim7]}(C'_R))[11 \sim$

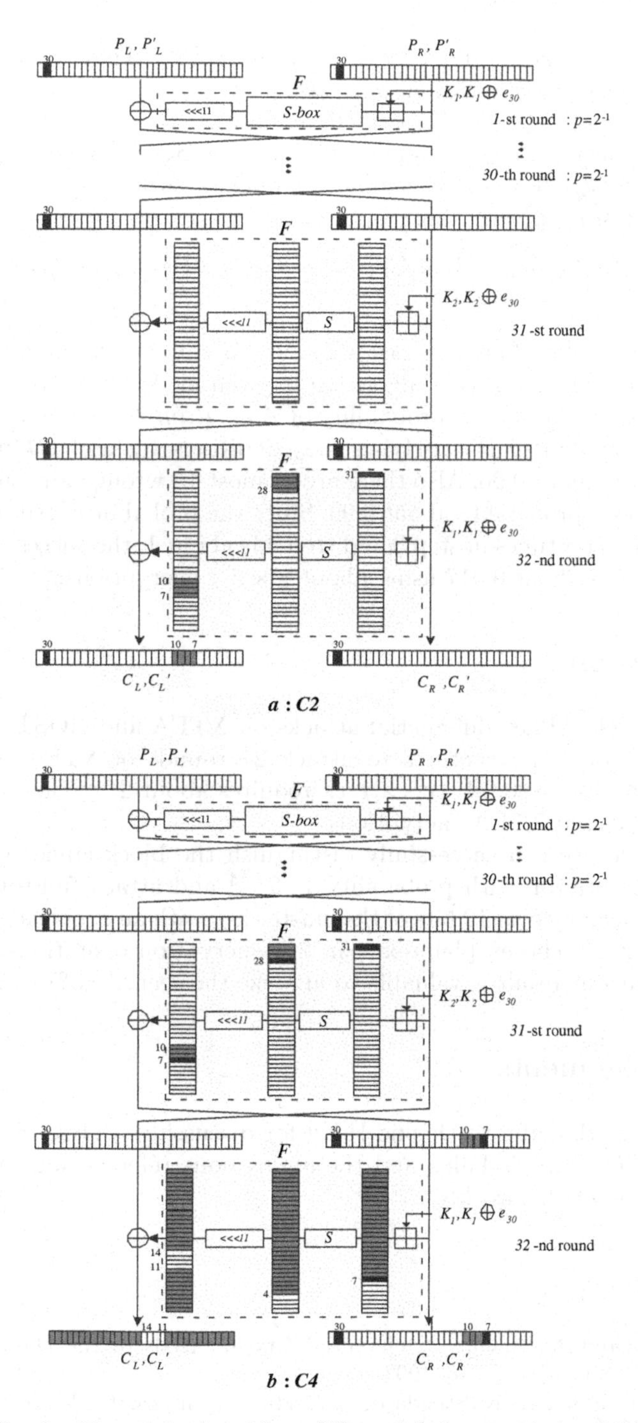

Fig. 8. 32-round related key differential characteristic of GOST

18] $\oplus C'_L[11 \sim 18]$ for any arbitrary candidate key $K_1[0 \sim 7]$, so we cannot find the right value $K_1[0 \sim 7]$ with high probability. That is the reason why we only consider recovering $K_1[8 \sim 11]$.

C4. $C_R \oplus C'_R \neq e_{30}$, $(C_R \oplus C'_R)[7] \neq 0$. (Refer to Fig. 8b.)
This case means that the input difference of S_8 in round 31 is nonzero and $(C_R \oplus C'_R)[7] \neq 0$. By similar arguments as for case *C3* we can recover $K_1[4 \sim 11]$ by checking

$$(F_{K_1[0\sim 11]}(C_R))[15 \sim 22] \oplus C_L[15 \sim 22] = (F_{K'_1[0\sim 11]}(C'_R))[15 \sim 22] \oplus C'_L[15 \sim 22].$$

Note that the key bits found in case *C1* and *C3* can also be found in case *C2* and *C4*, respectively. An attack algorithm is given in *Appendix A*.

Let us consider the success probability of *Algorithm* 2. If we choose 2^{35} pairs, there exist at least 4 pairs satisfying the conditions *C2* and *C4* respectively, with probability about 0.96. Also there are at most 15 wrong pairs surviving the filtering step with probability about 0.99. Since the probability that a wrong key is counted at most 3 times in step 3 and step 4 is about 1, the success probability of *Algorithm* 2 is about 0.917 using about 2×2^{35} encryptions.

5 Conclusion

We presented related key differential attacks on XTEA and GOST. In the case of XTEA, we use 121 structures to attack 27 rounds of XTEA with an expected success rate of 96.9%; this attack requires about $2^{20.5}$ chosen-plaintexts and $2^{115.15}$ 27-round XTEA encryptions.

Furthermore, we can successfully distinguish the block cipher GOST from a random permutation with probability $1-2^{-64}$ and attack full-round GOST. As a result, we can recover 12 bits of the master key with an expected success rate of 91.7% using 2^{35} chosen plaintexts, in 2^{36} encryption operations. Therefore, we believe that our result is valuable to analyze the security of GOST.

Acknowledgements

The authors are thankful to Deukjo Hong for discussing XTEA and very much appreciate Helena Handschuh's and the anonymous referees' aid in improving the presentation of this work.

References

[1] E. Biham and A. Shamir, "Differential Cryptanalysis of the Data Encryption Standard", Springer-Verlag, 1993. 311

[2] GOST, Gosudarstvennyi Standard 28147-89, "Cryptographic Protection for Data Processing Systems", Government Committee of the USSR for Standards, 1989. 299

[3] S. Hong, D. Hong, Y. Ko, D. Chang, W. Lee, and S. Lee, "Differential Cryptanalysis of TEA and XTEA", *Pre-Proceedings of the 6th Annual International Conference on Information Security and Cryptology (ICISC '03)*, Lecture Notes in Computer Science. Springer-Verlag, 2003. pp. 413-428. 299, 300, 301, 302

[4] J. Kelsey, B. Schneier, and D. Wagner, "Key Schedule Cryptanalysis of IDEA, G-DES, GOST, SAFER, and Triple-DES", *Advances in Cryptology - CRYPTO '96*, volume 1109 of *Lecture Notes of Computer Science*, Springer-Verlag, 1996, pp. 237-251. 300, 303

[5] J. Kelsey, B. Schneier, and D. Wagner, "Related-Key Cryptanalysis of 3-WAY, Biham-DES, CAST, DES-X, NewDES, RC2, and TEA", *Proceedings of International Conference on Information and Communications Security (ICICS '97)*, volume 1334 of *Lecture Notes of Computer Science*, Springer-Verlag, 1997, pp. 233-246. 299

[6] D. Moon, K. Hwang, W. Lee, S. Lee, and J. Lim, "Impossible Differential Cryptanalysis of Reduced Round XTEA and TEA", *Fast Software Encryption '02*, volume 2365 of *Lecture Notes of Computer Science*, Springer-Verlag, 2002, pp. 49-60. 300

[7] R. Needham and D. Wheeler, "eXtended Tiny Encryption Algorithm", October, 1997. 299

[8] H. Seki and T. Kaneko, "Differential Cryptanalysis of Reduced Rounds of GOST", *Seventh Annual Workshop on Selected Areas in Cryptography (SAC '00)*, volume 2012 of *Lecture Notes of Computer Science*, Springer-Verlag, 2001, pp. 315-323. 300, 308, 309, 310, 311

[9] A. Selçuk and A. Biçak "On Probability of Success in Linear and Differential Cryptanalysis", *Third International Conference, SCN 2002*, volume 2365 of *Lecture Notes of Computer Science*, Springer-Verlag, 2002, pp. 174-185. 311

[10] D. Wheeler and R. Needham, "TEA, a Tiny Encryption Algorithm", *Fast Software Encryption, Second International Workshop Proceedings*, volume 1008 of *Lecture Notes of Computer Science*, Springer-Verlag, 1995, pp. 97-110. 299

6 Appendix A

Assumption : The attacker knows that $K \oplus K'$ is equal to $(e_{30}, e_{30}, \cdots, e_{30})$
Input : (P_i, P_i'), $(i = 1, \cdots, 2^{35})$ where $P_i \oplus P_i' = (e_{30}, e_{30})$ as in Fig. 8
Output: 12-bit partial key K_1; $K_1[4 \sim 11]$ and $K_1[28 \sim 31]$

// Setup stage //
- Let$\mathcal{K} = \{k_1, k_2, \cdots, k_{24}\}$ and $\mathcal{K}' = \{k_1', k_2', \cdots, k_{28}'\}$ be the set of candidate keys for $K_1[28 \sim 31]$ and $K_1[4 \sim 11]$, respectively
- $\mathcal{D}, \mathcal{D}'$: empty set
- $ctr_1 = 0, \cdots, ctr_{24} = 0, ctr_1' = 0, \cdots, ctr_{28}' = 0$

// Filtering step //
1. For $i = 1, \ldots, 2^{35}$
   1.1. Request the ciphertexts $C_i = E_K(P_i)$ and $C_i' = E_{K'}(P_i)$
   If $C_i \oplus C_i'$ satisfies condition C2, $\mathcal{D} = \mathcal{D} \cup \{(C_i, C_i')\}$
   If $C_i \oplus C_i'$ satisfies condition C4, $\mathcal{D}' = \mathcal{D}' \cup \{(C_i, C_i')\}$

// Finding key $K_1[28 \sim 31]$ //
2. For each $(C_i, C_i') \in \mathcal{D}$
   /* For convenience let $C_i = (C_L, C_R)$ and $C_i' = (C_L', C_R')$ */
   2.1. For $j = 1, \ldots, 2^4$
   If $F_{k_j}(C_R[28 \sim 31]) \oplus C_L[7 \sim 10] = F_{k_j}(C_R'[28 \sim 31]) \oplus C_L'[7 \sim 10]$,
   $ctr_j += 1$
   If $F_{k_j}(C_R[28 \sim 31] + 1) \oplus C_L[7 \sim 10] = F_{k_j}(C_R'[28 \sim 31] + 1) \oplus C_L'[7 \sim 10]$,
   $ctr_j += 1$
   If $ctr_j \geq 4$, output k_j as $K_1[28 \sim 31]$ and goto 3

// Finding key $K_1[4 \sim 11]$ //
3. For each $(C_m, C_m') \in \mathcal{D}'$
   /* For convenience let $C_i = (C_L, C_R)$ and $C_i' = (C_L', C_R')$ */
   3.1. For $j = 1, \ldots, 2^8$
   3.2. For $i = 0, \ldots, 2^4 - 1$
   If $F_{k_j'||i}(C_R)[15 \sim 22] \oplus C_L[15 \sim 22] = F_{k_j'||i}(C_R')[15 \sim 22] \oplus C_L'[15 \sim 22]$,
   $ctr_j' += 1$
   /* Since k_j' denotes $K_1[4 \sim 11]$ and i denotes $K_1[0 \sim 3]$, $k_j'||i$ denotes $K_1[0 \sim 11]$ */
   3.3. If $ctr_j' \geq 4$, output k_j' as $K_1[4 \sim 11]$ and terminate this algorithm

---

Algorithm 2: 32-round related key differential attack on GOST.

On the Additive Differential Probability of Exclusive-Or

Helger Lipmaa[1], Johan Wallén[1], and Philippe Dumas[2]

[1] Laboratory for Theoretical Computer Science
Helsinki University of Technology
P.O.Box 5400, FIN-02015 HUT, Espoo, Finland
{helger,johan}@tcs.hut.fi
[2] Algorithms Project
INRIA Rocquencourt, 78153 Le Chesnay Cedex, France
Philippe.Dumas@inria.fr

Abstract. We study the differential probability $\mathrm{adp}^{\oplus}$ of exclusive-or when differences are expressed using addition modulo 2^N. This function is important when analysing symmetric primitives that mix exclusive-or and addition—especially when addition is used to add in the round keys. (Such primitives include IDEA, Mars, RC6 and Twofish.) We show that $\mathrm{adp}^{\oplus}$ can be viewed as a formal rational series with a linear representation in base 8. This gives a linear-time algorithm for computing $\mathrm{adp}^{\oplus}$, and enables us to compute several interesting properties like the fraction of impossible differentials, and the maximal differential probability for any given output difference. Finally, we compare our results with the dual results of Lipmaa and Moriai on the differential probability of addition modulo 2^N when differences are expressed using exclusive-or.

Keywords: Additive differential probability, differential cryptanalysis, rational series.

1 Introduction

Symmetric cryptographic primitives like block ciphers are typically constructed from a small set of simple building blocks like bitwise exclusive-or and addition modulo 2^N. Surprisingly little is known about how these two operations interact with respect to different cryptanalytic attacks, and some of the fundamental relations between them have been established only recently [LM01, Lip02, Wal03]. Our goal is to share light to this question by studying the interaction of these two operations in one concrete application: differential cryptanalysis [BS91], by studying the differential probability of exclusive-or when differences are expressed using addition modulo 2^N. This problem is dual to the one explored by Lipmaa and Moriai [LM01, Lip02]. We hope that our results will be helpful in evaluating the precise security of ciphers that mix addition and exclusive-or against differential cryptanalysis.

B. Roy and W. Meier (Eds.): FSE 2004, LNCS 3017, pp. 317–331, 2004.

Differential Cryptanalysis. Differential cryptanalysis studies the propagation of differences in functions. Let G, H be Abelian groups and let $f\colon G \to H$ be a function. The input difference $x - x^* \in G$ is said to propagate to the output difference $f(x) - f(x^*) \in H$ through f. A *differential* of f is a pair $(\alpha, \beta) \in G \times H$. This is usually denoted by $\alpha \to \beta$. If the difference between $x, x^* \in G$ is $x - x^* = \alpha$, the differential $\alpha \to \beta$ can be used to predict the corresponding output difference $f(x) - f(x^*)$. It is thus natural to measure the efficiency of a differential by its *differential probability*

$$\mathrm{dp}^f(\alpha \to \beta) = \Pr_{x \in G}[f(x + \alpha) - f(x) = \beta] \ .$$

When a cipher uses both bitwise exclusive-or and addition modulo 2^N, both operators are natural choices for expressing differences (depending on how the round keys are added). Depending on this choice, one must either study the differential properties of addition when differences are expressed using exclusive-or, or the dual differential probability of exclusive-or when differences are expressed using addition modulo 2^N. The differential probability of addition was studied in detail by [LM01, Lip02]. However, the dual differential probability of exclusive-or has remained open. This dual case is just as interesting in practise, since most of the popular block ciphers that mix addition and exclusive-or use addition—and not exclusive-or—for adding in the round keys. (Examples include IDEA [LMM91], Mars [BCD+98], RC6 [RRSY98] and Twofish [SKW+99].)

We will exclusively deal with the set $\{0, 1, \dots, 2^N - 1\}$ equipped with two group operations. On one hand, we use the usual addition modulo 2^N, which we denote by $+$. On the other hand, we identify $\{0, 1, \dots, 2^N - 1\}$ and the set $\mathbf{Z}_2^N$ of N-tuples of bits using the natural correspondence that identifies $x_{N-1}2^{N-1} + \dots + x_1 2 + x_0 \in \mathbf{Z}_{2^N}$ with $(x_{N-1}, \dots, x_1, x_0) \in \mathbf{Z}_2^N$. In this way the usual componentwise addition $\oplus$ in $\mathbf{Z}_2^N$ (or bitwise exclusive-or) carries over to a group operation in $\{0, 1, \dots, 2^N - 1\}$. We can thus especially view $\oplus$ as a function $\oplus\colon \mathbf{Z}_{2^N} \times \mathbf{Z}_{2^N} \to \mathbf{Z}_{2^N}$. We call the differential probability of the resulting mapping the *additive differential probability* of exclusive-or and denote it by $\mathrm{adp}^{\oplus}\colon \mathbf{Z}_{2^N}^3 \to [0, 1]$,

$$\mathrm{adp}^{\oplus}(\alpha, \beta \to \gamma) = \Pr_{x,y}[((x + \alpha) \oplus (y + \beta)) - (x \oplus y) = \gamma] \ . \tag{1}$$

The dual mapping, the *exclusive-or differential probability* of addition, denoted $\mathrm{xdp}^+\colon \mathbf{Z}_{2^N}^3 \to [0, 1]$, is given by

$$\mathrm{xdp}^+(\alpha, \beta \to \gamma) = \Pr_{x,y}[((x \oplus \alpha) + (y \oplus \beta)) \oplus (x + y) = \gamma] \ .$$

This dual mapping was studied in detail by Lipmaa and Moriai [LM01, Lip02], who gave a closed formula for xdp^+. Their formula in particular leads to an $\Theta(\log N)$-time algorithm for computing xdp^+ and the differential probability of some related mappings like the pseudo-Hadamard transform [Lip02].

Our contributions. In this paper, we present a detailed analysis of the mapping $\mathrm{adp}^{\oplus}: \mathbf{Z}_{2^N}^3 \rightarrow [0, 1]$. This concrete problem has been addressed (and in a rather ad hoc manner) in a few papers, including [Ber92], but it has never been addressed completely—probably because of its "apparent complexity" [Ber92]. We show that $\mathrm{adp}^{\oplus}$ can be expressed as a formal rational series in the sense of formal language theory with a linear representation in base 8. That is, there are eight square matrices A_i, a column vector C and a row vector L, such that if we write the differential $(\alpha, \beta \rightarrow \gamma)$ as an octal word $w = w_{N-1} \cdots w_1 w_0$ in a natural way,

$$\mathrm{adp}^{\oplus}(\alpha, \beta \rightarrow \gamma) = \mathrm{adp}^{\oplus}(w) = L A_{w_{N-1}} \cdots A_{w_1} A_{w_0} C \ .$$

This representation immediately gives a linear-time algorithm for computing $\mathrm{adp}^{\oplus}$. This should be be compared to the naïve $\Theta(2^{2N})$-time algorithm which seems to be the only previously known algorithm for $\mathrm{adp}^{\oplus}$. In addition, we derive some other properties, like the fraction $\frac{3}{7} + \frac{4}{7} \cdot \frac{1}{8^N}$ of differentials with nonzero probability, and determine the maximal differential probability $\max_{\alpha,\beta} \mathrm{adp}^{\oplus}(\alpha, \beta \rightarrow \gamma)$ for any given output difference γ. Finally, we show how our approach based on rational series easily can be adapted for studying the dual mapping xdp^{+}.

The paper is organised as follows. We first show that $\mathrm{adp}^{\oplus}$ is a rational series and derive a linear representation for it. This gives an efficient algorithm that computes $\mathrm{adp}^{\oplus}(w)$ in time $O(|w|)$. In Sect. 3, we discuss the distribution of $\mathrm{adp}^{\oplus}$ and differentials with maximal probability. Sect. 4 describes how similar methods can be used to analyse xdp^{+}. The appendix contains some omitted proofs.

2 Rational Series $\mathrm{adp}^{\oplus}$

Throughout this paper, we let N denote the default word length. We will consider $\mathrm{adp}^{\oplus}$ as a function of octal words by writing the differential $(\alpha, \beta \rightarrow \gamma)$ as the octal word $w = w_{N-1} \cdots w_0$, where $w_i = \alpha_i 4 + \beta_i 2 + \gamma_i$. This defines $\mathrm{adp}^{\oplus}$ as a function from the octal words of length N to the interval $[0, 1]$. As N varies in the set of nonnegative integers, we obtain a function from the set of all octal words to $[0, 1]$.

In the terminology of formal language theory, the additive differential probability $\mathrm{adp}^{\oplus}$ is a formal series over the monoid of octal words with coefficients in the field of real numbers. A remarkable subset of these series is the set of *rational series* [BR88]. One possible characterisation of such a rational series S is the following: there exists a square matrix A_k of size $q \times q$ for each letter k in the alphabet, a row matrix L of size $1 \times q$ and a column matrix C of size $q \times 1$ such that for each word $w = w_1 \cdots w_\ell$, the value of the series is $S(w) = L A_{w_1} \cdots A_{w_\ell} C$. The family $L, (A_k)_k, C$ is called a *linear representation* of dimension q of the rational series. In our case, the alphabet is the octal alphabet $\{0, 1, \ldots, 7\}$.

Theorem 1 (Linear representation of $\mathrm{adp}^{\oplus}$**).** *The formal series* $\mathrm{adp}^{\oplus}$ *has the* 8*-dimensional linear representation* L, $(A_k)_k$, C, *where* $L = \begin{pmatrix} 1 & 1 & 1 & 1 & 1 & 1 & 1 & 1 \end{pmatrix}$,

$C = (1\,0\,0\,0\,0\,0\,0\,0)^\top$,

$$A_0 = \frac{1}{4}\begin{pmatrix} 4&0&0&1&0&1&1&0 \\ 0&0&0&1&0&1&0&0 \\ 0&0&0&1&0&0&1&0 \\ 0&0&0&1&0&0&0&0 \\ 0&0&0&0&0&1&1&0 \\ 0&0&0&0&0&1&0&0 \\ 0&0&0&0&0&0&1&0 \\ 0&0&0&0&0&0&0&0 \end{pmatrix},$$

and A_k, $k \neq 0$, is obtained from A_0 by permuting row i with row $i \oplus k$ and column j with column $j \oplus k$: $(A_k)_{ij} = (A_0)_{i\oplus k, j\oplus k}$. (For completeness, the matrices $A_0, \ldots, A_7$ are given in Table 1.) Thus, $\mathrm{adp}^\oplus$ *is a rational series.*

For example, the differential $(\alpha, \beta \to \gamma) = (00110, 10100 \to 01110)$ corresponds to the octal word $w = 21750$ and $\mathrm{adp}^\oplus(\alpha, \beta \to \gamma) = \mathrm{adp}^\oplus(w) = LA_2A_1A_7A_5A_0C = \frac{5}{32}$. The linear representation immediately implies that $\mathrm{adp}^\oplus(w)$ can be computed using $O(|w|)$ arithmetic operations. Since the arithmetic operations can be carried out using $2|w|$-bit integer arithmetic, which can be implemented in constant time on a $|w|$-bit RAM model, we have

Corollary 1. *The additive differential probability* $\mathrm{adp}^\oplus(w)$ *can be computed in time $O(|w|)$ on a standard unit cost $|w|$-bit* RAM *model of computation.*

This can be compared with the $O(\log|w|)$-time algorithm for computing $\mathrm{xdp}^+(w)$ from [LM01].

As a side remark (we will not use this result later), note that the matrices $A_0, \ldots A_7$ in the linear representation for $\mathrm{adp}^\oplus$ are substochastic. Thus, we could view the linear representation as a inhomogeneous Markov chain by adding a dummy state and dummy state transitions.

The rest of this section is devoted to the technical proof of Theorem 1. To prove this result, we will first give a different formulation of $\mathrm{adp}^\oplus$. For $x, y \in \{0, \ldots, 2^N - 1\}$, let xy denote their componentwise product in $\mathbf{Z}_2^N$ (equivalently, the bitwise and of two N-bit strings). Let $\mathrm{borrow}(x, y) = x \oplus y \oplus (x - y)$

Table 1. All eight matrices A_i

$$A_0 = \frac{1}{4}\begin{pmatrix} 4&0&0&1&0&1&1&0 \\ 0&0&0&1&0&1&0&0 \\ 0&0&0&1&0&0&1&0 \\ 0&0&0&1&0&0&0&0 \\ 0&0&0&0&0&1&1&0 \\ 0&0&0&0&0&1&0&0 \\ 0&0&0&0&0&0&1&0 \\ 0&0&0&0&0&0&0&0 \end{pmatrix} \quad A_1 = \frac{1}{4}\begin{pmatrix} 0&0&1&0&1&0&0&0 \\ 0&4&1&0&1&0&0&1 \\ 0&0&1&0&0&0&0&0 \\ 0&0&1&0&0&0&0&1 \\ 0&0&0&0&1&0&0&0 \\ 0&0&0&0&1&0&0&1 \\ 0&0&0&0&0&0&0&0 \\ 0&0&0&0&0&0&0&1 \end{pmatrix} \quad A_2 = \frac{1}{4}\begin{pmatrix} 0&1&0&0&1&0&0&0 \\ 0&1&0&0&0&0&0&0 \\ 0&1&4&0&1&0&0&1 \\ 0&1&0&0&0&0&0&1 \\ 0&0&0&0&1&0&0&0 \\ 0&0&0&0&0&0&0&0 \\ 0&0&0&0&1&0&0&1 \\ 0&0&0&0&0&0&0&1 \end{pmatrix} \quad A_3 = \frac{1}{4}\begin{pmatrix} 1&0&0&0&0&0&0&0 \\ 1&0&0&0&0&1&0&0 \\ 1&0&0&0&0&0&1&0 \\ 1&0&0&4&0&1&1&0 \\ 0&0&0&0&0&0&0&0 \\ 0&0&0&0&0&1&0&0 \\ 0&0&0&0&0&0&1&0 \\ 0&0&0&0&0&1&1&0 \end{pmatrix}$$

$$A_4 = \frac{1}{4}\begin{pmatrix} 0&1&1&0&0&0&0&0 \\ 0&1&0&0&0&0&0&0 \\ 0&0&1&0&0&0&0&0 \\ 0&0&0&0&0&0&0&0 \\ 0&1&1&0&4&0&0&1 \\ 0&1&0&0&0&0&0&1 \\ 0&0&1&0&0&0&0&1 \\ 0&0&0&0&0&0&0&1 \end{pmatrix} \quad A_5 = \frac{1}{4}\begin{pmatrix} 1&0&0&0&0&0&0&0 \\ 1&0&0&1&0&0&0&0 \\ 0&0&0&0&0&0&0&0 \\ 0&0&0&1&0&0&0&0 \\ 1&0&0&0&0&0&1&0 \\ 1&0&0&1&0&4&1&0 \\ 0&0&0&0&0&0&1&0 \\ 0&0&0&1&0&0&1&0 \end{pmatrix} \quad A_6 = \frac{1}{4}\begin{pmatrix} 1&0&0&0&0&0&0&0 \\ 0&0&0&0&0&0&0&0 \\ 1&0&0&1&0&0&0&0 \\ 0&0&0&1&0&0&0&0 \\ 1&0&0&0&0&1&0&0 \\ 0&0&0&0&0&1&0&0 \\ 1&0&0&1&0&1&4&0 \\ 0&0&0&1&0&1&0&0 \end{pmatrix} \quad A_7 = \frac{1}{4}\begin{pmatrix} 0&0&0&0&0&0&0&0 \\ 0&1&0&0&0&0&0&0 \\ 0&0&1&0&0&0&0&0 \\ 0&1&1&0&0&0&0&0 \\ 0&0&0&0&1&0&0&0 \\ 0&1&0&0&1&0&0&0 \\ 0&0&1&0&1&0&0&0 \\ 0&1&1&0&1&0&0&4 \end{pmatrix}$$

denote the borrows, as an N-tuple of bits, in the subtraction $x - y$. Alternatively, $\text{borrow}(x, y)$ can be recursively defined by $\text{borrow}(x, y)_0 = 0$ and $\text{borrow}(x, y)_{i+1} = 1$ if and only if $x_i - \text{borrow}(x, y)_i < y_i$ as integers. This can be used to *define* $\text{borrow}(x, y)_N = 1$ if and only if $x_{N-1} - \text{borrow}(x, y)_{N-1} < y_{N-1}$ as integers. The borrows can be used to give an alternative formulation of $\text{adp}^{\oplus}$.

Lemma 1. *For all* $\alpha, \beta, \gamma \in \mathbf{Z}_{2^N}$,

$$\text{adp}^{\oplus}(w) = \Pr_{x,y}[a \oplus b \oplus c = \alpha \oplus \beta \oplus \gamma] ,$$

where $a = \text{borrow}(x, \alpha)$, $b = \text{borrow}(y, \beta)$ *and* $c = \text{borrow}(x \oplus y, (x - \alpha) \oplus (y - \beta))$.

Proof. By replacing x and y with $x - \alpha$ and $y - \beta$ in the definition (1) of $\text{adp}^{\oplus}$, we see that $\text{adp}^{\oplus}(\alpha, \beta \to \gamma) = \Pr_{x,y}[(x \oplus y) - ((x - \alpha) \oplus (y - \beta)) = \gamma]$. Since $(x \oplus y) - ((x - \alpha) \oplus (y - \beta)) = \gamma$ if and only if $\gamma = c \oplus x \oplus y \oplus (x - \alpha) \oplus (y - \beta) = a \oplus b \oplus c \oplus \alpha \oplus \beta$ if and only if $a \oplus b \oplus c = \alpha \oplus \beta \oplus \gamma$, the result follows. □

We furthermore need the following technical lemma.

Lemma 2. *For all* x, y, α, β, γ,

$$\begin{aligned} a_{i+1} &= (aa' \oplus \alpha \oplus a'x)_i , \\ b_{i+1} &= (bb' \oplus \beta \oplus b'y)_i \quad \text{and} \\ c_{i+1} &= [c \oplus a' \oplus b' \oplus c(a' \oplus b') \oplus (a' \oplus b')(x \oplus y)]_i , \end{aligned}$$

where $a = \text{borrow}(x, \alpha)$, $b = \text{borrow}(y, \beta)$, $c = \text{borrow}(x \oplus y, (x - \alpha) \oplus (y - \beta))$, $a' = a \oplus \alpha$ *and* $b' = b \oplus \beta$.

Proof. By the recursive definition of $\text{borrow}(x, y)$, $\text{borrow}(x, y)_{i+1} = 1$ if and only if $x_i < y_i + \text{borrow}(x, y)_i$ as integers. The latter event occurs if and only if either $y_i = \text{borrow}(x, y)_i$ and at least two of x_i, y_i and $\text{borrow}(x, y)_i$ are one, or $y_i \neq \text{borrow}(x, y)_i$ and at least two of x_i, y_i and $\text{borrow}(x, y)_i$ are zero. That is, $\text{borrow}(x, y)_{i+1} = 1$ if and only if $y_i \oplus \text{borrow}(x, y)_i \oplus \text{maj}(x_i, y_i, \text{borrow}(x, y)_i) = 1$, where $\text{maj}(u, v, w)$ denotes the majority of the bits u, v, w. Since $\text{maj}(u, v, w) = uv \oplus uw \oplus vw$, we have $\text{borrow}(x, y)_{i+1} = [y \oplus \text{borrow}(x, y) \oplus xy \oplus x\,\text{borrow}(x, y) \oplus y\,\text{borrow}(x, y)]_i$.

For a, we thus have $a_{i+1} = (\alpha \oplus a \oplus x\alpha \oplus xa \oplus \alpha a)_i = (a' \oplus a\,\alpha \oplus a'x)_i = [a' \oplus a\,(a' \oplus a) \oplus a'x]_i = (aa' \oplus \alpha \oplus a'x)_i$. The formula for b_{i+1} is completely analogous. For c, we have $c_{i+1} = [(x - \alpha) \oplus (y - \beta) \oplus c \oplus (x \oplus y)((x - \alpha) \oplus (y - \beta)) \oplus (x \oplus y)c \oplus ((x - \alpha) \oplus (y - \beta))c]_i = [x \oplus a' \oplus y \oplus b' \oplus c \oplus (x \oplus y)(x \oplus a' \oplus y \oplus b') \oplus (x \oplus y)c \oplus (x \oplus a' \oplus y \oplus b')c]_i = [c \oplus a' \oplus b' \oplus c(a' \oplus b') \oplus (a' \oplus b')(x \oplus y)]_i$. □

Proof (of Theorem 1). Let $(\alpha, \beta \to \gamma)$ be the differential associated with the word w. Denote $N = |w|$ and let x, y be uniformly distributed random variables in $\mathbf{Z}_{2^N}$. Denote $a = \text{borrow}(x, \alpha)$, $b = \text{borrow}(y, \beta)$ and $c = \text{borrow}(x \oplus y, (x - \alpha) \oplus (y - \beta))$. Let ξ be the octal word of borrow triples, $\xi_i = a_i 4 + b_i 2 + c_i$. We define ξ_N in the natural way using $\text{borrow}(u, v)_N = 1$ if and only if $u_{N-1} -$

$\mathrm{borrow}(u,v)_{N-1} < v_{N-1}$ as integers. For compactness, denote $\mathrm{xor}(w) = \alpha\oplus\beta\oplus\gamma$ and $\mathrm{xor}(\xi) = a \oplus b \oplus c$. Let $P(w,k)$ be the 8×1 substochastic matrix

$$P_j(w,k) = \Pr_{x,y}[\mathrm{xor}(\xi) \equiv \mathrm{xor}(w) \pmod{2^k}, \xi_k = j]$$

for $0 \leq k \leq N$. Let $M(w,k)$ be the 8×8 substochastic transition matrix

$$M_{ij}(w,k) = \Pr_{x,y}[\mathrm{xor}(\xi)_k = \mathrm{xor}(w)_k, \xi_{k+1} = i \mid \mathrm{xor}(\xi) \equiv \mathrm{xor}(w) \pmod{2^k}, \xi_k = j]$$

for $0 \leq k < N$. Then $P_i(w,k+1) = \sum_j M_{ij}(w,k)P_j(w,k)$ and thus $P(w,k+1) = M(w,k)P(w,k)$. Note furthermore that $P(w,0) = C$, since $a_0 = b_0 = c_0 = 0$, and that $LP(w,N) = \sum_j \Pr_{x,y}[\mathrm{xor}(\xi) \equiv \mathrm{xor}(w) \pmod{2^N}, \xi_N = j] = \Pr_{x,y}[\mathrm{xor}(\xi) \equiv \mathrm{xor}(w) \pmod{2^N}] = \mathrm{adp}^{\oplus}(w)$, where the last equality is due to Lemma 1. We will show that $M(w,k) = A_{w_k}$ for all k. By induction, it follows that $\mathrm{adp}^{\oplus}(w) = LP(w,N) = LM(w,N-1)\cdots M(w,0)C = LA_{w_{N-1}}\cdots A_{w_0}C$.

It remains to show that $M(w,k) = A_{w_k}$ for all k. Towards this end, let x, y be such that $\mathrm{xor}(\xi) \equiv \mathrm{xor}(w) \pmod{2^k}$ and $\xi_k = j$. We will count the number of ways we can choose (x_k, y_k) such that $\mathrm{xor}(\xi)_k = \mathrm{xor}(w)_k$ and $\xi_{k+1} = i$.

Denote $a' = a \oplus \alpha$, $b' = b \oplus \beta$ and $c' = c \oplus \gamma$. Note that $\mathrm{xor}(\xi)_k = \mathrm{xor}(w)_k$ if and only if $c'_k = (a' \oplus b')_k$. Under the assumption that $\mathrm{xor}(\xi)_k = \mathrm{xor}(w)_k$ we have $(cc' \oplus \gamma)_k = [c(a' \oplus b') \oplus c \oplus a' \oplus b']_k$. By Lemma 2, (x_k, y_k) must thus be a solution to $V\begin{pmatrix}x_k & y_k\end{pmatrix}^{\top} = U$ in $\mathbf{Z}_2$, where U and V are the matrices

$$U = \begin{pmatrix}(aa' \oplus \alpha)_k \oplus a_{k+1} \\ (bb' \oplus \beta)_k \oplus b_{k+1} \\ (cc' \oplus \gamma)_k \oplus c_{k+1}\end{pmatrix} \quad \text{and} \quad V = \begin{pmatrix}a'_k & 0 \\ 0 & b'_k \\ (a' \oplus b')_k & (a' \oplus b')_k\end{pmatrix}$$

over $\mathbf{Z}_2$. If this equation has a solution, it has exactly $2^{2-\mathrm{rank}(V)}$ solutions. But $\mathrm{rank}(V) = 0$ if and only if $a'_k = b'_k = 0$ (then there are 4 solutions) and $\mathrm{rank}(V) = 2$ otherwise (then there is 1 solution).

The equation has a solution (x_k, y_k) exactly when $\mathrm{rank}(V) = \mathrm{rank}(V\ U)$. From this and from the requirement that $c'_k = (a' \oplus b')_k$, we see that there are solutions exactly in the following cases.

- If $a'_k = b'_k = 0$, then $c'_k = 0$ and $\mathrm{rank}(V) = 0$. There are solutions (4 solutions) if and only if $a_{k+1} = \alpha_k$, $b_{k+1} = \beta_k$ and $c_{k+1} = \gamma_k$.
- If $a'_k = 0$ and $b'_k = 1$ then $c'_k = 1$ and $\mathrm{rank}(V) = 2$. There is a single solution if and only if $a_{k+1} = \alpha_k$.
- If $a'_k = 1$ and $b'_k = 0$, then $c'_k = 1$ and $\mathrm{rank}(V) = 2$. There is a single solution if and only if $b_{k+1} = \beta_k$.
- If $a'_k = 1$ and $b'_k = 1$ then $c'_k = 0$ and $\mathrm{rank}(V) = 2$. There is a single solution if and only if $c_{k+1} = \gamma_k$.

Since $j = \xi_k = a_k 4 + b_k 2 + c_k$ and $i = \xi_{k+1} = a_{k+1} 4 + b_{k+1} 2 + c_{k+1}$, the derivation so far can be summarised as

$$M_{ij}(w,k) = \begin{cases} 1\ , & j = (\alpha_k, \beta_k, \gamma_k)\ ,\ i = (\alpha_k, \beta_k, \gamma_k)\ , \\ 1/4\ , & j = (\alpha_k, \beta_k \oplus 1, \gamma_k \oplus 1)\ ,\ i = (\alpha_k, *, *)\ , \\ 1/4\ , & j = (\alpha_k \oplus 1, \beta_k, \gamma_k \oplus 1)\ ,\ i = (*, \beta_k, *)\ , \\ 1/4\ , & j = (\alpha_k \oplus 1, \beta_k \oplus 1, \gamma_k)\ ,\ i = (*, *, \gamma_k)\ , \\ 0\ , & \text{otherwise}\ , \end{cases}$$

where we have identified the integer $r_2 4 + r_1 2 + r_0$ with the binary tuple (r_2, r_1, r_0) and $*$ represents an arbitrary element of $\{0, 1\}$. It follows that $M(w,k) = A_0$ if $w_k = 0$ and $M_{i,j}(w,k) = M_{i \oplus w_k, j \oplus w_k}(0,k)$. That is, $M(w,k) = A_{w_k}$ for all w, k. This completes the proof. □

3 Distribution of $\mathrm{adp}^{\oplus}$ and Maximal Differentials

3.1 Distribution

We will use notation from formal languages to describe octal words (and thus differentials). In particular, we will use concatenation (xy), the corresponding powers ($x^0 = \lambda$ is the empty word and $x^{n+1} = xx^n$), union ($x + y$) and the Kleene star ($x^* = \sum_{n \geq 0} x^n$). Throughout this section, L, $(A_k)_k$, C is the linear representation of $\mathrm{adp}^{\oplus}$.

We will first consider the effect of tailing and leading zeros.

Corollary 2. *For all octal words w, $\mathrm{adp}^{\oplus}(w0^*) = \mathrm{adp}^{\oplus}(w)$.*

This trivial result follows from the observation that $A_0 C = C$.

Corollary 3. *Let w be a word and let $a = \begin{pmatrix} a_0 & \cdots & a_7 \end{pmatrix}^{\top} = A_{|w|-1} \cdots A_{w_0} C$. Let $\alpha = a_0$ and $\beta = a_3 + a_5 + a_6$. Let w' be a word of the form $w' = 0^* w$. Then $\mathrm{adp}^{\oplus}(w') = \alpha + \frac{\beta}{3} + \frac{8}{3} \cdot \beta \cdot 4^{-(|w'|-|w|)}$.*

Proof. Using a Jordan form $J = P^{-1} A_0 P$ of A_0, it is easy to see that

$$A_0^k = 4^{-k} \begin{pmatrix} 4^k & 0 & 0 & \frac{4^k-1}{3} & 0 & \frac{4^k-1}{3} & \frac{4^k-1}{3} & 0 \\ 0 & 0 & 0 & 1 & 0 & 1 & 0 & 0 \\ 0 & 0 & 0 & 1 & 0 & 0 & 1 & 0 \\ 0 & 0 & 0 & 1 & 0 & 0 & 0 & 0 \\ 0 & 0 & 0 & 0 & 0 & 1 & 1 & 0 \\ 0 & 0 & 0 & 0 & 0 & 1 & 0 & 0 \\ 0 & 0 & 0 & 0 & 0 & 0 & 1 & 0 \\ 0 & 0 & 0 & 0 & 0 & 0 & 0 & 0 \end{pmatrix} .$$

If we let $j = |w'| - |w|$, we see that $L A_0^j a = a_0 + \frac{4^j - 1}{3 \cdot 4^j}(a_3 + a_5 + a_6) + \frac{3}{4^j}(a_3 + a_5 + a_6) = \alpha + \frac{\beta}{3} + \frac{8}{3} \cdot \beta \cdot 4^{-(|w'|-|(|w))}$. □

This means that $\mathrm{adp}^{\oplus}(0^n w)$ decreases with n and $\mathrm{adp}^{\oplus}(0^n w) \to \alpha + \beta/3$ as $n \to \infty$. This can be compared to [LM01], where it was shown that $\mathrm{xdp}^{+}(00w) = \mathrm{xdp}^{+}(0w)$ for all w.

Theorem 2. *The additive differential probability* $\mathrm{adp}^{\oplus}(w)$ *is nonzero if and only if* w *has the form* $w = 0^*$ *or* $w = w'(3+5+6)0^*$ *for any octal word* w'.

Proof. Since $A_1C = A_2C = A_4C = A_7C = 0$, $\mathrm{adp}^{\oplus}(w'(1+2+4+7)0^*) = 0$. Conversely, let w be a word of the form $w = w'(3+5+6)0^*$. Let e_i be the canonical (column) basis vector with a 1 in the ith component and 0 in the others. By direct computation, the kernels are $\ker A_0 = \ker A_3 = \ker A_5 = \ker A_6 = \langle e_1, e_2, e_4, e_7 \rangle$ and $\ker A_1 = \ker A_2 = \ker A_4 = \ker A_7 = \langle e_0, e_3, e_5, e_6 \rangle$. For all i and $j \neq i$, $e_j \notin \ker A_i$, it can be seen that $A_i e_i = e_i$ and that $A_i e_j$ has the form $A_i e_j = (e_k + e_\ell + e_m + e_n)/4$, where $k \neq \ell$, $m \neq n$, $e_k, e_\ell \in \ker A_0$ and $e_m, e_n \in \ker A_1$. Since $C = e_0$, we see by induction that $A_{w_i} \cdots A_{w_0} C \notin \ker A_{w_{i+1}}$ for all i. Thus, $\mathrm{adp}^{\oplus}(w) \neq 0$.

Since the matrices A_k are nonnegative, an alternative proof is the following. Since

$$LA_{w_{N-1}} \cdots A_{w_0} C = \sum_{i_0,\ldots,i_N} L_{i_N} (A_{w_{N-1}})_{i_N, i_{N-1}} \cdots (A_{w_0})_{i_1, i_0} C_{i_0} ,$$

we see that $\mathrm{adp}^{\oplus}(w)$ is nonzero if and only if there are indexes $i_0, \ldots, i_N$ such that $L_{i_N} (A_{w_{N-1}})_{i_N, i_{N-1}} \cdots (A_{w_0})_{i_1, i_0} C_{i_0} \neq 0$. We construct a nondeterministic finite automaton with state set $\{0, \ldots, 7, \mathit{init}\}$ and input alphabet $\{0, \ldots, 7\}$ as follows. There is an empty transition from the initial state *init* to state i if and only if $L_i \neq 0$. There is a transition labelled x from state i to state j if and only if $(A_x)_{i,j} \neq 0$. The state i is accepting if and only if $C_i \neq 0$. Clearly, the automaton accepts the word w read from left to right if and only if $\mathrm{adp}^{\oplus}(w) \neq 0$. If we convert the automaton to a minimal deterministic automation, we obtain the following automation.

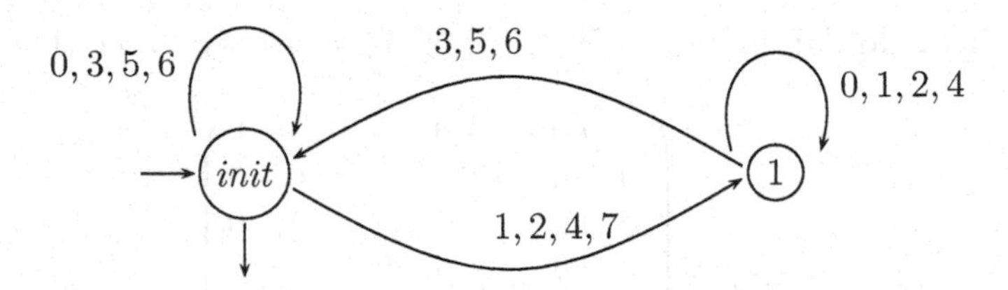

This automaton clearly accepts the language $0^* + (0+1+\cdots+7)^*(3+5+6)0^*$. □

A complete determination of the distribution of $\mathrm{adp}^{\oplus}$ falls out of the scope of this paper and we will restrict ourselves to some of the most important results. First, we turn to the fraction of *possible* differentials—that is, differentials with $\mathrm{adp}^{\oplus}(w) \neq 0$.

Corollary 4. *For all* $N \geq 0$, $\Pr_{|w|=N}[\mathrm{adp}^{\oplus}(w) \neq 0] = \frac{3}{7} + \frac{4}{7} \cdot \frac{1}{8^N}$.

Proof. According to Theorem 2, $\mathrm{adp}^{\oplus}(w) \neq 0$ if and only if w is the zero word or has form $w = w'\xi 0^k$, where w' is an arbitrary word of length $N - k - 1$ and $\xi \in \{3, 5, 6\}$. For a fixed value of k, we can choose w' and ξ in $3 \cdot 8^{N-k-1}$ ways. Thus, there are $1 + \sum_{k=0}^{N-1} 3 \cdot 8^{N-k-1} = \frac{4}{7} + \frac{3}{7} \cdot 8^N$ words with $\mathrm{adp}^{\oplus}(w) \neq 0$ in total. □

This result can be compared with [LM01, Theorem 2], which states that the corresponding probability for xdp^+ is $\Pr_{|w|=N}[\mathrm{xdp}^+(w) \neq 0] = \frac{4}{7} \cdot \left(\frac{7}{8}\right)^N$. This means, in particular, that $\Pr_{|w|=N}[\mathrm{adp}^{\oplus}(w) \neq 0] \to \frac{3}{7}$ while $\Pr_{|w|=N}[\mathrm{xdp}^+(w) \neq 0] \to 0$ as $N \to \infty$. Since the number of possible differentials is larger for $\mathrm{adp}^{\oplus}$ than for xdp^+, the average possible differential will obtain a smaller probability.

Next, if $w = (0+3+5+6)0^*$ then clearly $\mathrm{adp}^{\oplus}(w) = 1$. On the other hand, for any $\xi \in \{0, \dots, 7\}$, $\mathrm{adp}^{\oplus}(\xi w) \leq 1/2$. It follows that $\Pr_{|w|=N}[\mathrm{adp}^{\oplus}(w) = 1] = 4 \cdot 8^{-N}$, and $\Pr_{|w|=N}[\mathrm{adp}^{\oplus}(w) = k] = 0$ if $1/2 < k < 1$. One can further establish easily that $\mathrm{adp}^{\oplus}(w) = 1/2$ if and only if $w = \Sigma(0 + 3 + 5 + 6)0^*$, where $\Sigma = 0+1+\cdots+7$. The following straightforward lemma is useful in such calculations.

Lemma 3. *Let w be an octal word. Denote $\Sigma_0 = \{0, 3, 5, 6\}$, $\Sigma_1 = \{1, 2, 4, 7\}$ and $A_w = A_{w_{|w|-1}} \cdots A_{w_0}$. Then $\mathrm{adp}^{\oplus}(xw) = \mathrm{adp}^{\oplus}(0w)$ for all $x \in \Sigma_0$ and $\mathrm{adp}^{\oplus}(yw) = \mathrm{adp}^{\oplus}(1w)$ for all $y \in \Sigma_1$. Thus, $\mathrm{adp}^{\oplus}(w) = \mathrm{adp}^{\oplus}(xw) + \mathrm{adp}^{\oplus}(yw)$ and $\mathrm{adp}^{\oplus}(xzw) = \mathrm{adp}^{\oplus}(yzw) + (A_w C)_z$ for all $x, z \in \Sigma_b$ and $y \in \Sigma_{1-b}$.*

Proof. By direct calculation, we see that $LA_0 = LA_3 = LA_5 = LA_6 = \begin{pmatrix}1 & 0 & 0 & 1 & 0 & 1 & 1 & 0\end{pmatrix}$ and $LA_1 = LA_2 = LA_4 = LA_7 = \begin{pmatrix}0 & 1 & 1 & 0 & 1 & 0 & 0 & 1\end{pmatrix}$. Since $\mathrm{adp}^{\oplus}(vw) = LA_v A_w C$, we have $\mathrm{adp}^{\oplus}(xw) = \mathrm{adp}^{\oplus}(0w)$ and $\mathrm{adp}^{\oplus}(yw) = \mathrm{adp}^{\oplus}(1w)$ for all $x \in \Sigma_0$ and $y \in \Sigma_1$. Finally, if $x, z \in \Sigma_b$ and $y \in \Sigma_{1-b}$, we have $\mathrm{adp}^{\oplus}(w) = LA_w C = (LA_x + LA_y)A_w C = \mathrm{adp}^{\oplus}(xw) + \mathrm{adp}^{\oplus}(yw)$ and $\mathrm{adp}^{\oplus}(xzw) - \mathrm{adp}^{\oplus}(yzw) = (L(A_x - A_y)A_z)A_w C = e_z A_w C = (A_w C)_z$, where e_z is a row vector with a 1 in column z and 0 in the other columns. □

3.2 Maximal Differentials

Although many of the enumerative aspects of $\mathrm{adp}^{\oplus}$ seem infeasible, some optimisation problems are surprisingly simple. For all output differences γ, denote

$$\mathrm{adp}^{\oplus}_{2\max}(\gamma) = \max_{\alpha,\beta} \mathrm{adp}^{\oplus}(\alpha, \beta \to \gamma) \ .$$

For all γ, there is a simple differential with $\mathrm{adp}^{\oplus}(\alpha, \beta \to \gamma) = \mathrm{adp}^{\oplus}_{2\max}(\gamma)$.

Theorem 3. *For all output differences γ, $\mathrm{adp}^{\oplus}(0^N, \gamma \to \gamma) = \mathrm{adp}^{\oplus}_{2\max}(\gamma)$.*

The proof is omitted from the conference version.

4 Rational Series xdp^+

Lipmaa and Moriai [LM01, Lip02] used completely different techniques to analyse the exclusive-or differential probability xdp^+ of addition. We will now demonstrate the power of our approach by showing that it can easily adapt to analyse xdp^+ as well.

4.1 Linear Representation

As for $\mathrm{adp}^\oplus$, we write the differential $(\alpha, \beta \to \gamma)$ as the octal word $w = w_{N-1} \dots w_0$, where $w_i = \alpha_i 4 + \beta_i 2 + \gamma_i$. When N varies, we obtain a rational series xdp^+ with a linear representation of dimension 2.

Theorem 4 (Linear representation of xdp^+). *The formal series xdp^+ has the 2-dimensional linear representation L, $(X_k)_{k=0}^7$, C, where $L = \begin{pmatrix}1 & 1\end{pmatrix}$, $C = \begin{pmatrix}1 & 0\end{pmatrix}^\top$ and X_k is given by*

$$(X_k)_{ij} = \begin{cases} 1 - T(k_2 + k_1 + j) & \text{if } i = 0 \text{ and } k_2 \oplus k_1 \oplus k_0 = j \ , \\ T(k_2 + k_1 + j) & \text{if } i = 1 \text{ and } k_2 \oplus k_1 \oplus k_0 = j \ , \\ 0 & \text{otherwise} \end{cases}$$

for $i, j \in \{0, 1\}$, where $k = k_2 4 + k_1 2 + k_0$ and $T \colon \{0, 1, 2, 3\} \to \mathbf{R}$ is the mapping $T(0) = 0$, $T(1) = T(2) = \frac{1}{2}$ and $T(3) = 1$. (For completeness, all the matrices X_k are given in Table 2.) Thus, xdp^+ is a rational series.

For example, the differential $(\alpha, \beta \to \gamma) = (11100, 00110 \to 10110)$ corresponds to the word $w = 54730$ and $\mathrm{xdp}^+(\alpha, \beta \to \gamma) = \mathrm{xdp}^+(w) = L X_5 X_4 X_7 X_3 X_0 C = \frac{1}{4}$. The proof of this result is given in the appendix.

4.2 Words with a Given Probability

The simplicity of the linear representation of xdp^+ allows us to derive an explicit description of all words with a certain differential probability.

Theorem 5. *For all nonempty words w, $\mathrm{xdp}^+(w) \in \{0\} \cup \{2^{-k} \mid k \in \{0, 1, \dots, |w| - 1\}\}$. The differential probability $\mathrm{xdp}^+(w) = 0$ if and only if w has the form $w = w'(1 + 2 + 4 + 7)$, $w = w'(1 + 2 + 4 + 7)0w''$ or $w = w'(0 + 3 + 5 + 6)7w''$ for arbitrary words w', w'', and $\mathrm{xdp}^+(w) = 2^{-k}$ if and only if $\mathrm{xdp}^+(w) \neq 0$ and $|\{0 \leq i < N - 1 \mid w_i \neq 0, 7\}| = k$.*

Table 2. All the eight matrices X_k in Theorem 4

$$X_0 = \tfrac{1}{2}\begin{pmatrix}2 & 0\\0 & 0\end{pmatrix} \quad X_1 = \tfrac{1}{2}\begin{pmatrix}0 & 1\\0 & 1\end{pmatrix} \quad X_2 = \tfrac{1}{2}\begin{pmatrix}0 & 1\\0 & 1\end{pmatrix} \quad X_3 = \tfrac{1}{2}\begin{pmatrix}1 & 0\\1 & 0\end{pmatrix}$$

$$X_4 = \tfrac{1}{2}\begin{pmatrix}0 & 1\\0 & 1\end{pmatrix} \quad X_5 = \tfrac{1}{2}\begin{pmatrix}1 & 0\\1 & 0\end{pmatrix} \quad X_6 = \tfrac{1}{2}\begin{pmatrix}1 & 0\\1 & 0\end{pmatrix} \quad X_7 = \tfrac{1}{2}\begin{pmatrix}0 & 0\\0 & 2\end{pmatrix}$$

Proof. Let L, X_k and C be as in Theorem 4 and denote $e_0 = \begin{pmatrix}1 & 0\end{pmatrix}^\top$ and $e_1 = \begin{pmatrix}0 & 1\end{pmatrix}^\top$. Then the kernels of X_i are $\ker X_0 = \ker X_3 = \ker X_5 = \ker X_6 = \langle e_1 \rangle$ and $\ker X_1 = \ker X_2 = \ker X_4 = \ker X_7 = \langle e_0 \rangle$. By direct calculation, $X_0 e_0 = e_0$, $X_3 e_0 = X_5 e_0 = X_6 e_0 = \frac{1}{2}(e_0 + e_1)$, $X_1 e_1 = X_2 e_1 = X_4 e_1 = \frac{1}{2}(e_0 + e_1)$ and $X_7 e_1 = e_1$. Since $C = e_0$, we thus have $\mathrm{xdp}^+(w) = 0$ if and only if w has the form $w = w'(1+2+4+7)$, $w = w'(1+2+4+7)0w''$ or $w = w'(0+3+5+6)7w''$ for arbitrary words w', w''. Similarly, when w is such that $\mathrm{adp}^+(w) \neq 0$, we see that $X_{w_{n-1}} \cdots X_{w_0} C$ has the form $\begin{pmatrix}2^{-\ell} & 2^{-\ell}\end{pmatrix}^\top$, $\begin{pmatrix}2^{-\ell} & 0\end{pmatrix}^\top$ or $\begin{pmatrix}0 & 2^{-\ell}\end{pmatrix}^\top$, where $\ell = |\{w_i \mid w_i \notin \{0,7\}, 0 \leq i < n\}|$ for all n. Thus, $\mathrm{xdp}^+(w) = 2^{-k}$, where $k = |\{0 \leq i < N-1 \mid w_i \neq 0, 7\}|$. □

For example, if w is the word $w = 54730$, we see that $\mathrm{xdp}^+(w) \neq 0$ and $|\{0 \leq i < 4\} \mid w_i \neq 0,7\}| = 2$. Thus, $\mathrm{xdp}^+(w) = 2^{-2}$. This result immediately gives the closed formula from [LM01] and thus the $O(\log N)$-time algorithm.

4.3 Distribution

Based on the explicit description of all words with a certain differential probability, it is easy to determine the distribution of xdp^+. Let $\mathcal{A}(n,k)$, $\mathcal{B}(n,k)$ and $\mathcal{C}(n,k)$ denote the languages that consist of the words of length $n > 0$ with $\mathrm{xdp}^+(w) = 2^{-k}$, and $w_{n-1} = 0$, $w_{n-1} = 7$ and $w_{n-1} \neq 0,7$, respectively. The languages are clearly given recursively by

$$\begin{aligned}
\mathcal{A}(n,k) &= 0\mathcal{A}(n-1,k) + 0\mathcal{C}(n-1,k-1) \ , \\
\mathcal{B}(n,k) &= 7\mathcal{B}(n-1,k) + 7\mathcal{C}(n-1,k-1) \ , \\
\mathcal{C}(n,k) &= \Sigma_0\mathcal{A}(n-1,k) + \Sigma_1\mathcal{B}(n-1,k) + (\Sigma_0 + \Sigma_1)\mathcal{C}(n-1,k-1) \ ,
\end{aligned}$$

where $\Sigma_0 = 3+5+6$ and $\Sigma_1 = 1+2+4$. The base cases are $\mathcal{A}(1,0) = 0$, $\mathcal{B}(1,0) = \emptyset$ and $\mathcal{C}(1,0) = 3+5+6$. Let $A(z,u) = \sum_{n,k} |\mathcal{A}(n,k)| u^k z^n$, $B(z,u) = \sum_{n,k} |\mathcal{B}(n,k)| u^k z^n$ and $C(z,u) = \sum_{n,k} |\mathcal{C}(n,k)| u^k z^n$ be the corresponding ordinary generating functions. The recursive description of the languages immediately gives the the linear system

$$\begin{cases}
A(z,u) = zA(z,u) + uzC(z,u) + z \ , \\
B(z,u) = zB(z,u) + uzC(z,u) \ , \\
C(z,u) = 3zA(z,u) + 3zB(z,u) + 6uzC(z,u) + 3z \ .
\end{cases}$$

Denote $D(z,u) = A(z,u) + B(z,u) + C(z,u) + 1$. Then the coefficient of $u^k z^n$ in $D(z,u)$, $[u^k z^n]D(z,u)$, gives the number of words of length n with $\mathrm{xdp}^+(w) = 2^{-k}$ (the extra 1 comes from the case $n = 0$). By solving the linear system, we see that

$$D(z,u) = 1 + \frac{4z}{1-(1+6u)z} \ .$$

Since the coefficient of z^n in $D(z,u)$ for $n > 0$ is

$$[z^n]D(z,u) = 4[z^n]z\sum_{m=0}^{\infty}(1+6u)^m z^m = 4(1+6u)^{n-1} \ ,$$

Table 3. Short comparison of the functions $\mathrm{xdp}^{\oplus}$ and $\mathrm{adp}^{\oplus}$ and of their computational complexity

	xdp^{+}	$\mathrm{adp}^{\oplus}$
Possibility verification		
Algorithm complexity	$\Theta(1)$	$\Theta(\log N)$
Probability of possibility	$\frac{1}{2} \cdot \left(\frac{7}{8}\right)^{N-1}$	$\frac{3}{7} + \frac{4}{7} \cdot 8^{-N}$
Evaluation of a possible differential		
Algorithm complexity	$\Theta(\log N)$	$\Theta(N)$
Maximal differentials $\mathrm{adp}^{\oplus}_{2\max}$ and $\mathrm{xdp}^{+}_{2\max}$		
Finding max. differential (α, β)	$\Theta(\log N)$	$\Theta(1)$
Computing max. differential when (α, β) is known	$\Theta(\log N)$	$\Theta(N)$

we see that

$$[u^k z^n]D(z, u) = 4 \cdot 6^k \binom{n-1}{k}$$

for all $0 \le k < n$. The coefficient of z^n in $D(z, 1)$ for $n > 0$, $[z^n]D(z, 1) = 4[z^n]\frac{z}{1-7z} = 4 \cdot 7^{n-1}$ gives the number of words of length n with $\mathrm{xdp}^{+}(w) \neq 0$.

Theorem 6 ([LM01, Theorem 2]). *There are $4 \cdot 7^{n-1}$ words of length $n > 0$ with $\mathrm{xdp}^{+}(w) \neq 0$. Of these, $4 \cdot 6^k \binom{n-1}{k}$ have probability 2^{-k} for all $0 \le k < n$.*

5 Conclusions

We analysed the additive differential probability $\mathrm{adp}^{\oplus}$ of exclusive-or. We expect that our results combined with the work of Lipmaa and Moriai will facilitate advanced differential cryptanalysis of ciphers that mix addition and exclusive-or as well as the design of such ciphers. These results can also be used to guide the choice between addition and exclusive-or as the key-mixing operation.

In general, $\mathrm{adp}^{\oplus}$ is much more difficult to analyse than xdp^{+} (note especially the straightforward analysis of xdp^{+} in Sect. 4). On the other hand, it is easier to find the maximal differentials for $\mathrm{adp}^{\oplus}$, although the maximal differentials for xdp^{+} have higher probability: it can be seen that $\mathrm{adp}^{\oplus}_{2\max}(\gamma) \le \mathrm{xdp}^{+}_{2\max}(\gamma) = \max_{\alpha,\beta} \mathrm{xdp}^{+}(\alpha, \beta \to \gamma)$ for all γ. (See Fig. 1.) A short comparison of some of the properties of xdp^{+} and $\mathrm{adp}^{\oplus}$ is given in Table 3.

Maybe the main contribution of this paper is the formal series approach. In addition to the new results, we were able to give a simpler proof of the results of Lipmaa and Moriai on xdp^{+}. The results from [Wal03] can also be rephrased using our approach. We expect that our approach of using formal series has also other applications in cryptanalysis.

Acknowledgements

This work was partially supported by the Finnish Defence Forces Research Institute of Technology.

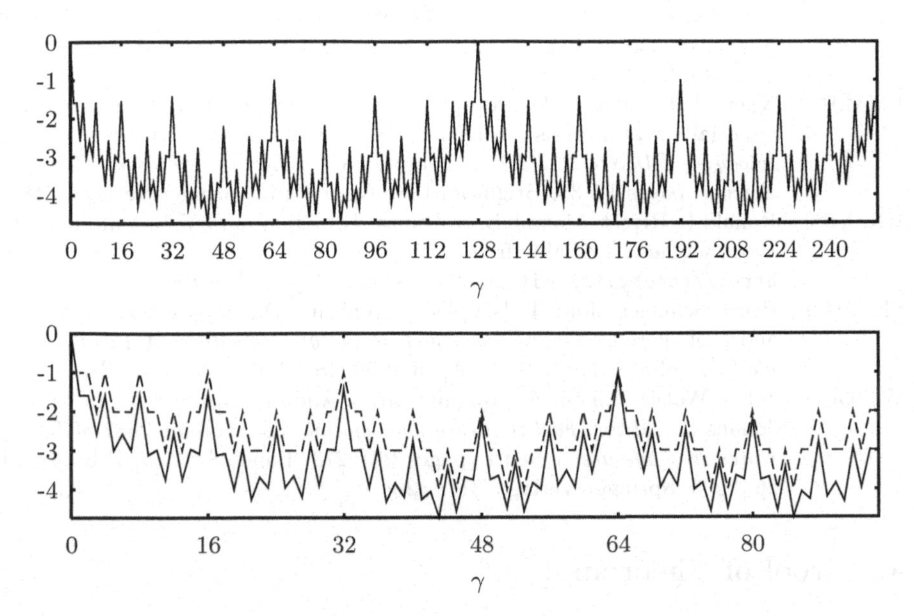

Fig. 1. Top: tabulation of values $\log_2 \mathrm{adp}^{\oplus}_{2\max}(\gamma)$, $0 \le \gamma \le 255$, for $N = 8$. Bottom: partial tabulation of values $\log_2 \mathrm{adp}^{\oplus}_{2\max}(\gamma)$ (solid) and $\log_2 \mathrm{xdp}^{+}_{2\max}(\gamma)$ (dashed), $0 \le \gamma \le 95$, for $N = 8$

References

[BCD+98] Carolynn Burwick, Don Coppersmith, Edward D'Avignon, Rosario Gennaro, Shai Halevi, Charanjit Jutla, Stephen M. Matyas Jr., Luke O'Connor, Mohammad Peyravian, David Safford, and Nevenko Zunic. MARS — A Candidate Cipher for AES. Available from `http://www.research.ibm.com/security/mars.html`, June 1998. 318

[Ber92] Thomas A. Berson. Differential Cryptanalysis Mod 2^{32} with Applications to MD5. In Rainer A. Rueppel, editor, *Advances in Cryptology — EUROCRYPT '92*, volume 658 of *Lecture Notes in Computer Science*, pages 71–80, Balatonfüred, Hungary, 24–28 May 1992. Springer-Verlag. ISBN 3-540-56413-6. 319

[BR88] Jean Berstel and Christophe Reutenauer. *Rational Series and Their Languages.* EATCS Monographs on Theoretical Computer Science. Springer-Verlag, 1988. 319

[BS91] Eli Biham and Adi Shamir. Differential Cryptanalysis of DES-like Cryptosystems. *Journal of Cryptology*, 4(1):3–72, 1991. 317

[Lip02] Helger Lipmaa. On Differential Properties of Pseudo-Hadamard Transform and Related Mappings. In Alfred Menezes and Palash Sarkar, editors, *INDOCRYPT 2002*, volume 2551 of *Lecture Notes in Computer Science*, pages 48–61, Hyderabad, India, 15–18 December 2002. Springer-Verlag. 317, 318, 326

[LM01] Helger Lipmaa and Shiho Moriai. Efficient Algorithms for Computing Differential Properties of Addition. In Mitsuru Matsui, editor, *Fast Soft-*

ware Encryption 2001, volume 2355 of *Lecture Notes in Computer Science*, pages 336–350, Yokohama, Japan, 2–4 April 2001. Springer-Verlag, 2002. 317, 318, 320, 324, 325, 326, 327, 328, 330

[LMM91] Xuejia Lai, James L. Massey, and Sean Murphy. Markov Ciphers and Differential Cryptanalysis. In Donald W. Davies, editor, *Advances in Cryptology — EUROCRYPT '91*, volume 547 of *Lecture Notes in Computer Science*, pages 17–38, Brighton, UK, April 1991. Springer-Verlag. 318

[RRSY98] Ronald L. Rivest, Matt J. B. Robshaw, R. Sidney, and Y. L. Yin. The RC6 Block Cipher. Available from `http://theory.lcs.mit.edu/~rivest/rc6.ps`, June 1998. 318

[SKW+99] Bruce Schneier, John Kelsey, Doug Whiting, David Wagner, Chris Hall, and Niels Ferguson. *The Twofish Encryption Algorithm: A 128-Bit Block Cipher*. John Wiley & Sons, April 1999. ISBN: 0471353817. 318

[Wal03] Johan Wallén. Linear Approximations of Addition Modulo 2^n. In Thomas Johansson, editor, *Fast Software Encryption 2003*, volume 2887 of *Lecture Notes in Computer Science*, pages 261–273, Lund, Sweden, February24–26 2003. Springer-Verlag. 317, 328

A Proof of Theorem 4

In order to prove Theorem 4, we introduce the following notation. Define the carry function carry: $\mathbf{Z}_2^N \times \mathbf{Z}_2^N \to \mathbf{Z}_2^N$ of addition modulo 2^N by $\operatorname{carry}(x,y) = (x+y) \oplus x \oplus y$. It is easy to see that

$$\operatorname{xdp}^+(\alpha,\beta \to \gamma) = \Pr_{x,y}[\operatorname{carry}(x,y) \oplus \operatorname{carry}(x \oplus \alpha, y \oplus \beta) = \alpha \oplus \beta \oplus \gamma] \ .$$

Denote $c = \operatorname{carry}(x,y)$ and $c^* = \operatorname{carry}(x \oplus \alpha, y \oplus \beta)$, where x, y, α and β are understood from context. Note that c_i can be recursively defined as $c_0 = 0$ and $c_{i+1} = 1$ if and only if at least two of x_i, y_i and c_i are 1. To simplify some of the formulae, denote $\operatorname{xor}(x,y,z) = x \oplus y \oplus z$ and $\Delta c = c \oplus c^*$. Then $\operatorname{xdp}^+(\alpha,\beta \to \gamma) = \Pr_{x,y}[\Delta c = \operatorname{xor}(\alpha,\beta,\gamma)]$. Let furthermore xy denote the componentwise product of x and y, $(xy)_i = x_i y_i$.

The linear representation of xdp^+ follows easily from the following result [LM01, Lemma 2].

Lemma 4. *Fix $\alpha, \beta \in \mathbf{Z}_2^n$ and $i \geq 0$. Then*

$$\Pr_{x,y}[\Delta c_{i+1} = 1 \mid \Delta c_i = r] = T(\alpha_i + \beta_i + r) \ ,$$

where T is as in Theorem 4.

This result follows easily from the recursive definition of the carry function and a case-by-case analysis.

Proof (of Theorem 4). Let $(\alpha, \beta \to \gamma)$ be the differential associated to the word w. Let x, y be uniformly distributed independent random variables over

$\mathbf{Z}_2^{|w|}$. For compactness, we denote $\mathrm{xor}(w) = \alpha \oplus \beta \oplus \gamma$. Let $P(w,k)$ be the 2×1 substochastic matrix given by

$$P_j(w,k) = \Pr_{x,y}[\Delta c \equiv \mathrm{xor}(w) \pmod{2^k}, \Delta c_k = j]$$

for $0 \leq k \leq |w|$ and let $M(w,k)$ be the 2×2 substochastic transition matrix

$$M_{ij}(w,k) = \Pr_{x,y}[\Delta c_k = \mathrm{xor}(w)_k, \Delta c_{k+1} = i \mid \Delta c \equiv \mathrm{xor}(w) \pmod{2^k}, \Delta c_k = j]$$

for $0 \leq k < |w|$. Since $P_i(w,k+1) = \sum_j M_{ij}(w,k) P_j(w,k)$, $P(w,k+1) = M(w,k)P(w,k)$. Note furthermore that $P(w,0) = C$ and that $\mathrm{xdp}^+(w) = \sum_j P_j(w,|w|) = LP(w,|w|)$. By Lemma 4, it is clear that

$$M_{ij}(w,k) = \begin{cases} 1 - T(\alpha_k + \beta_k + j) & \text{if } i = 0 \text{ and } \mathrm{xor}(w)_k = j\ , \\ T(\alpha_k + \beta_k + j) & \text{if } i = 1 \text{ and } \mathrm{xor}(w)_k = j \text{ and} \\ 0 & \text{otherwise}\ . \end{cases}$$

That is, $M(w,k) = X_{w_k}$ for all k. It follows by induction that $\mathrm{xdp}^+(w) = LX_{w_{|w|-1}} \cdots X_{w_0} C$. □

Two Power Analysis Attacks against One-Mask Methods

Mehdi-Laurent Akkar[1], Régis Bévan[2], and Louis Goubin[3]

[1] Texas Instruments
821 Avenue Jack-Kilby, BP 5, 06271 Villeneuve-Loubet Cedex, France
akkar@ti.com

[2] Oberthur Card Systems
25 rue Auguste Blanche, 92800 Puteaux, France
r.bevan@oberthurcs.com

[3] Axalto Cryptography Research & Advanced Security
36-38 rue de la Princesse, BP 45, 78431 Louveciennes Cedex, France
LGoubin@axalto.com

Abstract. In order to protect a cryptographic algorithm against Power Analysis attacks, a well-known method consists in hiding all the internal data with randomly chosen masks.
Following this idea, an AES implementation can be protected against Differential Power Analysis (DPA) by the "*Transformed Masking Method*", proposed by Akkar and Giraud at CHES'2001, requiring *two* distinct masks. At CHES'2002, Trichina, De Seta and Germani suggested the use of a *single* mask to improve the performances of the protected implementation. We show here that their countermeasure can still be defeated by usual first-order DPA techniques.
In another direction, Akkar and Goubin introduced at FSE'2003 a new countermeasure for protecting secret-key cryptographic algorithms against high-order differential power analysis (HO-DPA). As particular case, the "*Unique Masking Method*" is particularly well suited to the protection of DES implementations. However, we prove in this paper that this method is not sufficient, by exhibiting a (first-order) enhanced differential power analysis attack. We also show how to avoid this new attack.

Keywords: Tamper-resistant devices, Side-Channel attacks, Power Analysis, DPA, Transformed Masking Method, Unique Masking Method, DES, AES.

1 Introduction

The framework of Differential Power Analysis (also known as DPA) was introduced by P. Kocher, J. Jaffe and B. Jun in 1998 ([13]) and subsequently published in 1999 ([14]). The initial focus was on symmetrical cryptosystems such as DES (see [13, 16, 2]) and the AES candidates (see [4, 5, 8]), but public key cryptosystems have since also been shown to be vulnerable to the DPA attacks (see [17, 7, 11, 12, 19]).

B. Roy and W. Meier (Eds.): FSE 2004, LNCS 3017, pp. 332–347, 2004.

In software two main families of countermeasures against DPA are known:

- In [11, 12], L. Goubin and J. Patarin described a generic countermeasure consisting in splitting all the intermediate variables, using the secret sharing principle. This *duplication method* was also proposed shortly after by S. Chari *et al.* in [5] and [6].
- In [3], M.-L. Akkar and C. Giraud introduced the *transformed masking method* (TMM), an alternative countermeasure to the DPA. The basic idea is to perform all the computation such that all the data are XORed with a random mask. Moreover, the tables (*e.g.* the DES S-Boxes) are modified such that the output of a round is masked by the same mask as the input.

Both these methods have been proven secure against the initial DPA attacks, and are now widely used in real life implementations of many algorithms.

The TMM method can be used to protect AES implementations against DPA. Two masking values are then required to cope with the (non-linear) ByteSub operation. In a recent paper E. Trichina, D. De Seta, L. Germani [20] proposed the "*Simplified Adaptive Multiplicative Masking*" (SAMM), a variation of TMM with a single masking value, thus providing simpler and faster implementations for AES. Unfortunately, we will show in this paper that this method can be broken by usual DPA attacks.

Also suggested by P. Kocher, J. Jaffe and B. Jun [13, 14], and formalized by T. Messerges [15], Higher-Order Differential Power Analysis (HO-DPA) consists in studying correlations between the secret data and *several* points of the electric consumption curves (instead of *single* points for the basic DPA attack). To protect secret-key algorithms against this new class of attacks, M.-L. Akkar and L. Goubin recently proposed [1] a new countermeasure: the so-called "*Unique Masking Method*" (UMM).

In this paper, we describe an unexpected power-analysis attack, which can be applied to implementations of secret-key algorithms using the UMM method. More precisely, in the chosen-plaintext model, the attacker can recover the secret key by successively applying two classical DPA attacks on the second round.

The paper is organized as follows:

- In section 2, we recall basic notions about Differential Power Analysis (DPA), the Transformed Masking Method (TMM) and about the Simplified Adaptive Multiplicative Masking (SAMM).
- In section 3, we analyze the mathematical hypotheses on which the security of SAMM relies and point out a flaw in the design of the countermeasure.
- In section 4, we show how this flaw can be exploited by studying the power consumption of a real component.
- In section 5, we recall basic notions about Higher-Order Differential Power Analysis (HO-DPA) and about the Unique Masking Method (UMM).
- In section 6, we theoretically study the security of the UMM applied to DES and show how it could be cryptanalysed.
- In section 7, we give our perspectives and conclusions about the attacks presented in this paper.

2 Background

2.1 Differential Power Analysis

Differential Power Analysis (DPA) was introduced by Kocher, Jaffe and Jun in 1998 [13] and published in 1999 [14]. The basic idea is to make use of potential *correlations* between the data handled by the micro-controller and the electric consumption measured values. Since these correlations are often very low, *statistical* methods must be applied to deduce sufficient information from them.

The principle of DPA attacks consists in comparing consumption values measured on the *real* physical device (for instance a GSM chip or a smart card) with values computed in an *hypothetical model* of this device (the hypotheses being made among others on the nature of the implementation, and chiefly on a part of the secret key). By comparing these two sets of values, the attacker tries to recover all or part of the secret key.

The initial target of DPA attacks was limited to symmetric algorithms. Vulnerability of DES – first shown by Kocher, Jaffe and Jun [13, 14] – was further studied by Goubin and Patarin [11, 12], Messerges, Dabbish, Sloan [16] and Akkar, Bévan, Dischamp, Moyart [2]. Applications of these attacks were also largely taken into account during the AES selection process, notably by Biham, Shamir [4], Chari, Jutla, Rao, Rohatgi [5] and Daemen, Rijmen [8].

However public-key algorithms were also shown to be threatened: Goubin, Patarin [11, 12] and Messerges, Dabbish, Sloan [17] showed how to apply DPA against RSA, and the case of elliptic curve cryptosystems was analyzed by Coron [7], Okeya, Sakurai [19] and many others (see for instance [10] for a detailed bibliography).

In the basic DPA attack (see [13, 14] or [9]), also known as first-order DPA (or just DPA), the attacker records the power consumption signals and computes statistical properties of the signal for each individual moment in time of the computation. This attack does not require any knowledge about the individual electric consumption of each instruction, nor about the position in time of each of these instructions. It only relies on the following fundamental hypothesis (quoted from [12]):

Fundamental hypothesis (order 1): *There exists an intermediate variable, that appears during the computation of the algorithm, such that knowing a few key bits (in practice less than 32 bits) allows us to decide whether two inputs (respectively two outputs) give or not the same value for this variable.*

2.2 The Transformed Masking Method for AES

More details about this technique can be found in [3].

The idea is to mask the message at the beginning of the AES algorithm, and to recover the same mask at the end of each round. An important step for the AES is to securely perform the inversion step. For this, one need to compute $A_{i,j}^{-1} \oplus X_{i,j}$ from $A_{i,j} \oplus X_{i,j}$ where $A_{i,j}$ is the block (i, j) in an AES

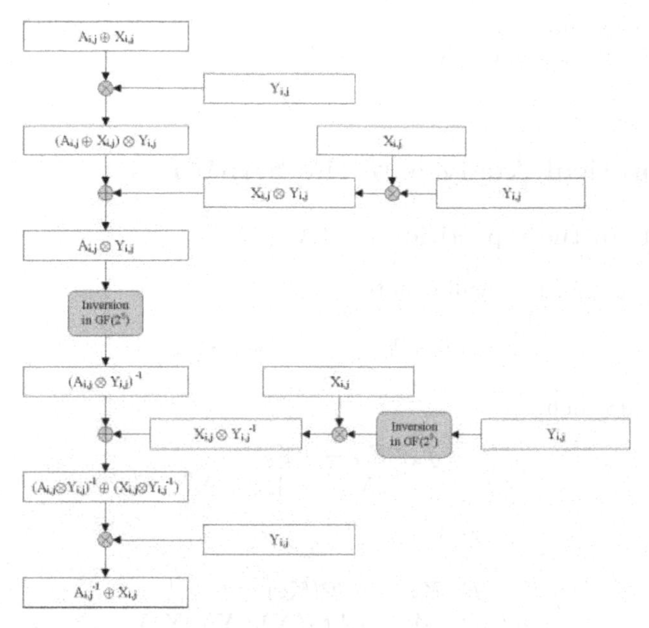

Fig. 1. Modified inversion in GF(2^8) with masking countermeasure

computation and $X_{i,j}$ the corresponding masking value. To perform this securely, Akkar and Giraud proposed to use the following operations (see Fig. 1):

1. Multiply the masked value A by a non zero random value Y to get $AY \oplus XY$
2. XOR with XY to get AY
3. Perform the inversion to get $A^{-1}Y^{-1}$
4. XOR with XY^{-1} to get $A^{-1}Y^{-1} \oplus XY^{-1}$
5. Multiply by Y to get $A^{-1} \oplus X$

2.3 The Simplified Adaptive Multiplicative Masking for AES

More details about this technique can be found in [20].

The general idea is not to use a different Y masking value to switch from an additive mask to a multiplicative one, but to use the same masking value X instead. The algorithm is the following.

1. Multiply the masked value A by X to get $AX \oplus X^2$
2. XOR with X^2 to get AX
3. Perform the inversion to get $A^{-1}X^{-1}$
4. XOR with 1 to get $A^{-1}X^{-1} \oplus 1$
5. Multiply by X to get $A^{-1} \oplus X$

One can notice that the particular value $AX \oplus X^2$ appears during the computation. The authors of [20] suggest that even if $AX \oplus X^2$ is not fully random, it is sufficiently random to serve the purpose. In the next section, we will precisely study the function $AX \oplus X^2$ and show that it introduces a weakness in the method.

3 Theoretical Analysis of the SAMM

3.1 Study of the repartition of $AX + X^2$

In the following part we will denote:

$$\mathbb{K} = \mathbb{F}_2 \ \ and \ \ \mathbb{K}_8 = \mathbb{F}_{2^8}$$

We will also define:

$$\begin{aligned} f_A\colon \mathbb{K}_8 &\longrightarrow \mathbb{K}_8 \\ X &\longrightarrow AX + X^2 \end{aligned}$$

and

$$\begin{aligned} F\colon \mathbb{K}_8 &\longrightarrow \mathcal{P}(\mathbb{K}_8) \\ A &\longrightarrow \{f_A(X),\ X \in \mathbb{K}_8\} \end{aligned}$$

where $\mathcal{P}(\mathbb{K}_8)$ represents the power set of $\mathbb{K}_8$.

Remark: In what follows, we will denote by $\#(A)$ the cardinality of A.

With these definitions we have the following result:

Theorem 1. *If $A \in \mathbb{K}_8$ then we have:*

$$\begin{aligned} \#F(A) &= 128 \text{ if } A \neq 0 \\ &= 256 \text{ if } A = 0 \end{aligned}$$

Proof.

- Case $A = 0$: $f_0(X) = X^2$ and is bijective on $\mathbb{K}_8$. Therefore we obviously have $\#F(0) = 256$.
- Case $A \neq 0$: $f_{A \neq 0}(X) = AX + X^2$, if given a Y, we want to solve the equation $f_A(X) = Y$, by defining $Z = X/A$ (because $A \neq 0$), we obtain the following equation to solve:

$$Z^2 + Z = Y/A^2$$

and it is well known that this equation has no solution if $Trace(Y/A^2) = 1$ and two solutions (if one gets one solution W the other is $W + 1$) if $Trace(Y/A^2) = 0$. Therefore it is easy to see that $\#F(A) = 128$.

This theorem shows that the simplified mask covers only one half of the 256 possible values on $\mathbb{K}_8$. This was already noticed in the article [20].

However, let us now study the distribution of $F(A_1)$ and $F(A_2)$, for two distinct values A_1 and A_2. The following proposition gives us a more precise result:

Theorem 2. *If* $(A_1, A_2) \in (\mathbb{K}_8 \setminus \{0\})^2$ *and* $A_1 \neq A_2$ *then:*

$$\#(F(A_1) \cap F(A_2)) = 64$$

Proof.

Let A_1 and A_2 be two elements of $\mathbb{K}_8 \setminus \{0\}$. We are looking for the values Y such as the two equations

$$\begin{cases} Y = A_1 X + X^2 \\ Y = A_2 X' + X'^2 \end{cases}$$

are simultaneously solvable or unsolvable. This is equivalent to:

$$Trace(Y/A_1) = Trace(Y/A_2)$$

By linearity of the trace operator we obtain:

$$Trace(Y/A_1 - Y/A_2) = 0$$

Let now consider $Z \in \mathbb{K}_8$. To Z corresponds a unique value Y such as $Y/A_1 - Y/A_2 = Z$, which is $Y = Z/(1/A_1 - 1/A_2)$. Since there exist 128 elements Z of trace 0, there exist 128 elements Y such that the previous system is simultaneously solvable or unsolvable.

Let us now consider:

$$\begin{cases} n_1 = \{Y \in \mathbb{K}_8 \text{ such that } Trace(Y/A_1) = 0 \text{ and } Trace(Y/A_2) = 1\} \\ n_2 = \{Y \in \mathbb{K}_8 \text{ such that } Trace(Y/A_2) = 0 \text{ and } Trace(Y/A_1) = 1\} \\ n_3 = \{Y \in \mathbb{K}_8 \text{ such that } Trace(Y/A_1) = 0 \text{ and } Trace(Y/A_2) = 0\} \\ n_4 = \{Y \in \mathbb{K}_8 \text{ such that } Trace(Y/A_2) = 1 \text{ and } Trace(Y/A_1) = 1\} \end{cases}$$

The last result gives us the following equation: $n_3 + n_4 = 128$

This equation, together with obvious trace considerations, gives the following system:

$$\begin{cases} n_3 + n_4 = 128 \\ n_1 + n_3 = 128 \\ n_2 + n_3 = 128 \\ n_2 + n_4 = 128 \end{cases}$$

Solving this system, we obtain: $n_1 = n_2 = n_3 = n_4 = 64$, thus achieving the proof of Theorem 2.

3.2 Consequences

Theorem 2 is really important because it proves that two distinct values (during the computation) are "projected" onto two sets of 128 values when the SAMM is implemented and that these two sets have 64 common values and 64 different ones. Therefore when an attacker performs a DPA on an implementation including the SAMM he will on average record the consumption of $F(A)$ instead of A but for two distinct values A and B, $F(A)$ and $F(B)$ are also distinct, allowing the attacker to distinguish the two cases. This give strong evidence that the attack is very likely to work. Moreover even if the attacker does not know the fact that SAMM is implemented, he will be able to recover the key because the attack is exactly the same.

4 Semi-real and Real Analysis of the SAMM

We have seen in the previous section that there is a theoretical flaw in the SAMM method. To go further, we have to check which results are obtained when using different models of power consumption. We also have to experiment on a real embedded device. More work about the consumption model of embedded device can be found in many papers: see for example [6, 2, 16].

4.1 Linear Model

The "linear model" considers that the power consumption of the card is proportional to the value of each bit of the manipulated value. For example the consumption of an 8-bit value

$$X = b_0 b_1 b_2 b_3 b_4 b_5 b_6 b_7$$

will be equal to

$$c_0 * b_0 + c_1 * b_1 + c_2 * b_2 + c_3 * b_3 + c_4 * b_4 + c_5 * b_5 + c_6 * b_6 + c_7 * b_7$$

where $\{c_i\}_{i \leq 8}$ is the average consumption of bit i. For example the Hamming weight (denoted by HW) model is a linear one with $c_i = c_j$ for all (i, j).

We have performed some experimentations in the following way. We have computed for every $A = 0..255$ the whole subset $F(A)$ and we have computed the average weight of each bit of the values in $F(A)$ to check if there was any bias in the results. The results are as follows:

- For 8 values of A (0x03,0x15,0x87,0x8C,0xCE,0xEB,0xED and 0xF6), one specific bit of the eight bits of $f_A(X)$ always vanishes, whatever the value of X is!
- For the 248 other values A, the probability that "the i-th bit of $f_A(X)$ is equal to zero" is equal to $\frac{1}{2}$. So an attacker is unable to predict one bit in order to compute a classical DPA attack.

If we now analyze the repartition of the HW of the values $F(A)$ we get the nine following sets:

Subset	#	HW 0	HW 1	HW 2	HW 3	HW 4	HW 5	HW 6	HW 7	HW 8
1	1	1	8	28	56	70	56	28	8	1
2	28	2	12	32	52	60	52	32	12	2
3	56	2	10	26	50	70	62	30	6	0
4	8	2	14	42	70	70	42	14	2	0
5	70	2	8	24	56	76	56	24	8	2
6	28	2	4	32	60	60	60	32	4	2
7	8	2	2	42	42	70	70	14	14	0
8	56	2	6	26	62	70	50	30	10	0
9	1	2	0	56	0	140	0	56	0	2

For example, line 8 in the table means that there are 56 values of A such as, there exists 2 values X for which $HW(f_A(X)) = 0$, 6 values X for which $HW(f_A(X)) = 1$, ..., and no element X such as $HW(f_A(X)) = 8$. Moreover, one can notice that the average Hamming weight of all the subsets is equal to 4 except for the subset 4 (which contains the 8 special values detailed in the previous paragraph), whose mean is equal to 3.5. The explicit values A of the nine subsets can be found in the appendix.

The conclusion is that even if only a small bias exists, if a component respects a linear model of consumption (such as the HW model), the masking method would probably be quite efficient in practice. However, one has to check what happens with a real embedded device: some experiments on an 8051-based 8-bit CPU are discussed in the following section.

4.2 Real Device Analysis of the SAMM

To obtain concrete results, we have made a comparison between the power consumption of a card manipulating a value A and a card manipulating the value $AX \oplus X^2$ with random X. For this we have recorded the consumption of a load operation on an 8 bits CPU based on an 8051 core.

The following curves (see Fig. 2 and Fig. 3) have been obtained by using the average consumption of 1024 power consumption traces. For the unmasked value we have used the record of 512 times the same value A. For the SAMM we have used the average of twice the 256 values $AX \oplus X^2$ with $X \in [0, 255]$. Then we have ordered the value per consumption to get an idea of the variance of the value. As seen before there are eight special values (the ones for which one bit is always 0) that we have excluded from the curves, indeed the consumption of these values was really different from the others[1].

As can be seen on these two curves, masked or unmasked, there is a quite important difference between different consumption values. That proves that,

[1] That shows that probably a SPA attack may be quite easy by focusing on these particular values !

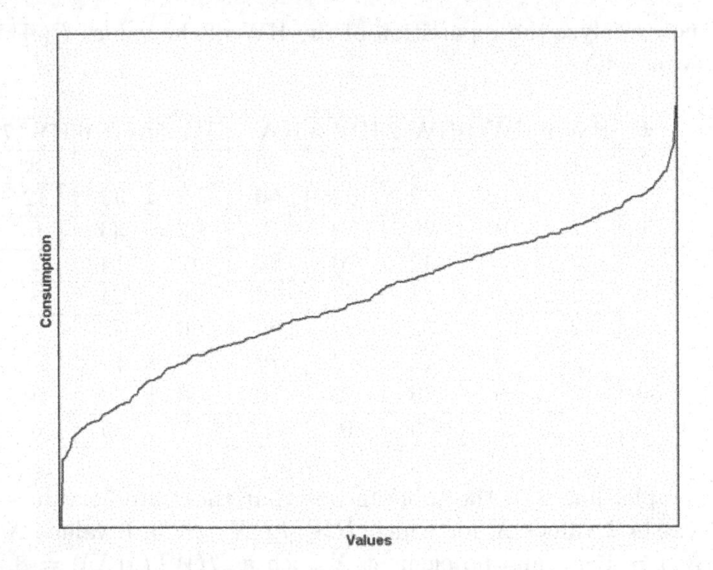

Fig. 2. Repartition of the consumption with unmasked values

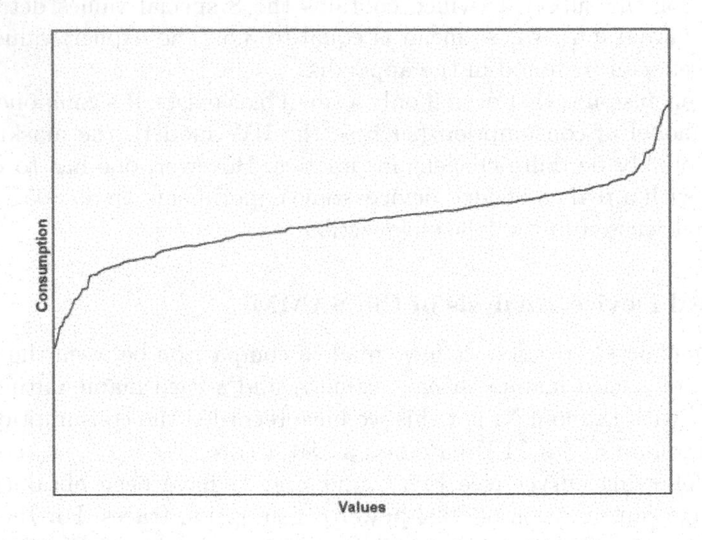

Fig. 3. Repartition of the consumption with SAMM values

in a real embedded device, even if it gets somewhat slower (the variance in the SAMM curve is smaller), a successful DPA attack on a classical AES implementation will also succeed on an AES implementation using the SAMM method.

If the attacker knows that the SAMM method is used, he can use an adapted DPA attack to retrieve the key with less measurements than the usual first order DPA. For every hypothesis K_j on the key byte, one computes the difference of the means of the following sets:

$$S_0 = \{\text{measurements for the messages } M_k | M_k \oplus K_j \in \text{subset } 4\}$$
$$S_1 = \{\text{measurements for the messages } M_k | M_k \oplus K_j \notin \text{subset } 4\}$$

Then the correct hypothesis is the one with the highest difference of means.

Remark. Due to obvious confidentiality reasons, details about the chip we used have to remain undisclosed.

5 The Unique Masking Method

5.1 High-Order Differential Power Analysis

As mentioned in section 2, Differential Power Analysis implies the use of a hypothetical model of the physical device which performs the cryptographic computations. If this model is able to predict a *single* value, for instance the electric consumption of the device for a single instant t, the differential power analysis is said to be of *first order*. If the model is able to predict *several* such values, the differential power analysis is said to be of *higher order*.

High-order differential power analysis (HO-DPA), suggested by Kocher, Jaffe and Jun [13, 14] (see also [9]), was formalized by Messerges in [15]. In the spirit of [12], Akkar and Goubin (see [1]) gave a necessary and sufficient condition for a DPA attack of order n to be applicable:

Fundamental hypothesis (order n): *There exists a set of n intermediate variables, that appear during the computation of the algorithm, such that knowing a few key bits (in practice less than 32 bits) allows to decide whether two inputs (respectively two outputs) give or not the same value for a known function of these n variables.*

In [1], Akkar and Goubin studied the impacts of HO-DPA attacks on implementations of cryptographic algorithms, and showed that usual countermeasures against DPA are not sufficient to avoid this new class of attacks. Moreover, they proposed a generic protection against higher-order attacks, illustrated in details for the DES case.

5.2 A Countermeasure against HO-DPA

The so-called "*Unique Masking Method*" (UMM) aims at providing a generic protection against differential power analysis of order n, whatever the value n may be. The two principles of this method is first to mask only the values that depend on less than 32 bits of the key in order to prevent DPA and secondly intermediate independent variables depending on less than 32 bits of the key must not be masked by the same value in order to thwart HO-DPA.

Let us describe the basic idea for the DES example. The unique mask is a random 32-bit value α. From this value, two sets of 8 S-boxes, denoted by $\tilde{S}_1$ and $\tilde{S}_2$, are defined by

$$\begin{cases} \forall x \in (\mathbb{F}_2)^{48},\ \tilde{S}_1(x) = S(x \oplus E(\alpha)) \\ \forall x \in (\mathbb{F}_2)^{48},\ \tilde{S}_2(x) = S(x) \oplus P^{-1}(\alpha) \end{cases}$$

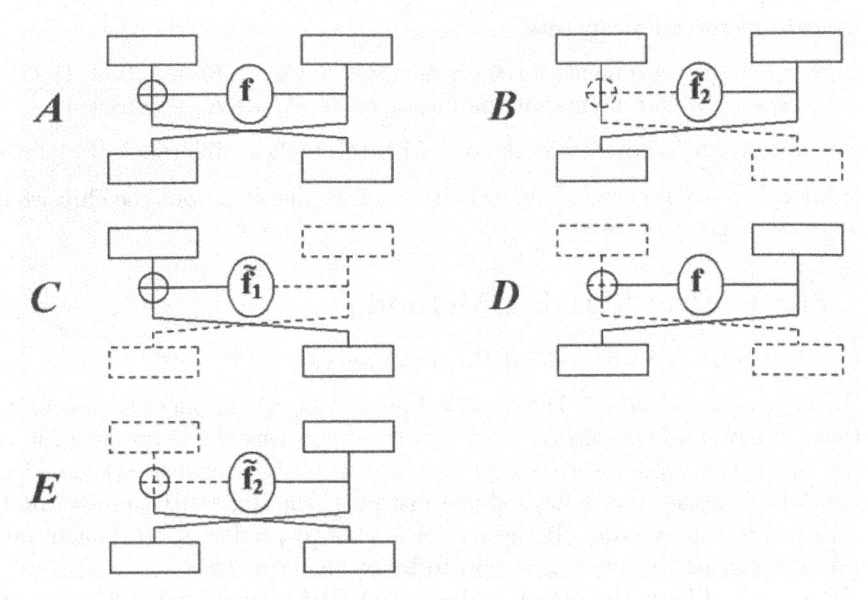

Fig. 4. The five possibilities for DES rounds

where S denotes the 8 usual DES S-boxes, E is the expansion function and P is the classical DES permutation just after the S-boxes.

Let f_{K_i} ($1 \leq i \leq 16$) be the functions involved in the Feistel scheme of DES (with the usual S-boxes), and $\tilde{f}_{1,K_i}$ (resp. $\tilde{f}_{2,K_i}$) the analogous function with the $\tilde{S}_1$ (resp. $\tilde{S}_2$) S-boxes. DES rounds can then be built from five possible frames, given in Figure 4 (solid lines correspond to unmasked data, dashed lines to masked data).

To be consistent with the DES computation, the sequence of rounds has to follow some rules, which can be summarized by a finite automaton, as shown in Figure 5. Initial states correspond to non-masked inputs (A or B), final states correspond to non-masked outputs (A or E). As an example, $BCDCDCEBCDCDCDCE$ is a valid sequence.

To provide a protection against differential power analysis attacks, all the values depending on less than 36 key bits are masked by α. This gives further constraints on the sequence of rounds: the three first ones have to be of the form BCD or BCE, and symmetrically the three last ones must be of the form BCE or DCE. For instance, the sequence $BCDCDCEBCDCDCDCE$ fulfills these conditions.

The unique masking method has several advantages: the structure of classical implementations can be kept unchanged, the only difference is the generation of random tables (see [1] for several practical methods for securely generating S-boxes from the mask α). Moreover performances remain acceptable. For instance, [1] reports on a DES implementation, including the unique masking

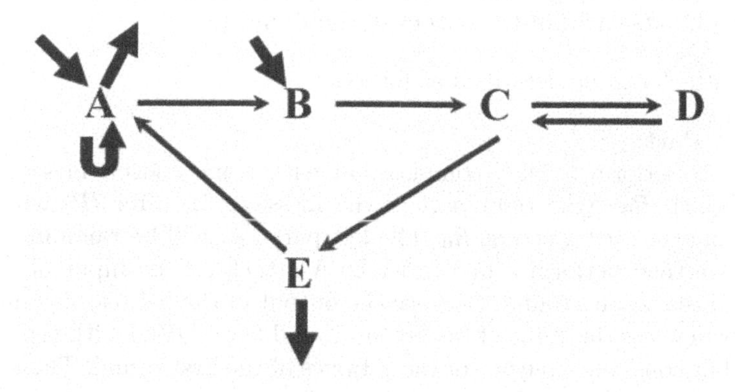

Fig. 5. Valid sequences for the rounds

method (together with SPA and DFA protections), which runs in 38 ms (at 10 MHz) on a ST19 chip.

6 Enhanced DPA of the Unique Masking Method

6.1 Basic Idea

For all the proposed sequences of rounds, the second round is always a "C"-type round. The output of the S-Boxes is unmasked and stay unmasked after being XORed with the left part of the message. After this XOR, the value goes through an "E"-type or "D"-type round for which the computation of the beginning of the f function (expansion function) is unmasked. For the second case (where the third round is of "D" type) even the S-Boxes output and the P permutation are unmasked.

The fact that the output of second round S-Boxes is unmasked will be the base of our attack.

6.2 Attacking the UMM

The attack is a chosen message attack. The main idea of the attack is to retrieve two intermediate values which are not protected against DPA, and then to get the key bits by solving an equation involving the two intermediate values. Before describing the attack, let us introduce some notations:

- IP denotes the initial permutation.
- E denotes the expansion function (from 32 to 48 bits).
- S denotes the S-Boxes.
- P denotes the P permutation after the S-Boxes
- $\oplus$ denotes the XOR operation
- K_i denotes the 48 bits subkey of round i.

- Finally, for a message M, L_i and R_i will denote the left part and the right part (32 bits each) of the output of the round i.

The attack can be described as follows:

- **First Part:**
  - We perform a DES computation with some chosen messages M_i for which $R_{0,i}$ (the right part of the message M_i after *IP*) will be set to an arbitrary constant R_0. The left part $L_{0,i}$ will be random.
  - We then perform a first-order DPA attack on the input of each S-Box of the second round. Because the output of the S-Boxes is unmasked we will guess the value of the second round key XORed with the – unknown but constant – output of the S-Boxes of the first round. The found value will be:

$$\delta = K_2 \oplus E(P(S(K_1 \oplus E(R_0))))$$

- **Second Part:**
  - We will perform a second first-order DPA with other messages with a different known constant value R_0', which will provide:

$$\delta' = K_2 \oplus E(P(S(K_1 \oplus E(R_0'))))$$

- **Final Part:**
  - By taking XOR of the two values found at last steps, we obtain the value:

$$\delta \oplus \delta' = (K_2 \oplus E(P(S(K_1 \oplus E(R_0))))) \oplus (K_2 \oplus E(P(S(K_1 \oplus E(R_0')))))$$

    The value K_2 vanishes and the linearity of functions E and P gives us the equation:

$$S(K_1 \oplus E(R_0)) \oplus S(K_1 \oplus E(R_0')) = g(\delta \oplus \delta')$$

  - Because we know R_0 and R_0', doing a exhaustive search on each 6-bits subkeys of K_1, will give us all the possible values for the subkey K_1.

On average, the differential properties of S will give us about 4 possibilities for each subkeys. Since there are 8 subkeys and we need to find the 8 bits which are not in K_1, this gives us $4^8 \times 2^8 = 2^{24}$ possibilities on the key. So an exhaustive search with one known plaintext/ciphertext pair will take a few seconds on a PC. If one does not have access to such a pair, another attack with a R_0'' constant value will decrease the possibilities for K_1 (practically, K_1 is completely known if R_0'' is not badly chosen). Then it is possible to perform classical DPA against K_2, once K_1 is known, to get directly the 56 bits of the DES key.
The reader has to notice that even if the attack exploits the correlation of two results, the attack consists just in applying twice a really usual first-order DPA attack. So the number of traces and the processing time is just twice those needed for a classical DPA against an unprotected DES.

6.3 Attack Scenarios and Countermeasures

- Our attack is feasible only if the attacker is able to keep constant the right part of the message after the initial permutation. If it is not possible the attacker has to "wait" for messages having the same R_0 part (1 message every 2^{32} messages on average). So the attack becomes quite long to perform.
- If we consider a scenario for which only the output is known, the attack becomes as difficult as the one for which the message could not be controlled. A chosen cipher text attack mainly applies against an authentication scheme, in which the device has to cipher a challenge from the outside.
- A solution consists in masking the output of the second round, which seems to make this attack unfeasible. One can use a different mask but the use of α_1 is not forbidden since the bits of R_1 and R_2 that are masked by the same value depends on 42 bits of the key. So we need one more function $\tilde{f}_{3,K_i}$ with the modified S-Boxes $\tilde{S}_3(x)$ such that $\forall x \in (\mathbb{F}_2)^{48}$, $\tilde{S}_3(x \oplus E(\alpha)) = S(x) \oplus P^{-1}(\alpha)$.

7 Conclusion

In this paper we presented two new power analysis attacks.

The first one applies to the Simplified Adaptive Multiplicative Masking (SAMM). Even if, in some models (HW Model or Linear Model), the SAMM seems to be quite secure, the SAMM is (in practice) vulnerable to usual (first-order) DPA attacks. Moreover we have seen that, from a theoretical point of view, this countermeasure is not correct because of the distribution of the values $AX \oplus X^2$, which is quite unbalanced. We thus recommend, to obtain DPA-resistant implementations of AES, using the original TMM method with two masks (correctly implemented due to the zero problem as described in [20, 1]) or, to use a dynamic inversion of the S-Box if the 256 bytes needed in RAM are available.

The second attack applies to the Unique Masking Method for DES, showing that in the chosen-message scenario, an enhanced first-order DPA attack is still possible. This is not due to the method itself but to a different model of the attacker, allowing her to have full access to the inputs of the algorithm. In the case of the DES, we have then shown that unique masks have to be extended to at least two rounds to protect the implementation against a chosen text attack. The drawbacks are an overhead on performances and an increase of required memory by one third (corresponding to the computation of one more modified S-box).

References

[1] M.-L. Akkar, L. Goubin, *A Generic Protection against High-Order Differential Power Analysis*. In Proceedings of FSE'2003, LNCS 2887, Springer-Verlag, 2003. 333, 341, 342, 345

[2] M.-L. Akkar, R. Bevan, P. Dischamp, D. Moyart, *Power Analysis: What is now Possible*. In Proceedings of ASIACRYPT'2000, LNCS 1976, pp. 489-502, Springer-Verlag, 2000. 332, 334, 338

[3] M.-L. Akkar, C. Giraud, *An Implementation of DES and AES Secure against Some Attacks*. In Proceedings of CHES'2001, LNCS 2162, pp. 309-318, Springer-Verlag, 2001. 333, 334

[4] E. Biham, A. Shamir, *Power Analysis of the Key Scheduling of the AES Candidates*. In Proceedings of the Second Advanced Encryption Standard (AES) Candidate Conference, March 1999. Available from `http://csrc.nist.gov/encryption/aes/round1/Conf2/aes2conf.htm` 332, 334

[5] S. Chari, C. S. Jutla, J. R. Rao, P. Rohatgi, *A Cautionary Note Regarding Evaluation of AES Candidates on Smart-Cards*. In Proceedings of the Second Advanced Encryption Standard (AES) Candidate Conference, March 1999. Available from `http://csrc.nist.gov/encryption/aes/round1/Conf2/aes2conf.htm` 332, 333, 334

[6] S. Chari, C. S. Jutla, J. R. Rao, P. Rohatgi, *Towards Sound Approaches to Counteract Power-Analysis Attacks*. In Proceedings of CRYPTO'99, LNCS 1666, pp. 398-412, Springer-Verlag, 1999. 333, 338

[7] J.-S. Coron, *Resistance Against Differential Power Analysis for Elliptic Curve Cryptosystems*. In Proceedings of CHES'99, LNCS 1717, pp. 292-302, Springer-Verlag, 1999. 332, 334

[8] J. Daemen, V. Rijmen, *Resistance Against Implementation Attacks: A Comparative Study of the AES Proposals*. In Proceedings of the Second Advanced Encryption Standard (AES) Candidate Conference, March 1999. Available from `http://csrc.nist.gov/encryption/aes/round1/Conf2/aes2conf.htm` 332, 334

[9] J. Daemen, M. Peters, G. Van Assche, *Bitslice Ciphers and Power Analysis Attacks*. In Proceedings of FSE'2000, LNCS 1978, Springer-Verlag, 2000. 334, 341

[10] L. Goubin, *A Refined Power-Analysis Attack on Elliptic Curve Cryptosystems*. In Proceedings of PKC'2003, LNCS 2567, pp. 199-211, Springer-Verlag, 2003. 334

[11] L. Goubin, J. Patarin, *Procédé de sécurisation d'un ensemble électronique de cryptographie à clé secrète contre les attaques par analyse physique*. European Patent, Bull CP8, February 4th, 1999, Publication Number: 2789535. 332, 333, 334

[12] L. Goubin, J. Patarin, *DES and Differential Power Analysis – The Duplication Method*. In Proceedings of CHES'99, LNCS 1717, pp. 158-172, Springer-Verlag, 1999. 332, 333, 334, 341

[13] P. Kocher, J. Jaffe, B. Jun, *Introduction to Differential Power Analysis and Related Attacks*. Technical Report, Cryptography Research Inc., 1998. Available from `http://www.cryptography.com/dpa/technical/index.html` 332, 333, 334, 341

[14] P. Kocher, J. Jaffe, B. Jun, *Differential Power Analysis*. In Proceedings of CRYPTO'99, LNCS 1666, pp. 388-397, Springer-Verlag, 1999. 332, 333, 334, 341

[15] T. S. Messerges, *Using Second-Order Power Analysis to Attack DPA Resistant software*. In Proceedings of CHES'2000, LNCS 1965, pp. 238-251, Springer-Verlag, 2000. 333, 341
[16] T. S. Messerges, E. A. Dabbish, R. H. Sloan, *Investigations of Power Analysis Attacks on Smartcards*. In Proceedings of the USENIX Workshop on Smartcard Technology, pp. 151-161, May 1999. Available from `http://www.eecs.uic.edu/~tmesserg/papers.html` 332, 334, 338
[17] T. S. Messerges, E. A. Dabbish, R. H. Sloan, *Power Analysis Attacks of Modular Exponentiation in Smartcards*. In Proceedings of CHES'99, LNCS 1717, pp. 144-157, Springer-Verlag, 1999. 332, 334
[18] I. Mironov, *(Not So) Random Shuffles of RC4*. In Proceedings of CRYPTO'2002, LNCS 2442, pp. 304-319, Springer-Verlag, 2002.
[19] K. Okeya, K. Sakurai, *Power Analysis Breaks Elliptic Curve Cryptosystem even Secure against the Timing Attack*. In Proceedings of INDOCRYPT'2000, LNCS 1977, pp. 178-190, Springer-Verlag, 2000. 332, 334
[20] E. Trichina, D. De Seta, L. Germani, *Simplified Adaptive Multiplicative Masking for AES*. In Proceedings of CHES'2002, LNCS 2523, pp. 187-197, Springer-Verlag, 2002. 333, 335, 337, 345

Appendix. Explicit subsets of values A, depending of the repartition of the Hamming weight of $f_A(X)$.

- Subset 1: 0x00
- Subset 2: 0x01, 0x06, 0x09, 0x22, 0x24, 0x2B, 0x46, 0x48, 0x56, 0x62, 0x65, 0x68, 0x75, 0x7B, 0x7C, 0x7D, 0x99, 0x9D, 0xA9, 0xB1, 0xC1, 0xCA, 0xCD, 0xEF, 0xF0, 0xF8, 0xFA, 0xFB
- Subset 3: 0x02, 0x04, 0x07, 0x0C, 0x0E, 0x12, 0x18, 0x19, 0x1C, 0x23, 0x29, 0x2A, 0x2D, 0x2E, 0x31, 0x37, 0x3E, 0x3F, 0x44, 0x49, 0x52, 0x54, 0x59, 0x5B, 0x5C, 0x5F, 0x60, 0x67, 0x6B, 0x78, 0x79, 0x81, 0x89, 0x8D, 0x8E, 0x8F, 0x90, 0x95, 0x96, 0x98, 0x9C, 0xA5, 0xB0, 0xB2, 0xB6, 0xB7, 0xB8, 0xBF, 0xC0, 0xC3, 0xD6, 0xD9, 0xE1, 0xE7, 0xEA, 0xF7
- Subset 4: 0x03, 0x15, 0x87, 0x8C, 0xCE, 0xEB, 0xED, 0xF6
- Subset 5: 0x05, 0x08, 0x0A, 0x0D, 0x11, 0x1A, 0x1D, 0x1E, 0x1F, 0x21, 0x27, 0x32, 0x34, 0x38, 0x3A, 0x3B, 0x3D, 0x43, 0x47, 0x4B, 0x4C, 0x4E, 0x4F, 0x51, 0x64, 0x69, 0x6D, 0x6E, 0x76, 0x77, 0x7E, 0x85, 0x88, 0x92, 0x97, 0x9B, 0x9E, 0xA0, 0xA1, 0xA2, 0xA4, 0xA8, 0xAA, 0xAC, 0xAE, 0xB3, 0xB9, 0xBB, 0xBE, 0xC4, 0xC5, 0xC7, 0xC8, 0xC9, 0xCF, 0xD0, 0xD1, 0xD2, 0xD5, 0xDC, 0xE5, 0xE8, 0xE9, 0xEC, 0xF3, 0xF4, 0xF5, 0xFD, 0xFE, 0xFF
- Subset 6: 0x0B, 0x0F, 0x10, 0x20, 0x26, 0x28, 0x36, 0x39, 0x40, 0x5E, 0x63, 0x6C, 0x6F, 0x83, 0x8A, 0x9F, 0xA7, 0xAD, 0xB5, 0xBC, 0xBD, 0xC2, 0xCC, 0xD8, 0xDB, 0xE4, 0xE6, 0xF9
- Subset 7: 0x13, 0x1B, 0x2C, 0x66, 0x72, 0x80, 0x84, 0xD3
- Subset 8: 0x14, 0x17, 0x25, 0x2F, 0x30, 0x33, 0x35, 0x3C, 0x41, 0x42, 0x45, 0x4A, 0x4D, 0x50, 0x53, 0x55, 0x57, 0x58, 0x5A, 0x5D, 0x61, 0x6A, 0x70, 0x71, 0x73, 0x74, 0x7A, 0x7F, 0x82, 0x86, 0x8B, 0x91, 0x93, 0x94, 0x9A, 0xA3, 0xA6, 0xAB, 0xAF, 0xB4, 0xBA, 0xC6, 0xCB, 0xD4, 0xD7, 0xDA, 0xDD, 0xDE, 0xDF, 0xE0, 0xE2, 0xE3, 0xEE, 0xF1, 0xF2, 0xFC
- Subset 9: 0x16

Nonce-Based Symmetric Encryption

Phillip Rogaway[1,2]

[1] Dept. of Computer Science, University of California
Davis, CA 95616, USA
[2] Dept. of Computer Science, Faculty of Science
Chiang Mai University, Chiang Mai 50200, Thailand
rogaway@cs.ucdavis.edu
www.cs.ucdavis.edu/~rogaway/

Abstract. Symmetric encryption schemes are usually formalized so as to make the encryption operation a probabilistic or state-dependent function $\mathcal{E}$ of the message M and the key K: the user supplies M and K and the encryption process does the rest, flipping coins or modifying internal state in order to produce a ciphertext C. Here we investigate an alternative syntax for an encryption scheme, where the encryption process $\mathcal{E}$ is a deterministic function that surfaces an *initialization vector* (IV). The user supplies a message M, key K, and initialization vector N, getting back the (one and only) associated ciphertext $C = \mathcal{E}_K^N(M)$. We concentrate on the case where the IV is guaranteed to be a *nonce*—something that takes on a new value with every message one encrypts. We explore definitions, constructions, and properties for nonce-based encryption. Symmetric encryption with a surfaced IV more directly captures real-word constructions like CBC mode, and encryption schemes constructed to be secure under nonce-based security notions may be less prone to misuse.

Keywords: Initialization vector, modes of operation, nonces, provable security, symmetric encryption.

1 Introduction

Ever since Goldwasser and Micali's landmark paper [7], formalizations of encryption schemes have usually made the encryption algorithm probabilistic or stateful. In this paper we investigate a different formalization for symmetric encryption: the encryption algorithm is made to be a deterministic function, but one of its argument is a user-supplied *initialization vector* (IV). Effectively, the user and not the encryption algorithm is made responsible for flipping coins or maintaining state. We are mostly interested in security properties that can be guaranteed as long as the IV is a *nonce*—a value, like a counter, used at most once within a session. Our formalization leads to what is effectively a stronger notion of privacy than the conventional formalization, and a stronger notion of authenticity as well. As a consequence, encryption schemes created so as to satisfy the given notions would seem to be less likely to be misused.

B. Roy and W. Meier (Eds.): FSE 2004, LNCS 3017, pp. 348–359, 2004.

Algorithm $\text{CBC.Encrypt}_K^N(M)$
if $|M| \notin \{n, 2n, 3n, \ldots\}$ **then return** $\star$
Parse M into $M_1 \cdots M_m$ where $|M_i| = n$
$C_0 \leftarrow N$
for $i \leftarrow 1$ **to** m **do**
  $C_i \leftarrow E_K(C_{i-1} \oplus M_i)$
return $C_1 \cdots C_m$

Algorithm $\text{CBC.Decrypt}_K^N(C)$
if $|C| \notin \{n, 2n, 3n, \ldots\}$ **then return** $\star$
Parse C into $C_1 \cdots C_m$ where $|M_i| = n$
$C_0 \leftarrow N$
for $i \in [1 .. m]$ **do**
  $M_i \leftarrow C_{i-1} \oplus E_K^{-1}(C_i)$
return $M_1 \cdots M_m$

Fig. 1. Scheme CBC. Encryption and decryption depend on a block cipher $E\colon \mathsf{Key} \times \{0,1\}^n \to \{0,1\}^n$. The key space for the encryption scheme is the same as the key space for the block cipher. The IV N is a string in $\{0,1\}^n$. The encryption scheme is the pair (CBC.Encrypt, CBC.Decrypt)

Algorithm $\text{CBC\$.Encrypt}_K(M)$
if $|M| \notin \{n, 2n, 3n, \ldots\}$ **then return** $\star$
Parse M into $M_1 \cdots M_m$ where $|M_i| = n$
$C_0 \xleftarrow{\$} \{0,1\}^n$
for $i \leftarrow 1$ **to** m **do**
  $C_i \leftarrow E_K(C_{i-1} \oplus M_i)$
return $C_0 C_1 \cdots C_m$

Algorithm $\text{CBC\$.Decrypt}_K(C)$
if $|C| \notin \{2n, 3n, 4n, \ldots\}$ **then return** $\star$
Parse C into $C_0 C_1 \cdots C_m$ where $|M_i| = n$
for $i \in [1 .. m]$ **do**
  $M_i \leftarrow C_{i-1} \oplus E_K^{-1}(C_i)$
return $M_1 \cdots M_m$

Fig. 2. Scheme CBC\$. The mechanism depends on a block cipher $E\colon \mathsf{Key} \times \{0,1\}^n \to \{0,1\}^n$. Scheme CBC\$ is a conventional probabilistic encryption scheme. It is just like scheme CBC except that the encryption routine chooses the IV $N = C_0$ internally and at random. The user cannot influence it. The value is now returned as part of the ciphertext

COMPARING CBC AND CBC\$ ENCRYPTION. Popular modes of operation for encryption have always surfaced an IV. For example, CBC using block cipher $E\colon \mathsf{Key} \times \{0,1\}^n \to \{0,1\}^n$ requires an initialization vector $N \in \{0,1\}^n$ to encrypt a message M (or decrypt a message C) under key $K \in \mathsf{Key}$. See Fig. 1 and note that the initialization vector N is an argument to both CBC.Encrypt and CBC.Decrypt. Given that the IV is manifestly present in the description of CBC mode, this would seem to be quite natural. Nonetheless, the approach is at odds with the customary formalization of symmetric encryption going back to [1, 7]. There one explicitly models some particular manner of generating the IV and folds this into the definition of the scheme. For example, one considers the scheme CBC\$ (i.e., CBC with a random IV) as defined in Fig. 2.

CONTRIBUTIONS. It is the purpose of this note to treat symmetric encryption schemes in a way that explicitly surfaces the IV. The approach was first taken in our earlier work on authenticated encryption [9, 10], where we adopted nonce-based definitions without significant comment. Here we more systematically investigate the explicit-IV notion of encryption, giving definitions, schemes, and basic results. We are mostly interested in the case when the IV is a nonce: a value used at most once within the scope of a given session.

This note aims to call attention to the explicit-IV approach and to nudge future work on practical encryption schemes into adopting the nonce-based framework.

STANDARDS. We believe that a nonce-based formalization is especially desirable when constructing an encryption scheme for a cryptographic standard: not knowing how the scheme will be used, standards would do well to achieve the strongest practical notion of security relative to the interface that they export. The viewpoint, then, is that conventional encryption modes like CBC, as defined in Fig. 1, are "deficient" insofar as they do not achieve a strong notion of security unless one assumes something very strong about their IVs. One would prefer an encryption mode that achieves a strong notion of security when one assumes very little about the IV. It is thus our view that, in the future, standards for privacy-only encryption would do well to achieve privacy in the ind$-sense that we will define in Section 3, while standards for authenticated encryption would do well to achieve, in addition, authenticity in the auth-sense that we define in Section 6.

FURTHER REASONS TO SURFACE THE IV. Another motivation for explicitly surfacing the IV in the definition of an encryption scheme is that books and systems often get wrong what it may or may not be. Books will say, for example, that it is fine for the IV in CBC encryption to be a counter, or the last block of encrypted ciphertext. Both statements are wrong, assuming that one intends to achieve a strong notion of privacy. Having definitions that expose the IV across the encryption and decryption interface facilitates answering *what* the IV may or may not be in order to achieve a given notion of security.

Yet another motivation for surfacing the IV is that it allows a particularly simple and strong notion of privacy: indistinguishability from random bits with respect to an adaptive chosen-plaintext-and-IV attack (ind$, to be defined later). This attack allows the adversary to select not only plaintexts but also the IVs that will be used to encrypt each of them, subject only to the constraint that no IV is reused. The model captures the possibility that the IVs may be chosen in an unfortunate way by the sender, possibly even influenced by the adversary, when we do not mandate any requirement on an IV beyond its non-reuse.

A SMALL WARNING. Nothing in this paper should be construed to suggest that the overall encryption process should become deterministic and stateless (and therefore not semantically secure). We are simply drawing the abstraction boundary a little differently, so that what is "inside" the scheme is deterministic, the coins or state being pushed "outside" of the scheme's formalization.

2 Syntax

DEFINITIONS. We begin by specifying the syntax for an encryption schemes that surfaces an IV. An *IV-based encryption scheme* is a pair of algorithms $\Pi = (\mathcal{E}, \mathcal{D})$ where $\mathcal{E}\colon \mathsf{Key} \times \mathsf{IV} \times \mathsf{Plaintext} \to \mathsf{Ciphertext}$ and $\mathcal{D}\colon \mathsf{Key} \times \mathsf{IV} \times \mathsf{Ciphertext} \to \mathsf{Plaintext} \cup \{\star\}$ are deterministic functions. These functions are called the *encryption function* and the *decryption function*, respectively. Here Key, IV, Plaintext, and Ciphertext are nonempty sets of strings, the first of which is finite or is otherwise endowed with a distribution (the understood distribution on a finite set being the uniform one). These sets are called the *key space*, the *IV space*, the *message space*, and the *ciphertext space*. We insist that Plaintext has the structure that if it contains a string M then it contains all string M' having the same length of M. We often write $\mathcal{E}_K^N(M)$ in place of $\mathcal{E}(K, N, M)$ and $\mathcal{D}_K^N(C)$ in place of $\mathcal{D}(K, N, C)$. We require that $\mathcal{D}_K^N(\mathcal{E}_K^N(M)) = M$ for any $K \in \mathsf{Key}$ and $N \in \mathsf{IV}$ and $M \in \mathsf{Plaintext}$. For simplicity, we assume that $|\mathcal{E}_K^N(M)|$ depends only on $|M|$ and, in particular, that $|\mathcal{E}_K^N(M)| = |M| + \tau$ for some constant τ associated to the encryption scheme. We call τ the *stretch* of the encryption scheme.

COMMENTS. (1) We will often use the word *nonce* instead of *IV* and write Nonce, the *nonce space*, instead of IV. We do this when we are thinking in terms of our nonce-based definitions for privacy (to follow). In such cases we call an IV-based encryption scheme a *nonce-based encryption scheme.* (2) We emphasize that $\mathcal{E}$ and $\mathcal{D}$ are deterministic and stateless functions: they may not flip coins or preserve state. (3) What we call the ciphertext $C = \mathcal{E}_K^N(M)$ is not expected to encode the IV, even though the IV is needed to decrypt. The IV may be communicated "out of band" to the receiver, maintained as shared state, or it may be manifest within the context of use, as when the IV is the sector index on a disk. (4) The encryption function $\mathcal{E}$ may be length-preserving, meaning that $|\mathcal{E}_K^N(M)| = |M|$ for all K, N, M. Indeed we will see that encryption schemes can achieve a strong notion of privacy yet have zero stretch. (5) We have allowed for the possibility that the decryption of a string returns the distinguished value $\star$, which is used to indicate that the ciphertext is invalid. While this possibility is not needed for basic notions of privacy, it is needed for defining authenticity. (6) We have not said that the sender or receiver are stateless and without benefit of coins, only that $\mathcal{E}$ and $\mathcal{D}$ are. For example, the sender might maintain a counter to use as the IV. It is simply that this state is outside of the functionality of $\mathcal{E}$.

3 Privacy

INDISTINGUISHABILITY FROM RANDOM BITS. Our preferred notion of privacy is "indistinguishability from random bits under an adaptive chosen-plaintext-and-IV attack". To formalize this, let adversary A be an algorithm with access to an oracle and let $\Pi = (\mathcal{E}, \mathcal{D})$ be an IV-based encryption scheme with key space

Key and IV space Nonce and stretch τ. We define

$$\mathbf{Adv}_{\Pi}^{\mathrm{ind\$}}(A) = \Pr\left[K \stackrel{\$}{\leftarrow} \mathsf{Key}:\ A^{\mathcal{E}_K(\cdot,\cdot)} \Rightarrow 1\right] - \Pr\left[A^{\$(\cdot,\cdot)} \Rightarrow 1\right]$$

The superscript ind\$ may alternatively be written as ind\$-cpa. The oracle $\mathcal{E}_K(\cdot,\cdot)$, on input (N, M), returns $\mathcal{E}_K^N(M)$. We sometimes refer to this as the *real* encryption oracle. The oracle $\$(\cdot,\cdot)$, on input (N, M), returns $|M| + \tau$ random bits. We sometimes refer to this as the *random-bits* oracle. Both oracles return $\star$ if $N \notin \mathsf{Nonce}$ or $M \notin \mathsf{Plaintext}$. When we write $A^{\mathcal{O}} \Rightarrow 1$ we are referring to the event that adversary A, running with its oracle $\mathcal{O}$, outputs the bit 1. We call an adversary A *nonce-respecting* if it never repeats a nonce: if A asks (N, M) it never subsequently asks (N, M') for any M'. This must hold regardless of A's coins and regardless of oracle responses. We assume that any ind\$-adversary is nonce-respecting.

CONVENTIONAL INDISTINGUISHABILITY. It is more customary to focus on a different kind of indistinguishability. Once again, let $\Pi = (\mathcal{E}, \mathcal{D})$ be a nonce-based encryption scheme with key space Key and nonce space Nonce. Let A be an adversary. Then define

$$\mathbf{Adv}_{\Pi}^{\mathrm{ind}}(A) = \Pr\left[K \stackrel{\$}{\leftarrow} \mathsf{Key}:\ A^{\mathcal{E}_K(\cdot,\cdot)} \Rightarrow 1\right] - \Pr\left[K \stackrel{\$}{\leftarrow} \mathsf{Key}:\ A^{\mathcal{E}_K(\cdot,0^{|\cdot|})} \Rightarrow 1\right]$$

The superscript ind\$ may alternatively be written as ind-cpa. The first oracle is a *real* encryption oracle, as before. The second oracle, on input (N, M), returns $\mathcal{E}_K(N, 0^{|M|})$. We call this a *fake* encryption oracle. Both oracles return $\star$ if $N \notin \mathsf{Nonce}$ or $M \notin \mathsf{Plaintext}$, and A is always assumed to be nonce-respecting.

RESOURCE-PARAMETERIZED DEFINITIONS. If Π is a scheme and A is an adversary and $\mathbf{Adv}_{\Pi}^{\mathrm{xxx}}(A)$ is a measure of adversarial advantage already defined, then we write $\mathbf{Adv}_{\Pi}^{\mathrm{xxx}}(\mathcal{R})$ to mean the maximal value of $\mathbf{Adv}_{\Pi}^{\mathrm{xxx}}(A)$ over all adversaries A that use resources bounded by $\mathcal{R}$. Here $\mathcal{R}$ is a list of variables specifying the resources of interest for the adversary in question. Adversarial resources to which we pay attention are: t—the running time of the adversary; q—the number of queries asked by the adversary; and σ—the aggregate length of these queries, plus the length of the adversary's output, measured in n-bit blocks, for some understood value n. When an adversary's query or output is a tuple of strings we count in σ the sum of the lengths of each component. Fractional blocks and the emptystring contribute 1. By convention, the running time of an algorithm includes the description size of that algorithm, relative to some standard encoding.

DISCUSSION: FAVORING IND\$ OVER IND. It is easy to verify that the ind\$-notion of security implies the ind-notion, and by a tight reduction, while ind does not imply ind\$ at all. (The same is true if one speaks of indistinguishability and indistinguishability from random bits in the context of conventional, probabilistic encryptions schemes.) Despite this, typical encryption schemes seem to achieve

Algorithm CBC1.Encrypt$_K^N(M)$	**Algorithm** CBC1.Decrypt$_K^N(C)$
if $\lvert M\rvert \notin \{n, 2n, 3n, \ldots\}$ **then return** $\star$	**if** $\lvert C\rvert \notin \{n, 2n, 3n, \ldots\}$ **then return** $\star$
Parse M into $M_1 \cdots M_m$ where $\lvert M_i\rvert = n$	Parse C into $C_1 \cdots C_m$ where $\lvert M_i\rvert = n$
$C_0 \leftarrow E_K(N)$	$C_0 \leftarrow E_K(N)$
for $i \leftarrow 1$ **to** m **do**	**for** $i \in [1..m]$ **do**
$\quad C_i \leftarrow E_K(C_{i-1} \oplus M_i)$	$\quad M_i \leftarrow C_{i-1} \oplus E_K^{-1}(C_i)$
return $C_1 \cdots C_m$	**return** $M_1 \cdots M_m$

Fig. 3. Scheme CBC1. The scheme is not secure

ind\$ if they achieve ind (again, the IV is *not* considered part of the ciphertext). Furthermore, it usually seems to be no extra trouble—indeed often it is slightly simpler—to directly demonstrate that some scheme achieves ind\$-security. Doing so is useful because an encryption scheme that satisfies ind\$ makes a more versatile tool: it can be used to directly provide a pseudorandom generator or a pseudorandom function. Finally, we find ind\$ seems to us conceptually simpler and easier to work with. For all of these reasons, we like ind\$ as the basic notion of security for building practical IV-based symmetric encryption schemes. (The counter-argument is that being indistinguishable from random bits is irrelevant to the goal of encryption—one would argue that it goes beyond the intuition about what secure encryption needs to provide. This is true, and yet it has often proven desirable to use definitions that reach beyond the minimal notions that satisfy one's intuition.)

4 Insecure Schemes

One can see right away that CBC encryption, as formalized in Fig. 1, is not ind-secure (and therefore it is not ind\$-secure, either). Here is the attack. The adversary is trying to distinguish a real encryption oracle from a fake encryption oracle. Let us write $\mathbf{0}$ for 0^n and $\mathbf{1}$ for $0^{n-1}1$. The adversary asks a first oracle query of $(N_1, M_1) = (\mathbf{0}, \mathbf{0})$, getting back a ciphertext C_1. Let it then ask a second oracle query of $(N_2, M_2) = (\mathbf{1}, \mathbf{1})$, getting back a ciphertext C_2. If $C_1 = C_2$ then the adversary outputs 1 (it believes it has a real encryption oracle) and otherwise the adversary outputs 0 (it knows that it has a fake encryption oracle). The adversary is extremely efficient and has advantage close to 1.

The attack above motivates a natural alternative to CBC: encipher the IV before using it, as shown in Fig. 3. We call the scheme CBC1. The key space Key for the encryption scheme remains the key space for the underlying block cipher and the nonce space Nonce is $\{0, 1\}^n$.

The scheme CBC1 still doesn't work. Let the adversary ask query $(N_1, M_1) = (\mathbf{0}, \mathbf{00})$, obtaining ciphertext $C_1^1 C_1^2$ (where C_1^1 and C_1^2 are n bits). Note that if the adversary was provided a real encryption oracle then $C_1^2 = E_K(C_1^1)$. So next the adversary asks $(N_2, M_2) = (C_1^1, C_1^2 \oplus C_1^1)$, getting result C_2^1. If $C_2^1 = C_1^2$

Algorithm $\text{CBC2.Encrypt}_{K1\,K2}^{N}(M)$	**Algorithm** $\text{CBC2.Decrypt}_{K1\,K2}^{N}(C)$
if $\|M\| \notin \{n, 2n, 3n, \ldots\}$ **then return** $\star$	**if** $\|C\| \notin \{n, 2n, 3n, \ldots\}$ **then return** $\star$
Parse M into $M_1 \cdots M_m$ where $\|M_i\| = n$	Parse C into $C_1 \cdots C_m$ where $\|M_i\| = n$
$C_0 \leftarrow E_{K1}(N)$	$C_0 \leftarrow E_{K1}(N)$
for $i \leftarrow 1$ **to** m **do**	**for** $i \in [1..m]$ **do**
$\quad C_i \leftarrow E_{K2}(C_{i-1} \oplus M_i)$	$\quad M_i \leftarrow C_{i-1} \oplus E_{K2}^{-1}(C_i)$
return $C_1 \cdots C_m$	**return** $M_1 \cdots M_m$

Fig. 4. Scheme CBC2. The scheme is now ind$-secure

then the adversary outputs 1 (it guesses that it has a real encryption oracle) and otherwise it outputs 0 (it is sure that it has a fake encryption oracle). The adversary is very efficient and is easily seen to have advantage close to 1.

5 Secure Schemes

Despite the two examples above, it is easy to construct encryption schemes that are secure in the ind$-sense. Consider first the scheme CBC2 shown in Fig. 4. The key space for the encryption scheme is $\mathsf{Key} \times \mathsf{Key}$, where Key is the key space for the underlying block cipher. The nonce space is $\{0,1\}^n$. The message space remains $(\{0,1\}^n)^+$.

The following result shows that CBC2 is a secure encryption scheme. We state the theorem in the information-theoretic setting. Passing to the complexity-theoretic case is standard. By $\mathcal{P}_n$ we mean the set of all permutations on $\{0,1\}^n$. These are block ciphers in the natural way. Thus by $\text{CBC2}[\mathcal{P}_n \times \mathcal{P}_n]$ we mean the scheme where E_{K1} and E_{K2} are random permutations from n bits to n bits.

Theorem 1. *Let* $n, \sigma \geq 1$. *Then* $\mathbf{Adv}^{\text{ind\$}}_{\text{CBC2}[\mathcal{P}_n \times \mathcal{P}_n]}(\sigma) \leq \sigma^2/2^n$ $\diamondsuit$

To avoid having two block-cipher keys one can modify the scheme using tricks like those from [4, 8]. However, it is not necessary to use a CBC-like scheme at all; simple forms of counter mode (CTR) work fine, and such modes have the advantage of being parallelizable and working directly on messages of any bit length. See Fig. 5 and Fig. 6 for two counter-based encryption schemes that are secure in the ind$-sense. The first has a nonce space of $\{0,1\}^{n/2}$ (assume that n is even) and the second has a nonce space of $\{0,1\}^n$ but uses one extra block-cipher call. When S is an n-bit string and i is a number we denote by $S+i$ the n-bit string which is obtained by treating S as a number (msb first, lsb last), adding i modulo 2^n to this number, and then turning the result back into an n-bit string (msb first, lsb last).

One should anticipate use of CTR1 only on strings of at most $2^{n/2}$ blocks, though the ind$-security of the scheme has already vanished by that point when the block cipher E is a PRP. Similarly, one should anticipate use of CTR2 only on strings of at most 2^n blocks, though the ind$-security of the scheme has long before vanished when the function E is a PRP.

Algorithm CTR1.Encrypt$_K^N(M)$	**Algorithm** CTR1.Decrypt$_K^N(C)$
$S \leftarrow N \parallel 0^{n/2}$	$S \leftarrow N \parallel 0^{n/2}$
$m \leftarrow \lceil \lvert M \rvert / n \rceil$	$m \leftarrow \lceil \lvert M \rvert / n \rceil$
$P \leftarrow E_K(S+0) \parallel \cdots \parallel E_K(S+m-1)$	$P \leftarrow E_K(S) \parallel \cdots \parallel E_K(S+m-1)$
$C \leftarrow M \oplus P$[first $\lvert M \rvert$ bits]	$M \leftarrow C \oplus P$[first $\lvert C \rvert$ bits]
return C	**return** M

Fig. 5. Scheme CTR1. The nonce space is $\mathsf{Nonce} = \{0,1\}^{n/2}$

Algorithm CTR2.Encrypt$_K^N(M)$	**Algorithm** CTR2.Decrypt$_K^N(C)$
$S \leftarrow E_K(N)$	$S \leftarrow E_K(N)$
$m \leftarrow \lceil \lvert M \rvert / n \rceil$	$m \leftarrow \lceil \lvert M \rvert / n \rceil$
$P \leftarrow E_K(S+0) \parallel \cdots \parallel E_K(S+m-1)$	$P \leftarrow E_K(S) \parallel \cdots \parallel E_K(S+m-1)$
$C \leftarrow M \oplus P$[first $\lvert M \rvert$ bits]	$M \leftarrow C \oplus P$[first $\lvert C \rvert$ bits]
return C	**return** M

Fig. 6. Scheme CTR2. The nonce space is $\mathsf{Nonce} = \{0,1\}^{n}$

Theorem 2. *Let* $n, \sigma \geq 1$. *Then* $\mathbf{Adv}^{\mathrm{ind\$}}_{\mathrm{CTR1}[\mathcal{P}_n \times \mathcal{P}_n]}(\sigma) \leq \sigma^2/2^n$ ◇

Theorem 3. *Let* $n, \sigma \geq 1$. *Then* $\mathbf{Adv}^{\mathrm{ind\$}}_{\mathrm{CTR2}[\mathcal{P}_n \times \mathcal{P}_n]}(\sigma) \leq \sigma^2/2^n$ ◇

Recall our conventions that when multiple strings are encoded into a single one, as in a query (N, M), one sums the length of each component in the resource bound σ. This explains the absence of a term like $q^2/2^n$ in the second bound (where q is the number of queries).

6 Stronger Notions of Security

One desirable property of a nonce-based encryption scheme is that an adversarial-produced ciphertext, coupled with its nonce, should be deemed *invalid* by the receiver unless, of course, it is a copy a prior ciphertext and its nonce. We recall the definition of this property and then look at some other strong properties for a nonce-based encryption scheme.

AUTHENTICITY. A notion of *authenticity of ciphertexts* for nonce-based encryption schemes was formalized in [9, 10] following [6, 3, 2]. Fix an encryption scheme $\Pi = (\mathcal{E}, \mathcal{D})$ with key space Key. Let A be a nonce-respecting adversary having an encryption oracle $\mathcal{E}_K$. We say that A *forges* if it outputs a pair (N, C) such that C was not the response to any $\mathcal{E}_K(N, M)$ query and $\mathcal{D}_K^N(C) \neq \star$. We write

$$\mathbf{Adv}^{\mathrm{auth}}_{\Pi}(A) = \Pr\left[K \stackrel{\$}{\leftarrow} \mathsf{Key} : A^{\mathcal{E}_K(\cdot,\cdot)} \text{ forges}\right]$$

and lift this to give resource-bounded definitions in the usual way.

CHOSEN-CIPHERTEXT SECURITY. We define indistinguishability from random bits under an adaptive chosen-plaintext-and-ciphertext-and-IV attack. The defining game is as with ind\$-cpa except that the adversary is given access to a decryption oracle as well. Queries may not be repeated, and one forbids the adversary from making a decryption query of (N, C) if the adversary already encrypted some (N, M) and got back an answer C; and one similarly forbids the adversary from encrypting (N, M) if the adversary already decrypted some (N, C) and got back an answer M. These restrictions must hold regardless of the adversary's coins and query responses. Only such an adversary is deemed to be *valid*.

In defining chosen-ciphertext security one restricts attention to valid, nonce-respecting adversaries. Be clear that the nonce-respecting condition applies only to encryption-queries; the adversary is free to repeat nonces in its decryption oracle. This reflects the understanding that the party encrypting a message is the one that is responsible for providing fresh nonces; the receiver may be stateless.

Let adversary A be a valid, nonce-respecting adversary and let $\Pi = (\mathcal{E}, \mathcal{D})$ be a nonce-based encryption scheme with key space Key. We define

$$\mathbf{Adv}_{\Pi}^{\text{ind\$-cca}}(A) = \Pr\left[K \stackrel{\$}{\leftarrow} \mathsf{Key}:\ A^{\mathcal{E}_K(\cdot,\cdot)\,\mathcal{D}_K(\cdot,\cdot)} \Rightarrow 1\right] - \Pr\left[A^{\$(\cdot,\cdot)\,\mathcal{D}_K(\cdot,\cdot)} \Rightarrow 1\right]$$

The notion can be modified to ind-cca in the natural way.

NONMALLEABILITY. The notion of nonmalleability [5] can likewise be adapted to the nonce-based setting. The adversary's goal will be to create a ciphertext (perhaps by modifying ciphertexts that have already been seen) whose underlying plaintext is something about which the adversary can say something interesting. More specifically, the adversary will output a tuple (N, C, f, Y) where N is a nonce and C is a ciphertext and $f\colon \{0,1\}^* \cup \{\star\} \to \{0,1\}^*$ is a function (encoded as a program) and Y is a string. The output should be interpreted as the adversary guessing that the value of $f(M)$ is Y, where $M = \mathcal{D}_K^N(C)$. The formalization captures the idea that the adversary should be right about this guess no more often than that which is inherent for the game.

We define nonmalleability under chosen-ciphertext attack (meaning chosen-plaintext-and-ciphertext-and-IV attack). Fix an encryption scheme $\Pi = (\mathcal{E}, \mathcal{D})$ having key space Key. Consider a valid, nonce-respecting adversary A with access to oracles $\mathcal{E}_K(\cdot,\cdot)$ and $\mathcal{D}_K(\cdot,\cdot)$. At the end of the adversary's execution, let $\mathit{Dec}(N, C)$ be $\{M\}$ if the adversary asked some query $\mathcal{E}_K(N, M)$ and this returned C or the adversary asked some query $\mathcal{D}_K(N, C)$ and this returned M. If the adversary asked no such query, let $\mathit{Dec}(N, C) = \star$. One can regard $\mathit{Dec}(N, C)$ as a "fake" decryption of C for nonce N: if the adversary trivially knows the decryption to be M then the value is M; otherwise, the value is the "guess" that the ciphertext is invalid. Then define $\mathbf{Adv}_{\Pi}^{\text{nm-cca}}(A)$ as

$$\Pr\left[K \stackrel{\$}{\leftarrow} \mathsf{Key};\ (N, C, f, Y) \stackrel{\$}{\leftarrow} A^{\mathcal{E}_K(\cdot,\cdot)\,\mathcal{D}_K(\cdot,\cdot)};\ M \leftarrow \mathcal{D}_K(N, C):\ f(M) = Y\right] -$$

$$\Pr\left[K \stackrel{\$}{\leftarrow} \mathsf{Key};\ (N, C, f, Y) \stackrel{\$}{\leftarrow} A^{\mathcal{E}_K(\cdot,\cdot)\,\mathcal{D}_K(\cdot,\cdot)};\ M \leftarrow \mathit{Dec}(N, C):\ f(M) = Y\right].$$

The corresponding notion for nonmalleability under a chosen-plaintext attack (nm-cpa) is obtained by insisting that the adversary asks no decryption queries.

Though the above notions might not look like nonmalleability, really they are: the case of creating a ciphertext C whose plaintext M is related to the plaintext M' of some other ciphertext C' is just a special case.

IMPLICATIONS AND SEPARATIONS. As with probabilistic encryption, one can work out the complete set of implications and separations between the defined notions of nonce-based encryption. The most useful relations are that ind$ plus auth implies both ind$-cca and nm-cca. The intuition is clear: the auth-condition effectively renders useless the decryption oracle, since it almost always returns an answer that the adversary can anticipate. We omit further details.

ACHIEVING IND$+AUTH BY GENERIC COMPOSITION. None of the ind$-secure encryption schemes given so far (CBC$, CTR1, CTR2) achieve the auth-notion of authenticity (nor do they achieve ind-cca or nm-cca). We now explain the most natural way to modify an ind$-secure encryption scheme so as to achieve authenticity (while preserving ind$-security, of course).

Let $\Pi = (\mathcal{E}, \mathcal{D})$ be a nonce-based encryption scheme having nonce space $\mathsf{Nonce} = \{0,1\}^n$ and key space Key. Think of Π as being ind$-secure. Let $F\colon \mathsf{Key}' \times \mathsf{Dom} \to \{0,1\}^n$ be a function. Think of it as being good as a pseudorandom function. We want to combine Π and F to give an encryption scheme $\bar{\Pi} = (\bar{\mathcal{E}}, \bar{\mathcal{D}})$ that will be ind$-secure and auth-secure. The simplest possibilities are as follows.

- *Encrypt-then-PRF.* Let $\bar{\mathcal{E}}^N_{K\,K'}(M) = C \parallel T$ where $T = F_{K'}(N \parallel C)$ and $C = \mathcal{E}^N_K(M)$. Decryption (including the test for authenticity) proceeds in the natural way.
- *PRF-then-Encrypt.* Let $\bar{\mathcal{E}}^N_{K\,K'}(M) = \mathcal{E}^N_K(M \parallel T)$ where $T = F_{K'}(N \parallel M)$. Decryption (including the test for authenticity) proceeds in the natural way.

The definition above assumes that the encryption scheme Π and the PRF F have appropriately matching domains.

The situation is different from conventional probabilistic encryption [2]; for nonce-based encryption, *both* encrypt-then-PRF and PRF-then-encrypt work correctly. The proofs are straightforward; see [10] for the slightly more complex setting in which "associated data" is present as well.

7 Directions

With the syntax of an encryption having been modified to surface the IV, a number of weaker notions of security for IV-based encryption make sense. For example, to capture the requirement that "the IVs are to be some fixed sequence of distinct values" have the adversary provide a deterministic algorithm F that gives distinct n-bit strings $F(1), F(2), \ldots, F(2^n)$. Require indistinguishability from random bits with respect to the resulting scheme.

In this paper we have only treated symmetric encryption. Public-key encryption schemes traditionally do *not* surface an IV. But they do use random bits, and it makes just as much sense to consider nonce-based public-key encryption schemes as it does to consider nonce-based symmetric encryption schemes. This provides an approach to effectively weakening the requirement for randomness on the sender.

Acknowledgements

Some of the ideas of this note were developed in the course of preparing some lectures for Helsinki University of Technology, Finland (April 2002); thanks to Helger Lipmaa for inviting me to give those lectures. Kind thanks also to Mihir Bellare, John Black, and Tom Shrimpton for their useful comments and suggestions. This work was supported under NSF CCR-0208842 and a gift from CISCO Systems.

References

[1] M. Bellare, A. Desai, E. Jokipii and P. Rogaway. A concrete security treatment of symmetric encryption: Analysis of the DES modes of operation. *Proceedings of 38th Annual Symposium on Foundations of Computer Science* (FOCS 97), IEEE, 1997. 349

[2] M. Bellare and C. Namprempre. Authenticated encryption: Relations among notions and analysis of the generic composition paradigm. *Advances in Cryptology—Asiacrypt '00*, Lecture Notes in Computer Science, vol. 1976, T. Okamoto, ed., Springer-Verlag, 2000. 355, 357

[3] M. Bellare and P. Rogaway. Encode-then-encipher encryption: How to exploit nonces or redundancy in plaintexts for efficient cryptography. *Advances in Cryptology—Asiacrypt '00*. Lecture Notes in Computer Science, vol. 1976, T. Okamoto, ed., Springer-Verlag, 2000. 355

[4] J. Black and P. Rogaway. CBC MACs for arbitrary-length messages: The three-key constructions. *Advances in Cryptology—CRYPTO '00*, Lecture Notes in Computer Science, vol. 1880, M. Bellare, ed., Springer-Verlag, pp. 197–215, Aug 2000. 354

[5] D. Dolev, C. Dwork, and M. Naor. Non-malleable cryptography. *SIAM J. Computing*, vol. 30, no. 2, pp. 391–437, 2000. 356

[6] J. Katz and M. Yung. Unforgeable encryption and chosen ciphertext secure modes of operation. *Fast Software Encryption* (FSE 2000), Lecture Notes in Computer Science, vol 1978, B. Schneier, ed., Springer, pp. 284–299, 2001. 355

[7] S. Goldwasser and S. Micali. Probabilistic encryption. *Journal of Computer and System Sciences*, vol. 28, pp. 270–299, April 1984. 348, 349

[8] T. Iwata and K. Kurosawa. One-key CBC MAC. *Fast Software Encryption* (FSE 2003). Lecture Notes in Computer Science (to appear), 2003. 354

[9] P. Rogaway, M. Bellare, J. Black, and T. Krovetz. OCB: A block-cipher mode of operation for efficient authenticated encryption. *Proceedings of the 8th ACM Conference on Computer and Communications Security* (CCS '01), ACM Press, pp. 196–205, 2001. 349, 355

[10] P. Rogaway. Authenticated-encryption with associated-data. *Proceedings of the 9th ACM Conference on Computer and Communications Security* (CCS '02), ACM Press, pp. 98–107, 2002. 349, 355, 357

Ciphers Secure against Related-Key Attacks

Stefan Lucks

University of Mannheim, Germany
http://th.informatik.uni-mannheim.de/people/lucks/index.html

Abstract. In a related-key attack, the adversary is allowed to transform the secret key and request encryptions of plaintexts under the transformed key. This paper studies the security of PRF- and PRP-constructions against related-key attacks. For adversaries who can only transform a part of the key, we propose a construction and prove its security, assuming a conventionally secure block cipher is given. By the terms of concrete security, this is an improvement over a recent result by Bellare and Kohno [2]. Further, based on some technical observations, we present two novel constructions for related-key secure PRFs, and we prove their security under number-theoretical infeasibility assumptions.

Keywords: related-key attacks, provable security, pseudorandom functions, block ciphers, concrete security

1 Introduction

In a related-key scenario, the adversary can partially control the key. It remains secret to the adversary (i.e., she can't *read* it), but she can *choose key transformations*, *modify* the key accordingly, and request *encryptions under the modified keys*. The well-known DES complementation property can be viewed as a vulnerability against a related-key DES-distinguisher.

One motivation to study related-key attacks is to evaluate the security of secret-key cryptosystems, namely the security of block ciphers and their "key schedules ", see Knudsen [11] and Biham [3]. Kelsey, Schneier and Wagner [9, 10] presented related-key attacks against several block ciphers, including three-key triple-DES. Today, related-key attacks are a well established tool to evaluate the security of block ciphers, e.g. in the context of the AES [4, 5, 7]. *Another motivation* is the existence of cryptographic schemes, whose security depends on the related-key security of some underlying primitive. Two examples are tweakable block ciphers by Liskov, Rivest and Wagner [13] and RMAC by Jaulmes, Joux and Valette [8]. Knudsen and Kohno [12] pointed out that the triple-DES based variant of RMAC (which had been proposed for standardisation [6]) can be attacked by exploiting the related-key insecurity of triple-DES.

Recently, Bellare and Kohno [1, 2] investigated related-key attacks from a theoretical point of view. They presented an approach to formally handle the notion of related-key attacks. As it turned out, the security of a scheme against related-key attacks greatly depends on the adversary's capabilities, namely on the set of key transformations available to her.

B. Roy and W. Meier (Eds.): FSE 2004, LNCS 3017, pp. 359–370, 2004.

1.1 Focus of this Paper and Overview

In the current paper, we follow the approach from [1, 2], presenting some improved *possibility results*, i.e., constructions for block ciphers (PRPs) and pseudorandom function generators (PRFs), which are provably secure against related-key (RK) adversaries. We first concentrate on "partially-transforming " adversaries, where a part of the secret key is unaffected and remains constant. Then we deal with stronger "T^+-transforming " adversaries, where the adversary can add a known (or rather chosen) difference to the secret key. See Section 1.2 for the exact definitions. Finally, we provide a short summary. Our main results are:

- For some applications of RK-secure PRFs or PRPs, it would suffice to use a cipher being secure against *partially-transforming* RK adversaries [2]. Section 2 introduces a new construction for secure PRPs provably secure against partially-transforming adversaries. A similar construction and a proof of security can be found in [2][1]. The concrete complexity (i.e., the upper bound on the adversary's advantage) shown in [2] turns out to be rather weak, though. The construction in Section 2 allows to prove a better bound.
- In Section 3, we explore equivalent constructions for related-key secure PRFs, and we consider the composition of conventionally secure and related-key secure PRFs. Our observations may be useful as a tool for finding PRFs provably secure against more general related-key adversaries, instead of only partially-transforming ones.
- Section 4 describes two new PRF-constructions. Based on certain number-theoretical assumptions, we prove the security of these constructions against T^+-transforming adversaries. This is a step towards solving a challenge posed in [1, 2]. To the best of the author's knowledge, these constructions are the only PRFs so far with a standard-model proof of security against group-induced transformations. Note though, that the assumptions we make are new and not well-studied.
- Sections 5 and 6 conclude the paper with a remark on using a hash function as a tool to ensure related-key security, and with a summary.

1.2 Notation and Definitions

We write PRF for a Pseudo-Random Function generator and PRP for a Pseudo-Random Permutation generator (= block cipher). Let K, D and R be finite sets. We write Perm(D) for the *set of permutations* over D. I.e., $p : D \to D$ is in Perm(D) if and only if $p^{-1} : D \to D$ exists with $p^{-1}(p(d)) = d$ for all $d \in D$. We view a function $F : K \times D \to R$ as a *family of functions* $F(k, \cdot) = F_k(\cdot)$ indexed by $k \in K$. If additionally $D = R$ and $F_k \in$ Perm(D) for all $k \in K$, then F is a *family of permutations*, also called a *block cipher*. We write F_k^{-1} for the inverse of F_k, i.e., for the decryption function. Perm(K, D) denotes the *set of all block ciphers* $E : K \times D \to D$. Below, $E : K \times D \to D$ denotes a block cipher encryption function and E^{-1} denotes its inverse.

Recall the advantage of an adversary in a (conventional) chosen plaintext attack (cf. e.g. [14]): Given E and an adversary $A(\langle\text{CP-oracle}\rangle)$ with access to a chosen plaintext

[1] ..., but it was not included in the Eurocrypt version [1] of that paper.

oracle, the PRP-advantage of A when attacking E is the unsigned difference for A to distinguish the *real case* from a *random case*:

$$\mathrm{Adv}_E^{\mathrm{prp}}(A) = \Big| \Pr[k \in_{\mathrm{R}} K : A(E_k(\cdot)) = 1] - \Pr[g \in_{\mathrm{R}} \mathrm{Perm}(D) : A(g(\cdot)) = 1] \Big|.$$

Let $k \in_{\mathrm{R}} K$ be a secret key. A *related-key oracle* $E_{\mathrm{rk}(\cdot,k)}(\cdot)$ is an oracle with two inputs, a key transformation $t : K \to K$, and an element $d \in D$. Given a query (t, d), the related-key oracle responds $E_{t(k)}(d)$ in the real case, which is to be distinguished from a random case.

Definition 1 (Security of a PRP under RK attacks).
Let the block cipher E and the set of transformations T be given. The adversary $A(\langle \mathrm{RK} - \mathrm{oracle} \rangle)$ *with access to a related-key oracle is a* T-transforming *adversary,*[2] *if she is allowed to choose queries* $(t, d) \in T \times D$ *as oracle queries. The* PRP-RK-advantage *of a T-transforming adversary A when attacking E is*

$$Adv_{T,E}^{\mathrm{prp-rk}} = \Big| \Pr[k \in_{\mathrm{R}} K : A(E_{\mathrm{rk}(\cdot,k)}(\cdot)) = 1] - \Pr[k \in_{\mathrm{R}} K, G \in_{\mathrm{R}} Perm(K, D) : A(G_{\mathrm{rk}(\cdot,k)}(\cdot)) = 1] \Big|.$$

Here, the *real case* is the experiment "randomly choose $k \in K$ and, on a query (t, d), respond the value $E_{t(k)}(d)$ ". The *random case* is: "Randomly choose $k \in K$ and $G \in \mathrm{Perm}(K, D)$, i.e., a family of $|K|$ independent random permutations. Respond $G_{t(k)}(d)$ to oracle queries (t, d). " The *attack game* for A means to distinguish the real from the random case.

Similarly, we define the **security of a PRF under RK attacks**.

We concentrate on the following types of key transformations:

Group-induced transformations: Let $(K, \diamond)$ be a group. We define
$$T^{\diamond} := \{f : K \to K \mid \exists \delta \in K : f(k) = k \diamond \delta\}.$$
In Section 4, we focus on T^+, where "+ " denotes addition mod $|K|$.

Partial transformations: Set $K = K_1 \times K_2$ for non-empty sets K_1 and K_2. T is a set of partial transformations, if T can be rewritten as
$$T = \{t \mid \exists t' \in T' : t(k_1, k_2) = (k_1, t'(k_2))\},$$
where T' is a set of functions $K_2 \to K_2$.

Collision free sets of transformations: T is *collision-free*, if, for all $k, k' \in K$, there exists at most one $t \in T$ with $t(k) = k'$. This is relevant in the context of protocol design, such the previously mentioned RMAC and tweakable block ciphers. Sets of group-induced transformations are collision free. Sets of partial transformations can be collision-free.

[2] [1, 2] call this a "T-restricted adversary ". This could be misleading, since RK adversaries appear to be *enhanced* and not restricted, in comparison to conventional adversaries.

2 Secure PRPs and Partial Transformations

Set $K^* := \{0,1\}^{m+n}$ and consider a set T of partial transformations:

$$T \subseteq \{t \in \{K^* \to K^*\} \mid \exists \tau : \{0,1\}^n \to \{0,1\}^n : t(x,y) = (x, \tau(y))\}. \quad (1)$$

Let $E : \{0,1\}^m \times \{0,1\}^n \to \{0,1\}^n$ be a block cipher and consider

$$E^0 : \{0,1\}^{m+n} \times \{0,1\}^n \to \{0,1\}^n, \quad E^0_{(X,Y)}(M) = E_X(M).$$

The adversary has no control over the key X in use. So if E is conventionally secure, shouldn't E^0 be secure against T-transforming adversaries? Consider the following adversary: Choose transformations $\sigma, \tau : \{0,1\}^n \to \{0,1\}^n$ with $\sigma(Y) \neq \tau(Y)$ being likely for random Y. Ask for the encryptions of a random plaintext M under $(X, \sigma(Y))$ and $(X, \tau(Y))$. In the *real case* (encryption using E_0), you get the same answer both times. In the *random case*, if $\sigma(Y) \neq \tau(Y)$ then M is encrypted under two independent random permutations – and the two answers are probably different. By comparing the two answers, the adversary can win her attack game.

So we need a different construction. Assume E (as above) being conventionally secure and consider

$$E' : \{0,1\}^{m+n} \times \{0,1\}^n \to \{0,1\}^n, \quad E'_{(X,Y)}(M) = E_X(Y \oplus E_X(M)).$$

Theorem 1 (Security of E' [2]). *Let K^*, T, E, and E' be as above. Let A' be a T-transforming adversary. We limit the oracle-queries $(t_i, x_{i,j})$ made by A' as follows: r is the number of different transformations t_i and q is the highest number of different queries $(t_i, x_{i,j})$ for any transformation t_i. (Formally r and q are defined as $r = \left|\{t_i \in T \mid \exists \text{ query } (t_i, \cdot)\}\right|$ and $q = \max_{t_i}\left|\{x_{i,j} \in \{0,1\}^n \mid \exists \text{ query } (t_i, x_{i,j})\}\right|$.[3])*

For any such RK-adversary A' attacking E', we can construct a chosen plaintext adversary A attacking E with

$$Adv_E^{\mathrm{prp}}(A) \geq Adv_{T,E'}^{\mathrm{prp-rk}}(A') - \frac{16r^2q^2 + rq'(q'-1)}{2^{n+1}},$$

where $q' = q * \max_{k,k' \in \{0,1\}^{m+n}} |\{$ transformations $t \in T$ with $t(k) = k'\}|$, and A needs the same running time as A'.

Theorem 1 describes the concrete security of E', depending on the security of E. As usual with concrete security analysis, Theorem 1 should provide a *practically relevant security assurance* for security architects. Intuitively: *The difference between $Adv_E^{\mathrm{prp}}(A)$ and $Adv_{T,E'}^{\mathrm{prp-rk}}(A')$ is low (or rather negligible).* Unfortunately, this only holds for large n. E.g., with E=AES and thus $n = 128$, the difference may exceed $16r^2q^2/2^{n+1}$, even if T is collision-free. Assume the AES to be practically secure against chosen plaintext attacks. This means that the advantage of any "reasonable-time" adversary A against E is $\mathrm{Adv}_E^{\mathrm{prp}}(A) = \epsilon \approx 0$. Allow for $r = q = 2^{31}$. Since $n = 128$, a "reasonable-time" RK-adversary A' *can exist*, who distinguishes E' from random with $\mathrm{Adv}_{T,E'}^{\mathrm{prp-rk}}(A') > 1/2 + \epsilon$.

[3] Hence, the actual number of oracle queries A makes is between $(r + q - 1)$ and rq.

The number of oracle queries made by A' can be as low as $p+q-1 < 2^{32}$. Therefore, E' can be insecure in practice, in spite of Theorem 1 and the (assumed) security of E=AES. We don't claim A' exists – but we would like to prove its *nonexistence*.

Thus, it is practically interesting to find an improved bound, either for construction E', or an alternative construction. Below, we consider

$$E'' : \{0,1\}^{2n} \times \{0,1\}^n \to \{0,1\}^n, \quad E''_{(X,Y)}(M) = E_{E_X(Y)}(M),$$

where $E : \{0,1\}^n \times \{0,1\}^n \to \{0,1\}^n$ is conventionally secure, as above. Thus, we have $K^* = \{0,1\}^{2n}$, and T is a set of partial transformations, as before (Eq. 1). For simplicity, we additionally require T to be *collision-free*.

Theorem 2 (Security of E''). *Let $K^* = \{0,1\}^{2n}$, T a collision-free set of partial transformations. A'' is a T-transforming adversary for E''. Count the transformations in A''-queries by $r = |\{t_i \in T \mid \exists \text{ query } (t_i, \cdot)\}|$. Then a chosen plaintext adversary A for E exists, making no more oracle queries than A, with the same running time as A'' and the advantage*

$$Adv_E^{\text{prp}}(A) \geq \frac{Adv_{T,E''}^{\text{prp-rk}}(A'')}{r+1}.$$

Proof. Assume the *nonexistence* of an adversary A for E with the advantage $a \geq \text{Adv}_{T,E''}^{\text{prp-rk}}(A'')/(r+1)$.

Observe that the oracle queries $(t_i, d_{i,j})$ from A'' can be viewed as accessing r different oracles, each implementing a permutation. So in the real case, A'' is querying the r-tuple

$$P = (E_{E_X(t_1(Y))}, \ldots, E_{E_X(t_r(Y))})$$

of permutations over $\{0,1\}^n$. An oracle query $(t_i, d_{i,j})$ is equivalent to asking the i-th permutation $p_i = E_{E_X(t_i(Y))}$ for $p_i(d_{i,j})$.[4]

Due to the collision-freeness of T, we have $t_i(Y) \neq t_j(Y)$ for $t_i \neq t_j$, thus $E_X(t_i(Y)) \neq E_X(t_j(Y))$. Hence, the r permutations in P are defined by r different keys $E_X(t_1(Y))$, $\ldots$, $E_X(t_r(Y))$. But in the random case, the tuple of permutations can actually be viewed as r *independent random permutations* E_i^* over $\{0,1\}^n$. We write this tuple as

$$P_r = (E_1^*, \ldots, E_{r-1}^*, E_r^*).$$

The attack game of A'' is equivalent to distinguishing the r-tuple P of permutations from P_r. In doing so, the advantage of A'' is $\text{Adv}_{T,E''}^{\text{prp-rk}}(A'')$.

There are other ways to respond to oracle queries $(t_i, d_{i,j})$, different from both the real and the random case. Let E^* be a random permutation, and replace $E_{E_X(t_i(Y))}(M)$ by $E_{E^*(t_i(Y))}(M)$. This way, we get r independent random values $Z_i = E^*(t_i(Y))$, and a new r-tuple of permutations

$$P_0 = (E_{Z_1}, \ldots, E_{Z_r}).$$

[4] This does not restrict the order in which A makes her oracle queries. After making an oracle query $(t_i, \cdot)$, and having seen the answer, A'' may of course freely choose some queries $(t_{i'}, \cdot)$ for arbitrary values $i' \in \{1, \ldots, i\}$.

Distinguishing P_0 from P is equivalent to distinguishing E_X from E^*, which is exactly the attack game for A. From the assumption on A, we conclude that A'' can only distinguish P from P_0 with an advantage $< a$.

What is the advantage of A'' in distinguishing P_0 from P_r? Consider the r-tuples

$$\begin{aligned} P_1 &= (E_{Z_1}, \ldots, E_{Z_{r-2}}, E_{Z_{r-1}}, E_r^*), \\ P_2 &= (E_{Z_1}, \ldots, E_{Z_{r-2}}, E_{r-1}^*, E_r^*), \\ &\vdots \\ P_r &= (E_1^*, \ldots, E_{r-2}^*, E_{r-1}^*, E_r^*). \end{aligned}$$

If, for any $i \in \{1, \ldots, r\}$, A'' could distinguish P_{i-1} from P_i with an advantage a, then A'' could as well distinguish E_{Z_i} from E_i^* in the same running time. Since Z_i is just a random value, and E_i^* a random function, independent from the other values and functions here, distinguishing E_{Z_i} from E_i^* is (again) equivalent to winning the attack game for A. Thus, the advantage of A'' to distinguish P_{i-1} from P_i must be less than a.

Finally, we put things together: A'' can only distinguish P from P_0 with an advantage less than a, A'' can only distinguish P_0 from P_1 with an advantage less than a, ..., A'' can only distinguish P_{r-1} from P_r with an advantage less than a. Consequently, the advantage for A in distinguishing P from P_r must be strictly smaller than $(r+1)a$. See the picture below.

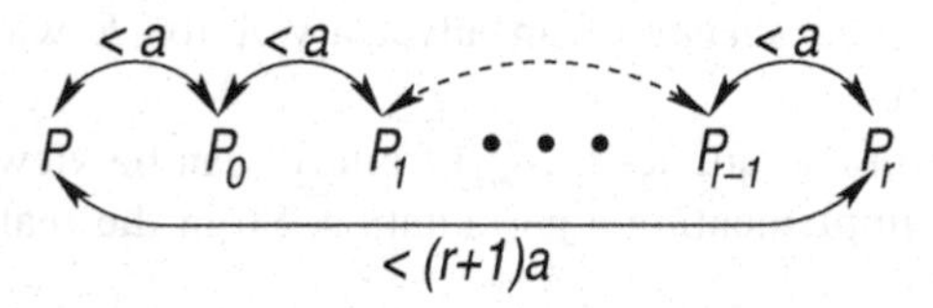

By the definition of a, we know that A'' can distinguish P from P_r with the advantage $(r+1)a$. This contradicts the assumption on A. □

Theorem 2 implies that if E is practically secure and r is not overwhelmingly large, then E'' is secure, too. As above, consider E=AES (with a key size of 128 bit) and assume the AES to be practically secure against chosen plaintext attacks. This means that the advantage of any "reasonable-time " adversary A against E is $\mathrm{Adv}_E^{\mathrm{prp}}(A) = \epsilon \approx 0$. Restrict A'' to less than 2^{32} oracle queries, thus $r < 2^{32}$. In this case, attacking E'' can be at most 2^{32}-times better (i.e. lead to an advantage 2^{32}-times as large), compared to an attack on the AES in the same running time.

We argue that the bound in Theorem 2 is sharp, and hence our result is close to optimal: The attack scenario on E'' allows the adversary to see encryptions under r different keys. Consider an exhaustive key-search attack against E'', trying to find any of the 2^r keys and compare it with an exhaustive key-search attack against E. The chances of successfully attacking E'' are 2^r-times better than the chances of successfully attacking E.

3 Equivalence and Composition of PRF Constructions

In this section, we make some technical observations. While rather simple, these observations may nevertheless be useful both for understanding the phenomenon of related-key security, and for designing ciphers provably secure against related-key attacks.

Let F be a function $F : K \times D \to R$ (which equivalently is a family of functions $D \to R$). For F, we consider a set of transformations $T \subseteq \{K \to K\}$. Let $D = D' \times D''$ (where even $|D'| = 1$ or $|D''| = 1$ is allowed). We can rewrite F as a function F' with

$$F' : \overbrace{(K \times D')}^{K'} \times D'' \to R.$$

Equivalently, F' is a family of functions $D'' \to R$. We consider the set T' of transformations:

$$T' = \{t' : K' \to K' \mid \exists t \in T, d' \in D' : t'(k, x) = (t(k), d')\}.$$

Theorem 3 (Equivalence of F and F').

1. *Let A be a T-transforming RK adversary for F. A T'-transforming RK adversary A' for F' exists, with the same running time, the same number of oracle queries and the same advantage.*
2. *Let A' be a T'-transforming RK adversary for F'. A T-transforming RK adversary A for F exists, with the same running time, the same number of oracle queries and the same advantage.*

Proof. Consider claim 1 and the T-transforming RK adversary A for F. A's queries are of the form $(t, (d', d'')) \in T \times (D' \times D'')$. Our T'-transforming adversary A' for F' is identical to T, except that each query $(t, (d', d''))$ is replaced by the equivalent query $((t, d'), d'') \in K' \times D''$. Thus, T' makes exactly the same number of oracle queries, needs the same running time and wins the attack game with the same advantage as T. Proving claim 2 is similar. □

In the context of Theorem 3, we even allowed $|D''| = 1$. In this case, $F' : K' \times D'' \to R$ can, of course, be rewritten as $F' : K' \to R'$. This apparently trivial case is worth investigating. By means of some function $F'' : R' \times D \to R$, we define a composed function

$$F : K' \times D \to R, \quad F_k(d) = F''_{F'(k)}(d).$$

Theorem 4 (Security of composed function F). *Let A be a T-transforming adversary for F. We can construct a T-transforming adversary A' for F', and a chosen ciphertext adversary A'' for F'', such that neither the running time of A', nor the running time of A'' exceed the running time of A, and the following condition holds:*

$$Adv^{\text{prf-rk}}_{T,F}(A) \le Adv^{\text{prf-rk}}_{T,F'}(A') + Adv^{\text{prf}}_{F''}(A''). \tag{2}$$

Neither A' nor A'' makes more oracle queries than A.

Proof. Let $k \in K'$ be a random key, unknown to the adversary A. A distinguishes between the events Real and Random

- Real: All responses to oracle queries $(\delta, d) \in K' \times D$ are generated as

$$F''_{F'(k+\delta)}(d).$$

- Random: Let $F^* : K' \times D \to R$ be a random function. All responses to oracle queries $(\delta, d) \in K' \times D$ are generated as $F^*(\delta, d)$.

We introduce a third event, K'-Random:

- K'-Random: Let $F^{**} : K' \rightarrow R'$ be a random function. All responses to oracle queries $(\delta, d) \in K' \times D$ are generated as

$$F''_{F^{**}(\delta)}(d).$$

Distinguishing Real from K'-Random means to distinguish K' from a random function. This is exactly the task A' is supposed to do. Thus, we can turn A into A' without increasing either running time or number of queries.

Similarly, we observe that if we can distinguish K'-Random from Random, we can mount a chosen plaintext attack against F'' and thus turn A into A'', again with the same running time and number of queries.

What about Condition 2? See the picture below.

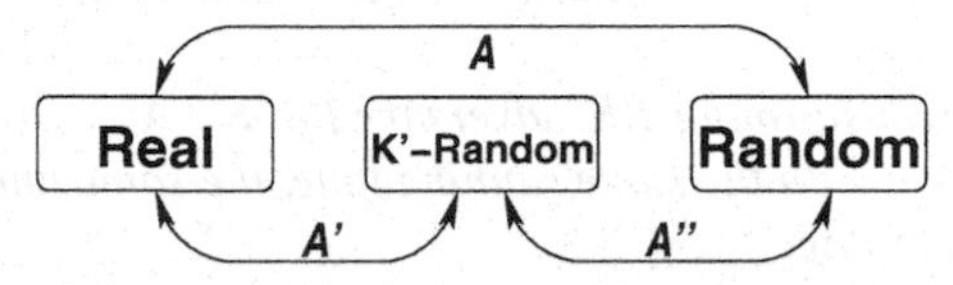

With A' and A'' as described above, condition 2 holds. □

In short, Theorem 4 implies that if F' is practically secure against T-transforming adversaries and F'' is practically secure against chosen ciphertext adversaries, then F must be practically secure against T-transforming adversaries. This provides us with a *tool* for finding RK-secure PRFs, or proving their existence under reasonable assumptions:

Let T, K', D and R be given. We are searching for

$$F : K' \times D \rightarrow R,$$

practically secure w.r.t. T-transforming adversaries. It is sufficient to choose an appropriate R' and a conventionally secure PRF $F'' : R' \times D \rightarrow R$, and then search for a function

$$F' : K' \rightarrow R',$$

practically secure w.r.t. T-transforming adversaries.

The idea is that finding F' may be less difficult than finding F directly. In the next section, we concentrate on finding appropriate functions F'.

4 PRFs and Group-Induced Key Transformations

In this section, we describe two PRFs and prove their security against T^+-transforming adversaries under certain assumptions from algorithmic number-theory. This is a step towards solving a "challenging problem " posed by Bellare and Kohno [1, 2]. Note though, that our assumptions are non-standard and have not much been studied much, so far. It remains an open problem, to describe some PRFs or PRPs and reduce their security against T^+-transforming adversaries to some cryptographic standard assumption, such as Decisional Diffie-Hellman, Quadratic Residousity, or others.

4.1 The RSA-Based PRF F'_{RSA}

Let N be the product of two large random primes. We define the function

$$F'_{\text{RSA}} : \mathbb{Z}_N \to \mathbb{Z}_N, \quad F'_{\text{RSA}}(k) := k^N \bmod N.$$

To evaluate the security of F'_{RSA}, we define an appropriate problem:

Definition 2. *Let N be the product of two large random primes. Let R be a random value in $\mathbb{Z}_N$. Define*

$$f(x) = (x + R)^N \bmod N. \tag{3}$$

Interactive Dependent RSA Problem (IDRP): *Distinguish f from a random function $\mathbb{Z}_N \to \mathbb{Z}_N$. The distinguisher is given N and oracle access to the function (but neither R nor the factors of N).*

Interactive Dependent RSA Assumption: *The IDRP is infeasible.*

Some remarks on the IDRP:

1. We can make the above scheme and the IDRP more "RSA-like " by choosing any large RSA-exponent e (that means, e and $\varphi(N)$ have no common divisors) and rewriting Equation 3 by $f(x) = (x + R)^e \bmod N$. If e is small, however, this variant of the IDRP is feasible [15].
2. The IDRP can be seen as a generalisation of Pointcheval's Dependent-RSA problem [15]: For independent random $x, y \in \mathbb{Z}_N$, distinguish the random pair (x, y) from the pair $(x^e \bmod N, (x + 1)^e \bmod N)$.
   Given an efficient algorithm to solve the Dependent-RSA problem, we could efficiently solve the IDRP.

Theorem 5 (Security of F'_{RSA}). *Under the Interactive Dependent RSA Assumption, no efficient T^+-transforming adversary with significant advantage for F'_{RSA} can exist.*

Proof. The proof is quite straightforward. Assume F'_{RSA} to be insecure. Then an efficient T^+-transforming adversary A for F'_{RSA} wins the following attack game with significant advantage:

- choose $\delta \in \mathbb{Z}_N$, define a key transformation $t_\delta(k) = k + \delta \bmod N$,
- ask for $F'_{\text{RSA}}(t_\delta(k)) = F'_{\text{RSA}}(k + \delta) = (k + \delta)^N \bmod N$ (with k unknown),
- and, after repeating the above two steps a couple of times, distinguish the results from the outputs of a random function.

This attack game is equivalent to solving the IDRP. □

4.2 The Diffie-Hellman based PRF F'_{DH}

In [1, 2], Bellare and Kohno consider two PRF-constructions which are provably secure against chosen plaintext adversaries under the Decisional Diffie-Hellman assumption. Both turn out to be insecure against additive-transforming adversaries. How can we define a Diffie-Hellman based PRF, with plausible hope for security against additive-transforming adversaries?

Let P, P_2, P_3 and P_4 be primes, $P = 2P_2 + 1$, $P_2 = 2P_3 + 1$. $P_3 = 2P_4 + 1$. Let g be an element of order P_2 in $\mathbb{Z}_P^*$. Let g_2 be an element of order P_3 in $\mathbb{Z}_{P_2}^*$. Let g_3 be an element of order P_4 in $\mathbb{Z}_{P_3}^*$. As before, the key transformations are additions (below, we will formally define the set T^+ in this context). We consider the following functions:

- The function $F_1'(k) = g^k \bmod P$ is weak, since $g^{k+\delta} = g^k * g^\delta \bmod P$. Thus, given $g^k = F_1'(k+0)$ and δ, we can compare a response from the RK oracle with $F_1'(k+0) = g^k * g^\delta \bmod P$. This is used in [1, 2] for straightforward RK attacks certain Diffie-Hellman based PRFs.
- Similarly, the function $F_2'(k) = g^{(g_2^k)} \bmod P$ is also weak, since $g^{(g_2^{k+\delta})} = g^{(g_2^k)(g_2^\delta)} = \left(g^{(g_2^k)}\right)^{(g_2^\delta)} \bmod P$.
- The function

$$F_{\mathrm{DH}}'(k) = g^{\left(g_2^{(g_3^k)}\right)} \bmod P$$

looks like a promising candidate.

For $F_{\mathrm{DH}}'(k)$, the set of keys is $\mathbb{Z}_{P_4}$. Consequently, our set T^+ of key transformations is defined by the addition modulo P_4.

Definition 3. *Let P, P_4, g, g_2, and g_3 be defined as above. Let r be a random value in $\mathbb{Z}_{P_4}^*$. Define*

$$f(x) = g^{\left(g_2^{(g_3^{x+r})}\right)} \bmod P.$$

Define $R = \{z \in \mathbb{Z}_P \mid \exists k \in \mathbb{Z}_{P_4} : z = F_{\mathrm{DH}}'(k)\}$.

Diffie-Hellman Random Function Assumption (DHRFA): *It is infeasible, to distinguish f from a random function $\mathbb{Z}_{P_4} \to R$.*

Theorem 6 (Security of F_{DH}').
Under the DHRFA, there exists no efficient T^+-transforming adversary for F_{DH}' with significant advantage.

The *proof* of Theorem 6 is similar to the proof of Theorem 5 and omitted here.

5 Using a Hash Function

As Ross Anderson pointed out at the FSE workshop in Delhi, a common engineering technique to ensure related-key security is to combine a block cipher E with a hash function H, defining a new block cipher

$$E_X^H(M) = E_{H(X)}(M).$$

This is a reasonable construction. In fact, for many types of related-key adversaries – including those considered in the current paper – it is straightforward to prove the security of E^H in the *random oracle model*, assuming E to be conventionally secure.

This approach has the following drawbacks, however:

- For implementing E^H, we need to implement two cryptographic primitives E and H.
- The security of E^H depends on the security of E *and* on the security of H. If, e.g., E is conventionally secure but H fails to meet its security requirements, E^H can be insecure.
- A random oracle proof of security for E^H does reveal the security requirements for H.
  On the other hand, it may be possible to prove the security of E^H against certain kinds of related-key adversaries in the *standard model*, making some nonstandard assumptions on H.

6 Summary

This paper presented new constructions for related-key secure PRFs.

For one construction, a tight security bound against partially-transforming adversaries has been shown, improving the concrete complexity of previous constructions. The proof assumes some block cipher to be secure in the conventional sense (i.e., without related keys).

Two other constructions are shown secure against more general adversaries, however under certain non-standard number-theoretical assumptions.

References

[1] M. Bellare, T. Kohno. A theoretical treatment of related-key attacks: RKA-PRPs RKA-PRFs and applications. E. Biham, editor, *Eurocrypt 2003*, Springer Lecture Notes in Computer Science # 2654, pp. 491–506. 359, 360, 361, 366, 367, 368, 369

[2] M. Bellare, T. Kohno. A theoretical treatment of related-key attacks: RKA-PRPs RKA-PRFs and applications. March 18, 2003. Full version of [1]. `http://www.cs.ucsd.edu/users/tkohno/papers/RKA/` (URL checked: Jan. 14, 2004). 359, 360, 361, 362, 366, 367, 368

[3] E. Biham. New types of cryptanalytic attacks using related keys. T. Helleseth, editor, *Eurocrypt 93*, Springer Lecture Notes in Computer Science # 765, pp. 398–409. 359

[4] J. Daemen, V. Rijmen. *AES proposal: Rijndael.* 359

[5] J. Daemen, V. Rijmen. *The design of Rijndael*, Springer-Verlag, 2002. 359

[6] Morris Dworkin. DRAFT Recommendation for block cipher modes of operation: the RMAC authentication mode. *NIST Special Publication 800-38b.* October 18, 2002. 359

[7] N. Ferguson, J. Kelsey, S. Lucks, B. Schneier, M. Stay, D. Wagner, D. Whiting. Improved cryptanalysis of Rijndael. B. Schneier, editor, *Fast Software Encryption 2000*, Springer Lecture Notes in Computer Science # 1978, pp. 213–230. 359

[8] E. Jaulmes, A. Joux, F. Valette. On the security of randomized CBC-MAC beyond the birthday paradox limit: A new construction. J. Daemen, V. Rijmen, editors, *Fast Software Encryption 2002*, Springer Lecture Notes in Computer Science. 359

[9] J. Kelsey, B. Schneier, D. Wagner. Key-schedule cryptanalysis of IDEA, G-DES, GOST, SAFER, and Triple-DES. N. Koblitz, editor, *Crypto '96*, Springer Lecture Notes in Computer Science # 1109, pp. 237–251. 359

[10] J. Kelsey, B. Schneier, D. Wagner. Related-key cryptanalysis of 3-Way, Biham-DES, CAST, DES-X, NewDES, RC2, and TEA. Y. Han, T. Okamoto, S. Quing, editors, *Information and Communications Security '97*, Springer Lecture Notes in Computer Science # 1334, pp. 233–246. 359

[11] L. Knudsen. Cryptanalysis of LOKI91. J. Seberry, Y. Zheng, editors, *Auscrypt '92*, Springer Lecture Notes in Computer Science # 718, pp. 196–208. 359

[12] L. Knudsen, T. Kohno. Analysis of RMAC. T. Johansson, editor, *Fast Software Encryption 2003*, Springer Lecture Notes in Computer Science # 2887, pp. 182–191. 359

[13] M. Liskov, R. Rivest, D. Wagner. Tweakable block ciphers. M. Yung, editor, *Crypto '02*, Springer Lecture Notes in Computer Science # 2422, pp. 31–46. 359

[14] M. Naor, O. Reingold. On the construction of pseudo-random permutations: Luby-Rackoff revisited. *J. of Cryptology*, vol 12, 1999, pp. 29-66. 360

[15] D. Pointcheval. New public key cryptosystems based on the dependent-RSA problems. Jacques Stern, editor, *Eurocrypt 1999*, Springer Lecture Notes in Computer Science # 1592, pp. 239-254. 367

Cryptographic Hash-Function Basics: Definitions, Implications, and Separations for Preimage Resistance, Second-Preimage Resistance, and Collision Resistance

Phillip Rogaway[1,2] and Thomas Shrimpton[3]

[1] Dept. of Computer Science, University of California
Davis, CA 95616, USA
rogaway@cs.ucdavis.edu
www.cs.ucdavis.edu/~rogaway

[2] Dept. of Computer Science, Fac of Science, Chiang Mai University, 50200 Thailand

[3] Dept. of Electrical and Computer Engineering, University of California
Davis, CA 95616, USA
teshrim@ucdavis.edu
www.ece.ucdavis.edu/~teshrim

Abstract. We consider basic notions of security for cryptographic hash functions: collision resistance, preimage resistance, and second-preimage resistance. We give seven different definitions that correspond to these three underlying ideas, and then we work out all of the implications and separations among these seven definitions within the concrete-security, provable-security framework. Because our results are concrete, we can show two types of implications, *conventional* and *provisional*, where the strength of the latter depends on the amount of compression achieved by the hash function. We also distinguish two types of separations, *conditional* and *unconditional*. When constructing counterexamples for our separations, we are careful to preserve specified hash-function domains and ranges; this rules out some pathological counterexamples and makes the separations more meaningful in practice. Four of our definitions are standard while three appear to be new; some of our relations and separations have appeared, others have not. Here we give a modern treatment that acts to catalog, in one place and with carefully-considered nomenclature, the most basic security notions for cryptographic hash functions.

Keywords: collision resistance, cryptographic hash functions, preimage resistance, provable security, second-preimage resistance.

1 Introduction

This paper casts some new light on an old topic: the basic security properties of cryptographic hash functions. We provide definitions for various notions of collision-resistance, preimage resistance, and second-preimage resistance, and

B. Roy and W. Meier (Eds.): FSE 2004, LNCS 3017, pp. 371–388, 2004.

then we work out all of the relationships among the definitions. We adopt a concrete-security, provable-security viewpoint, using reductions and definitions as the basic currency of our investigation.

INFORMAL TREATMENTS OF HASH FUNCTIONS. Informal treatments of cryptographic hash functions can lead to a lot of ambiguity, with informal notions that might be formalized in very different ways and claims that might correspondingly be true or false. Consider, for example, the following quotes, taken from our favorite reference on cryptography [9, pp. 323–330]:

> *preimage-resistance* — for essentially all pre-specified outputs, it is computationally infeasible to find any input which hashes to that output, i.e., to find any preimage x' such that $h(x') = y$ when given any y for which a corresponding input is not known.
>
> *2nd-preimage resistance* — it is computationally infeasible to find any second input which has the same output as any specified input, i.e., given x, to find a 2nd-preimage $x' \neq x$ such that $h(x) = h(x')$.
>
> *collision resistance* — it is computationally infeasible to find any two distinct inputs x, x' which hash to the same output, i.e., such that $h(x) = h(x')$.
>
> **Fact** Collision resistance implies 2nd-preimage resistance of hash functions.
>
> **Note** (*collision resistance does not guarantee preimage resistance*)

In trying to formalize and verify such statements, certain aspects of the English are problematic and other aspects aren't. Consider the first statement above. Our community understands quite well how to deal with the term *computationally infeasible*. But how is it meant to specify the output y? (What, exactly, do "essentially all" and "pre-specified outputs" mean?) Is hash function h to be a fixed function or a random element from a set of functions? Similarly, for the second quote, is it really meant that the specified point x can be *any* domain point (e.g., it is not chosen at random)? As for the bottom two claims, we shall see that the first is true under two formalizations we give for 2nd-preimage resistance and false under a third; the second statement is true for all hash functions under two formalizations of preimage resistance, while under a third the strength of this separation depends on the extent to which the hash function is compressing. [1]

SCOPE. In this paper we are going to examine seven different notions of security for a hash function family $H\colon \mathcal{K} \times \mathcal{M} \to \{0,1\}^n$. For a more complete discussion of nomenclature, see Appendix A and reference [9].

[1] We emphasize that it is most definitely *not* our intent here to criticize one of the most useful books on cryptography; we only use it to help illustrate that there are many ways to go when formalizing notions of hash-function security, and how one chooses to formalize things matters for making even the most basic of claims.

Name	Find	Experiment	Some Aliases
Pre	preimage	random key, random challenge	OWF
ePre	preimage	random key, fixed challenge	
aPre	preimage	fixed key, random challenge	
Sec	2nd-preimage	random key, random challenge	weak CR
eSec	2nd-preimage	random key, fixed challenge	UOWHF
aSec	2nd-preimage	fixed key, random challenge	
Coll	collision	random key (no challenge)	strong CR, collision-free

How did we arrive at exactly these seven notions? We set out to be exhaustive. For *two* of our goals—finding a preimage and finding a second preimage—it makes sense to think of *three* different settings: the key and the challenge being random; the key being random and the challenge being fixed; or the key being fixed and the challenge being random. It makes no sense to think of the key and the challenge as both being fixed, for a trivial adversary would then succeed. For the final goal—finding a collision—there is no challenge and one is compelled to think of the key as being random, for a trivial adversary would prevail if the key were fixed. We thus have $2 \cdot 3 + 1 = 7$ sensible notions, which we name Pre, ePre, aPre, Sec, eSec, aSec, and Coll. The leading "a" in the name of a notion is meant to suggest *always*: if a hash function is secure for any fixed key, then it is "always" secure. The leading "e" in the name of a notion is meant to suggest *everywhere*: if a hash function is secure for any fixed challenge, then it is "everywhere" secure. Notions Coll, Pre, Sec, eSec are standard; variants ePre, aPre, and aSec would seem to be new.

COMMENTS. The aPre and aSec notions may be useful for designing higher-level protocols that employ hash functions that are to be instantiated with SHA1-like objects. Consider a protocol that uses an object like SHA1 but says it is using a collision-resistant hash function, and proves security under such an assumption. There is a problem here, because there is no natural way to think of SHA1 as being a random element drawn from some family of hash functions. If the protocol could instead have used an aSec-secure hash-function family, doing the proof from that assumption, then instantiating with SHA1 would seem to raise no analogous, foundational issues. In short, assuming that your hash function is aSec- or aPre-secure serves to eliminate the mismatch of using a standard cryptographic hash function after having done proofs that depend on using a random element from a hash-function family.

CONTRIBUTIONS. Despite the numerous papers that construct, attack, and use cryptographic hash functions, and despite a couple of investigations of cryptographic hash functions whose purpose was close to ours [15, 16], the area seems to have more than its share of conflicting terminology, informal notions, and assertions of implications and separations that are not supported by convincing proofs or counterexamples. Our goal has been to help straighten out some of the basics. See Appendix A for an abbreviated exposition of related work.

We begin by giving formal definitions for our seven notions of hash-function security. Our definitions are concrete (no asymptotics) and treat a hash function H as a family of functions, $H\colon \mathcal{K} \times \mathcal{M} \to \{0,1\}^n$.

After defining the different notions of security we work out all of the relationships among them. Between each pair of notions xxx and yyy we provide either an implication or a separation. Informally, saying that xxx *implies* yyy means that if H is secure in the xxx-sense then it is also secure in the yyy-sense. To separate notions, we say, informally, that xxx *nonimplies* yyy if H can be secure in the xxx-sense without being secure in the yyy-sense.[2] Our implications and separations are quantitative, so we provide both an implication *and* a separation for the cases where this makes sense. Since we are providing implications and separations, we adopt the strongest feasible notions of each, in order to strengthen our results.

We actually give two kinds of implications. We do this because, in some cases, the strength of an implication crucially depends on the amount of compression achieved by the hash function. For these *provisional* implications, if the hash function is substantially compressing (e.g., mapping 256 bits to 128 bits) then the implication is a strong one, but if the hash function compresses little or not at all, then the implication effectively vanishes. It is a matter of interpretation whether such a provisional implication is an implication with a minor "technical" condition, or if a provisional implication is fundamentally not an implication at all. A *conventional* implication is an ordinary one; the strength of the implication does not depend on how much the hash function compresses.

We will also use two kinds of separations, but here the distinction is less dramatic, as both flavors of separations are strong. The difference between a *conventional* separation and an *unconditional* separation lies in whether or not one must effectively assume the existence of an xxx-secure hash function in order to show that xxx nonimplies yyy.

When we give separations, we are careful to impose the hash-function domain and range first; we don't allow these to be chosen so as to make for convenient counterexamples. This makes the problem of constructing counterexamples harder, but it also make the results more meaningful. For example, if a protocol designer wants to know if collision-resistance implies preimage-resistance for a 160-bit hash function H, what good is a counterexample that uses H to make a 161-bit hash function H' that is collision resistant but not preimage-resistant? It would not engender any confidence that collision-resistance fails to imply preimage-resistance when all hash functions of interest have 160-bit outputs.

Some of the counterexamples we use may appear to be unnatural, or to exhibit behavior unlike "real world" hash functions. This is not a concern; our goal is to demonstrate when one notion does not imply another by constructing counterexamples that respect imposed domain and range lengths; there is no need for the examples to look natural.

---

[2] We say "nonimplies" rather than "does not imply" because a separation is not the negation of an implication; a separation is effectively stronger and more constructive than that.

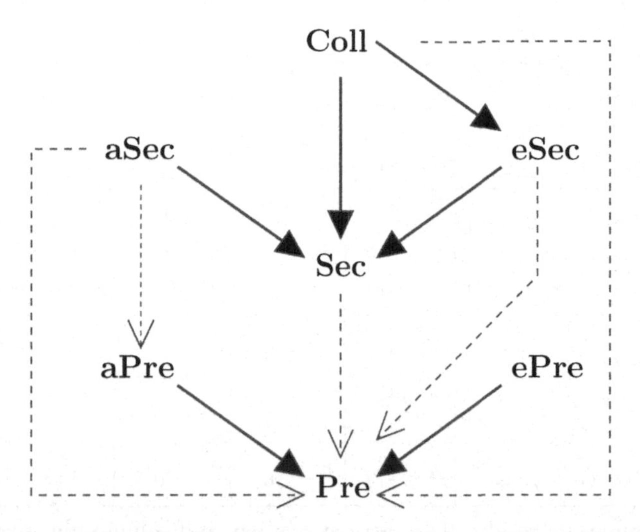

Fig. 1. Summary of the relationships among our seven notions of hash-function security. Solid arrows represent conventional implications, dotted arrows represent provisional implications (their strength depends on the relative size of the domain and range), and the lack of an arrow represents a separation

Our findings are summarized in Fig. 1, which shows when one notion implies the other (drawn with a solid arrow), when one notion provisionally implies the other (drawn with a dotted arrow), and when one notion nonimplies the other (we use the absence of an arrow and do not bother to distinguish between the two types of nonimplications). In Fig. 2 we give a more detailed summary of the results of this paper.

2 Preliminaries

We write $M \stackrel{\$}{\leftarrow} \mathcal{S}$ for the experiment of choosing a random element from the distribution $\mathcal{S}$ and calling it M. When $\mathcal{S}$ is a finite set it is given the uniform distribution. The concatenation of strings M and M' is denoted by $M \parallel M'$ or MM'. When $M = M_1 \cdots M_m \in \{0,1\}^m$ is an m-bit string and $1 \leq a \leq b \leq m$ we write $M[a..b]$ for $M_a \cdots M_b$. The bitwise complement of a string M is written $\overline{M}$. The empty string is denoted by ε. When a is an integer we write $\langle a \rangle_r$ for the r-bit string that represents a.

A *hash-function family* is a function $H\colon \mathcal{K} \times \mathcal{M} \to \mathcal{Y}$ where $\mathcal{K}$ and $\mathcal{Y}$ are finite nonempty sets and $\mathcal{M}$ and $\mathcal{Y}$ are sets of strings. We insist that $\mathcal{Y} = \{0,1\}^n$ for some $n > 0$. The number n is called the *hash length* of H. We also insist that if $M \in \mathcal{M}$ then $\{0,1\}^{|M|} \subseteq \mathcal{M}$ (the assumption is convenient and any reasonable hash function would certainly have this property). Often we will write the first argument to H as a subscript, so that $H_K(M) = H(K, M)$ for all $M \in \mathcal{M}$.

When $H\colon \mathcal{K} \times \mathcal{M} \to \mathcal{Y}$ and $\{0,1\}^m \subseteq \mathcal{M}$ we denote by $\mathrm{Time}_{H,m}$ the minimum, over all programs P_H that compute H, of the length of P_H plus the worst-case running time of P_H over all inputs (K, M) where $K \in \mathcal{K}$ and $M \in \{0,1\}^m$;

	Pre	ePre	aPre	Sec	eSec	aSec	Coll
Pre	$\to$	$\not\to$ to δ_3 (d)	$\not\to$ to δ_4 (e)	$\not\to$ (h)	$\not\to$ (h)	$\not\to$ (h)	$\not\to$ (h)
ePre	$\to$ (l)	$\to$	$\not\to$ to δ_4 (e)	$\not\to$ (h)	$\not\to$ (h)	$\not\to$ (h)	$\not\to$ (h)
aPre	$\to$ (l)	$\not\to$ to δ_3 (d)	$\to$	$\not\to$ (h)	$\not\to$ (h)	$\not\to$ (h)	$\not\to$ (h)
Sec	$\to$ to δ_1 (a) $\not\to$ to δ_2 (b)	$\not\to$ to δ_3 (d)	$\not\to$ to δ_4 (e)	$\to$	$\not\to$ to δ_5 (i)	$\not\to$ to δ_4 (e)	$\not\to$ to δ_5 (i)
eSec	$\to$ to δ_1 (a) $\not\to$ to δ_2 (c)	$\not\to$ (f)	$\not\to$ to δ_4 (e)	$\to$ (l)	$\to$	$\not\to$ to δ_4 (e)	$\not\to$ to δ_5 (j) $\not\to$ (k)
aSec	$\to$ to δ_1 (a) $\not\to$ to δ_2 (b)	$\not\to$ to δ_3 (d)	$\to$ to δ_1 (a) $\not\to$ to δ_2 (b)	$\to$ (l)	$\not\to$ to δ_5 (i)	$\to$	$\not\to$ to δ_5 (i)
Coll	$\to$ to δ_1 (a)	$\not\to$ (g)	$\not\to$ to δ_4 (e)	$\to$ (l)	$\to$ (l)	$\not\to$ to δ_4 (e)	$\to$

Fig. 2. Summary of results. The entry at row xxx and column yyy gives the relationships we establish between notions xxx and yyy. Here $\delta_1 = 2^{n-m}$, $\delta_2 = 1 - 2^{n-m-1}$, $\delta_3 = 2^{-m}$, $\delta_4 = 1/|\mathcal{K}|$, and $\delta_5 = 2^{1-m}$. The hash functions $H1, \ldots, H6$ and $G1, G2, G3$ are specified in Fig. 3. The annotations (a)-(j) mean: (a) see Theorem 1; (b) by $G1$, see Proposition 2; (c) by $G3$, see Proposition 3; (d) by $H1$, see Theorem 5; (e) by $H2$, see Theorem 5 (f) by $H6$, see Theorem 4; (g) by $H6$, see Theorem 3; (h) by $H3$, see Theorem 5; (i) by $H4$, see Theorem 5; (j) by $G2$, see Proposition 4; (k) by $H5$, see Theorem 2; (l) see Proposition 1

plus the the minimum, over all programs P_K that sample from $\mathcal{K}$, of the time to compute the sample plus the size of P_K. We insist that P_H read its input, so that $\mathrm{Time}_{H,m}$ will always be at least m. Some underlying RAM model of computation must be fixed.

An *adversary* is an algorithm that takes any number of inputs. Some of these inputs may be long strings and so we establish the convention that the adversary can read the ith bit of argument j by writing (i, j), in binary, on distinguished query tape. The resulting bit is returned to the adversary in unit time. If A is an adversary and $\mathbf{Adv}_H^{\mathrm{xxx}}(A)$ is a measure of adversarial advantage already defined then we write $\mathbf{Adv}_H^{\mathrm{xxx}}(\mathcal{R})$ to mean the maximal value of $\mathbf{Adv}_H^{\mathrm{xxx}}(A)$ over all adversaries A that use resources bounded by $\mathcal{R}$. In this paper it is sufficient to consider only the resource t, the running time of the adversary. By convention, the running time is the actual worst case running time of A (relative to some fixed RAM model) plus the description size of A (relative to some fixed encoding of algorithms).

3 Definitions of Hash-Function Security

PREIMAGE RESISTANCE. One would like to speak of the difficulty with which an adversary is able to find a preimage for a point in the range of a hash function. Several definitions make sense for this intuition of inverting.

$$H1_K(M) = \begin{cases} 0^n \text{ if } M = 0^m \\ H_K(M) \text{ otherwise} \end{cases}$$

$$H2_K(M) = \begin{cases} 0^n \text{ if } K = K_0 \\ H_K(M) \text{ otherwise} \end{cases}$$

$$H3^b_K(M) = H_K(M[1..m-1] \parallel b)$$

$$H4_K(M) = \begin{cases} 0^n \text{ if } M = 0^m \text{ or } M = 1^m \\ H_K(M) \text{ otherwise} \end{cases}$$

$$H5^c_K(M) = \begin{cases} H_K(0^{m-n} \parallel H_K(c)) \text{ if } M = 1^{m-n} \parallel H_K(c) & (1) \\ H_K(M) \text{ otherwise} & (2) \end{cases}$$

$$H6_K(M) = \begin{cases} 0^n \text{ if } M = 0^m & (1) \\ H_K(M) \text{ if } M \neq 0^m \text{ and } H_K(M) \neq 0^n & (2) \\ H_K(0^m) \text{ otherwise} & (3) \end{cases}$$

$$G1_K(M) = \begin{cases} M[1..n] \text{ if } M[n+1..m] = 0^{m-n} \\ 0^n \text{ otherwise} \end{cases}$$

$$G2_K(M) = \begin{cases} 1^{n-m} \parallel K \text{ if } M \in \{K, \overline{K}\} \\ 0^{n-m} \parallel M \text{ otherwise} \end{cases}$$

$$G3_K(M) = \begin{cases} \langle i \rangle_n \text{ if } M = \langle (K+i) \bmod 2^m \rangle_m \text{ for some } i \in [1..2^n - 1] \\ 0^n \text{ otherwise} \end{cases}$$

Fig. 3. Given a hash function H: $\mathcal{K} \times \{0,1\}^m \to \{0,1\}^n$ we construct hash functions $H1, \ldots, H6$: $\mathcal{K} \times \{0,1\}^m \to \{0,1\}^n$ for our conditional separations. The value $K_0 \in \mathcal{K}$ is fixed and arbitrary. The hash functions $G1$: $\{\varepsilon\} \times \{0,1\}^m \to \{0,1\}^n$, $G2$: $\{0,1\}^m \times \{0,1\}^m \to \{0,1\}^n$, $G3$: $\{1, \ldots, 2^m - 1\} \times \{0,1\}^m \to \{0,1\}^n$, are used in our unconditional separations

Definition 1 [Types of preimage resistance] Let $H = \mathcal{K} \times \mathcal{M} \to \mathcal{Y}$ be a hash-function family and let m be a number such that $\{0,1\}^m \subseteq \mathcal{M}$. Let A be an adversary. Then define:

$$\mathbf{Adv}_H^{\mathrm{Pre}\,[m]}(A) = \Pr\left[K \xleftarrow{\$} \mathcal{K};\, M \xleftarrow{\$} \{0,1\}^m;\, Y \leftarrow H_K(M);\, M' \xleftarrow{\$} A(K,Y) : H_K(M') = Y\right]$$

$$\mathbf{Adv}_H^{\mathrm{ePre}}(A) = \max_{Y \in \mathcal{Y}} \left\{ \Pr\left[K \xleftarrow{\$} \mathcal{K};\, M \xleftarrow{\$} A(K) :\ H_K(M) = Y\right] \right\}$$

$$\mathbf{Adv}_H^{\mathrm{aPre}\,[m]}(A) = \max_{K \in \mathcal{K}} \left\{ \Pr\left[M \xleftarrow{\$} \{0,1\}^m;\, Y \leftarrow H_K(M);\, M' \xleftarrow{\$} A(Y) : H_K(M') = Y\right] \right\} \qquad \bullet$$

The first definition, *preimage resistance* (Pre), is the usual way to define when a hash-function family is a *one-way function*. (Of course the notion is different from a function f: $\mathcal{M} \to \mathcal{Y}$ being a one-way function, as these are syntactically different objects.) The second definition, *everywhere preimage-resistance* (ePre), most directly captures the intuition that it is infeasible to find the preimage of range points: for *whatever* range point is selected, it is computationally hard to find its preimage. The final definition, *always preimage-resistance* (aPre), strengthens the first definition in the way needed to say that a function like SHA1 is one-way: one regards SHA1 as one function from a family of hash functions (keyed, for example, by the initial chaining value) and we wish to say that for this particular function from the family it remains hard to find a preimage of a random point.

SECOND-PREIMAGE RESISTANCE. It is likewise possible to formalize multiple definitions that might be understood as technical meaning for second-preimage resistance. In all cases a domain point M and a description of a hash function H_K are known to the adversary, whose job it is to find an M' different from M such that $H(K,M) = H(K,M')$. Such an M and M' are called *partners*.

Definition 2 [Types of second-preimage resistance] Let $H\colon \mathcal{K} \times \mathcal{M} \to \mathcal{Y}$ be a hash-function family and let m be a number such that $\{0,1\}^m \subseteq \mathcal{M}$. Let A be an adversary. Then define:

$$\mathbf{Adv}_H^{\mathrm{Sec}\,[m]}(A) = \Pr\Big[K \stackrel{\$}{\leftarrow} \mathcal{K};\ M \stackrel{\$}{\leftarrow} \{0,1\}^m;\ M' \stackrel{\$}{\leftarrow} A(K,M) : \\ (M \neq M') \wedge (H_K(M) = H_K(M'))\Big]$$

$$\mathbf{Adv}_H^{\mathrm{eSec}\,[m]}(A) = \max_{M \in \{0,1\}^m} \Big\{\Pr\Big[K \stackrel{\$}{\leftarrow} \mathcal{K};\ M' \stackrel{\$}{\leftarrow} A(K) : \\ (M \neq M') \wedge (H_K(M) = H_K(M'))\Big]\Big\}$$

$$\mathbf{Adv}_H^{\mathrm{aSec}\,[m]}(A) = \max_{K \in \mathcal{K}} \Big\{\Pr\Big[M \stackrel{\$}{\leftarrow} \{0,1\}^m;\ M' \stackrel{\$}{\leftarrow} A(M) : \\ (M \neq M') \wedge (H_K(M) = H_K(M'))\Big]\Big\} \quad \bullet$$

The first definition, *second-preimage resistance* (Sec), is the standard one. The second definition, *everywhere second-preimage resistance* (eSec), most directly formalizes that it is hard to find a partner for any particular domain point. This notion is also called a *universal one-way hash-function family* (UOWHF) and it was first defined by Naor and Yung [12]. The final definition, *always second-preimage resistance* (aSec), strengthens the first in the way needed to say that a function like SHA1 is second-preimage resistant: one regards SHA1 as one function from a family of hash functions and we wish to say that for this particular function it is remains hard to find a partner for a random point.

COLLISION RESISTANCE. Finally, we would like to speak of the difficulty with which an adversary is able to find two distinct points in the domain of a hash function that hash to the same range point.

Definition 3 [Collision resistance] Let $H\colon \mathcal{K} \times \mathcal{M} \to \mathcal{Y}$ be a hash-function family and let A be an adversary. Then we define:

$$\mathbf{Adv}_H^{\mathrm{Coll}}(A) = \Pr\Big[K \stackrel{\$}{\leftarrow} \mathcal{K};\ (M,M') \stackrel{\$}{\leftarrow} A(K) : (M \neq M') \wedge (H_K(M) = H_K(M'))\Big] \quad \bullet$$

It does not make sense to think of strengthening this definition by maximizing over all $K \in \mathcal{K}$: for any fixed function $h\colon \mathcal{M} \to \mathcal{Y}$ with $|\mathcal{M}| > |\mathcal{Y}|$ there is is an efficient algorithm that outputs an M and M' that collide under h. While this program might be hard to find in practice, there is no known sense in which this can be formalized.

4 Equivalent Formalizations with a Two-Stage Adversary

Four of our definitions (ePre, aPre, eSec, aSec) maximize over some quantity that one may imagine the adversary to know. In each of these cases it possible to modify the definition so as to have the adversary itself choose this value. That is, in a "first phase" of the adversary's execution it chooses the quantity in question, and then a random choice is made by the environment, and then the adversary continues from where it left off, but now given this randomly chosen value. The corresponding definitions are then as follows:

Definition 4 [Equivalent versions of ePre, aPre, eSec, aSec] Let $H = \mathcal{K} \times \mathcal{M} \to \mathcal{Y}$ be a hash-function family and let m be a number such that $\{0,1\}^m \subseteq \mathcal{M}$. Let A be an adversary. Then define:

$$\mathbf{Adv}_H^{\mathrm{ePre}}(A) = \Pr\left[(Y,S) \xleftarrow{\$} A();\ K \xleftarrow{\$} \mathcal{K};\ M \xleftarrow{\$} A(K,S):\ H_K(M) = Y\right]$$

$$\mathbf{Adv}_H^{\mathrm{aPre}\,[m]}(A) = \Pr\Big[(K,S) \xleftarrow{\$} A();\ M \xleftarrow{\$} \{0,1\}^m;\ Y \leftarrow H_K(M);\ M' \xleftarrow{\$} A(Y,S): \quad H_K(M') = Y\Big]$$

$$\mathbf{Adv}_H^{\mathrm{eSec}\,[m]}(A) = \Pr\Big[(M,S) \xleftarrow{\$} A();\ K \xleftarrow{\$} \mathcal{K};\ M' \xleftarrow{\$} A(K,S): \quad (M \neq M') \wedge (H_K(M) = H_K(M'))\Big]$$

$$\mathbf{Adv}_H^{\mathrm{aSec}\,[m]}(A) = \Pr\Big[(K,S) \xleftarrow{\$} A();\ M \xleftarrow{\$} \{0,1\}^m;\ M' \xleftarrow{\$} A(M,S): \quad (M \neq M') \wedge (H_K(M) = H_K(M'))\Big] \quad \bullet$$

In the two-stage definition of $\mathbf{Adv}_H^{\mathrm{eSec}\,[m]}(A)$ we insist that the message M output by A is of length m bits, that is $M \in \{0,1\}^m$. Each of these four definitions are extended to their resource-parameterized version in the usual way.

The two-stage definitions above are easily seen to be equivalent to their one-stage counterparts. Saying here that definitions xxx and yyy are *equivalent* means that there is a constant C such that $\mathbf{Adv}_H^{\mathrm{xxx}\,[m]}(t) \le \mathbf{Adv}_H^{\mathrm{yyy}\,[m]}(C(t + m + n))$ and $\mathbf{Adv}_H^{\mathrm{yyy}\,[m]}(t) \le \mathbf{Adv}_H^{\mathrm{xxx}\,[m]}(C(t + m + n))$. Omit mention of $+m$ and $[m]$ in the definition for everywhere preimage resistance since this does not depend on m. Since the exact interpretation of time t was model-dependent anyway, two measures of adversarial advantage that are equivalent need not be distinguished.

We give an example of the equivalence of one-stage and two-stage adversaries, explaining why eSec and eSec2 are equivalent, where eSec2 temporarily denotes the version of eSec defined in Definition 4 (and eSec refers to what is given in

Definition 2). Let A attack hash function H in the eSec sense. For every fixed M there is a two-stage adversary $A2$ that does as well as A at finding a partner for M. Specifically, let $A2$ be an adversary with the value M "hardwired in" to it. Adversary $A2$ prints out M and when it resumes it behaves like A. Similarly, let $A2$ be a two-stage adversary attacking H in the eSec2 sense. Consider the random coins used by $A2$ during its first stage and choose specific coins that maximize the probability that $A2$ will subsequently succeed. For these coins there is a specific pair (M, S) that $A2$ returns. Let A be a (one-stage) adversary that on input (K, M) runs exactly as $A2$ would on input (K, S).

5 Implications

DEFINITIONS OF IMPLICATIONS. In this section we investigate which of our notions of security (Pre, aPre, ePre, Sec, aSec, eSec, and Coll) imply which others. First we explain our notion of an implication.

Definition 5 [Implications] Fix $\mathcal{K}$, $\mathcal{M}$, m, and n where $\{0,1\}^m \subseteq \mathcal{M}$. Suppose that xxx and yyy are labels for which $\mathbf{Adv}_H^{\text{xxx}\cdot}$ and $\mathbf{Adv}_H^{\text{yyy}\cdot}$ have been defined for any $H\colon \mathcal{K} \times \mathcal{M} \to \{0,1\}^n$.

- *Conventional implication.* We say that xxx **implies** yyy, written xxx $\to$ yyy, if $\mathbf{Adv}_H^{\text{yyy}\cdot}(t) \le c\,\mathbf{Adv}_H^{\text{xxx}\cdot}(t')$ for all hash functions $H\colon \mathcal{K} \times \mathcal{M} \to \{0,1\}^n$ where c is an absolute constant and $t' = t + c\,\text{Time}_{H,m}$.
- *Provisional implication.* We say that xxx **implies** yyy **to** ϵ, written xxx $\to$ yyy to ϵ, if $\mathbf{Adv}_H^{\text{yyy}\cdot}(t) \le c\,\mathbf{Adv}_H^{\text{xxx}\cdot}(t') + \epsilon$ for all hash functions $H\colon \mathcal{K}\times\mathcal{M} \to \{0,1\}^n$ where c is an absolute constant and $t' = t+c\,\text{Time}_{H,m}$.

•

In the definition above, and later, the $\cdot$ is a placeholder which is either $[m]$ (for Pre, aPre, Sec, aSec, eSec) or empty (for ePre, Coll).

Conventional implications are what one expects: xxx $\to$ yyy means that if a hash function is secure in the xxx-sense, then it is secure in the yyy-sense. Whether or not a provisional implication carries the usual semantics of the word *implication* depends on the value of ϵ. Below we will demonstrate provisional implications with a value of $\epsilon = 2^{n-m}$ and so the interpretation of such a result is that we have demonstrated a "real" implication for hash functions that are substantially compressing (e.g., if the hash function maps 256 bits to 128 bits) while we have given a *non-result* if the hash function is length-preserving, length-increasing, or it compresses just a little.

CONVENTIONAL IMPLICATIONS. The conventional implications among our notions are straightforward, so we quickly dispense with those, omitting the proofs. In particular, the following are easily verified.

Proposition 1. [Conventional implications] Fix $\mathcal{K}$, $\mathcal{M}$, m, such that $\{0,1\}^m \subseteq \mathcal{M}$, and $n > 0$. Let Coll, Pre, aPre, ePre, Sec, aSec, eSec be the corresponding security notions. Then:

(1) Coll $\rightarrow$ Sec
(2) Coll $\rightarrow$ eSec
(3) aSec $\rightarrow$ Sec
(4) eSec $\rightarrow$ Sec
(5) aPre $\rightarrow$ Pre
(6) ePre $\rightarrow$ Pre •

In addition to the above, of course xxx $\rightarrow$ xxx for each notion xxx that we have given.

PROVISIONAL IMPLICATIONS. We now give five provisional implications. The value of ϵ implicit in these claims depends on the relative difference of the domain length m and the hash length n. Intuitively, one can follow paths through the graph in Figure 1, composing implications to produce the five provisional implications. The formal proof of these five results appears in the full version [14].

Theorem 1. [Provisional implications] Fix $\mathcal{K}$, $\mathcal{M}$, m, such that $\{0,1\}^m \subseteq \mathcal{M}$, and $n > 0$. Let Coll, Pre, aPre, Sec, aSec, eSec be the corresponding security notions. Then:
(1) Sec $\rightarrow$ Pre to 2^{n-m}
(2) aSec $\rightarrow$ Pre to 2^{n-m}
(3) eSec $\rightarrow$ Pre to 2^{n-m}
(4) Coll $\rightarrow$ Pre to 2^{n-m}
(5) aSec $\rightarrow$ aPre to 2^{n-m} •

6 Separations

DEFINITIONS. We now investigate separations among our seven security notions. We emphasize that asserting a separation—which we will also call a *nonimplication*—is *not* the assertion of a lack of an implication (though it does effectively imply this for any practical hash function). In fact, we will show that both a separation and an implication can exist between two notions, the relative strength of the separation/implication being determined by the amount of compression performed by the hash function. Intuitively, xxx nonimplies yyy if it is possible for something to be xxx-secure but not yyy-secure. We provide two variants of this idea. The first notion, a *conventional* nonimplication, says that if H is a hash function that is secure in the xxx-sense then H can be converted into a hash function H' having the same domain and range that is still secure in the xxx-sense but that is now completely *insecure* in the yyy-sense. The second notion, an *unconditional* nonimplication, says that there is a hash function H that is secure in the xxx-sense but completely insecure in the yyy-sense. Thus the first kind of separation effectively assumes an xxx-secure hash function in order to separate xxx from yyy, while the second kind of separation does not need to do this.[3]

Definition 6 **[Separations]** Fix $\mathcal{K}$, $\mathcal{M}$, m, and n where $\{0,1\}^m \subseteq \mathcal{M}$. Suppose that xxx and yyy be labels for which $\mathbf{Adv}_H^{\text{xxx}\cdot}$ and $\mathbf{Adv}_H^{\text{yyy}\cdot}$ have been defined for any $H\colon \mathcal{K}\times\mathcal{M}\to\{0,1\}^n$.

- *Conventional separation.* We say that xxx **nonimplies** yyy **to** ϵ, in the conventional sense, written xxx $\not\rightarrow$ yyy to ϵ, if for any $H\colon \mathcal{K}\times\mathcal{M}\to\{0,1\}^n$ there exists an $H'\colon \mathcal{K}\times\mathcal{M}\to\{0,1\}^n$ such that $\mathbf{Adv}_{H'}^{\text{xxx}\cdot}(t) \le c\,\mathbf{Adv}_H^{\text{xxx}\cdot}(t')+\epsilon$ and yet $\mathbf{Adv}_{H'}^{\text{yyy}\cdot}(t') = 1$ where c is an absolute constant and $t' = t + c\,\mathrm{Time}_{H,m}$.
- *Unconditional separation.* We say that xxx **nonimplies** yyy **to** ϵ, in the unconditional sense, written xxx $\not\twoheadrightarrow$ yyy to ϵ, if there exists an $H\colon \mathcal{K}\times\mathcal{M}\to\{0,1\}^n$ such that $\mathbf{Adv}_H^{\text{xxx}\cdot}(t) \le \epsilon$ for all t and yet $\mathbf{Adv}_H^{\text{yyy}\cdot}(t') = 1$ where $t' = c\,\mathrm{Time}_{H,m}$ for some absolute constant c. •

When $\epsilon = 0$ above we say that we have a *strong* separation and we omit saying "to ϵ" in speaking of it. When $\epsilon > 0$ above we say that we have a *provisional* separation. The degree to which a provisional separation should be regarded as a "real" separation depends on the value ϵ.

SOME PROVISIONAL SEPARATIONS. The following separations depend on the relative values of the domain size m and the range size n. As an example, if the hash-function family H is length-preserving, meaning $H\colon \mathcal{K}\times\{0,1\}^n\to\{0,1\}^n$, then it being second preimage resistant won't imply it being preimage resistant: just consider the identify function, which is perfectly second preimage resistant (no domain point has a partner) but trivially breakable in the sense of finding preimages. This counterexample is well-known. We now generalize and extend this counterexample, giving a "gap" of $1-2^{n-m-1}$ for three of our pairs of notions. Thus we have a strong separation when $m=n$ and a rapidly weakening separation as m exceeds n by more and more. Taken together with Proposition 1 we see that this behavior is not an artifact of the proof: as m exceeds n, the 2^{n-m}-implication we have given effectively takes over.

Proposition 2. **[Separations, part 1a]** Fix $m \ge n > 0$ and let Sec, Pre, aSec, aPre be the corresponding security notions. Then:

(1) Sec $\not\twoheadrightarrow$ Pre to $1-2^{n-m-1}$
(2) aSec $\not\twoheadrightarrow$ Pre to $1-2^{n-m-1}$
(3) aSec $\not\twoheadrightarrow$ aPre to $1-2^{n-m-1}$ •

The proof is given in the full version of this paper [14].

Proposition 3. **[Separations, part 1b]** Fix $m \ge n > 0$, and let Pre and eSec be the corresponding security notions. Then eSec $\not\twoheadrightarrow$ Pre to $1-2^{n-m-1}$.

[3] That unconditional separations are (sometimes) possible in this domain is a consequence of the fact that, for some values of the domain and range, secure hash functions trivially exist (e.g., the identity function $H_K(M)=M$ is collision-free).

The proof is given in the full version of this paper [14].

ADDITIONAL SEPARATIONS. We now give some further nonimplications. Unlike those just given, these nonimplications do not have a corresponding provisional implication. Here, the separation is the whole story of the relationship between the notions, and the strength of the separation is not dependent on the amount of compression performed by the hash function.

Theorem 2. [Separations, part 2A] Fix $m > n > 0$ and let eSec and Coll be the corresponding security notions. Then eSec $\not\to$ Coll. •

The proof is in Appendix B. Because of the structure of the counterexample used in Theorem 2, we give the following proposition for completeness.

Proposition 4. Fix $n > 0$ and $m \leq n$, and let eSec and Coll be the corresponding security notions. Then eSec $\not\to$ Coll to $2^{-(m+1)}$. •

The proof is given in the full version of this paper [14].

Theorem 3. [Separations, part 2B] Fix m, n such that $n > 0$, and let Coll and ePre be the corresponding security notions. Then Coll $\not\to$ ePre. •

The proof is given in the full version of this paper [14].

Theorem 4. [Separations, part 2C] Fix m, n such that $n > 0$, and let eSec and ePre be the corresponding security notions. Then eSec $\not\to$ ePre. •

The proof is given in the full version of this paper [14].

The remaining 28 separations are not as hard to show those given so far, so we present them as one theorem and without proof. The specific constructions $H1, H2, H3, H4$ are those given in Fig. 3.

Theorem 5. [Separations, part 3] Fix m, n such that $n > 0$, and let Coll, Pre, aPre, ePre, Sec, aSec, eSec be the corresponding security notions. Let $H\colon \mathcal{K} \times \{0,1\}^m \to \{0,1\}^n$ be a hash function and define $H1, \ldots, H6$ from it according to Fig. 3. Then:

(1) Pre $\not\to$ ePre to 2^{-m}: $\mathbf{Adv}_{H1}^{\mathrm{Pre}}(t) \leq 1/2^m + \mathbf{Adv}_{H}^{\mathrm{Pre}}(t)$ and $\mathbf{Adv}_{H1}^{\mathrm{ePre}}(t') = 1$

(2) Pre $\not\to$ aPre to $1/|\mathcal{K}|$: $\mathbf{Adv}_{H2}^{\mathrm{Pre}}(t) \leq 1/|\mathcal{K}| + \mathbf{Adv}_{H}^{\mathrm{Pre}}(t)$ and $\mathbf{Adv}_{H2}^{\mathrm{aPre}}(t') = 1$

(3) Pre $\not\to$ Sec: $\mathbf{Adv}_{H3}^{\mathrm{Pre}}(t) \leq 2 \cdot \mathbf{Adv}_{H}^{\mathrm{Pre}}(t)$ and $\mathbf{Adv}_{H3}^{\mathrm{Sec}}(t') = 1$

(4) Pre $\not\to$ eSec: $\mathbf{Adv}_{H3}^{\mathrm{Pre}}(t) \leq 2 \cdot \mathbf{Adv}_{H}^{\mathrm{Pre}}(t)$ and $\mathbf{Adv}_{H3}^{\mathrm{eSec}}(t') = 1$

(5) Pre $\not\to$ aSec: $\mathbf{Adv}_{H3}^{\mathrm{Pre}}(t) \leq 2 \cdot \mathbf{Adv}_{H}^{\mathrm{Pre}}(t)$ and $\mathbf{Adv}_{H3}^{\mathrm{aSec}}(t') = 1$

(6) Pre $\not\to$ Coll: $\mathbf{Adv}_{H3}^{\mathrm{Pre}}(t) \leq 2 \cdot \mathbf{Adv}_{H}^{\mathrm{Pre}}(t)$ and $\mathbf{Adv}_{H3}^{\mathrm{Coll}}(t') = 1$

(7) ePre $\not\to$ aPre to $1/|\mathcal{K}|$: $\mathbf{Adv}_{H2}^{\mathrm{ePre}}(t) \leq 1/|\mathcal{K}| + \mathbf{Adv}_{H}^{\mathrm{ePre}}(t)$ and $\mathbf{Adv}_{H2}^{\mathrm{aPre}\,[m]}(t') = 1$

(8) ePre $\not\to$ Sec: $\mathbf{Adv}_{H3}^{\mathrm{ePre}}(t) \leq 2 \cdot \mathbf{Adv}_{H}^{\mathrm{ePre}}(t)$ and $\mathbf{Adv}_{H3}^{\mathrm{Sec}\,[m]}(t') = 1$

(9) ePre $\not\to$ eSec: $\mathbf{Adv}_{H3}^{\mathrm{ePre}}(t) \leq 2 \cdot \mathbf{Adv}_{H}^{\mathrm{ePre}}(t)$ and $\mathbf{Adv}_{H3}^{\mathrm{eSec}\,[m]}(t') = 1$

(10) ePre $\not\to$ aSec: $\mathbf{Adv}_{H3}^{\mathrm{ePre}}(t) \leq 2 \cdot \mathbf{Adv}_{H}^{\mathrm{ePre}}(t)$ and $\mathbf{Adv}_{H3}^{\mathrm{aSec}\,[m]}(t') = 1$

(11) ePre $\not\to$ Coll: $\mathbf{Adv}_{H3}^{\mathrm{ePre}}(t) \le 2 \cdot \mathbf{Adv}_{H}^{\mathrm{ePre}}(t)$ and $\mathbf{Adv}_{H3}^{\mathrm{Coll}}(t') = 1$

(12) aPre $\not\to$ ePre to 2^{-m}: $\mathbf{Adv}_{H1}^{\mathrm{aPre}\,[m]}(t) \le 1/2^m + \mathbf{Adv}_{H}^{\mathrm{aPre}\,[m]}(t)$ and $\mathbf{Adv}_{H1}^{\mathrm{ePre}}(t') = 1$

(13) aPre $\not\to$ Sec: $\mathbf{Adv}_{H3}^{\mathrm{aPre}\,[m]}(t) \le 2 \cdot \mathbf{Adv}_{H}^{\mathrm{aPre}\,[m]}(t)$ and $\mathbf{Adv}_{H3}^{\mathrm{Sec}\,[m]}(t') = 1$

(14) aPre $\not\to$ eSec: $\mathbf{Adv}_{H3}^{\mathrm{aPre}\,[m]}(t) \le 2 \cdot \mathbf{Adv}_{H}^{\mathrm{aPre}\,[m]}(t)$ and $\mathbf{Adv}_{H3}^{\mathrm{eSec}\,[m]}(t') = 1$

(15) aPre $\not\to$ aSec: $\mathbf{Adv}_{H3}^{\mathrm{aPre}\,[m]}(t) \le 2 \cdot \mathbf{Adv}_{H}^{\mathrm{aPre}\,[m]}(t)$ and $\mathbf{Adv}_{H3}^{\mathrm{aSec}\,[m]}(t') = 1$

(16) aPre $\not\to$ Coll: $\mathbf{Adv}_{H3}^{\mathrm{aPre}\,[m]}(t) \le 2 \cdot \mathbf{Adv}_{H}^{\mathrm{aPre}\,[m]}(t)$ and $\mathbf{Adv}_{H3}^{\mathrm{Coll}}(t') = 1$

(17) Sec $\not\to$ ePre to 2^{-m}: $\mathbf{Adv}_{H1}^{\mathrm{Sec}\,[m]}(t) \le 1/2^m + \mathbf{Adv}_{H}^{\mathrm{Sec}\,[m]}(t)$ and $\mathbf{Adv}_{H1}^{\mathrm{ePre}}(t') = 1$

(18) Sec $\not\to$ aPre to $1/|\mathcal{K}|$: $\mathbf{Adv}_{H2}^{\mathrm{Sec}\,[m]}(t) \le 1/|\mathcal{K}| + \mathbf{Adv}_{H}^{\mathrm{Sec}\,[m]}(t)$ and $\mathbf{Adv}_{H2}^{\mathrm{aPre}\,[m]}(t') = 1$

(19) Sec $\not\to$ eSec to 2^{-m+1}: $\mathbf{Adv}_{H4}^{\mathrm{Sec}\,[m]}(t) \le 1/2^{m-1} + \mathbf{Adv}_{H}^{\mathrm{Sec}\,[m]}(t)$ and $\mathbf{Adv}_{H4}^{\mathrm{eSec}\,[m]}(t') = 1$

(20) Sec $\not\to$ aSec to 2^{-m}: $\mathbf{Adv}_{H2}^{\mathrm{Sec}\,[m]}(t) \le 1/|\mathcal{K}| + \mathbf{Adv}_{H}^{\mathrm{Sec}\,[m]}(t)$ and $\mathbf{Adv}_{H2}^{\mathrm{aSec}\,[m]}(t') = 1$

(21) Sec $\not\to$ Coll to 2^{-m+1}: $\mathbf{Adv}_{H4}^{\mathrm{Sec}\,[m]}(t) \le 1/2^{m-1} + \mathbf{Adv}_{H}^{\mathrm{Sec}\,[m]}(t)$ and $\mathbf{Adv}_{H4}^{\mathrm{Coll}}(t') = 1$

(22) eSec $\not\to$ aPre to $1/|\mathcal{K}|$: $\mathbf{Adv}_{H2}^{\mathrm{eSec}\,[m]}(t) \le 1/|\mathcal{K}| + \mathbf{Adv}_{H}^{\mathrm{eSec}\,[m]}(t)$ and $\mathbf{Adv}_{H2}^{\mathrm{aPre}\,[m]}(t') = 1$

(23) eSec $\not\to$ aSec to $1/|\mathcal{K}|$: $\mathbf{Adv}_{H2}^{\mathrm{eSec}\,[m]}(t) \le 1/|\mathcal{K}| + \mathbf{Adv}_{H}^{\mathrm{eSec}\,[m]}(t)$ and $\mathbf{Adv}_{H2}^{\mathrm{aSec}\,[m]}(t') = 1$

(24) aSec $\not\to$ ePre to 2^{-m}: $\mathbf{Adv}_{H1}^{\mathrm{aSec}\,[m]}(t) \le 1/2^m + \mathbf{Adv}_{H}^{\mathrm{aSec}\,[m]}(t)$ and $\mathbf{Adv}_{H1}^{\mathrm{ePre}}(t') = 1$

(25) aSec $\not\to$ eSec to 2^{-m}: $\mathbf{Adv}_{H4}^{\mathrm{aSec}\,[m]}(t) \le 1/2^{m-1} + \mathbf{Adv}_{H}^{\mathrm{aSec}\,[m]}(t)$ and $\mathbf{Adv}_{H4}^{\mathrm{eSec}\,[m]}(t') = 1$

(26) aSec $\not\to$ Coll to 2^{-m+1}: $\mathbf{Adv}_{H4}^{\mathrm{aSec}\,[m]}(t) \le 1/2^{m-1} + \mathbf{Adv}_{H}^{\mathrm{aSec}\,[m]}(t)$ and $\mathbf{Adv}_{H4}^{\mathrm{Coll}}(t') = 1$

(27) Coll $\not\to$ aPre to $1/|\mathcal{K}|$: $\mathbf{Adv}_{H2}^{\mathrm{Coll}}(t) \le 1/|\mathcal{K}| + \mathbf{Adv}_{H}^{\mathrm{Coll}}(t)$ and $\mathbf{Adv}_{H2}^{\mathrm{aPre}}(t') = 1$

(28) Coll $\not\to$ aSec to $1/|\mathcal{K}|$: $\mathbf{Adv}_{H2}^{\mathrm{Coll}}(t) \le 1/|\mathcal{K}| + \mathbf{Adv}_{H}^{\mathrm{Coll}}(t)$ and $\mathbf{Adv}_{H2}^{\mathrm{aSec}}(t') = 1$

where $t' = c\,\mathrm{Time}_{H,m}$ for some absolute constant c. •

Acknowledgements

Thanks to Mihir Bellare, Daniel Brown, and to various anonymous reviewers who provided useful comments on an earlier draft of this paper.

This work was supported by NSF 0085961, NSF 0208842, and a gift from Cisco Systems. Many thanks to the NSF and Cisco for their support. Work on this paper was carried out while the authors were at Chiang Mai University, Chulalongkorn University, and UC Davis.

References

[1] R. Anderson. The classification of hash functions. In *IMA Conference in Cryptography and Coding IV*, pages 83–94, December 1993. 386

[2] M. Bellare, A. Desai, D. Pointcheval, and P. Rogaway. Relations among notions of security for public-key encryption schemes. In H. Krawczyk, editor, *Advances in Cryptology – CRYPTO '98*, volume 1462 of *Lecture Notes in Computer Science*, pages 232–249. Springer-Verlag, 1998. 387

[3] M. Bellare and P. Rogaway. Collision-resistant hashing: Towards making UOWHFs practical. In *Advances in Cryptology – CRYPTO 97*, volume 1294 of *Lecture Notes in Computer Science*, pages 470–484, 1997. 386

[4] J. Black, P. Rogaway, and T. Shrimpton. Black-box analysis of the block-cipher-based hash-function constructions from PGV. In *Advances in Cryptology – CRYPTO '02*, volume 2442 of *Lecture Notes in Computer Science*. Springer-Verlag, 2002. 386

[5] D. Brown and D. Johnson. Formal security proofs for a signature scheme with partial message recovery. *Lecture Notes in Computer Science*, 2020:126–144, 2001. 386

[6] I. Damgård. Collision free hash fucntions and public key signature schemes. In *Advances in Cryptology – EUROCRYPT '87*, volume 304 of *Lecture Notes in Computer Science*. Springer-Verlag, 1988. 386

[7] I. Damgård. A design principle for hash functions. In G. Brassard, editor, *Advances in Cryptology – CRYPTO '89*, volume 435 of *Lecture Notes in Computer Science*. Springer-Verlag, 1990. 386

[8] S. Goldwasser and S. Micali. Probabilistic encryption. *Journal of Computer and System Sciences*, 28:270–299, April 1984. 387

[9] A. Menezes, P. van Oorschot, and S. Vanstone. *Handbook of Applied Cryptography*. CRC Press, 1996. 372

[10] R. Merkle. One way hash functions and DES. In G. Brassard, editor, *Advances in Cryptology – CRYPTO '89*, volume 435 of *Lecture Notes in Computer Science*. Springer-Verlag, 1990. 386

[11] I. Mironov. Hash functions: From Merkle-Damgård to Shoup. In *Advances in Cryptology – EUROCRYPT '01*, Lecture Notes in Computer Science. Springer-Verlag, 2001. 386

[12] M. Naor and M. Yung. Universal one-way hash functions and their cryptographic applications. In *Proceedings of the Twenty-first ACM Symposium on Theory of Computing*, pages 33–43, 1989. 378, 386

[13] B. Preneel. *Cryptographic hash functions*. Katholieke Universiteit Leuven (Belgium), 1993. 386

[14] P. Rogaway and T. Shrimpton. Cryptographic hash-function basics: Definitions, implications and separations for preimage resistance, second-preimage resistance, and collision resistance. Full version of this paper,www.cs.ucdavis.edu/~rogaway, 2004. 381, 382, 383

[15] D. Stinson. Some observations on the theory of cryptographic hash functions. Technical Report 2001/020, University of Waterloo, 2001. 373, 386

[16] Y. Zheng, T. Matsumoto, and H. Imai. Connections among several versions of one-way hash functions. In *Special Issue on Cryptography and Information Security, Proceedings of IEICE of Japan*, 1990. 373, 386, 387

A Brief History

It is beyond the scope of the current work to give a full survey of the many hash-function security-notions in the literature, formal an informal, and the many relationships that have (and have not) been shown among them. We touch upon some of the more prominent work that we know.

The term *universal one-way hash function*(UOWHF) was introduced by Naor and Yung [12] to name their asymptotic definition of second-preimage resistance. Along with Damgård [7, 6], who introduced the notion of *collision freeness*, these papers were the first to put notions of hash-function security on a solid formal footing by suggesting to study keyed family of hash functions. This was a necessary step for developing a meaningful formalization of collision-resistance. Contemporaneously, Merkle [10] describes notions of hash-function security: *weak collision resistance* and *strong collision resistance*, which refer to second-preimage and collision resistance, respectively. Damgård also notes that a compressing collision-free hash function has one-wayness properties (our pre notion), and points out some subtleties in this implication.

Merkle and Damgård [10, 7] each show that if one properly iterates a collision-resistant function with a fixed domain, then one can construct a collision-resistant hash-function with an enlarged domain. This iterative method is now called the Merkle-Damgård construction.

Preneel [13] describes *one-way hash functions* (those which are both preimage-resistant and second-preimage resistant) and *collision-resistant hash functions* (those which are preimage, second-preimage and collision resistant). He identifies four types of attacks and studies hash functions constructed from block ciphers.

Bellare and Rogaway [3] give concrete-security definitions for hash-function security and study second-preimage resistance and collision resistance. Their *target collision-resistance*(TCR) coincides with a UOWHF (eSec) and their *any collision-resistance*(ACR) coincides with Coll-security.

Brown and Johnson [5] define a *strong hash* that, if properly formalized in the concrete setting, would include our ePre notion.

Mironov [11] investigates a class of asymptotic definitions that bridge between conventional collision resistance and UOWHF. He also looks at which members of that class are preserved by the Merkle-Damgård constructions.

Anderson [1] discusses some unconventional notions of security for hash functions that might arise when one considers how hash functions might interact with higher-level protocols.

Black, Rogaway, and Shrimpton [4] use a concrete definition of preimage resistance that requires inversion of a uniformly selected range point.

Two papers set out on a program somewhat similar to ours [15] and [16]. Stinson [15] considers hash function security from the perspective that the notions of primary interest are those related to producing digital signatures. He considers four problems (zero-preimage, preimage, second-preimage, collision) and describes notions of security based on them. He considers in some depth the relationship between the preimage problem and the collision problem.

Zheng, Matsumoto and Imai [16] examine some asymptotic formalizations of the notions of second-preimage resistance and collision resistance. In particular, they suggest five classes of second-preimage resistant hash functions and three classes of collision resistant hash functions, and then consider the relationships among these classes.

Our focus on provable security follows a line that begins with Goldwasser and Micali [8]. In defining several related notions of security and then working out all relations between them, we follow work like that of Bellare, Desai, Pointcheval, and Rogaway [2].

B Proof of Theorem 2

Let H: $\mathcal{K} \times \{0,1\}^m \to \{0,1\}^n$ be a hash function family and let $H5$: $\mathcal{K} \times \{0,1\}^m \to \{0,1\}^n$ be the function defined in Fig. 3. We show that

$$\mathbf{Adv}_{H5}^{\mathrm{eSec}\,[m]}(t) \le 2\,\mathbf{Adv}_{H}^{\mathrm{eSec}\,[m]}(t') \quad \text{and} \quad \mathbf{Adv}_{H5}^{\mathrm{Coll}}(t') = 1$$

where $t' \le t + \ell \mathrm{Time}_{H,m}$ for some absolute constant ℓ.

Let $\Pr_K$ denote probability taken over $K \in \mathcal{K}$. Given H we define for every $c \in \{0,1\}^m$ an n-bit string Y_c and a real number δ_c as follows. Let Y_c be the lexicographically first string that maximizes $\delta_c = \Pr_K[H_K(c) = Y_c]$. Over all pairs c, c' we select the lexicographically first pair c, c' (when considered as the $2n$-bit string $c \parallel c'$) such that $c \ne c'$ and $Y_c = Y_{c'}$ and δ_c is maximized (ie, $\Pr_K[H_K(c) = H_K(c')]$ is maximized). Now let $H5 = H5^c$ be defined according to Fig. 3.

We begin by exhibiting an adversary T that gains $\mathbf{Adv}_{H5}^{\mathrm{Coll}}(T) = 1$ and runs in time ℓm for some absolute constant ℓ. On input $K \in \mathcal{K}$, let T output $M = 1^{m-n} \parallel H_K(c)$ and $M' = 0^{m-n} \parallel H_K(c)$.

Now we show that if H is strong in the eSec-sense then so is $H5$. Let A be a two-stage adversary that gains advantage $\delta_m = \mathbf{Adv}_{H5}^{\mathrm{eSec}\,[m]}(A)$ and runs in time t. Let second-preimage-finding adversaries B and C be constructed as follows:

Algorithm B
  [Stage 1] On input ():
    Run $(M, S) \leftarrow A()$
    return (M, S)
  [Stage 2] On input (K, S):
    Run $M' \leftarrow A(K, S)$
    if $M \ne M'$ **and** $M \ne 1^{m-n} \parallel H_K(c)$
      then return M'
    else return $0^{m-n} \parallel H_K(c)$

Algorithm C
  [Stage 1] On input ():
    return (c, ε)
  [Stage 2] On input (K, S)
    return c'

The central claim of the proof is as follows:

Claim: $\mathbf{Adv}_{H5}^{\mathrm{eSec}\,[m]}(A) \le \mathbf{Adv}_{H}^{\mathrm{eSec}\,[m]}(B) + \mathbf{Adv}_{H}^{\mathrm{eSec}\,[m]}(C)$

Let us prove this claim. Recall that the job of A is to find an M and an M' such that $M \ne M'$ and $H5(M) = H5(M')$. Referring to the line numbers in Fig. 3, we say that u-v is a collision if M caused $H5$ to output on line $u \in \{1,2\}$ and $M' \neq M$ caused $H5$ to output on line $v \in \{1,2\}$, and $H5(M) = H5(M')$. We analyze the three possible u-v collisions that A can create. (Note that 1-1 is not a collision, since then $M = M'$.)

[Case 2-2] Assume A wins by causing a 2-2 collision. In this case $M \neq M'$ and $M \neq 1^{m-n} \parallel H_K(c)$ and $M' \neq 1^{m-n} \parallel H_K(c)$. Thus $H_K(M) = H_K(M')$ and so B finds a collision under H. We have then that $\Pr_K[A \text{ wins by a 2-2 collision}] \leq \mathbf{Adv}_H^{\mathrm{eSec}\,[m]}(B)$.

[Case 1-2] Assume that A wins by creating a 1-2 collision. Then $M \neq M'$ and $M = 1^{m-n} \parallel H_K(c)$. We claim that in this case adversary C wins. To see this, note that $\Pr[M \xleftarrow{\$} A();\, K \xleftarrow{\$} \mathcal{K} \colon M = 1^{m-n} \parallel H_K(c)] = \Pr_K[H_K(c) = Y]$ for some fixed $Y \in \{0,1\}^n$. By the way we chose c and c' we have $\Pr_K[H_K(c) = Y] \leq \Pr_K[H_K(c) = Y_c] = \Pr_K[H_K(c) = Y_{c'}] = \Pr_K[H_K(c) = H_K(c')]$; hence $\Pr[M \xleftarrow{\$} A();\, K \xleftarrow{\$} \mathcal{K} \colon M = 1^{m-n} \parallel H_K(c)] \leq \Pr_K[H_K(c) = H_K(c')]$. The conclusion is that $\Pr_K[A \text{ wins by a 1-2 collision}] \leq \Pr[M \xleftarrow{\$} A();\, K \xleftarrow{\$} \mathcal{K} \colon M = 1^{m-n} \parallel H_K(c)] \leq \mathbf{Adv}_H^{\mathrm{eSec}\,[m]}(C)$.

[Case 2-1] Assume that A wins by creating a 2-1 collision. Then $M \neq M'$ and $M' = 1^{m-n} \parallel H_K(c)$, and so $H_K(M) = H_K(0^{m-n} \parallel H_K(c))$. We claim that in this case either adversary B wins, or C does. Let BAD be the event that $M = 0^{m-n} \parallel H_K(c)$. If $M \neq 0^{m-n} \parallel H_K(c)$ then clearly B wins, so $\Pr_K[A \text{ wins by a 2-1 collision} \wedge \overline{\mathsf{BAD}}] \leq \mathbf{Adv}_H^{\mathrm{eSec}\,[m]}(B)$. If $M = 0^{m-n} \parallel H_K(c)$ then we have that $\Pr_K[A \text{ wins by a 2-1 collision} \wedge \mathsf{BAD}] \leq \Pr[M \xleftarrow{\$} A();\, K \xleftarrow{\$} \mathcal{K} \colon M = 0^{m-n} \parallel H_K(c)] \leq \mathbf{Adv}_H^{\mathrm{eSec}\,[m]}(C)$ by an argument nearly identical to that given for Case 1-2,.

Pulling together all of the cases yields the following:

$$
\begin{aligned}
\mathbf{Adv}_{H5}^{\mathrm{eSec}\,[m]}(A) &= \Pr_K[A \text{ wins by a 2-2 collision}] \Pr_K[\text{2-2 collision}] \\
&\quad + \Pr_K[A \text{ wins by a 1-2 collision}] \Pr_K[\text{1-2 collision}] \\
&\quad + \Pr_K[A \text{ wins by a 2-1 collision} \wedge \overline{\mathsf{BAD}}] \Pr_K[\text{2-1 collision} \wedge \overline{\mathsf{BAD}}] \\
&\quad + \Pr_K[A \text{ wins by a 2-1 collision} \wedge \mathsf{BAD}] \Pr_K[\text{2-1 collision} \wedge \mathsf{BAD}] \\
&\leq \mathbf{Adv}_H^{\mathrm{eSec}\,[m]}(B) \Pr_K[\text{2-2 collision}] + \mathbf{Adv}_H^{\mathrm{eSec}\,[m]}(C) \Pr_K[\text{1-2 collision}] \\
&\quad + \mathbf{Adv}_H^{\mathrm{eSec}\,[m]}(B) \Pr_K[\text{2-1 collision} \wedge \overline{\mathsf{BAD}}] \\
&\quad + \mathbf{Adv}_H^{\mathrm{eSec}\,[m]}(C) \Pr_K[\text{2-1 collision} \wedge \mathsf{BAD}] \\
&\leq \mathbf{Adv}_H^{\mathrm{eSec}\,[m]}(B) + \mathbf{Adv}_H^{\mathrm{eSec}\,[m]}(C)
\end{aligned}
$$

where the last inequality is because of convexity. This completes the proof of the claim.

Finally, since the running time of B is $t + \mathrm{Time}_{H,m} + \ell m$ for some absolute constant ℓ, and this is greater than the running time of C, we are done.

The EAX Mode of Operation

Mihir Bellare[1], Phillip Rogaway[2,3], and David Wagner[4]

[1] Dept. of Computer Science & Engineering, University of California at San Diego
9500 Gilman Drive, La Jolla, California 92093, USA
mihir@cs.ucsd.edu
www-cse.ucsd.edu/users/mihir

[2] Department of Computer Science, University of California at Davis
Davis, California 95616, USA
rogaway@cs.ucdavis.edu
www.cs.ucdavis.edu/~rogaway/

[3] Department of Computer Science, Faculty of Science, Chiang Mai University
Chiang Mai 50200, Thailand

[4] Department of Electrical Engineering and Computer Science
University of California at Berkeley
Berkeley, California 94720, USA
daw@cs.berkeley.edu
www.cs.berkeley.edu/~daw/

Abstract. We propose a block-cipher mode of operation, EAX, for solving the problem of authenticated-encryption with associated-data (AEAD). Given a nonce N, a message M, and a header H, our mode protects the privacy of M and the authenticity of both M and H. Strings N, M, and H are *arbitrary* bit strings, and the mode uses $2\lceil |M|/n \rceil + \lceil |H|/n \rceil + \lceil |N|/n \rceil$ block-cipher calls when these strings are nonempty and n is the block length of the underlying block cipher. Among EAX's characteristics are that it is on-line (the length of a message isn't needed to begin processing it) and a fixed header can be pre-processed, effectively removing the per-message cost of binding it to the ciphertext.

Keywords: Authenticated encryption, CCM, EAX, message authentication, CBC MAC, modes of operation, OMAC, provable security.

1 Introduction

An authenticated encryption (AE) scheme is a symmetric-key mechanism by which a message M is a transformed into a ciphertext CT with the goal that CT protect both the privacy *and* the authenticity of M. The last few years has seen the emergence of AE as a recognized cryptographic goal. With this has come the development of new authenticated-encryption schemes and the analysis of old ones. This paper offers up a new authenticated-encryption scheme, EAX, and provides a thorough analysis of it. To understand why we are defining a new AE scheme, we need to give some background.

B. Roy and W. Meier (Eds.): FSE 2004, LNCS 3017, pp. 389–407, 2004.

Flavors of authenticated encryption. It useful to distinguish two kinds of AE schemes. In a *two-pass* scheme we make two passes through the data, one aimed at providing privacy and the other, authenticity. One way of making a two-pass AE scheme is by *generic composition*, wherein one pass constitutes a (privacy-only) symmetric-encryption scheme, while the other pass is a message authentication code (MAC). The encryption scheme and the MAC each use their own key. Analyses of some generic composition methods can be found in [6, 20, 5].

In a *one-pass* AE scheme we make a single pass through the data, simultaneously doing what is needed to engender both privacy and authenticity. Typically, the computational cost is about half that of a two-pass scheme. Such schemes emerged only recently. They include IAPM, OCB, and XCBC [17, 25, 12].

Soon after the emergence of one-pass AE schemes it was realized that often not all the data should be privacy-protected. Changes were needed to the basic definitions and mechanisms in order to support the possibility that some information, like a packet header, must *not* be encrypted. Thus was born the notion of *authenticated-encryption with associated-data* (AEAD), first formally defined in [24]. The non-secret data is called the *associated data* or the *header*. Like an AE schemes, an AEAD scheme might make one pass or two.

Standardizing a two-pass AEAD scheme. Traditionally, it has been the designers of applications and network protocols who were responsible for combining privacy and authenticity mechanisms in order to make a two-pass AEAD scheme. This has not worked well. It turns out that there are numerous ways to go wrong in trying to make a secure AEAD scheme, and many protocols, products, and standards have done just that. (For example, see [11] for a wrong one-pass scheme, see [5] for weaknesses in the AEAD mechanism of SSH, and [6, 20] for attacks on some methods of popular use.)

Nowadays, some standards bodies (including NIST, IETF, and IEEE 802.11) would like to standardize on an AEAD scheme. Indeed IEEE 802.11 has already done so. This is a good direction. Standardized AEAD might help minimize errors in mis-combining cryptographic mechanisms.

So far, standards bodies have been unwilling to standardize on any of the one-pass schemes due to pending patents covering them. There is, accordingly, an established desire for standardizing on a two-pass AEAD scheme. The two-pass scheme should be *as good as possible* subject to the limitation of falling within the two-pass framework.

Generic-composition would seem to be the obvious answer. But defining a generic-composition AEAD scheme is not an approach that has moved forward within any of the standards bodies. There would seem to be a number of reasons. One reason is a relatively minor inefficiency—the fact that generic composition methods must use two keys. Probably a bigger issue is that the architectural advantage of generic composition brings with it an "excessive" degree of choice—after deciding on a generic composition method, one still needs two lower-level specifications, namely a symmetric encryption scheme and a MAC, for each of which numerous block-cipher based choices exist. Standards bodies want something self-contained, as well as being a patent-avoiding, block-cipher based, single-key mechanism.

So far, there has been exactly one proposal for such a method (though see the "contemporaneous work" section below). It is called CCM [26], and is due to Whiting, Housley, and Ferguson [26]. CCM has enjoyed rapid success, and is now the required mechanism for IEEE 802.11 wireless LANs as well as 802.15.4 wireless personal area networks. NIST has indicated that it plans to put out a "Recommendation" based on CCM.

Our contributions. It is our view that CCM has a good deal of pointless complexity and inefficiency. It is the first contribution of this paper to explain these limitations. It is the second and main contribution of this paper to provide a new AEAD scheme, EAX, that avoids these limitations.

CCM limitations. A description of CCM, together with a detailed description of its shortcomings, can be found in the full version of this paper [8]. Some of the points we make and elaborate on there are the following. CCM is not on-line, meaning one needs to know the lengths of both the plaintext and the associated data before one can proceed with encryption. This may be inconvenient or inefficient. CCM does not allow pre-processing of static associated data. (If, for example, we have an unchanging header attached to every packet being authenticated, we would like that the cost of authenticating this header be paid only once, meaning header authentication should have no significant cost after a single pre-computation. CCM fails to have this property.) CCM's parameterization is more complex than necessary, including, in addition to the block cipher and tag length, a message-length parameter. CCM's nonce length is restricted in such a way that it may not provide adequate security when nonces are chosen randomly. Finally, CCM implementations could suffer performance hits because the algorithm can disrupt word alignment in the associated data.

EAX and its attributes. EAX is a nonce-using AEAD scheme employing no tool beyond the block cipher $E\colon \mathsf{Key} \times \{0,1\}^n \rightarrow \{0,1\}^n$ on which it is based. We expect that E will often be instantiated by AES, but we make no restrictions in this direction. (In particular we do not require that $n = 128$.) Nothing is assumed about the nonces except that they are non-repeating. EAX provides both privacy, in the sense of indistinguishability from random bits, and authenticity, in the sense of an adversary's inability to produce a new but valid $\langle$nonce, header, ciphertext$\rangle$ triple. EAX is simple, avoiding complicated length-annotation. It is a conventional two-pass AEAD scheme, making a separate privacy pass and authenticity pass, using no known intellectual property.

EAX is flexible in the functionality it provides. It supports arbitrary-length messages: the message space is $\{0,1\}^*$. The key space for EAX is the key space Key of the underlying block cipher. EAX supports arbitrary nonces, meaning the nonce space is $\{0,1\}^*$. Any tag length $\tau \in [0 .. n]$ is possible, to allow each user to select how much security she wants from the authenticity guarantees. The only user-selectable parameters are the block cipher E and that tag length τ.

EAX has desirable performance attributes. Message expansion is minimal: the length of the ciphertext (which, following the conventions of [25], excludes the nonce) is only τ bits more than the length of the plaintext. Implementations can profitably pre-process static associated data. (If an unchanging header is attached to every packet, authenticating this header has no significant cost after a single pre-computation.) Key-setup is efficient: all block-cipher calls use the same underlying key, so that we do not incur the cost of key scheduling more than once. For both encryption and decryption, EAX uses only the forward direction of the block cipher, so that hardware implementations do not need to implement the decryption functionality of the block cipher. The scheme is on-line for both the plaintext M and the associated data H, which means that one can process streaming data on-the-fly, using constant memory, not knowing when the stream will stop.

PROVABLE SECURITY. We prove that EAX is secure assuming that the block cipher that it uses is a secure pseudorandom permutation (PRP). Security for EAX means indistinguishability from random bits *and* authenticity of ciphertexts. The combination implies other desirable goals, like nonmalleability and indistinguishability under a chosen-ciphertext attack.

The proof of security for EAX is surprisingly complex. The key-collapse of EAX2 destroys a fundamental abstraction boundary. Our security proof relies on a result about the security of a tweakable extension of OMAC (Lemma 3) in which an adversary can obtain not only a tag for a message of its choice, but also an associated key-stream.

PRAGMATICS. The main reason there is any interest in two-pass schemes, as we have already discussed, is that one-pass schemes would seem to be subject to patents. Motivated by this, standardization bodies have expressed the intent of standardizing on a conventional, two-pass scheme, even understanding the factor-of-two performance hit. The merit of this judgment is debatable, but the pragmatic reality is that there has emerged a desire for a conventional scheme, like EAX, that is *as good as possible* subject to the two-pass constraint. Lack of a scheme like EAX will simply lead to an inferior scheme being standardized, which is to the disadvantage of the user community. Accordingly, EAX addresses a real and practical design problem. We took up work on this design problem at the suggestion of the co-Chair of the IRTF (Internet Research Task Force), which supports the standardization efforts of the IETF. We believe that EAX has the potential for widespread adoption and use.

AFTERWARDS. One non-goal of EAX was to be parallelizable. Another recent two-pass design, CWC [19], is parallelizable. It pays for this advantage with a somewhat complex algorithm, based on Carter-Wegman hashing using polynomial evaluation over a prime field. More recent still is GCM [22], a parallelizable, two-pass design based on multiplication in the finite field with 2^{128} elements.

Other recent AEAD mechanisms include Helix [10] and SOBER-128 [13]. These are stream ciphers that aim to provide authenticity. The provable-security methodology does not apply to these objects since they are built directly rather than from lower level primitives.

2 Preliminaries

All strings in this paper are over the binary alphabet $\{0,1\}$. For $\mathcal{L}$ a set of strings and $n \geq 0$ a number, we let $\mathcal{L}^n$ and $\mathcal{L}^*$ have their usual meanings. The concatenation of strings X and Y is denoted $X \,\|\, Y$ or simply XY. The string of length 0, called the *empty string*, is denoted ε. If $X \in \{0,1\}^*$ we let $|X|$ denote its length, in bits. If $X \in \{0,1\}^*$ and $\ell \leq |X|$ then the first ℓ bits of X are denoted X [first ℓ bits]. The set $\text{Byte} = \{0,1\}^8$ contains all the strings of length 8, and a string $X \in \text{Byte}^*$ is called a *byte string* or an *octet string*. If $X \in \text{Byte}^*$ we let $\|X\|_8 = |X|/8$ denote its length in bytes. For $\ell \geq 1$ a number, we write $\text{Byte}^{<\ell}$ for all byte strings having fewer than ℓ bytes. If $X \in \text{Byte}^*$ and $\ell \leq \|X\|_n$ then the first ℓ bytes of X are denoted X [first ℓ bytes]. When $X \in \{0,1\}^n$ is a nonempty string and $t \in \mathbb{N}$ is a number we let $X + t$ be the n-bit string that results from regarding X as a nonnegative number x (binary notation, most-significant-bit first), adding x to t, taking the result modulo 2^n, and converting this number back into an n-bit string. If $t \in [0..2^n - 1]$ we let $[t]_n$ denote the encoding of t into an n-bit binary string (msb first, lsb last). If X and P are strings then we let $X \oplus\!\!\rightarrow P$ (the *xor-at-the-end* operator) denote the string of length $\ell = \max\{|X|, |P|\}$ bits that is obtained by prepending $\big||X| - |P|\big|$ zero-bits to the shorter string and then xoring this with the other string. (In other words, xor the shorter string into the *end* of the longer string.) A *block cipher* is a function $E\colon \mathsf{Key} \times \{0,1\}^n \to \{0,1\}^n$ where Key is a finite, nonempty set and $n \geq 1$ is a number and $E_K(\cdot) = E(K, \cdot)$ is a permutation on $\{0,1\}^n$. The number n is called the *block length*. Throughout this note we fix such a block cipher E.

In Figure 1 we define the algorithms CBC, CTR, pad, OMAC (no superscript), and OMAC$^\bullet$ (with superscript). The algorithms CBC (the CBC MAC) and CTR (counter-mode encryption) are standard. Algorithm pad is used only to define OMAC. Algorithm OMAC [14] is a pseudorandom function (PRF) that is a one-key variant of the algorithm XCBC [9]. Algorithm OMAC$^\bullet$ is like OMAC but takes an extra argument, the integer t. This algorithm is a "tweakable" PRF [21], tweaked in the most simple way possible.

We explain the notation used in the definition of OMAC. The value of iL (line 40: i an integer in $\{2,4\}$ and $L \in \{0,1\}^n$) is the n-bit string that is obtained by multiplying L by the n-bit string that represents the number i. The multiplication is done in the finite field $\text{GF}(2^n)$ using a canonical polynomial to represent field points. The canonical polynomial we select is the lexicographically first polynomial among the irreducible polynomials of degree n that have a minimum number of nonzero coefficients. For $n = 128$ the indicated polynomial is $^{128} +^{7} +^{2} + +1$. In that case, $2L = L{\ll}1$ if the first bit of L is 0 and $2L = (L{\ll}1) \oplus 0^{120}10000111$ otherwise, where $L{\ll}1$ means the left shift of L by one position (the first bit vanishing and a zero entering into the last bit). The

Algorithm $\mathrm{CBC}_K(M)$	**Algorithm** $\mathrm{CTR}_K^{\mathcal{N}}(M)$
10 Let $M_1 \cdots M_m \leftarrow M$ where $\lvert M_i \rvert = n$	20 $m \leftarrow \lceil \lvert M \rvert / n \rceil$
11 $C_0 \leftarrow 0^n$	21 $S \leftarrow E_K(\mathcal{N}) \parallel \cdots \parallel E_K(\mathcal{N}+m-1)$
12 **for** $i \leftarrow 1$ **to** m **do**	22 $C \leftarrow M \oplus S$ [first $\lvert M \rvert$ bits]
13 $C_i \leftarrow E_K(M_i \oplus C_{i-1})$	23 **return** C
14 **return** C_m	

Algorithm $\mathsf{pad}(M;\ B, P)$	**Algorithm** $\mathrm{OMAC}_K(M)$
30 **if** $\lvert M \rvert \in \{n, 2n, 3n, \ldots\}$	40 $L \leftarrow E_K(0^n)$; $B \leftarrow 2L$; $P \leftarrow 4L$
31 **then return** $M \oplus\!\!\rightarrow B$,	41 **return** $\mathrm{CBC}_K(\mathsf{pad}(M;\ B, P))$
32 **else return** $(M \parallel 10^{n-1-(\lvert M \rvert \bmod n)}) \oplus\!\!\rightarrow P$	**Algorithm** $\mathrm{OMAC}_K^t(M)$
	50 **return** $\mathrm{OMAC}_K([t]_n \parallel M)$

Fig. 1. Basic building blocks. The block cipher $E\colon \mathsf{Key} \times \{0,1\}^n \rightarrow \{0,1\}^n$ is fixed and $K \in \mathsf{Key}$. For CBC, $M \in (\{0,1\}^n)^+$. For CTR, $M \in \{0,1\}^*$ and $\mathcal{N} \in \{0,1\}^n$. For pad, $M \in \{0,1\}^*$ and $B, P \in \{0,1\}^n$ and the operation $\oplus\!\!\rightarrow$ xors the shorter string into the end of longer one. For OMAC, $M \in \{0,1\}^*$ and $t \in [0..2^n-1]$ and the multiplication of a number by a string L is done in $\mathrm{GF}(2^n)$

value of $4L$ is simply $2(2L)$. We warn that to avoid side-channel attacks one must implement the doubling operation in a constant-time manner.

We have made a small modification to the OMAC algorithm as it was originally presented, changing one of its two constants. Specifically, the constant 4 at line 40 was the constant 1/2 (the multiplicative inverse of 2) in the original definition of OMAC [14]. The OMAC authors indicate that they will promulgate this modification [15], which slightly simplifies implementations.

3 The EAX Algorithm

ALGORITHM. Fix a block cipher $E\colon \mathsf{Key} \times \{0,1\}^n \rightarrow \{0,1\}^n$ and a tag length $\tau \in [0..n]$. These parameters should be fixed at the beginning of a particular session that will use EAX mode. Typically, the parameters would be agreed to in an authenticated manner between the sender and the receiver, or they would be fixed for all time for some particular application. Given these parameters, EAX provides a nonce-based AEAD scheme $\mathrm{EAX}[E, \tau]$ whose encryption algorithm has signature $\mathsf{Key} \times \mathsf{Nonce} \times \mathsf{Header} \times \mathsf{Plaintext} \rightarrow \mathsf{Ciphertext}$ and whose decryption algorithm has signature $\mathsf{Key} \times \mathsf{Nonce} \times \mathsf{Header} \times \mathsf{Ciphertext} \rightarrow \mathsf{Plaintext} \cup \{\textsc{Invalid}\}$ where Nonce, Header, Plaintext, and Ciphertext are all $\{0,1\}^*$. The EAX algorithm is specified in Figure 2 and a picture illustrating EAX encryption is given in Figure 3. We now discuss various features of our algorithm and choices underlying the design.

Algorithm EAX.Encrypt$_K^{N\,H}$ (M)	**Algorithm** EAX.Decrypt$_K^{N\,H}$ (CT)
10 $\mathcal{N} \leftarrow \text{OMAC}_K^0(N)$	20 **if** $\|CT\| < \tau$ **then return** INVALID
11 $\mathcal{H} \leftarrow \text{OMAC}_K^1(H)$	21 Let $C \parallel T \leftarrow CT$ where $\|T\| = \tau$
12 $C \leftarrow \text{CTR}_K^{\mathcal{N}}(M)$	22 $\mathcal{N} \leftarrow \text{OMAC}_K^0(N)$
13 $\mathcal{C} \leftarrow \text{OMAC}_K^2(C)$	23 $\mathcal{H} \leftarrow \text{OMAC}_K^1(H)$
14 $Tag \leftarrow \mathcal{N} \oplus \mathcal{C} \oplus \mathcal{H}$	24 $\mathcal{C} \leftarrow \text{OMAC}_K^2(C)$
15 $T \leftarrow Tag$ [first τ bits]	25 $Tag' \leftarrow \mathcal{N} \oplus \mathcal{C} \oplus \mathcal{H}$
16 **return** $CT \leftarrow C \parallel T$	26 $T' \leftarrow Tag'$ [first τ bits]
	27 **if** $T \neq T'$ **then return** INVALID
	28 $M \leftarrow \text{CTR}_K^{\mathcal{N}}(C)$
	29 **return** M

Fig. 2. Encryption and decryption under EAX mode. The plaintext is M, the ciphertext is CT, the key is K, the nonce is N, and the header is H. The mode depends on a block cipher E (that CTR and OMAC implicitly use) and a tag length τ

NO ENCODINGS. We have avoided any nontrivial encoding of multiple strings into a single one.[1] Some other approaches that we considered required a PRF to be applied to what was logically a tuple, like (N, H, C). Doing this raises encoding issues we did not want to deal with because, ultimately, there would seem to be no simple, efficient, compelling, on-line way to encode multiple strings into a single one. Alternatively, one could avoid encodings and consider a new kind of primitive, a multi-argument PRF. But this would be a non-standard tool and we didn't want to use any non-standard tools. All in all, it seemed best to find a way to sidestep the need to do encodings.

WHY NOT GENERIC COMPOSITION? Why have we specified a block-cipher based (BC-based) AEAD scheme instead of following the generic-composition approach of combining a (privacy-only) encryption method and a message authentication code? In fact, there are reasonable arguments in favor of generic composition, based on aesthetic or architectural sensibilities. One can argue that generic composition better separates conceptually independent elements (privacy and authenticity) and, correspondingly, allows greater implementation flexibility [6, 20]. Correctness becomes much simpler and clearer as well. All the same, BC-based AEAD modes have some important advantages of their own. They make it easier for implementors to use a scheme without knowing a lot of cryptography, presenting a simpler abstraction boundary. They make it easier to obtain interoperably. They reduce the risk that implementors will choose insecure parameters. They can save on key bits and key-setup time, as generic-composition methods invariably require a pair of separate keys.

[1] One could view the prefixing of $[t]_n$ to M in the definition of $\text{OMAC}_K^t(M)$ as an encoding, but $[t]_n$ is a constant, fixed-length string, and the aim here is just to "tweak" the PRF. This is very different from needing to encode arbitrary-length strings into a single string.

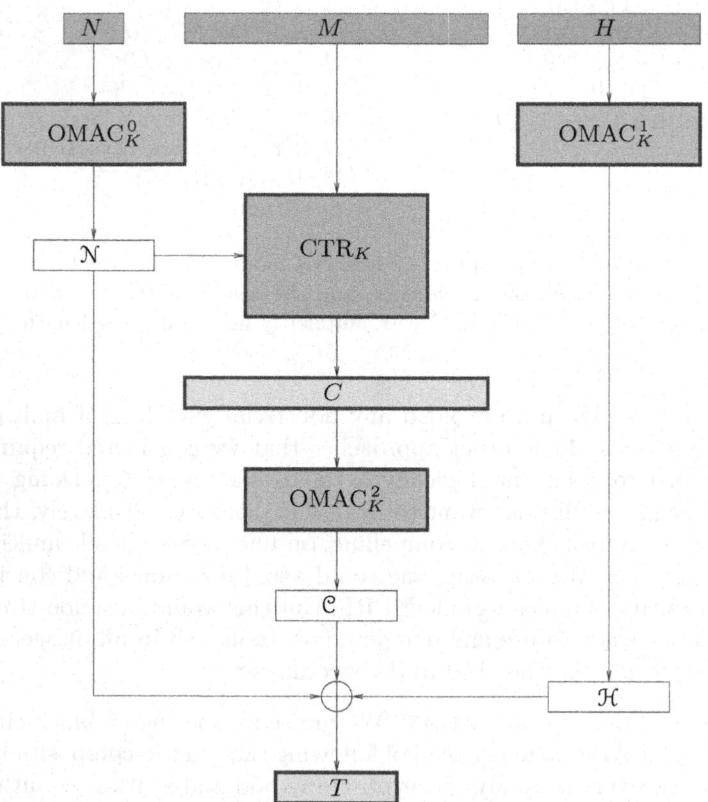

Fig. 3. Encryption under EAX. The message is M, the key is K, and the header is H. The ciphertext is $CT = C \parallel T$.

EAX can be viewed as having been derived from a generic-composition scheme we call EAX2, described in Section 4. Specifically, one instantiates EAX2 using CTR mode (counter mode) and OMAC, and then collapses the two keys into one. If one favors generic composition, EAX2 is a nice algorithm for it.

ON-LINE. We say that an algorithm is *on-line* if it is able to process a stream of data as it arrives, with constant memory, not knowing in advance when the stream will end. Observe then that on-line methods should not require knowledge of the length of a message until the message is finished. A failure to be on-line has been regarded as a significant defect for an encryption scheme or a MAC. EAX is on-line.

Now it is true that in many contexts where one would be encrypting a string one *does* know the length of the string in advance. For example, many protocols

	CCM	EAX
Functionality	AE with AD	AE with AD
Built from	Block cipher E with 128-bit blocksize	Block cipher E with n-bit blocksize
Parameters	Block cipher E Tag length $\boldsymbol{\tau}\in\{4,6,8,10,12,14,16\}$ Length of message length field $\boldsymbol{\lambda}\in[2..8]$	Block cipher E Tag length $\tau \in [0..n]$
Message space	Parameterized: 7 choices: $\boldsymbol{\lambda} \in [2..8]$. Each possible message space a subset of BYTE^*, from $\text{BYTE}^{2^{16}-1}$ to $\text{BYTE}^{<2^{64}-1}$	$\{0,1\}^*$
Nonce space	Parameterized, with a value of $15-\boldsymbol{\lambda}$ bytes. From 56 bits to 104 bits	$\{0,1\}^*$
Key space	One block-cipher key	One block-cipher key
Ciphertext expansion	$\boldsymbol{\tau}$ bytes	τ bits
Block-cipher calls	$2\left\lceil\frac{\lvert M\rvert}{128}\right\rceil+\left\lceil\frac{\lvert H\rvert}{128}\right\rceil+2+\delta$, for $\delta\in\{0,1\}$	$2\left\lceil\frac{\lvert M\rvert}{n}\right\rceil+\left\lceil\frac{\lvert H\rvert}{n}\right\rceil+\left\lceil\frac{\lvert N\rvert}{n}\right\rceil$
Block-cipher calls with static header	$2\left\lceil\frac{\lvert M\rvert}{128}\right\rceil+\left\lceil\frac{\lvert H\rvert}{128}\right\rceil+2+\delta$, for $\delta\in\{0,1\}$	$2\left\lceil\frac{\lvert M\rvert}{n}\right\rceil+\left\lceil\frac{\lvert N\rvert}{n}\right\rceil$
Key setup	Block cipher subkeys	Block cipher subkeys 3 block-cipher calls
IV requirements	Non-repeating nonce	Non-repeating nonce
Parallelizable?	No	No
On-line?	No	Yes
Preprocessing (/msg)	Limited (key stream)	Limited (key stream, header)
Memory rqmts	Small constant	Small constant
Provable security?	Yes (if E is a good PRP) Bound of $\Theta(\sigma^2/2^{128})$	Yes (if E is a good PRP) Bound of $\Theta(\sigma^2/2^n)$
Patent-encumbered?	No	No

Fig. 4. A comparison of basic characteristics of CCM and EAX. The count on block-cipher calls for EAX ignores key-setup costs. We denote by τ the length of the EAX tag in bits, and by $\boldsymbol{\tau}$ (boldface) the length of the CCM tag in bytes

will already have "packaged up" the string length at a lower level. In effect, such strings have been represented in the computing system as sequence of bytes and a count of those bytes. But there are also contexts where one does *not* know the length of a message in advance of getting an indication that it is over. For examples, a printable string is often represented in computer systems as a sequence of non-zero bytes followed by a terminal zero-byte. Certainly one should be able to efficiently encrypt a string which has been represented in this way.

ABILITY TO PROCESS STATIC AD. In many scenarios the associated data H will be static over the course of a communications session. For example, the associated data may include information such as the IP address of the sender, the receiver, and fixed cryptographic parameters associated to this session. In such a case one would like that the amount of time to compute $\text{Encrypt}_K^{N\,H}(M)$ and $\text{Decrypt}_K^{N\,H}(C)$ should be independent of $|H|$, disregarding the work done in

Algorithm EAX2.Encrypt$^{N\,H}_{K1,K2}$ (M)	**Algorithm** EAX2.Decrypt$^{N\,H}_{K1,K2}$ (CT)
10 $\mathcal{N} \leftarrow F^0_{K1}(N)$	20 **if** $\lvert CT\rvert < \tau$ **then return** INVALID
11 $\mathcal{H} \leftarrow F^1_{K1}(H)$	21 Let $C \parallel T \leftarrow CT$ where $\lvert T\rvert = \tau$
12 $C \leftarrow \mathcal{E}^{\mathcal{N}}_{K2}(M)$	22 $\mathcal{N} \leftarrow F^0_{K1}(N)$
13 $\mathcal{C} \leftarrow F^2_{K1}(C)$	23 $\mathcal{H} \leftarrow F^1_{K1}(H)$
14 $Tag \leftarrow \mathcal{N} \oplus \mathcal{C} \oplus \mathcal{H}$	24 $\mathcal{C} \leftarrow F^2_{K1}(C)$
15 $T \leftarrow Tag$ [first τ bits]	25 $Tag' \leftarrow \mathcal{N} \oplus \mathcal{C} \oplus \mathcal{H}$
16 **return** $CT \leftarrow C \parallel T$	26 $T' \leftarrow Tag'$ [first τ bits]
	27 **if** $T \neq T'$ **then return** INVALID
	28 $M \leftarrow \mathcal{D}^{\mathcal{N}}_{K2}(C)$
	29 **return** M

Fig. 5. Encryption and decryption under EAX2. The mode is built from a PRF $F\colon \mathsf{Key1} \times \{0,1\}^* \rightarrow \{0,1\}^n$ and an IV-based encryption scheme $\Pi = (\mathcal{E}, \mathcal{D})$ having key space $\mathsf{Key2}$ and message space $\{0,1\}^*$. The plaintext is M and the key is $(K1, K2)$ and the header is H. By F^i_K we mean the function where $F^i_K(M) = F_K([i]_n \parallel M)$

a preprocessing step. The significance of this goal was already explained in [24]. EAX achieves this goal.

ADDITIONAL FEATURES. Invalid messages can be rejected at half the cost of decryption. This is one of the benefits of following what is basically an encrypt-then-authenticate approach as opposed to an authenticate-then-encrypt approach.

To obtain a MAC as efficient as the PRF underlying EAX define $\mathrm{MAC}_K(H) = \mathrm{Encrypt}^{0^n\,H}_K(\varepsilon)$.

COMPARISON WITH CCM. Figure 4 compares CCM and EAX along a few relevant dimensions. A description of CCM and an extended comparison can be found in the full version of this paper [8].

4 EAX2 Algorithm

To understand the the proof of security of EAX and the approach taken for its design, we introduce EAX2, a generic composition method. EAX is EAX2 for the particular case of CTR encryption and OMAC authentication, but then collapsed to a single key.

EAX2 COMPOSITION. Let $F\colon \mathsf{Key1} \times \{0,1\}^* \rightarrow \{0,1\}^n$ be a PRF, where $n \geq 2$. Let $\Pi = (\mathcal{E}, \mathcal{D})$ be an IV-based encryption scheme having key space $\mathsf{Key2}$ and IV space $\{0,1\}^n$. This means that $\mathcal{E}\colon \mathsf{Key2} \times \{0,1\}^n \times \{0,1\}^* \rightarrow \{0,1\}^*$ and $\mathcal{D}\colon \mathsf{Key2} \times \{0,1\}^n \times \{0,1\}^* \rightarrow \{0,1\}^*$ and $\mathsf{Key2}$ is a set of keys and for every $K \in \mathsf{Key2}$ and $\mathcal{N} \in \{0,1\}^n$ and $M \in \{0,1\}^*$, if $C = \mathcal{E}^{\mathcal{N}}_K(M)$ then $\mathcal{D}^{\mathcal{N}}_K(C) = M$. Let $\tau \leq n$ be a number. Now given F and Π and τ we define an AEAD scheme $\text{EAX2}[\Pi, F, \tau] = (\text{EAX2.Encrypt}, \text{EAX2.Decrypt})$ as follows. Set $F_K^t(M) = F_K([t]_n \parallel M)$. Set $\mathsf{Key} = \mathsf{Key1} \times \mathsf{Key2}$. Then the encryption algorithm $\text{EAX2.Encrypt}\colon \mathsf{Key} \times \{0,1\}^* \times \{0,1\}^* \to \{0,1\}^*$ and the decryption algorithm $\text{EAX2.Decrypt}\colon \mathsf{Key} \times \{0,1\}^* \times \{0,1\}^* \to \{0,1\}^* \cup \{\text{INVALID}\}$ are defined in Figure 5. Scheme $\text{EAX2}[\Pi, F, \tau]$ is provably secure under natural assumptions about Π and F. See Section 6.

EAX1 COMPOSITION. Let EAX1 be the single-key variant of EAX2 where one insists that $\mathsf{Key} = \mathsf{Key1} = \mathsf{Key2}$ and where one keys F, $\mathcal{E}$, and $\mathcal{D}$ with a single key $K \in \mathsf{Key}$. One associates to F and Π the scheme $\text{EAX1}[\Pi, F, \tau]$ that is defined as with EAX2 but where the one key K keys everything. Notice that $\text{EAX}[E, \tau] = \text{EAX1}[\text{CTR}[E], \text{OMAC}[E], \tau]$. This is a useful way to look at EAX.

5 Definitions

AEAD SCHEMES. A *set of keys* is a nonempty set having a distribution (the uniform distribution when the set is finite). A (nonce-based) *authenticated-encryption with associated-data* (AEAD) scheme is a pair of algorithms $\mathbf{\Pi} = (\mathbf{E}, \mathbf{D})$ where $\mathbf{E}$ is a deterministic *encryption* algorithm $\mathbf{E}\colon \mathsf{Key} \times \mathsf{Nonce} \times \mathsf{Header} \times \mathsf{Plaintext} \to \mathsf{Ciphertext}$ and a $\mathbf{D}$ is a deterministic *decryption* algorithm $\mathbf{D}\colon \mathsf{Key} \times \mathsf{Nonce} \times \mathsf{Header} \times \mathsf{Ciphertext} \to \mathsf{Plaintext} \cup \{\text{INVALID}\}$. The *key space* Key is a set of keys while the *nonce space* Nonce and the *header space* Header (also called the space of *associated data*) are nonempty sets of strings. We write $\mathbf{E}_K^{N\,H}(M)$ for $\mathbf{E}(K, N, H, M)$ and $\mathbf{D}_K^{N\,H}(CT)$ for $\mathbf{D}(K, N, H, CT)$. We require that $\mathbf{D}_K^{N\,H}(\mathbf{E}_K^{N\,H}(M)) = M$ for all $K \in \mathsf{Key}$ and $N \in \mathsf{Nonce}$ and $H \in \mathsf{Header}$ and $M \in \mathsf{Plaintext}$. In this note we assume, for notational simplicity, that Nonce, Header, Plaintext, and Ciphertext are all $\{0,1\}^*$ and that $|\mathbf{E}_K^{N\,H}(M)| = |M|$. An adversary is a program with access to one or more oracles.

NONCE-RESPECTING. Suppose A is an adversary with access to an *encryption oracle* $\mathbf{E}_K^{\cdot\,\cdot}(\cdot)$. This oracle, on input (N, H, M), returns $\mathbf{E}_K^{N\,H}(M)$. Let $(N_1, H_1, M_1), \ldots, (N_q, H_q, M_q)$ denote its oracle queries. The adversary is said to be *nonce-respecting* if $N_1, \ldots, N_q$ are always distinct, regardless of oracle responses and regardless of A's internal coins.

PRIVACY OF AEAD SCHEMES. We consider adversaries with access to an encryption oracle $\mathbf{E}_K^{\cdot\,\cdot}(\cdot)$. We assume that any privacy-attacking adversary is nonce-respecting. The advantage of such an adversary A in violating the privacy of AEAD scheme $\mathbf{\Pi} = (\mathbf{E}, \mathbf{D})$ having key space Key is

$$\mathbf{Adv}_{\mathbf{\Pi}}^{\mathbf{priv}}(A) = \Pr\left[K \xleftarrow{\$} \mathsf{Key} :\ A^{\mathbf{E}_K^{\cdot\,\cdot}(\cdot)} = 1\right] - \Pr\left[K \xleftarrow{\$} \mathsf{Key} :\ A^{\$^{\cdot\,\cdot}(\cdot)} = 1\right]$$

where $\$^{\cdot\,\cdot}(\cdot)$ denotes the oracle that on input (N, H, M) returns a random string of length $|M|$.

AUTHENTICITY OF AEAD SCHEMES. This time we provide the adversary with two oracles, an encryption oracle $\mathbf{E}_K^{\cdot\,\cdot}(\cdot)$ as above and also a *verification oracle*

$\widehat{\mathbf{D}}_K^{\cdot\cdot}(\cdot)$. The latter oracle takes input (N, H, CT) and returns 1 if $\mathbf{D}_K^{N\,H}(CT) \in$ Plaintext and returns 0 if $\mathbf{D}_K^{N\,H}(CT) = \text{INVALID}$. The adversary is assumed to satisfy three conditions, and these must hold regardless of the responses to its oracle queries and regardless of A's internal coins:

- Adversary A must be nonce-respecting. (The condition is understood to apply only to the adversary's encryption oracle. Thus a nonce used in an encryption-oracle query may be used in a verification-oracle query.)
- Adversary A may never make a verification-oracle query (N, H, CT) such that the encryption oracle previously returned CT in response to a query (N, H, M).
- Adversary A must call its verification-oracle exactly once, and may not subsequently call its encryption oracle. (That is, it makes a sequence of encryption-oracle queries, then a verification-oracle query, and then halts.)

We say that such an adversary *forges* if its verification oracle returns 1 in response to the single query made to it. The advantage of such an adversary A in violating the authenticity of AEAD scheme $\mathbf{\Pi} = (\mathbf{E}, \mathbf{D})$ having key space Key is

$$\mathbf{Adv}^{\mathbf{auth}}_{\mathbf{\Pi}}(A) \;=\; \Pr\left[K \stackrel{\$}{\leftarrow} \mathsf{Key}:\; A^{\mathbf{E}_K^{\cdot\cdot}(\cdot),\, \widehat{\mathbf{D}}_K^{\cdot\cdot}(\cdot)} \text{ forges}\right].$$

IV-BASED ENCRYPTION. An *IV-based encryption scheme* (an IVE scheme) is a pair of algorithms $\Pi = (\mathcal{E}, \mathcal{D})$ where $\mathcal{E}\colon \mathsf{Key} \times \mathsf{IV} \times \mathsf{Plaintext} \to \mathsf{Ciphertext}$ is a deterministic *encryption* algorithm and $\mathcal{D}\colon \mathsf{Key} \times \mathsf{IV} \times \mathsf{Ciphertext} \to \mathsf{Plaintext} \cup \{\text{INVALID}\}$ is a deterministic *decryption* algorithm. The *key space* Key is a set of keys and the *plaintext space* Plaintext and *ciphertext space* Ciphertext and *IV space* IV are all nonempty sets of strings. We write $\mathcal{E}_K^R(M)$ for $\mathcal{E}(K, R, M)$ and $\mathcal{D}_K^R(C)$ for $\mathcal{D}(K, R, C)$. We require that $\mathcal{D}_K^R(\mathcal{E}_K^R(M)) = M$ for all $K \in \mathsf{Key}$ and $R \in \mathsf{IV}$ and $M \in \mathsf{Plaintext}$. We assume, as before, that $\mathsf{Plaintext} = \mathsf{Ciphertext} = \{0,1\}^*$ and that $|\mathcal{E}_K^R(M)| = |M|$. We also assume that $\mathsf{IV} = \{0,1\}^n$ for some $n \geq 1$ called the *IV length*.

PRIVACY OF IVE SCHEMES WITH RANDOM IVS. Let $\Pi = (\mathcal{E}, \mathcal{D})$ be an IVE scheme with key space Key and IV space $\mathsf{IV} = \{0,1\}^n$. Let $\mathcal{E}^\$$ be the probabilistic algorithm defined from $\mathcal{E}$ that, on input K and M, chooses an IV R at random from $\{0,1\}^n$, computes $C \leftarrow \mathcal{E}_K^R(M)$, and then returns C along with the chosen IV:

```
Algorithm E$_K(M)     //The probabilistic encryption scheme built from IVE scheme E
R <-$ {0,1}^n ; C <- E^R_K(M); return R || C
```

Then we define the advantage of an adversary A in violating the privacy of Π (as an encryption scheme using random IV) by

$$\mathbf{Adv}^{\mathrm{priv}}_{\Pi}(A) \;=\; \Pr\left[K \stackrel{\$}{\leftarrow} \mathsf{Key}:\; A^{\mathcal{E}_K^\$(\cdot)} = 1\right] - \Pr\left[K \stackrel{\$}{\leftarrow} \mathsf{Key}:\; A^{\$(\cdot)} = 1\right]$$

where $\$(\cdot)$ denotes the oracle that on input M returns a random string of length $n + |M|$. This is just the ind\$-privacy of the randomized symmetric encryption scheme associated to Π. We comment that we have used a superscript of "priv" for an IVE scheme and "**priv**" (bold font) for an AEAD scheme.

PSEUDORANDOM FUNCTIONS. A *family of functions*, or a *pseudorandom function* (PRF), is a map $F\colon \mathsf{Key} \times D \to \{0,1\}^n$ where Key is a set of keys and D is a nonempty set of strings. We call n the *output length* of F. We write F_K for the function $F(K,\cdot)$ and we write $f \xleftarrow{\$} F$ to mean $K \xleftarrow{\$} \mathsf{Key};\ f \leftarrow F_K$. We denote by $\mathcal{R}_n^*$ the set of all functions with domain $\{0,1\}^*$ and range $\{0,1\}^n$; by $\mathcal{R}_n^n$ the set of all functions with domain $\{0,1\}^n$ and range $\{0,1\}^n$; and by $\mathcal{R}_n^I$ the set of all functions with domain I and range $\{0,1\}^n$. We identify a function with its key, making $\mathcal{R}_n^n$, $\mathcal{R}_n^*$ and $\mathcal{R}_n^I$ pseudorandom functions. The advantage of adversary A in violating the pseudorandomness of the family of functions $F\colon \mathsf{Key} \times \{0,1\}^* \to \{0,1\}^n$ is

$$\mathbf{Adv}_F^{\mathrm{prf}}(A) \;=\; \Pr\left[K \xleftarrow{\$} \mathsf{Key} :\; A^{F_K(\cdot)} = 1\right] - \Pr\left[\rho \xleftarrow{\$} \mathcal{R}_n^* :\; A^{\rho(\cdot)} = 1\right]$$

A family of functions $E\colon \mathsf{Key} \times D \to \{0,1\}^n$ is a *block cipher* if $D = \{0,1\}^n$ and each E_K is a permutation. We let $\mathcal{P}_n$ denote all the permutations on $\{0,1\}^n$ and define

$$\mathbf{Adv}_E^{\mathrm{prp}}(A) \;=\; \Pr\left[K \xleftarrow{\$} \mathsf{Key} :\; A^{E_K(\cdot)} = 1\right] - \Pr\left[\pi \xleftarrow{\$} \mathcal{P}_n :\; A^{\pi(\cdot)} = 1\right]$$

RESOURCES. If xxx is an advantage notion for which $\mathbf{Adv}_\Pi^{\mathrm{xxx}}(A)$ has been defined we write $\mathbf{Adv}_\Pi^{\mathrm{xxx}}(R)$ for the maximal value of $\mathbf{Adv}_\Pi^{\mathrm{xxx}}(A)$ over all adversaries A that use resources at most R. When counting the resource usage of an adversary, one maximizes over all possible oracle responses, including those that could not be returned by any experiment we have specified for adversarial advantage. Resources of interest are: t—the running time; q—the total number of oracle queries; q_{e}—the number of oracle queries to the adversary's first oracle; q_{v}—the number of oracle queries to the adversary's second oracle; and σ—the data complexity. The running time t of an algorithm is its actual running time (relative to some fixed RAM model of computation) plus its description size (relative to some standard encoding of algorithms). The data complexity σ is defined as the sum of the lengths of all strings encoded in the adversary's oracle queries, plus the total number of all of these strings.[2] In this paper the length of strings is measured in n-bit blocks, for some understood value n. The number of blocks in a string M is defined as $\|M\|_n = \max\{1, \lceil |M|/n \rceil\}$, so that the empty string counts as one block. As an example, an adversary that asks queries (N_1, H_1, M_1), (N_2, H_2, M_2) to its first oracle and query (N, H, M) to its second oracle has data complexity $\|N_1\|_n + \|H_1\|_n + \|M_1\|_n + \|N_2\|_n + \|H_2\|_n + \|M_2\|_n + \|N\|_n + \|H\|_n + \|M\|_n + 9$. The name of a resource measure (t, t', q, etc.) will be enough to make clear what resource it refers to.

[2] There is a certain amount of arbitrariness in this convention, but it is reasonable and simplifies subsequent accounting.

When we use big-O notation it is understood that the constant hidden inside the notation may depend on n. We write $\widetilde{O}(f(x))$ for $O(f(x)\lg(f(x)))$. When F is a function we write $\mathrm{Time}_F(\sigma))$ for the maximal amount of time to compute the function F over inputs of total length σ. When $\Pi = (\mathcal{E}, \mathcal{D})$ is an AEAD scheme or an IVE scheme with key space Key we write $\mathrm{Time}_{\mathcal{E}}(\sigma)$ for the time to compute a random element $K \stackrel{\$}{\leftarrow} \mathsf{Key}$ plus the maximal amount of time to compute the function $\mathcal{E}_K$ on arguments of total length σ.

6 Security Results

We first obtain results about the security of EAX2 and then prove a result about the security of a tweakable-OMAC extension. These results are applied to derive results about the security of EAX. The notation and security measures referred to below are defined in Section 5.

SECURITY OF EAX2. We begin by considering the EAX2$[\Pi, F, \tau]$ scheme with F being equal to $\mathcal{R}_n^n$, the set of all functions with domain $\{0,1\}^n$ and range $\{0,1\}^n$. In other words, we are considering the case where F_{K1} is a random function with domain $\{0,1\}^n$ and range $\{0,1\}^n$. First we show that EAX2$[\Pi, \mathcal{R}_n^n, \tau]$ inherits the privacy of the underlying IVE scheme Π. The proof of the following is in the full version of this paper [8].

Lemma 1. [Privacy of EAX2 with a random PRF] *Let Π be an IVE scheme with IV space $\{0,1\}^n$ and let $\tau \in [0..n]$. Then*

$$\mathbf{Adv}^{\mathrm{priv}}_{\mathrm{EAX2}[\Pi,\mathcal{R}_n^n,\tau]}(t,q,\sigma) \le \mathbf{Adv}^{\mathrm{priv}}_{\Pi}(t',q,\sigma)$$

where $t' = t + \widetilde{O}(\sigma)$. □

We now turn to authenticity. The following shows that EAX2$[\Pi, \mathcal{R}_n^n, \tau]$ provides authenticity under the assumption that the underlying IVE scheme Π provides privacy. The proof is in the full version of this paper [8].

Lemma 2. [Authenticity of EAX2 with a random PRF] *Let Π be an IVE scheme with IV space $\{0,1\}^n$ and let $\tau \in [0..n]$. Then*

$$\mathbf{Adv}^{\mathrm{auth}}_{\mathrm{EAX2}[\Pi,\mathcal{R}_n^n,\tau]}(t,q,\sigma) \le \mathbf{Adv}^{\mathrm{priv}}_{\Pi}(t',q,\sigma) + 2^{-\tau}$$

where $t' = t + \widetilde{O}(\sigma)$. □

Our definition of authenticity allows the adversary only one query to its verification oracle, meaning only one forgery attempt. A standard argument says that the advantage of an adversary making q_{v} verification queries can grow by a factor of at most q_{v}. As per the above this means it is at most $q_{\mathrm{v}} \cdot [2^{-\tau} + \mathbf{Adv}^{\mathrm{priv}}_{\Pi}(t',q,\sigma)]$. We believe that in fact the bound is better than this, namely that it is $q_{\mathrm{v}}2^{-\tau} + \mathbf{Adv}^{\mathrm{priv}}_{\Pi}(t',q,\sigma)$. However, we do not have a proof of this stronger bound.

The above allows us to obtain results about the security of the general EAX2$[\Pi, F, \tau]$ scheme based on assumptions about the security of the component schemes. The proof of the following is in the full version of this paper [8].

Theorem 1. [Security of EAX2] *Let* $F\colon \mathsf{Key1} \times \{0,1\}^* \to \{0,1\}^n$ *be a family of functions, let* $\Pi = (\mathcal{E}, \mathcal{D})$ *be an IVE scheme with IV space* $\{0,1\}^n$ *and let* $\tau \in [0..n]$. *Then*

$$\mathbf{Adv}^{\mathbf{auth}}_{\mathrm{EAX2}[\Pi,F,\tau]}(t, q, \sigma) \le \mathbf{Adv}^{\mathrm{priv}}_{\Pi}(t_2, q, \sigma) + \mathbf{Adv}^{\mathrm{prf}}_{F}(t_1, 3q+3, \sigma) + 2^{-\tau} \quad (1)$$

$$\mathbf{Adv}^{\mathbf{priv}}_{\mathrm{EAX2}[\Pi,F,\tau]}(t, q, \sigma) \le \mathbf{Adv}^{\mathrm{priv}}_{\Pi}(t_2, q, \sigma) + \mathbf{Adv}^{\mathrm{prf}}_{F}(t_3, 3q, \sigma) \quad (2)$$

where $t_1 = t + \mathrm{Time}_{\mathcal{E}}(\sigma) + \widetilde{O}(\sigma)$ *and* $t_2 = t + \widetilde{O}(\sigma + nq)$ *and* $t_3 = t + \mathrm{Time}_{\mathcal{E}}(\sigma) + \widetilde{O}(\sigma)$. □

We remark that although "birthday" terms of the form $\sigma^2/2^n$ or $q^2/2^n$ do not appear explicitly in the bounds above, they may appear when we bound the $\mathbf{Adv}^{\mathrm{priv}}_{\Pi}(\cdot,\cdot,\cdot)$ and $\mathbf{Adv}^{\mathrm{prf}}_{F}(\cdot,\cdot,\cdot)$ in terms of their arguments.

Security of a Tweakable-OMAC Extension. This section develops the core result underlying why key-reuse "works" across OMAC and CTR modes. To do this, we consider the following extension of the tweakable-OMAC construction. Fix $n \ge 1$ and let $t \in \{0,1,2\}$ and $\rho \in \mathcal{R}^n_n$ and $M \in \{0,1\}^*$ and $s \in \mathbb{N}$. Then define

```
Algorithm 𝕆𝕄𝔸ℂ_ρ(t, M, s)
10  R ← OMAC^t_ρ(M)
11  for j ← 0 to s − 1 do S_j ← ρ(R + j)
12  return R S_0 S_1 ··· S_{s−1}
```

Thus an $\mathbb{OMAC}_\rho$ oracle, when asked (t, M, s), returns not only $R = \mathrm{OMAC}^t_\rho(M)$ but also a key stream $S_0 S_1 \ldots S_s$ formed using CTR-mode and start-index R. We emphasize that the key stream is formed using the *same* function ρ (that is, the same key) that underlies the OMAC computation. Note too that we have limited the tweak t to a small set, $\{0,1,2\}$.

We imagine providing an adversary A with one of two kinds of oracles. The first is an oracle $\mathbb{OMAC}_\rho(\cdot,\cdot,\cdot)$ for a randomly chosen $\rho \in \mathcal{R}^n_n$. The second is an oracle $\$_n(\cdot,\cdot,\cdot)$ that, on input (t, M, s), returns $n(s+1)$ random bits. Either way, we assume that the adversary is *length-committing*: if the adversary asks a query (t, M, s) it does not ask any subsequent query (t, M, s'). As the adversary runs, it asks some sequence of queries $(t_1, M_1, s_1), \ldots, (t_q, M_q, s_q)$. The resources of interest to us are the sum of the block lengths of the messages being MACed, $\sigma_1 = \sum \|M_i\|_n$, and the total number $\sigma_2 = \sum s_i$ of key-stream blocks that the adversary requests. We claim that a reasonable adversary will have little advantage in telling apart the two oracles, and we bound its distinguishing probability in terms of the resources σ_1 and σ_2 that it expends. Recall that for oracles X and Y and an adversary A we measure A's ability to distinguish between oracles X and Y by the number $\mathbf{Adv}^{\mathrm{dist}}_{X,Y}(A) = \Pr[A^X = 1] - \Pr[A^Y = 1]$. The proof of the following is in the full version of this paper [8].

Lemma 3. [Pseudorandomness of $\mathbb{OMAC}$] *Fix $n \geq 2$. Then, for length-committing adversaries,*

$$\mathbf{Adv}^{\mathrm{dist}}_{\mathbb{OMAC}[\mathcal{R}^n_n],\$_n}(\sigma_1, \sigma_2) \leq \frac{(\sigma_1 + \sigma_2 + 3)^2}{2^n} \qquad \square$$

SECURITY OF EAX. We are now ready to consider the security of EAX. The proof of the following is in the full version of this paper [8].

Theorem 2. [Security of EAX] *Let $n \geq 2$ and $\tau \in [0..n]$. Then*

$$\mathbf{Adv}^{\mathbf{priv}}_{\mathrm{EAX}[\mathcal{R}^n_n,\tau]}(\sigma) \leq \frac{9\,\sigma^2}{2^n}$$

$$\mathbf{Adv}^{\mathbf{auth}}_{\mathrm{EAX}[\mathcal{R}^n_n,\tau]}(\sigma) \leq \frac{10.5\,\sigma^2}{2^n} + \frac{1}{2^\tau} \qquad \square$$

Finally, we may, in the customary way, pass to the corresponding complexity-theoretic result where we start with an arbitrary block cipher E.

Corollary 1. [Security of EAX] *Let $n \geq 2$ and $E\colon \mathsf{Key} \times \{0,1\}^n \times \{0,1\}^n$ be a block cipher and let $\tau \in [0..n]$. Then*

$$\mathbf{Adv}^{\mathbf{priv}}_{\mathrm{EAX}[E,\tau]}(t, \sigma) \leq \frac{9.5\,\sigma^2}{2^n} + \mathbf{Adv}^{\mathrm{prp}}_{E}(t', \sigma)$$

$$\mathbf{Adv}^{\mathbf{auth}}_{\mathrm{EAX}[E,\tau]}(t, \sigma) \leq \frac{11\,\sigma^2}{2^n} + \frac{1}{2^\tau} + \mathbf{Adv}^{\mathrm{prp}}_{E}(t', \sigma)$$

where $t' = t + O(\sigma)$. $\square$

We omit the proof, which is completely standard.

References

[1] M. Bellare, A. Desai, E. Jokipii, and P. Rogaway. A concrete security treatment of symmetric encryption: Analysis of the DES modes of operation. *Proceedings of the 38th Symposium on Foundations of Computer Science*, IEEE, 1997.

[2] M. Bellare, R. Guérin, and P. Rogaway. XOR MACs: New methods for message authentication using finite pseudorandom functions. *Advances in Cryptology – CRYPTO '95*, Lecture Notes in Computer Science, vol. 963, D. Coppersmith ed., Springer-Verlag, 1995.

[3] M. Bellare, O. Goldreich, and H. Krawczyk. Stateless evaluation of pseudorandom functions: Security beyond the birthday barrier. *Advances in Cryptology – CRYPTO '96*, Lecture Notes in Computer Science, vol. 1109, N. Koblitz ed., Springer-Verlag, 1996.

[4] M. Bellare, J. Kilian, and P. Rogaway. The security of the cipher block chaining message authentication code. *Journal of Computer and System Sciences* (JCSS), vol. 61, no. 3, pp. 362–399, Dec 2000.

[5] M. Bellare, T. Kohno, and C. Namprempre. Authenticated encryption in SSH: provably fixing the SSH binary packet protocol. *Proceedings of the 9th Annual Conference on Computer and Communications Security*, ACM, 2002. 390

[6] M. Bellare and C. Namprempre. Authenticated encryption: Relations among notions and analysis of the generic composition paradigm. *Advances in Cryptology – ASIACRYPT '00*, Lecture Notes in Computer Science, vol. 1976, T. Okamoto ed., Springer-Verlag, 2000. 390, 395

[7] M. Bellare and P. Rogaway. Encode-then-encipher encryption: How to exploit nonces or redundancy in plaintexts for efficient encryption. *Advances in Cryptology – ASIACRYPT '00*, Lecture Notes in Computer Science, vol. 1976, T. Okamoto ed., Springer-Verlag, 2000.

[8] M. Bellare, P. Rogaway and D. Wagner. The EAX mode of operation (A two-Pass authenticated-encryption scheme optimized for simplicity and efficiency). Full version of this paper, available via http://www.cs.ucdavis.edu/~rogaway 391, 398, 402, 403, 404

[9] J. Black and P. Rogaway. CBC MACs for arbitrary-length messages: The three-key constructions. *Advances in Cryptology – CRYPTO '00*, Lecture Notes in Computer Science, vol. 1880, M. Bellare ed., Springer-Verlag, 2000. 393

[10] N. Ferguson, D. Whiting, B. Schneier, J. Kelsey, S. Lucks, and T. Kohno. Helix: Fast encryption and authentication in a single cryptographic primitive. *Fast Software Encryption* (FSE 2003), Lecture Notes in Computer Science, vol. 2887, Springer-Verlag, pp. 330–346, 2003. 392

[11] V. Gligor and P. Donescu. Integrity-aware PCBC encryption. Security Protocols, 7th International Workshop. Lecture Notes in Computer Science, vol. 1796, Springer-Verlag, pp. 153–171, 1999. 390

[12] V. Gligor and P. Donescu. Fast encryption and authentication: XCBC encryption and XECB authentication modes. Presented at the 2nd NIST Workshop on AES Modes of Operation, Santa Barbara, CA, August 24, 2001. 390

[13] P. Hawkes and G. Rose. Primitive specification for SOBER-128. Cryptology ePrint Archive Report 2003/48. April 2003. 392

[14] T. Iwata and K. Kurosawa. OMAC: One-key CBC MAC. *Fast Software Encryption* (FSE 2003), Lecture Notes in Computer Science, vol. 2887, Springer-Verlag, pp. 129–153, 2003. 393, 394

[15] T. Iwata and K. Kurosawa. Personal communications, January 2002. 394

[16] J. Jonsson. On the security of CTR + CBC-MAC. *Proceedings of Selected Areas of Cryptography (SAC)*, 2002.

[17] C. Jutla. Encryption modes with almost free message integrity. *Advances in Cryptology – EUROCRYPT '01*, Lecture Notes in Computer Science, vol. 2045 , B. Pfitzmann ed., Springer-Verlag, 2001. 390

[18] J. Katz and M. Yung. Unforgeable encryption and adaptively secure modes of operation. *Fast Software Encryption '00*, Lecture Notes in Computer Science, vol. 1978, B. Schneier ed., Springer-Verlag, 2000.

[19] T. Kohno, J. Viega, and D. Whiting. A high-performance conventional authenticated encryption mode. These proceedings. 392

[20] H. Krawczyk. The order of encryption and authentication for protecting communications (or: how Secure is SSL?). *Advances in Cryptology – CRYPTO '01*, Lecture Notes in Computer Science, vol. 2139, J. Kilian ed., Springer-Verlag, 2001. 390, 395

[21] M. Liskov, R. Rivest, and D. Wagner. *Advances in Cryptology – CRYPTO '02*, Lecture Notes in Computer Science, vol. 2442, pp. 31–46, Springer-Verlag, 2002. 393

[22] D. McGrew and J. Viega. Flexible and efficient message authentication in hardware and software. Manuscript, 2003. Available from http://www.zork.org/ 392

[23] E. Petrank and C. Rackoff. CBC MAC for real-time data sources. *Journal of Cryptology*, vol. 13, no. 3 pp. 315–338, 2000.

[24] P. Rogaway. Authenticated-encryption with associated-data. *Proceedings of the 9th Annual Conference on Computer and Communications Security* (CCS-9), pp. 98–107, ACM, 2002. 390, 398

[25] P. Rogaway, M. Bellare, and J. Black. OCB: A block-cipher mode of operation for efficient authenticated encryption. *ACM Transactions on Information and System Security* (TISSEC), vol. 6, no. 3, pp. 365–403, Aug. 2003. 390, 392

[26] D. Whiting, R. Housley, and N. Ferguson. Counter with CBC-MAC (CCM). June 2002. Available at http://csrc.nist.gov/encryption/modes/proposedmodes/ 391, 406

A Definition of CCM

Since CCM [26] was a major motivation for our work, we recall its definition, writing it in a new form. First some notation. Write string constants in hexadecimal, as in 0xFFFE. When $X \in \{0,1\}^\ell$ is a nonempty string and $i \in \mathbb{N}$ is a number we let $X + i$ be the ℓ-bit string that results from regarding X as a nonnegative number x (binary notation, msb first), adding x to i, taking the result modulo 2^n, and converting this number back into an ℓ-bit string. Now CCM depends on three parameters:

- E — the *block cipher* — where $E\colon \mathsf{Key} \times \{0,1\}^{128} \to \{0,1\}^{128}$
- τ — the *tag length* — where $\tau \in \{4, 6, 8, 10, 12, 14, 16\}$
- λ — the *length-of-the-message-length-field* — where $\lambda \in \{2, 3, 4, 5, 6, 7, 8\}$

Once parameters (E, τ, λ) have been fixed, where $E\colon \mathsf{Key} \times \{0,1\}^{128} \to \{0,1\}^{128}$ is a block cipher, CCM is the AE scheme specified in Figure 6. The nonce space is $\mathsf{Nonce} = \textsc{Byte}^{15-\lambda}$ and the header space is $\mathsf{Header} = \textsc{Byte}^{<2^{64}}$ and the message space is $\mathsf{Plaintext} = \textsc{Byte}^{<2^{8\lambda}}$. There is a tradeoff between the length of nonces, $\eta = |N| = 15 - \lambda$ bytes, and the longest permitted message, $256^\lambda - 1$ bytes.

```
Algorithm CCM.Encrypt_K^{N H}(M)
100 B ← 0 || if H = ε then 0 else 1 endif || [τ/2 − 1]_3 || [λ − 1]_3 ||
101       N || [||M||_n]_{8λ} ||
102       if H = ε then ε elseif ||H||_n < 62580 then [||H||_n]_16 elseif ||H||_n < 2^32
103       then 0xFFFE || [||H||_n]_32 else 0xFFFF || [||H||_n]_64 endif ||
104       H ||
105       if H = ε then ε elseif ||H||_n < 62580 then [0]_n^{(14−||H||_n) mod 16}
106       elseif ||H||_n < 2^32 then [0]_n^{(10−||H||_n) mod 16} else [0]_n^{(6−||H||_n) mod 16} endif
107       || M ||
108       [0]_n^{(−||M||_n) mod 16}
109 U ← CBC_K(B)
110 A_0 ← [λ − 1]_8 || N || [0]_n^{15−λ}
111 V || C ← CTR_K^{A_0}(U || M) where |V| = 128
112 T ← V [first τ bytes]
113 return CT ← C || T

Algorithm CCM.Decrypt_K^{N H}(CT)
200 if ||CT||_n < τ then return INVALID
201 Partition CT into C || T where ||T||_n = τ
202 if ||C||_n > 2^λ − 1 then return INVALID

210 A_0 ← [λ − 1]_8 || N || [0]_n^{15−λ}
211 M ← CTR_K^{A_0+1}(C)

220 B ← 0 || if H = ε then 0 else 1 endif || [τ/2 − 1]_3 || [λ − 1]_3 ||
221       N || [||M||_n]_{8λ} ||
222       if H = ε then ε elseif ||H||_n < 62580 then [||H||_n]_16 elseif ||H||_n < 2^32
223       then 0xFFFE || [||H||_n]_32 else 0xFFFF || [||H||_n]_64 endif
224       || H ||
225       if H = ε then ε elseif ||H||_n < 62580 then [0]_n^{(14−||H||_n) mod 16}
226       elseif ||H||_n < 2^32 then [0]_n^{(10−||H||_n) mod 16} else [0]_n^{(6−||H||_n) mod 16} endif
227       || M ||
228       [0]_n^{(−||M||_n) mod 16}
230 U ← CBC_K(B)
231 V ← E_K(A_0) ⊕ U
232 T' ← V [first τ bytes]
233 if T ≠ T' then return INVALID
234 return M
```

Fig. 6. Encryption and decryption under $\mathrm{CCM}[E, \tau, \lambda]$

CWC: A High-Performance Conventional Authenticated Encryption Mode

Tadayoshi Kohno[1], John Viega[2], and Doug Whiting[3]

[1] UC San Diego
tkohno@cs.ucsd.edu
[2] Virginia Tech
viega@securesoftware.com
[3] Hifn, Inc. dwhiting@hifn.com

Abstract. We introduce CWC, a new block cipher mode of operation for protecting both the privacy and the authenticity of encapsulated data. CWC is the first such mode having all five of the following properties: provable security, parallelizability, high performance in hardware, high performance in software, and no intellectual property concerns. We believe that having all five of these properties makes CWC a powerful tool for use in many performance-critical cryptographic applications. CWC is also the first appropriate solution for some applications; e.g., standardization bodies like the IETF and NIST prefer patent-free modes, and CWC is the first such mode capable of processing data at 10Gbps in hardware, which will be important for future IPsec (and other) network devices. As part of our design, we also introduce a new parallelizable universal hash function optimized for performance in both hardware and software.

1 Introduction

An *authenticated encryption associated data* (AEAD) scheme is a symmetric encryption scheme designed to protect both the privacy and the authenticity of encapsulated data. There has recently been a strong push toward producing block cipher-based AEAD schemes [13, 10, 12, 24, 28, 23, 5]. Despite this push, among the previous works there does not exist any AEAD scheme simultaneously having all of the following properties: provable security, parallelizability, high performance in hardware, high performance in software, and free from intellectual property concerns. Even though not all applications will require all five of the these properties, almost all applications will require at least one of the them, and may very likely have to be able to interoperate with an application requiring a different property. We thus view finding an appropriate scheme having all five of these properties as a very important research goal.

Finding an appropriate balance between all five of the aforementioned properties is, however, not easy because the most natural approaches to addressing some of the properties are actually disadvantageous with respect to other properties. We believe we have overcome these challenges and, in doing so, introduce a new mode of operation called CWC, or Carter-Wegman Counter mode.

B. Roy and W. Meier (Eds.): FSE 2004, LNCS 3017, pp. 408–426, 2004.

Motivating example. One of the primary motivations for such a block cipher-based AEAD scheme is IPsec. From a pragmatic perspective, we note that many vendors and standardization bodies prefer patent-free modes over patented modes (the elegant OCB mode was apparently rejected from the IEEE 802.11 working group because of patent concerns). And, from a hardware performance perspective, we note that because none of the existing patent-free AEAD schemes are parallelizable, it to impossible to make existing patent-free AEAD schemes run faster than about 2Gbps using conventional ASIC technology and a single processing unit. Nevertheless, future network devices will be expected to run at 10Gbps. CWC addresses these issues, being both patent-free and capable of processing data at 10Gbps using conventional ASIC technology.

The CWC solution. Our new mode of operation, called CWC, has all five of the properties mentioned above. It is provably secure. Moreover, our provable security-based analyses helped guide our research and helped us reject other schemes with similar performance properties but with slightly worse provable security bounds. CWC is also parallelizable, which means that we can make CWC-AES run at 10Gbps using conventional ASIC technology. CWC is also fast in software. Our current implementation of CWC-AES runs at about the same speed as the other patent-free modes on 32-bit architectures (Table 1), and we anticipate significant performance gains on 32-bit CPUs when using more sophisticated implementation techniques (Section 6), and we also see significantly better performance on 64-bit architectures. Of course, we do remark that the patented modes like OCB are capable of running even faster in software, which would make them very attractive were they not also encumbered in intellectual property issues.

Like the other two unpatented block cipher-based AEAD modes, CCM [28] and EAX [5], CWC avoids patents by using two inter-related but mostly independent modules: one module to "encrypt" the data and one module to "authenticate" the data. Adopting the terminology used in [5], it is because of the two-module structure that we call CWC a "conventional" block cipher-based AEAD scheme. Although CWC uses two modules, it can be implemented efficiently in a single pass. By using the conventional approach, CCM, EAX, and CWC are very much like composition-based AEAD scheme [4, 15], or AEAD schemes composed from existing encryption schemes and MACs. Unlike composition-based AEAD schemes, however, by designing CWC directly from a block cipher, we eliminate redundant steps and fine-tune CWC for efficiency, again keeping in mind both our hardware and software goals. For example, we use only one block cipher key, which saves expensive memory access in hardware.

The encryption core of CWC is essentially counter (CTR) mode encryption, which is well-known to be efficient and parallelizable. Finding an appropriate algorithm for the authentication core of CWC proved to be more of a challenge. For authentication, we decided to base our design on the Carter-Wegman [27] universal hash function approach for message authentication. Part of the difficulty in the design came down to choosing the right type of universal hash function,

Table 1. Software performance (in clocks per byte) for the three patent-free block cipher-based AEAD modes on a Pentium III. Values are averaged over 50 000 samples

	Linux/gcc-3.2.2					Windows 2000/VS6.0				
	Payload lengths (bytes)					Payload lengths (bytes)				
Mode	128	256	512	2048	8192	128	256	512	2048	8192
CWC-AES	105.5	88.4	78.9	72.2	70.5	84.7	70.2	62.2	56.5	55.0
CCM-AES	97.9	87.1	82.0	78.0	77.1	64.8	56.7	52.5	49.5	48.7
EAX-AES	114.1	94.9	86.1	79.1	77.5	75.2	61.8	55.3	50.4	49.1

with the right parameters. Since polynomial evaluation can be parallelized (if the polynomial is in x, one can split it into i polynomials in x^i), we chose to use a universal hash function consisting of evaluating a polynomial modulo the prime $2^{127}-1$. We note the our hash function is similar to Bernstein's hash127 [6] except that Bernstein's hash function was optimized for software performance at the expense of hardware performance. To address this issue, we use larger coefficients than Bernstein uses. We believe our hardware- and software-optimized universal hash function to be of independent interest.

Notation. As part of our research, we first created a general approach for combining CTR mode encryption with a universal hash function in order to provide authenticated encryption. We shall refer to this general approach as CWC (note no change in font), and shall use CWC-BC to refer to a CWC instantiation with a 128-bit block cipher BC as the underlying block cipher and with the universal hash function described briefly above. We shall use CWC as shorthand for CWC-BC and use CWC-AES to mean CWC-BC with AES [8] as the underlying block cipher. Other instantiations of the general CWC approach are possible, e.g., for legacy 64-bit block ciphers. Since we are primarily targeting new applications, and since a mode using a 128-bit block cipher will never be asked to interoperate with a mode using a 64-bit block cipher, we focus this paper only on our 128-bit CWC instantiation.

When we say that an AEAD scheme's encryption algorithm takes a pair (A, M) as input and produces a ciphertext as output, we mean that the AEAD scheme is designed to protect the privacy of M and the authenticity of both A and M. This will be made more formal in the body.

Performance. Let (A, M) be some input to the CWC encryption algorithm. The CWC encryption algorithm derives a universal hash subkey from the block cipher key. Assuming that the universal hash subkey is maintained across invocations, encrypting (A, M) takes $\lceil |M|/128 \rceil + 2$ block cipher invocations. The polynomial used in CWC's universal hashing step will have degree $d = \lceil |A|/96 \rceil + \lceil |M|/96 \rceil$. There are several ways to evaluate this polynomial (details in Section 6). As noted above, we could evaluate it in parallel. Serially, assuming no precomputation, we could evaluate this polynomial using d 127x127-bit multiplies. As another example, assuming n precomputed powers of the hash subkey, which are cheap to maintain in software for reasonable n, we could evaluate the polynomial using $d - m$ 96x127-bit multiplies and m 127x127-bit multiplies, where $m = \lceil (d+1)/n \rceil - 1$.

In hardware using conventional ASIC technology at 0.13 micron, it takes approximately 300 Kgates to reach 10 Gbps throughput for CWC-AES. This is around twice as much as OCB, but avoids IP negotiation overhead and royalty payments to three parties. Table 1 relates the software performance, on a Pentium III, of CWC-AES to the two other patent-free AEAD modes CCM and EAX; the patented modes such as OCB are not included in this table, but are about twice as fast as the times given for the patent-free modes. The implementations used to compute Table 1 were written in C by Brian Gladman [9] and all use 128-bit AES keys; the current CWC-AES implementation does not use the above-mentioned precomputation approach for evaluating the polynomial. Table 1 shows that the current implementations of the three modes have comparable performance in software, the relative "best" depending on the OS/compiler and the length of the message. Using the above-mentioned precomputation approach and switching to assembly, we anticipate reducing the cost of CWC's universal hashing step to around 8 cpb, thereby significantly improving the performance of CWC-AES in software compared to CCM-AES and EAX-AES (since the authentication portions of CCM-AES and EAX-AES are limited by the speed of AES but the authentication portion of CWC-AES is limited by the speed of the universal hash function). For comparison, Bernstein's related hash127, which also evaluates a polynomial modulo $2^{127} - 1$ but whose specific structure makes it less attractive in hardware, runs around 4 cpb on a Pentium III when written in assembly and using the precomputation approach. On 64-bit G5s, our initial implementation of the hash function runs at around 6 cpb, thus showing that CWC-AES is very attractive on 64-bit architectures (when running the G5 in 32-bit mode, our implementation runs at around 15 cpb).

We do not claim that CWC-AES will be particularly efficient on low-end CPUs such as 8-bit smartcards. However, our goal was not to develop an AEAD scheme for such low-end processors.

The patent issue. The patent issue is a very peculiar one. While it may initially sound odd to let patents influence research, we note that it is also not uncommon, especially in other sciences. Indeed, we view this line of research as discovering the most appropriate solution given real-world constraints. And, just like performance constraints, intellectual property constraints are very real.

Background and related work. The notion of an *authenticated encryption (AE) scheme* was formalized by Katz and Yung [13] and by Bellare and Namprempre [4] and the notion of an *authenticated encryption with associated data (AEAD) scheme* was formalized by Rogaway [23]. Bellare and Namprempre [4] and Krawczyk [15] explored ways to combine standard encryption schemes with MACs to achieve authenticated encryption. A number of dedicated AE and

AEAD schemes also exist, including RPC [13], XECB [10], IAPM [12], OCB [24], CCM [28], and EAX [5]. CWC is similar to the combination of McGrew's UST [20] and TMMH [19], where one of the main advantages of CWC over UST+TMMH is CWC's small key size, which, as the author of UST and TMMH noted, can be a bottleneck for UST+TMMH in hardware at high speeds. The integrity portion of CWC builds on top of the Carter-Wegman universal hashing approach to message authentication [27]. The specific hash function CWC uses is similar to Bernstein's hash127 [6], but is better suited for hardware. Shoup [26] and Nevelsteen and Preneel [22] also worked on software optimizations for universal hash functions. Rogaway and Wagner released a critique of CCM [25]. For each issue raised in [25], we find that we have addressed the issue (e.g., we designed CWC to be on-line) or we disagree with the issue (e.g., we feel that it is sufficient for new modes of operation to handle arbitrary octet-length, as opposed to arbitrary bit-length, messages; we stress, however, that, if desired, it is easy to modify CWC to handle arbitrary bit-length messages, see Section 5). CWC recently served as the starting point for GCM [21], another promising new conventional authenticated encryption mode.

2 Preliminaries

Notation. If x is a string then $|x|$ denotes its length in bits. Let ε denote the empty string. If x and y are two equal-length strings, then $x \oplus y$ denotes the XOR of x and y. If x and y are strings, then $x \| y$ denotes their concatenation. If N is a non-negative integer and l is an integer such that $0 \leq N < 2^l$, then $\mathsf{tostr}(N, l)$ denotes the encoding of N as an l-bit string in big-endian format. If x is a string, then $\mathsf{toint}(x)$ denotes the integer corresponding to string x in big-endian format (the most significant bit is *not* interpreted as a sign bit). For example, $\mathsf{toint}(10000010) = 2^7 + 2 = 130$. If b is a bit and n a non-negative integer, then b^n denote b concatenated with itself n times; e.g., 10^7 is the string 10000000. Let $x \leftarrow y$ denote the assignment of y to x. If X is a set, let $x \stackrel{\$}{\leftarrow} X$ denote the process of uniformly selecting at random an element from X and assigning it to x. If f is a randomized algorithm, let $x \stackrel{\$}{\leftarrow} f(y)$ denote the process of running f with input y and a uniformly selected random tape. When we refer to the time of an algorithm or experiment, we include the size of the code (in some fixed encoding). There is also an implicit big-$\mathcal{O}$ surrounding all such time references.

Authenticated encryption schemes with associated data. We use Rogaway's notion of an *authenticated encryption with associated data (AEAD) scheme* or *mode* [23]. An AEAD scheme $\mathcal{SE} = (\mathcal{K}_e, \mathcal{E}, \mathcal{D})$ consists of three algorithms and is defined over some key space $\mathsf{KeySp}_{\mathcal{SE}}$, some nonce space $\mathsf{NonceSp}_{\mathcal{SE}} = \{0,1\}^n$, n a positive integer, some associated data (header) space $\mathsf{AdSp}_{\mathcal{SE}} \subseteq \{0,1\}^*$, and some payload message space $\mathsf{MsgSp}_{\mathcal{SE}} \subseteq \{0,1\}^*$. We require that membership in $\mathsf{MsgSp}_{\mathcal{SE}}$ and $\mathsf{AdSp}_{\mathcal{SE}}$ can be efficiently tested and that if M, M' are two strings such that $M \in \mathsf{MsgSp}_{\mathcal{SE}}$ and $|M'| = |M|$, then $M' \in \mathsf{MsgSp}_{\mathcal{SE}}$.

The randomized key generation algorithm $\mathcal{K}_e$ returns a key $K \in \mathsf{KeySp}_{\mathcal{SE}}$; we denote this process as $K \stackrel{\$}{\leftarrow} \mathcal{K}_e$. The deterministic encryption algorithm $\mathcal{E}$ takes as input a key $K \in \mathsf{KeySp}_{\mathcal{SE}}$, a nonce $N \in \mathsf{NonceSp}_{\mathcal{SE}}$, a header (or associated data) $A \in \mathsf{AdSp}_{\mathcal{SE}}$, and a payload message $M \in \mathsf{MsgSp}_{\mathcal{SE}}$, and returns a ciphertext $C \in \{0,1\}^*$; we denote this process as $C \leftarrow \mathcal{E}_K^{N,A}(M)$ or $C \leftarrow \mathcal{E}_K(N, A, M)$. The deterministic decryption algorithm $\mathcal{D}$ takes as input a key $K \in \mathsf{KeySp}_{\mathcal{SE}}$, a nonce $N \in \mathsf{NonceSp}_{\mathcal{SE}}$, a header $A \in \mathsf{AdSp}_{\mathcal{SE}}$, and a string $C \in \{0,1\}^*$ and outputs a message $M \in \mathsf{MsgSp}_{\mathcal{SE}}$ or the special symbol $\mathsf{INVALID}$ on error; we denote this process as $M \leftarrow \mathcal{D}_K^{N,A}(C)$. We require that $\mathcal{D}_K^{N,A}(\mathcal{E}_K^{N,A}(M)) = M$ for all $K \in \mathsf{KeySp}_{\mathcal{SE}}$, $N \in \mathsf{NonceSp}_{\mathcal{SE}}$, $A \in \mathsf{AdSp}_{\mathcal{SE}}$, and $M \in \mathsf{MsgSp}_{\mathcal{SE}}$. Let $l(\cdot)$ denote the *length function* of $\mathcal{SE}$; i.e., for all keys K, nonces N, headers A, and messages M, $|\mathcal{E}_K^{N,A}(M)| = l(|M|)$.

Under the correct usage of an AEAD scheme, after a random key is selected, the application should never invoke the encryption algorithm twice with the same nonce value until a new key is randomly selected. In order to ensure that a nonce does not repeat, implementations typically use nonces that contain counters. We use the notion of a nonce, rather than simply a counter, because the notion of a nonce is more general and allows the developer the freedom to structure the nonce as he or she desires.

Block ciphers. A block cipher $E : \{0,1\}^k \times \{0,1\}^L \to \{0,1\}^L$ is a function from k-bit keys and L-bit blocks to L-bit blocks. We use $E_K(\cdot)$, $K \in \{0,1\}^k$, to denote the function $E(K, \cdot)$ and we use $f \stackrel{\$}{\leftarrow} E$ as short hand for $K \stackrel{\$}{\leftarrow} \{0,1\}^k$; $f \leftarrow E_K$. Block ciphers are families of permutations; namely, for each key $K \in \{0,1\}^k$, E_K is a permutation on $\{0,1\}^L$. We call k the key length of E and we call L the block length.

We adopt the notion of security for block ciphers introduced in [17] and adopted for the concrete setting in [2]. Let $E : \{0,1\}^k \times \{0,1\}^L \to \{0,1\}^L$ be a block cipher and let $\mathsf{Perm}(L)$ denote the set of all permutations on $\{0,1\}^L$. Let A be an adversary with access to an oracle and that returns a bit. Then

$$\mathbf{Adv}_F^{\mathrm{prp}}(A) = \Pr\left[f \stackrel{\$}{\leftarrow} E \,:\, A^{f(\cdot)} = 1 \right] - \Pr\left[g \stackrel{\$}{\leftarrow} \mathsf{Perm}(L) \,:\, A^{g(\cdot)} = 1 \right]$$

denotes the PRP-advantage of A in distinguishing a random instance of E from a random permutation. Intuitively, we say that E is a secure PRP, or a secure block cipher, if the PRP-advantages of all adversaries using reasonable resources is small. Modern block ciphers, such as AES [8], are believed to be secure PRPs.

3 The CWC Mode of Operation

We now describe our new AEAD scheme. Let $\mathsf{BC} : \{0,1\}^{\mathsf{kl}} \times \{0,1\}^{128} \to \{0,1\}^{128}$ be a 128-bit block cipher. Let $\mathsf{tl} \leq 128$ is the desired tag length in bits. Then the CWC mode of operation using BC with tag length tl, $\mathsf{CWC\text{-}BC\text{-}tl} =$

$(\mathcal{K}, \mathsf{CWC\text{-}ENC}, \mathsf{CWC\text{-}DEC})$, is defined as follows. The message spaces are:

$$\begin{aligned}\mathsf{MsgSp}_{\mathsf{CWC\text{-}BC\text{-}tl}} &= \{\, x \in (\{0,1\}^8)^* \; : \; |x| \leq \mathsf{MaxMsgLen} \,\} \\ \mathsf{AdSp}_{\mathsf{CWC\text{-}BC\text{-}tl}} &= \{\, x \in (\{0,1\}^8)^* \; : \; |x| \leq \mathsf{MaxAdLen} \,\} \\ \mathsf{KeySp}_{\mathsf{CWC\text{-}BC\text{-}tl}} &= \{0,1\}^{\mathsf{kl}} \\ \mathsf{NonceSp}_{\mathsf{CWC\text{-}BC\text{-}tl}} &= \{0,1\}^{88}\end{aligned}$$

where $\mathsf{MaxMsgLen}$ and $\mathsf{MaxAdLen}$ are both $128 \cdot (2^{32} - 1)$. That is, the payload and associated data spaces for CWC-BC-tl consist of all strings of octets that are at most $2^{32} - 1$ blocks long.

The CWC-BC-tl key generation, encryption, and decryption algorithms are defined as follows:

Algorithm $\mathcal{K}$
  $K \xleftarrow{\$} \{0,1\}^{\mathsf{kl}}$
  Return K

Algorithm $\mathsf{CWC\text{-}ENC}_K(N, A, M)$
  $\sigma \leftarrow \mathsf{CWC\text{-}CTR}_K(N, M)$
  $\tau \leftarrow \mathsf{CWC\text{-}MAC}_K(N, A, \sigma)$
  Return $\sigma \| \tau$

Algorithm $\mathsf{CWC\text{-}DEC}_K(N, A, C)$
  If $|C| < \mathsf{tl}$ then return INVALID
  Parse C as $\sigma \| \tau$ where $|\tau| = \mathsf{tl}$
  If $A \notin \mathsf{AdSp}_{\mathsf{CWC\text{-}BC\text{-}tl}}$ or $\sigma \notin \mathsf{MsgSp}_{\mathsf{CWC\text{-}BC\text{-}tl}}$ then
    return INVALID
  If $\tau \neq \mathsf{CWC\text{-}MAC}_K(N, A, \sigma)$ then
    return INVALID
  $M \leftarrow \mathsf{CWC\text{-}CTR}_K(N, \sigma)$
  Return M

The remaining algorithms (CWC-CTR, CWC-MAC, CWC-HASH) are defined below. The CWC-CTR algorithm handles generating the encryption and decryption keystreams, CWC-MAC handles the generation of an authentication tag, and uses CWC-HASH as the underlying universal hash function.

Algorithm $\mathsf{CWC\text{-}CTR}_K(N, M)$
  $\alpha \leftarrow \lceil |M|/128 \rceil$
  For $i = 1$ to α do
    $s_i \leftarrow \mathsf{BC}_K(10^7 \| N \| \mathsf{tostr}(i, 32))$
  $x \leftarrow$ first $|M|$ bits of $s_1 \| s_2 \| \cdots \| s_\alpha$
  $\sigma \leftarrow x \oplus M$
  Return σ

Algorithm $\mathsf{CWC\text{-}MAC}_K(N, A, \sigma)$
  $R \leftarrow \mathsf{BC}_K(\mathsf{CWC\text{-}HASH}_K(A, \sigma))$
  $\tau \leftarrow \mathsf{BC}_K(10^7 \| N \| 0^{32}) \oplus R$
  Return first tl bits of τ

Algorithm $\mathsf{CWC\text{-}HASH}_K(A, \sigma)$
  $Z \leftarrow$ last 127 bits of $\mathsf{BC}_K(110^{126})$
  $K_h \leftarrow \mathsf{toint}(Z)$
  $l \leftarrow$ min int such that 96 divides $|A \| 0^l|$
  $l' \leftarrow$ min int such that 96 divides $|\sigma \| 0^{l'}|$
  $X \leftarrow A \| 0^l \| \sigma \| 0^{l'}$; $\beta \leftarrow |X|/96$
  Break X into chunks $X_1, X_2, \ldots, X_\beta$
  For $i = 1$ to β do
    $Y_i \leftarrow \mathsf{toint}(X_i)$
  $l_\sigma \leftarrow |\sigma|/8$; $l_A \leftarrow |A|/8$
  $Y_{\beta+1} \leftarrow 2^{64} \cdot l_A + l_\sigma$
  $R \leftarrow Y_1 K_h^\beta + \cdots + Y_\beta K_h + Y_{\beta+1} \mod 2^{127} - 1$
  Return $\mathsf{tostr}(R, 128)$

4 Theorem Statements

The CWC scheme is a provably secure AEAD scheme assuming that the underlying block cipher, e.g., AES, is a secure pseudorandom permutation. This is a quite reasonable assumption since most modern block ciphers, including AES, are believed to be pseudorandom. Furthermore, all provably-secure block cipher modes of operation that we are aware of make at least the same assumptions we make, and some modes, such as OCB [24], require the stronger, albeit still reasonable, assumption of super-pseudorandomness.

The specific results for CWC appear as Theorem 1 and Theorem 2 below, and are proven in the full version of this paper [14]. In [14] we also present results for the general CWC construction, from which Theorems 1 and 2 follow.

4.1 Privacy

We first show that if BC is a secure block cipher, then CWC-BC-tl will preserve privacy under chosen-plaintext attacks. For our notion of privacy for AEAD schemes, we use the strong definition of indistinguishability from [23]. Let $\mathcal{SE} = (\mathcal{K}_e, \mathcal{E}, \mathcal{D})$ be an AEAD scheme with length function $l(\cdot)$. Let $\$(\cdot,\cdot,\cdot)$ be an oracle that, on input $(N, A, M) \in \mathsf{NonceSp}_{\mathcal{SE}} \times \mathsf{AdSp}_{\mathcal{SE}} \times \mathsf{MsgSp}_{\mathcal{SE}}$, returns a random string of length $l(|M|)$. Let B be an adversary with access to an oracle and that returns a bit. Then

$$\mathbf{Adv}^{\mathrm{priv}}_{\mathcal{SE}}(B) = \Pr\left[\, K \xleftarrow{\$} \mathcal{K}_e \;:\; B^{\mathcal{E}_K(\cdot,\cdot,\cdot)} = 1 \,\right] - \Pr\left[\, B^{\$(\cdot,\cdot,\cdot)} = 1 \,\right]$$

is the IND\$-CPA-*advantage* of B in breaking the privacy of $\mathcal{SE}$ under chosen-plaintext attacks; i.e., $\mathbf{Adv}^{\mathrm{priv}}_{\mathcal{SE}}(B)$ is the advantage of B in distinguishing between ciphertexts from $\mathcal{E}_K(\cdot,\cdot,\cdot)$ and random strings. An adversary B is *nonce-respecting* if it never queries its oracle with the same nonce twice. Intuitively, a scheme $\mathcal{SE}$ preserves privacy under chosen plaintext attacks if the IND\$-CPA-advantage of all nonce-respecting adversaries using reasonable resources is small.

Theorem 1. *[Privacy of* **CWC**.*] Let CWC-BC-tl be as in Section 3. Then given a nonce-respecting* IND\$-CPA *adversary A against CWC-BC-tl one can construct a* PRP *adversary C_A against BC such that if A makes at most q oracle queries totaling at most μ bits of payload message data, then*

$$\mathbf{Adv}^{priv}_{\mathit{CWC\text{-}BC\text{-}tl}}(A) \le \mathbf{Adv}^{prp}_{\mathit{BC}}(C_A) + \frac{(\mu/128 + 3q + 1)^2}{2^{129}} \;. \tag{1}$$

Furthermore, the experiment for C_A takes the same time as the experiment for A and C_A makes at most $\mu/128 + 3q + 1$ oracle queries.

Let us elaborate on why Theorem 1 implies that CWC-BC will preserve privacy under chosen-plaintext attacks. Assume BC is a secure block cipher. This means that $\mathbf{Adv}^{\mathrm{prp}}_{\mathsf{BC}}(C)$ must be small for all adversaries C using reasonable resources and, in particular, this means that, for C_A as described in the theorem statement, $\mathbf{Adv}^{\mathrm{prp}}_{\mathsf{BC}}(C_A)$ must be small assuming that A uses reasonable resources. And if $\mathbf{Adv}^{\mathrm{prp}}_{\mathsf{BC}}(C_A)$ is small and μ, q are small, then, because of the above equations, $\mathbf{Adv}^{\mathrm{priv}}_{\mathsf{CWC\text{-}BC\text{-}tl}}(A)$ must also be small as well. I.e., any adversary A using reasonable resources will only be able to break the privacy of CWC-BC-tl with some small probability.

As a concrete example, let us consider limiting the number of applications of CWC-BC-tl between rekeyings to some reasonable value such as $q = 2^{32}$, and let us limit the total number of payload bits between rekeyings to $\mu = 2^{50}$. Then Equation 1 becomes

$$\mathbf{Adv}^{\text{priv}}_{\text{CWC-BC-tl}}(A) \leq \mathbf{Adv}^{\text{prp}}_{\text{BC}}(C_A) + \frac{1}{2^{42}}$$

which means that, assuming that the underlying block cipher is a secure PRP, an attacker will not be able to break the privacy of CWC-BC-tl with advantage much greater than 2^{-42}.

4.2 Integrity/Authenticity

We now present our results showing that if BC is a secure block cipher, then CWC-BC-tl will protect the authenticity of encapsulated data. We use the strong notion of authenticity for AEAD schemes from [23]. Let $\mathcal{SE} = (\mathcal{K}_e, \mathcal{E}, \mathcal{D})$ be an AEAD scheme. Let F be a forging adversary and consider an experiment in which we first pick a random key $K \xleftarrow{\$} \mathcal{K}_e$ and then run F with oracle access to $\mathcal{E}_K(\cdot,\cdot,\cdot)$. We say that F *forges* if F returns a pair (N, A, C) such that $\mathcal{D}^{N,A}_K(C) \neq \mathsf{INVALID}$ but F did not make a query (N, A, M) to $\mathcal{E}_K(\cdot,\cdot,\cdot)$ that resulted in a response C. Then

$$\mathbf{Adv}^{\text{auth}}_{\mathcal{SE}}(F) = \Pr\left[\, K \xleftarrow{\$} \mathcal{K}_e \;:\; F^{\mathcal{E}_K(\cdot,\cdot,\cdot)} \text{ forges} \,\right]$$

is the AUTH-*advantage* of F in breaking the integrity/authenticity of $\mathcal{SE}$. Intuitively, the scheme $\mathcal{SE}$ preserves integrity/authenticity if the AUTH-advantage of all nonce-respecting adversaries using reasonable resources is small.

Theorem 2. *[Integrity/authenticity of* **CWC**.*] Let CWC-BC-*tl *be as specified in Section 3. (Recall that BC is a 128-bit block cipher and that the tag length* tl *is* ≤ 128.*) Consider a nonce-respecting* AUTH *adversary A against CWC-BC-*tl. *Assume the execution environment allows A to query its oracle with associated data that are at most* $n \leq \mathsf{MaxAdLen}$ *bits long and with messages that are at most* $m \leq \mathsf{MaxMsgLen}$ *bits long. Assume A makes at most* $q-1$ *oracle queries and the total length of all the payload data (both in these* $q-1$ *oracle queries and the forgery attempt) is at most* μ. *Then given A we can construct a* PRP *adversary* C_A *against BC such that*

$$\mathbf{Adv}^{auth}_{\text{CWC-BC-tl}}(A) \leq \mathbf{Adv}^{prp}_{BC}(C_A) + \frac{(\mu/128 + 3q + 1)^2}{2^{129}} + \frac{n+m}{2^{133}} + \frac{1}{2^{125}} + \frac{1}{2^{\text{tl}}}\,. \quad (2)$$

Furthermore, the experiment for C_A *takes the same time as the experiment for A and* C_A *makes at most* $\mu/128 + 3q + 1$ *oracle queries.*

Let us elaborate on why Theorem 2 implies that CWC-BC will preserve authenticity. Assume BC is a secure block cipher. This means that $\mathbf{Adv}^{\text{prp}}_{\text{BC}}(C)$ must be small for all adversaries C using reasonable resources and, in particular, this means that, for C_A as described in the theorem statement, $\mathbf{Adv}^{\text{prp}}_{\text{BC}}(C_A)$ must be small assuming that A uses reasonable resources. And if $\mathbf{Adv}^{\text{prp}}_{\text{BC}}(C_A)$ is small and μ, q, m and n are small, then, because of the above equations, $\mathbf{Adv}^{\text{auth}}_{\text{CWC-BC-tl}}(A)$ must also be small as well. I.e., any adversary A using reasonable resources will only be able to break the authenticity of CWC-BC-tl with some small probability.

Let us consider some concrete examples. Let n = MaxAdLen and m = MaxMsgLen, which is the maximum possible allowed by the CWC-BC construction. Then Equation 2 becomes

$$\mathbf{Adv}^{\mathrm{auth}}_{\mathsf{CWC\text{-}BC\text{-}tl}}(A) \leq \mathbf{Adv}^{\mathrm{prp}}_{\mathsf{BC}}(C_A) + \frac{(\mu/128 + 3q + 1)^2}{2^{129}} + \frac{1}{2^{93}} + \frac{1}{2^{\mathsf{tl}}} .$$

If we set $q = 2^{32}$ and $\mu = 2^{50}$ as before, and if we take $\mathsf{tl} \geq 43$, then the above equation becomes

$$\mathbf{Adv}^{\mathrm{auth}}_{\mathsf{CWC\text{-}BC\text{-}tl}}(A) \leq \mathbf{Adv}^{\mathrm{prp}}_{\mathsf{BC}}(C_A) + \frac{1}{2^{41}}$$

which means that, assuming that the underlying block cipher is a secure PRP, an attacker will not be able to break the unforgeability of CWC-BC-tl with probability much greater than 2^{-41}.

Remark 1. **[Chosen-ciphertext privacy.]** Since CWC-BC-tl preserves privacy under chosen-plaintext attacks (Theorem 1) *and* provides integrity (Theorem 2) assuming that BC is a secure pseudorandom permutation, it also provides privacy under chosen-ciphertext attacks under the same assumption about BC. See [4, 23] for a discussion of the relationship between chosen-plaintext privacy, integrity, and chosen-ciphertext privacy; this relationship was also used, for example, by the designers of OCB [24].

5 Design Decisions

Finding an appropriate balance between provable security, hardware efficiency, and software efficiency, while simultaneously avoiding existing intellectual property issues, proved to be one the the biggest challenges of this research project. In this section we discuss how our diverse set of goals affected our design decisions.

The CWC-HASH universal hash function. We found that the best way to simultaneously achieve our parallelizability, hardware, and software goals was to base the authentication portion of CWC on the Carter-Wegman [27] universal hash function approach to message authentication. This is because universal hash functions, and especially the one we created for CWC, can be implemented in a multitude of ways, thus allowing different platforms and applications to implement CWC-HASH in the way most appropriate for them. For example, hardware implementations will like parallelize the computation of CWC-HASH by splitting it into multiple polynomials in K_h^i for some i. In more detail, if the polynomial is

$$Y_1 K_h^{\beta} + Y_2 K_h^{\beta-1} + Y_3 K_h^{\beta-2} + Y_4 K_h^{\beta-3} + \cdots + Y_\beta K_h + Y_{\beta+1} \bmod 2^{127} - 1 .$$

then, setting $i = 2$, and $y = K_h^2 \bmod 2^{127} - 1$, and assuming β is odd for illustration purposes, we can rewrite the above polynomial as

$$\Big(Y_1 y^m + Y_3 y^{m-1} + \cdots + Y_\beta\Big)x + \Big(Y_2 y^m + Y_4 y^{m-1} + \cdots + Y_{\beta+1}\Big) \bmod 2^{127} - 1 ,$$

After splitting the polynomial, hardware implementations will then likely compute each polynomial using Horner's rule (e.g., the polynomial $aK_h^{2i} + bK_h^i + c$ would be evaluated as $(((a)K_h^i + b)K_h^i) + c)$. Software implementations on modern CPUs, for which memory is cheap, will likely precompute a number of powers of K_h and evaluate the CWC-HASH polynomial directly, or almost directly, using a hybrid between a precomputation approach and Horner's rule. We consider a number of possible implementation strategies in more detail in Section 6.

CWC-HASH is an instantiation of the classic polynomial universal hash approach to message authentication [27], and is closely related to Bernstein's hash127 [6], which also evaluates a polynomial modulo $2^{127}-1$. Although hash127 is very fast in software, its structure makes it less suitable for use on high-speed hardware. In particular, Bernstein's choice of 32-bit coefficients, while great for software implementations with precomputed powers of K_h, means that hardware implementations using Horner's rule will be "wasting work." Specifically, even with 32-bit coefficients, incorporating each new coefficient using Horner's rule will require a 127x127-bit multiply because the accumulated value will be 127 bits long. By defining the CWC-HASH coefficients to be 96-bits long, we increase the performance of Horner's rule implementations by a factor of three. (Of course, we could have gone even further and made the coefficients 126 bits long, but doing so would have required considerable additional complexity to perform bit and byte shifting within the coefficients.) An alternative approach for increasing the performance of a serial implementation of Horner's rule would be to reduce the size of the CWC-HASH subkey K_h to 96 bits. We discuss why we rejected this option in more detail later, but remark here that there are already more efficient strategies than Horner's rule for implementing CWC-HASH in software, and that in a parallelized approach the values K_h^i, $i \geq 2$, will most often be full 127-bit values even if K_h is only 96-bits long.

On using a single key. From a security perspective, it would have been perfectly acceptable, and in fact more traditional, to make the CWC-CTR block cipher key and the two CWC-MAC block cipher keys independent. Like others [28, 5], however, we acknowledge that there are several important reasons for sharing keys between the encryption and authentication portions of modes such as CWC. One of the most important reasons is simplicity of key management. Indeed, fetching key material can be a major bottleneck in high-speed hardware, and minimizing key material is thus important. This fact is also why we derive the hash subkey from the block cipher key rather than use an independent hash subkey. We could, of course, have defined a mode that derived a number of essentially independent block cipher and hash keys from a single block cipher key, but doing so would either have required more memory or more computation and, because we have proofs that our construction works, would have been unnecessary.

Sharing the block cipher key in the way described above and deriving the hash subkey from the block cipher key did, however, mean that we had to be very careful with our proofs of security. To facilitate our proofs, we took extra care in our design to ensure that there would never be a collision in the plaintext inputs to the block cipher between the different usages of the block cipher. For example, by defining CWC-HASH to produce a 127-bit value as output, we know that the first application of BC to $\text{CWC-HASH}_K(A, \sigma)$ in CWC-MAC will always have its first bit set to 0. To avoid a collision with the input to the keystream generator, the block cipher inputs in CWC-CTR always have the first two bits set to 10. When using the block cipher to create the hash subkey K_h, the first two bits of the input are set to 11.

On the choice of parameters. Part of this effort involved specifying the appropriate parameters for the CWC encryption mode. Example parameters include the nonce length and the way the nonce is encoded in the input to the block cipher. We chose to fix these parameters for interoperability purposes, but note that our general approach in [14] does not have theses parameters fixed. We chose to set the nonce length to 88 bits in order to handle future IPsec sequence numbers. And we chose to set the block counter length to 32 bits in order to allow CWC to be used with IPsec jumbograms and other large packets. We also chose to use big-endian byte ordering for consistency purposes and to maintain compatibility with McGrew's ICM Internet-Draft [18] and the IETF, which strongly favors big-endian byte-ordering.

Handling arbitrary bit-length messages. Since we do not believe that many applications will actually require the ability to encrypt arbitrary bit-length messages, we do not define CWC to take arbitrary bit-length messages as input. That said, we did design CWC in such a way that it will be easy to modify the specification to take arbitrary bit-length messages without affecting interoperability with existing implementations when octet-strings are communicated. For example, one could augment the computation of $Y_{\beta+1}$ in CWC-HASH as follows:

$$r_A \leftarrow |A| \bmod 8\,;\ r_\sigma \leftarrow |\sigma| \bmod 8\,;\ Y_{\beta+1} \leftarrow 2^{120} \cdot r_A + 2^{112} \cdot r_\sigma + 2^{64} \cdot l_A + l_\sigma\,.$$

Of course, a cleaner approach for handling arbitrary bit-length messages would be to compute $l_A \leftarrow |A|$ and $l_\sigma \leftarrow |\sigma|$ in CWC-HASH. We do not define CWC this way because we do not consider it a good trade-off to define a mode for arbitrary bit-length messages at the expense of octet-oriented systems.

64-bit block ciphers. We did not define CWC for use with 64-bit block ciphers because we are targeting future high-speed cryptographic applications. Nevertheless, the general CWC approach in [14] can be instantiated with 64-bit block ciphers. A 64-bit instantiation may, however, require several uncomfortable tradeoffs; e.g., in the length of the nonce.

Some possible alternatives. Here we discuss some other possible alternatives to CWC and why we rejected these alternatives. First, as noted earlier, it is possible to improve the performance in some situations by using shorter hash subkeys K_h, say of length 96 bits. Such an alternative will not increase the performance in high-speed hardware implementations that will parallelize the computation of CWC-HASH by evaluating a polynomial in (at least) K_h^2. A 96-bit hash subkey would have increased Horner's rule performance in software, but would still be comparable in speed to a software-based approach using precomputed powers of K_h (see Section 6), so reducing the size of K_h to 96 bits would not provide a significant advantage in software either. In [14] we also consider what happens to our provable security bounds when the length of the hash subkey is reduced to less than 96 bits.

There are a number of possible approaches for reducing the number of block cipher applications in the CWC-MAC algorithm by one. For example, one could use $\mathsf{BC}_K(h'_K(N, A, \sigma))$ as the tag, where h' is a modified version of CWC-HASH designed to hash 3-tuples instead of pairs of strings. One could also use something like $\mathsf{BC}_K(N) + Y_1 K_h^{\beta+2} + \cdots + Y_\beta K_h^3 + l_A K_h^2 + l_\sigma K_h \bmod 2^{127} - 1$ as the tag. In [14] we consider these and other alternatives and discuss why we chose to define CWC the way that we did instead of using an option with one fewer block cipher invocation. In the case of the two alternatives mentioned in this paragraph, we note that we rejected them because we were able to prove better bounds on the security of CWC as currently defined.

Motivated by EAX2 [5], one possible alternative to CWC might be to use $\mathsf{BC}_K(1110^5 \| N)$ both as the value to encrypt R in CWC-MAC and as the initial counter to CTR mode-encrypt M (with the first two bits of the counter always set to 10). Other EAX2-motivated constructions also exist. For example, the tag might be set to $\mathsf{BC}_K(h(X_0 \| N)) \oplus \mathsf{BC}_K(h(X_1 \| A)) \oplus \mathsf{BC}_K(h(X_2 \| \sigma))$, where X_0, X_1, X_2 are strings, none of which is a prefix of the other, and h is a parallelizable universal hash function, like CWC-HASH but hashing only single strings (as opposed to pairs of strings). Compared to CWC, these alternatives have the ability to take longer nonces as input, and, from a functional perspective, can be applied to strings up to 2^{126} blocks long. But we do not view this as a reason to prefer these alternatives over CWC. From a practical perspective, we do not foresee applications needing nonces longer than 11 octets, or needing to encrypt messages longer than $2^{32} - 1$ blocks. Moreover, from a security perspective, applications should not encrypt too many packets between rekeyings, implying that even 11 octet nonces are more than sufficient. We do comment, however, that we believe the alternatives discussed in this paragraph are still more attractive than EAX because, like CWC but unlike EAX, these alternatives are parallelizable.

We chose not to base the authentication portion of our new mode on XOR-MAC [3] or PMAC [7] because of patent concerns and our software performance requirements and we chose not to base the authentication portion on software-efficient MACs such as HMAC [1] because of our hardware parallelizability requirement.

6 Performance

Hardware. Since one of our main goals was to achieve high performance in hardware and, in particular, to provide a solution for future 10 Gbps IPsec (and other) network devices, let us focus first on hardware costs. As noted in the introduction, using 0.13 micron CMOS ASIC technology, it should take approximately 300 Kgates to achieve 10 Gbps throughput for CWC-AES. This estimate, which is applicable to AES with all key lengths, includes four AES counter-mode encryption engines, each running at 200 MHz and requiring about 25Kgates each. In addition, there are two 32x128-bit multiply/accumulate engines, each running at 200 MHz with a latency of four clocks, one each for the even and odd polynomial coefficients. Of course, simply keeping these engines "fed" may be quite a feat in itself, but that is generally true of any 10 Gbps path. Also, there may well be better methods to structure an implementation, depending on the particular ASIC vendor library and technology, but, regardless of the implementation strategy, 10 Gbps is quite achievable because of the inherent parallelism of CWC.

Since OCB is CWC's main competitor for high-speed environments, it is worth comparing CWC with OCB instantiated with AES (we do not compare CWC with CCM and EAX here since the latter two are not parallelizable). We first note that CWC-AES saves some gates because we only have to implement AES encryption in hardware. However, at 10 Gbps, OCB still probably requires only about half the silicon area of CWC-AES. The main question for many hardware designers is thus whether the extra silicon area for CWC-AES costs more than three royalty payments, as well as negotiation costs and overhead. With respect to negotiation costs and royalty payments, we note that despite significant demands, to date the relevant parties have not all offered publicly available IP fee schedules. Given this fact, and given today's silicon costs, we believe that the extra silicon for CWC-AES is probably cheaper overall than the negotiation costs and IP fees required for OCB.

Software. CWC-AES can also be implemented efficiently in software. Table 1 shows timing information for CWC-AES, as well as CCM-AES and EAX-AES, on a 1.133GHz mobile Pentium III dual-booting RedHat Linux 9 (kernel 2.4.20-8) and Windows 2000 SP2. The numbers in the table are the clocks per byte for different message lengths averaged over 50 000 runs and include the entire time for setting up (e.g., expanding the AES key-schedule) and encrypting. All implementations were in C and written by Brian Gladman [9] and use 128-bit AES keys. The Linux compiler was gcc version 3.2.2; the Windows compiler was Visual Studio 6.0. To be fair, we note that OCB does run at about twice the speeds given in Table 1.

From Table 1 we conclude that the three patent-free modes, as currently implemented by Gladman, share similar software performances. The "best" performing one appears to depend on OS/compiler and the length of the message being processed. On Linux, it appears that CWC-AES performs slightly better than EAX-AES for all message lengths that we tested, and better than CCM-AES for the longer messages, whereas Gladman's CCM-AES and EAX-AES implementations slightly outperform his CWC-AES implementation on Windows for all the message lengths that we tested.

Note, however, that all the implementations used to compute Table 1 were written in C. Furthermore, the current CWC-AES code does not make use of all of the optimization techniques (and in particular precomputation) that we describe below. By switching to assembly and using the additional optimization techniques, we anticipate the speed for CWC-HASH to drop to better than 8 clocks per byte, whereas the speed for the CBC-MAC portion of CCM-AES and EAX-AES will be limited by the speed of AES (the best reported speed for AES on a Pentium III is 14.1 cpb, due to a proprietary library by Helger Lipmaa; Gladman's free hand-optimized Windows assembly implementation runs at 17.5 cpb [16]). Returning to the speed of CWC-HASH, for reference we note that Bernstein's related hash127 [6] runs around 4 cpb on a Pentium III when written in assembly and using the precomputation approach. Bernstein's hash127 also works by evaluating a polynomial modulo $2^{127}-1$; the main difference is that the coefficients for hash127 are 32 bits long, whereas the coefficients for CWC-HASH are 96 bits long (recall Section 5, which discusses why we use 96-bit coefficients). We also note that the performance of CWC-HASH will increase dramatically on 64-bit architectures with larger multiplies; an initial implementation on a G5 using 64-bit integer operations runs at around 6 cpb (when running the G5 in 32-bit mode, the hash function runs at around 15 cpb).

Since the implementation of CWC-HASH is more complicated than the implementation of the CWC-CTR portion of CWC, we devote the rest of this section to discussing CWC-HASH.

Precomputation. As noted in Section 5, there are two general approaches to implementing CWC-HASH in software. The first is to use Horner's rule. The second is to evaluate the polynomial directly, which can be faster if one precomputes powers of the hash key K_h at setup time (here the powers of K_h can be viewed as an expanded key-schedule). In particular, as noted in Section 5, evaluating the polynomial using Horner's rule requires a 127x127-bit multiply for each coefficient, whereas evaluating the polynomial directly using precomputed powers of K_h requires a 96x127-bit multiply for each coefficient. (We discuss elsewhere why we did not make the hash subkey 96-bits, which could have sped up a serial Horner's rule implementation.) The advantage with precomputation was first observed by Bernstein in the context of hash127 [6].

The above description of the precomputation approach assumed that if the polynomial is $Y_1 K_h^{\gamma-1} + \cdots + Y_{\gamma-1} K_h + Y_\gamma$ (i.e., the polynomial has γ coefficients), then we had precomputed the powers of K_h^i for all $i \in \{1, \ldots, \gamma-1\}$. The precomputation approach extends naturally to the case where we have precomputed the powers K_h^j, $j \in \{1, \ldots, n\}$, for some $n \leq \gamma-1$. For simplicity, first assume that we know the polynomial has a multiple of n coefficients. For such a polynomial, one processes the first n coefficients (to get $Y_1 K_h^{n-1} + \ldots + Y_{n-1} K_h + Y_n$), then multiplies the intermediate result by K_h^n (to get $Y_1 K_h^{2n-1} + \ldots + Y_{n-1} K_h^{n+1} + Y_n K_h^n$). After that, one can continue processing data with the same precomputed values (to get $Y_1 K_h^{2n-1} + \ldots + Y_{2n-1} K_h + Y_{2n}$), and so on. Note that each chunk of n coefficients takes $(n-1)$ 96x127-bit multiplies, and all but the last chunk takes an additional 127x127-bit multiply. Now assume that the number of coefficients m in the polynomial is not necessarily a multiple of n. If m is known in advance, one could first process m mod n coefficients, multiply by K_h^n, then process in n-coefficient chunks as before. Alternately, as long as the end of the message is known n coefficients in advance, one could process n-coefficients chunks, and then finish off the final m mod n coefficients using Horner's rule. Or, if the number of coefficients in the polynomial is not known until the final coefficient is reached, one could process the message in n-coefficient chunks and then multiply by a precomputed power of K_h^{-1} once the end of the message hash been reached.

Naturally, precomputation requires extra memory, but that is usually cheap and plentiful in a software-based environment. Using 32-bit multiplies, the precomputation approach requires 12 32-bit multiplies per 96-bit coefficient, as well as 17 adds, all of which may carry. In assembly, most of these carry operations can be implemented for free, or close to it by using a special variant of the add instruction that adds in the operand as well as the value of the carry from the previous add operation. But when implemented in C, they will generally compile to code that requires a conditional branch and an extra addition. An implementation using Horner's rule requires an additional four multiplies and three additions with carry per coefficient, adding about 33% overhead, since the multiplies dominate the additions. A 64-bit platform only requires four multiplies and four adds (which may all carry), no matter the implementation strategy taken, which explains why implementations of CWC-HASH for 64-bit architectures are much faster.

Exploiting the parallelism of some instruction sets. On most 32-bit platforms, it turns out that the integer execution unit is not the fastest way to implement CWC-HASH. Many platforms have multimedia instructions that can be used to speed up the implementation. As another alternative, Bernstein demonstrated that, on most platforms, the floating point unit can be used to implement this class of universal hash functions far more efficiently than can be done in the integer unit. This is particularly true on the x86 platform where, in contrast to using the standard registers, two floating point multiples can be started in close proximity without introducing a pipeline stall. That is, the x86 can effectively perform two floating-point operations in parallel. The disadvantage of using floating-point registers is that the operands for the individual multiplies need to be small, so that the operations can be done without loss of precision. On the x86, Bernstein multiplies 24-bit values, allowing the sums of product terms to fit into double precision values with 53 bits of precision without loss of information. Bernstein details many ways to optimize this sort of calculation in [6].

As noted before, there are only two main differences between the structure of the polynomials of Bernstein's hash127 and CWC-HASH. The first is that Bernstein uses signed coefficients, whereas CWC-HASH uses unsigned coefficients; this should not have an impact on efficiency. The other difference is that Bernstein uses 32-bit coefficients, whereas CWC-HASH uses 96-bit coefficients. While both solutions average one multiplication per byte when using integer math, Bernstein's solution requires only .75 additions per byte, whereas CWC-HASH requires 1.42 additions per byte, nearly twice as many. Using 32-bit multiplies to build a 96x127 multiplier (assuming precomputation), CWC-HASH should therefore perform no worse than at half the speed of hash127. When using 24-bit floating point coefficients to build a multiply (without applying any non-obvious optimizations), hash127 requires 12 multiplies and 16 adds per 32-bit word. CWC can get by with 8 multiples per word and 12.67 additions per word. This is because a 96-bit coefficient fits exactly into four 24-bit values, meaning we can use a 6x4 multiply for every three words. With 32-bit coefficients, we need to use two 24-bit values to represent each coefficient, resulting in a single 6x2 multiply that needs to be performed for each word.

Gladman's C implementation of CWC-HASH uses floating point arithmetic, but uses Horner's rule instead of performing precomputation to achieve extra speed. Nothing about the CWC hash indicates that it should run any worse than half the speed of hash127, if implemented in a similar manner, in assembly, and using the floating point registers and precomputation. This upper-bound paints an encouraging picture for CWC performance, because hash127 on a Pentium III runs around 4 cpb when implemented in assembly and using the floating point registers and precomputation. This indicates that a well-optimized software version of CWC-HASH should run no slower than 8 cycles per byte on the same machine.

Finally, it may be possible to further improve the performance of CWC-HASH. For example, literature from the gaming community [11] indicates that one can use both integer and floating point registers in parallel. Although we have not tested this approach, it seems reasonable to conclude that one might be able to interleave integer operations, and thereby obtain additional speedups.

Acknowledgements

We thank Peter Gutmann, David McGrew, Fabian Monrose, Avi Rubin, Adam Stubblefield, and David Wagner for their comments. Additionally, we thank Brian Gladman for helping to validate our test vectors and for working with us to obtain timing information. T. Kohno was supported by a National Defense Science and Engineering Fellowship.

References

[1] M. Bellare, R. Canetti, and H. Krawczyk. Keying hash functions for message authentication. In N. Koblitz, editor, *CRYPTO '96*, volume 1109 of *LNCS*, pages 1–15. Springer-Verlag, Aug. 1996. 420

[2] M. Bellare, A. Desai, E. Jokipii, and P. Rogaway. A concrete security treatment of symmetric encryption. In *Proc. of the 38th FOCS*, pages 394–403. IEEE Computer Society Press, 1997. 413

[3] M. Bellare, R. Guérin, and P. Rogaway. XOR MACs: New methods for message authentication using finite pseudorandom functions. In D. Coppersmith, editor, *CRYPTO '95*, volume 963 of *LNCS*, pages 15–28. Springer-Verlag, Aug. 1995. 420

[4] M. Bellare and C. Namprempre. Authenticated encryption: Relations among notions and analysis of the generic composition paradigm. In T. Okamoto, editor, *ASIACRYPT 2000*, volume 1976 of *LNCS*, pages 531–545. Springer-Verlag, Dec. 2000. 409, 411, 417

[5] M. Bellare, P. Rogaway, and D. Wagner. The EAX mode of operation. In W. Meier and B. Roy, editors, *FSE 2004*, LNCS. Springer-Verlag, 2004. 408, 409, 412, 418, 420

[6] D. Bernstein. Floating-point arithmetic and message authentication, 2000. Available at `http://cr.yp.to/papers.html##hash127`. 410, 412, 418, 422, 423

[7] J. Black and P. Rogaway. A block-cipher mode of operation for parallelizable message authentication. In L. Knudsen, editor, *EUROCRYPT 2002*, volume 2332 of *LNCS*. Springer-Verlag, 2002. 420

[8] J. Daemen and V. Rijmen. *The Design of Rijndael*. Springer-Verlag, 2002. 410, 413

[9] B. Gladman. AES and combined encryption/authentication modes, 2003. Available at `http://fp.gladman.plus.com/AES/index.htm`. 411, 421

[10] V. Gligor and P. Donescu. Fast encryption and authentication: XCBC encryption and XECB authentication modes. In M. Matsui, editor, *FSE 2001*, LNCS. Springer-Verlag, 2001. 408, 412

[11] C. Hecker. Perspective texture mapping, part V: It's about time. *Game Developer*, Apr. 1996. Available at `http://www.d6.com/users/checker/pdfs/gdmtex5.pdf`. 424

[12] C. Jutla. Encryption modes with almost free message integrity. In B. Pfitzmann, editor, *EUROCRYPT 2001*, volume 2045 of *LNCS*, pages 529–544. Springer-Verlag, May 2001. 408, 412

[13] J. Katz and M. Yung. Unforgeable encryption and chosen ciphertext secure modes of operation. In B. Schneier, editor, *FSE 2000*, volume 1978 of *LNCS*, pages 284–299. Springer-Verlag, Apr. 2000. 408, 411, 412

[14] T. Kohno, J. Viega, and D. Whiting. CWC: A high-performance conventional authenticated encryption mode, 2003. Full version of this paper, available at `http://eprint.iacr.org/2003/106/`. 415, 419, 420

[15] H. Krawczyk. The order of encryption and authentication for protecting communications (or: How secure is SSL?). In J. Kilian, editor, *CRYPTO 2001*, volume 2139 of *LNCS*, pages 310–331. Springer-Verlag, Aug. 2001. 409, 411

[16] H. Lipmaa. AES/Rijndael: speed, 2003. Available at `http://www.tcs.hut.fi/~helger/aes/rijndael.html`. 422

[17] M. Luby and C. Rackoff. How to construct pseudorandom permutations from pseudorandom functions. *SIAM J. Computation*, 17(2), Apr. 1988. 413

[18] D. McGrew. Integer counter mode, Oct. 2002. Available at `http://www.ietf.org/internet-drafts/draft-irtf-cfrg-icm-01.txt`. 419

[19] D. McGrew. The truncated multi-modular hash function (TMMH), version two, Oct. 2002. Available at `http://www.ietf.org/internet-drafts/draft-irtf-cfrg-tmmh-00.txt`. 412

[20] D. McGrew. The universal security transform, Oct. 2002. Available at http://www.ietf.org/internet-drafts/draft-irtf-cfrg-ust-01.txt. 412

[21] D. McGrew and J. Viega. Galois/counter mode. Submission to NIST. Available at http://csrc.nist.gov/CryptoToolkit/modes/proposedmodes/, 2004. 412

[22] W. Nevelsteen and B. Preneel. In J. Stern, editor, *EUROCRYPT '99*, volume 1592 of *LNCS*, pages 24–41. Springer-Verlag, 1999. 412

[23] P. Rogaway. Authenticated encryption with associated data. In *Proc. of the 9th CCS*, Nov. 2002. 408, 411, 412, 415, 416, 417

[24] P. Rogaway, M. Bellare, J. Black, and T. Krovetz. OCB: A block-cipher mode of operation for efficient authenticated encryption. In *Proc. of the 8th CCS*, pages 196–205. ACM Press, 2001. 408, 412, 415, 417

[25] P. Rogaway and D. Wagner. A critique of CCM, Apr. 2003. Available at http://eprint.iacr.org/2003/070/. 412

[26] V. Shoup. On fast and provably secure message authentication based on universal hashing. In N. Koblitz, editor, *CRYPTO '96*, volume 1109 of *LNCS*, pages 313–328. Springer-Verlag, Aug. 1996. 412

[27] M. Wegman and L. Carter. New hash functions and their use in authentication and set equality. *Journal of Computer and System Sciences*, 22:265–279, 1981. 409, 412, 417, 418

[28] D. Whiting, N. Ferguson, and R. Housley. Counter with CBC-MAC (CCM). Submission to NIST. Available at http://csrc.nist.gov/CryptoToolkit/modes/proposedmodes/, 2002. 408, 409, 412, 418

New Security Proofs for the 3GPP Confidentiality and Integrity Algorithms

Tetsu Iwata[1] and Tadayoshi Kohno[2]

[1] Dept. of Computer and Information Sciences,
Ibaraki University, 4–12–1 Nakanarusawa, Hitachi, Ibaraki 316-8511, Japan
iwata@cis.ibaraki.ac.jp

[2] Dept. of Computer Science and Engineering,
University of California at San Diego
9500 Gilman Drive, La Jolla, California 92093, USA
tkohno@cs.ucsd.edu

Abstract. This paper analyses the 3GPP confidentiality and integrity schemes adopted by Universal Mobile Telecommunication System, an emerging standard for third generation wireless communications. The schemes, known as $f8$ and $f9$, are based on the block cipher KASUMI. Although previous works claim security proofs for $f8$ and $f9'$, where $f9'$ is a generalized versions of $f9$, it was recently shown that these proofs are incorrect. Moreover, Iwata and Kurosawa (2003) showed that it is *impossible* to prove $f8$ and $f9'$ secure under the standard PRP assumption on the underlying block cipher. We address this issue here, showing that it is possible to prove $f8'$ and $f9'$ secure if we make the assumption that the underlying block cipher is a secure PRP-RKA against a certain class of related-key attacks; here $f8'$ is a generalized version of $f8$. Our results clarify the assumptions necessary in order for $f8$ and $f9$ to be secure and, since no related-key attacks are known against the full eight rounds of KASUMI, lead us to believe that the confidentiality and integrity mechanisms used in real 3GPP applications are secure.

1 Introduction

Background. Within the security architecture of the 3rd Generation Partnership Project (3GPP) system there are two standardized constructions: A confidentiality scheme $f8$, and an integrity scheme $f9$ [1]. 3GPP is the body standardizing the next generation of mobile telephony. Both $f8$ and $f9$ are modes of operations based on the block cipher KASUMI [2]. $f8$ is a symmetric encryption scheme which is a variant of the Output Feedback (OFB) mode with full feedback, and $f9$ is a Message Authentication Code (MAC) which is a variant of the CBC MAC.

Provable Security. Provable security is a standard security goal for block cipher modes of operations. Indeed, many of the block cipher modes of operations

B. Roy and W. Meier (Eds.): FSE 2004, LNCS 3017, pp. 427–445, 2004.

are provably secure assuming that the underlying block cipher is a secure pseudorandom permutation, or a super-pseudorandom permutation [21]. For example, we have: CTR mode [3] and CBC encryption mode [3] for symmetric encryption schemes, PMAC [8] and OMAC [14] for message authentication codes, and IAPM [17], OCB mode [22], CCM mode [23, 16], EAX mode [6] and CWC mode [20] for authenticated encryption schemes.

Therefore, it is natural to ask whether $f8$ and $f9$ are provably secure if the underlying block cipher is a secure pseudorandom permutation. Making this assumption, it was claimed that $f8$ is a secure symmetric encryption scheme in the sense of left-or-right indistinguishability [18] and that $f9'$ is a secure MAC [12], where $f9'$ is a generalized version of $f9$. However, these claims were disproven [15]. One of the remarkable aspects of $f8$ and $f9$ is the use of a nonzero constant called a "key modifier," or KM. In the $f8$ and $f9$ schemes, KASUMI is keyed with K and $K \oplus \mathrm{KM}$. The paper [15] constructs a secure pseudorandom permutation F with the following property: For any key K, the encryption function with key K is the decryption function with $K \oplus \mathrm{KM}$. That is, $F_K(\cdot) = F^{-1}_{K \oplus \mathrm{KM}}(\cdot)$. Then it was shown that $f8$ and $f9'$ are insecure if F is used as the underlying block cipher. This result shows that it is *impossible* to prove the security of $f8$ and $f9'$ even if the underlying block cipher is a secure pseudorandom permutation.

Our Contribution. Given the results in [15], it is logical to ask if there are assumptions under which $f8$ and $f9$ are actually secure and, if so, what those assumptions are. The answers to these questions would give us greater insights into the security of these two modes. Because of the constructions' use of keys related by fixed xor differences, the natural conjecture is that if the constructions are actually secure, then the minimum assumption on the block cipher must be that the block cipher is secure against some class of xor-restricted related-key attacks, as introduced in [7] and formalized in [5].

We prove that the above hypotheses are in fact correct and, in doing so, we clarify what assumptions are actually necessary in order for the $f8$ and $f9$ modes to be secure. In more detail, we first consider a generalized version of $f8$, which we call $f8'$. $f8'$ is a nonce-based symmetric encryption scheme, and is the natural nonce-based extension of the original $f8$. We then show that $f8'$ is a secure nonce-based deterministic symmetric encryption mode in the sense of indistinguishability from random strings if the underlying block cipher is secure against related-key attacks in which an adversary is able to obtain chosen-plaintext samples of the underlying block cipher using two keys related by a fixed known xor difference.

We next consider a generalized version of $f9$, which we call $f9'$. $f9'$ is a deterministic MAC, and is a natural extension of $f9$ that gives the user, or adversary, more liberty in controlling the input to the underlying CBC MAC core. We then show that $f9'$ is a secure pseudorandom function, which provably implies a secure MAC, if the underlying block cipher resists related-key attacks in which an adversary is able to obtain chosen-plaintext samples of the underlying block cipher using two keys related by a fixed known xor difference.

Since both $f8'$ and $f9'$ are generalized versions of $f8$ and $f9$, and, since the best known related-key attack against KASUMI breaks only six out of eight rounds [9], our results show that unless a novel new attack is discovered against KASUMI, the 3GPP confidentiality and integrity mechanisms are actually secure. We view this as an important practical corollary of our research since the 3GPP constructions are destined for use in future mobile telephony applications. Additionally, because our proofs explicitly quantify what properties of the underlying block cipher are necessary in order for $f8'$ and $f9'$ to be secure, our results can help others decide whether it is safe to instantiate the generalized 3GPP modes with block ciphers other than KASUMI. Of course, because the assumptions we make are stronger than the standard pseudorandomness assumptions, as proven necessary in [15], unless there is a significant reason to do otherwise, we suggest that future systems use more conventional modes such as CTR mode and OMAC.

For our proofs, rather than trying to find and re-use correct portions of the analyses in [18] and [12], we chose instead to prove the security of $f8'$ and $f9'$ directly. We did this in order to ensure the correctness of our results and to avoid presenting proofs covered with patches. We discuss some of problems with the previous analyses in more detail in Appendices A.1 and B.1.

Related Works. Initial security evaluation of KASUMI, $f8$ and $f9$ can be found in [11]. Knudsen and Mitchell analyzed the security of $f9'$ against forgery and key recovery attacks [19].

2 Preliminaries

Notation. If x is a string then $|x|$ denotes its length in bits. If x and y are two equal-length strings, then $x \oplus y$ denotes the xor of x and y. If x and y are strings, then $x \| y$ denotes their concatenation. Let $x \leftarrow y$ denote the assignment of y to x. If X is a set, let $x \stackrel{R}{\leftarrow} X$ denote the process of uniformly selecting at random an element from X and assigning it to x. If $F : \{0,1\}^k \times \{0,1\}^n \rightarrow \{0,1\}^m$ is a family of functions from $\{0,1\}^n$ to $\{0,1\}^m$ indexed by keys $\{0,1\}^k$, then we use the notation $F_K(D)$ as shorthand for $F(K,D)$. We say F is a family of permutations, i.e., a block cipher, if $n = m$ and $F_K(\cdot)$ is a permutation on $\{0,1\}^n$ for each $K \in \{0,1\}^k$. Let $\mathrm{Rand}(n,m)$ denote the set of all functions from $\{0,1\}^n$ to $\{0,1\}^m$. When we refer to the time of an algorithm or experiment in the provable security sections of this paper, we include the size of the code (in some fixed encoding). There is also an implicit big-$\mathcal{O}$ surrounding all such time references.

PRP-RKAs. The PRP-RKA notion was introduced in [5], and is based on the pseudorandomness notions introduced in [21] and later made concrete in [4]. The notion was designed to model block ciphers secure against related-key attacks [7].

Let $\mathrm{Perm}(k,n)$ denote the set of all block ciphers with domain $\{0,1\}^n$ and keys $\{0,1\}^k$. The notation $G \stackrel{R}{\leftarrow} \mathrm{Perm}(k,n)$ thus corresponds to selecting a random block-cipher, and comes down to defining G via

$$\text{For each } K \in \{0,1\}^k \text{ do: } \ G_K \stackrel{R}{\leftarrow} \mathrm{Perm}(n) \ ,$$

where $\mathrm{Perm}(n)$ is the set of all permutations on $\{0,1\}^n$.

Given a family of functions $F : \{0,1\}^k \times \{0,1\}^n \to \{0,1\}^n$ and a key $K \in \{0,1\}^k$, we define the related-key oracle $F_{\mathrm{RK}(\cdot,K)}(\cdot)$ as an oracle that takes two arguments, a function $\phi : \{0,1\}^k \to \{0,1\}^k$ and an element $M \in \{0,1\}^n$, and that returns $F_{\phi(K)}(M)$, or the encipherment of M under the key $\phi(K)$. In this context, we shall refer to ϕ as a related-key-deriving (RKD) function.

The PRP-RKA notion, which we now describe, is parameterized by a set of RKD functions Φ. Let $E : \{0,1\}^k \times \{0,1\}^n \to \{0,1\}^n$ be a family of functions and let Φ be a set of RKD functions over $\{0,1\}^k$. Let $\mathcal{A}$ be an adversary with access to a related-key oracle, and restricted to queries of the form (ϕ, x) in which $\phi \in \Phi$ and $x \in \{0,1\}^n$, and let $\mathcal{A}$ return a bit. Then

$$\begin{aligned}\mathbf{Adv}^{\text{prp-rka}}_{\Phi,E}(\mathcal{A}) \stackrel{\text{def}}{=} \Big| & \Pr(K \stackrel{R}{\leftarrow} \{0,1\}^k : \mathcal{A}^{E_{\mathrm{RK}(\cdot,K)}(\cdot)} = 1) \\ & - \Pr(K \stackrel{R}{\leftarrow} \{0,1\}^k ; G \stackrel{R}{\leftarrow} \mathrm{Perm}(k,n) : \mathcal{A}^{G_{\mathrm{RK}(\cdot,K)}(\cdot)} = 1) \Big|\end{aligned}$$

is defined as the *PRP-RKA-advantage* of $\mathcal{A}$ in a Φ-restricted related-key attack (RKA) on E. Intuitively, we say that E is a *secure PRP-RKA under Φ-restricted related-key attacks* if the PRP-RKA-advantage of all adversaries using reasonable resources is small.

In this work we are primarily interested in keys that are related by some xor difference. For any $\Delta \in \{0,1\}^k$ we let $\mathrm{XOR}_\Delta : \{0,1\}^k \to \{0,1\}^k$ denote the function which on input $K \in \{0,1\}^k$ returns $K \oplus \Delta$. We define $\Phi_k^{\oplus}$ as $\Phi_k^{\oplus} \stackrel{\text{def}}{=} \{ \mathrm{XOR}_\Delta : \Delta \in \{0,1\}^k \}$. We briefly remark that modern block ciphers, e.g., AES [10], are designed to be secure PRP-RKAs under $\Phi_k^{\oplus}$-restricted related-key attacks. Additionally, the best-known $\Phi_k^{\oplus}$-restricted related-key attack against the block cipher KASUMI, which was designed for use with the 3GPP modes, only breaks six out of eight rounds [9].

3 Specifications of $f8$, $f8'$, $f9$ and $f9'$

3.1 3GPP Confidentiality Algorithm $f8$ [1]

$f8$ is a symmetric encryption scheme standardized by 3GPP[1]. It uses a block cipher $\mathrm{KASUMI} : \{0,1\}^{128} \times \{0,1\}^{64} \to \{0,1\}^{64}$ as the underlying primitive. The $f8$ key generation algorithm returns a random 128-bit key K. The $f8$ encryption algorithm takes a 128-bit key K, a 32-bit counter COUNT, a 5-bit radio bearer

[1] The original specification [1] refers $f8$ as a symmetric synchronous stream cipher. The specification presented here is fully compatible with the original one.

identifier BEARER, a 1-bit direction identifier DIRECTION, and a message $M \in \{0,1\}^*$ to return a ciphertext C, which is the same length as M. Also, it uses a 128-bit constant $\mathrm{KM} = (01)^{64}$ (or 0x55...55 in hexadecimal) called the key modifier. In more detail, the encryption algorithm is defined as follows:

Algorithm $\mathsf{f8\text{-}Encrypt}_K(\mathrm{COUNT}, \mathrm{BEARER}, \mathrm{DIRECTION}, M)$
- $m \leftarrow \lceil |M|/64 \rceil$
- $Y[0] \leftarrow 0^{64}$
- $A \leftarrow \mathrm{COUNT} \| \mathrm{BEARER} \| \mathrm{DIRECTION} \| 0^{26}$
- $A \leftarrow \mathrm{KASUMI}_{K \oplus \mathrm{KM}}(A)$
- For $i = 1$ to m do:
  - $X[i] \leftarrow A \oplus [i-1]_{64} \oplus Y[i-1]$
  - $Y[i] \leftarrow \mathrm{KASUMI}_K(X[i])$
- $C \leftarrow M \oplus$ (the leftmost $|M|$ bits of $Y[1] \| \cdots \| Y[m]$)
- Return C

In the above description, $[i-1]_{64}$ denotes the 64-bit binary representation of $i-1$. The decryption algorithm, which takes COUNT, BEARER, DIRECTION, and a ciphertext C as input and returns a plaintext M, is defined in the natural way.

Since we analyze and prove results about a variant of $f8$ whose encryption algorithm takes a nonce as input in lieu of COUNT, BEARER, and DIRECTION, we do not describe the specifics of how COUNT, BEARER, and DIRECTION are used in real 3GPP applications. We do note that 3GPP applications will never invoke the $f8$ encryption algorithm twice with the same (COUNT, BEARER, DIRECTION) triple, which means that our nonce-based variant is appropriate.

3.2 A Generalized Version of $f8$: $f8'$

$f8'$ is a nonce-based deterministic symmetric encryption scheme, which is a generalized (and weakened) version of $f8$. It uses a block cipher $E : \{0,1\}^k \times \{0,1\}^n \to \{0,1\}^n$ as the underlying primitive. Let $f8'[E, \Delta]$ be $f8'$, where E is used as the underlying primitive and Δ is a non-zero k-bit key modifier. The $f8'$ key generation algorithm returns a random k-bit key K. The $f8'[E, \Delta]$ encryption algorithm, which we call $\mathsf{f8'\text{-}Encrypt}$, takes an n-bit nonce N instead of COUNT, BEARER and DIRECTION. That is, the encryption algorithm takes a k-bit key K, an n-bit nonce N, and a message $M \in \{0,1\}^*$ to return a ciphertext C, which is the same length as M. Then the encryption algorithm proceeds as follows:

Algorithm $\mathsf{f8'\text{-}Encrypt}_K(N, M)$
$m \leftarrow \lceil |M|/n \rceil$
$Y[0] \leftarrow 0^n$
$A \leftarrow N$
$A \leftarrow E_{K \oplus \Delta}(A)$
For $i = 1$ to m do:
$X[i] \leftarrow A \oplus [i-1]_n \oplus Y[i-1]$
$Y[i] \leftarrow E_K(X[i])$
$C \leftarrow M \oplus$ (the leftmost $|M|$ bits of $Y[1] \| \cdots \| Y[m]$)
Return C

In the above description, $[i-1]_n$ denotes n-bit binary representation of $i-1$. Decryption is done in an obvious way.

Notice that we treat COUNT, BEARER and DIRECTION as a nonce. That is, as we will define in Section 4, we allow the adversary to choose these values. Consequently, $f8'$ can be considered a weakened version of $f8$ since it gives the an adversary the ability to control the entire initial value of A, rather than only a subset of the bits as would be the case for an adversary attacking $f8$.

3.3 3GPP Integrity Algorithm $f9$ [1]

$f9$ is a message authentication code standardized by 3GPP. It uses KASUMI as the underlying primitive. The $f9$ key generation algorithm returns a random 128-bit key K. The $f9$ tagging algorithm takes a 128-bit key K, a 32-bit counter COUNT, a 32-bit random number FRESH, a 1-bit direction identifier DIRECTION, and a message $M \in \{0,1\}^*$ and returns a 32-bit tag T. It uses a 128-bit constant $\mathrm{KM} = (10)^{64}$ (or `0xAA...AA` in hexadecimal), called the key modifier.

Let $M = M[1] \| \cdots \| M[m]$ be a message, where each $M[i]$ $(1 \le i \le m-1)$ is 64 bits. The last block $M[m]$ may have fewer than 64 bits. We define $\mathsf{pad}_{64}(\mathrm{COUNT}, \mathrm{FRESH}, \mathrm{DIRECTION}, M)$ as follows: It concatenates COUNT, FRESH, M and DIRECTION, and then appends a single "1" bit, followed by between 0 and 63 "0" bits so that the total length is a multiple of 64 bits. More precisely,

$$\mathsf{pad}_{64}(\mathrm{COUNT}, \mathrm{FRESH}, \mathrm{DIRECTION}, M) = \mathrm{COUNT} \| \mathrm{FRESH} \| M \| \mathrm{DIRECTION} \| 1 \| 0^{63-(|M|+1 \bmod 64)} .$$

Then the tagging algorithm is defined as follows:

Algorithm $\text{f9-Tag}_K(\text{COUNT}, \text{FRESH}, \text{DIRECTION}, M)$
$M \leftarrow \text{pad}_{64}(\text{COUNT}, \text{FRESH}, \text{DIRECTION}, M)$
Break M into 64-bit blocks $M[1] \| \cdots \| M[m]$
$Y[0] \leftarrow 0^{64}$
For $i = 1$ to m do:
$X[i] \leftarrow M[i] \oplus Y[i-1]$
$Y[i] \leftarrow \text{KASUMI}_K(X[i])$
$T \leftarrow \text{KASUMI}_{K \oplus \text{KM}}(Y[1] \oplus \cdots \oplus Y[m])$
$T \leftarrow$ the leftmost 32 bits of T
Return T

The $f9$ verification algorithm is defined in the natural way.

As with $f8$, since we analyze and prove the security of a generalized version of $f9$, we do not describe how COUNT, FRESH, and DIRECTION are used in real 3GPP applications.

3.4 A Generalized Version of $f9$: $f9'$ [12, 19, 15]

The message authentication code $f9'$ is a generalized (and weakened) version of $f9$ that gives the user (or adversary) almost complete control over the input the underlying CBC MAC core. It uses a block cipher $E : \{0,1\}^k \times \{0,1\}^n \to \{0,1\}^n$ as the underlying primitive. Let $f9'[E, \Delta, l]$ be $f9'$, where E is used as the underlying block cipher, Δ is a non-zero k-bit key modifier, and the tag length is l, where $1 \le l \le n$. The key generation algorithm returns a random k-bit key K. The tagging algorithm, which we call f9′-Tag, takes a k-bit key K and a message $M \in \{0,1\}^*$ as input and returns an l-bit tag T.

Let $M = M[1] \| \cdots \| M[m]$ be a message, where each $M[i]$ $(1 \le i \le m-1)$ is n bits. The last block $M[m]$ may have fewer than n bits. In $f9'$, we use pad'_n instead of pad_{64}. $\text{pad}'_n(M)$ works as follows: It simply appends a single "1" bit, followed by between 0 and $n-1$ "0" bits so that the total length is a multiple of n bits. More precisely,

$$\text{pad}'_n(M) = M \| 1 \| 0^{n-1-(|M| \bmod n)} \enspace .$$

Thus, we simply ignore COUNT, FRESH, and DIRECTION. Equivalently, we consider COUNT, FRESH, and DIRECTION as a part of the message. The rest of the tagging algorithm is the same as with $f9$. In pseudocode,

Algorithm $\text{f9}'\text{-Tag}_K(M)$
$M \leftarrow \text{pad}'_n(M)$
Break M into n-bit blocks $M[1] \| \cdots \| M[m]$
$Y[0] \leftarrow 0^n$
For $i = 1$ to m do:
$X[i] \leftarrow M[i] \oplus Y[i-1]$
$Y[i] \leftarrow E_K(X[i])$
$T \leftarrow E_{K \oplus \Delta}(Y[1] \oplus \cdots \oplus Y[m])$
$T \leftarrow$ the leftmost l bits of T
Return T

The verification algorithm is defined in the natural way.

As we will define in Section 5, our adversary is allowed to choose COUNT, FRESH, and DIRECTION since $f9'$ treats them as a part of the message. In this sense, $f9'$ can be considered as a weakened version of $f9$.

4 Security of $f8'$

Definitions. Before proving the security of $f8'$, we must first formally define what we mean by a nonce-based encryption scheme, and what it means for such an encryption scheme to be secure.

Mathematically, a nonce-based symmetric encryption scheme $\mathcal{SE} = (\mathcal{K}, \mathcal{E}, \mathcal{D})$ consists of three algorithms and is defined for some nonce length n. The randomized key generation algorithm $\mathcal{K}$ takes no input and returns a random key K. The stateless and deterministic encryption algorithm takes a key K, an nonce $N \in \{0,1\}^n$, and a message $M \in \{0,1\}^*$ as input and returns a ciphertext C such that $|C| = |M|$; we write $C \leftarrow \mathcal{E}_K(N, M)$. The stateless and deterministic decryption algorithm takes a key K, a nonce $N \in \{0,1\}^n$, and a ciphertext $C \in \{0,1\}^*$ as input and returns a message M such that $|M| = |C|$; we write $M \leftarrow \mathcal{D}_K(N, C)$. For consistency, we require that for all keys K, nonces N, and messages M, $\mathcal{D}_K(N, \mathcal{E}_K(N, M)) = M$.

We adopt the strong notion of privacy for nonce-based encryption schemes from [22]. This notion, which we call indistinguishability from random strings, provably implies the more standard notions given in [3]. Let $\$(\cdot,\cdot)$ denote an oracle that on input a pair of strings (N, M) returns a random string of length $|M|$. If $\mathcal{A}$ is an adversary with access to an oracle, then

$$\mathbf{Adv}^{\mathrm{priv}}_{\mathcal{SE}}(\mathcal{A}) \stackrel{\mathrm{def}}{=} \left| \Pr(K \stackrel{R}{\leftarrow} \mathcal{K} : \mathcal{A}^{\mathcal{E}_K(\cdot,\cdot)} = 1) - \Pr(\mathcal{A}^{\$(\cdot,\cdot)} = 1) \right|$$

is defined as the *PRIV-advantage* of $\mathcal{A}$ in distinguishing the outputs of the encryption algorithm with a randomly selected key from random strings. We say that $\mathcal{A}$ is nonce-respecting if it never queries its oracle twice with the same nonce value. Intuitively, we say that an encryption scheme *preserves privacy under chosen-plaintext attacks* if the PRIV-advantage of all nonce-respecting adversaries $\mathcal{A}$ using reasonable resources is small.

Provable Security Results. Let $p8'[n]$ be a variant of $f8'$ that uses random functions on n-bits instead of E_K and $E_{K \oplus \Delta}$. Specifically, the key generation algorithm for $p8'[n]$ returns two randomly selected functions R_1, R_2 from $\mathrm{Rand}(n, n)$. The encryption algorithm for $p8'[n]$, $\mathsf{p8'\text{-}Encrypt}$, takes R_1 and R_2 as "keys" and uses them instead of E_K and $E_{K \oplus \Delta}$. The decryption algorithm is defined in the natural way.

We first upper-bound the advantage of an adversary in breaking the privacy of $p8'[n]$. Let (N_i, M_i) denote a privacy adversary's i-th oracle query. If the adversary makes exactly q oracle queries, then we define the total number of blocks for the adversary's queries as $\sigma = \sum_{i=1}^{q} \lceil |M_i|/n \rceil$.

Lemma 4.1. *Let $p8'[n]$ be as described above and let $\mathcal{A}$ be a nonce-respecting privacy adversary which asks at most q queries totaling at most σ blocks. Then*

$$\mathbf{Adv}^{\mathrm{priv}}_{p8'[n]}(\mathcal{A}) \leq \frac{\sigma^2}{2^n} \ . \tag{1}$$

A proof sketch is given in Appendix A, and a proof is given in the full version of this paper [13].

We now present our main result for $f8'$ (Theorem 4.1 below). At a high level, our theorem shows that if a block cipher E is secure against Φ-restricted related key attacks, where Φ is a small subset of $\Phi_k^{\oplus}$, then the construction $f8'[E, \Delta]$ based on E will be a provably secure encryption scheme. In more detail, our theorem states that given any adversary $\mathcal{A}$ attacking the privacy of $f8'[E, \Delta]$ and making at most q oracle queries totaling at most σ blocks, we can construct a Φ-restricted PRP-RKA adversary $\mathcal{B}$ attacking E such that $\mathcal{B}$ uses similar resources as $\mathcal{A}$ and $\mathcal{B}$ has advantage $\mathbf{Adv}^{\mathrm{prp\text{-}rka}}_{\Phi,E}(\mathcal{B}) \geq \mathbf{Adv}^{\mathrm{priv}}_{f8'[E,\Delta]}(\mathcal{A}) - (3\sigma^2 + q^2)/2^{n+1}$. If we assume that E is secure against Φ-restricted related-key attacks and that $\mathcal{A}$ (and therefore $\mathcal{B}$) uses reasonable resources, then $\mathbf{Adv}^{\mathrm{prp\text{-}rka}}_{\Phi,E}(\mathcal{B})$ must be small by definition, and thus $\mathbf{Adv}^{\mathrm{priv}}_{f8'[E,\Delta]}(\mathcal{A})$ must also be small. This means that under these assumptions on E, $f8'[E, \Delta]$ is provably secure.

Since many block ciphers, including AES and KASUMI, are believed to resist $\Phi_k^{\oplus}$-restricted related-key attacks, and since Φ is a small subset of $\Phi_k^{\oplus}$, this theorem means that $f8'$ constructions built from these block ciphers will be provably secure. Additionally, because Φ is a small subset of $\Phi_k^{\oplus}$, the $f8'$ construction actually requires a much weaker assumption on the underlying block cipher than resistance to the full class of $\Phi_k^{\oplus}$-restricted related-key attacks, meaning that it is more likely for the underlying block cipher to resist Φ-restricted related-key attacks than $\Phi_k^{\oplus}$-restricted related-key attacks. Of course, our results also suggest that if a block cipher is known to be insecure under Φ-restricted related-key attacks, that block cipher should not be used in the $f8'$ construction.

Since $f8'$ is a weakened version of the KASUMI-based $f8$ encryption scheme, and since KASUMI is currently believed to resist $\Phi_k^{\oplus}$-restricted related-key attacks, our result shows that $f8$ as designed for use in the 3GPP protocols is secure.

Our main theorem statement for $f8'$ is given below.

Theorem 4.1 (Main Theorem for $f8'$). *Let $E : \{0,1\}^k \times \{0,1\}^n \to \{0,1\}^n$ be a block cipher and let Δ be a non-zero k-bit constant. Let $f8'[E, \Delta]$ be as described in Sec. 3.2. Let* id *be the identity function on $\{0,1\}^k$ and let $\Phi = \{\mathrm{id}, \mathrm{XOR}_{\Delta}\} \subseteq \Phi_k^{\oplus}$ be a set of RKD functions over $\{0,1\}^k$. If $\mathcal{A}$ is a nonce-respecting privacy adversary which asks at most q queries totaling at most σ blocks, then we can construct a Φ-restricted PRP-RKA adversary $\mathcal{B}$ against E such that*

$$\mathbf{Adv}^{\mathrm{priv}}_{f8'[E,\Delta]}(\mathcal{A}) \leq \frac{3\sigma^2 + q^2}{2^{n+1}} + \mathbf{Adv}^{\mathrm{prp\text{-}rka}}_{\Phi,E}(\mathcal{B}) \ . \tag{2}$$

Furthermore, $\mathcal{B}$ makes at most $\sigma + q$ oracle queries and uses the same time as $\mathcal{A}$.

Proof. Let f8′-Encrypt denote the encryption algorithm for $f8'[E,\Delta]$ and let p8′-Encrypt denote the encryption algorithm for $p8'[n]$. Expanding the definition of $\mathbf{Adv}^{\mathrm{priv}}_{f8'[E,\Delta]}(\mathcal{A})$, we get:

$$\begin{aligned}\mathbf{Adv}^{\mathrm{priv}}_{f8'[E,\Delta]}(\mathcal{A}) &= \left|\Pr(K \xleftarrow{R} \{0,1\}^k : \mathcal{A}^{\text{f8}'\text{-Encrypt}_K(\cdot,\cdot)} = 1) - \Pr(\mathcal{A}^{\$(\cdot,\cdot)} = 1)\right| \\ &\le \mathbf{Adv}^{\mathrm{priv}}_{p8'[n]}(\mathcal{A}) + \left|\Pr(K \xleftarrow{R} \{0,1\}^k : \mathcal{A}^{\text{f8}'\text{-Encrypt}_K(\cdot,\cdot)} = 1)\right. \\ &\qquad \left. - \Pr(R_1,R_2 \xleftarrow{R} \mathrm{Rand}(n,n) : \mathcal{A}^{\text{p8}'\text{-Encrypt}_{R_1,R_2}(\cdot,\cdot)} = 1)\right| .\end{aligned}$$

Applying Lemma 4.1 we get

$$\begin{aligned}\mathbf{Adv}^{\mathrm{priv}}_{f8'[E,\Delta]}(\mathcal{A}) \le \frac{\sigma^2}{2^n} &+ \left|\Pr(K \xleftarrow{R} \{0,1\}^k : \mathcal{A}^{\text{f8}'\text{-Encrypt}_K(\cdot,\cdot)} = 1)\right. \\ &\left. - \Pr(R_1,R_2 \xleftarrow{R} \mathrm{Rand}(n,n) : \mathcal{A}^{\text{p8}'\text{-Encrypt}_{R_1,R_2}(\cdot,\cdot)} = 1)\right| .\end{aligned}$$

Let $\mathcal{B}$ be a Φ-restricted related-key adversary against E that runs $\mathcal{A}$ and that returns the same bit that $\mathcal{A}$ returns. Let $F_{\mathrm{RK}(\cdot,K)}(\cdot)$ denote $\mathcal{B}$'s related-key oracle. When $\mathcal{A}$ makes an oracle query (N,M) to its oracle, $\mathcal{B}$ essentially computes the f8′-Encrypt algorithm, except that it uses its related-key oracle in place of E_K and $E_{K\oplus\Delta}$. In pseudocode,

```
Algorithm B^{F_RK(·,K)(·)}
Run A, replying to A's oracle queries (N, M) as follows:
    m ← ⌈|M|/n⌉
    Y[0] ← 0^n
    A ← N
    A ← F_RK(XOR_Δ,K)(A)
    For i = 1 to m do:
        X[i] ← A ⊕ [i − 1]_n ⊕ Y[i − 1]
        Y[i] ← F_RK(id,K)(X[i])
    C ← M ⊕ (the leftmost |M| bits of Y[1]‖ ··· ‖Y[m])
    Return C to A
When A outputs b:
    output b
```

We now observe that

$$\Pr(K \xleftarrow{R} \{0,1\}^k : \mathcal{A}^{\text{f8}'\text{-Encrypt}_K(\cdot,\cdot)} = 1) = \Pr(K \xleftarrow{R} \{0,1\}^k : \mathcal{B}^{E_{\mathrm{RK}(\cdot,K)}(\cdot)} = 1)$$

since $\mathcal{B}$, when given related-key oracle access to E with a randomly selected key K, responds to $\mathcal{A}$ exactly as the f8′-Encrypt$_K(\cdot,\cdot)$ oracle would respond with a randomly selected key K.

Let $\mathrm{Rand}(k,n,n)$ denote the set of all functions from $\{0,1\}^k \times \{0,1\}^n$ to $\{0,1\}^n$. Then the equation

$$\begin{aligned}\Pr(R_1,R_2 \xleftarrow{R} \mathrm{Rand}(n,n) &: \mathcal{A}^{\text{p8}'\text{-Encrypt}_{R_1,R_2}(\cdot,\cdot)} = 1) \\ &= \Pr(K \xleftarrow{R} \{0,1\}^k \,;\, G \xleftarrow{R} \mathrm{Rand}(k,n,n) : \mathcal{B}^{G_{\mathrm{RK}(\cdot,K)}(\cdot)} = 1)\end{aligned}$$

follows from the fact that when G is randomly selected from $\mathrm{Rand}(k,n,n)$, regardless of the key K and since we assume $\Delta \neq 0^k$, G_K and $G_{K\oplus\Delta}$ are both randomly selected functions from $\mathrm{Rand}(n,n)$.

Combining the above equations, we have that

$$\begin{aligned}
\mathbf{Adv}^{\mathrm{priv}}_{f8'[E,\Delta]}(\mathcal{A}) \leq \frac{\sigma^2}{2^n} + \Big| &\Pr(K \xleftarrow{R} \{0,1\}^k : \mathcal{B}^{E_{\mathrm{RK}(\cdot,K)}(\cdot)} = 1) \\
&- \Pr(K \xleftarrow{R} \{0,1\}^k \,;\, G \xleftarrow{R} \mathrm{Rand}(k,n,n) : \mathcal{B}^{G_{\mathrm{RK}(\cdot,K)}(\cdot)} = 1)\Big| \\
= \frac{\sigma^2}{2^n} + \Big| &\Pr(K \xleftarrow{R} \{0,1\}^k : \mathcal{B}^{E_{\mathrm{RK}(\cdot,K)}(\cdot)} = 1) \\
&- \Pr(K \xleftarrow{R} \{0,1\}^k \,;\, H \xleftarrow{R} \mathrm{Perm}(k,n) : \mathcal{B}^{H_{\mathrm{RK}(\cdot,K)}(\cdot)} = 1) \\
&+ \Pr(K \xleftarrow{R} \{0,1\}^k \,;\, H \xleftarrow{R} \mathrm{Perm}(k,n) : \mathcal{B}^{H_{\mathrm{RK}(\cdot,K)}(\cdot)} = 1) \\
&- \Pr(K \xleftarrow{R} \{0,1\}^k \,;\, G \xleftarrow{R} \mathrm{Rand}(k,n,n) : \mathcal{B}^{G_{\mathrm{RK}(\cdot,K)}(\cdot)} = 1)\Big| \,.
\end{aligned}$$

Using the PRP-RKA definition and applying a variant of the PRF/PRP switching lemma from [5], we get

$$\mathbf{Adv}^{\mathrm{priv}}_{f8'[E,\Delta]}(\mathcal{A}) \leq \mathbf{Adv}^{\mathrm{prp\text{-}rka}}_{\Phi,E}(\mathcal{B}) + \frac{\sigma(\sigma-1)}{2^{n+1}} + \frac{q(q-1)}{2^{n+1}} + \frac{\sigma^2}{2^n} \,.$$

For the application of the PRF/PRP switching lemma, we note that $\mathcal{B}$ queries its related-key oracle with the RKD function id at most σ times and the RKD function XOR_Δ at most q times. Rearranging the above equation and simplifying gives (2), as desired.

5 Security of $f9'$

Definitions. Before proving the security of $f9'$, we must first formally define what we mean by a MAC, and what it means for a MAC to be secure.

Mathematically, a message authentication scheme or MAC $\mathcal{MA} = (\mathcal{K}, \mathcal{T}, \mathcal{V})$ consists of three algorithms and is defined for some tag length l. The randomized key generation algorithm $\mathcal{K}$ takes no input and returns a random key K. The stateless and deterministic tagging algorithm takes a key K and a message $M \in \{0,1\}^*$ as input and returns a tag $T \in \{0,1\}^l$; we write $T \leftarrow \mathcal{T}_K(M)$. The stateless and deterministic verification algorithm takes a key K, a message $M \in \{0,1\}^*$, and a candidate tag $T \in \{0,1\}^l$ as input and returns a bit b; we write $b \leftarrow \mathcal{V}_K(M,T)$. For consistency, we require that for all keys K and messages M, $\mathcal{V}_K(M, \mathcal{T}_K(M)) = 1$.

For security, we adopt a strong notion of security for MACs, namely pseudorandomness (PRF). In [4] it was proven that if a MAC is secure PRF, then it is also unforgeable. If $\mathcal{A}$ is an adversary with access to an oracle, then

$$\mathbf{Adv}^{\mathrm{prf}}_{\mathcal{MA}}(\mathcal{A}) \stackrel{\mathrm{def}}{=} \Big|\Pr(K \xleftarrow{R} \mathcal{K} : \mathcal{A}^{\mathcal{T}_K(\cdot)} = 1) - \Pr(g \xleftarrow{R} \mathrm{Rand}(*,l) : \mathcal{A}^{g(\cdot)} = 1)\Big|$$

is defined as the *PRF-advantage* of $\mathcal{A}$ in distinguishing the outputs of the tagging algorithm with a randomly selected key from the outputs of a random function with the same domain and range. Intuitively, we say that a message authentication code is *pseudorandom* or secure if the PRF-advantage of all adversaries $\mathcal{A}$ using reasonable resources is small.

Provable Security Results. Let $p9'[n]$ be a variant of $f9'$ that always outputs a full n-bit tag and that uses random functions on n-bits instead of E_K and $E_{K\oplus\Delta}$. Specifically, the key generation algorithm for $p9'[n]$ returns two randomly selected functions R_1, R_2 from $\mathrm{Rand}(n,n)$. The tagging algorithm for $p9'[n]$, $\mathsf{p9'}$-Tag, takes R_1 and R_2 as "keys" and uses them instead of E_K and $E_{K\oplus\Delta}$. The verification algorithm is defined in the natural way.

We first upper-bound the advantage of an adversary in attacking the pseudorandomness of $p9'[n]$. Let M_i denote an adversary's i-th oracle query. If an adversary makes exactly q oracle queries, then we define the total number of blocks for the adversary's queries as $\sigma = \sum_{i=1}^{q} \lceil |M_i|/n \rceil$.

Lemma 5.1. *Let $p9'[n]$ be as described above and let $\mathcal{A}$ be an adversary which asks at most q queries totaling at most σ blocks. Then*

$$\mathbf{Adv}^{\mathrm{prf}}_{p9'[n]}(\mathcal{A}) \le \frac{\sigma^2 + q^2}{2^{n+1}} \ . \tag{3}$$

A proof sketch is given in Appendix B, and a proof is given in the full version of this paper [13].

We now present our main result for $f9'$ (Theorem 5.1), which we interpret as follows: our theorem shows that if a block cipher E is secure against Φ-restricted related-key attacks, where Φ is a small subset of $\Phi_k^{\oplus}$, then the construction $f9'[E,\Delta,l]$ based on E will be a provably secure message authentication code. In more detail, we show that given any adversary $\mathcal{A}$ attacking $f9'[E,\Delta,l]$ and making at most q oracle queries totaling at most σ blocks, we can construct a Φ-restricted PRP-RKA adversary $\mathcal{B}$ against E such that $\mathcal{B}$ uses similar resources as $\mathcal{A}$ and $\mathcal{B}$ has advantage $\mathbf{Adv}^{\mathrm{prp\text{-}rka}}_{\Phi,E}(\mathcal{B}) \ge \mathbf{Adv}^{\mathrm{prf}}_{f9'[E,\Delta,l]}(\mathcal{A}) - (3q^2 + 2\sigma^2 + 2\sigma q)/2^{n+1}$. If we assume that E is secure against Φ-restricted related-key attacks and that $\mathcal{A}$ (and therefore $\mathcal{B}$) uses reasonable resources, then $\mathbf{Adv}^{\mathrm{prp\text{-}rka}}_{\Phi,E}(\mathcal{B})$ must be small by definition. Therefore $\mathbf{Adv}^{\mathrm{prf}}_{f9'[E,\Delta,l]}(\mathcal{A})$ must be small as well, proving that under these assumptions on E, $f9'[E,\Delta,l]$ is secure.

Since many block ciphers, including AES and KASUMI, are believed to resist $\Phi_k^{\oplus}$-restricted related-key attacks, and since Φ is a small subset of $\Phi_k^{\oplus}$, this theorem means that $f9'$ constructions built from these block ciphers will be provably secure. Furthermore, because $f9'$ is a weakened version of the KASUMI-based $f9$ message authentication code, our result shows that $f9$ as designed for use in the 3GPP protocols is secure.

The precise theorem statement is as follows:

Theorem 5.1 (Main Theorem for $f9'$). *Let $E : \{0,1\}^k \times \{0,1\}^n \to \{0,1\}^n$ be a block cipher, let Δ be a non-zero k-bit constant, and let l, $1 \le l \le n$, be* *a constant. Let $f9'[E, \Delta, l]$ be as described in Sec. 3.4. Let* id *be the identity function on $\{0,1\}^k$ and let $\Phi = \{\mathrm{id}, \mathrm{XOR}_\Delta\} \subseteq \Phi_k^\oplus$ be a set of RKD functions over $\{0,1\}^k$. If $\mathcal{A}$ is a PRF adversary which asks at most q queries totaling at most σ blocks, then we can construct a Φ-restricted PRP-RKA adversary $\mathcal{B}$ against E such that*

$$\mathbf{Adv}^{\mathrm{prf}}_{f9'[E,\Delta,l]}(\mathcal{A}) \leq \frac{3q^2 + 2\sigma^2 + 2\sigma q}{2^{n+1}} + \mathbf{Adv}^{\mathrm{prp\text{-}rka}}_{\Phi,E}(\mathcal{B}) \ . \tag{4}$$

Furthermore, $\mathcal{B}$ makes at most $\sigma + 2q$ oracle queries and uses the same time as $\mathcal{A}$.

Proof. We first note that given any PRF adversary $\mathcal{A}$ against $f9'[E, \Delta, l]$, we can construct a PRF adversary $\mathcal{C}$ against $f9'[E, \Delta, n]$ such that the following equation holds

$$\mathbf{Adv}^{\mathrm{prf}}_{f9'[E,\Delta,l]}(\mathcal{A}) \leq \mathbf{Adv}^{\mathrm{prf}}_{f9'[E,\Delta,n]}(\mathcal{C}) \ . \tag{5}$$

This standard result follows from the fact that the extra bits provided to the adversary can only improve its chance of success.

Our approach to upper-bounding $\mathbf{Adv}^{\mathrm{prf}}_{f9'[E,\Delta,n]}(\mathcal{C})$ is similar to the approach we used to upper-bound $\mathbf{Adv}^{\mathrm{priv}}_{f8'[E,\Delta]}(\mathcal{A})$ in the proof of Theorem 5.1. Let $\mathsf{f9'\text{-}Tag}$ denote the tagging algorithm for $f9'[E, \Delta, n]$ and let $\mathsf{p9'\text{-}Tag}$ denote the tagging algorithm for $p9'[n]$. Expanding the definition of $\mathbf{Adv}^{\mathrm{prf}}_{f9'[E,\Delta,n]}(\mathcal{C})$ and applying Lemma 5.1, we get:

$$\begin{aligned}
\mathbf{Adv}^{\mathrm{prf}}_{f9'[E,\Delta,n]}(\mathcal{C}) &= \Big| \Pr(K \stackrel{R}{\leftarrow} \{0,1\}^k : \mathcal{C}^{\mathsf{f9'\text{-}Tag}_K(\cdot)} = 1) \\
&\qquad - \Pr(g \stackrel{R}{\leftarrow} \mathrm{Rand}(*, n) : \mathcal{C}^{g(\cdot)} = 1) \Big| \\
&\leq \frac{\sigma^2 + q^2}{2^{n+1}} + \Big| \Pr(K \stackrel{R}{\leftarrow} \{0,1\}^k : \mathcal{C}^{\mathsf{f9'\text{-}Tag}_K(\cdot)} = 1) \\
&\qquad - \Pr(R_1, R_2 \stackrel{R}{\leftarrow} \mathrm{Rand}(n, n) : \mathcal{C}^{\mathsf{p9'\text{-}Tag}_{R_1,R_2}(\cdot)} = 1) \Big| \ .
\end{aligned}$$

As with the proof of Lemma 5.1, let $\mathcal{B}$ be a Φ-restricted related-key adversary against E that runs $\mathcal{C}$ and that returns the same bit that $\mathcal{C}$ returns. Let $F_{\mathrm{RK}(\cdot,K)}(\cdot)$ denote $\mathcal{B}$'s related-key oracle. This time, when $\mathcal{C}$ makes an oracle query (N, M) to its oracle, $\mathcal{B}$ essentially computes the $\mathsf{f9'\text{-}Tag}$ algorithm, except that it uses its related-key oracle in place of E_K and $E_{K \oplus \Delta}$. In pseudocode,

Algorithm $\mathcal{B}^{F_{\mathrm{RK}(\cdot,K)}(\cdot)}$
Run $\mathcal{C}$, replying to $\mathcal{C}$'s oracle queries M as follows:
$M \leftarrow \mathsf{pad'}_n(M)$
Break M into n-bit blocks $M[1] \| \cdots \| M[m]$
$Y[0] \leftarrow 0^n$
For $i = 1$ to m do:
$X[i] \leftarrow M[i] \oplus Y[i-1]$
$Y[i] \leftarrow F_{\mathrm{RK}(\mathrm{id},K)}(X[i])$
$T \leftarrow F_{\mathrm{RK}(\mathrm{XOR}_\Delta,K)}(Y[1] \oplus \cdots \oplus Y[m])$
Return T to $\mathcal{C}$
When $\mathcal{C}$ outputs b:
output b

We first observe that when $\mathcal{B}$ is given related-key oracle access to E with key K, it replies to $\mathcal{C}$'s oracle queries exactly as $\mathsf{f9'}\text{-}\mathsf{Tag}_K(\cdot)$ does. This means that the following equation holds:

$$\Pr(K \stackrel{R}{\leftarrow} \{0,1\}^k : \mathcal{C}^{\mathsf{f9'}\text{-}\mathsf{Tag}_K(\cdot)} = 1) = \Pr(K \stackrel{R}{\leftarrow} \{0,1\}^k : \mathcal{B}^{E_{\mathrm{RK}(\cdot,K)}(\cdot)} = 1) \ .$$

We also observe that when $\mathcal{B}$ is given related-key oracle access to G with key K, where G is a randomly selected function family from $\mathrm{Rand}(k,n,n)$, the functions $G_K(\cdot)$ and $G_{K\oplus\Delta}(\cdot)$ are both randomly selected from $\mathrm{Rand}(n,n)$. This means that $\mathcal{B}$ replies to $\mathcal{C}$'s oracle queries exactly as $\mathsf{p9'}\text{-}\mathsf{Tag}_{R_1,R_2}(\cdot)$ would with two randomly selected functions R_1, R_2 from $\mathrm{Rand}(n,n)$. Consequently, the following equation holds:

$$\begin{aligned}\Pr(R_1, R_2 \stackrel{R}{\leftarrow} \mathrm{Rand}(n,n) : \mathcal{C}^{\mathsf{p9'}\text{-}\mathsf{Tag}_{R_1,R_2}(\cdot)} = 1)& \\ = \Pr(K \stackrel{R}{\leftarrow} \{0,1\}^k \,;\, G \stackrel{R}{\leftarrow} \mathrm{Rand}(k,n,n) : \mathcal{B}^{G_{\mathrm{RK}(\cdot,K)}(\cdot)} = 1)&\end{aligned}$$

Combining these equations, we have that

$$\begin{aligned}\mathbf{Adv}^{\mathrm{prf}}_{f9'[E,\Delta,n]}(\mathcal{C}) \le \frac{\sigma^2+q^2}{2^{n+1}} + \Big| &\Pr(K \stackrel{R}{\leftarrow} \{0,1\}^k : \mathcal{B}^{E_{\mathrm{RK}(\cdot,K)}(\cdot)} = 1) \\ &- \Pr(K \stackrel{R}{\leftarrow} \{0,1\}^k \,;\, H \stackrel{R}{\leftarrow} \mathrm{Perm}(k,n) : \mathcal{B}^{H_{\mathrm{RK}(\cdot,K)}(\cdot)} = 1) \\ &+ \Pr(K \stackrel{R}{\leftarrow} \{0,1\}^k \,;\, H \stackrel{R}{\leftarrow} \mathrm{Perm}(k,n) : \mathcal{B}^{H_{\mathrm{RK}(\cdot,K)}(\cdot)} = 1) \\ &- \Pr(K \stackrel{R}{\leftarrow} \{0,1\}^k \,;\, G \stackrel{R}{\leftarrow} \mathrm{Rand}(k,n,n) : \mathcal{B}^{G_{\mathrm{RK}(\cdot,K)}(\cdot)} = 1)\Big| \ .\end{aligned}$$

Applying the PRP-RKA definition and a variant of the PRF/PRP switching lemma from [5], we get

$$\begin{aligned}\mathbf{Adv}^{\mathrm{prf}}_{f9'[E,\Delta,n]}(\mathcal{C}) \le \mathbf{Adv}^{\mathrm{prp\text{-}rka}}_{\Phi,E}(\mathcal{B}) &+ \frac{(\sigma+q)\cdot(\sigma+q-1)}{2^{n+1}} \\ + \frac{q\cdot(q-1)}{2^{n+1}} + \frac{\sigma^2+q^2}{2^{n+1}} &\ .\end{aligned}$$

For the application of the PRF/PRP switching lemma, we note that $\mathcal{B}$ queries its related-key oracle with the RKD function id at most $\sigma + q$ times and the RKD function XOR_Δ at most q times. Combining the above with equation (5) and simplifying gives the theorem statement.

Acknowledgements

T. Kohno was supported by a National Defense Science and Engineering Fellowship.

References

[1] 3GPP TS 35.201 v 3.1.1. Specification of the 3GPP confidentiality and integrity algorithms, Document 1: $f8$ and $f9$ specification. Available at http://www.3gpp.org/tb/other/algorithms.htm. 427, 430, 432

[2] 3GPP TS 35.202 v 3.1.1. Specification of the 3GPP confidentiality and integrity algorithms, Document 2: KASUMI specification. Available at http://www.3gpp.org/tb/other/algorithms.htm. 427

[3] M. Bellare, A. Desai, E. Jokipii, and P. Rogaway. A concrete security treatment of symmetric encryption. Proceedings of *The 38th Annual Symposium on Foundations of Computer Science, FOCS '97,* pp. 394–405, IEEE, 1997. 428, 434

[4] M. Bellare, J. Kilian, and P. Rogaway. The security of the cipher block chaining message authentication code. *JCSS,* vol. 61, no. 3, pp. 362–399, 2000. Earlier version in Y. Desmedt, editor, *Advances in Cryptology – CRYPTO '94*, volume 839 of *Lecture Notes in Computer Science*, pages 341–358. Springer-Verlag, Berlin Germany, 1994. 429, 437

[5] M. Bellare, and T. Kohno. A theoretical treatment of related-key attacks: RKA-PRPs, RKA-PRFs, and applications. In E. Biham, editor, *Advances in Cryptology – EUROCRYPT 2003*, volume 2656 of *Lecture Notes in Computer Science*, pages 491–506. Springer-Verlag, Berlin Germany, 2003. 428, 429, 437, 440

[6] M. Bellare, P. Rogaway, and D. Wagner. The EAX mode of operation. In W. Meier and B. Roy, editors, *Fast Software Encryption, FSE 2004*, Springer-Verlag, 2004. 428

[7] E. Biham. New types of cryptanalytic attacks using related keys. In T. Helleseth, editor, *Advances in Cryptology – EUROCRYPT '93*, volume 765 of *Lecture Notes in Computer Science*, pages 398–409. Springer-Verlag, Berlin Germany, 1993. 428, 429

[8] J. Black and P. Rogaway. A block-cipher mode of operation for parallelizable message authentication. In L. R. Knudsen, editor, *Advances in Cryptology – EUROCRYPT 2002*, volume 2332 of *Lecture Notes in Computer Science*, pages 384–397. Springer-Verlag, Berlin Germany, 2002. 428

[9] M. Blunden and A. Escott. Related key attacks on reduced round KASUMI. In M. Matsui, editor, *Fast Software Encryption, FSE 2001*, volume 2355 of *Lecture Notes in Computer Science*, pages 277–285. Springer-Verlag, Berlin Germany, 2002. 429, 430

[10] J. Daemen and V. Rijmen. *The Design of Rijndael.* Springer-Verlag, Berlin Germany, 2002. 430

[11] Evaluation report (version 2.0). Specification of the 3GPP confidentiality and integrity algorithms, Report on the evaluation of 3GPP confidentiality and integrity algorithms. Available at http://www.3gpp.org/tb/other/algorithms.htm. 429

[12] D. Hong, J-S. Kang, B. Preneel and H. Ryu. A concrete security analysis for 3GPP-MAC. In T. Johansson, editor, *Fast Software Encryption, FSE 2003*, volume 2887 of *Lecture Notes in Computer Science*, pages 154–169. Springer-Verlag, Berlin Germany, 2003. 428, 429, 433, 443, 444, 445

[13] T. Iwata and T. Kohno. New security proofs for the 3GPP confidentiality and integrity algorithms. Full version of this paper, available at http://eprint.iacr.org/, 2004. 435, 438, 442, 443, 445
[14] T. Iwata and K. Kurosawa. OMAC: One-Key CBC MAC. In T. Johansson, editor, *Fast Software Encryption, FSE 2003*, volume 2887 of *Lecture Notes in Computer Science*, pages 129–153. Springer-Verlag, Berlin Germany, 2003. 428
[15] T. Iwata and K. Kurosawa. On the correctness of security proofs for the 3GPP confidentiality and integrity algorithms. In K. G. Paterson, editor, *Cryptography and Coding, Ninth IMA International Conference*, volume 2898 of *Lecture Notes in Computer Science*, pages 306–318. Springer-Verlag, Berlin Germany, 2003. 428, 429, 433
[16] J. Jonsson. On the Security of CTR + CBC-MAC. In K. Nyberg and H. M. Heys, editors, *Selected Areas in Cryptography, 9th Annual Workshop (SAC 2002)*, volume 2595 of *Lecture Notes in Computer Science*, pages 76–93. Springer-Verlag, Berlin Germany, 2002. 428
[17] C. S. Jutla. Encryption modes with almost free message integrity. In B. Pfitzmann, editor, *Advances in Cryptology – EUROCRYPT 2001*, volume 2045 of *Lecture Notes in Computer Science*, pages 529–544. Springer-Verlag, Berlin Germany, 2001. 428
[18] J-S. Kang, S-U. Shin, D. Hong and O. Yi. Provable security of KASUMI and 3GPP encryption mode $f8$. In C. Boyd, editor, *Advances in Cryptology – ASIACRYPT 2001*, volume 2248 of *Lecture Notes in Computer Science*, pages 255–271. Springer-Verlag, Berlin Germany, 2001. 428, 429, 443
[19] L. R. Knudsen and C. J. Mitchell. Analysis of 3gpp-MAC and two-key 3gpp-MAC. *Discrete Applied Mathematics*, vol. 128, no. 1, pp. 181–191, 2003. 429, 433
[20] T. Kohno, J. Viega, and D. Whiting. CWC: A high-performance conventional authenticated encryption mode. In W. Meier and B. Roy, editors, *Fast Software Encryption, FSE 2004*, Springer-Verlag, 2004. 428
[21] M. Luby and C. Rackoff. How to construct pseudorandom permutations from pseudorandom functions. *SIAM J. Comput.*, vol. 17, no. 2, pp. 373–386, April 1988. 428, 429
[22] P. Rogaway, M. Bellare, J. Black, and T. Krovetz. OCB: a block-cipher mode of operation for efficient authenticated encryption. *Proceedings of ACM Conference on Computer and Communications Security, ACM CCS 2001*, ACM, 2001. 428, 434, 443
[23] D. Whiting, R. Housley, and N. Ferguson. Counter with CBC-MAC (CCM). Submission to NIST. Available at http://csrc.nist.gov/CryptoToolkit/modes/. 428

A Proof Sketch of Lemma 4.1

We sketch the proof of Lemma 4.1 here, leaving the details to [13]. The adversary has an oracle which is either $\mathsf{p8'}\text{-}\mathsf{Encrypt}_{R_1,R_2}(\cdot,\cdot)$ or $\$(\cdot,\cdot)$. Let (N_i, M_i) denote the adversary's i-th oracle query, and let C_i denote the answer from the oracle. Assume that the length of M_i (and C_i) is m_i blocks, where $m_i \geq 1$. We write $M_i = M_i[1]\|\cdots\|M_i[m_i]$ and $C_i = C_i[1]\|\cdots\|C_i[m_i]$.

We define a bad query-answer pair and a bad event.

Bad Query-Answer Pair. We say that (N_i, M_i, C_i) is a bad query-answer pair if some string is repeated in the multiset

$$\left\{[0]_n, M_i[1] \oplus C_i[1] \oplus [1]_n, \ldots, M_i[m_i - 1] \oplus C_i[m_i - 1] \oplus [m_i - 1]_n\right\} .$$

For the i-th query-answer pair (N_i, M_i, C_i), the input sequence of R_2 is

$$\left\{A_i \oplus [0]_n, A_i \oplus M_i[1] \oplus C_i[1] \oplus [1]_n, \ldots, A_i \oplus M_i[m_i - 1] \oplus C_i[m_i - 1] \oplus [m_i - 1]_n\right\}$$

where $A_i = R_1(N_i)$. Thus, for a bad query-answer pair, there is a collision among the input sequence of R_2.

Bad Event. Let $i < j$. For (N_i, M_i, C_i) and (N_j, M_j, C_j), we say that a bad event occurs if

$$\left\{A_i \oplus [0]_n, A_i \oplus M_i[1] \oplus C_i[1] \oplus [1]_n, \ldots, A_i \oplus M_i[m_i - 1] \oplus C_i[m_i - 1] \oplus [m_i - 1]_n\right\}$$

and

$$\left\{A_j \oplus [0]_n, A_j \oplus M_j[1] \oplus C_j[1] \oplus [1]_n, \ldots, A_j \oplus M_j[m_j - 1] \oplus C_i[m_j - 1] \oplus [m_j - 1]_n\right\}$$

have some common element, where $A_i = R_1(N_i)$ and $A_j = R_1(N_j)$. If a bad event occurs, there is a collision between some input to R_2 at the i-th query and some input to R_2 at the j-th query. This implies $\mathsf{p8'}\text{-}\mathsf{Encrypt}_{R_1,R_2}(\cdot,\cdot)$ does not behave like $\$(\cdot,\cdot)$.

Intuitively, we show that if all the query-answer pairs are not bad and the bad event does not occur, then the adversary cannot distinguish between $\mathsf{p8'}\text{-}\mathsf{Encrypt}_{R_1,R_2}(\cdot,\cdot)$ and $\$(\cdot,\cdot)$. The proof is completed by upper bounding the probability of some bad query-answer pair occurs, or some bad event occurs.

A.1 Discussion of the Previous Work [18]

[18, p. 269, Lemma 7] might be seen to correspond to Lemma 4.1. However, there is a problem with the definition of their encryption scheme. Their encryption scheme, which we call $p8''[n]$, is described as follows: The key generation algorithm for $p8''[n]$ returns a randomly selected permutation P_1 from $\mathrm{Perm}(n)$. The encryption algorithm for $p8''[n]$ takes P_1 as a "key" and uses P_1 and P_2 instead of E_K and $E_{K\oplus\Delta}$, but it is not defined how P_2 is derive from P_1. We note that [12, p. 166, Lemma 2] has a similar problem, which is described in Appendix B.1.

We also adopt the strong notion of privacy, indistinguishability from random strings [22]. This security notion is strictly stronger than the left-or-right indistinguishability used in [18, p. 269, Lemma 7].

In [13], we present the full security proof for $p8'[n]$ in order to achieve this strong security notion and to establish self contained security proof.

B Proof Sketch of Lemma 5.1

To prove Lemma 5.1, we define $p9'$-E$[n]$, a variant of $p9'[n]$. The tagging algorithm for $p9'$-E$[n]$ takes only messages of length multiple of n, and it does not perform the final encryption. Specifically, the key generation algorithm for $p9'$-E$[n]$ returns a randomly selected function R_1 from $\mathrm{Rand}(n,n)$. The tagging algorithm for $p9'$-E$[n]$, $\mathsf{p9'\text{-}E\text{-}Tag}$, takes R_1 as a "key" and a message M such that $|M| = mn$ for some $m \geq 1$. In pseudocode:

Algorithm $\mathsf{p9'\text{-}E\text{-}Tag}_{R_1}(M)$
 Break M into n-bit blocks $M[1] \| \cdots \| M[m]$
 $Y[0] \leftarrow 0^n$
 For $i = 1$ to m do:
  $X[i] \leftarrow M[i] \oplus Y[i-1]$
  $Y[i] \leftarrow R_1(X[i])$
 Return $Y[1] \oplus \cdots \oplus Y[m]$

The verification algorithm is defined in the natural way.

Let $M_1, \ldots, M_q$ be any fixed and distinct bit strings such that $|M_i| = m_i n$, where $m_i \geq 1$. Then Lemma 5.1 is proved by deriving the upper bound of the collision probability among the output of $\mathsf{p9'\text{-}E\text{-}Tag}_{R_1}(M_i)$. We show the following lemma.

Lemma B.1. *Let $m_1, \ldots, m_q$, $M_1, \ldots, M_q$ be as described above. Then*

$$\Pr(R_1 \stackrel{R}{\leftarrow} \mathit{Rand}(n,n) : 1 \leq {}^{\exists}i < {}^{\exists}j \leq q, \mathsf{p9'\text{-}E\text{-}Tag}_{R_1}(M_i) = \mathsf{p9'\text{-}E\text{-}Tag}_{R_1}(M_j))$$

is at most $(\sigma^2 + q^2)/2^{n+1}$, where $\sigma = m_1 + \cdots + m_q$.

Given the above lemma, the proof of Lemma 5.1 is completed from the well known fact that applying a random function to the output of an almost-universal hash function family is a PRF.

B.1 Discussion of the Previous Work [12]

[12, p. 162, Lemma 1] corresponds to our Lemma 5.1. Then one might wonder if the relevant portion can be re-used. However, in the proof of [12, p. 162, Lemma 1], there is a flaw in the analysis of Game 5. We use our notation. Let $q = 2$ in Lemma B.1. Then [12, p. 166] says

$$\Pr(R_1 \stackrel{R}{\leftarrow} \mathrm{Rand}(n,n) : \mathsf{p9'\text{-}E\text{-}Tag}_{R_1}(M_1) = \mathsf{p9'\text{-}E\text{-}Tag}_{R_1}(M_2)) = \frac{1}{2^n} \ ,$$

since $Y_1[1]$ is a random string in $\{0,1\}^n$, where $Y_1[1] = R_1(M_1[1])$. However, if $M_1[1] = M_2[1]$, then we have $Y_1[1] = Y_2[1]$, where $Y_2[1] = R_1(M_2[1])$, and their randomness disappears. This part needs to be fixed, which is done in Lemma B.1.

Also, [12, p. 166, Lemma 2] doesn't hold. There is a problem with the definition of their MAC. Their MAC, which we call $p9''[n]$, is described as follows: the key generation algorithm for $p9''[n]$ returns a randomly selected permutation P_1 from $\mathrm{Perm}(n)$. The tagging algorithm for $p9''[n]$ takes P_1 as a "key" and uses P_1 and P_2 instead of E_K and $E_{K \oplus \Delta}$, and outputs a full n-bit tag, where $P_2 \in \mathrm{Perm}(n) \setminus \{P_1\}$ is determined from P_1 by some means. The verification algorithm is defined in the natural way. Then [12, p. 159] says the security of $p9''[n]$ does not depend on how P_2 is derived from P_1, which is not correct. For example if P_2 is chosen as $P_2 = P_1^{-1}$, then it is easy to make a forgery.

In [13], we present the full security proof for $p9'[n]$ in order to avoid presenting proof covered with patches, and to establish self contained security proof.

Cryptanalysis of a Message Authentication Code due to Cary and Venkatesan

Simon R. Blackburn and Kenneth G. Paterson*

Department of Mathematics
Royal Holloway, University of London
Egham, Surrey, TW20 0EX, U.K.
{simon.blackburn,kenny.paterson}@rhul.ac.uk

Abstract. A cryptanalysis is given of a MAC proposal presented at CRYPTO 2003 by Cary and Venkatesan. A nice feature of the Cary-Venkatesan MAC is that a lower bound on its security can be proved when a certain block cipher is modelled as an ideal cipher. Our attacks find collisions for the MAC and yield MAC forgeries, both faster than a straightforward application of the birthday paradox would suggest. For the suggested parameter sizes (where the MAC is 128 bits long) we give a method to find collisions using about $2^{48.5}$ MAC queries, and to forge MACs using about 2^{55} MAC queries. We emphasise that our results do not contradict the lower bounds on security proved by Cary and Venkatesan. Rather, they establish an upper bound on the MAC's security that is substantially lower than one would expect for a 128-bit MAC.

Keywords: Message authentication, MAC, matrix groups, cryptanalysis, birthday paradox.

1 Introduction

This paper is concerned with a proposal for a Message Authentication Code (MAC) presented at CRYPTO 2003 by Cary and Venkatesan [1]. Their idea is to take a MAC construction of Jakubowski and Venkatesan [3] based on linear operations over a finite field, and alter it by replacing finite field operations by operations in the ring of integers modulo some power of 2 (as the latter operations are more efficient on the current generation of processors). Cary and Venkatesan [1] have proved a lower bound on the security of their MAC. This paper presents two attacks on the MAC, and so establishes a corresponding upper bound on the MAC's security. The first attack shows that an adversary with access to a MAC oracle is able to find collisions of the MAC considerably faster than a straightforward application of the birthday paradox would suggest. For an introduction to MACs and birthday attacks on them, see [4]. The second attack does more: it derives most of the secret key material for the MAC (which enables MACs to be forged). This second attack works by exploiting certain

* This author supported by the Nuffield Foundation NUF-NAL 02.

B. Roy and W. Meier (Eds.): FSE 2004, LNCS 3017, pp. 446–453, 2004.

collisions in the MAC; these collisions are found almost as efficiently as in the first attack. Thus the proposal of Cary and Venkatesan [1], while efficient and offering some interesting provable security properties, does not offer the level of security that a MAC of its output length should aspire to.

The next section describes the Cary–Venkatesan MAC. Sections 3 and 4 describe our two attacks on this MAC. The concluding section explains how our attacks impact on the practical level of security offered by the MAC when the suggested parameter sizes are used.

2 The Cary–Venkatesan MAC

Let ℓ, k and t be integers. The Cary–Venkatesan MAC operates on blocks consisting of t words $x_1, x_2, \ldots, x_t$ each word being of length ℓ bits. We regard the words x_i as ℓ-bit integers. The MAC has a $t(\ell-1)+k$-bit secret key. This key is made up of odd ℓ-bit integers $a_1, a_2, \ldots, a_t$ together with a k-bit string K.

The MAC consists of two parts, a *compression function* H and a *block cipher* E. The compression function H takes as input the vector $\underline{a} = (a_1, a_2, \ldots, a_t)$ and a block $x = x_1 x_2 \ldots x_t$ where $x_i \in \{0, 1, \ldots, 2^\ell - 1\}$; it returns a 4ℓ-bit string $h = H_{\underline{a}}(x)$. The block cipher operates on 4ℓ-bit blocks. It takes as input the key K and the output h of the compression function; the cipher returns the 4ℓ-bit value $E_K(h)$ and this is the output of the MAC.

Cary and Venkatesan allow the block cipher E to be any secure block cipher acting on 4ℓ-bit blocks with a k-bit key. They model E as an ideal cipher and concentrate their efforts on designing an efficient compression function H of the following form.

Let $A_1, A_2, \ldots, A_{t-1}$ be fixed 2×2 matrices, and let z_0 and σ_0 be fixed column vectors of length 2; suppose all the entries of these matrices and vectors lie in the ring $\mathbb{Z}_{2^\ell}$ of integers modulo 2^ℓ. (These matrices and vectors are public, and some suggested examples are given in [1].) Vectors $v_1, v_2, \ldots, v_t \in (\mathbb{Z}_{2^\ell})^2$ are calculated as follows. Let $i \in \{1, 2, \ldots, t\}$ be fixed. Multiply the ℓ-bit integers a_i and x_i, to produce a 2ℓ-bit integer; this product is then broken into two ℓ-bit integers, and the result v_i is regarded as an element of $(\mathbb{Z}_{2^\ell})^2$. The way in which the product $a_i x_i$ is split to form v_i is not specified in [1]; we assume that a natural choice of $v_i = \left[a_i x_i \bmod 2^\ell,\ a_i x_i \operatorname{div} 2^\ell\right]^T$ is used. (Another natural choice would be $v_i = \left[a_i x_i \operatorname{div} 2^\ell,\ a_i x_i \bmod 2^\ell\right]^T$. Our results are unaffected if this choice is used instead.)

The output h of H is defined to be the pair (z, σ), where

$$z = z_0 + v_1 + A_1 v_2 + A_1 A_2 v_3 + \cdots + A_1 A_2 \cdots A_{t-1} v_t$$

and where

$$\sigma = \sigma_0 + v_1 + v_2 + \cdots + v_t.$$

Here all operations are over $\mathbb{Z}_{2^\ell}$.

Cary and Venkatesan propose two variants of their MAC: a way of chaining the compression function so that it can compress more than one block into 4ℓ-bits (by making the 'initial values' z_0 and σ_0 used in the next block depend on z and σ above), and a method for doubling the length of output of the compression function (by computing the compression function above twice on the same block, using different keys, and then concatenating their outputs). Our attacks below can be adapted to apply to these variants as well, although we will not discuss the straightforward modifications that are needed.

3 The First Attack

For MACs such as the one considered here, which consist of a relatively weak keyed compression function followed by a block cipher encryption, it is generally assumed that it is computationally infeasible to invert the block cipher E without knowledge of the secret key K. (If the block cipher can be inverted efficiently, the output of the compression function H is available to the cryptanalyst. The keys used in the compression function can then usually be derived from MACs of a few chosen messages. Once these keys are known, MACs of a wide variety of messages may be forged. This shows that, in practice, the security level offered by a MAC of this type cannot be greater than the length of the block cipher key. This is certainly the case with the proposal of [1].)

The final cipher E is often modelled as an ideal cipher, namely a set of random permutations indexed by the key K. An adversary has access to an oracle that adds MACs to messages; the adversary aims to generate a valid MAC for any message that has not been sent as a query to the oracle. In this ideal cipher model it is intuitively clear (and indeed formally provable) that finding two messages that are compressed by H to the same value h (finding a *collision*) is a prerequisite to breaking the MAC. Since the block length of the cipher E in the proposal of [1] is 4ℓ bits, the birthday paradox implies that a collision will be found with reasonable probability after approximately $2^{2\ell}$ oracle queries on random messages. A good scheme in this model should therefore have the property that it is impossible for an adversary to produce collisions with reasonable probability unless the number of oracle queries is close to this upper bound. However, we show that when the MAC proposed in [1] is used, an adversary can produce collisions using significantly fewer than $2^{2\ell}$ messages.

The basic idea of our collision finding attack is to construct a large set of messages with the property that the compression function H maps each message into a small subset of inputs to the block cipher (irrespective of the key $a_1, a_2, \ldots, a_t$ used). Because the block cipher is a permutation, there is a collision in the MAC output if and only if there is a collision at the input to the block cipher, i.e. at the output of the compression function H.

Let r be an integer that is as large as possible subject to the conditions that $0 \leq r < \ell$ and that $2^{t(\ell-r)}$ is at least $n = \lceil \sqrt{2^{4\ell-r}} \rceil$. (For most choices of parameters, $r = \ell - 1$ or $r = \ell - 2$ will suffice.)

There are $2^{t(\ell-r)}$ messages $x = x_1x_2\cdots x_t$ with the property that 2^r divides x_i for all $i \in \{1,2,\ldots,t\}$. We denote this set of messages by X and let Y be a subset of X consisting of n such messages: a subset of this size exists, by our choice of r. We obtain the MACs of all the messages in Y from the MAC oracle. Define B to be the set of all pairs $(z,\sigma) \in ((\mathbb{Z}_{2^\ell})^2)^2$ such that the first component of $\sigma - \sigma_0$ is divisible by 2^r. Our condition on the elements x_i for messages in X implies that the first component of each vector v_i is divisible by 2^r, and so the same is true for the vector $\sigma - \sigma_0$. Hence the image of X under H lies in B, whatever the value of the secret key $a_1, a_2, \ldots, a_t$. Now,

$$|B| = 2^{4\ell-r} < n^2.$$

Since we have requested the MACs of the n messages in the set $Y \subseteq X$, the birthday paradox implies that we will find a collision with reasonable probability (in fact, with probability about 0.63). Note that n is considerably less than $2^{2\ell}$. Indeed, when $r = \ell - 1$ or $r = \ell - 2$, we have that n is approximately $2^{3\ell/2}$. So we have found collisions for the function H, and hence for the MAC, considerably faster than a straightforward use of the birthday paradox would imply.

We have assumed in this analysis that the image of X under H is uniformly distributed in B. A non-uniform distribution only enhances the probability of collisions.

4 The Second Attack

Our first attack found collisions efficiently. However, it is not clear how knowledge of these collisions could be used to forge MACs. We now present a second method for finding collisions, almost as efficient as the method above, with the extra feature that collisions may be used to find the key words a_i, and hence to forge MACs.

We begin by choosing two integer parameters d and s, which affect the efficiency and probability of success of the attack. The value of d is the number of collisions we need to find in the compression function, and is chosen as follows. Let V be a $\mathbb{Z}_2$-vector space of dimension $t-1$. Then d is chosen so that the probability of a set of d randomly chosen vectors from V forming a spanning set is large (say at least 0.5). A choice of $d = t$ or $d = t+1$ will suffice in most situations. The parameter s relates to the size of a set Y' which is analogous to the set Y in our first attack. We choose s to be a positive integer such that 2^{ts} is just greater than $w = \lceil \sqrt{2dt(2^{3\ell+s})} \rceil$. For most parameter sets, this means that s is small ($s = 2$ or $s = 3$, say). Our choice of the values of d and s will become clear in the description of the attack below.

The attack proceeds in 3 stages. In stage 1, we find many collisions in the compression function H by requesting MACs of messages of a special form. In stage 2, we use data gathered from these collisions to form a system of linear equations in the unknown key words a_i. Solving these equations and performing a moderate amount of additional computation allows us to find the key $\underline{a}$ for the compression function H. In stage 3, we use knowledge of $\underline{a}$ to quickly find a collision in H without querying the MAC. This immediately leads to a MAC forgery.

Stage 1: We ask for the MACs of a subset Y' taken from the set of messages $X' = \{x_1 x_2 \cdots x_t \mid 0 \leq x_i < 2^s, \ \forall i \in \{1, 2, \ldots, t\}\}$. There are $2^{ts} \geq w$ messages in X' and we take $|Y'| = w$. Define B' to be the set of all pairs $(z, \sigma) \in ((\mathbb{Z}_{2^\ell})^2)^2$ such that the second component of $\sigma - \sigma_0$ lies between 0 and $t2^s$. Our condition on the elements x_i for messages in X' implies that $x_i a_i < 2^{s+\ell}$ for all i. So the second component of v_i is less than 2^s and hence the second component of $\sigma - \sigma_0$ is less than $t2^s$. Thus the image of X' under H is contained in B', whatever the value of the secret key $a_1, a_2, \ldots, a_t$.

Notice that B' is of size $u = t2^{3\ell+s}$ and $Y' \subset X'$ is of size $w = \lceil\sqrt{2dt(2^{3\ell+s})}\rceil$. Our choice of parameters implies that, by the birthday paradox, we will find collisions in the MAC function, and hence in the compression function, for the set Y'. In fact, we have chosen d, s and w so that $|Y'|$ is a little larger than is needed for the birthday paradox to apply. This is so that we are likely to find many collisions. Indeed, applying results of [2], we have that for large w and u, the frequency distribution $Q(x)$ of the number of collisions x can be approximated by a Poisson distribution $P_\lambda(x)$ with parameter $\lambda = w^2/2u$. The error in approximating $Q(x)$ by $P_\lambda(x)$ can be bounded using results of [2]. We have:

$$|Q(x) - P_\lambda(x)| \leq \frac{5}{w}x^2 + \frac{3w}{u}x + \frac{w^2}{3u^2}. \tag{1}$$

For our choice of parameters and for $x \leq d$, the error term in eqn. (1) can be bounded by $8d^{3/2}/u^{1/2}$. In our situation, d will be much smaller than u. Thus the approximation by a Poisson distribution will be excellent when $x \leq d$. For our choice of parameters, we have $\lambda = d$. So in the case of interest to us (where d is approximately equal to t and so is reasonably large), the Poisson distribution has mean d, and median approximately d. Putting all of this together, we see that the probability that we find d or more collisions in Y' is approximately equal to 0.5. The above analysis can of course be refined, but suffices for our purposes.

Stage 2: We now assume that at least d collisions of the above type have been found. We proceed by examining the first component of σ for each of these collisions. Each gives an equation of the form

$$x_1^{(j)} a_1 + x_2^{(j)} a_2 + \cdots + x_t^{(j)} a_t = {x'_1}^{(j)} a_1 + {x'_2}^{(j)} a_2 + \cdots + {x'_t}^{(j)} a_t \mod 2^\ell,$$

for $j \in \{1, 2, \ldots, d\}$. Writing $y_i^{(j)} = x_i^{(j)} - {x'_i}^{(j)}$, we obtain a system of d equations in t unknowns $a_1, \ldots, a_t \in \mathbb{Z}_{2^\ell}$:

$$y_1^{(j)} a_1 + y_2^{(j)} a_2 + \cdots + y_t^{(j)} a_t = 0 \bmod 2^\ell, \quad 1 \leq j \leq d. \tag{2}$$

Define vectors $y^{(j)} \in (\mathbb{Z}_{2^\ell})^t$ by $y^{(j)} = (y_1^{(j)}, y_2^{(j)}, \ldots, y_t^{(j)})$. The number of solutions to the system (2) depends on the linear independence properties of the vectors $y^{(j)}$ considered modulo 2. Let $z^{(j)}$ denote $y^{(j)}$ mod 2 and let V denote the dimension $t-1$ subspace of $\mathbb{Z}_2^t$ that consists of all vectors of even parity. Because the a_i are odd and the equations (2) hold, we have that $z^{(j)} \in V$ for $1 \leq j \leq d$. It is then elementary to show that the system (2) has a unique solution up to a $\mathbb{Z}_{2^\ell}$ scalar multiple if and only if the d vectors $z^{(j)}$ span the space V.

The probability that the d vectors $z^{(j)}$ span V is at least $1 - 1/2^{d-t+1}$, assuming the vectors to be random. (To see this, notice that the vectors fail to span V if and only if they all lie in some subspace U of co-dimension 1 in V. There are exactly $2^{t-1} - 1$ such subspaces U. Assuming the vectors $z^{(j)}$ to be randomly distributed in V, the probability that they all lie in any given U is equal to 2^{-d}. Hence the probability that the vectors do not span V is at most $(2^{t-1}-1)\cdot 2^{-d} \leq 1/2^{d-t+1}$.) This probability is close to 1 as soon as d is slightly greater than $t-1$. Given that the vectors $z^{(j)}$ do span V, a standard Gaussian elimination procedure over $\mathbb{Z}_{2^\ell}$ can be used to produce $b_1, b_2, \ldots, b_t \in \mathbb{Z}_{2^\ell}$ such that there exists an odd constant c with the property that $a_i = cb_i \bmod 2^\ell$ for all $i \in \{1, 2, \ldots, t\}$.

Stage 1 of our attack has given us d pairs of messages that collide under the compression function. To find c, we simply try each of the $2^{\ell-1}$ possibilities in turn and check whether the compression function with key $a_i = cb_i$ produces collisions for these pairs of messages. It is highly likely that a single value of c will produce all the correct collisions; this value will be the correct value of c. Thus we have recovered the value of the key words a_i.

Stage 3: Finally, we produce a MAC forgery as follows. We search for collisions in the compression function as in Sect. 3; however, since we now know the key to the compression function, we do not need to query the MAC oracle to obtain these collisions. After about $2^{3\ell/2}$ trials, we find a collision in the compression function: $H(x) = H(x')$ for distinct messages x and x'. We then query the MAC oracle on the message x. The resulting MAC will be valid for the message x', and so we have forged a MAC as required.

To summarise, we have forged a MAC after making

$$w + 1 = \lceil \sqrt{2dt(2^{3\ell+s})} \rceil + 1$$

oracle queries, and a comparatively small amount of additional effort (which mainly consists of storing the oracle outputs, together with computing the compression function about $2^{3\ell/2}$ times). The probability that our attack works is approximately $(1 - 1/2^{d-t+1})/2$.

The probability of success can be made arbitrarily close to 1, firstly by increasing the value of d and secondly by taking a larger number of MAC queries to increase the probability that d pairs of collisions will result. We omit the routine details of these enhancements.

5 Consequences for Suggested Parameter Sizes

In [1, Sect. 5], Cary and Venkatesan give details of an implementation of their MAC for the parameters $\ell = 32$ and $t = 50$. They are able to prove that the resulting MAC offers 54 bits of security, which can be interpreted as meaning that collisions for the MAC should not be found until after at least 2^{27} MAC queries have been made.

Our attack in Sect. 3 shows that, for these parameters, MAC collisions can be found using about $2^{48.5}$ MAC queries. (The attack will set $r = 31$; then the space B will be of size 2^{97} and $2^{48.5}$ MAC queries will be needed to obtain a collision with probability 0.63.). Taking $d = t = 50$, our attack in Sect. 4 uses $s = 2$ and finds MAC forgeries with probability about 0.25 using approximately 2^{55} MAC queries.

We have conducted experiments on cut down versions of the MAC. Using $t = 50$ and $8 \leq \ell \leq 14$, collisions in the compression function occurred as frequently as our analysis predicts. While we have not implemented our attacks for $\ell = 32$, we see no reason why our attack should not scale as predicted.

For both of our attacks, the complexity is significantly less than the 2^{64} queries implied by a standard application of the birthday paradox, though a good deal greater than the level of security that has been established for the MAC. It seems fair to say that the MAC proposal of [1] does not offer the security levels that one would expect of a strong MAC algorithm with a 128 bit output. We expect that more sophisticated attacks than the ones presented here may reduce the complexity of finding and exploiting collisions further.

In response to the attacks presented in this paper, the authors of [1] have suggested a new scheme, namely that the output of their compression function should first be passed through SHA-1 and then 54 bits of the result be taken as output. The intention of this construction is to match the proved security level of the original compression function with the length of the MAC. (The security level of this construction can be no greater than the 54 bits proved for the original MAC.) Note that the attacks presented in this paper do not carry over to this new scheme, since the overwhelming majority of collisions in the MAC will be caused by collisions in the final stage rather than by collisions in the compression function. Of course, a proof of security for this modified MAC can no longer rely on modelling a block cipher as a random permutation. Instead, this assumption would need to be replaced by an assumption concerning the collision properties of SHA-1 (or perhaps by modelling SHA-1 as a random function).

Acknowledgements

The authors would like to thank Sean Murphy for his help with background to the birthday paradox arguments in Sect. 4, Bart Preneel for bringing reference [2] to our attention, and Chris Mitchell for his comments on an earlier draft of this paper. The authors would also like to thank the referees of the paper, for their constructive comments and suggestions.

References

[1] M. Cary and R. Venkatesan, "A message authentication code based on unimodular matrix groups", in D. Boneh, editor, *Advances in Cryptology – Proc. CRYPTO 2003*, Lecture Notes in Computer Science Volume 2729, (Springer, Berlin, 2003), 500-512. 446, 447, 448, 452

[2] M. Girault, R. Cohen and M. Campana, "A generalized birthday attack", in C. G. Günther, editor, *Advances in Cryptology – Proc. EUROCRYPT'88*, Lecture Notes in Computer Science Volume 330, (Springer, Berlin, 1988), 129-156. 450, 452

[3] M. H. Jakubowski and R. Venkatesan, "The chain and sum primitive and its applications to MACs and stream ciphers", in K. Nyberg, editor, *Advances in Cryptology – Proc. EUROCRYPT '98*, Lecture Notes in Computer Science Volume 1403, (Springer, Berlin, 1998), 281-293. 446

[4] A. Menezes, P. C. van Oorschot and S. Vanstone, *Handbook of Applied Cryptography*, CRC Press, Boca Raton, 1997. 446

Fast Software-Based Attacks on SecurID

Scott Contini[1] and Yiqun Lisa Yin[2]

[1] Macquarie University
Computing Department
NSW 2109 Australia
scontini@comp.mq.edu.au,
[2] Princeton University
EE Department
Princeton, NJ 08540, USA
yyin@princeton.edu

Abstract. SecurID is a widely used hardware token for strengthening authentication in a corporate environment. Recently, Biryukov, Lano, and Preneel presented an attack on the alleged SecurID hash function [1]. They showed that *vanishing differentials* – collisions of the hash function – occur quite frequently, and that such differentials allow an attacker to recover the secret key in the token much faster than exhaustive search. Based on simulation results, they estimated that the running time of their attack would be about 2^{48} full hash operations when using only a single 2-bit vanishing differential.
In this paper, we present techniques to improve the [1] attack. Our theoretical analysis and implementation experiments show that the running time of our improved attack is about 2^{45} hash operations. We then investigate into the use of extra information that an attacker would typically have: multiple vanishing differentials or knowledge that other vanishing differentials do not occur in a nearby time period. When using the extra information, we believe that key recovery can always be accomplished within about 2^{40} hash operations.

1 Introduction

The SecurID, developed by RSA Security, is a hardware token used for strengthening authentication when logging in to remote systems, since passwords by themselves tend to be easily guessable and subject to dictionary attacks. The SecurID adds an "extra factor" of authentication: one must not only prove themselves by getting their password correct, but also by demonstrating that they have the SecurID token assigned to them. The latter is done by entering the 6- or 8-digit code that is being displayed on the token at the time of login.

Each token has within it a 64-bit secret key and an internal clock. Every minute, or every half-minute in some tokens, the secret key and the current time are sent through a cryptographic hash function. The output of the hash function determines the next two authenticator codes, which are displayed on the LCD

B. Roy and W. Meier (Eds.): FSE 2004, LNCS 3017, pp. 454–471, 2004.

screen. The secret key is also known to the "ACE/server", so that the same authenticator can independently be computed and verified at the remote end.

If ever a user loses their token, they must report it so that the current token can be deactivated and replaced with a new one. Thus, the user bears some responsibility in maintaining the security of the system. On the other hand, if the user were to temporarily leave his token in a place where it could be observed by others and then later recover it, then it should not be the case that the security of the device could be entirely breached, assuming the device is well-designed.

The scenario just described was considered in a recent publication by Biryukov, Lano, and Preneel [1], where they showed that the hash function that is alleged to be used by SecurID [4] (ASHF) has weak properties that could allow one to find the key much faster than exhaustive search. The attack they describe requires recording all outputs of the SecurID using a PC camera with OCR software, and then later searching the outputs for indication of a *vanishing differential* – two closely related input times that result in the same output hash. If one is discovered, the attacker then has a good chance of finding the internal secret key using a search algorithm that they estimated to be equivalent to 2^{48} hash function operations. On a 2.4 GHz PC, 2^{48} hash operations take about 111 years. [1] It would require over 1300 of these PC's to find the key in a month.

In this paper, we present three techniques to significantly speed up the filtering, which is the bottleneck of their attack. Our theoretical analysis and implementation experiments show that the time complexity can be reduced to about 2^{45} hash operations when using only a single vanishing differential.

We then investigate into the use of extra information that an attacker would ordinarily have, in order to speed up the attack further. This information consists of either multiple vanishing differentials, or knowledge that no other vanishing differentials occur in a nearby time period of the observed one. In either case, the running time can be reduced significantly. Our preliminary analysis suggests that after a vanishing differential is observed, the attacker would nearly always be able to perform the key search algorithm in 2^{40} hash operations or less. On a typical PC, this can be done in about 5 months, making the computing power requirements for the search attainable by almost any individual.

The success probability of all attacks (including [1]) depend upon how long the attacker must wait for a vanishing differential to occur. Simulations have shown that in any one-week period, 1% of the SecurID cards will have a vanishing differential; in any one-year period, 35% of the tokens will have a vanishing differential. According to these statistics, we mention two realistic scenarios in which the token could be compromised. In the first scenario, a user may be on vacation for one week and left his token behind in a place where others could observe it, in which case there is a small but definitely *non-negligible* chance that a collision would happen. In the second scenario, the success is much more likely. Since the cost of SecurID tokens is very expensive, tokens are often reassigned to new users when a previous owner leaves a company [5]. This is a bad idea,

[1] Requires some optimisations to Wiener's code, such as re-ordering bytes to eliminate bswaps.

since the original user would have a high chance of being able to find the internal key, assuming he recorded many of the outputs while it was in his possession. In light of our new results, token reassignment becomes a very serious risk.

2 The SecurID Hash Function

We provide a high level description of the alleged SecurID hash function, following the same notation as in [1] wherever possible. More detailed descriptions can be found in [1, 4].

The function can be modeled as a keyed hash function $y = H(k, t)$, where k is a 64-bit secret key stored on the SecurID token, t is a 24-bit time obtained from the clock every 30 or 60 seconds, and y is two 6- or 8-digit codes. The function consists of the following steps:

– an expansion function that expands t into a 64-bit "plaintext",
– an initial key-dependent permutation,
– four key-dependent rounds, each of which has 64 subrounds,
– an exclusive-or of the output of each round onto the key,
– a final key-dependent permutation (same algorithm as the initial one), and
– a key-dependent conversion from hexadecimal to decimal.

Throughout the paper, we use the following notation to represent bits, nibbles, and bytes in a word: a 64-bit word b, consisting of bytes $B_0, ..., B_7$, nibbles $\mathtt{B}_0, ..., \mathtt{B}_{15}$, and bits $b_0 b_1 ... b_{63}$. The nibble $\mathtt{B}_0$ corresponds to the most significant nibble of byte 0 and the bit b_0 corresponds to the most significant bit. The other values are as one would expect.

For our analysis, only the time expansion, key-dependent permutation, and the key-dependent rounds are of interest. In the next three sections, we will describe them in more detail.

2.1 Time Expansion

The time t is a 24-bit number representing twice the number of minutes since January 1, 1986 GMT. So the least significant bit is always 0, and if the token outputs codes every minute, then the expansion function will clear the 2nd least significant bit as well. Let the result be represented by the bytes $T_0 T_1 T_2$ where T_0 is the most significant. The expansion is of the form $T_0 T_1 T_2 T_2 T_0 T_1 T_2 T_2$. Note that the least significant byte is replicated 4 times, and the other two bytes are replicated 2 times each.

2.2 Key-Dependent Permutation

We give a more insightful description of how the ASHF key-dependent permutation really works. The original code, obtained by Wiener [4] (apparently by reverse engineering the ACE/server code), is quite cryptic. Our description is different, but produces an equivalent output to his code.

The key-dependent permutation uses the key nibbles $K_0 \ldots K_{15}$ in order to select bits of the data for output into a permuted_data array. The data bits will be taken 4 at a time, copied to the permuted_data array from right to left (i.e. higher indexes are filled in first), and then removed from the original data array. Every time 4 bits are removed from the original data array, the size shrinks by 4. Indexes within that array are always modulo the number of bits remaining.

A pointer m is first initialised to the index K_0. The first 4 bits that are taken are those right before the index of m. For example, if K_0 is 0x2, then bits 62, 63, 0, and 1 are taken. As these bits are removed from the array, the index m is adjusted accordingly so that it continues to point at the same bit it pointed to before the 4 bits were removed. The pointer m is then increased by a value of K_1, and the 4 bits prior to this are taken, as before. The process is repeated until all bits have been taken.

Note that once the algorithm gets down to the final 3 or less key and data nibbles, the number of data bits remaining is at most 12 yet the number of choices for each key nibble is 16. Hence, multiple keys will result in the same permutation, which we call "redundancy of the key with respect to the permutation." This was used in the attack [2], and to a lesser extent in [1].

2.3 Key-Dependent Rounds

Each of the four key-dependent rounds takes as inputs a 64-bit key k and a 64-bit value b^0, and outputs a 64-bit value b^{64}. The key k is then exclusive-ored with the output b^{64} to produce the new key to be used in the next round.

One round consists of 64 subrounds. For $i = 1, ..., 64$, subround i transforms b^{i-1} into b^i using a single key bit k_{i-1}. Depending on whether the key k_{i-1} is equal to b_0^{i-1}, the value b^{i-1} is transformed according to two different functions, denoted by R and S. The details of R and S are not so important for our research, with the exception of two properties:

1. Both the R and the S functions are byte-oriented, that is, they update each of the eight bytes in b^i separately. After the update, only bytes B_0 and B_4 are modified, and the other six bytes remain the same.
2. The way R and S are used causes the hash function to have easy-to-find collisions after a small number of subrounds within the *first* round.

At the end of each subround, all the bits are rotated one bit position to the left. So, up to subround $N \leq 25$ of the first round, only $2N + 14$ data bits have been involved in the computation. This property is used in the Biryukov, Lano, and Preneel attack.

3 The Attack of Biryukov, Lano, and Preneel

The attack of Biryukov, Lano, and Preneel [1] can determine the full 64-bit secret key when given a single collision of the hash function. Suppose that two input times t and t' get expanded and permuted to become 64-bit words b and b', and the two words collide in subround N of the first round. The collision from the pair (t, t') is called a *vanishing differential.* In their key recovery attack, the attacker first guesses the subround N, and then uses a *filtering algorithm* for each N to search the set of candidate keys that make such a vanishing differential possible. According to their simulations, one only needs to do up to $N = 12$ to have a 50% chance of finding the key. [2] A summary of their description for $N = 1$ is given below. For simplicity, assume that a 2-bit vanishing differential is used, though this need not be the case.

A one-time cost precomputation table is needed before the filtering starts. The table contains entries the form

$$(k_0, B_0, B_4, B'_0, B'_4).$$

where k_0 represents a key bit, (B_0, B_4) represent data bytes of b after the initial keyed permutation, and (B'_0, B'_4) represent data bytes of b' after the permutation. The exact entries in the table are those where (B_0, B_4) differs from (B'_0, B'_4) in exactly 2-bits known as the "difference bits," and for which a vanishing differential occurs during the first subround. Since none of the other key bits or data bytes are involved in the first subround, whether a vanishing differential can happen or not for $N = 1$ is completely characterised by this table.

For each entry in the table, the filtering proceeds in two phases, each of which contains two steps.

- *First Phase.* (process the first half of the key bits)
  - *Step One.* Guess key bits $k_1, ..., k_{27}$. Together with k_0, 28 key bits are set, which determines 28 bits of b and b' after the initial key-dependent permutation. Since these bits overlap with the entries in the table in nibbles $\mathtt{B}_9$ and $\mathtt{B}'_9$, a key value that does not produce the correct nibbles for both b and b' is filtered out.
  - *Second Step.* Continue to guess key bits $k_{28}, ..., k_{31}$. Filtering is done using overlaps in nibbles $\mathtt{B}_8$ and $\mathtt{B}'_8$.
- *Second Phase.* (process the second half of the key bits)
  - *First Step.* Continue to guess key bits $k_{32}, ..., k_{59}$. Filtering is done using overlaps in nibbles $\mathtt{B}_1$ and $\mathtt{B}'_1$.
  - *Second Step.* Continue to guess key bits $k_{60}, ..., k_{63}$. Filtering is done using overlaps in nibbles $\mathtt{B}_0$ and $\mathtt{B}'_0$.

Finally, each candidate key that passes the filtering is tested by performing a full hash function to see if it is the correct key. For general N, the two phases of filtering each involve $\lceil \frac{7+N}{4} \rceil$ data nibbles, so the phases each have $\lceil \frac{7+N}{4} \rceil$ steps.

[2] Our own simulations suggest that one needs to search up to $N = 16$. The discrepancy is due to differences in the way the attack is viewed, which we elaborate on in Section 7.1. For larger values of N, the cost of the precomputation stage becomes prohibitive.

4 Analysis of the Biryukov, Lano, and Preneel Attack

Biryukov, Lano, and Preneel estimated the time complexity of their attack through simulation. They provided results for $N = 1$: step 1 of phase 1 reduced the number of possibilities to 2^{27}, step 2 of phase 1 further reduced the count to to 2^{25}, step 1 of phase 2 increased the count to 2^{45}, and step 2 of phase 2 resulted in 2^{41} true candidates. For larger values of N, they expect that the complexity of the attack would be lower due to stronger filtering.

Here we analyse their algorithm, giving some mathematical justification for the simulation results they observed and also showing that their conjecture of the filtering improving for larger N appears to be correct. In our analysis, we sometimes treat probabilities as if they are independent, which is not always true, but it is assumed that it provides a reasonable approximation.

Some properties of the precomputed tables are used in the analysis. For a given value of N, the table entries are of the following form:

- legal values for the key bits in indices $0, \dots, N-1$,
- legal values for the plaintext pairs after the initial permutation in bit indices $32, 33, \dots, 38+N$ which we label as (W_4, W_4') (we use the subscript 4 because the words begins at byte B_4), and
- legal values for the plaintext pairs after the initial permutation in bit indices $0, 1, \dots, 6+N$ which we label as (W_0, W_0') (the word begins at byte B_0).

The words W_0, W_0', W_4, W_4' each consist of $7+N$ bits and the number of key bits is N. By "legal values" we mean that the combination of plaintext bits after the initial permutation and key bits will cause the difference to vanish in subround N. We also have one other requirement, which was previously overlooked (including in an earlier version of this research): the values of the two bits in b (or b') where the differences are located must be the same, due to the way the time expansion works. This reduces the number of table entries and results in a speedup to the filtering. Although this is one of our three main filtering speedups, we apply it to the analysis of the original [1] algorithm in order to keep things as clean as possible.

Analysis of Final Number of Candidates: Analysing the final step is equivalent to determining the true number of candidates that need to be tested with the full SecurID hash function. The expected number of true candidates can easily be determined since anything that matches an entry in the precomputed table will result in a vanishing differential. In other words, the entries in the table are not only a necessary set of cases for a vanishing differential to occur, but also sufficient.

For each entry in the precomputed table, we have:

- Only a portion of about $1/\binom{64}{2}$ of the 2^{64} keys will permute the 2 difference bits into the locations corresponding to what is in that table entry.

– With probability $\frac{1}{2}$, the value of the two difference bits will match those in the table (recall, the 2 bits in b must be the same, and the corresponding bits in b' are the complement).
– With probability $\frac{1}{2^{2N+12}}$, the remaining permuted data bits will match the table entry.
– With probability $\frac{1}{2^N}$ the guessed key bits will match the entry of the table.

Hence, the expected number of final candidates is:

$$\text{table size} \times 2^{64} \times \frac{1}{\binom{64}{2}} \times \frac{1}{2} \times \frac{1}{2^{2N+12}} \times \frac{1}{2^N} \ . \tag{1}$$

Run Time Analysis of Phase 2, Step 1: Phase 2, step 1 of the Biryukov, Lano, and Preneel attack is typically the dominant cost. To analyse it, we must first determine the number of candidates passing phase 1.

Define C_0 to be the number of unique table entries of the form $(k_0, \ldots, k_{N-1}, W_4, W_4')$ where $W_4 = W_4'$, C_1 similarly except $W_4 \oplus W_4'$ having hamming weight 1, and C_2 similarly except $W_4 \oplus W_4'$ having hamming weight 2.

Among the of 2^{32} key bits considered in phase 1, a fraction of $\binom{57-N}{2}/\binom{64}{2}$ will put no difference in the tuple (W_4, W_4'). Of those, only a fraction of $\frac{C_0}{2^{7+N}}$ will match one of the C_0 unique entries in the table for W_4 (which is the same as W_4'). With probability $\frac{1}{2^N}$, the guessed key bits will match those in the table as well. Thus, the expected number of 32-bit keys resulting in no difference in (W_4, W_4') that pass phase 1 is:

$$2^{32} \times \frac{\binom{57-N}{2}}{\binom{64}{2}} \times \frac{C_0}{2^{7+N}} \times \frac{1}{2^N} = 2^{19-2N} \times \frac{3192 - 113N + N^2}{63} \times C_0 \ .$$

For 1-bit differences, the equation is

$$2^{32} \times \frac{\binom{57-N}{1}}{\binom{64}{2}} \times \frac{1}{2} \times \frac{C_1}{2^{6+N}} \times \frac{1}{2^N} = 2^{20-2N} \times \frac{57 - N}{63} \times C_1 \ .$$

For 2-bit differences, the equation is

$$2^{32} \times \frac{1}{\binom{64}{2}} \times \frac{1}{2} \times \frac{C_2}{2^{5+N}} \times \frac{1}{2^N} = 2^{21-2N} \times \frac{C_2}{63} \ .$$

The $\frac{1}{2}$ in this last equation accounts for whether the two difference bits in the first plaintext match the table entry (the bits must be the same). Thus, the expected number of candidates to pass the phase 1 is

$$T = \frac{2^{19-2N}}{63} \times \left[(3192 - 113N + N^2)C_0 + (114 - 2N)C_1 + 4C_2\right] \ . \tag{2}$$

The first step in phase 2 involves guessing enough key bits so that the resulting permuted data array just begins to overlap with W_0 and W_0'. The exact

Table 1. Computing the running time estimates of algorithm [1] for $N = 1..6$

N	Table size	C_0	C_1	C_2	T	Time for phase 2, step 1	Time for testing final candidates	Total time
1	12	5	2	0	$2^{25.0}$	$2^{47.0}$	$2^{40.6}$	$2^{47.0}$
2	152	11	64	44	$2^{24.3}$	$2^{43.2}$	$2^{41.3}$	$2^{43.5}$
3	1130	64	362	128	$2^{24.8}$	$2^{43.6}$	$2^{41.2}$	$2^{43.9}$
4	7292	453	1750	712	$2^{25.5}$	$2^{44.3}$	$2^{40.9}$	$2^{44.4}$
5	48212	2775	10614	3864	$2^{26.0}$	$2^{44.9}$	$2^{40.6}$	$2^{44.9}$
6	276788	15076	52716	19520	$2^{26.4}$	$2^{41.4}$	$2^{40.1}$	$2^{41.9}$

number of key bits guessed in this step is $4 \times \lfloor \frac{29-N}{4} \rfloor$. Under the assumption that the permutation is 5% of the time required to do the full SecurID hash, the running time is equivalent to

$$T \times 2^{4 \times \lfloor \frac{29-N}{4} \rfloor} \times \frac{4 \times \lfloor \frac{29-N}{4} \rfloor}{64} \times 2 \times 0.05 \times s \tag{3}$$

full hash operations, where s is the speedup factor that can be obtained by taking advantage of the redundancy in the key with respect to the permutation. The value of s is $\frac{96}{256}$ for $N = 1$, $\frac{12}{16}$ for $N = 2..5$, and 1 for all other values.

We remark that in some cases, there is a chance that the second step of phase 2 may be a bit more time consuming than the first. A sufficient but not necessary condition for step 1 to be the most time consuming is if the fraction of values that remain is less than $\frac{\lfloor \frac{29-N}{4} \rfloor}{16}$ of the values considered. This is usually the case. We shall ignore the exceptional cases for now, but will deal with them when we present our filtering speedups.

Combined Analysis: The running time of algorithm [1] for a particular value of N is expected to be the approximately the sum of equations 3 and 1. For $N = 1..6$, these running times are given in Table 1. Again, we reiterate that the table sizes are different from [1] because of an extra condition due to the time expansion, which also gives a small improvement in the running time. The analysis for $N = 1$ closely matches the simulated results from [1].[3]

Even though the number of candidates T after the first phase are approximately the same as N goes from 1 to 2 and also from 5 to 6, the running times of the phase 2, step 1 drop significantly. This is because one less nibble of the key is being guessed, and an extra filtering step is being added. In general, we see the pattern that larger values of N are contributing less and less to the sum of the running times, which agrees with the conjecture from [1]. The total running time for $N = 1$ to 6 is $2^{47.7}$ and larger values of N would appear to add minimally to this total. For vanishing differentials that involve $\geq$ 4-bits, which happens about one third of the time, preliminary analysis suggests that the run time is better.

[3] A small discrepancy for T exists due to the fact that their simulations involved a precomputed table about twice as big ours.

5 Faster Filtering

Table 1 illustrates that the trick to speeding up the key recovery attack in [1] is faster filtering. We have found three ways in which their third filtering can be sped up:

1. Only include entries in the precomputed table that actually can be derived from the time expansion. In particular, the values of the two bits in b (or b') where the differences are located must be the same.
2. In the original filter, a separate permutation is computed for each trial key. This is inefficient, since most of the permuted bits from one particular permutation will overlap with those from many other permutations. Thus, we can amortise the cost of the permutation computations.
3. We can detect ahead of time when a large portion of keys will result in "bad" permutations in steps 1 of both phase 1 and phase 2, and the filtering process can skip past chunks of these bad permutations.

The first technique was already applied to the analyses in the previous section. Without this improvement, the running time would have been about 50% worse.

The second technique is aimed at reducing the numerator of the factor $\frac{4\times\lfloor\frac{29-N}{4}\rfloor}{64} = \frac{\lfloor\frac{29-N}{4}\rfloor}{16}$ in equation 3. To do this, we view the key as a 64-bit counter, where k_0 is the most significant bit and k_{63} is the least. In phase 2, step 1 of the filter, the bits $k_0, \ldots, k_{31}$ are fixed and so are some of the least significant bits (the exact number depends upon N), so we can exclude these for now. The keys are tried in order via a recursive procedure that handles one key nibble at a time. At the jth recursive branch, each of the possibilities for nibble $\mathtt{K}_{7+j}$ are tried. The part of the permutation for that nibble is computed, and then the $j+1^{\text{st}}$ recursive branch is taken. The level of recursion stops when key nibble $\mathtt{K}_{7+\lfloor\frac{29-N}{4}\rfloor}$ is reached. Thus, the $\lfloor\frac{29-N}{4}\rfloor$ from equation 3 gets replaced with the average cost per permutation trial, which is $\sum_{i=0}^{\lfloor\frac{29-N}{4}\rfloor-1} 2^{-4i} \approx 1.07$. Observe that when $N = 1$, this results in a factor of $\frac{7}{1.07} \approx 6.5$ speedup. This trick alone knocks more than 2 bits off the running time.

The third speedup is dependent upon the second. It will apply in both phases of the filtering. During the process of trying a permutation, there will be large chunks of bad trial keys that can be identified immediately and skipped. In particular, whenever a difference bit is placed outside of words (W_0, W_0') and (W_4, W_4'), the key can be skipped because the difference is not in a legal position. Moreover, any other key with the same most significant bits (up to the key nibble that placed the difference bit) will also result in illegal values, implying that the entire recursive branch can be skipped. Heuristically, one would expect that the number of keys that get tested for filtering in phase 2, step 1 to be about a fraction of about $\binom{14+2N}{2}/\binom{64}{2}$ of the number for the attack in [1]. However, this over simplifies the analysis. A more proper analysis can be done similar to our analysis in the previous section.

Table 2. Running times using our improved filter, for $N = 1..6$

N	Time for phase 2, step 1	Time for testing final candidates	Total time
1	$2^{38.7}$	$2^{40.6}$	$2^{40.9}$
2	$2^{36.4}$	$2^{41.3}$	$2^{41.3}$
3	$2^{37.1}$	$2^{41.2}$	$2^{41.3}$
4	$2^{37.9}$	$2^{40.9}$	$2^{41.1}$
5	$2^{38.6}$	$2^{40.6}$	$2^{40.9}$
6	$2^{35.7}$	$2^{40.1}$	$2^{40.2}$

The combined speedups give the run times in Table 2. In all cases, phase 2, step 1 has become faster than the time for testing the final candidates. The running time for $N = 1..6$ is $2^{43.6}$, so we conjecture that the run time for N up to 16 is no more than $16/6 \times 2^{43.6} \approx 2^{45}$. We remark that the run times for the third speedup ignore the overhead time for rejecting keys in phase 2 where the difference bit gets put outside of (W_0, W_0'), but such overhead time is expected to make little difference. We have also ignored the time for other filtering steps of the algorithm. Of those, only step 2 of phase 2 is expected to have comparable cost to step 1 of phase 2. In fact, it can be more costly, especially when $N \equiv 2 \bmod 4$. However, there are several possible speedups for this step, particularly when N is small (this restriction is for practical reasons) where the run time becomes most relevant. Such speedups involve using additional preocomputed lookup tables to determine valid keys from the remaining data bits and testing whether the hamming weight of the remaining data bits matches that of the precomputed table entries before blindly trying keys. Therefore, it seems fair to assume that the testing of final candidates will always be the dominant cost in the modified algorithm.

Although it appears that we cannot do much better using only a single vanishing differential, we can improve the situation if we use other information that an attacker would have. In later sections we will show that we can improve the time greatly if we take advantage of multiple vanishing differentials, or if we take advantage of knowledge that no other vanishing differentials occur within a small time period of the observed one.

6 Software Implementation

The attack of Biryukov, Lano, and Preneel was specially designed to keep RAM usage low - only one of the precomputed table entries needs to be in program memory at a time. We tested our ideas only for $N = 1$ and 2-bit differences, and since the table size is small, we took the freedom of implementing a slight variant of their attack which kept the whole precomputed table in memory at once.

We programmed all filtering steps of both phases and the three main filtering speedups. In addition, we programmed an extra "table lookup" speedup that would improve the running time by a factor of 8 for $N = 1$. The extra speedup is only applicable for small values of N due to the memory requirements. Thus, the running time is expected to be 8 times faster than the $2^{38.7}$ listed in Table 2. On our 2.4 GHz PC, this translates to about 8 days of effort.

Our code did the search in numerical order, when the key is viewed as a counter as described in Section 5. The only thing we did not do was testing the final candidates using the real function. Instead, we just stopped when we arrived at the target key. So our implementation was designed to test and time the filtering only, in order to confirm that filtering is significantly faster than testing of the final candidates.

At the time of writing, we have not done the full key search yet. However, we have done a search that starts out knowing the correct first nibble of the key. The key we were searching for is `356b48b3ae15c271` which yields a vanishing differential when times `0x1c3ba8` and `0x1c3aa8` are sent in. We were able to find the key in 13.8 hours. If we assume that the full search will take at most 2^4 times longer, the full running time would be 9.2 days, which is on target of expectations.

7 Multiple Vanishing Differentials

There are two scenarios for multiple vanishing differentials: when they have the same difference and when they have different differences. The former is more likely to occur, but in either case we can speed up the attack.

7.1 Multiple Vanishing Differentials with the Same Difference

According to computer simulations, about 45% of the keys that had a collision over a two month period will actually have at least 2 collisions. There is a simple explanation for this, and a way to use the observation to speed up the key search even more.

Consider a vanishing differential which comes from times $t = T_0T_1T_2$ and $t' = T_0'T_1'T_2'$. As we saw earlier, the only bits that determine whether the vanishing differential will occur at a particular subround are those that get permuted into words W_0, W_0', W_4, and W_4'. Suppose we flip one of the bits in T_2 and T_2' (the same bit in each). This bit will be replicated four times in the time expansion. If, after the permutation, none of those bits end up in W_0, W_0', W_4, or W_4', then we will witness another vanishing differential. The new vanishing differential will follow the same difference path and disappear in the same subround. Thus, new information is learned that can be used to speed up the key search, which we explain below. In the case that another vanishing differential does *not* occur, information is also learned which can improve the search, which is detailed in Section 8.

Following the above thought process, it is evident that:

- Flipping time bits in T_1, T_1' or T_0, T_0' will only replicate the flipped bit twice in the expansion. Since there are only two bits that are not allowed to be

Table 3. Number of final candidates assuming the attacker became aware of z-bits that do not get permuted into words W_0, W_0', W_4, or W_4'

N	Number of final cands using only a single collision	Number of final cands with $z = 2$	Number of final cands with $z = 4$	Number of final cands with $z = 8$
1	$2^{40.6}$	$2^{39.8}$	$2^{38.9}$	$2^{37.0}$
2	$2^{41.3}$	$2^{40.3}$	$2^{39.3}$	$2^{37.2}$
3	$2^{41.2}$	$2^{40.1}$	$2^{39.0}$	$2^{36.6}$
4	$2^{40.9}$	$2^{39.7}$	$2^{38.4}$	$2^{35.7}$
5	$2^{40.6}$	$2^{39.2}$	$2^{37.8}$	$2^{34.8}$
6	$2^{40.1}$	$2^{38.6}$	$2^{37.0}$	$2^{33.6}$

in W_0, W_0', W_4, and W_4', the collision is more likely to occur. On the other hand, the time between the collisions is increased, since these are more significant time bits.

- Multiple vanishing differentials are more likely to occur when the first collision happened in a small number of subrounds. This is because the words W_0, W_0', W_4, and W_4' are smaller, giving more places where the flipped bits can land without interfering with the collision.[4]
- The converse of these observations is that when multiple vanishing differentials occur, it is most often the case that the collisions all happened in the same subround and followed the same difference path. Moreover, the collisions usually happen within a few subrounds.

By simply eying the time data that caused the multiple vanishing differentials, one can determine with close to 100% accuracy whether this situation has happened. The signs of it are: 1) Same input difference for all vanishing differentials, 2) All input times differ in only a few bits, and 3) It is the same bits that differ in all cases. An example is given in Appendix B.

The attacker learns $z \geq 2$ bits which cannot be permuted to words W_0, W_0', W_4, or W_4'. This new knowledge can be combined with our third filtering speedup to skip past more bad keys. The expected number of final key candidates to be tested becomes a fraction of $\binom{50-2N}{z}/\binom{64}{z}$ of the values given in Table 2. See Table 3 for a summary of these figures when $z = 2$, $z = 4$, and $z = 8$. The times can be further reduced using information about where certain related plaintexts did not cause a vanishing differential: see Section 8.

[4] This is the reason for the apparent discrepancy between our research claiming that one needs to precompute up to $N = 16$ in order to have a $\geq 50\%$ of find the key and [1] claiming 12. In our view, the attacker has a single token and will perform a key search once a single vanishing differential has occurred. In their view, the attacker has several tokens for a fixed period of time, and the attacker selects a vanishing differential randomly among all vanishing differentials that have occurred [3]. Since their view includes multiple vanishing differentials, the expected number of subrounds is less.

7.2 Multiple Vanishing Differentials with Different Differences

Given two vanishing differentials with different differences, the number of candidate keys can be reduced significantly by constructing more effective filters in each step. Denote the two pairs of vanishing differentials V_1 and V_2, and their N values N_1 and N_2.

We first make a guess of (N_1, N_2). The number of guesses will be quadratic in the number of subrounds tested up to. The following is a simplified sketch for the new filtering algorithm.

- *First Phase.* Take V_1 and guess the first 32 bits of the key. For each 32-bit key that produces a valid (W_4, W_4'), test it against V_2 to see if it also produces a valid (W_4, W_4').
- *Second Phase.* For 32-bit keys that pass phase 1, do the same thing to guess the second 32 bits of the key.

The main idea here is to do double filtering within each stage so that the number of candidate keys is further reduced in comparison to when only a single vanishing differential is used.

When $N_1 = N_2 = 1$, the probability that a 32-bit key passes phase 1 (see Table 1) is $2^{25.0}/2^{32} = 2^{-7.0}$ (assuming using the original filter of [1] - it is even more reduced using our improved filter), and the probability that a 64-bit key passes both phases is $2^{40.6}/2^{64} = 2^{-23.4}$. If the two vanishing differentials are indeed *independent*, we would expect the number of keys to pass the first phase to be

$$2^{32} \times 2^{-7.0} \times 2^{-7.0} = 2^{18}$$

and the number of keys to pass both phases to be

$$2^{64} \times 2^{-23.4} \times 2^{-23.4} = 2^{17.2}.$$

Experimental results will reveal whether these figures are attainable in practice, but even if they are not, a big speed up is still expected. The situation should be better in the cases where differences with hamming weights ≥ 4 are involved.

We should mention the caveat that the chances of success using the above technique are lower, since we need *both* difference pairs to disappear within 16 subrounds. On the other hand, the cost of trying this algorithm for two difference pairs is expected to be substantially cheaper than trying the previous algorithms for only one. Therefore, the double filtering should add negligible overhead to the search in the cases that it fails, and would greatly speedup the search when it is successful.

8 Using Non-Vanishing Differentials with a Vanishing Differential

In Section 7.1, we argued that even if only a single vanishing differential occurs over some time period, the search can still be sped up if one takes advantage of knowing where related differentials do not vanish. Here, we give the details.

Assume a vanishing differential occurred at times t and t', but no vanishing differential occurred among the time pairs $(t \oplus 2^i, t' \oplus 2^i)$ for $i = 2, \ldots, j$. We start with $i \geq 2$ because in the most typical case, where authenticators are displayed every minute, the least two significant bits of the time are 0 (see Section 2.1). For the values $2 \leq i \leq 7$, the difference is replicated 4 times in the time expansion, and for $i \geq 8$, it is replicated twice.

For each value of i, we learn a set of 2 or 4 bits for which at least one in each set must be permuted into the words W_0, W_0', W_4, or W_4'. Let us label these sets as $U_2, \ldots, U_j$. For simplicity, we will take $j = 13$, which corresponds to no other vanishing differential within a window of 2.8 days before or after the observed one. So, we are interested in the probability of at least one bit in each of these sets getting permuted into words W_0, W_0', W_4, or W_4'.

We say a set U_i is *represented* with $c_i \geq 1$ bits if exactly c_i bits from U_i get permuted into W_0, W_0', W_4, or W_4'. The number of ways $2N + 14$ bits can be selected to end up in W_0, W_0', W_4, or W_4' is $\binom{64}{2N+14}$. The number of ways that exactly c_i bits are represented in the selection for $2 \leq i \leq 13$ is

$$\prod_{i=2}^{7} \binom{4}{c_i} \times \prod_{i=8}^{13} \binom{2}{c_i} \times \binom{28}{2N + 14 - \sum_{i=2}^{13} c_i}.$$

The first product tells the number of ways of selecting c_i bits from each set that has 4 bits, the second product is the same except for among sets with 2 bits, and the third product is the number of ways of selecting the remaining bits from the 28 bits that are not among any of the U_i. Thus, our desired probability is:

$$\sum_{\text{all valid } (c_2, \ldots, c_{13})} \frac{\prod_{i=2}^{7} \binom{4}{c_i} \times \prod_{i=8}^{13} \binom{2}{c_i} \times \binom{28}{2N+14-\sum_{i=2}^{13} c_i}}{\binom{64}{2N+14}} \tag{4}$$

where *valid* $(c_2, \ldots, c_{13})$ means that each value is at least 1, but the sum of all values is no more than $2N + 14$.

We have computed these probabilities using the Magma [6] computer algebra package. The probabilities, and corresponding running time for the testing of final candidates are given in Table 4. Monte Carlo experiments have been done to double-check the accuracy of these results. The fact that the probabilities are so small for low values of N is consistent with the argument in Section 7.1 that when a collision happens early, other collisions are likely to follow soon after.

One should not assume that the times for the testing the final candidates given in Table 4 are the dominant cost in applying this strategy. Unlike the filtering speedups given in Sections 5 and 7.1, the use of non-vanishing differentials seem to require more overhead in checking the conditions. So although we do not have an exact running time, we confidently surmise that the use of non-vanishing differentials will reduce the time down below 2^{40} hash operations.

Table 4. Assuming no more vanishing differentials occur within 2.8 days before or after of a given vanishing differential, the final testing of candidates can be improved by the amounts given in this table

N	Fraction of keys having property	Time for testing final candidates
1	$2^{-14.3}$	$2^{26.3}$
2	$2^{-11.7}$	$2^{29.6}$
3	$2^{-9.7}$	$2^{31.5}$
4	$2^{-8.1}$	$2^{32.8}$
5	$2^{-6.7}$	$2^{33.9}$
6	$2^{-5.7}$	$2^{34.4}$

9 Conclusion

The design of the alleged SecurID hash function appears to have several problems. The most serious appears to be collisions that happen far too frequently and very early within the computation. The involvement of only a small fraction of bits in the subrounds exacerbates the problem. Moreover, the redundancy of the key with respect to the initial permutation adds an extra avenue of attack. Altogether, ASHF is substantially weaker than one would expect from a modern day hash function.

Our research has shown that the key recovery attack in [1] can be sped up by more than a factor of 8, giving an improved attack with time complexity about 2^{45} hash operations. In practice, the attacker can actually obtain more information than just a single collision. We have shown that, with this extra information, the time complexity can be further reduced to about 2^{40} hash operations, making the attack doable by anyone with a modern PC.

Acknowledgements

We are grateful to Joe Lano for his insights and helpful comments, and for his hospitality while the first author of this document visited Brussels.

References

[1] A. Biryukov, J. Lano, B. Preneel. *Cryptanalysis of the Alleged SecurID Hash Function*, In Proceedings of SAC 2003, to appear in LNCS. A longer version of this paper is available online from http://eprint.iacr.org/2003/162. 454, 455, 456, 457, 459, 461, 462, 465, 466, 468, 469

[2] S. Contini, *The Effect of a Single Vanishing Differential in ASHF*, sci.crypt post, 6 Sep, 2003. 457

[3] J. Lano, private communication, 28 Oct, 2003. 465

[4] I.C. Wiener, *Sample SecurID Token Emulator with Token Secret Import*, post to BugTraq, http://archives.neohapsis.com/archives/bugtraq/2000-12/0428.html, 21 Dec, 2000. 455, 456
[5] *Tips on Reassigning SecurID Cards and Requesting New SecurID Cards*, AMS Newsletter, March 2002, Issue No. 117. Available at http://www.utoronto.ca/ams/news/117/html/117-5.htm . 455
[6] *The Magma Computer Algebra Package*. Information available at http://magma.maths.usyd.edu.au/magma/ . 467

A Analysing Precomputed Tables

Using computer experiments, we were able to exhaustively search for valid entries in the precomputed table up to $N = 6$ for 2-bit vanishing differentials and up to $N = 4$ for 4-bit differentials at this point. It was predicted in [1] that the size of the table gets larger by a factor of 8 as N grows and it may take up to 2^{44} steps and 500GB memory to precompute the table for $N = 12$.

Here we make an attempt to derive the entries in the table analytically when $N = 1$. If we could extend the method to $N > 1$, we may be able to enumerate the entries analytically without expensive precomputation and storage.

We start with Equation (6) in [1]. Note that we are trying to find constraints for the values in the subround $i-1$. So for simplicity, we will omit the superscript $i-1$ from now on, and Equation (6) becames the following.

$$B'_4 = ((((B_0 >>> 1) - 1) >>> 1) - 1) \oplus B_4, \qquad (5)$$
$$B'_0 = 100 - B_4 \ .$$

We first note that B_0 and B'_0 have to be different in the msb. Therefore, there is at least one bit difference in (B_0, B'_0). The other bit difference can be placed either in the remaining 7 bits of (B_0, B'_0) or any of the 8 bits in (B_4, B'_4).

Rewriting Equation 5, we have

$$B_0 = (((B_4 \oplus B'_4) + 1) <<< 1) + 1) <<< 1.$$

Since there are at most one bit difference in (B_4, B'_4), it can only take on 9 possible values: 0 (for no bit difference) or 2^i (for one bit difference in bit i). Below, for each possible value of (B_4, B'_4), we enumerate the possible values of (B_0, B'_0). During the enumeartion, we also take into consideration the additional requirement that the two bits in b where the differences occur must be the same (See Section 4).

- If $B_4 \oplus B'_4 = 0$, then B_0 =0x06. Since there is no bit difference in (B_4, B'_4), we know that B_0 and B'_0 differ in two bits – one of them must be the msb, and the other can be any of the remaining bits.

$B_4 \oplus B'_4$	B_0	B'_0	k_0
0x00	0x06	0x87, 84, 82, 8e, 96, a6, c6	0

Table 5. Example of 16 vanishing differentials that happened within 1.3 days, using key `b5 a9 f4 8c 16 23 a6 1a`

First plaintext	Second plaintext
1e 80 8c 8c 1e 80 8c 8c	1e 90 8c 8c 1e 90 8c 8c
1e 81 8c 8c 1e 81 8c 8c	1e 91 8c 8c 1e 91 8c 8c
1e 82 8c 8c 1e 82 8c 8c	1e 92 8c 8c 1e 92 8c 8c
1e 83 8c 8c 1e 83 8c 8c	1e 93 8c 8c 1e 93 8c 8c
1e 84 8c 8c 1e 84 8c 8c	1e 94 8c 8c 1e 94 8c 8c
1e 85 8c 8c 1e 85 8c 8c	1e 95 8c 8c 1e 95 8c 8c
1e 86 8c 8c 1e 86 8c 8c	1e 96 8c 8c 1e 96 8c 8c
1e 87 8c 8c 1e 87 8c 8c	1e 97 8c 8c 1e 97 8c 8c
1e 88 8c 8c 1e 88 8c 8c	1e 98 8c 8c 1e 98 8c 8c
1e 89 8c 8c 1e 89 8c 8c	1e 99 8c 8c 1e 99 8c 8c
1e 8a 8c 8c 1e 8a 8c 8c	1e 9a 8c 8c 1e 9a 8c 8c
1e 8b 8c 8c 1e 8b 8c 8c	1e 9b 8c 8c 1e 9b 8c 8c
1e 8c 8c 8c 1e 8c 8c 8c	1e 9c 8c 8c 1e 9c 8c 8c
1e 8d 8c 8c 1e 8d 8c 8c	1e 9d 8c 8c 1e 9d 8c 8c
1e 8e 8c 8c 1e 8e 8c 8c	1e 9e 8c 8c 1e 9e 8c 8c
1e 8f 8c 8c 1e 8f 8c 8c	1e 9f 8c 8c 1e 9f 8c 8c

The additional requirement rules out two possible values of B'_0 (0x84, 0x82), leaving 5 possible combinations.

- If $B_4 \oplus B'_4 = 2^i$, then there is only one bit difference in (B_0, B'_0), which is the msb. In this case, there are only one choice for B'_0 for each B_0.

$B_4 \oplus B'_4$	B_0	B'_0	k_0
0x01	0x0a	0x8a	0
0x02	0x0e	0x8e	0
0x04	0x16	0x96	0
0x08	0x26	0xa6	0
0x10	0x46	0xc6	0
0x20	0x86	0x06	1
0x40	0x07	0x87	0
0x80	0x08	0x88	0

The additional requirement rules out every combination above except the first one (B_0 =0x0a and B'_0 =0x8a).

Combining the above two cases, we have $5 + 1 = 6$ pairs of (B_0, B'_0), each of which giving a valid tuple $(k_0, B_0, B_4, B'_0, B'_4)$, where k_0 is the msb of B_0.

Finally, note that if (k_0, a, b, c, d) is a valid tuple, than (k_0, c, d, a, b) is also a valid typle. For example, if (0, 0x06, 0xdd, 0x87, 0xdd) is valid, then (0, 0x87, 0xdd, 0x06, 0xdd) is also valid. Therefore, the table consists of a total of $2 \times 6 = 12$ entries. These entries match the results from our simulation.

B Example of Multiple Vanishing Differentials

Table 5 is an example where 16 vanishing differentials happened within 1.3 days. All had the same difference path, which collided at $N = 2$. One can see that only the 4 least significant bits of time byte T_1 differ. Since each of these bits are duplicated twice, the expected running time of the last steps is given by $z = 8$ in Table 3. Taking into consideration $N = 2$, the total time is expected to be on the order of 2^{38} operations.

A MAC Forgery Attack on SOBER-128

Dai Watanabe[1] and Soichi Furuya[1]

Systems Development Laboratory, Hitachi, Ltd.
292 Yoshida-cho, Totsuka-ku, Yokohama, 244-0817, Japan
{daidai, soichi}@sdl.hitachi.co.jp

Abstract. SOBER-128 is a stream cipher designed by Rose and Hawkes in 2003. It can be also uses for generating Message Authentication Codes (MACs). The developers claimed that it is difficult to forge the MAC generated by SOBER-128, though, the security model defined in the proposal paper is not realistic. In this paper, we examine the security of the MAC generation function of SOBER-128 under the security notion given by Bellare and Namprempre. As a result, we show the MAC generation function of SOBER-128 is vulnerable against differential cryptanalysis. The success probability of this attack is estimated at 2^{-6}.

Keywords: Stream cipher, Message Authentication Code, Differential cryptanalysis, SOBER

1 Introduction

The desire to use a known cryptographic module for various applications exists from long ago. The first trial is realized as modes of operation of a block cipher. The OFB-mode and the counter-mode are the usages of a block cipher for a random number generation and CBC-MAC provides a message authentication mechanism. On the other hand, Anderson and Biham presented the construction of a block cipher from a pseudo-random number generator (PRNG) and a hash function [1]. The security of these modes of operation are provable under the assumption that the underlying primitives are ideal cryptographic functions.

Daemen considered the construction of elemental cryptographic functions, a block cipher, a PRNG , and a hash function, from unreliably weak functions, e.g., a round function of a block cipher [6]. The security of these constructions are not certain, but their processing speeds are often significantly faster than that of modes of operation. Daemen and Clapp proposed a cryptographic module PANAMA in 1998 [7]. PANAMA can be used as a PRNG and as a hash function. Ferguson *et al.* proposed an authenticated encryption algorithm Helix in 2003 [11]. Helix can be also used as a PRNG and for a MAC generation.

SOBER [18] is a stream cipher developed by Rose in 1998. SOBER adopts a linear feedback shift register (LFSR) defined over $GF(2^8)$ as the update function of the internal state. This construction enables an efficient software implementation in 8-bit processors so that LFSRs defined over an extension field are employed not only by SOBER, but also by SSC2 [20], SNOW [8, 10], and so on.

B. Roy and W. Meier (Eds.): FSE 2004, LNCS 3017, pp. 472–482, 2004.

Several variants of SOBER have been developed to strengthen its security or to be suitable for 16-bit and 32-bit processors. SOBER-t16 [12], t32 [13], Turing [14], SOBER-128 [15] are the published algorithms. SOBER-t16 and t32 were submitted to NESSIE project, but were rejected because some security flaws are reported [4, 5, 9]. A weakness of the initialization of Turing has been also reported [16].

SOBER-128 is the latest algorithm in SOBER family, and is the modified version of SOBER-t32. In addition, SOBER-128 can be used not only as a stream cipher, but also as a MAC generation. However, the security of MAC generation algorithm of SOBER-128 has not been evaluated, so is still unclear.

In this paper we examine the security of the MAC generation function of SOBER-128. We show that MAC generation algorithm of SOBER-128 is vulnerable against differential cryptanalysis and the forgery of the MAC generated by SOBER-128 is successful with high probability, about 2^{-4}.

The attack that we present in this paper is out of the expectation of the designers. Our attack follows the security notion defined by Bellare and Namprempre [2].

The rest of this paper is organized as follows. Firstly we define the supposed attacker in this paper in Sect. 2. Secondly we briefly describe the MAC generation algorithm of SOBER-128 in Sect. 3. Then we show how to forge MAC generated by SOBER-128 in Sect. 4 and discuss the weakness of SOBER-128 in Sect. 5. At last we summarize the result in Sect. 6.

2 The Model of the Attack

The designers mentioned that it is necessary for the recovery of the internal state (or the secret key) to forge a MAC generated by SOBER-128 and resolved the difficulty of a MAC forgery into that of the recovery of the internal state. In the attack to recover the internal state, the attacker can generate a MAC for a message with a secret key and an IV only once. Under the security notional description, the attacker can make queries only to a (probabilistic) MAC generation oracle.

In this paper, we consider a MAC forgery attack without any knowledge about the secret internal state. The attacker we supposed is a malicious person in a public communication channel. In our attack, there are a sender of messages, a receiver, and the attacker. The sender sends pairs of messages and the associated tags $(M, MAC(M, K))$ to the receiver. The attacker intercepts them, changes only the messages, then sends $(M', MAC(M, K))$ to the receiver. The attack is successful if the receiver does not detect the manipulation of a message.

The security notion of this attack was given by Bellare and Namprempre [2]. Under the notion, the attacker can make queries not only to the MAC generation oracle, but also to the MAC verification oracle. For each query, the MAC verification oracle returns 1 if the attached tag is correct, and return 0 otherwise. In real communication, the MAC verification oracle is the receiver and he outputs a requirement of retransmission to the sender if he detects the manipulation.

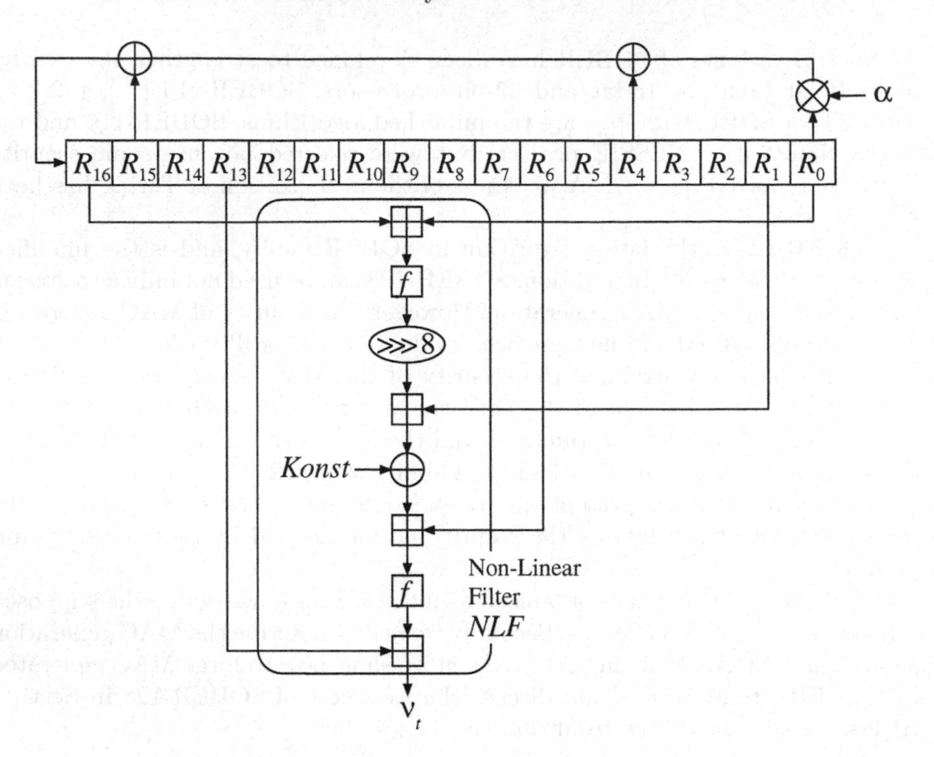

Fig. 1. The structure of non-linear filter NLF

There is no difference between these two attacks if the oracles are deterministic as a MAC verification oracle can be constructed from a MAC generation oracle. However, they are essentially different if the oracle is probabilistic (i.e. the oracle has a nonce as an input). In our model, the attacker can make a query for each oracle. On the other hand, the attacker whom the designers supposed can make only a query to a MAC generation oracle. The difference is critical to apply our attack given in Sect. 4.

3 The Algorithm of SOBER-128

SOBER-128 is a filtering generator which takes a secret key and an initial vector (IV) as inputs. The lengths of inputs are less than 128 bits. The update function of the internal state is an LFSR whose feedback polynomial is defined over GF(2^{32}). The non-linear filtering function consists of additions modulo 2^{32}, XORing, circular shift, and 8×32-bit substitution box (S-box) S (See Figure 1 for detail).

3.1 The Linear Feedback Shift Register

Before giving the definition of the LFSR, we firstly define the bit representation of elements of a finite field $\mathrm{GF}(2^w)$. The extension field over GF(2) is defined by the residue field $\mathrm{GF}(2)[z]/(\varphi(z))$, where $\varphi(z)$ is the irreducible polynomial. An element of $\mathrm{GF}(2^w)$ is a polynomial whose degree is less than w. The bit representation of a polynomial $a_{w-1}z^{w-1} + a_{w-2}z^{w-2} + \cdots + a_0$ is given by $a_{w-1}z^{w-1}||a_{w-2}z^{w-2}|| \cdots ||a_0$, where the notation $||$ means a bit-concatenation.

In the proposal of SOBER-128, the finite field $\mathrm{GF}(2^{32})$ is defined by an extension field of a subfield $\mathrm{GF}(2^8)$. Firstly the subfield $\mathrm{GF}(2^8)$ is given by the residue field $\mathrm{GF}(2)[z]/(x^8 + z^6 + z^3 + z^2 + 1)$. Next, $\mathrm{GF}(2^{32})$ is defined by the residue field $\mathrm{GF}(2^8)[y]/(g(y))$, where $g(y) = \mathtt{0xd0} \cdot y^3 + \mathtt{0x2b} \cdot y^2 + \mathtt{0x43} \cdot y + \mathtt{0x67}$. The coefficients in the irreducible polynomial g are the hexadecimal expressions of the elements of $\mathrm{GF}(2^8)$.

Now we can define the feedback polynomial $p(x)$ of the LFSR of SOBER-128:

$$p(x) = x^{17} + x^{15} + x^4 + \alpha, \tag{1}$$

where $\alpha = \mathtt{0x00000100}$ (hexadecimal expression) $= y$.

The registers of SOBER-128 is denoted by $R = (R[0], R[1], \ldots, R[16])$, where each $R[i]$ is a register of 32-bit length. We use subscript, like R_t, if the time t should be clarified. The update function defined by above feedback polynomial is calculated as follows:

$$\begin{aligned} \mathtt{tmp} &= R[0], \\ R[i] &= R[i+1], \quad 0 \leq i < 16, \\ R[16] &= R[14] \oplus R[4] \oplus \alpha \cdot \mathtt{tmp}. \end{aligned} \tag{2}$$

Furthermore, we can reduce the calculation of the multiplication with a constant in $\mathrm{GF}(2^{32})$ on 32-bit processor. The multiplication consists of only two operations, referring a pre-computed table Multab indexed by the most significant byte and a shift operation:

$$\alpha \cdot x = (x << 8) \;\hat{}\; \mathrm{Multab}[(x >> 24)\&\mathtt{0xFF}], \tag{3}$$

where ^ is the XORing operation and <<, >> are the left and right shift operations respectively.

3.2 Non-linear Filtering *NLF*

The description of the non-linear filtering function NLF is not necessary for explaining our attack, but we give the brief description.

NLF is defined by the following equation (See Figure 1.):

$$NLF(R) = f(((R[0] \bullet R[16])) \bullet \; 8 \bullet R[1] \oplus Konst) \bullet R[6]) \bullet R[13],$$

where $\bullet$ is the addition modulo 2^{32} and $\bullet$ is the rotation to right in a 32-bit register. 8×32-bit non-linear substitution f is defined by

$$f(a) = S_1[a_H]||(S_2[a_H] \oplus a_L),$$

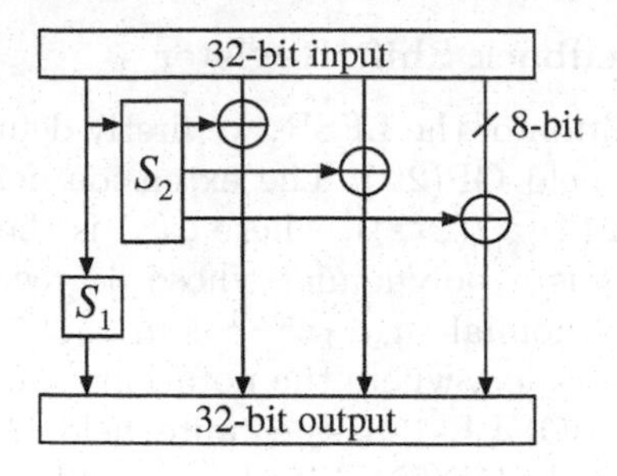

Fig. 2. The structure of function f

where a_H and a_L are the most significant 8 bits and the least significant 24 bits of the input a respectively (See Figure 2.). S_1 and S_2 in Figure 2 are the substitution boxes that output 8 bits and 24 bits values respectively. Please refer the proposal document of SOBER-128 [15] for more detail about the substitution table. $Konst$ is dependent on the initial values, and does not change after the initialization. Hence we can consider it is a key-dependent constant.

3.3 The Message Feedback Function *PFF*

The message feedback function PFF is used for injecting a message to the LFSR in MAC generation. PFF is given by the following equation:

$$PFF(R[4], p, Konst) = f((f(R[4] \boxplus p) \ggg 8) \boxplus Konst).$$

The value of the register $R[4]$ is replaced by the output of PFF.

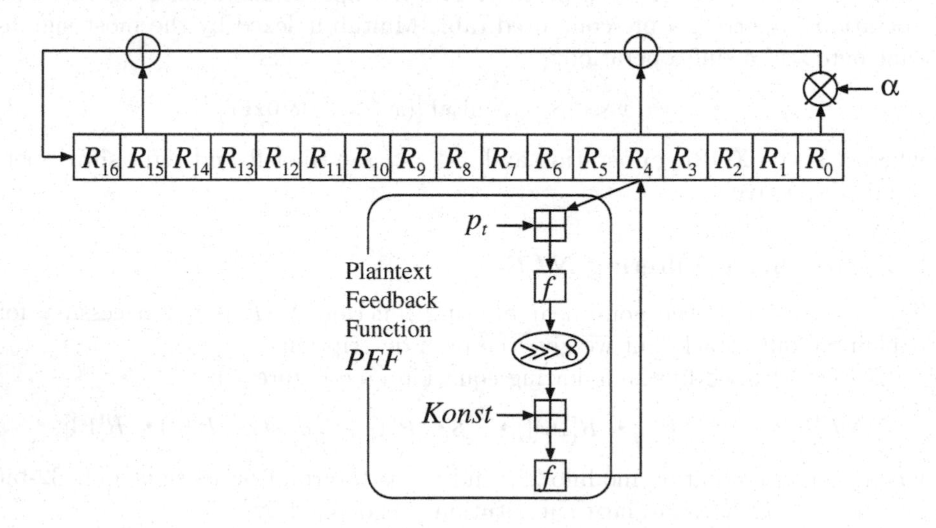

Fig. 3. The structure of plaintext feedback function PFF

3.4 MAC Generation

The MAC generation of SOBER-128 is divided into two phases. The first phase is called MAC accumulation and the second phase is called MAC finalization. Before starting MAC generation, the internal state is initialized with a secret key and an IV.

In the MAC accumulation phase, the message words are injected into the LFSR by using the message feedback function PFF:

$$R_t[4] \leftarrow PFF(R_t[4], p_t, Konst) = f((f(R_t[4] \bullet\ p_t) \bullet\quad 8) \bullet\ Konst). \quad (4)$$

In the MAC finalization phase, the internal state is mixed by XORing the constant $INITKONST$ to $R[15]$, applying the non-linear update function *Diffuse* 18 times. The non-linear function *Diffuse* is defined as follows:

Step 1. Updating the LFSR.
Step 2. Applying NLF to the register and replacing $R[4]$ with the value $R[4] \oplus \nu$, where ν is the output of NLF.

After the mixing, SOBER-128 outputs MAC of arbitrary length. The last step is same as the random number generation using NLF.

4 The MAC Forgery

Our attack aims to find a collision of messages with a same secret key and a same IV. For that, we apply differential cryptanalysis developed by Biham and Shamir [3]. Differential cryptanalysis observes the differential propagation of the target encryption function. In differential cryptanalysis, the attacker firstly encrypts amount of plaintext pairs with a fixed differential. and observes the distribution of the differentials of ciphertext. If the output differentials do not distribute uniformly, the attacker can recover some information of the secret key from the deviation of the distribution.

The basic idea of our attack is finding a pair of message sequence P and P', which yields same internal state value at a certain time T with a secret key and an IV. In a practical sense, the purpose of the attack is finding the differential sequence $\{\Delta p_t\}_t$ with which any message sequence pair $P = \{p_t\}_t$ and $P' = \{p_t \oplus \Delta p_t\}_t$ make a collision with high probability.

The search can be divided into two part. The search for the differential propagation of PFF with high probability and the construction of the differential elimination equation in the LFSR. In the following two subsection, each topic is discussed in detail.

4.1 The Differential Propagation in the Message Feedback Function PFF

Firstly, we discuss about the differential propagation in the message feedback function PFF.

We denote the differential propagation in the function f by $\Delta \xrightarrow{f} \Delta'$. The differential characteristic of non-linear permutation f is very large because the transformation is not uniform. Only the most significant byte is transformed non-linearly. so that any differential Δ whose most significant byte is zero does not influenced by f, i.e., $\Delta \xrightarrow{f} \Delta$ holds with probability 1. Hence, if the byte-wise expression of the differential of the register $R[4]$ is given by $(0, \delta_1, \delta_2, 0)$ and the differential of the message injection p is given by $(0, \delta_1', \delta_2', 0)$, the following differential propagation of PFF holds with high probability:

$$\begin{aligned}((0, \delta_1, \delta_2, 0), (0, \delta_1', \delta_2', 0)) &\xrightarrow{\boxplus} (0, \delta_1'', \delta_2'', 0) \\ &\xrightarrow{f} (0, \delta_1'', \delta_2'', 0) \\ &\xrightarrow{\ggg} (0, 0, \delta_1'', \delta_2'') \\ &\xrightarrow{\boxplus} (0, 0, \delta_1'', \delta_2'') \\ &\xrightarrow{f} (0, 0, \delta_1'', \delta_2''). \end{aligned} \tag{5}$$

4.2 How to Eliminate Differentials in the LFSR

Next we discuss how to eliminate differentials in the LFSR. To simplify the discussion, we replace the message feedback function PFF by XORing the message directly to $R_t[4]$. In fact, Eq. 5 indicates that any differential of $R_t[4]$ can be entirely overwritten by addition to a message block. Hence the discussion with the simplified message injection is useful. To avoid confusion, we denote the message block at time t by q_t and the differential by Δq_t respectively.

The internal state initialized by a secret key K and an initial vector I is denoted by R_0. We denote by $R_0(K, I)$ if the initial values should be clarified.

We denote the expression matrix of the LFSR of SOBER-128 defined over GF(2^{32}) by A. In the following discussion, we treat the internal state (registers) of the LFSR as a vector space over GF(2^{32}). Eq. 5 shows that the injection of the differential to the LFSR is identified by addition of the vector D_t given by

$$D_t = {}^t(0, 0, 0, 0, \Delta q_t, 0, \ldots, 0).$$

The corresponding internal state to message sequence $q_t \oplus \Delta q_t$ is given by

$$\begin{aligned} R_T' \oplus D_T &= A \cdot (R_{T-1}' \oplus D_{T-1}) \oplus D_T \\ &= A^2 \cdot (R_{T-2}' \oplus D_{T-2}) \oplus A \cdot D_{T-1} \oplus D_T \\ &\vdots \\ &= A^T \cdot R_0 \oplus \sum_{t=0}^{T} A^t \cdot D_{T-t}. \end{aligned} \tag{6}$$

Eq. 6 indicates that the differential vector of the internal state becomes zero at time T iff the following equation is satisfied.

$$\sum_{t=0}^{T} A^t \cdot D_{T-t} = 0. \tag{7}$$

The differential vector $D_t = {}^t(0, 0, 0, 0, \Delta q_t, 0, \ldots, 0)$ is divided into the elemental vector $e_4 = {}^t(0, 0, 0, 0, 1, 0, \ldots, 0)$ and the coefficient Δq_i:

$$D_t = \Delta q_t \cdot e_4.$$

Multiplying a constant is commutative with any matrix, hence we can transform Eq. 7 as follows:

$$\begin{aligned} 0 &= \sum_{t=0}^{T} A^t \cdot D_{T-t} \\ &= \sum_{t=0}^{T} A^t (\Delta q_{T-t} \cdot e_4) \\ &= (\sum_{t=0}^{T} \Delta q_{T-t} A^t) e_4. \end{aligned} \tag{8}$$

A differential sequence which satisfies this equation is given by Cayley-Hamilton relation. The characteristic polynomial of matrix A is given by Eq. 1 so that A satisfies the following relation:

$$A^{17} + A^{15} + A^4 + \alpha E = 0, \tag{9}$$

where E is the identity matrix of dimension 17. Therefore the following differential sequence Δq_t satisfies Eq. 8:

$$\Delta q_0 = \Delta q_2 = \Delta q_{13} = a, \quad \Delta q_{17} = \alpha \cdot a. \tag{10}$$

On the other hand, we examined exhaustive search for bite-wise truncated differential of message sequences $\{\Delta q_t\}_t$ which satisfy Eq. 9 and $T \leq 20$ on PC. The experiment concludes that Eq. 10 gives the best differential set, i.e. both the time T and the number of non-zero differentials are smallest.

Now we get back to the message injection by PFF. There are two different purpose to inject a message differential Δp_t into the LFSR. One is for injecting a certain differential value into a register, And the other is for eliminating a differential in $R_t[4]$. For each case, $\Delta p_t = \Delta q_t \bullet \quad 8$ and $\Delta p_t = \Delta q_t$ yields the same result as in the discussion of the simplified message injection.

Table 1 shows the propagation of the differential given in Eq. 10 in the internal state, where $b = \alpha \cdot a$.

Eq. 3 implies a significant characteristic of α. If the byte-wise expression of the differential Δq_0 is given by $(0, x, y, z)$, then $\alpha \cdot \Delta q_0 = (x, y, z, 0)$ holds for every x, y, z. For example, the 32-bit differentials corresponding to the truncated differentials a and b in Table 1 are given by 0x00000001 and 0x00000100 respectively.

4.3 The Success Probability of the MAC Forgery

In the differential propagation of the message feedback function PFF given by Eq. 5, the differential changes probabilistically only at the addition over $\mathrm{GF}(2^{32})$.

Table 1. The differential propagation in the LFSR

Time	Δp_t	Δq_t	Differential in the LFSR
00	$a \bullet 8$	a	0000*a*000000000000
01	0	0	000*a*000000000000*a*
02	$a \bullet 8$	a	00*a*0*a*0000000000*a*0
03	0	0	0*a*0*a*0000000000*a*00
04	0	0	*a*0*a*0000000000*a*000
05	0	0	0*a*0000000000*a*000*b*
06	0	0	*a*0000000000*a*000*b*0
07	0	0	0000000000*a*000*b*00
08	0	0	000000000*a*000*b*000
09	0	0	00000000*a*000*b*0000
10	0	0	0000000*a*000*b*00000
11	0	0	000000*a*000*b*000000
12	0	0	00000*a*000*b*0000000
13	a	a	00000000*b*00000000
14	0	0	0000000*b*000000000
15	0	0	000000*b*0000000000
16	0	0	00000*b*00000000000
17	b	b	00000000000000000

Lipmaa and Moriai presented the efficient algorithm for calculating the differential probability of an addition modulo 2^n for arbitrary n [17]. For example, if the differential $a =$ 0x00000001 in Eq. 10 is chosen, the differential probabilities of PFF at each time $t = 0, 2, 13, 17$ are 2^{-2}, 2^{-2}, 2^{-1}, 2^{-1} respectively. Therefore the success probability of the MAC forgery of SOBER-128 is about 2^{-6}.

We examined the attack with a sample code provided by the designers. The success probability of the experimental attack is about $2^{-5.5}$.

5 Discussion

Generally, the supposed attack in MAC forgery is adaptive chosen plaintext attack. However, the attack applied in the previous section is known plaintext attack. Any information about IVs and the secret key is not necessary as the attack is applicable even if the internal state is perfectly random. Besides, though SOBER-128 has authenticated encryption mechanism, encrypting messages does not strengthen the security of message authentication function because altering ciphertext is equal to altering plaintext for the encryption by a stream cipher.

In this section, we examine some idea to improve the security of SOBER-128 and consider its efficiency.

The vulnerability of MAC generation algorithm of SOBER-128 is mainly derived from the fact that the substitution f is not uniformly non-linear. From the viewpoint of differential propagation, only the transformation of the most significant byte is non-linear. Besides, a rotation has no diffusion function. Hence at least 4 times iterations of these functions is necessary to apply the non-linear transformation to all bytes. This reduces the performance of SOBER-128.

Another idea to improve the security is injecting the output of PFF to plural registers for increasing the complexity. However, the transformation in Eq. 8 holds for any vector whose elements are defined by $v_i = x_i \cdot v \quad (\mathrm{GF}(2^{32}))$. Hence this change of algorithm does not improve the security.

6 Conclusion

In this paper, we show that the MAC generation algorithm of SOBER-128 is vulnerable under a certain practical assumption. We assume that the attacker intercepts the message and change some bits of the message, and sends it to the receiver. Under this assumption, the attacker can forge a message with probability 2^{-6}.

References

[1] R. Anderson and E. Biham, "The Practical and Provably Secure Block Ciphers: BEAR and LION," *Fast Software Encryption, FSE'96*, Springer-Verlag, LNCS 1039, pp. 113–120, 1996. 472

[2] M. Bellare and C. Namprempre, "Authenticated Encryption: Relations among Notions and Analysis of the Generic Composition Paradigm," *Advances in Cryptology, Asiacrypto 2000*, Springer-Verlag, LNCS 1976, pp. 531–545, 2000. 473

[3] E. Biham and A. Shamir, *Differential Cryptanalysis of the Data Encryption Standard*, Springer-Verlag, 1993. 477

[4] S. Babbage and J. Lano, "Probabilistic Factors in the Sober-t Stream Ciphers," *Third Open NESSIE Workshop*, proceedings, 2002. 473

[5] C. De Cannière, J. Lano, B. Preneel, and J. Vandewalle, "Distinguishing Attacks on Sober-t32," *Third Open NESSIE Workshop*, proceedings, 2002. 473

[6] J. Daemen, "Cipher and hash function design strategies based on linear and differential cryptanalysis," *Doctoral Dissertation*, K. U.Leuven, 1995. 472

[7] J. Daemen and C. Clapp, "Fast Hashing and Stream Encryption with PANAMA," *Fast Software Encryption, FSE'98*, Springer-Verlag, LNCS 1372, pp. 60–74, 1998. 472

[8] P. Ekdahl and T. Johansson, "SNOW – a new stream cipher, " *NESSIE project submission*, 2000, available at `http://www.cryptonessie.org/`. 472

[9] P. Ekdahl and T. Johansson, "Distinguishing Attacks on SOBER-t16 and t32," *Fast Software Encryption, FSE 2002*, Springer-Verlag, LNCS 2365, pp. 210–224, 2002. 473

[10] P. Ekdahl and T. Johansson, "A new version of the stream cipher SNOW, " *Selected Areas in Cryptography, SAC 2002*, Springer-Verlag, LNCS 2595, pp. 47–61, 2002. 472

[11] N. Ferguson, D. Whiting, B. Schneier, J. Kelsey, S. Lucks, and T. Kohno, "Helix, Fast Encryption and Authentication in a Single Cryptographic Primitive," *Fast Software Encryption, FSE 2003*, pre-proceedings, pp. 345–362, 2003. 472

[12] P. Hawkes and G. Rose, "Primitive Specification and Supporting Documentation for SOBER-t16 Submission to NESSIE," *First Open NESSIE Workshop*, proceedings, 2000. 473

[13] P. Hawkes and G. Rose, "Primitive Specification and Supporting Documentation for SOBER-t32 Submission to NESSIE," *First Open NESSIE Workshop*, proceedings, 2000. 473

[14] P. Hawkes and G. Rose, "Turing, A Fast Stream Cipher," *Fast Software Encryption, FSE 2003*, pre-proceedings, pp. 307–324, 2003. 473

[15] P. Hawkes and G. Rose, "Primitive Specification for SOBER-128," IACR ePrint Archive, `http://eprint.iacr.org/2003/81/`, 2003. 473, 476

[16] A. Joux and F. Muller, "A Chosen IV Attack against Turing," *Selected Areas in Cryptography, SAC 2003*, pre-proceedings, 2003. 473

[17] H. Lipmaa and S. Moriai, "Efficient Algorithms for Computing Differential Properties of Addition," *Fast Software Encryption, FSE 2001*, Springer-Verlag, LNCS 2355, pp. 336–350, 2001. 480

[18] G. Rose, "A Stream Cipher based on Linear Feedback over GF(2^8)," "Proc. Australian Conference on Information Security and Privacy," Springer-Verlag, 1998. 472

[19] R. Rueppel, *Analysis and Design of Stream Ciphers*, Springer-Verlag, 1986.

[20] "The Software-Oriented Stream Cipher SSC2," *Fast Software Encryption, FSE 2000*, Springer-Verlag, LNCS 1978, pp. 31–48, 2000. 472

On Linear Approximation of Modulo Sum

Alexander Maximov

Department of Information Technology
Lund University
Box 118, SE–22100 Lund, Sweden
movax@it.lth.se

Abstract. The general case for a linear approximation of the form "$X_1 + \cdots + X_k \mod 2^n$" $\rightarrow$ "$X_1 \oplus \cdots \oplus X_k \oplus N$" is investigated, where the variables and operations are n-bit based, and the noise variable N is introduced due to the approximation. An efficient and practical algorithm of complexity $O(n \cdot 2^{3(k-1)})$ to calculate the probability $\Pr\{N\}$ is given, and in some cases it can be reduced to $O(2^{k-2})$.

1 Introduction

Linear approximations of nonlinear blocks in a cipher is a common tool for cryptanalysis. One of the most typical approximations is the substitution of the arithmetical sum modulo 2^n ($\boxplus$) with the XOR-operation ($\oplus$) of the input variables. We introduce a noise variable N and write: $X_1 \boxplus \cdots \boxplus X_k = X_1 \oplus \cdots \oplus X_k \oplus N$. For a distinguishing attack the bias of a linear combination of noise variables can be calculated if their distributions are known. For the considered approximation the distribution of N can be calculated in two ways:

I. for $X_1 = 0 \ldots 2^n - 1$ $\leftarrow O(2^{k \cdot n})$

$\ddots$

for $X_k = 0 \ldots 2^n - 1$
$\text{Dist}_N[(X_1 \boxplus \cdots \boxplus X_k) \oplus$
$(X_1 \oplus \cdots \oplus X_k)]$++;

or

II. for $C = 0 \ldots 2^n - 1$ $\leftarrow O(c \cdot 2^n)$
Dist_N[C]=ProbOfN(C);

where the function ProbOfN(C) calculates the corresponding probability (see Section 2). Note that we deal with integer-valued distribution tables, i.e., $\Pr\{N = C\} = \text{Dist}_N[C]/2^{k \cdot n}$.

2 The Function ProbOfN(C)

Let $C = \overline{c_n \ldots c_2 0}$ (note that $\Pr\{N = \overline{c_n \ldots c_2 1}\} = 0$). Then:

$$\text{ProbOfN}(C) = (1\,1 \ldots 1) \times \prod_{i=n}^{2} \mathbf{T}_{c_i} \times \mathbf{S_0},$$

B. Roy and W. Meier (Eds.): FSE 2004, LNCS 3017, pp. 483–484, 2004.

where $\mathbf{T_0}$, $\mathbf{T_1}$, and $\mathbf{S_0}$ are fixed matrices. The algorithm to construct the matrices $\mathbf{T_0}$, $\mathbf{T_1}$, and $\mathbf{S_0}$ is given below.

Initialization:

$\mathbf{S_0} = (0)$ - is of size $(2^{k-1} \times 1)$

$\mathbf{T_0} = \mathbf{T_1} = (0)$ - is of size $(2^{k-1} \times 2^{k-1})$

Algorithm 1: $\mathbf{S_0}$ – *construction*

1. for $X = 0$ to $2^k - 1$
2. $\quad \mathbf{S_0}[\lfloor \frac{\#X}{2} \rfloor] + = 1$

Algorithm 2: $\mathbf{T_0}$, $\mathbf{T_1}$ – *construction*

1. for $C = 0$ to $2^{k-2} - 1$
2. $\quad$ for $X = 0$ to $2^k - 1$
3. $\quad\quad \mathbf{T_0}[C + \lfloor \frac{\#X}{2} \rfloor][2C] + +,$
4. $\quad\quad \mathbf{T_1}[C + \lfloor \frac{\#X+1}{2} \rfloor][2C + 1] + +;$

where $\#X$ is the *Hamming weight* of X.

3 Example

Assume that $n = 5$ and $k = 3$, i.e., $N = (X_1 \boxplus X_2 \boxplus X_3) \oplus (X_1 \oplus X_2 \oplus X_3)$. Then:

$$\mathbf{T_0} = \begin{pmatrix} 4\,0\,0\,0 \\ 4\,0\,4\,0 \\ 0\,0\,4\,0 \\ 0\,0\,0\,0 \end{pmatrix} \mathbf{T_1} = \begin{pmatrix} 0\,1\,0\,0 \\ 0\,6\,0\,1 \\ 0\,1\,0\,6 \\ 0\,0\,0\,1 \end{pmatrix} \mathbf{S_0} = \begin{pmatrix} 4 \\ 4 \\ 0 \\ 0 \end{pmatrix}.$$

Let $C = \overline{10110}$, then ProbOfN(C)= $(1\,1\,1\,1) \times \mathbf{T_1} \times \mathbf{T_0} \times \mathbf{T_1} \times \mathbf{T_1} \times \mathbf{S_0}$, and $\Rightarrow \Pr\{N = \overline{10110}\} = 1536/2^{3 \cdot 5} = 0.046875$.

4 Optimization Ideas

If n is not very large, say $n = 32$ bits, then optimization can be done in the following way. Represent $C = \overline{AB0}$, where $A = \overline{c_{32} \ldots c_{16}}$ and $B = \overline{c_{15} \ldots c_2}$. Then create two tables of vectors: $R_{Left}[A] = (1\ \ 1 \ldots 1) \times \prod_{i=32}^{16} \mathbf{T}_{c_i}$ and $R_{Right}[B] = \prod_{i=15}^{2} \mathbf{T}_{c_i} \times \mathbf{S_0}$, for all A and B. Then the probability $\Pr\{N = C\}$ is just a scalar product $R_{Left}[\overline{c_{32} \cdots c_{16}}] \times R_{Right}[\overline{c_{15} \cdots c_2}]$, and the time complexity is $O(2^{k-2})$. This idea of partitioning can be extended to larger n as well.

Author Index

Lecture Notes in Computer Science

For information about Vols. 1–3005

please contact your bookseller or Springer-Verlag

Vol. 3120: J. Shawe-Taylor, Y. Singer (Eds.), Learning Theory. X, 648 pages. 2004.

Vol. 3105: S. Göbel, U. Spierling, A. Hoffmann, I. Iurgel, O. Schneider, J. Dechau, A. Feix (Eds.), Technologies for Interactive Digital Storytelling and Entertainment. XVI, 304 pages. 2004.

Vol. 3099: J. Cortadella, W. Reisig (Eds.), Applications and Theory of Petri Nets 2004. XI, 505 pages. 2004.

Vol. 3098: J. Desel, W. Reisig, G. Rozenberg (Eds.), Advanced Course on Petri Nets. VIII, 849 pages. 2004.

Vol. 3096: G. Melnik, H. Holz (Eds.), Advances in Learning Software Organizations. X, 173 pages. 2004.

Vol. 3094: A. Nürnberger, M. Detyniecki (Eds.), Adaptive Multimedia Retrieval. VIII, 229 pages. 2004.

Vol. 3093: S.K. Katsikas, S. Gritzalis, J. Lopez (Eds.), Public Key Infrastructure. XIII, 380 pages. 2004.

Vol. 3092: J. Eckstein, H. Baumeister (Eds.), Extreme Programming and Agile Processes in Software Engineering. XVI, 358 pages. 2004.

Vol. 3091: V. van Oostrom (Ed.), Rewriting Techniques and Applications. X, 313 pages. 2004.

Vol. 3089: M. Jakobsson, M. Yung, J. Zhou (Eds.), Applied Cryptography and Network Security. XIV, 510 pages. 2004.

Vol. 3086: M. Odersky (Ed.), ECOOP 2004 – Object-Oriented Programming. XIII, 611 pages. 2004.

Vol. 3085: S. Berardi, M. Coppo, F. Damiani (Eds.), Types for Proofs and Programs. X, 409 pages. 2004.

Vol. 3084: A. Persson, J. Stirna (Eds.), Advanced Information Systems Engineering. XIV, 596 pages. 2004.

Vol. 3083: W. Emmerich, A.L. Wolf (Eds.), Component Deployment. X, 249 pages. 2004.

Vol. 3079: Z. Mammeri, P. Lorenz (Eds.), High Speed Networks and Multimedia Communications. XVIII, 1103 pages. 2004.

Vol. 3078: S. Cotin, D.N. Metaxas (Eds.), Medical Simulation. XVI, 296 pages. 2004.

Vol. 3077: F. Roli, J. Kittler, T. Windeatt (Eds.), Multiple Classifier Systems. XII, 386 pages. 2004.

Vol. 3076: D. Buell (Ed.), Algorithmic Number Theory. XI, 451 pages. 2004.

Vol. 3074: B. Kuijpers, P. Revesz (Eds.), Constraint Databases and Applications. XII, 181 pages. 2004.

Vol. 3073: H. Chen, R. Moore, D.D. Zeng, J. Leavitt (Eds.), Intelligence and Security Informatics. XV, 536 pages. 2004.

Vol. 3072: D. Zhang, A.K. Jain (Eds.), Biometric Authentication. XVII, 800 pages. 2004.

Vol. 3070: L. Rutkowski, J. Siekmann, R. Tadeusiewicz, L.A. Zadeh (Eds.), Artificial Intelligence and Soft Computing - ICAISC 2004. XXV, 1208 pages. 2004. (Subseries LNAI).

Vol. 3068: E. André, L. Dybkj{\}ae r, W. Minker, P. Heisterkamp (Eds.), Affective Dialogue Systems. XII, 324 pages. 2004. (Subseries LNAI).

Vol. 3067: M. Dastani, J. Dix, A. El Fallah-Seghrouchni (Eds.), Programming Multi-Agent Systems. X, 221 pages. 2004. (Subseries LNAI).

Vol. 3066: S. Tsumoto, R. S lowiński, J. Komorowski, J.W. Grzymala-Busse (Eds.), Rough Sets and Current Trends in Computing. XX, 853 pages. 2004. (Subseries LNAI).

Vol. 3065: A. Lomuscio, D. Nute (Eds.), Deontic Logic in Computer Science. X, 275 pages. 2004. (Subseries LNAI).

Vol. 3064: D. Bienstock, G. Nemhauser (Eds.), Integer Programming and Combinatorial Optimization. XI, 445 pages. 2004.

Vol. 3063: A. Llamosí, A. Strohmeier (Eds.), Reliable Software Technologies - Ada-Europe 2004. XIII, 333 pages. 2004.

Vol. 3062: J.L. Pfaltz, M. Nagl, B. Böhlen (Eds.), Applications of Graph Transformations with Industrial Relevance. XV, 500 pages. 2004.

Vol. 3061: F.F. Ramas, H. Unger, V. Larios (Eds.), Advanced Distributed Systems. VIII, 285 pages. 2004.

Vol. 3060: A.Y. Tawfik, S.D. Goodwin (Eds.), Advances in Artificial Intelligence. XIII, 582 pages. 2004. (Subseries LNAI).

Vol. 3059: C.C. Ribeiro, S.L. Martins (Eds.), Experimental and Efficient Algorithms. X, 586 pages. 2004.

Vol. 3058: N. Sebe, M.S. Lew, T.S. Huang (Eds.), Computer Vision in Human-Computer Interaction. X, 233 pages. 2004.

Vol. 3057: B. Jayaraman (Ed.), Practical Aspects of Declarative Languages. VIII, 255 pages. 2004.

Vol. 3056: H. Dai, R. Srikant, C. Zhang (Eds.), Advances in Knowledge Discovery and Data Mining. XIX, 713 pages. 2004. (Subseries LNAI).

Vol. 3055: H. Christiansen, M.-S. Hacid, T. Andreasen, H.L. Larsen (Eds.), Flexible Query Answering Systems. X, 500 pages. 2004. (Subseries LNAI).

Vol. 3054: I. Crnkovic, J.A. Stafford, H.W. Schmidt, K. Wallnau (Eds.), Component-Based Software Engineering. XI, 311 pages. 2004.

Vol. 3053: C. Bussler, J. Davies, D. Fensel, R. Studer (Eds.), The Semantic Web: Research and Applications. XIII, 490 pages. 2004.

Vol. 3052: W. Zimmermann, B. Thalheim (Eds.), Abstract State Machines 2004. Advances in Theory and Practice. XII, 235 pages. 2004.

Vol. 3051: R. Berghammer, B. Möller, G. Struth (Eds.), Relational and Kleene-Algebraic Methods in Computer Science. X, 279 pages. 2004.

Vol. 3050: J. Domingo-Ferrer, V. Torra (Eds.), Privacy in Statistical Databases. IX, 367 pages. 2004.

Vol. 3049: M. Bruynooghe, K.-K. Lau (Eds.), Program Development in Computational Logic. VIII, 539 pages. 2004.

Vol. 3047: F. Oquendo, B. Warboys, R. Morrison (Eds.), Software Architecture. X, 279 pages. 2004.

Vol. 3046: A. Laganà, M.L. Gavrilova, V. Kumar, Y. Mun, C.K. Tan, O. Gervasi (Eds.), Computational Science and Its Applications – ICCSA 2004. LIII, 1016 pages. 2004.

Vol. 3045: A. Laganà, M.L. Gavrilova, V. Kumar, Y. Mun, C.K. Tan, O. Gervasi (Eds.), Computational Science and Its Applications – ICCSA 2004. LIII, 1040 pages. 2004.

Vol. 3044: A. Laganà, M.L. Gavrilova, V. Kumar, Y. Mun, C.K. Tan, O. Gervasi (Eds.), Computational Science and Its Applications – ICCSA 2004. LIII, 1140 pages. 2004.

Vol. 3043: A. Laganà, M.L. Gavrilova, V. Kumar, Y. Mun, C.K. Tan, O. Gervasi (Eds.), Computational Science and Its Applications – ICCSA 2004. LIII, 1180 pages. 2004.

Vol. 3042: N. Mitrou, K. Kontovasilis, G.N. Rouskas, I. Iliadis, L. Merakos (Eds.), NETWORKING 2004, Networking Technologies, Services, and Protocols; Performance of Computer and Communication Networks; Mobile and Wireless Communications. XXXIII, 1519 pages. 2004.

Vol. 3040: R. Conejo, M. Urretavizcaya, J.-L. Pérez-de-la-Cruz (Eds.), Current Topics in Artificial Intelligence. XIV, 689 pages. 2004. (Subseries LNAI).

Vol. 3039: M. Bubak, G.D.v. Albada, P.M. Sloot, J.J. Dongarra (Eds.), Computational Science - ICCS 2004. LXVI, 1271 pages. 2004.

Vol. 3038: M. Bubak, G.D.v. Albada, P.M. Sloot, J.J. Dongarra (Eds.), Computational Science - ICCS 2004. LXVI, 1311 pages. 2004.

Vol. 3037: M. Bubak, G.D.v. Albada, P.M. Sloot, J.J. Dongarra (Eds.), Computational Science - ICCS 2004. LXVI, 745 pages. 2004.

Vol. 3036: M. Bubak, G.D.v. Albada, P.M. Sloot, J.J. Dongarra (Eds.), Computational Science - ICCS 2004. LXVI, 713 pages. 2004.

Vol. 3035: M.A. Wimmer (Ed.), Knowledge Management in Electronic Government. XII, 326 pages. 2004. (Subseries LNAI).

Vol. 3034: J. Favela, E. Menasalvas, E. Chávez (Eds.), Advances in Web Intelligence. XIII, 227 pages. 2004. (Subseries LNAI).

Vol. 3033: M. Li, X.-H. Sun, Q. Deng, J. Ni (Eds.), Grid and Cooperative Computing. XXXVIII, 1076 pages. 2004.

Vol. 3032: M. Li, X.-H. Sun, Q. Deng, J. Ni (Eds.), Grid and Cooperative Computing. XXXVII, 1112 pages. 2004.

Vol. 3031: A. Butz, A. Krüger, P. Olivier (Eds.), Smart Graphics. X, 165 pages. 2004.

Vol. 3030: P. Giorgini, B. Henderson-Sellers, M. Winikoff (Eds.), Agent-Oriented Information Systems. XIV, 207 pages. 2004. (Subseries LNAI).

Vol. 3029: B. Orchard, C. Yang, M. Ali (Eds.), Innovations in Applied Artificial Intelligence. XXI, 1272 pages. 2004. (Subseries LNAI).

Vol. 3028: D. Neuenschwander, Probabilistic and Statistical Methods in Cryptology. X, 158 pages. 2004.

Vol. 3027: C. Cachin, J. Camenisch (Eds.), Advances in Cryptology - EUROCRYPT 2004. XI, 628 pages. 2004.

Vol. 3026: C. Ramamoorthy, R. Lee, K.W. Lee (Eds.), Software Engineering Research and Applications. XV, 377 pages. 2004.

Vol. 3025: G.A. Vouros, T. Panayiotopoulos (Eds.), Methods and Applications of Artificial Intelligence. XV, 546 pages. 2004. (Subseries LNAI).

Vol. 3024: T. Pajdla, J. Matas (Eds.), Computer Vision - ECCV 2004. XXVIII, 621 pages. 2004.

Vol. 3023: T. Pajdla, J. Matas (Eds.), Computer Vision - ECCV 2004. XXVIII, 611 pages. 2004.

Vol. 3022: T. Pajdla, J. Matas (Eds.), Computer Vision - ECCV 2004. XXVIII, 621 pages. 2004.

Vol. 3021: T. Pajdla, J. Matas (Eds.), Computer Vision - ECCV 2004. XXVIII, 633 pages. 2004.

Vol. 3019: R. Wyrzykowski, J.J. Dongarra, M. Paprzycki, J. Wasniewski (Eds.), Parallel Processing and Applied Mathematics. XIX, 1174 pages. 2004.

Vol. 3018: M. Bruynooghe (Ed.), Logic Based Program Synthesis and Transformation. X, 233 pages. 2004.

Vol. 3017: B. Roy, W. Meier (Eds.), Fast Software Encryption. XI, 485 pages. 2004.

Vol. 3016: C. Lengauer, D. Batory, C. Consel, M. Odersky (Eds.), Domain-Specific Program Generation. XII, 325 pages. 2004.

Vol. 3015: C. Barakat, I. Pratt (Eds.), Passive and Active Network Measurement. XI, 300 pages. 2004.

Vol. 3014: F. van der Linden (Ed.), Software Product-Family Engineering. IX, 486 pages. 2004.

Vol. 3012: K. Kurumatani, S.-H. Chen, A. Ohuchi (Eds.), Multi-Agnets for Mass User Support. X, 217 pages. 2004. (Subseries LNAI).

Vol. 3011: J.-C. Régin, M. Rueher (Eds.), Integration of AI and OR Techniques in Constraint Programming for Combinatorial Optimization Problems. XI, 415 pages. 2004.

Vol. 3010: K.R. Apt, F. Fages, F. Rossi, P. Szeredi, J. Váncza (Eds.), Recent Advances in Constraints. VIII, 285 pages. 2004. (Subseries LNAI).

Vol. 3009: F. Bomarius, H. Iida (Eds.), Product Focused Software Process Improvement. XIV, 584 pages. 2004.

Vol. 3008: S. Heuel, Uncertain Projective Geometry. XVII, 205 pages. 2004.

Vol. 3007: J.X. Yu, X. Lin, H. Lu, Y. Zhang (Eds.), Advanced Web Technologies and Applications. XXII, 936 pages. 2004.

Vol. 3006: M. Matsui, R. Zuccherato (Eds.), Selected Areas in Cryptography. XI, 361 pages. 2004.